Hydrocarbon Migration and Its Near-Surface Expression

Outgrowth of the AAPG Hedberg Research Conference
Vancouver, British Columbia, April 24–28, 1994

Edited by

Dietmar Schumacher

Michael A. Abrams

AAPG Memoir 66

Published by
The American Association of Petroleum Geologists
Tulsa, Oklahoma, U.S.A.
Printed in the U.S.A.

Association Editor: Kevin T. Biddle
Science Director: Richard Steinmetz
Publications Manager: Kenneth M. Wolgemuth
Special Projects Editor: Anne H. Thomas
Production: Kathy and Dana Walker, Editorial Technologies, Renton, Washington

This and other AAPG publications are available from

AAPG Bookstore
P.O. Box 979
Tulsa, OK 74101-0979
U.S.A.
Tel (918) 584-2555
 or (800) 364-AAPG (U.S.A.—book orders only)
Fax (918) 584-0469
 or (800) 898-2274 (U.S.A.—book orders only)

Australian Mineral Foundation
AMF Bookshop
63 Conyngham Street
Glenside, South Australia 5065
Australia
Tel (08) 379-0444
Fax (08) 379-4634

Geological Society Publishing House
Unit 7, Brassmill
Enterprise Centre
Brassmill Lane
Bath BA1 3JN
United Kingdom
Tel 0225-445046
Fax 0225-442836

Canadian Society of Petroleum Geologists
#505, 206 7th Avenue S.W.
Calgary, Alberta T2P 0W7
Canada
Tel (403) 264-5610

Foreword

Utilization of Hydrocarbon Seep Information

Geochemical prospecting for petroleum is the search for chemically identifiable surface or near-surface occurrences of hydrocarbons as clues to the location of oil or gas accumulations. It extends through a range from observation of clearly visible oil and gas seepages at one extreme to the identification of minute traces of hydrocarbons determinable only by highly sophisticated analytical methods at the other. There is no question in principle about the value of the method for petroleum exploration if properly applied. Historically, most of the world's major petroleum-bearing areas and many of its largest oil and gas fields were first called to attention because of visible oil and gas seepages. The mere presence of higher hydrocarbons in a region is encouraging in that it usually proves that conditions in that region have been suitable for at least some petroleum generation. Often seepages are in close proximity to commercial oil and gas pools, but the absence of seepage does not at all negate prospects because it may only indicate that there has been little escape from such pools due to good sealing rocks.

Oil and gas are mobile fluids and rocks are generally permeable. Surface oil and gas seeps primarily reflect avenues of migration (or escape) from deeper and sometimes laterally distant locations. Moreover, because avenues of migration (or escape) from deeper accumulations vary considerably in the degree to which they are sealed, the quantitative size of a seep has little relationship to the size of the accumulation. Some small accumulations are marked by strong visible seepages, whereas some of the largest accumulations are so well sealed that they show no visible seepages and only microscopic seepages or none at all. The value of seepages, visible or microscopic, is thus largely a matter of the accuracy with which they can be interpreted geologically. In some case (e.g., Burgan field) a well drilled vertically at the site of seepage would have discovered the field. In other cases where escape of hydrocarbons has been along low dipping fault planes or low dipping carrier beds, surface seepages may be many miles laterally from vertical superposition over the oil or gas accumulation. Again, the value of the information on the seepage, visible or microscopic, is always there, but it is only the geologic interpretation that allows cashing in on its value.

On land, most visible seepages have already been recorded and the nature of the relationship to subsurface petroleum accumulations has been at least studied if not always successfully determined. The main task now for geochemical prospecting is the identification of the invisible or less clearly manifested "seepages" that can be determined only by detailed chemical analysis of fluids in surface and near-surface rocks. The problems are not whether there is any value to the data but rather are (1) the techniques for identification, (2) the geologic interpretation, and (3) the quality of the interpretation good enough to justify the cost.

Offshore, the situation is slightly different. Visual observation of offshore seepages has been impeded by the water cover, and reliance must be placed mainly on chemical analysis of the water column and the interstitial waters filling the pores of the blanket of young sediment covering the sea floor. Again, there seems to me no question of the innate value of the geochemical information, positive or negative. And again, the problems are with the techniques of identification and geologic interpretation, and whether the interpretation is good enough to justify costs. There is nothing wrong with the concept; it is only a question of our ability to collect the data adequately and to interpret the results correctly, at a reasonable cost.

A geochemical survey should be thought of not as a black magic means of spotting the location of oil and gas pools but only as a simple common sense method of gathering data on hydrocarbon occurrences too dilute to make visible seeps or impregnations—data which if collected reliably, interpreted wisely, and used intelligently along with all other lines of evidence will always be helpful in petroleum exploration of any area.

Hollis D. Hedberg
March 1981

AAPG
Wishes to thank the following
for their generous contributions
to
Hydrocarbon Migration
and Its Near-Surface Expression

Association of Petroleum Geochemical Explorationists

Conoco Inc.

Dietmar Schumacher

Exploration Technologies, Inc.

Exxon Ventures (CIS), Inc.

Geo-Microbial Technologies, Inc.

Geoscience & Technology, Inc.

Petrobras

Phillips Petroleum Company

Shell Offshore, Inc.

The University of Saskatchewan

Zonge Engineering and Research Organization, Inc.

Contributions are applied against the production
costs of publication, thus directly reducing the book's purchase
price and making the volume available
to a greater audience.

Contents

About the Editors

Dietmar ("Deet") Schumacher is currently a Research Professor with the Earth Sciences and Resources Institute (ESRI) of the University of Utah in Salt Lake City. He received his B.S. and M.S. degrees in geology from the University of Wisconsin and his Ph.D. from the University of Missouri. Deet taught geology at the University of Arizona for 7 years before joining Phillips Petroleum as a research geologist in 1977. He held a variety of positions at Phillips, including Research Supervisor for petroleum geology and Senior Geological Specialist. Deet then joined Pennzoil in 1982 and served as manager of geology/geochemistry before transferring to assignments with Pennzoil International, Pennzoil Offshore, and Pennzoil's Technology Group. In 1994, Deet accepted a position as Research Professor at ESRI. He is presently an Associate Editor of the AAPG *Bulletin* and a past president of both the Houston Geological Society and the Association of Petroleum Geochemical Explorationists. Deet has had a

long-standing interest in exploration and development applications of petroleum geochemistry, particularly surface exploration methods. It is this interest that resulted in his convening (with Michael Abrams) the AAPG Hedberg Research Conference "Near-Surface Expression of Hydrocarbon Migration" and the editing of this volume.

Michael Abrams is presently a Senior Exploration Geochemist with Exxon Ventures (CIS), Inc. Michael has had assignments in exploration, production, and research all over the world during his 15 years with the Exxon Corporation. He was introduced to surface geochemistry as a marine geologist early in his career as an explorationist. His first assignment with Exxon was to design a research program to investigate acoustic anomalies in offshore Alaska which were thought to be due to leakage of subsurface hydrocarbons. Michael has since championed surface geochemistry as a viable exploration tool in frontier basins, both within Exxon and through his publications. Michael attended the University of Rochester, George Washington University, and the University of Southern California.

Preface

Over the past 60 years, numerous direct and indirect hydrocarbon exploration methods have been developed. The application of these surface prospecting methods to oil and gas exploration has resulted in varied success and considerable controversy. Few question that hydrocarbons migrate to the near surface in amounts that are detectable, but many are skeptical of how such information can be integrated into more conventional exploration and development programs. Our understanding of the process of hydrocarbon migration from source or reservoir to the near surface is poorly understood and severely limits the interpretation of surface geochemical data. The past decade has seen a renewed interest in this topic which, when coupled with developments in analytical and interpretive methods, has produced a new body of data and insights in this area.

This publication is a direct outgrowth of the AAPG Hedberg Research Conference held in April 1994 entitled "Near-Surface Expression of Hydrocarbon Migration." The purpose of this research conference was to gather international experts from industry and academia to critically examine the process of hydrocarbon migration and its varied near-surface expressions. The wide range of topics discussed is reflected by the papers selected for inclusion in this volume: near-surface manifestations of hydrocabon migration, hydrocarbon-induced alteration of soils and sediments, migration mechanisms, hydrocarbon flux measurements, sampling and analytical techniques, survey design and interpretation, physical and geological implications of hydrocarbon leakage, and finally, exploration case studies.

Conference participants engaged in lively discussion, and despite a lack of consensus on a number of topics, there was general agreement on the following conclusions:

- ✓ Hydrocarbon accumulations are dynamic; seals are imperfect.
- ✓ All petroleum basins have some type of near-surface hydrocarbon leakage.
- ✓ Surface expression of leakage is not always detectable by conventional means.
- ✓ Hydrocarbon seepage can be active or passive, and it can be visible (macroseepage) or only chemically detectable (microseepage).
- ✓ Seepage expression, whether active or passive, is a function of many factors other than the mere presence or absence of active hydrocarbon generation and migration.
- ✓ Migration occurs mainly vertically, but it can also occur over long distances laterally.
- ✓ Hydrocarbons can move vertically through thousands of meters of strata without observable faults or fractures in a relatively short time (weeks to years).
- ✓ Relationships between surface geochemical anomalies and subsurface accumulation can be complex; proper interpretation requires integration of seepage data with geological, geophysical, and hydrological data.
- ✓ Hydrocarbon migration mechanisms are still poorly understood. Present evidence favors effusion as the process of macroseepage and bouyancy of microbubbles as the mechanism for microseepage.

It is our hope that the information and ideas presented in this volume will assist the formulation of more effective exploration and development strategies by providing a better understanding of hydrocarbon migration and its near-surface effects. Only through a fuller understanding of these processes can surface exploration technology achieve its full potential, a goal increasingly important as our industry strives to improve exploration efficiency during these times of economic uncertainty.

Dietmar Schumacher
Michael Abrams

Acknowledgments

We sincerely appreciate the assistance given by the following individuals who generously provided critical reviews of the manuscripts published in this volume:

Jim Allan* (Imperial Oil Resources)
Peter Blanchette (Surface Exploration)
Norman Carlson (Zonge Engineering)
Timothy Collett (USGS)
James Corthay (Exxon Exploration Co.)
Steve Creaney (Exxon Exploration Co.)
Wally Dow (DGSI)
Doug Elmore (University of Oklahoma)
John Geissman (University of New Mexico)
Joel Gevirtz (Interscience)
H. Robert Hopkins (Geosat Committee)
Jeff Hulen (ESRI, University of Utah)
Alan James (Exxon Production Research)
Dirk Kettel (Consultant)
Ron Klusman (Colorado School of Mines)
Alan Kornacki (Shell Offshore)

Steve May (Exxon Production Research)
Marty Gorbaty (Exxon Corporate Research)
Paul Philp (University of Oklahoma)
Neil Piggott (British Petroleum)
Bill Powell (Exxon Production Research)
Leigh Price (USGS)
Melodye Rooney (Mobil Research)
Roger Sassen (GERG, Texas A&M)
Donald Saunders (Recon Exploration)
Len Srnka (Exxon Production Research)
Lori Summa (Exxon Production Research)
Ken Sundberg (Phillips Petroleum)
Jane Thrasher (British Petroleum)
Allan Tripp (University of Utah)
W. A. Young (Exxon Production Research).

The conscientious and painstaking reviews of the manuscripts by these colleagues have greatly improved the scientific content and value of this volume.

We also wish to thank the editorial and production staff of AAPG, particularly Anne Thomas and Kathy Walker.

Dietmar Schumacher

Michael Abrams

*Jim Allan died in June 1996 after a difficult battle with cancer. Jim's contribution to this publication , as well as his contributions to the science of geochemistry, will always be remembered and appreciated.

Abrams, M. A., 1996, Distribution of subsurface hydrocarbon seepage in near-surface marine sediments, *in* D. Schumacher and M. A. Abrams, eds., Hydrocarbon migration and its near-surface expression: AAPG Memoir 66, p. 1–14.

Chapter 1

Distribution of Subsurface Hydrocarbon Seepage in Near-Surface Marine Sediments

Michael A. Abrams

Exxon Ventures (CIS), Inc.
Houston, Texas, U.S.A.

Abstract

Hydrocarbon seeps in surficial marine sediments are of two types: active and passive. Active seeps occur where gas bubbles, pockmarks, or bright spots are visible on seismic profiles and where chemosynthetic communities are present in conjunction with large concentrations of migrated hydrocarbons (macroseeps). These generally occur where generation and migration of hydrocarbons from source rocks are ongoing today (at maximum burial) or where significant migration pathways have developed from recent tectonic activity. Passive seeps occur where concentrations of migrated hydrocarbons are usually low (microseeps) with few or no geophysical anomalies. These occur typically in areas where generation and expulsion is relict (no longer at maximum burial) or regional seals prevent significant vertical migration.

The type of seepage controls the distribution of migrated hydrocarbons in the near-surface sediments and should dictate the sampling equipment and approach used to detect seeps. Active seeps are usually detected near the water–sediment interface, in the water column or at the sea surface, and at relatively large distances from major leak points. Most conventional sediment and water samplers can capture active seeps. The Gulf of Mexico, Santa Barbara Channel, and parts of the North Sea have active hydrocarbon seeps.

Passive seeps can only be detected relatively far below the water–sediment interface and require samples to be collected near leak points. Sampling equipment must penetrate the zone of maximum disturbance or any shallow migration barriers. In areas where surficial sediments are coarse grained or compacted, conventional gravity corers will not work. Other options for subsurface sampling include vibracores, jet cores, and rotary cores. Precise location of samples (site-specific) using seismic profiles to locate leak points is critical to detect passive hydrocarbon seeps. The Beaufort and Bering seas, offshore Alaska, and parts of the North Sea contain passive seeps.

INTRODUCTION

Surface geochemical techniques were first applied by Laubmeyer and Sokolov almost 60 years ago with both success and failure (Laubmeyer, 1933; Sokolov, 1935; Rosaire, 1940). The concept that hydrocarbons migrate from subsurface accumulations and mature source rocks to near-surface sediments is well documented (Jones and Horvitz, 1939, 1969, 1980, 1981, 1985a,b; Drozd, 1983; Price, 1986; Kennicutt and Brooks, 1988). Few question that hydrocarbons migrate to near-surface marine sediments in detectable amounts, but many doubt that such information can be integrated into conventional exploration or development programs. Many papers detail direct and indirect geochemical methods that purport to demonstrate surface geochemical anomalies above known subsurface accumulations. Nevertheless, it is probable that an equal number of largely unpublished studies fail to find such a correlation.

Why do some or all surface geochemical techniques work in some areas and not in others? The key to understanding and properly interpreting surface geochemical data lies in understanding how subsurface hydrocarbons leak and migrate into surficial sediments. Subsurface hydrocarbon migration is poorly understood. Several recent papers have attempted to discuss the physical processes of subsurface hydrocarbon movement to the near surface (Price, 1986; Sweeney, 1988; Kettel, 1990;

Klusman, 1993). It is doubtful that geochemists and geologists will ever fully understand all the mechanisms by which subsurface hydrocarbons migrate to near-surface sediments. However, the distribution of hydrocarbons migrating from depth to the surface and into the water/air column does provide insight when interpreting surface geochemical data.

In this paper, I relate the distribution of subsurface hydrocarbon leakage within and above near-surface sediments to the type of seep activity. Where seepage is very active, such as in the Gulf of Mexico, hydrocarbon seeps are easily discernible by almost any technique. Where seepage is passive, such as offshore Alaska, hydrocarbon seeps are difficult to nearly impossible to detect. The type of seep strongly controls the distribution of migrating hydrocarbons in near-surface sediments, the water column, and the ocean surface, and this dictates the sampling equipment and approach required to detect seeps.

SEEPAGE ACTIVITY

Surface hydrocarbon seep activity is defined here as the relative rate of seepage and its associated effects. It has two distinct end-members:

1. Active—Areas where subsurface hydrocarbons actively seep in large concentrations within and above the surface marine sediments are called *active seeps*. When active seeps contain anomalous low molecular weight (C_1 to C_{5+}) and high molecular weight hydrocarbons that may visibly stain sediment cores, they are generally referred to as *macroseeps*. Certain deep-water active seeps also support chemosynthetic communities (Kennicutt et al., 1988; MacDonald et al., 1989) and are often associated with gas hydrates (Kvenvolden and McMenamin, 1980). Active seeps are easily detected as acoustic anomalies (e.g., wipe-out zones and bottom simulating reflector) on conventional and high-resolution seismic profiles. Also, active seeps may be seen as gas bubble traces and pockmarks on subbottom profiler and side-scan sonar records (Hovland and Judd, 1988). Active seeps are widespread and easily detected by most sampling techniques at the seabed, in the water column, or at the sea surface (as slicks).

Active seeps typically occur in basins that are now actively generating hydrocarbons or that contain excellent migration pathways. Examples of active seeps are found in the Gulf of Mexico (Reitsema et al., 1978; Anderson et al., 1983; Brooks and Carey, 1986), offshore California (Fischer and Stevenson, 1973; Kvenvolden and Field, 1981; Kennicutt and Brooks, 1988), parts of the North Sea (Gevirtz et al., 1983; Faber and Stahl, 1984; Sweeney, 1988), and offshore Indonesia (Thompson et al., 1991).

2. Passive—Areas where subsurface hydrocarbons are not actively seeping are called *passive seeps*. These seep zones usually contain low molecular weight hydrocarbon concentrations above normal background levels (*microseeps*), but they can also contain high molecular

weight hydrocarbons (*macroseeps*). Acoustic anomalies may be found on conventional and high-resolution seismic profiles, but water column anomalies are rarely present. Anomalous hydrocarbon concentrations are usually only detected very close to major leak points and require sampling relatively far below the water–sediment interface.

In basins where hydrocarbon generation is relict or where migration is sporadic or inhibited by a major migration barrier, seepage is passive. Areas with passive seeps include parts of offshore Alaska (Abrams, 1992), the northwest shelf of Australia, Antarctica (Whiticar et al., 1985), central Sumatra (Thompson et al., 1991), and certain regions of the North Sea (Faber and Stahl, 1984).

In each end-member, distinctive physical characteristics control the distribution of the migrating hydrocarbons in surface sediments. The vertical and horizontal distribution of anomalous hydrocarbons should determine the sampling program and equipment. This is a key component in interpreting surface geochemical data. In many cases, the failure to recognize positive indications of thermally generated hydrocarbons is not due to the absence of subsurface generation and subsequent accumulation but is more a function of sampling equipment/procedures and analytical procedures used to detect hydrocarbon seepage.

NEAR-SURFACE EXPRESSION OF SUBSURFACE LEAKAGE

Horizontal Distribution

Seeps vary in shape, size, and form. They cause anomalous (significantly higher than *in situ* background) hydrocarbon concentration patterns to form over petroleum accumulations. These patterns align themselves into halo (doughnut), apical (focal), crescent, or linear (straight-line) shapes. The hydrocarbon concentration pattern may be related to structural features such as sea floor highs, subsurface highs, diapirs, faults, unconformities, and spill points (which may result from structural failure of trapping components). Most of the surveys discussed in the following sections have relied on low molecular weight hydrocarbons or secondary alterations that result from hydrocarbon leakage (mineralogic changes) (Klusman, 1989).

Halo and Apical Anomalies

Horvitz (1981) observed halo microseepage patterns in offshore Louisiana. Relatively low hydrocarbon concentrations occur in the sediments over the main part of the subsurface accumulation, with relatively high values in the sediment above the edges. Low values were also observed in the background areas away from the accumulation. Horvitz noted that, in selected cases, hydrocarbon anomalies did not appear directly above the

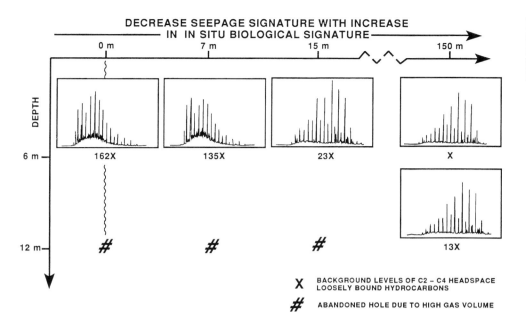

DECREASE SEEPAGE SIGNATURE WITH INCREASE IN IN SITU BIOLOGICAL SIGNATURE

X BACKGROUND LEVELS OF C2 – C4 HEADSPACE LOOSELY BOUND HYDROCARBONS

ABANDONED HOLE DUE TO HIGH GAS VOLUME

Figure 1—Distribution of anomalous C_{15+} saturate hydrocarbons along a leaky fault, offshore Alaska, showing passive seepage.

subsurface accumulations. Hitchon (1974) suggests that fluid flow within the near-surface region may affect hydrocarbon movement and should be taken into consideration when interpreting surface geochemical data. These zones of anomalous hydrocarbon concentrations that rim the edges or are directly above the accumulations result from vertical microseepage, according to Price (1986). Offshore areas, such as the Gulf of Mexico and offshore California, that show halos and apical anomalies also display macroseepage (high molecular weight hydrocarbons). Macroseeps are usually associated with faults and fractures.

Fault-Related Anomalies

Studies by Fisher and Stevenson (1973), Voytov et al. (1972), Reitsema et al. (1978), and Abrams (1992) all show examples where hydrocarbon concentrations are greatest near faults. Jet core samples collected on and off a leaky fault in the Bering Sea demonstrated the site-specific nature of passive hydrocarbon seepage (Abrams, 1992). Core samples collected on or near (within 75 m) the fault scarp detected anomalous low and high molecular weight thermogenic hydrocarbons, whereas core samples collected more than 75 m from the fault scarp displayed typical background levels of low molecular weight hydrocarbons and contained biological C_{15+} signatures (Figure 1). No geophysical or geochemical anomalies were found in the water column or on the ocean surface.

Similar results have been noted in areas of active seepage. The boundary between thermogenic seepage and masked *in situ* biological compounds is much less distinct in areas of active seepage (Figure 2). Studies in the Gulf of Mexico by Texas A&M (Ian MacDonald, personal communication, 1995) and Exxon (unpublished proprietary surveys) detected macroseepage up to several hundred meters away from fault scarps. This may have been partially due to leaky surficial sediments (relatively coarse grained and unconsolidated), the fracture system, or sub-

seismic-scale faulting undetected by high-resolution seismic profiles (usually associated with areas of diapiric uplift).

Other Anomalies

Local shallow (less than 30 m) stratigraphy and lithology can strongly control the distribution and type of hydrocarbons. Examination of several deep (4–6 m) gravity cores and conventional and high-resolution seismic profiles demonstrate localized subsurface barriers that appear to prevent movement into sediments less than 6 m deep (Figure 3). Low molecular weight hydrocarbons (C_1 to C_{5+}) and carbon dioxide move up the major fault system into near-surface sediments where local barriers prevent movement into the very near-surface sediments. Hydrocarbon and carbon dioxide anomalies were found only where the localized barriers were not present. Similar observations have been made in other surface geochemical surveys, especially in areas where hydrates and permafrost are present (unpublished Exxon studies). Clayton (1992) comments that although gas may appear to leak from distinct fractures within surficial sediments, it is actually more dispersed within the shallow sediments.

No Anomalies

Finally, many unpublished surveys show no detectable geochemical anomalies above or near relatively significant subsurface hydrocarbon accumulations. One such example in offshore Alaska included an ocean bottom sniffer and two gravity core geochemical surveys above a fault-dependent accumulation of 200–300 million bbl of oil (Figure 4). No migrated hydrocarbons were detected in any of the detailed geochemical data, which included headspace, cuttings, adsorbed, and water column low molecular weight hydrocarbons; total scanning spectrofluorescence; capillary gas chromatography; and gas chromatography–mass spectrometry data.

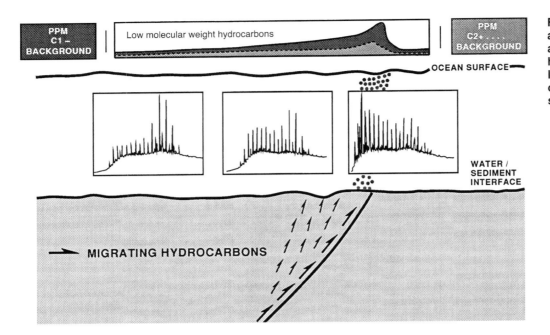

Figure 2—Distribution of anomalous C_{15+} saturate and low molecular weight hydrocarbons along a leaky fault, offshore Gulf of Mexico, showing active seepage.

Vertical Distribution

Hydrocarbon samples collected from above and within surficial marine sediments at several different depth intervals demonstrate significant differences in hydrocarbon concentrations and type. These differences can be explained by physical and biological processes within the shallow surficial sediments, by analytical procedures, and by the seepage rate.

Free Low Molecular Weight Hydrocarbons

Profiles of total free hydrocarbon gas, which includes headspace plus cuttings low molecular weight hydrocarbon (C_1 to C_{5+}) gas, often display increases with depth (Whelan et al., 1975; Zhizhchenko, 1978; Bernard, 1978; Kvenvolden and Field, 1981; Carlson et al., 1985; Whiticar et al., 1985; Faber et al., 1990; Abrams, 1992). Sediment samples collected from jet and gravity core boreholes in areas of passive seepage demonstrate total hydrocarbon gas concentrations increasing to two times background levels (estimated at 1 m below the water–sediment interface based on 1000 samples) at 2 m, 16 times background at 6 m, and 48 times background at 46 m (Figure 5). Similar results have been shown in the offshore Gulf of Mexico by Anderson et al. (1983) in deep gravity cores near active seeps. Total gas concentrations increased over 100 times from 1 to 10 m (Figure 6). Note that the magnitude of concentrations and changes are very different in the active and passive seep zones.

Total hydrocarbon gas concentrations are usually dominated by the methane, which is 90–99% of the total gas. Examination of wet gas concentrations only (ethane, propane, butane, and pentane obtained from headspace and cuttings) also often reveal increases with depth (Figure 6).

Water Column Versus Surficial Sediments

Brooks et al. (1979) examined both sediment and water column profiles near several active seeps in the offshore Gulf of Mexico. Both profiles indicated that anomalous hydrocarbons were present, but the type of hydrocarbons were quite different. The water column hydrocarbons of the near-bottom waters appeared to be biogenic in origin, based on the low concentrations of ethane and propane relative to methane. In contrast, the sediment sample hydrocarbons appeared to be thermogenic, based on the high concentrations of ethane and propane relative to methane. Brooks et al. (1979) explained that the differences result from molecular fractionation, that is, higher molecular weight hydrocarbons are preferentially retained in the sediments relative to methane. The seepage could be from microbial sources in which small quantities of ethane and propane are concentrated by preferential retention. Conversely, the seepage could be from thermogenic sources at depth undergoing molecular fractionation, which would account for the higher C_1 to C_2 + C_3 ratios.

Similar results have been noted near passive seeps. Water column surveys in several offshore Alaskan basins did not detect thermogenic hydrocarbons, whereas deep core samples collected in areas of subsurface hydrocarbon accumulations contained thermally derived high molecular weight hydrocarbons (C_{10+}).

Adsorbed Low Molecular Weight Hydrocarbons

Adsorbed hydrocarbons, also called bound, sorbed, or acid extraction (see analytical section for details), are generally relatively consistent in downcore profiles near active and passive seeps. Faber et al. (1990) examined adsorbed and cuttings (loosely bound) hydrocarbon

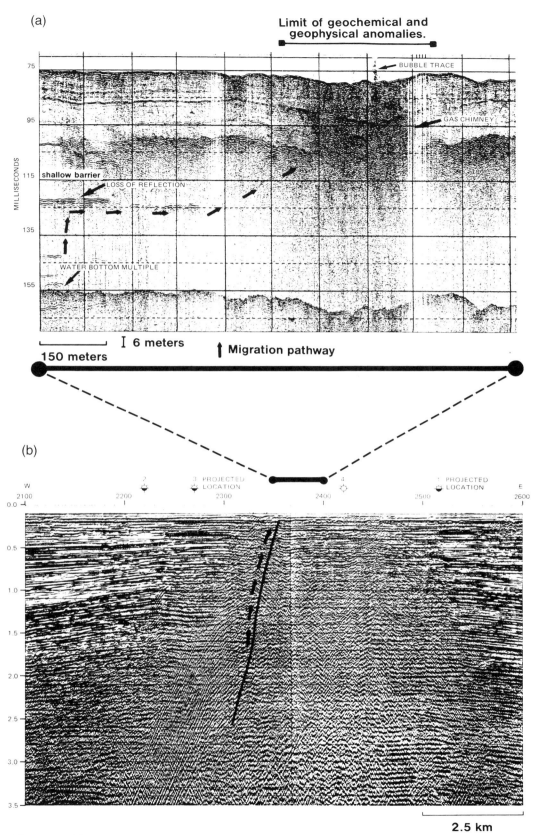

Figure 3—(a) Local-scale and (b) regional-scale seismic profiles showing anomalous hydrocarbons offsetting major leakage.

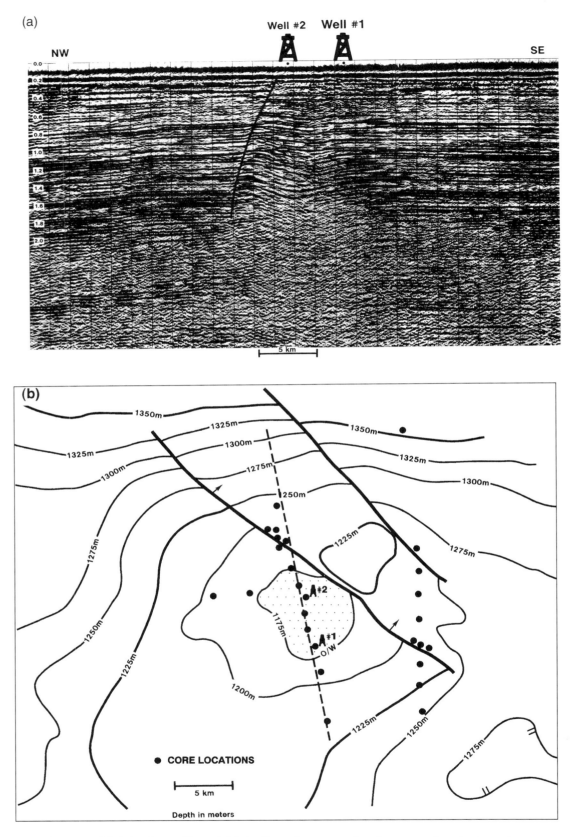

Figure 4—(a) Seismic profile and (b) depth contour map showing no surface hydrocarbon anomaly above a hydrocarbon accumulation, offshore Alaska.

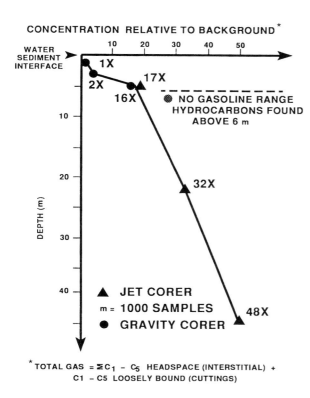

Figure 5—Concentration of low molecular weight hydrocarbons (C_1 to C_{5+}) versus depth, Bering Sea shelf, offshore Alaska, showing passive seepage. (From Abrams, 1993.)

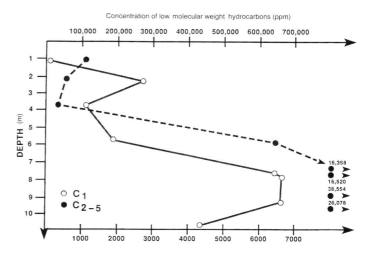

Figure 6—Concentration of low molecular weight hydrocarbons (C_1 to C_{5+}) versus depth, offshore Gulf of Mexico, showing active seepage. (From Anderson et al., 1983.)

gases in a 5-m core from the Gulf of Mexico (Figure 7). Methane cuttings gas concentrations increased dramatically below 2.5 m. The ethane-plus hydrocarbons (C_{2+}) were absent in the shallow part (less than 2.3 m) of the core and present in only minor amounts in the deeper part of the core. Gas concentrations for the adsorbed methane and ethane-plus hydrocarbons showed little variation in concentration over the same intervals.

In addition, the cuttings methane carbon isotopic ratio was noticeably heavier in the shallow part of the core (less than 2.0 m), whereas the adsorbed methane carbon isotopic ratios were more consistent throughout the core. Abrams (1989) had similar results in shallow gravity cores. Most likely, the heavier isotopic ratios in the cuttings methane results from bacterial oxidation where residual methane is enriched in ^{13}C. This is generally not the case for adsorbed methane carbon isotopic ratios.

High Molecular Weight Hydrocarbons

High molecular weight hydrocarbons (C_{10+}) also increase with depth. Near passive seeps, migrating high molecular weight hydrocarbons were only detected 10 and 6 m below the water–sediment interface in two different offshore Alaskan basins. Stained samples collected below the 10-m threshold had molecular signatures similar to the subsurface hydrocarbon accumulations and

interpreted source rocks. Submersible dives in areas of active seepage, such as in the Santa Barbara Channel and Gulf of Mexico, demonstrate high molecular weight hydrocarbons seeps on and above the water–sediment interface. Visual examination of stained cores in areas of active seepage show high molecular weight hydrocarbons throughout the cores but the highest concentrations in major fractures.

Transient Nature of Leakage

Surface hydrocarbon concentrations can vary significantly with time. Many of these variations may be related to diurnal cycles or to changes within the subsurface accumulation. Surface hydrocarbon seeps and anomalies have been shown to appear and disappear in relatively short times (weeks to months). Studies by Glotov (1992) indicate that solar and lunar cycles and seasonal soil condition changes greatly affect measured low molecular weight hydrocarbons. Sivaborvon (1974) reported that a surface geochemical anomaly increased in intensity over Hilbig field in Bastrop County, Texas, during repressurization for secondary recovery studies. Coleman et al. (1977) demonstrated leakage of gas from 1000-m-deep underground storage reservoirs into shallow-water wells within a 1-year period. Horvitz (1985b) collected surface sediment cores in 1946 over the Hastings field in Brazoria County, Texas, and the halo pattern of anomalous hydrocarbon values outlined the producing area. Then in 1968, Horvitz recollected shallow-sediment samples after the field was depleted. The 1968 survey showed no significant pattern of anomalous hydrocarbon values. Similar postproduction offshore surveys are scarce due to the high cost of marine sampling programs. All these observations clearly indicate that seepage and migration are dynamic (ongoing).

Figure 7—Adsorbed hydrocarbon profiles for C_1, C_{2+}, and $\delta^{13}C_1$ with depth, offshore Gulf of Mexico. (Adapted from Faber et al., 1990.)

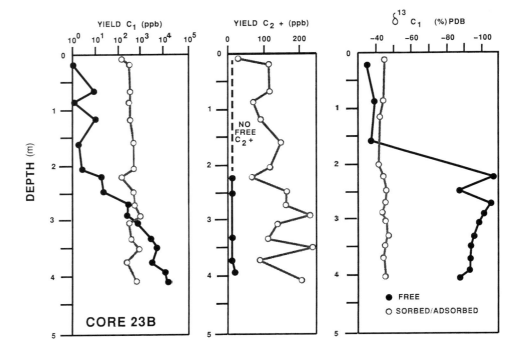

SURFACE GEOCHEMICAL PROGRAM FOR SEEPAGE DETECTION

Sampling Procedures

The sampling program for detection of seepage should fit the level of seepage activity and be designed to answer the key exploration questions. Surface geochemical samples have historically been collected according to a *grid pattern*, with a pre-set distance between sample sites, or as *site-specific*, in which the sites are based on specific geologic features. The data collection and subsequent evaluation differ for each collection process.

Geochemical data collected on a grid survey can be plotted on a map and contoured. Types of hydrocarbon geochemical data mapped generally include key low molecular weight hydrocarbon (methane, ethane, propane, and *n*-butane) concentrations, ratios of methane to ethane-plus hydrocarbons, and ratios of saturate to unsaturate low molecular weight hydrocarbons. Nonhydrocarbon data can also be contoured (mineralogical, geobotanical, and geothermal measurements). Contour patterns are examined to determine if a recognizable pattern is present above a subsurface feature.

Site-specific surveys use geological and geophysical data to look for evidence of shallow leakage, such as acoustic anomalies, gas bubbles, hydrates, and chemosynthetic communities. Instead of collecting core samples on a random grid, samples are collected at the defined leakage areas in the hope of collecting sufficient migrated thermal hydrocarbons for detailed chemical characterization.

Grid pattern surveys may be effective where seeps are very active but are not effective with passive seepage.

Grid surveys usually fail to detect subsurface accumulations and subsurface hydrocarbon generation in areas of passive seepage. Site-specific surveys are effective with both active and passive seepage and are thus the preferred approach.

The key exploration questions to be addressed should also play a major role in determining the type of survey. Grid surveys rarely detect major macroseepage because the sampling is random and it most likely misses major leak points. If the goal is to obtain high molecular weight hydrocarbons for oil source characterization, then use a site-specific survey in all situations but especially in areas of passive seepage.

Sampling Equipment

The type of sampling equipment used should be suited to the seepage activity (Figure 8). Many devices are used to collect seepage samples. Deep sediment cores (greater than 10 m) can be obtained from rotary/jet core equipment. Rotary/jet core devices are depth-limited only to the extent of unconsolidated sediments or permit restrictions. Shallow-sediment cores (less than 10 m) can be obtained from gravity-driven devices such as a piston or open barrel corer. Gravity core recoveries are limited by equipment design and sediment lithologies (Abrams, 1982). Gravity corers rarely recover sediments from 4–5 m below the water-sediment interface in most shelfal sediments. Near-bottom waters can be sampled by towed devices such as sniffers (Sackett, 1977) or by cabled devices such as Niskin and Nansen bottles, or more elaborately by manned or unmanned submersibles.

Airborne samplers and detectors vary in the type and sensitivity of slick detection. They include laser fluores-

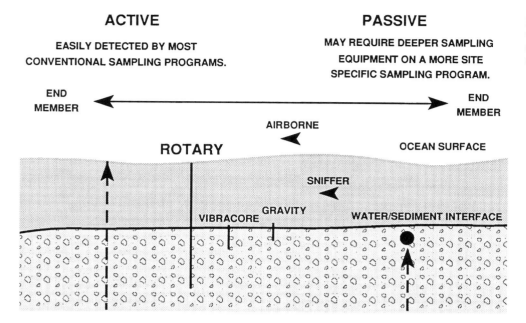

Figure 8—Sampling procedures and equipment used for seepage activity ranging from active to passive.

cence (such as ALF), satellite imagery (MacDonald et al., 1993), radar, air samplers (such as AIRTRACE), and visual inspection for slicks from slow-traveling aircraft. It should be realized that not all slicks result from seabed seepage; some are caused by pollution (oil spills). These remote airborne techniques do not provide a physical sample of the hydrocarbons to permit authentication of the slick as a natural seep. This represents a serious shortcoming unless these methods are followed up by surface "truthing."

Shallow and deep sediment samplers, water column samplers, and airborne techniques can detect low and high molecular weight hydrocarbon anomalies near active seeps, but passive seeps require deep sediment samples for detection of these anomalies.

Analytical Procedures

Low Molecular Weight Hydrocarbons

Several analytical procedures analyze low molecular weight hydrocarbons in surficial marine sediments. The most common is *headspace gas* analysis, also called interstitial gas. Gas is sampled through a silicone septum on the top of the can, which contains a specified amount of sediment. The can is shaken and sometimes heated prior to sampling to release interstitial gases contained within the pore spaces. *Cuttings gas* analysis, also known as loosely bound gas, uses a blender to break up a specified amount of sediment and release the loosely bound gases contained in the unconsolidated sediments. The last and most controversial technique is *adsorbed gas* analysis. In this process, the coarse-grain fraction is removed by wet sieving 300–1000 g of a bulk sediment sample. The fine-grain portion (63 μm and smaller) is heated in phosphoric acid in a partial vacuum to remove the "bound" hydrocarbons (Horvitz, 1985a).

Adsorbed and *total free* (headspace and cuttings) low molecular weight hydrocarbons display different vertical and horizontal distribution patterns (Faber et al., 1990). Abrams (1989) examined methane carbon isotopes from both headspace and adsorbed methane. He found that the adsorbed methane carbon isotopes matched the subsurface hydrocarbons, whereas the free methane carbon isotopic ratios ranged from thermogenic to biogenic. Horvitz was the first to report this phenomenon (Horvitz, 1982). Horvitz's studies concentrated in shallow soils above subsurface hydrocarbon accumulations where large volumes of *in situ* biological gas overwhelmed the migrated hydrocarbon signature. Horvitz believed that biologically generated hydrocarbons are retained in the interstitial portion of soils and marine sediments, whereas thermally formed low molecular weight hydrocarbons preferentially adsorb on the clays in the sediment.

An alternative hypothesis is that the biologically derived interstitial methane may actually result from methanogenic bacteria effectively feeding on seeping hydrocarbons. Faber et al. (1990) proposed that free and adsorbed gases do not exchange, thus minimizing masking effects of the more abundant biogenic gas on the adsorbed thermogenic hydrocarbons. Studies by Thompson (1987) indicate that the adsorbed sediment gas is trapped in carbonate minerals, probably as fluid inclusions. Similar conclusions were reached by Pflaum (1989) during studies of bound (adsorbed) hydrocarbons. Horvitz and Ma (1988) ground the coarse-grained fraction and extracted hydrocarbons using the standard adsorbed extraction process. Results were similar to the fine-grained fraction. Both studies suggest that the clay fraction is not the adsorbing medium, as Horvitz originally believed, but that the hydrocarbons may actually be trapped in carbonate fluid inclusions, as Thompson (1987) suggested. The carbonate could be authigenic, thus formed as a direct result of the seepage. Recent studies in the Gulf of Mexico demonstrate large carbonate buildups that are most likely a direct result of active hydrocarbon seepage (Roberts et al., 1989).

Areas of active seepage generally show large concentrations of both headspace and cuttings low molecular weight hydrocarbons. Areas of passive seepage generally show low headspace and cuttings low molecular weight hydrocarbon concentrations, which may not be detectable above background levels. Adsorbed hydrocarbon concentrations are generally much lower than headspace and cuttings concentrations. Furthermore, adsorbed hydrocarbon concentrations are also lower in passive seep zones than in active seep zones.

High Molecular Weight Hydrocarbons

Mature high molecular weight hydrocarbons (C_{10+}) are conclusive evidence of a working source system within a basin. Detection of high molecular weight hydrocarbon anomalies within surficial marine sediments requires many of the same analytical procedures currently used in conventional oil and rock analyses: chemical extraction, gas chromatography, and gas chromatography–mass spectrometry. Unfortunately, *in situ* organic material can often complicate analytical results. Deciphering the migrated seep molecular signal from the *in situ* sedimentary molecular signal can be difficult, particularly in passive seepage areas. The *in situ* organic characteristics, which may be from recent deposition or reworked source rocks, can overprint any migrated thermogenic hydrocarbons from depth. This is demonstrated in an example from a passive seepage area shown in Figure 7. The migrated hydrocarbon signature degrades quickly as the hydrocarbon concentrations decrease away from the fault. Similar studies in active seepage areas such as the Gulf of Mexico (A. G. Requejo, personal communication, 1994) indicate at least 350–500 ppm of extract threshold is required to see clearly through the sedimentary background. Another way to circumvent *in situ* biological overprinting is to look at compounds that cannot be produced biologically. Requejo suggests looking at selected aromatic compounds that are associated more with thermal processes.

Fluorescence spectroscopy uses ultraviolet excitation to cause fluorescence of organic compounds containing one or more aromatic functional groups. Fluorescence spectroscopy is not normally used in conventional exploration geochemistry, but it has been used for many years to detect anomalous hydrocarbons in marine waters and sediments. Kartsev et al. (1959) described fixed- and single-wavelength scanning fluorescence for geochemical prospecting. Whelan et al. (1977) used fixed emission at 265 nm and scanned excitation wavelengths at 240–500 nm to characterize aromatic mixtures that may have been related to reservoir leakage. Horvitz (1980) looked at the ratio of 320-nm and 365-nm emissions for excitation at 265 nm (R_1) to characterize oil versus gas. Texas A&M (Brooks et al., 1983) used an emission and excitation spectrum for wavelengths between 200 and 800 nm in total scanning spectrofluorescence (TSF). TSF gen-

erates three-dimensional contours to represent the varying emissions graphically.

Published fixed-wavelength and TSF distribution patterns indicate that fluorescence spectroscopy can provide surface distribution patterns and TSF signatures very similar to subsurface oil accumulations and oils (Brooks et al., 1983) in areas of active seepage. My own unpublished studies indicate that fixed- and single-wavelength and TSF data are often misleading in areas of passive seeps. *In situ* organic compounds from anthropogenic sources (relict hydrocarbons) often indicate that migrated hydrocarbon seepage is present with spectrofluorescence when conventional extraction and gas chromatography do not reveal thermal hydrocarbon compounds.

Alteration of Migrated Hydrocarbons

Seeping petroleum is usually degraded by bacteria. At temperatures below 70°C, both oil and gas offer a rich nutrient source for bacteria. Degradation is carried out primarily by aerobic bacteria, but anaerobes are also involved in feeding off the by-products of the aerobes. Collection of geochemical samples below the zone of aerobic activity helps to minimize degradation problems. Areas of active seepage often have few alteration problems due to the active movement of hydrocarbons, whereas areas of passive seepage typically show severe degradation.

DISCUSSION

The vertical and horizontal distribution of hydrocarbons in surficial marine sediments can be used to explain the wide range of observations. The variation in concentration and type of hydrocarbons is directly related to seepage activity. Areas of active seepage, such as offshore Gulf of Mexico, have relatively large concentrations of migrated hydrocarbons throughout the shallow marine sediments, within the water column, and on the ocean surface. Detection of active seepage is usually relatively easy because of the presence of large concentrations of hydrocarbons. In contrast, areas of passive seepage, such as offshore Alaska, rarely show anomalous hydrocarbons above the zone of maximum disturbance (see discussion below) or shallow migration barriers, and never in the water column or ocean surface. There are several possible explanations for why passive seepage is more difficult to detect:

1. **Lower concentration of hydrocarbons**—The lower concentration of hydrocarbons may be due to relict hydrocarbon generation, and movement of hydrocarbons may be related to tectonic activity.
2. **Alteration in very shallow surficial sediments**— The *zone of maximum disturbance* (Figure 9), first introduced by Abrams (1989, 1992), is where selected shallow-sediment processes, such as aerobic bacterial activity, pore water flushing, and loss of volatile hydrocarbons, alters the composition of

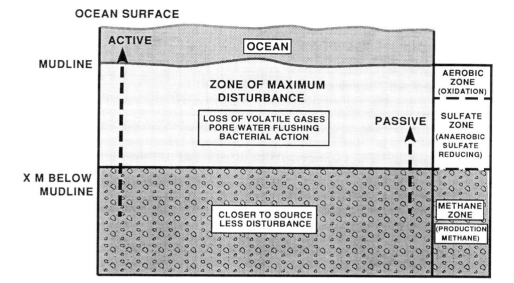

Figure 9—Schematic diagram of vertical variability in shallow-marine sediments showing the zone of maximum disturbance.

migrated hydrocarbons. Migrated subsurface hydrocarbons collected within this zone are usually altered by these *in situ* processes so that lower levels of seepage are difficult to detect.

3. **Local generation of low molecular weight hydrocarbons**—The generation of methane and higher homologs within the sulfate-reducing zone (Figure 9) by anaerobic respiration can mask migrated hydrocarbons. In addition, recent studies by Vogel et al. (1982) and Min'ko (1991) indicate that relatively low concentrations of C_2 to C_4 saturate and unsaturate hydrocarbons may be formed by low-temperature processes mediated by microorganisms. These *in situ* generated, nonmigrated hydrocarbons can often be mistaken for migrated hydrocarbons. Interpretations should be made only for hydrocarbon concentrations that are significantly above background concentrations.

4. **Migration barriers**—Some physical barriers, faults, and fractures may prevent detectable hydrocarbons from migrating into surface sediments (Figure 10). Deep permafrost hydrate layers act as top seals in the North Slope region of Alaska (Collett and Cunningham, 1994). Faults and fractures can act as migration conduits, but do not always reach the water–sediment interface.

Active and passive seepage as defined here are two distinct end-members. Many seeps actually represent a spectrum of seepage behavior between these two end-members. Thus, the interpreter should know that not all seeps are alike. Techniques and concepts developed to detect and collect seeps in one area may not work in others. In addition, seeps seldom reflect the producibility, size, and type of hydrocarbon accumulations at depth. Understanding the basin setting and geologic history relative to the surface geochemical data offers insight into the appropriate interpretation.

CONCLUSIONS

Chemically identifiable subsurface leakage into near-surface marine sediments falls into a spectrum between two end-members:

1. Active seepage is defined as ongoing seepage of migrated hydrocarbons in areas with active hydrocarbon generation and/or excellent migration pathways or poor seals.
2. Passive seepage is defined as relict seepage of migrated hydrocarbons in areas with passive hydrocarbon generation and/or poor migration pathways or excellent seals.

The type of seepage activity controls the near surface seepage distribution in marine sediments. In areas of active seepage, migrated subsurface hydrocarbons can be detected in near-surface sediments, in the water column, and on the ocean surface at the leak points and at relatively large distances from major leak points. In areas of passive seepage, migrated subsurface hydrocarbons can be detected only in near-surface marine sediments below the zone of maximum disturbance and at close proximity to major leak points.

The sampling program should reflect seepage activity, geologic setting, and key questions to be addressed by the survey. In areas of active seepage, shallow boring, water column, and airborne detection techniques are effective, whereas in areas of passive seepage, deep coring equipment is required to detect seepage. The sampling program should also reflect the type of study: regional versus prospect.

A lack of surface seepage does not necessarily mean that subsurface hydrocarbon accumulation or generation is not present but rather that the leakage could not be detected with conventional surface exploration tools.

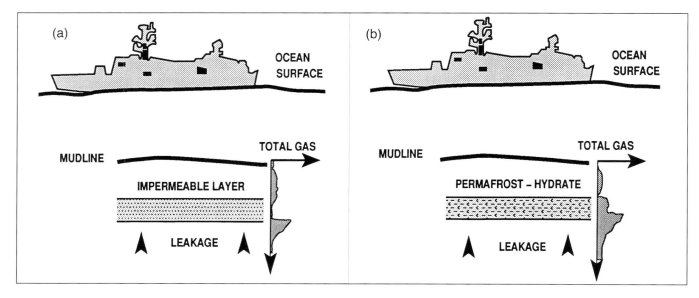

Figure 10—Schematic cross sections showing shallow migration barrier effect in (a) a low-porosity, low-permeability zone and (b) a permafrost hydrate layer.

Acknowledgments—The author gratefully acknowledges Exxon Exploration Company, Exxon Company USA, Exxon Production Research Company, and Exxon Company International for funding the many programs that provided the data for this paper. Informative discussions and critical review of the manuscript by W. A. Young, Neil Piggott, and Deet Schumacher are also gratefully acknowledged. I also thank the late Leo Horvitz with whom I had many invaluable discussions. Leo's studies pioneered surface geochemistry as a viable exploration tool. The figures were drafted by Judy Watson.

REFERENCES CITED

Abrams, M. A., 1982, Modifications for increasing recovery and penetration in an open barrel gravity corer: OCEANS 82 Conference Records 82CH1827.5, p. 661–666.

Abrams, M. A., 1989, Interpretation of methane carbon isotopes extracted from surficial marine sediments for detection of subsurface hydrocarbons: Association of Petroleum Geochemical Explorationists Bulletin, v. 5, p. 139–166.

Abrams, M. A., 1992, Geophysical and geochemical evidence for subsurface hydrocarbon leakage in the Bering sea, Alaska: Marine and Petroleum Geology Bulletin, v. 9, p. 208–221.

Andersen, R. K., R. S. Scalan, P. L. Parker, and E. W. Behrens, 1983, Seep oil and gas in Gulf of Mexico slope sediments: Science, v. 222, p. 619–621.

Bernard, B. D., 1978, Light hydrocarbons in marine sediments: Ph.D. dissertation, Texas A&M University, College Station, Texas, 144 p.

Brooks, J. M., and B. D. Carey, 1986, Offshore surface geochemical exploration: Oil and Gas Journal, v. 84, p. 66–72.

Brooks, J. M., B. B. Bernard, W. M. Sackett, and J. R. Schwarz, 1979, Natural gas seepage on the south Texas shelf: Proceedings of the Eleventh Annual Offshore Technology Conference, Houston, Texas, OTC-3411, p. 471–478.

Brooks, J. M., M. C. Kennicutt II, L. A. Bernard, G. J. Genoux, and B. D. Carey, 1983, Applications of total scanning fluorescence to exploration geochemistry: Offshore Technology Paper, OTC-4624, p. 393–400.

Carlson, P. R., M. S. Golan-Bac, H. A. Karl, and K. A. Kvenvolden, 1985, Seismic and geochemical evidence for shallow gas in sediment on Navarin continental margin, Bering Sea: AAPG Bulletin, v. 69, p. 422–436.

Clayton, C., 1992, Gas migration mechanisms from accumulation to surface: Shallow Gas Newsletter, no. 6.

Coleman, D. D., W. F. Meents, C.-L. Liu, and R. A. Keough, 1977, Isotopic identification of leakage gas from underground storage reservoirs: a progress report: Illinois State Geological Survey, Illinois Petroleum, no. 111, p 10.

Collett, T. S., and K. I. Cunningham, 1994, Geologic controls on gas migration within northern Alaska (abs.), *in* Near-surface expressions of hydrocarbon migration: AAPG Hedberg Research Conference Abstracts, April 24–28, Vancouver, British Columbia.

Faber, E., and W. Stahl, 1984, Geochemical surface exploration for hydrocarbons in North Sea: AAPG Bulletin, v. 68, p. 363–386.

Faber, E. W., J. Stahl, M. J. Whiticar, J. Lietz, and J. M. Brooks, 1990, Thermal hydrocarbons in Gulf Coast sediments, *in* Gulf Coast oils and gases: Proceedings of the Ninth Annual Research Conference, SEPM and Mineralogist Foundation, New Orleans, October 1, p. 297–307.

Fischer, P. J., and A. J. Stevenson, 1973, Natural hydrocarbon seeps along the northern shelf of the Santa Barbara Basin, California: Offshore Technology Conference, OTC-1738, p. 159–166.

Gevirtz, J. L., B. D. Carey, and S. R. Blanco, 1983, Surface geochemical exploration in the North Sea, *in* J. Brooks, ed., Petroleum geochemistry and exploration of Europe: Geological Society of London, Special Publication, no. 12, p. 35–50.

Glotov, V. E., 1992, Geochemical prospecting for oil and gas fields in the NE USSR (CIS): Journal of Petroleum Geology, v. 15-3, p. 345–358.

Hitchon, B., 1974, Application of geochemistry to the search for crude oil and natural gas, *in* A. A. Levinson, ed., Introduction to exploration geochemistry: Calgary, Alberta, Applied Publishing Ltd., p. 509–545.

Hovland, M., and A. G. Judd, 1988, Seabed pockmarks and seepage: London, Graham and Trotman, p. 293.

Horvitz, L., 1939, On geochemical prospecting: Geophysics, v. 4, p. 210–228.

Horvitz, L., 1969, Hydrocarbon prospecting after thirty years, *in* W. B. Heroy, ed., Unconventional methods in exploration for petroleum and natural gas: Dallas, Texas, Southern Methodist University Press, p. 205–218.

Horvitz, L., 1980, Near-surface evidence of hydrocarbon movement from depth, *in* W. H. Roberts III and R. J. Cordell, eds., Problems of petroleum migration: AAPG Studies in Geology 10, p. 241–269.

Horvitz, L., 1981, Hydrocarbon prospecting after forty years, *in* B. M. Gottleib, ed., Unconventional methods in exploration for petroleum and natural gas II: Dallas, Texas, Southern Methodist University Press, p. 83–95.

Horvitz, L., 1982, Upward migration of hydrocarbons from gas and oil deposits: Paper presented at American Chemical Society Division of Geochemistry, 183rd ACS National Meeting, March 28–April 2, Las Vegas, Nevada.

Horvitz, L., 1985a, Geochemical exploration for petroleum: Science, v. 229, p. 812–827.

Horvitz, L., 1985b, Near-surface hydrocarbons and non-hydrocarbon gases in petroleum exploration, *in* R. W. Klusman, ed., Surface and near-surface geochemical methods in petroleum exploration: Association of Petroleum Geochemical Explorationists, Special Publication 1, p. D1–D52.

Horvitz, E. P., and S. Ma, 1988, Hydrocarbons in near-surface sand, a geochemical survey of the Dolphin field in North Dakota: Association of Petroleum Geochemical Explorationists Bulletin, v. 4, n. 1, p. 30–46.

Laubmeyer, G., 1933, A new geophysical prospecting method, especially for deposits of hydrocarbons: Petroleum, v. 29, no. 18, p. 1–4.

Jones, V. T., and R. J. Drozd, 1983, Predictions of oil and gas potential by near-surface geochemistry: AAPG Bulletin, v. 67, p. 932–952.

Kartsev, A. A., Z. A. Tabasaranskii, M. I. Subbota, and G. A. Mogilevskii, 1959, Geochemical methods for prospecting and exploration for petroleum and natural gas (translated by P. A. Witherspoon and W. D. Romney, eds.): Los Angeles, University of California Press, 349 p.

Kennicutt, M. C., and J. M. Brooks, 1988, Relation between shallow sediment bitumens and deeper reservoired hydrocarbons, offshore Santa Maria Basin, California, U.S.A.: Applied Geochemistry, v 3, p. 573–582.

Kennicutt, M. C., J. M. Brooks, R. R. Bidogare, and G. J. Denoux, 1988, Gulf of Mexico hydrocarbon seep communities—I. regional distribution of hydrocarbon seepage and associated fauna: Deep Sea Research, v. 35, no. 9, p. 1639–1651.

Kettel, D., 1990, Physical and geological implications in surface geochemcial exploration: Bulletin of Swiss Association Petroleum Geology Engineering, v. 55, p. 27–40.

Klusman, R. W., 1989, Surface and near-surface geochemistry in petroleum exploration: 1989 Symposium of the Association of Petroleum Geochemical Explorationists, Denver, Colorado, July 26–27.

Klusman, R. W., 1993, Soil gas and related methods for natural resource exploration: New York, John Wiley and Sons, 483 p.

Kvenvolden, K. A., and M. A. McMenamin, 1980, Hydrates of natural gas: a review of their geologic occurrence: USGS Circular 825, 11 p.

Kvenvolden, K. A., and M. F. Field, 1981, Thermogenic hydrocarbons in unconsolidated sediment of Eel River Basin, offshore northern California: AAPG Bulletin, v. 65, no. 9, p. 1642–1646.

MacDonald, I. R., G. S. Boland, J. S. Baker, J. M. Brooks, M. C. Kennicutt, and R. R. Bidigare, 1989, Gulf of Mexico hydrocarbon seep communities, II: spatial distribution of seep organisms and hydrocarbons at Bush Hill: Marine Biology, v. 101, p. 235–247.

MacDonald, I. R., N. L. Guinasso, Jr., S. G. Ackleson, J. F. Amos, R. Duckworth, R. Sassen, and J. M. Brooks, 1993, Natural oil slicks in the Gulf of Mexico are visible from space: Journal of Geophysical Research, 98-C9, p. 16351–16364.

Min'ko, O. I., 1991, Generation of hydrocarbon gas by planetary soil cover: Geochemistry International, v. 28, no. 8, p. 2–12.

Pflaum, R. C., 1989, Gaseous hydrocarbons bound in marine sediments: Ph.D. dissertation, Texas A&M University, College Station, Texas, p. 155.

Price, L. C., 1986, A critical overview and proposed working model of surface geochemical exploration, *in* M. J. Davidson, ed., Unconventional methods in exploration for petroleum and natural gas—IV: Dallas, Texas, Southern Methodist University Press, p 245–304.

Roberts, H. H., R. Sassen, R. Carney, and P. Aharon, 1989, Carbonate buildups on the continental slope off central Louisiana: Offshore Technology Conference Proceedings, OTC-5953, p. 655–662.

Rosaire, E. E., 1940, Geochemical prospecting for petroleum: AAPG Bulletin, v. 24, p. 1418–1426.

Reitsema, R. H., F. A. Linberg, and A. J. Kaltenback, 1978, Light hydrocarbons in Gulf of Mexico water: sources and relation to structural highs: Journal of Geochemical Exploration, v. 10, p. 139–151.

Sackett, W. M., 1977, Use of hydrocarbon sniffing in offshore exploration: Journal of Geochemical Exploration, v. 7, p. 243–254.

Sivaborvon, V., 1974, Re-study of hydrocarbon distribution around Hilbig oil field, Bastrop County, Texas: Master's thesis, University of Texas, Austin, Texas, p. 79.

Sokolov, V. A., 1935, Summary of the experimental work of the gas survey: Neftyanoye Khozyaystvo, v. 27, no. 5, p. 28–34.

Sweeney, R. E., 1988, Petroleum related hydrocarbon seepage in recent North Sea sediments: Chemical Geology, v. 71, p. 53–71.

Thompson, M., C. Reminton, J. Purnomo, and D. Macgregor, 1991, Detection of liquid hydrocarbon seepage in Indonesian offshore frontier basins using airborne laser fluorescence (ALF): the results of a Pertamina/BP joint study: Proceedings of the Twentieth Annual Convention, Indonesian Petroleum Association, October, IPA 91-11.15, p. 664–689.

Thompson, R. ,1987, Relationship between mineralogy and adsorbed hydrocarbons in soils and sediments: ISEM-SMU Conference on vertical migration, Ft. Burgwin, New Mexico, October 15–18.

Vogel, T. M., R. S. Oremland, and K. A. Kvenvolden, 1982, Low-temperature formation of hydrocarbon gases in San Fransisco Bay sediment: Chemical Geology, v. 37, no. 3/4, p. 289–298.

Voytov, G. I., R. G. Grechukhina, and V. S. Lebedev, 1972, Chemical and isotopic composition of gas and water in southern Dagestan: Geochemistry, v. 205, p. 1217–1220.

Whelan III, T., J. M. Coleman, and J. N. Shuayada, 1975, The geochemistry of recent Mississippi river delta: gas concentrations and sediment stability: Proceedings of the 7th Offshore Technology Conference, v. 3, p. 71–84.

Whelan III, T., D. B. Purvis, G. Hart, and J. L. Albright, 1977, Seismic, fluorimetric, and chromatographic studies of reservoir leakage: Proceedings of the 9th Offshore Technology Conference, paper no. 2937, p. 453–458.

Whiticar, M. J., E. Suess, and H. Wehner, 1985, Thermogenic hydrocarbons in surface sediments of the Bransfield Strait, Antartica Peninsula: Nature, v. 314, no. 7, p. 87–90.

Zhizhchenko, B. P., 1978, Generation of hydrocarbon gases in recent marine sediments: International Geology Review, v. 21, no. 4, p. 163–170.

Kaluza, M. J., and E. H. Doyle, 1996, Detecting fluid migration in shallow sediments: continental slope environment, Gulf of Mexico, *in* D. Schumacher and M. A. Abrams, eds., Hydrocarbon migration and its near-surface expression: AAPG Memoir 66, p. 15–26.

Chapter 2

Detecting Fluid Migration in Shallow Sediments: Continental Slope Environment, Gulf of Mexico

Michael J. Kaluza

Fugro-McClelland Marine Geosciences, Inc.
Houston, Texas, U.S.A.

Earl H. Doyle

Shell Offshore, Inc.
Houston, Texas, U.S.A.

Abstract

The detection of shallow gas features on the northern Gulf of Mexico continental slope has been aided with a unique positively buoyant deep-towed subbottom profiler and side-scan sonar system. The tool provides high-quality, high-resolution seismic displays of the shallow (upper 75 m) stratigraphy and seafloor images (400-m swath) capable of resolving geologic features that may constrain exploration drilling or engineering development of potential petroleum reserves.

From the more than 20,000 km (~10,000 nmi) of deep-tow data collected, numerous encounters of shallow gas features have been made. Gas and other fluid vents have been seen on the seafloor in association with seafloor and shallow buried fault systems. In some cases, vents have been identified by distinctive seafloor topography expressed as hills and mounds and by seafloor depressions, craters, and blister-like features. No distinctive topographic irregularities occur at other seafloor vent areas, which are identified primarily by the seismic character of the records. Shallow subsurface gas has been identified by the amorphous and wiped-out character of stratified sequences on subbottom profiler data and by high-amplitude "bright spot" reflections within sediment packages. Gas flow, both vertically along fault planes and laterally along permeable sediment layers, can be identified from these types of data.

INTRODUCTION

The northern Gulf of Mexico continental slope offers the petroleum industry a boost in domestic oil exploration opportunities. Improved methods in 3-D seismic data acquisition and processing techniques have helped explorationists identify potential petroleum reserves in the large subsurface basins and canyons created along the seaward extents of large salt bodies. Just as these new methods have helped explorationists, new developments in collecting high-resolution shallow seismic data, necessary for assessing shallow soil and geologic conditions for production design, have helped engineering geologists.

Conventional surface-towed high-resolution seismic equipment, commonly used on the continental shelf, cannot provide high-quality data when operated much beyond the shelf-slope break. Identification of potential geologic constraints, such as seafloor faults, shallow gas, fluid expulsion events, and slope instability features, may go totally undetected using these conventional tools. A better method for collecting high-quality subbottom profiler and side-scan sonar images of the seafloor, using a positively buoyant deep-towed vehicle, has been developed and refined over the past decade. This paper describes the geologic features identified by this unique system that have been interpreted to represent shallow gas. We discuss only shallow geologic features seen in clay-rich sediments on the continental slope of the Gulf of Mexico.

DEEP-TOW SYSTEM

The deep-tow system used to collect the data examples shown in this paper is capable of collecting high-resolution subbottom profiler and side-scan sonar data in water depths ranging from about 300 to 3000 m (~1000–10,000 ft). It was developed by EDO Western Corporation in the late 1970s and was put into routine deep-water seismic surveying operations in the Gulf of Mexico in 1986 after

Figure 1—The deep-tow side-scan sonar and subbottom profiler tow vehicle shown during deployment exercises. The tow vehicle measures about 4 m in length and 1 m in diameter.

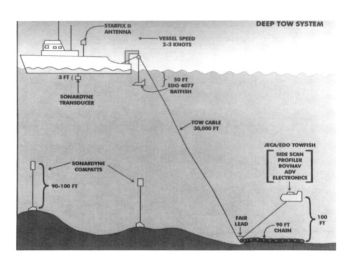

Figure 2—Schematic drawing of deep-tow operations (revised from Prior et al., 1988; Doyle et al., 1992). The positively buoyant tow vehicle is pulled by a 9000-m coaxial cable and is maintained at a constant height above the seabed by coupling a tow chain to a fair lead assembly ~30 m up the coaxial cable. Side-scan sonar and subbottom transmissions are multiplexed and transmitted up and down the coaxial tow cable. The Sonardyne Compatt network is deployed and calibrated before the survey begins. All tow vehicle navigation is multiplexed and transmitted up the tow cable, along with the seismic information, and interfaced with the ship's Starfix navigation.

first being used off the Atlantic east coast from 1981 to 1984 (Prior and Doyle, 1984). Since then, more than 20,000 km (~10,000 nmi) of high-quality shallow seismic data have been collected with this type of positively buoyant deep-tow system. The operational aspects of this tool have been discussed by Prior and Doyle (1984) and Prior et al. (1988).

The system consists of a positively buoyant tow vehicle coupled to the seismic vessel by a 9000-m (~30,000-ft) coaxial electrical towing cable (Figure 1). The tow body contains a subbottom profiler transducer with a frequency that is interchangeable from 3.5 k to 7.0 kHz. Dual 100-kHz side-scan sonar transducers provide planar images of the sea floor 200 m (~600 ft) out on either side of the deep-tow vehicle (400 m full swath). The images portrayed on the seismic recorders, and archived on optical disk, are analogous to low-oblique aerial photographs. The subbottom profiler transducer, typically set at 3.5 kHz frequency, provides penetration of the seafloor to as much as 75 m (250 ft) below the mud line. Resolution of individual seismic horizons is as good as 0.3 m (1 ft).

Since the deep-tow vehicle is a positively buoyant tow body, it must be counterweighted via a tow chain connected 30 m (100 ft) up the coaxial tow cable. A fair lead assembly with a weak-link swivel attachment joins the chain

with the electrical coaxial tow cable. The tow chain touches the sea floor, while the tow body always floats about 30 m (100 ft) above it, regardless of what the sea floor topographic changes may be. This technique provides a stable platform for the seismic transducers, maintaining the tow body at a consistent and optimum height above the sea floor at all times. The schematic drawing in Figure 2 shows the towing assembly for deep-tow operations.

This stable tow technique alleviates sidelobe distortion and seismic attenuation through long water column distances, as well as heave, pitch, roll, and yaw distortions to data quality common with negatively or neutrally buoyant tow vehicles requiring constant cable length changes in complex topography. In addition, towing the seismic tool near the sea bottom puts the transducers (and receivers) in a much quieter environment, thus the signal-to-noise ratio is very good. The result is high-quality, high-resolution sea floor and shallow seismic reflection data from which subtle geologic features and conditions can be assessed.

Positioning of the tow vehicle is done by using a long baseline acoustic seabed array of medium frequency transponders (Kelland, 1988; Prior et al., 1988). An acoustic transceiver within the deep-tow vehicle interrogates each transponder as the deep-tow vehicle traverses the array area. Distance measurements are relayed up the coaxial cable and integrated with the ship's satellite navigation systems (Starfix and DGPS). This provides accurate positioning for the deep-tow vehicle as well as for geologic features seen on the subbottom profiler and side-scan sonar records.

DEEP-WATER GEOLOGIC SETTING

A large inventory of deep-tow subbottom profiler and side-scan sonar data have been collected over the past 8 years across the northern Gulf of Mexico continental slope. These data have primarily been acquired to meet federal permitting requirements for petroleum exploration drilling and for engineering assessments of platforms and pipelines.

Many prospective oil and gas fields are situated near shallow salt diapirs. The thick Jurassic salt mass underlying much of the northern Gulf of Mexico is quite plastic and can be squeezed and extruded by the weight of the overlying sediment mass. The result is diapirs and salt spires being pushed upward and seaward through the sediment overburden (Martin, 1978). The salt mass is generally closer to the present sea floor along the upper and middle continental slope than on the continental shelf. The extruded salt is thus more evident on the present sea floor in the form of pronounced hills, ridges, basins, and troughs.

The salt movement causes stress on the ductile overburden, eventually breaking the sediment layers and forming faults. Salt evacuation can cause collapsed graben structures on the sea floor that generally trend for several thousands of feet over the uplifted and subsequently collapsed area. Fault displacement may continue with the continued flow of the salt, creating a growth-like character to the fault, which may reach the sea floor and continually or episodically cause further surface displacement and renewal of closed fault planes.

Many of the growth faults associated with diapiric activity show considerable sea floor expression. Since many oil and gas plays are closely associated with salt diapirs and the associated intraslope basins, the resulting faults become natural conduits for gas and other fluids to migrate upward. Fluid expulsion features consisting of gas, oil, mud, or brine have been identified in numerous locations along the continental slope in close association with many near-surface and deep-seated faults and diapiric structures (Roberts et al., 1990a; Prior and Doyle, 1993).

Fluids can continue to migrate along fault planes to the sea floor, or the fluids may migrate laterally away from the fault through permeable layers. When less dense sediment layers are interbedded between dense layers, migrating fluids may become partially blocked and the amount of flow may be choked from further upward migration. The fluids will then follow the path of least resistance along the permeable layer. Laterally migrating gas concentration typically shows higher amplitude reflection character in the permeable layer and especially along the upper interface of the layer with the overlying denser layer.

Pockets of gas-rich sediments are sometimes seen on geophysical records and associated with landslide deposits. The pockets may be evident, but no distinguishable path for migration may be seen. These pockets of gas-rich sediment appear to exist only where they are found and probably formed *in situ*. In many cases, gas migrating along fault planes or concentrated in permeable layers can occur in such small quantities that it may not be detectable using conventional seismic methods (lower frequency sound sources). Fault planes may be distinguishable on these data, but small volumes of migrating gas may not. The higher frequency subbottom profiler signal, however, may be distorted or wiped out or show high-amplitude "bright spot" responses when even minimal gas is present.

The shallow gas in deep water can be in any natural state of matter. Gas can be entrained in sediments in a liquid phase or in higher concentrations as a bubble or gaseous phase, or it can be occluded in a solid phase. Gas hydrate mounds form when proper temperature, pressure, and chemical composition for the gas is achieved (Neurauter and Bryant, 1989). Gas hydrates are solid, ice-like clathrate structures in which gas is compressed and concentrated between water molecules forming a crystalline water lattice. Up to 70 times the volume of gas can be concentrated in a solid hydrate for the same area encompassed by free gas (Sloan, 1990).

At other times, methane gas may enter into a chemical reaction in the shallow clay sediments to form solid carbonate rock as a by-product. These are known as authigenic carbonates and are quite common along the upper continental slope of the Gulf of Mexico, especially over diapirically uplifted areas and along sea floor fault displacements where petroleum seepage occurs. The basic process of carbonate precipitation involves the microbial oxidation of hydrocarbons at or near the water–sea floor interface to produce carbon dioxide and bicarbonate. Under proper conditions, this process speeds up the production of calcium and magnesium carbonates (Roberts et al., 1987, 1988, 1989, 1990b). Both hydrate and authigenic carbonate mounds lend their existence to the fluid migration of methane gas.

SEISMIC CHARACTER OF SHALLOW FLUID VENT FEATURES

This section provides insight into the varying types of vent-associated features recognized from deep-tow surveys. Venting and accumulated fluid flow have numerous effects on the shallow sediments and on sea floor character. Not all vents appear the same and not all vent features cause positive or negative relief. The fluids being vented through the sea floor may not necessarily always be gas. Any liquid or gaseous substance entrapped in the shallow sediment under pressures greater than hydrostatic will flow when a path is given. Exactly what a fluid vent will look like depends on the sediment environment, the amount of fluid in the

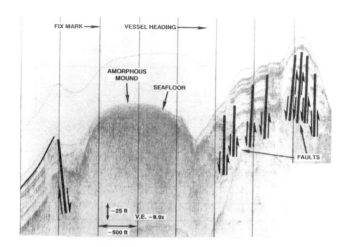

Figure 3—Deep-tow subbottom profiler record (3.5 kHz frequency) showing a rounded, dome-shaped, amorphous mound on top of a diapiric ridge and closely associated with faults (from Prior and Doyle, 1993). The mound is associated with fluid mud flows exiting active seeps to the sea floor, with the faults possibly supplying the conduits for escape from depth. Vertical scale is 7.6 m between timing lines (10 msec) and ~150 m between fix marks.

system, the time frame of vent activity, oceanographic conditions, and numerous other factors.

Many of the examples shown here are only inferred to be caused by gas expulsion or other fluid migration, while several have been verified by sampling techniques and manned submersible visual inspection. However, the seismic similarities of sites that have been verified provide confidence in our inferences made to nonsampled sites. We first describe the sea floor vent features and fol-

low with data examples of inferred shallow subsurface gas where there are no indications of fluids venting to the sea floor.

Sea Floor Mounds

Numerous sea floor mounds have been identified using the deep-tow tool. Typically, the mounds can be categorized by their shape, surface expression, and size. The mounds are assessed based on the environment they exist in to evaluate whether the mounds represent hard carbonate accumulation, soft free-venting gassy mud mounds, dormant vents, or hydrate mounds. Figures 3 through 6 illustrate several mounds sampled by the deep-tow subbottom profiler signal. The shapes of the mounds vary from well-rounded and smooth, to flat-topped, and to irregular and nonuniform. Mounds may rise only a few feet from the sea floor or may be as much as 150 m (500 ft) in elevation.

Sea floor mounds inferred to represent hydrate concentrations have generally been uniform in size, about 300–500 m (1000–1500 ft) in diameter, and have elevations ranging about 15–30 m (50–100 ft) above the sea floor (Figures 6, 7). They are generally rounded to oblong in shape (Figures 7, 8) and have a slightly irregular reflection character over their tops. Faults can be seen radiating from many inferred hydrate mounds. Hydrate mounds have been visited by manned submersibles and are usually covered by a thin drape of soft clay (Figure 9). Chemosynthetic communities are often associated with hydrate mounds (Brooks et al., 1987; Roberts et al., 1990a). Other round mounds with more subtle sea floor characteristics (Figure 3) may also represent hydrate accumulations. A slightly thicker drape sediment often overlies these solid hydrates.

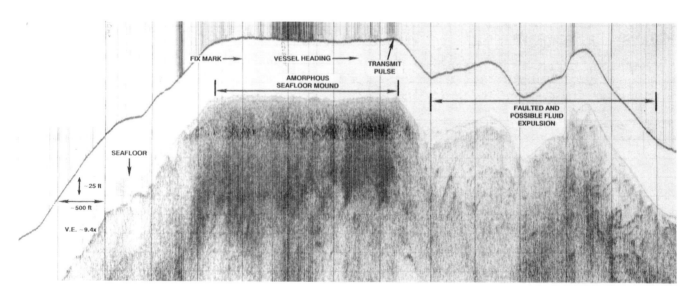

Figure 4—Deep-tow subbottom profiler record (3.5 kHz frequency) showing a flat-topped, circular, amorphous sea floor mound located within 0.6 km of the rounded mound shown in Figure 3. It also has active fluid expulsion and mud flows exiting the mound. Vertical scale is 7.6 m between timing lines (10 msec) and ~150 m between fix marks.

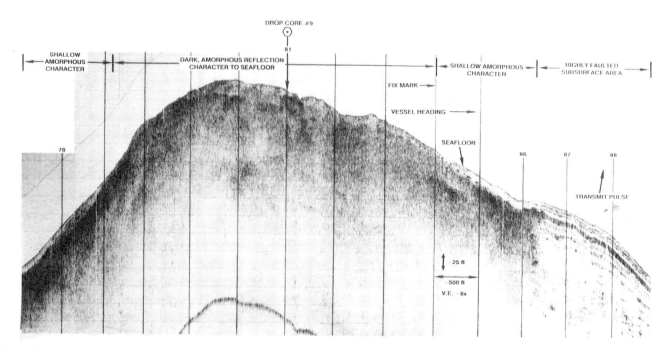

Figure 5—Deep-tow subbottom profiler record (3.5 kHz frequency) showing a nonuniform, slightly irregular, amorphous sea floor hill. Drop core #9 sampled ~2.75 m of a clay matrix with authigenic carbonate rock fragments that were stained and dripping with oil. Vertical scale is 7.6 m between timing lines (10 msec) and ~150 m between fix marks.

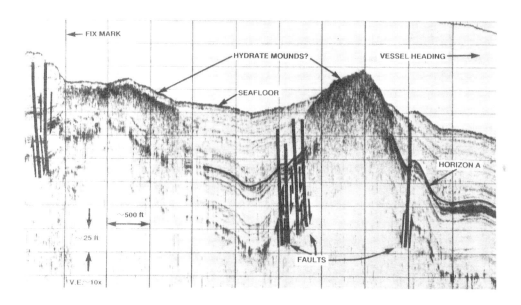

Figure 6—Deep-tow subbottom profiler record (3.5 kHz frequency) showing a nonuniform, slightly higher amplitude mound possibly consisting of hydrates. This mound is the same as the larger mound shown in Figure 7. Vertical scale is 7.6 m between timing lines (10 msec) and ~150 m between fix marks.

Other sea floor mounds with the same general shape do not always represent hydrates. Some mounds identified in the deep-tow surveys have flat-topped surfaces (Figure 4) and are generally about 300–500 m (1000–1500 ft) in diameter. Several visited on manned submersible dives were seen to have active vents expelling free gas or fluids into the sea water (Figure 10). Others have authigenic carbonate rock capping their surfaces (Figure 5). The sea floor hill in Figure 5 was sampled with a drop core that retrieved authigenic carbonate rock fragments

that were literally dripping with crude oil. Other mounds visited with manned submersibles showed thin-capped carbonate sheets with no apparent fluid seepage or biological growth. These mounds are thought to represent dead or dormant seep mounds. The thin carbonate cap may be caused by hydrate accumulation below the rock preventing the growth of carbonates deeper into the sediment. This may explain the flattened, thin formation of authigenic rock and the distinctive absence of larger boulder rock outcrops.

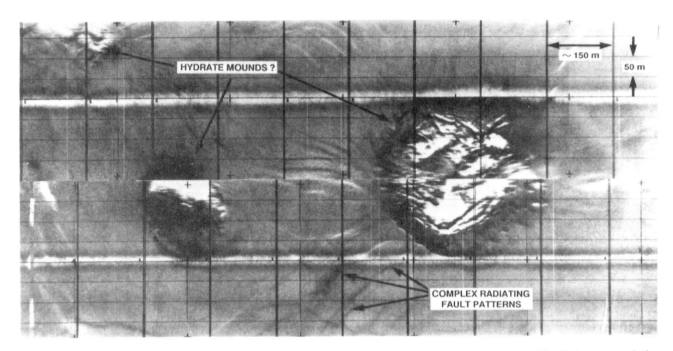

Figure 7—Mosaic of two deep-tow side-scan sonar lines over sea floor mounds that may represent hydrate accumulation. Timing lines are ~50 m apart and fix marks are ~150 m apart.

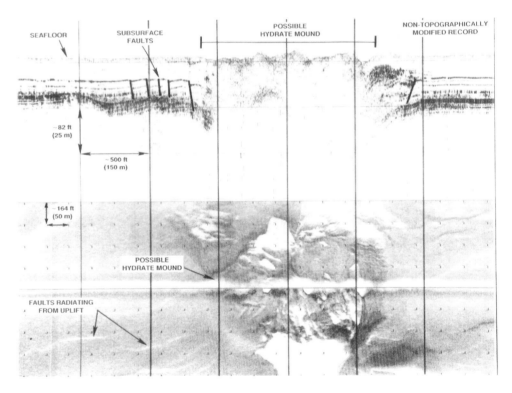

Figure 8—Split-trace subbottom profiler (top) and side-scan sonar (bottom) record over a possible sea floor hydrate mound. This comparison shows the complex faulting that circumvents and radiates from the uplift typical of hydrate mounds. Vertical scale on the subbottom profiler record is ~25 m, and tick marks on the side-scan sonar record are ~50 m apart. Fix marks are ~150 m apart.

Active venting can occur on seep mounds of any shape. The example in Figure 4 was acquired in an area where fluid debris flow material was seen exiting the mound and running down the side of a major ridge (Figure 11). This site was visited using a manned submersible (H. H. Roberts, personal communication, 1992, 1993). Small,

active cratered vents were seen on top of this mound (Figure 10) expelling gas and fine sediment that was cascading down the slope of the mound. In many cases, these seep areas are encircled with brightly colored bacterial mats. Bacterial mats have been described as commonly associated with active gas vents (Roberts et al., 1990b).

Figure 9—Manned submersible dive photograph over a known hydrate mound showing chemosynthetic tube worms. (Photo courtesy of Harry Roberts.)

Figure 10—Manned submersible dive photograph showing active gas seepage from the seabed. (Photo courtesy of Harry Roberts.)

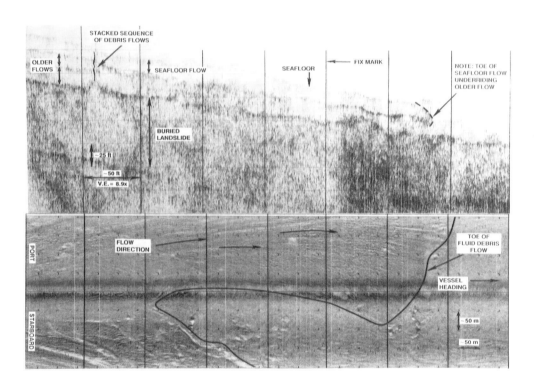

Figure 11—Split-trace subbottom profiler (top) and side-scan sonar (bottom) record showing fluid debris flow material moving downslope from the active seep mound shown in Figure 4. Vertical scale on the subbottom profiler record is ~25 m, and tick marks on the sidescan sonar record are ~50 m apart. Fix marks are ~150 m apart.

Sea Floor Depressions and Craters

Active gas vents do not always form hills. Many large depressions and several craters have been seen in the deep-tow survey data from which active fluid expulsion has been evident. The sizes of the depressions vary widely, from relatively minor shallow depressions having only a few meters of lateral sea floor expression (Figure 12) to large craters as much as 300–500 m (1000–1500 ft) in diameter (Figure 13) with tens of meters of vertical relief. The smaller depressions or lows suggest slow (low-energy) venting of gas, either continual or episodic, which

suspends fine, soft sea floor material that is subsequently carried away by bottom currents.

Catastrophic blowouts can occur to produce such large-scale craters as shown in Figures 13 and 14. This particular crater (Prior et al., 1989) may have been the result of high-pressure surges from an underlying hydrocarbon reservoir. The crater actually lies slightly off-center from the top of a small hill (Figure 15). Concentric, peripheral faults surround the crater (Figures 13, 14, 15), suggesting that intrusive forces were imparted on the original hill from below producing tensional forces on the overlying strata. The buildup of force apparently could

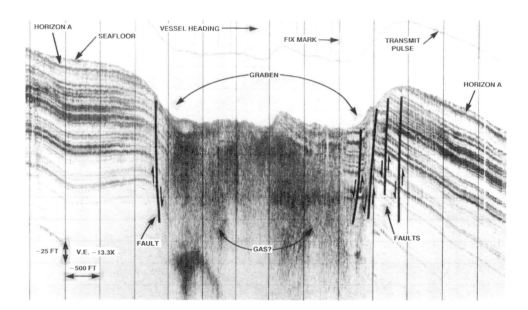

Figure 12—Deep-tow subbottom profiler record (3.5 kHz frequency) showing a low-relief sea floor depression bounded by faults and with possible fluid or gas expulsion. Vertical scale is 7.6 m between timing lines (10 msec) and ~150 m between fix marks.

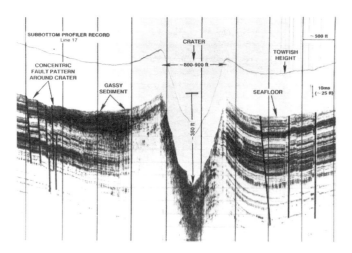

Figure 13—Deep-tow subbottom profiler record (3.5 kHz frequency) over a large sea floor crater possibly caused by a catastrophic blowout event (from Prior et al., 1989). Vertical scale is 7.6 m between timing lines (10 msec) and ~150 m between fix marks.

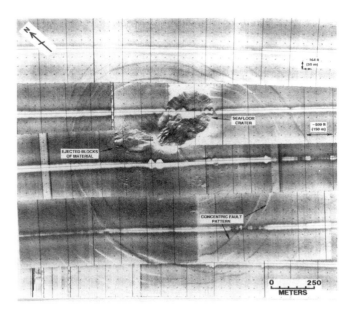

Figure 14—Composite side-scan sonar mosaic over the sea floor crater shown in Figure 13. Note the debris field and concentric faults surrounding the crater (from Prior et al., 1989).

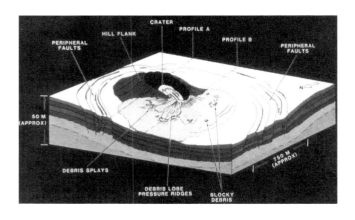

Figure 15—Interpretive three-dimensional schematic drawing of the sea floor crater shown in Figures 13 and 14 (from Prior et al., 1989).

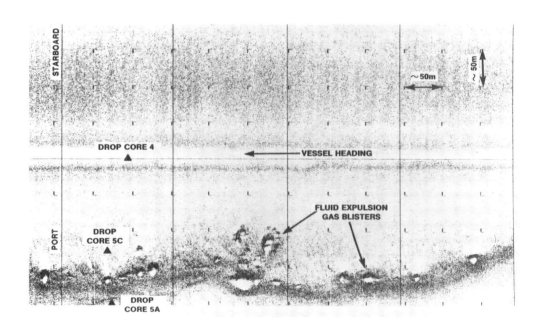

Figure 16—Deep-tow side-scan sonar record (100 kHz frequency) showing fluid expulsion along a fault plane. Drop cores taken above, below, and into the sea floor fault plane showed increased signs of gas along the fault. Tick marks are ~50 m apart, and fix marks are ~150 m apart.

not be retained by the overburden, and a potential blowout ensued. Ejected material along the present sea floor suggests that, when the active venting did occur, it was either catastrophic or had enough back pressure to expel large blocks of debris.

An alternative interpretation for this crater is that the hill may have formed by the growth of gas hydrates in near-surface sediments, expanding the sediments and forming the concentric fault pattern. Subsequent degradation of the hydrate structure would have released the concentrated gas, which then could have suspended the surrounding sediment to be carried away by near sea bottom currents. This would not, however, explain how the large blocks of debris came to rest downslope of the crater. A higher energy event would have been needed to move such large blocks that still remain partially intact.

Gas Blistering and Sea Floor "Staining"

Sea floor gas vents may be clusters of small expulsion features that acoustically contrast with the normal surrounding sea floor. Active vents along fault planes have been seen as blister-like expressions (Figure 16) on the side-scan sonar data, and they can acoustically void or wipe-out seismic character on subbottom profiler data (Figure 17). The blistering probably indicates continuous or intermittent vents that were active at least during the time of the survey. Most blistered areas are in close association with sea floor or very shallow fault displacements. Drop cores taken above, below, and through the fault shown in Figure 16 showed high concentrations of methane gas directly at the base of the fault but relatively little methane concentrations above and downslope of the base of sea floor expression.

Many fluid expulsion vents can only be identified as darker reflecting backscatter on the side-scan sonar data (Figure 18). This backscatter can look like a coffee stain on

the record with no other apparent reason for the venting. No vertical abnormalities may be present on the sea floor, such as depressions, mounds, or blisters. The subsurface, penetrated by subbottom profiler soundings, may provide the only indication that gas is present in the shallow sediments. Below the darker reflecting sea floor zone, the shallow sediments may appear amorphous, chaotic, or higher in amplitude than adjacent reflections (Figure 19). A vent may occur well away from any fault plane, but the source can generally be traced back to some subsurface displacement.

Sub–Sea Floor Gas Features

Shallow gas venting may never reach the sea floor. Impermeable stratigraphic sequences may block gas migration and seal further upward migration. If the seal is competent and fluids continue to accumulate, sub–sea floor mounds may form, causing a bulge in the overlying stratigraphy (Figure 20). Sometimes similar features appear in buried mass movement deposits and may only represent remnant, rotated blocks of intact sediments. It can be difficult at times to distinguish between gas zones and remnant landslide blocks. However, if faults happen to cut through landslide features and no venting is associated with the fault plane, then it can be assumed that the landslide is not gas saturated, or at least not gas saturated near the fault plane.

Gas may migrate up fault planes and find permeable stratigraphic units through they can migrate laterally. If the stratigraphic units are porous enough, the gas may travel considerable distances from the conduit. Typically, the top of the gas-enriched, porous stratigraphic unit will show a slightly higher amplification (Figure 21). Lateral migration of the gas may extend for several kilometers from the original vertical vent.

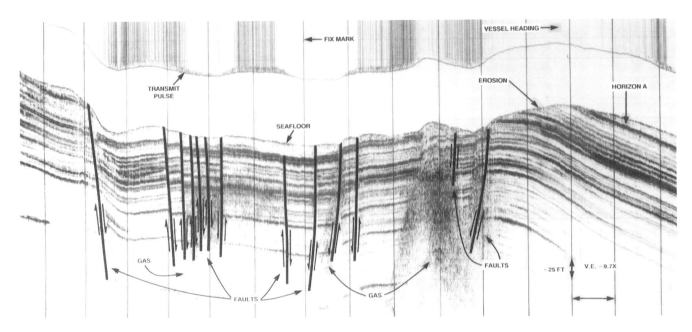

Figure 17—Deep-tow subbottom profiler record (3.5 kHz frequency) showing the acoustically wiped-out character of the seismic signal where gas or fluids are using fault planes as conduits. Vertical scale is 7.6 m between timing lines (10 msec) and ~150 m between fix marks.

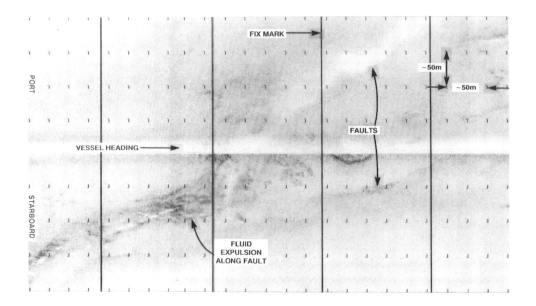

Figure 18—Deep-tow side-scan sonar record (100 kHz frequency) showing darker staining of sea floor where fluids are escaping near fault planes. Tick marks are ~50 m apart, and fix marks are ~150 m apart.

CONCLUSIONS

Shallow gas and fluid migration events are common along the upper and middle continental slope of the northern Gulf of Mexico. This is especially true where subsurface salt has deformed the overlying sediments and created faults within the overlying strata. Identifying shallow gas features in deep-water areas is made easier when high-resolution seismic tools can be operated near the sea bottom and the acquired data is of high quality and resolution. The positively buoyant deep-tow side-scan sonar and subbottom profiler system used in the Gulf of Mexico over the past decade has proved to be one of the best sources for acquiring such high-quality data.

Shallow gas zones and associated features with varying seismic character have been recognized both on the sea floor and in shallow sediments. These include the following:

- Sea floor seep mounds, hydrate hills, and authigenic carbonate rock outcrops
- Sea floor depressions and craters

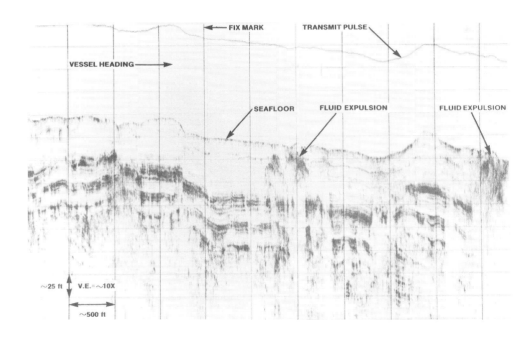

Figure 19—Deep-tow sub-bottom profiler record (3.5 kHz frequency) showing probable buried fluid expulsion zones associated with faulting where there is little or no indication of venting to the sea floor. Vertical scale is 7.6 m between timing lines (10 msec) and ~150 m between fix marks.

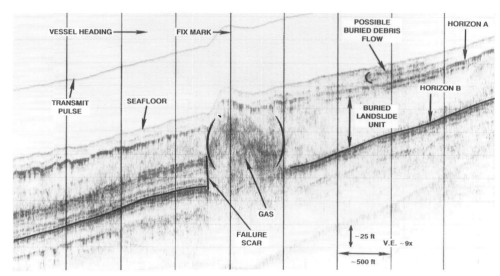

Figure 20—Deep-tow sub-bottom profiler record (3.5 kHz frequency) showing possible gas entrapped in a buried landslide unit. Vertical scale is 7.6 m between timing lines (10 msec) and ~150 m between fix marks.

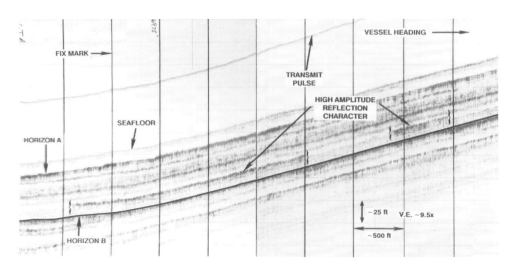

Figure 21—Deep-tow sub-bottom profiler record (3.5 kHz frequency) showing slightly higher amplification of a seismic event thought to be caused by gas enrichment. Vertical scale is 7.6 m between timing lines (10 msec) and ~150 m between fix marks.

- Blister-like vents associated with faulting
- Darker reflectivity or "stained" sea floor reflecting zones
- Subsurface chaotic, amorphous, and wipe-out seismic reflection character
- High-amplitude reflecting zones and amplified seismic horizons

The accumulation of more than 20,000 km (10,000 nmi) of deep-tow subbottom profiler and side-scan sonar data from the Gulf of Mexico continental slope allows many interesting and interrelated geologic features and processes to be recognized. Gas and other fluid-related features have been recorded, tested, and visually inspected so that a significant library of data examples have been accumulated for reference in future studies. The varying reflection patterns seen at shallow gas locations can help geoscientists evaluate and assess future data to make intelligent and reliable interpretations.

Acknowledgments—The authors would like to thank Shell Offshore Inc. (SOI) for permitting us to show these deep-water data examples. We would also like to thank Harry Roberts of the Coastal Studies Institute, Louisiana State University, for supplying the submersible dive photographs used in this paper. Data collection for the deep-tow operations were handled for SOI by John E. Chance & Associates, Inc. (JECA) of Lafayette, Louisiana. We would like to thank Jay Northcutt of JECA for supervising and coordinating all field activities associated with the collection of these data. Onboard quality control, in-office data interpretations, geologic assessments, and report writing were handled by Fugro-McClelland Marine Geosciences, Inc. (FMMGI) who have been Shell's consultant on all deep-tow projects. We would like to thank FMMGI's staff of geologists for their excellent work and data example archiving.

REFERENCES CITED

Brooks, J. M., M. C. Kennicutt, C. R. Fisher, S. A. Macko, K. Cole, J. J. Childress, R. R. Bridigare, and R. D. Vetter, 1987, Deep-sea hydrocarbon seep communities: evidence for energy and nutritional carbon sources: Science, v. 238, p. 1138–1142.

Doyle, E. H., M. J. Kaluza, and H. H. Roberts, 1992, Use of manned submersibles to investigate slumps in deep-water Gulf of Mexico, *in* Civil engineering in the oceans: Proceedings, International Conference Committees, Ocean Engineering & Wave and Wave Forces ASCE, Texas A&M University, College Station, Texas, November 2–5.

Kelland, N. C., 1988, Accurate acoustic position monitoring of deep-water geophysical towfish: Proceedings, 20th Annual Offshore Technology Conference, OTC 5780, Houston, Texas, May 2–5.

Martin, R. G., 1978, Northern and eastern Gulf of Mexico continental margin: stratigraphic and structural framework, *in* A. H. Bouma, G. T. Moore, and J. M. Coleman, eds., Framework, facies, and oil-trapping characteristics of the upper continental margin: AAPG Studies in Geology 7, p. 21–00

Neurauter, T. N., and W. R. Bryant, 1989, Gas hydrates and their association with mud diapir/mud volcanoes on the Louisiana continental slope: Proceedings, 21st Annual Offshore Technology Conference, OTC 5944, Houston, Texas, May 1–4, p. 599–607.

Prior, D. B., and E. H. Doyle, 1984, Geological hazard surveying for exploratory drilling in water depths of 2000 m: Proceedings, 16th Annual Offshore Technology Conference, OTC 4747, Houston, May 7–9, p. 311–315.

Prior, D. B., and E. H. Doyle, 1993, Submarine landslides—the value of high-resolution geophysical geohazard surveys for engineering: Paper presented at U.K. Academy of Engineering Conference, London, November.

Prior, D. B., E. H. Doyle, M. J. Kaluza, M. W. Woods, and J. W. Roth, 1988, Technical advances in high-resolution hazard surveying, deep water, Gulf of Mexico: Proceedings, 20th Annual Offshore Technology Conference, OTC 5758, Houston, Texas, May 2–5, p. 109–117.

Prior, D. B., E. H. Doyle, and M. J. Kaluza, 1989, Evidence for sediment eruption on deep sea floor, Gulf of Mexico: Science, v. 23, p. 517–519.

Roberts, H. H., R. Sassen, and P. Aharon, 1987, Carbonates of the Louisiana continental slope: Proceedings, 19th Annual Offshore Technology Conference, OTC 5463, Houston, Texas, April 27–30, p. 373–382.

Roberts, H. H., R. Sassen, and P. Aharon, 1988, Petroleum-derived authigenic carbonates of the Louisiana continental slope: Proceedings, Oceans '88 Conference, Baltimore, Maryland, October 31–November 2, v. 1, p. 101–105.

Roberts, H. H., R. Sassen, J. Carney, and P. Aharon, 1989, Carbonate buildups on the continental slope off Louisiana: Proceedings, 21st Annual Offshore Technology Conference, OTC 5953, Houston, Texas, May 1–4, p. 655–662.

Roberts, H. H., R. Sassen, R. Carney, and P. Aharon, 1990a, The role of hydrocarbons in creating sediment and small-scale topography on the Louisiana continental slope: Ninth Annual Research Conference Proceedings, GCS-SEPM Foundation, New Orleans, Louisiana, December 4–7, p. 311–324.

Roberts, H. H., P. Aharon, J. Carney, J. Larkin, and R. Sassen, 1990b, Sea floor responses to hydrocarbon seeps, Louisiana continental slope: Geo-Marine Letters, v. 10, p. 232–243.

Sloan, Jr., E. D., 1990, Clathrate hydrates of natural gases: Monticello, New York, Marcel Dekker, 641 p.

MacDonald, I. R., J. F. Reilly, Jr., S. E. Best, R. Venkataramaiah, R. Sassen, N. L. Guinasso, Jr., and J. Amos, 1996, Remote sensing inventory of active oil seeps and chemosynthetic communities in the northern Gulf of Mexico, *in* D. Schumacher and M. A. Abrams, eds., Hydrocarbon migration and its near-surface expression: AAPG Memoir 66, p. 27–37.

Chapter 3

Remote Sensing Inventory of Active Oil Seeps and Chemosynthetic Communities in the Northern Gulf of Mexico

I. R. MacDonald

Geochemical and Environmental Research Group
Texas A&M University
College Station, Texas, U.S.A.

J. F. Reilly, Jr.

Enserch Exploration, Inc.,
Dallas, Texas, U.S.A.

Present address:
National Aeronautics and Space Administration
Johnson Space Center
Houston, Texas, U.S.A.

S. E. Best

R. Venkataramaiah

R. Sassen

N. L. Guinasso, Jr.

Geochemical and Environmental Research Group
Texas A&M University
College Station, Texas, U.S.A.

J. Amos

Earth Satellite Corporation
Rockville, Maryland, U.S.A.

Abstract

We compiled locations of probable oil slicks from interpretation of a Space Shuttle photograph, a Landsat Thematic Mapper scene, three European Radar Satellite scenes, and collections of floating oil and observations of sea floor seeps from submarines. These locations were ranked according to recurrence of evidence for natural oil seepage among the various data sets. As a result, we have verified 43 biological communities that depend on hydrocarbon seeps and 63 locations where remote sensing data indicate that sea floor sources are capable of producing perennial oil slicks.

Monitoring individual seeps over time contributes to understanding the natural loading of hydrocarbons in the marine environment. The seeps also form a natural test bed for development of sensors and techniques to detect oil floating on the sea. Remote detection of natural seepage extends the probable range of chemosynthetic communities dependent on hydrocarbon seepage in the northern Gulf of Mexico. This technique for detecting areas of macroseepage is potentially applicable to hydrocarbon basins in which oil production is in an earlier stage of development than the Gulf of Mexico.

INTRODUCTION

Accurately gauging the scale, distribution, and physical properties of natural oil seepage in the ocean is important because detection of seeps can offer a cost-effective means for locating offshore oil reserves. Marine oil seepage is a source of surfactant in the upper 1 mm of the ocean that may affect climate by altering transfer rates of gas, water vapor, and momentum. Remote sensing provides an approach for addressing this issue. The difficulty lies in accurate interpretation of remote sensing targets that are generated by actual oil seeps, as opposed to other natural or artificial processes. We have used satellite and airborne remote sensing, as well as sea truth observations from ships and research submarines, to study natural oil seeps in the northern Gulf of Mexico. The combination of these data types allowed us to confirm the remote sensing characteristics of marine oil seeps in this region. We were then able to extend our interpretation of the total remote sensing data set and to compile a catalog of seeps across a significant extent of the continental slope.

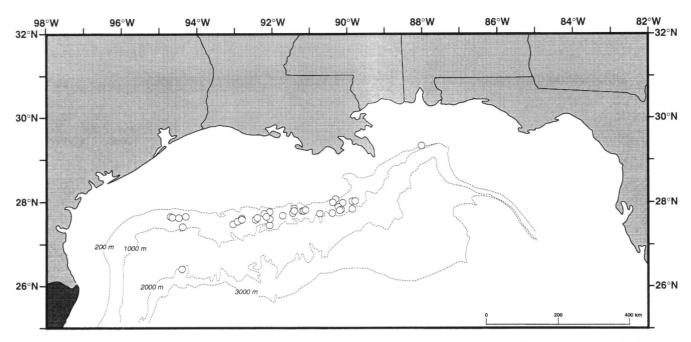

Figure 1—Map of Gulf of Mexico showing the distribution of known locations where communities of chemosynthetic megafauna (animals that depend on natural hydrocarbon seepage) have been collected, photographed, or observed.

Location

The northern continental slope of the Gulf of Mexico is a major hydrocarbon basin (Worrall and Snelson, 1989). Natural hydrocarbon seepage in this region has been documented by a variety of evidence, including historical records of floating and beached oil (Geyer and Giammona, 1980), extensive occurrence of oil-stained marine sediments (Brooks et al., 1986; Kennicutt et al., 1988b), and widespread communities of chemosynthetic tube worms, mussels, and clams, which depend on chemically reduced compounds associated with hydrocarbon seepage (Childress et al., 1986; Brooks et al., 1987; Kennicutt et al., 1988a; MacDonald et al., 1989). Seepage produces distinctive alteration to discrete areas of the sea floor; the geology of these sites suggests that seepage tends to persist at specific locations for at least hundreds of years (Behrens, 1988). The biological communities that form at these sites display predictable patterns of local distribution and diversity which are diagnostic of the location and style of seepage (MacDonald et al., 1990b). Figure 1 shows a map of locations where chemosynthetic communities are known to exist on the sea floor. Because oil and gas seeps can also be abundant where chemosynthetic communities are absent (e.g., offshore southern California), detailed site surveys are necessary to confirm the presence of a chemosynthetic community associated with a natural oil seep.

Visibility of Floating Oil

Oil seepage in the Gulf of Mexico produces persistent slicks on the surface of the water that can be detected from space (MacDonald et al., 1993). Oil floating on the sea can be detected by both active and passive methods of remote sensing. Synthetic aperture radar (SAR) and laser fluorescence are two active methods that have been used to detect natural oil slicks (Estes et al., 1985). Passive methods include microwave radiometry, multispectral scanning (MSS), ultraviolet (UV) fluorescence, and detection of enhanced sunglint. The spectral signature of oil in the UV range is most useful for thin layers, while thicker oil sheen tends to have a strong signature in the infrared radiation (IR) range (Rogne et al., 1993). Efforts to detect natural oil seeps by use of MSS undertaken by NASA, the U.S. Coast Guard, and the Marine Spill Response Corporation have not always proven successful (MacDonald, unpublished data). Generally, best results have been obtained by using SAR or sunglint to detect areas where floating oil has dampened the fine-scale surface roughness of the sea and created patches of increased specular reflectance. Sunglint has been photographed from the space shuttle and imaged by the Landsat Thematic Mapper for these purposes

Oil escaping from the sea floor arrives at the sea surface as a continuous rain of droplets, generally less than 1 cm in diameter. Each droplet of oil blossoms into a small patch of rainbow-colored sheen as it hits the surface. The sheen spreads rapidly, changing from rainbow-colored to silver gray, and within seconds merges into a continuous layer that is invisible to the naked eye. The downwind drift of these layers produces linear patches of water in which centimeter-scale capillary waves are strongly dampened. Specular reflection from these patches is enhanced, producing distinctive "slicks" (Figure 2). Although theory suggests that an invisible

layer of oil could have a monomolecular thickness (<<0.01 μm), experiments with known volumes of oil suggest that the effective thickness is 0.01–0.1 μm (R. Duckworth, personal communication, 1993).

In calm weather, the effect of floating oil is immediately visible even though there is no change in surface color. The smell of evaporating oil is also strong, so it is usually evident when the ship is in an area of freshly surfacing oil. The photographs in Figure 3 were take from the rail of a ship with a 28-mm lens fitted with a polarizing filter. Figure 3a shows unoiled water, while Figure 3b shows oiled water in a slick produced by natural seepage. In both cases, the wind is from the upper left, blowing about 5 m/sec.

The effect on light scattering evident in the two photographs can be quantified. We digitized the two photographs, then examined the power spectra of the sea surface for a column of pixels from the center of each photograph. Plots of these spectra (Figure 4) show that the oil film has dampened the high-frequency capillary waves and shifted the power spectrum to the longer wavelengths. The total area under the curve has also increased as a result of specular reflection directing more of the light toward the camera.

This same phenomenon applies to images of sunglint captured by the Space Shuttle and the Landsat Thematic Mapper (TM) sensor and to images of radar backscatter captured by the European Radar Satellite (ERS-1). Slicks form over natural oil seeps as a continual supply of oil arrives at the surface and drifts away from the source under the influence of wind and current. This results in distinctive, elongated slicks, the length of which is related to the rate of flow, the evaporation rate, the rate of advection, and the minimum thickness at which the slick will be detected by a given sensor (MacDonald et al., 1993). Figure 5 shows a complex of natural slicks imaged by ERS-1.

MATERIALS AND METHODS

Sea floor observations of the biological, chemical, and geological characteristics of hydrocarbon seeps in the northern Gulf were reviewed to compile an inventory of sites where chemosynthetic fauna have been collected by a trawl or submarine or have been definitively photographed from a submarine, remotely operated vehicle, or camera sled. As noted, collections and other observations of oil floating on the surface of the ocean above seeps predate offshore production by at least 40 years (Geyer and Giammona, 1980). Our study relied on collections made from surface ships beginning in 1991. Close to the source of a natural oil slick, individual droplets were collected by use of 5- by 30-cm strips of hydrophobic fabric that were clamped to a pole and gently laid on the sea surface. Sampling was done off the windward bow of a moving ship, and poles and clamps were rinsed with solvent between samples. Samples were stored chilled before being sonicated and extracted with methylene chloride for analysis. The hydrocarbon composition of the slick samples was determined by gas chromatography with flame-ionized detection and by total scanning fluorescence (Brooks et al., 1986). Locations of surface collections, biological communities, and geochemical anomalies were also added as layers in a geographic information system (GIS).

The remote sensing data base comprised images collected from three sensor systems on board three satellites (Table 1). The Space Shuttle photograph was scanned and digitally resampled to conform to the position of known coastal features and offshore installations. Suspected slicks in these images were outlined as closed polygons. The set of slick polygons for each image was then stored as a layer in the GIS by use of MAPIX® software.

(a)

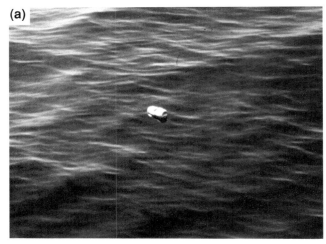

(b)

Figure 3—Two photographs taken from the side of a ship in calm weather showing the effect of oil seepage. The slicks show as areas of dampened surface roughness. (a) Unoiled water; (b) oiled water in same region. Surface reflectance was enhanced with a polarizing filter; scale is provided by floating beverage cans.

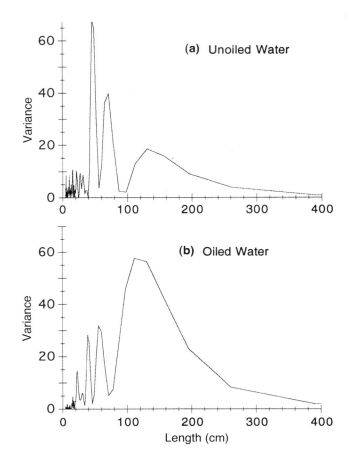

Figure 4—Graphs showing power spectra obtained for a single column of pixels from digitized versions of the two photographs in Figure 3. (a) Unoiled water; (b) oiled water. Length scale is in centimeters (approximate) as determined from the floating beverage cans.

Evaluation of Remote Sensing Images

The objective of remote sensing evaluation was to combine observations from the data sets described above, then review these observations to identify probable sites of sea floor seepage. This was accomplished in two stages. First, we compared all possible image pairs to determine co-locations of slick polygons. Overlapping or adjacent polygons were classified by the following ranking scheme:

Rank 1: Overlap of the upwind ends of slicks in two or more images
Rank 2: Overlap of slick ends regardless of wind direction
Rank 3: Ends of slicks within ~2 km of each other.
Rank 4: Overlap of any portion of slick polygons in two or more images.

Second, we compared the geographic coordinates of the slick polygons to collection sites of floating freshly surfaced oil, to known locations of chemosynthetic communities, and to oily soil samples collected in the same lease block with a surface slick or community of chemosynthetic animals.

Sea-Truthing Remote Sensing Images

Three locations were selected for dives by a research submarine because they coincided with remotely detected oil slicks. The purpose of the dives was to try to locate a chemosynthetic community or other definitive evidence for long-term seepage on the sea floor below each surface target. In particular, our concern was to explore potential new sites that were in greater water depths than previously known sites. All dives were made with the research submarine *Johnson Sea-Link*, which has a maximum depth rating of 1000 m. *Johnson Sea-Link* was supported by the mother ships R/V *Seward Johnson* and R/V *Edwin Link*. Dives consisted of about 2 hours of bottom time, during which the submarine was first directed to

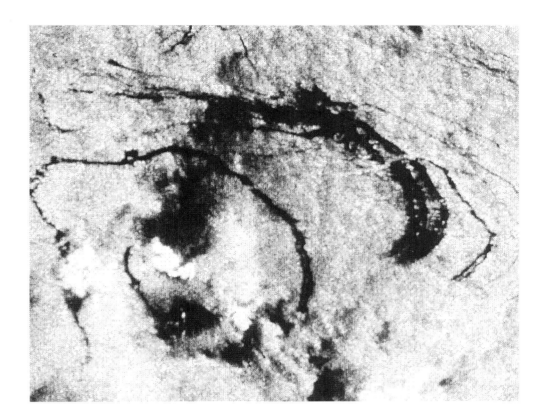

Figure 5—Remote sensing image of a complex of natural slicks taken by ERS-1 SAR on 9 September 1993. The boat wake that bisects several slicks fails to disturb the irregular target on the left, which is therefore interpreted not to be oil.

Table 1—Remote Sensing Systems Used in This Study

European Radar Satellite (ERS-1, SAR image)

Orbit	Frame	Date
9518	3051	8 May 1993
9475	3051	8 May 1993
11250	3051	9 Sept. 1993

Landsat (*Landsat 5*, TM image)

Path	Row	Collection Time	Date
22	41	1105 CST	31 July 1991

Space Shuttle (*Atlantis*, 120-mm lens, hand-held)

Mission	Roll	Frame	Collection Time	Date
STS-30	151	028	1535–1538 CST	5 May 1989

the appropriate location by a surface ship, then carried out autonomous exploration based on what was visible to scientists in the submarine.

RESULTS

Chemosynthetic Communities

A review of published and unpublished sources determined that there were 43 sites where chemosynthetic fauna have been collected or observed (Figure 1, Table 2). These locations were distributed zonally from about long. 95° to 88° W and in a bathymetric range from about 300 to 2200 m. Within this distribution, distinct discontinuities occurred that were centered at long. 94° and 89° W. The easternmost community was found in the Viosca Knoll 826 lease block (long. 88° W). Interestingly, reports of chemosynthetic fauna at this site (the clam *Calyptogena ponderosa* and the scallop-like *Acesta bullisi*) date from trawl collections made with the R/V *Oregon* in 1955 (Volkes, 1963; Boss, 1968). Although this was not recognized at the time of publication, these are some of the earliest reports of chemosynthetic fauna in the literature. The deepest site is the community in the 2250-m-deep Alaminos Canyon 645 lease block (Brooks et al., 1990). The distinct lack of known communities within the 1000–2000 m bathymetric interval was primarily due to lack of exploration.

Perennial Oil Slicks

By analyzing the remote sensing and ancillary data, we found 63 locations where multiple data sets indicated the presence of perennial oil slicks (Figure 6, Table 3). At six of these sites, we collected freshly surfaced oil at locations that coincided with a target in one of the remote sensing images. Oil-stained sediments were present in five cases and chemosynthetic fauna in three. At 39 of the 63 sites listed in Table 3, we made our primary determination based on overlap of the evident upwind ends of slicks in two or more of the remote sensing images (Figure 7). Floating oil was collected at two of these sites and oil-stained sediments were collected at an additional ten sites

Table 2—Sites Where Chemosynthetic Megafauna Have Been Collected or Definitively Photographed

Fauna[a]	Latitude N	Longitude W	MMS Lease Block[b]	Depth (m)	Observation Method[c]	Data Source[d]
VM	26°21.20'	94°29.80'	AC0645	2200	Sub	1
M	27°23.50'	94°29.45'	EB0602	1111	Trl	2
PG	27°27.55'	93°08.60'	GB0500	734	Trl	2
VC	27°30.05'	93°02.01'	GB0458	757	Trl	2
M	27°31.50'	92°10.50'	GB0476	750	Sub	3
MC	27°33.40'	92°32.40'	GB0424	570	Sub	3
V	27°35.00'	92°30.00'	GB0425	600	Sub	3
VC	27°34.50'	92°55.95'	GB0416	580	Sub	3
VC	27°36.00'	94°46.00'	EB0376	776	Sub	3
PG	27°36.15'	94°35.40'	EB0380	793	Trl	2
MC	27°36.50'	92°28.94'	GB0382	570	Sub	3
VC	27°36.60'	94°47.35'	EB0375	773	Trl	2
VC	27°36.82'	92°15.25'	GB0386	585	Sub, Trl	2, 3
VC	27°37.15'	92°14.40'	GB0387	781	Sub, Trl	2, 3
V	27°37.75'	91°49.15'	GC0310	780	Trl	2
VC	27°38.00'	92°17.50'	GB0342	425	Trl	2
C	27°39.15'	94°24.30'	EB0339	780	Trl	2
VC	27°39.60'	90°48.90'	GC0287	994	Sub, Trl	2
C	27°40.45'	90°29.10'	GC0293	1042	Trl	2
VC	27°40.50'	92°18.00'	GB0297	589	Trl	2
VMC	27°40.88'	91°32.10'	GC0272	720	Sub, Trl	2, 3, 4
VC	27°42.65'	92°10.45'	GB0300	719	Trl	2
V	27°43.10'	91°30.15'	GC0229	825	Trl	2
VM	27°43.30'	91°16.30'	GC0233	650	Sub	5
VMC	27°43.70'	91°17.55'	GC0233	813	Trl	2
VM	27°44.08'	91°15.27'	GC0234	600	Sub	3, 6
VM	27°44.30'	91°19.10'	GC0232	807	Sub	3
VM	27°44.80'	91°13.30'	GC0234	550	Sub	3, 7
VC	27°45.00'	90°16.31'	GC0210	715	Sub	3
C	27°45.50'	89°58.30'	GC0216	963	Sub, Phsl	8, 2
VMC	27°46.33'	90°15.00'	GC0210	796	Sub	3
VM	27°46.65'	91°30.35'	GC0184/5	580	Sub, Trl	2, 3, 9
VM	27°46.75'	90°14.70'	GC0166	767	Sub, Trl	2, 3
VM	27°49.16'	91°31.95'	GC0140	290	Sub	10
V	27°50.00'	90°19.00'	GC0121	767	Sub	3
VM	27°53.56'	90°07.07'	GC0081	682	Phsl	11
VC	27°54.40'	90°11.90'	GC0079	685	Trl	2
VM	27°55.50'	90°27.50'	GC0030	504	Sub	3
VPG	27°56.65'	89°58.05'	GC0040	685	Trl	2
C	27°57.10'	89°54.30'	MC0969	658	Trl	2
V	27°57.25'	89°57.50'	EW1010	597	Sub, Trl	2, 3
V	27°58.70'	90°23.40'	EW1001	430	Sub, Trl	2, 3
VC	29°11.00'	88°00.00'	VK0826	545	Sub, ROV, Trl	3, 4, 12

[a]Chemosynthetic megafauna found at site: V = vestimentiferan tube worms, M = seep mytilids, C = vesicomyid or lucinid clams, PG = pogonophoran tube worms.

[b]Lease block designators follow Minerals Management Service (MMS) standard abbreviations.

[c]Trl = Trawl, Sub = submarine, ROV = remotely operated vehicle, Phsl = photosled.

[d]Sources: 1 = Brooks et al. (1990), 2 = Kennicutt et al. (1988a,b), 3 = GERG unpubl. data, 4 = Callender et al. (1990), 5 = MacDonald et al. (1990c), 6 = MacDonald et al. (1990a), 7 = MacDonald et al. (1990b), 8 = Rosman et al. (1987), 9 = MacDonald et al. (1989), 10 = Roberts et al. (1990), 11 = Boland 1986, 12 = Boss (1968), Gallaway et al. (1990), Volkes (1963). Data sources give precedence to first observations published in open literature.

within this group. Among the remaining 24 locations listed in Table 3, four are distinguished as strong candidates for perennial seeps because of positive indicators in the supporting data sets. The remaining sites fit into our scheme of overlapping targets in multiple remote sensing images, but they lack supporting evidence.

Observations from a Research Submarine

Sites selected for dives were located in the Green Canyon 232, 287, and 321 lease blocks (Figure 7, Table 3). We were able to find chemosynthetic fauna at the first two of these sites but not at the third. The surface slick in Green Canyon 232 was spotted from an airplane and was clearly visible to shipboard observers before it was noted in the satellite data. During two dives to depths of 700 m, we noted dense aggregations of tube worms and seep mussels similar to assemblages at the Bush Hill site (MacDonald et al., 1989). One of us (Guinasso) inadvertently caused a copious discharge of oil and gas from the sea floor by attempting to collect some of the mussels. The mussels were clustered within a field of authigenic carbonate rubble. Probing the mussel cluster with a scoop sampler disturbed the carbonate substratum and initiated a burst of gas bubbles followed by a flow of oil droplets that continued unabated for over 30 min. Arrival of the oil droplets at the surface was noted by the shipboard observers at a location about 100 m from the submarine's sea floor position.

The surface slick at Green Canyon 287 was noted in the TM and ERS-1 data sets. The site is on a plateau that rises from a base depth of 1500 m to a crest depth of 760 m. A salt body has intruded almost to the sea floor, and salt movement has produced a series of shallow faults trending northwest-southeast normal to the salt (see Reilly et al., this volume). The *Johnson Sea-Link* bottomed at a depth of 800 m and followed a southward course across the flank of the uplift block. This track took the submarine across a series of northwest-southeast trending valleys. Clam shells, carbonate outcrops encrusted with tube worms, and occasional gas seeps were observed in the floors of these valleys. Although clam shells were broadly distributed, the abundance of chemosynthetic fauna in the communities at this site was much less than at the Green Canyon 232 site. Although gas venting was evident at several locations, macroseepage of oil was not observed. Extractable hydrocarbons from sediment samples at the site were heavily biodegraded (MacDonald, unpublished data).

Surface slicks at the Green Canyon 321 site were evident in all of the remote sensing data sets. We traversed the sea floor below these slicks repeatedly during more than 6 hours of submarine exploration that covered depths from 750 to 900 m. Occasional indications of oil seepage, such as mats of *Beggiatoa* bacteria and outcrops of authigenic carbonate, were observed, but neither significant aggregations of chemosynthetic fauna nor other definitive sea floor indications of macroseepage were seen.

DISCUSSION

Exploration by submarines and related methods of direct observation have provide most of our knowledge about the locations of macro- and megaseeps in the northern Gulf of Mexico. The distribution of chemosynthetic communities discovered in this fashion generally conforms to the distribution of oil fields in the northern Gulf offshore (Sassen et al., 1993a); however, there are two sources of bias in this distribution. First, exploration has been limited by the depth certification of the submarines available for sea floor observations. Most exploration by submarine has been carried out with the *Johnson Sea-Link* and the *NR-1*. Consequently, observations below 1000 m are restricted to the Alaminos Canyon region, where submarine *Alvin* conducted two dive series. Second, many of the primary sites that have been explored by submarines were chosen on the basis of geophysical and geochemical data collected in connection with energy prospecting.

In addition to these biases, the exploration efficiency of submarines is low—they scan a restricted field of view for brief intervals of bottom time along short distances. Submarines, ROVs, and other sea floor sampling devices provide invaluable insight into the geology and biology of hydrocarbon seeps, and they can be used to delineate the geographic-bathymetric envelope in which seeps occur. However, they cannot provide a comprehensive inventory of the number of seeps.

Remote sensing provides a different perspective on the occurrence of seeps. Although the coverage is potentially broader and far less subjective than submarine exploration, this technique also has limitations. Imaging oil slicks (i.e., patches of the sea surface where floating oil dampens capillary waves) has proven the most effective method for detecting the thin layers of oil produced by seeps. Also, this phenomenon is susceptible to false targets and to disruption by rough weather. Several slicks imaged in the Space Shuttle photograph appear to be due to something other than oil seepage because they do not reappear in subsequent images of the same area. The area covered by the 9 September 1993 ERS-1 image was also imaged in March 1993. The wind and seas during March were much higher than in September, and the SAR completely failed to detect oil slicks.

The coincidence of remotely sensed seeps with chemosynthetic communities is good, but far from universal. Prolonged seepage promotes processes that tend to retain oil in the sediments (Behrens, 1988; Sassen et al., 1993b). The most vigorous seeps observed have been active mud vents (Roberts and Neurauter, 1990; Lee et al., 1993), where few chemosynthetic fauna have established themselves. Formation of a dense chemosynthetic community appears to require protracted seepage and microbial activity to prepare the substratum (Sassen et al., 1994). We suggest that when the distributions of seep communities and of surface slicks caused by seepage are considered together, one sees a continuum of the seepage process over time.

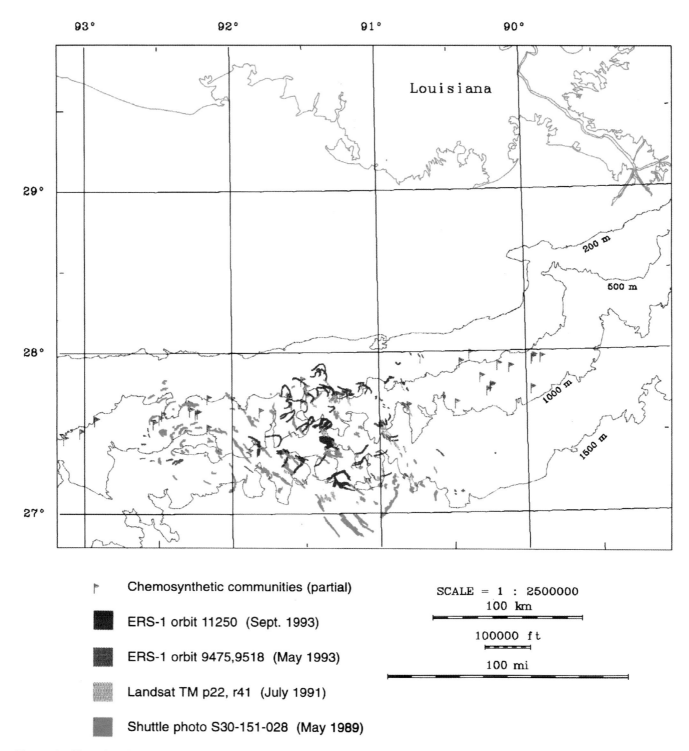

Chemosynthetic communities (partial)

ERS-1 orbit 11250 (Sept. 1993)

ERS-1 orbit 9475,9518 (May 1993)

Landsat TM p22, r41 (July 1991)

Shuttle photo S30-151-028 (May 1989)

SCALE = 1 : 2500000
100 km

100000 ft

100 mi

Figure 6—Map showing locations of chemosynthetic communities compiled by use of the geographic information system (GIS). Suspected oil slicks were manually traced in each image. Resulting polygons are shown for a Space Shuttle photograph (violet), a TM scene (yellow), a pair of ERS-1 images collected in May 1993 (blue), and an ERS-1 image collected in September 1993 (red). Locations of chemosynthetic communities are shown as green asterisks.

Table 3—Oil Slicks Detected in Remote Sensing Images, Northern Gulf of Mexico

Latitude[a] N	Longitude[a] W	Lease Block[b]	Depth[c] (m)	Remote Sensing Overlap[d]	Sea Floor Manifestation	Geochemical Anomaly	Floating Oil Collected
27° 44.83'	91° 13.33'	GC0234	831	No overlap	Chemo fauna	Oily core	Floating oil
27° 43.71'	91° 36.57'	GC0226	825	No overlap	—	Oily core	Floating oil
27° 40.53'	92° 18.00'	GB0297	589	No overlap	Chemo fauna	Oily core	Floating oil
27° 31.55'	92° 10.58'	GB0476	891	No overlap	Chemo fauna	—	Floating oil
27° 21.99'	92° 23.01'	GB0648	1058	No overlap	—	Oily core	Floating oil
27° 13.02'	91° 03.12'	GC0766	1523	No overlap	—	Oily core	Floating oil
27° 57.88'	90° 43.23'	EB0995	281	Up-wind ends	—	—	—
27° 55.63'	91° 52.86'	SM0205	184	Up-wind ends	—	Oily core	—
27° 48.73'	90° 51.03'	GC0154	715	Up-wind ends	—	Oily core	—
27° 48.65'	91° 07.20'	GC0148	551	Up-wind ends	—	—	—
27° 45.04'	91° 29.48'	GC0229	825	Up-wind ends	—	—	—
27° 42.74'	91° 19.14'	GC0232	807	Up-wind ends	—	Oily core	—
27° 40.52'	90° 49.46'	GC0287	994	Up-wind ends	Chemo fauna	—	—
27° 38.93'	90° 48.39'	GC0331	1037	Up-wind ends	—	—	—
27° 37.26'	91° 17.38'	GC0321	881	Up-wind ends	—	Oily core	—
27° 33.29'	90° 58.55'	GC0415	1049	Up-wind ends	—	—	—
27° 32.55'	91° 00.00'	GC0415	1049	Up-wind ends	—	—	—
27° 32.10'	90° 59.75'	GC0415	1049	Up-wind ends	—	—	—
27° 32.04'	91° 24.65'	GC0451	1027	Up-wind ends	—	—	—
27° 31.96'	91° 23.18'	GC0451	1027	Up-wind ends	—	Oily core	—
27° 31.41'	91° 22.31'	GC0451	1027	Up-wind ends	—	—	—
27° 27.12'	90° 55.64'	GC0504	1181	Up-wind ends	—	—	—
27° 24.89'	91° 22.67'	GC0539	1288	Up-wind ends	—	—	—
27° 24.53'	91° 21.86'	GC0540	1307	Up-wind ends	—	—	—
27° 24.12'	91° 39.70'	GC0577	1274	Up-wind ends	—	—	—
27° 23.88'	91° 21.05'	GC0584	1375	Up-wind ends	—	—	—
27° 22.70'	91° 45.02'	GC0576	1302	Up-wind ends	—	—	—
27° 22.56'	91° 10.25'	GC0588	1502	Up-wind ends	—	—	Floating oil
27° 22.32'	91° 10.45'	GC0587	1558	Up-wind ends	—	—	—
27° 21.39'	91° 03.09'	GC0590	1404	Up-wind ends	—	Oily core	—
27° 21.31'	90° 02.97'	GC0611	1341	Up-wind ends	—	Oily core	—
27° 17.97'	91° 11.95'	GC0675	1622	Up-wind ends	—	—	—
27° 14.35'	91° 36.36'	GC0711	1412	Up-wind ends	—	—	—
27° 13.98'	90° 54.12'	GC0725	1455	Up-wind ends	—	—	—
27° 13.62'	91° 01.21'	GC0723	1421	Up-wind ends	—	Oily core	—
27° 13.56'	90° 47.80'	GC0727	1374	Up-wind ends	—	Oily core	—
27° 12.41'	91° 02.04'	GC0766	1523	Up-wind ends	—	Oily core	—
27° 05.60'	91° 21.69'	GC0892	1999	Up-wind ends	—	—	Floating oil
27° 05.09'	91° 23.75'	GC0891	2006	Up-wind ends	—	—	—
27° 44.24'	91° 18.83'	GC0232	807	Ends undeter.	Oil seep obsd.	Oily core	Floating oil
27° 42.74'	91° 19.14'	GC0232	807	Ends undeter.	Chemo fauna	Oily core	—
27° 31.35'	91° 24.25'	GC0451	1027	Ends undeter.	—	—	—
27° 15.05'	91° 35.83'	GC0711	1412	Ends undeter.	—	—	—
27° 13.76'	91° 02.02'	GC0722	1414	Ends undeter.	—	Oily core	Floating oil
27° 13.57'	90° 48.89'	GC0727	1374	Ends undeter.	—	Oily core	—
27° 48.51'	90° 51.07'	GC0154	715	Ends <2 km	—	—	—
27° 39.64'	91° 22.03'	GC0319	909	Ends <2 km	—	—	—
27° 24.73'	91° 18.44'	GC0541	1374	Ends <2 km	—	—	—
27° 10.42'	91° 12.36'	GC0807	1676	Ends <2 km	—	Oily core	—
27° 47.63'	91° 07.23'	GC0148	551	Random overlap	—	—	—
27° 46.67'	91° 30.39'	GC0185	718	Random overlap	Oil seep obsd.	Oily core	Floating oil
27° 44.47'	91° 12.75'	GC0235	841	Random overlap	—	Oily core	—
27° 43.62'	91° 25.19'	GC0230	825	Random overlap	—	—	—
27° 40.93'	91° 36.46'	GC0271	851	Random overlap	—	Oily core	—
27° 32.79'	90° 58.34'	GC0416	1072	Random overlap	—	—	—
27° 32.60'	91° 32.57'	GC0404	947	Random overlap	—	Oily core	—
27° 28.39'	90° 56.73'	GC0504	1181	Random overlap	—	—	—
27° 27.09'	90° 54.70'	GC0505	1172	Random overlap	—	—	—
27° 26.18'	91° 50.52'	GC0530	1176	Random overlap	—	—	—
27° 22.43'	91° 21.62'	GC0584	1375	Random overlap	—	—	—
27° 20.55'	91° 23.44'	GC0627	1395	Random overlap	—	—	—
27° 18.90'	91° 38.93'	GC0666	1323	Random overlap	—	—	—
27° 18.13'	91° 10.80'	GC0675	1622	Random overlap	—	—	—

[a]Latitude and longitude are for point of overlap.
[b]Lease block designators follow standard MMS format.
[c]Depths given are mean depths for lease block.

[d]Slicks listed as having no overlap were confirmed by coincidence with sea truth collections or observation from submarine.

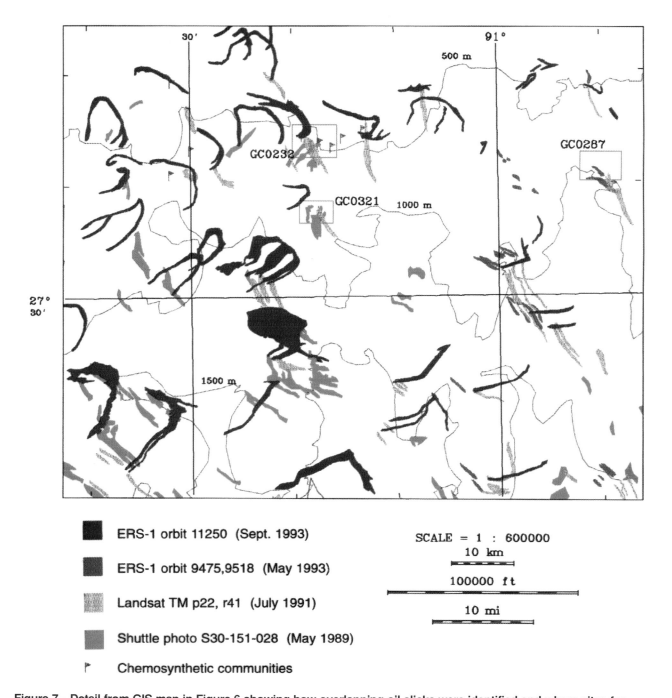

ERS-1 orbit 11250 (Sept. 1993)

ERS-1 orbit 9475,9518 (May 1993)

Landsat TM p22, r41 (July 1991)

Shuttle photo S30-151-028 (May 1989)

Chemosynthetic communities

SCALE = 1 : 600000
10 km

100000 ft

10 mi

Figure 7—Detail from GIS map in Figure 6 showing how overlapping oil slicks were identified and where sites for submarine dives were located.

To use remote sensing in the search for marine seeps, we need to interpret convincingly a variety of data types, including images from different orbital platforms, field observations, and geological records. The approach we have taken high-grades certain targets in a straight-forward manner while retaining other less certain sites for further investigation. By publishing this exhaustive listing of confirmed and probable seeps, we hope to stimulate further exploration that will test our interpretations.

Acknowledgments—We thank BP International Ltd. and Earth Satellite Corp. for supplying data and analyses in support of this paper. We thank the membership of the GOSAP and GEOSAT committees for enlightening discussion and assistance. This work was funded in part by the Minerals Management Service, Gulf of Mexico Regional OCS Office contract no. 14-12-0001-30555 to Texas A&M Research Foundation, GERG, and by the NASA EOCAP Program contract no. NAS13-444 to Earth Satellite Corp. Use of the sub-

marine Johnson Sea-Link was partially funded by the University of North Carolina at Wilmington, National Undersea Research Center. The Marine Spill Response Corporation has provided additional support for presentation of these results at meetings and symposia and for field work.

REFERENCES CITED

Behrens, E. W., 1988, Geology of a continental slope oil seep, northern Gulf of Mexico: AAPG Bulletin, v. 72, no. 2, p. 105–114.

Boland, G. S., 1986, Discovery of co-occurring bivalve *Acesta* sp. and chemosynthetic tube worms *Lamellibrachia*: Nature (London), v. 323, p. 759.

Boss, K. J., 1968, New species of Vesicomyidae from the Gulf of Darien, Caribbean Sea (Bivalvia, Mollusca): Bulletin of Marine Science, v. 18, no. 3, p. 731–748.

Brooks, J. M., H. B. Cox, W. R. Bryant, M. C. Kennicutt II, R. G. Mann, and T. J. McDonald, 1986, Association of gas hydrates and oil seepage in the Gulf of Mexico: Advances in Organic Geochemistry, v. 10, p. 221–234.

Brooks, J. M., M. C. Kennicutt II, C. R. Fisher, S. A. Macko, K. Cole, J. J. Childress, R. R. Bidigare, and R. D. Vetter, 1987, Deep-sea hydrocarbon seep communities: evidence for energy and nutritional carbon sources: Science, v. 238, p. 1138–1142.

Brooks, J. M., D. A. Wiesenburg, H. Roberts, R. S. Carney, I. R. MacDonald, C. R. Fisher, J. Guinasso N.L., W. W. Sager, S. J. McDonald, J. Burke R., P. Ahron, and T. J. Bright, 1990, Salt, seeps and symbiosis in the Gulf of Mexico: Eos, v. 71, p. 1772–1773.

Callender, W. R., G. M. Staff, E. N. Powell, and I. R. MacDonald, 1990, Gulf of Mexico hydrocarbon seep communitiers, V. Biofacies and shell orientation of autochthonous shell beds below storm wave base: Palaios, v. 5, p. 2–14.

Childress, J. J., C. R. Fisher, J. M. Brooks, M. C. Kennicutt II, R. Bidigare, and A. Anderson, 1986, A methanotrophic marine mulluscan (Bivalvia, Mytilidae) symbiosis: mussels fueled by gas: Science, v. 233, p. 1306–1308.

Estes, J. E., R. E. Crippen, and J. L. Star, 1985, Natural oil seep detection in the Santa Barbara Channel, California, with Shuttle Imaging Radar: Geology, April 13, p. 282–284.

Gallaway, B. J., L. R. Martin, and G. F. Hubbard, 1990, Characterization of the chemosynthetic fauna at Viosca Knoll Block 826: LGL Ecological Research Associates, 29 p.

Geyer, R. A., and C. P. Giammona, 1980, Naturally occurring hydrocarbons in the Gulf of Mexico and Caribbean Sea, *in* Marine environmental pollution, I. Hydrocarbons: New York, Elsevier Scientific, p. 37–106.

Kennicutt II, M. C., J. M. Brooks, R. R. Bidigare, and G. J. Denoux, 1988a, Gulf of Mexico hydrocarbon seep communities, I. Regional distribution of hydrocarbon seepage and associated fauna: Deep Sea Research, v. 35, p. 1639–1651.

Kennicutt II, M. C., J. M. Brooks, and G. J. Denoux, 1988b, Leakage of deep, reservoired petroleum to the near surface on the Gulf of Mexico continental slope: Marine Chemistry, v. 24, p. 39–59.

Lee, C. S., W. W. Sager, I. R. MacDonald, and J. F. Reilly, 1993, Active mud vents on a hydrocarbon seep mound, continental slope, Gulf of Mexico: Poster Session, Fall Meeting of American Geophysical Union, San Francisco, December 5–10.

MacDonald, I. R., G. S. Boland, J. S. Baker, J. M. Brooks, M. C. Kennicutt II, and R. R. Bidigare, 1989, Gulf of Mexico chemosynthetic communities II: spatial distribution of seep organisms and hydrocarbons at Bush Hill: Marine Biology, v. 101, p. 235–247.

MacDonald, I. R., W. R. Callender, R. A. Burke, and S. J. McDonald, 1990a, Fine-scale distribution of methanotrophic mussels at a Louisiana slope cold seep: Progress in Oceanography, v. 25, p. 15–24.

MacDonald, I. R., N. L. Guinasso, Jr., J. F. Reilly, Jr., J. M. Brooks, W. R. Callender, and S. G. Gabrielle, 1990b, Gulf of Mexico hydrocarbon seep communities, VI. Patterns of community structure and habitat: Geo-Marine Letters, v. 10, p. 244–252.

MacDonald, I. R., J. F. Reilly, Jr., N. L. Guinasso, Jr., J. M. Brooks, R. S. Carney, W. A. Bryant, and T. J. Bright, 1990c, Chemosynthetic mussels at a brine-filled pockmark in the northern Gulf of Mexico: Science, v. 248, p. 1096–1099.

MacDonald, I. R., N. L. Guinasso, S. G. Ackleson, J. F. Amos, R. Duckworth, and J. M. Brooks, 1993, Natural oil slicks in the Gulf of Mexico visible from space: Journal of Geophysical Research, v. 98-C9, p. 16351–16364.

Roberts, H. H., and T. W. Neurauter, 1990, Direct observations of a large active mud vent on the Louisiana continental slope (abs.): AAPG Bulletin, v. 74, p. 1508.

Rogne, T., I. MacDonald, A. Smith, M. C. Kennicutt II, and C. Giammona, 1993, Multispectral remote sensing data and sea truth data from the Tenyo Maru oil spill: Photogrammetric Engineering & Remote Sensing, v. 59, no. 3, p. 391–397.

Rosman, I., G. S. Boland, and J. S. Baker, 1987, Epifaunal aggregations of *Vesicomyidae* on the continental slope off Louisiana: Deep-Sea Research, v. 34, p. 1811–1820.

Sassen, R., J. M. Brooks, I. R. MacDonald, M. C. Kennicutt II, N. L. Guinasso, Jr., and A. G. Requejo, 1993a, Association of oil seeps and chemosynthetic communities with oil discoveries, upper continental slope, Gulf of Mexico: Gulf Coast Association of Geological Societies, v. 43, p. 349–355.

Sassen, R., H. H. Roberts, P. Aharon, J. Larkin, E. W. Chinn, and R. Carney, 1993b, Chemosynthetic bacterial mats at cold hydrocarbon seeps, Gulf of Mexico continental slope: Organic Geochemistry, v. 20, no. 1, p. 77–89.

Sassen, R., I. R. MacDonald, A. G. Requejo, M. C. Kennicutt II, J. M. Brooks, and N. L. Guinasso, Jr., 1994, Organic geochemistry of cold hydrocarbon vents in the Gulf of Mexico, offshore Texas and Louisiana: Geo-Marine Letters, v. 14, p. 110–119.

Volkes, H. E., 1963, Studies on Tertiary and Recent giant Limidae: Tulane Studies in Geology, v. 1, no. 2, p. 75–92.

Worrall, D. M., and S. Snelson, 1989, Evolution of the northern Gulf of Mexico, with emphasis on Cenozoic growth faulting and the role of salt: GSA Decade of North American Geology, v. A, p. 97–138.

Reilly, Jr., J. F., I. R. MacDonald, E. K., Biegert, and J. M. Brooks, 1996, Geologic controls on the distribution of chemosynthetic communities in the Gulf of Mexico, *in* D. Schumacher and M. A. Abrams, eds., Hydrocarbon migration and its near-surface expression: AAPG Memoir 66, p. 39–62.

Chapter 4

Geologic Controls on the Distribution of Chemosynthetic Communities in the Gulf of Mexico

James F. Reilly, Jr.

Enserch Exploration, Inc.,
Dallas, Texas, U.S.A.

Present address:
National Aeronautics and Space Administration
Johnson Space Center
Houston, Texas, U.S.A.

Ian R. MacDonald

Geochemical and Environmental Research Group
Texas A&M University
College Station, Texas, U.S.A.

E. K. Biegert

GEOSAT Committee
Norman, Oklahoma, U.S.A.

James M. Brooks

Geochemical and Environmental Research Group
Texas A&M University
College Station, Texas, U.S.A.

Abstract

Communities of chemoautotrophic organisms have been observed at multiple sites on the continental slope of the Gulf of Mexico where natural seepage of hydrocarbons has been recognized. Effects of this seepage are readily seen as modification of the sea floor in remote sensing data, direct observation, and sampling. Though faunal distribution at these sites is clustered, indicating an external control on community siting, the mechanism affecting the uneven distribution of organisms within a seep site is poorly understood. Possible geologic controls are likely to have influenced the development of environments conducive to colonization in four regions on the continental slope where chemoautotrophic fauna have been documented or where indications of active seepage and sea floor modification are known.

Distribution mapping of organisms from submersible observations were merged with surface and near-surface structure mapped from 3-D CDP-processed seismic data. Where organisms are present, a first-order correlation of community occurrence with surface expression of faulting was noted. Complex communities containing vestimentiferan tubeworms were found in areas where deeply rooted faulting occurred in response to simple shear. Biomass is concentrated along the surface traces of antithetic faults where extensive outcrops of authigenic calcium carbonate on the sea floor and evidence of active continuous seepage occur. Other sites contain simple communities of a single species lacking a vestimentiferan component. These communities, containing methanotrophic mytilids or calyptogenid or lucinid clams, exist in areas dominated by Coulomb shear resulting from halokinesis or mass failure. Other sites contain insignificant biomass but have extensive hydrocarbon seepage. Structure at these sites is dominated by shallow-piercement halokinesis reflecting spatial and temporal discontinuities in the seepage history, or recent initiation of seepage. A correlation is proposed between the occurrence of significant chemosynthetic communities and structural forcing mechanisms that developed the seep substrate.

Figure 1—Regional map showing location of survey sites included in this study. The box outlines the orbital imagery coverage from MacDonald et al. (1993).

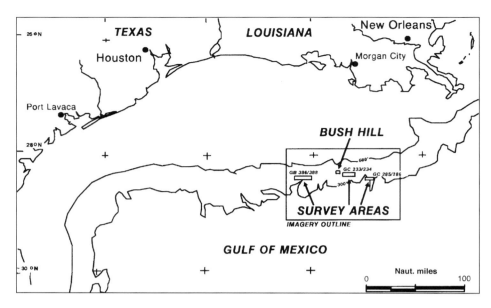

INTRODUCTION

Communities of chemoautolithotrophic tubeworms (*Lamellibrachia* sp. and *Escarpia* sp.) and bivalves (*Calyptogena ponderosa, Vesicomya cordata,* and *Pseudomiltha* sp.) have been documented at approximately 45 sites on the continental slope of the Gulf of Mexico (Kennicutt et al., 1985, 1988a; Brooks et al., 1987a,b, 1990; MacDonald et al., 1990a,b). These communities are taxonomically similar to the hydrothermal vent communities discovered along the East Pacific Rise. However, the Gulf communities differ from those communities in that the compounds forming the chemosynthetic subtrate (CH_4, HS^-, and H_2S) are delivered by natural hydrocarbon seepage rather than geothermal discharge.

Complex high-density communities consisting of several genera have been documented on the upper slope (MacDonald et al., 1989, 1990c; Brooks et al., 1989). These complex communities generally contain locally dense aggregations of the vestimentiferan tubeworms in close association with mussels and/or clams within discrete, localized sites. Seep fluids collected near the sediment–seawater interface in the vicinity of these communities contain significant concentrations of thermogenic, high molecular weight hydrocarbons (Brooks et al., 1989; MacDonald et al., 1989). Simple communities, consisting mainly of a single genus, occur where the seep fluid is dominated by methane, as in the case of the methanotrophic mussels or where the high molecular weight hydrocarbons have been highly degraded leaving high concentrations of sulfide compounds favored by the clams (Sassen et al., 1994). The variability in the geochemical substrate indicates subsurface controls on seep fluid type and flux.

This paper compares the distribution of chemosynthetic communities at several sites to subsurface structure as mapped from 3-D common depth point (CDP) processed seismic data. Study sites include complex communities and low-density, poorly developed occurrences.

Identification of relationships between large-scale geologic structures associated with seepage and fine-scale community occurrence may offer insight to the sequence of geologic and biological processes. Case studies are an important component for understanding the basinwide characteristics of oil and gas seeps and the organisms dependent on them.

GEOLOGIC SETTING AND BACKGROUND

Natural hydrocarbon seepage in the Gulf of Mexico is a widespread phenomenon with both mesoscale and microscale characteristics (Geyer and Giammona, 1980; Anderson et al., 1983; Brooks et al., 1985; MacDonald et al., 1994). Remote sensing shows that extensive slicks are present throughout much of the Central Gulf, indicating that seepage is an ubiquitous process (MacDonald et al., this volume). Sediment sampling and observation of active seepage are limited by being highly localized in their area of investigation (Anderson and Bryant, 1989; Brooks et al., 1985; Behrens, 1988; Roberts et al., 1992). High-frequency geophysical techniques image relatively large areas of the sea floor where seep conditions are present, but distortion of the geophysical signal by low concentrations of hydrocarbons or the byproducts of biodegradation in the sediments make it difficult to distinguish more subtle features of seep areas, such as biogenic versus thermogenic hydrocarbons (Anderson and Bryant, 1989; Neurauter and Bryant, 1990).

Piston core sediment samples acquired during 1986–1987 from sites characterized by attenuated seismic return confirmed the presence of thermogenic hydrocarbons at 23 sites, with free oil in the sediments at 6 sites. Otter trawls recovered chemoautotrophs at 19 sites, with significant recoveries of calyptogenid or vesicomyid clams at 12 sites, seep mytilids at 3 sites, and vestimen-

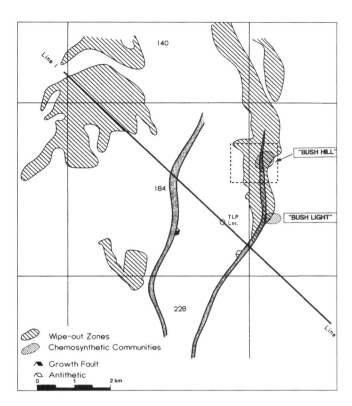

Figure 2—Surface expression map of the region containing the Bush Hill community illustrating areas of near-surface attenuation of high-frequency seismic signal, generalized traces of surface-cutting growth and antithetic faults (dark gray), and locations of the chemosynthetic communities.

tiferans at 15 sites (Kennicutt et al., 1988a). Anderson and Bryant (1989) have described the median size for areas exhibiting seismic characteristics of near-sea floor hydrocarbon saturation on the upper slope as about 0.6 km². Submersible surveys would subsequently find that biological communities consisted of locally dense patches on the order of tens of meters or less clustered in areas of less than 0.5 km² (MacDonald et al., 1989, 1990c).

No simple relationship exists between modification of the seismic signal and the presence of near-surface hydrocarbon concentration or biological communities. Furthermore, biological and geochemical studies are incapable of discerning the mechanisms of seep formation and preservation. By studying these data in concert, however, it is possible to form robust inferences about both structural control on the fine-scale distribution of seepage and the persistence of active migration in recent time.

MATERIALS, METHODS, AND STUDY AREA

Site selection for this study was based on data coverage and community development (Figure 1). The presence and continuity of natural slicks on the sea surface in

the vicinity of the study sites were observed in a time series of orbital imagery obtained from Landsat, *ERS-1*, and the space shuttle *Atlantis* (MacDonald et al., this volume). The outline of this coverage is shown in Figure 1.

The primary data sets at each site are 3-D CDP-processed seismic data recorded and processed to near-minimum phase and filtered to deliver a dominant frequency content of 10–60 Hz. Mapping of the sea floor and subsea floor structure was carried out on Landmark Graphics Corp. workstations. Regional bathymetric control was obtained from the USGS global relief data set. Additional site-specific bathymetry was supplied from high-resolution echosounder surveys conducted as part of geophysical hazard surveys, submersible surveys, and engineering surveys. Ground truth for selected terrains was supplied from sediment samples collected from surface ships. Community distribution and diversity at each site were documented from samples collected and photographic and video records made during cruises of the U.S. Navy's *NR-1* and Harbor Branch Oceanographic Institution's *Johnson Sea-Link I and II* submersibles.

SITE DESCRIPTIONS

Green Canyon Block 185 (Bush Hill)

The "Bush Hill" community in Green Canyon Block 185 is located in about 560 m water depth at lat. 27° 47' N, long. 91° 30.4' W. It is notable because thermogenic hydrates were first described here in the Gulf of Mexico (Brooks et al., 1984). Trawl recoveries in 1984 and submersible investigations of the site beginning in 1985 documented the first known occurrence of chemoautotrophic organisms similar in taxonomy to the hydrothermal vent organisms existing in a hydrocarbon seep environment (Kennicutt, et al., 1985). Over the past decade, the Bush Hill community has been extensively documented, becoming the *de facto* type locality for complex chemosynthetic communities in the Gulf of Mexico (Kennicutt et al., 1988a; Brooks et al., 1984, 1987a,b, 1989, 1991; MacDonald et al., 1989, 1990c).

This community is a dense, diverse assemblage of chemosynthetic organisms dominated by vestimentiferan tubeworms. It is located on the crest of a "hydrate hill" that has an areal extent of about 500 m by 500 m within a much larger area of gas-saturated sediments, as interpreted from shallow seismic data (Figure 2). Sediment samples obtained from the vicinity of the community characteristically contain a significant amount of thermogenic petroleum (Kennicutt et al., 1988b). Discharges of free streams of gas and occasionally liquid petroleum are commonly observed during submersible operations at the site. A perennial slick is commonly seen at this site and is can be seen extending south of the site in satellite imagery.

Both this community and a second community ("Bush Light") to the south are located on the sea floor trace of a north-south trending normal fault developed in response

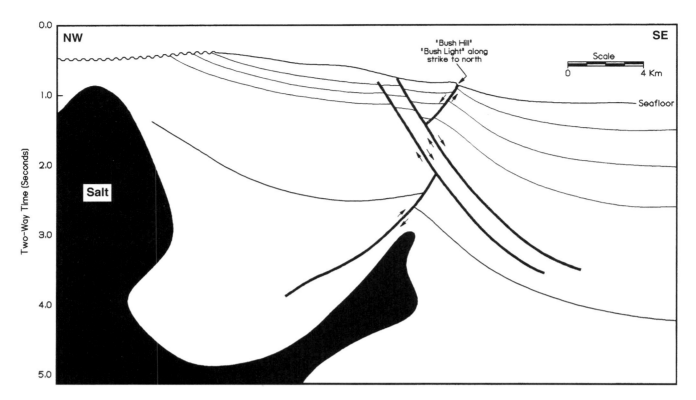

Figure 3—Schematic drawing of a seismic profile (line 1 on Figure 2) showing the subsurface structure beneath the Bush Hill community. Recent movement of the salt diapir to the northwest is evidenced by erosional truncation of the shallow horizons.

to mass movement over a salt-supported anticline. The line drawing of a CDP seismic section in Figure 3 (modified from Cook and D'Onfro, 1991) illustrates the west-dipping fault plane beneath the communities as antithetic to an east-dipping normal master fault that intersects the sea floor about 3 km west of the communities. A notable lack of both near-surface hydrocarbon indicators and chemosynthetic organisms along the western master fault suggests that fluid migration is preferentially captured by the antithetic system and concentrated at localized efflux sites. Continuous seepage at these sites has generated an environment suitable for colonization. This apparent linkage of faunal density with areas of enhanced seep flux was also noted in the community by MacDonald et al. (1989), who described a distinct clustering within the Bush Hill community near surface expressions of sea floor fractures.

Formation of the mound is a subject of some debate. Neurauter and Bryant (1990) suggest that the mound was formed as the result of heave or diapiric processes rather than sedimentary accretion. An alternative model by C. S. Lee (personal communication, 1995) proposes that the mound was constructively developed by vertical accretion in response to continuous seepage and subsequent hydrate formation. Recent observations of hydrate extrusion and detachment (MacDonald et al., 1994) support Lee's model. The presence of apparently continuous seepage at highly localized sites requires preservation of

conduits along the fault plane. Salt mobilization at intermediate depths, as proposed by Cook and D'Onfro (1991) and Roberts et al., (1992), combined with sediment compaction and rotation would provide the mechanisms for uninterrupted creep along the fault surfaces.

Green Canyon Blocks 233/234

The Green Canyon 233/234 site, located about 280 km southwest of the Mississippi River Delta, is notable in being the first documented occurrence of natural seepage of thermogenic hydrocarbons on the continental slope of the Gulf of Mexico (Anderson et al., 1983). Chemoautotrophic organisms occur within three distinctly different sites inside a zone of acoustic attenuation, or "wipe-out," noted on high-frequency geophysical data covering about 85% of the study area. To determine the spatial distribution of chemosynthetic organisms in the wipe-out area, a survey of the region was conducted in May 1989 with the submarine *NR-1*. Within this area, six transects covering 45 km were surveyed, with multicellular chemosynthetic organisms observed along 4.7 km, or 10.4% of the area. The majority of the biomass noted was concentrated within three zones: the complex tubeworm community in the northeast quadrant of Block 234, the "Mussel Beach" seep mytilid community in the southwest quadrant of Block 234, and the "NR-1 Brine Pool" mytilid community in the southeast quadrant of Block 233.

The bathymetry of the study area region is highly variable, ranging from low-relief gentle slopes (1°) north of the study sites to areas having slopes up to 12° at the head of a submarine canyon trending south-southeast just south of the sites. In the northeast quadrant of Block 234 near lat. 27° 44' N, long. 91° 13' W in about 560 m water depth, a densely populated complex community exists (Figure 4a) that consists of numerous large tubeworm clusters with an auxiliary bivalve population (Brooks et al., 1989, 1990). About 4 km west at lat. 27° 44.1' N, long. 91° 32.3' W in 640 m water depth, the Mussel Beach community comprised almost exclusively of seep mytilids (Figure 4b) is located in a zone of low-relief sea floor troughs (MacDonald et al., 1990a). Farther south and west, in the southeast quadrant of Block 233 at lat. 27° 43.4' N, long. 91° 16.8' W in 650 m water depth, a unique mytilid community is present as a 3–4-m-wide ring surrounding a stable brine pool (MacDonald et al., 1990b) (see Figure 9).

Visual documentation, sidescan sonar, vertical incidence sonar, and detailed bathymetric records of each of these sites were obtained from submersible surveys carried out in 1989, 1990, and 1991. Subsurface structural control is provided by 3-D CDP-processed seismic data acquired during 1985. Sea floor sediment samples were obtained by piston core acquisition between 1984 and 1991.

Complex Community

The complex community site located near lat. 27° 44' 34"N, long. 91° 13' 15" W in the northeast quadrant of Block 234 exhibits a diverse biota consisting of multiple clusters of vestimentiferan tubeworms with subordinate colonies of methanotrophic mytilids (Figure 4a). This area has been documented as an active thermogenic hydrocarbon seep site by Anderson et al. (1983) and Behrens (1988) and as a chemosynthetic community site by Brooks et al. (1989). Bathymetrically, the area containing the community is located to the east of a prominent ridge and at the transition between gentle sea floor gradients to the north and more steeply dipping gradients to the south.

Sea floor topography in the vicinity of the community is dominated by a series of surface cutting faults that exhibit a bimodal, rectilinear distribution oriented approximately east-west and northwest-southeast forming a series of grabens and half-grabens within a shallow depression on the eastern side of a structural nose. Surface traces of these faults have been labeled I through VIII from north to south (Figure 5). Distribution mapping of chemoautotrophic fauna from submersible surveys conducted during 1988, 1989, and 1994 show apparent concentrations of the communities in the north-central region of the study area and little or no settlement along the surface traces of the southern faults.

Sidescan sonar recorded by submarine *NR-1* around the community revealed a highly reflective, irregular surface over generally positive topographic features. Visual observations indicate that these features are outcrops of authigenic carbonate, some of which have several meters

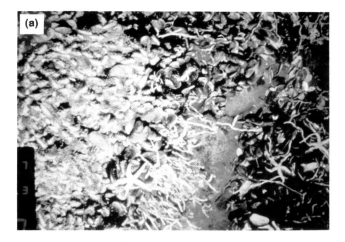

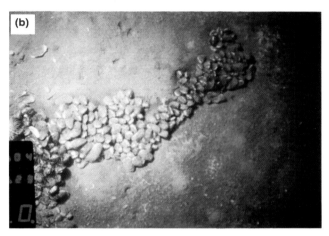

Figure 4—Photographs of community organisms at the Green Canyon 234 site. (a) Vestimentiferans and mytilids within the complex community site; image size is about 2.5 × 3.5 m. (b) Mytilid cluster occupying a shallow depression in the Mussel Beach community; image size is about 2 × 3 m.

of relief. When compared with the surface traces of faults I–VIII, the carbonate outcrop pattern noted in the sidescan sonar records generally mimics the fault fabric.

The subsurface structure beneath this community is illustrated by a north-south oriented 3-D CDP seismic profile (line CC-1, Figure 6). Annotation of observed chemosynthetic occurrence illustrates the location of the communities relative to these faults. At shallow depths, the structure is dominated by a series of down-to-the-north normal faults with a subordinate set of antithetic down-to-the-south faults. Faults I, II, IV, and V are antithetic to faults III and VI. Although fault VI shows relatively minor offset at the sea floor, it forms the primary, or master, fault to faults I–IV. On the southern boundary of the community site, faults VII, VIIa, and VIII are seen as the updip failure surface of the recent slump block that truncates faults IV and VI to the southeast. Truncation of the normal faults by the low-angle slump fault indicates

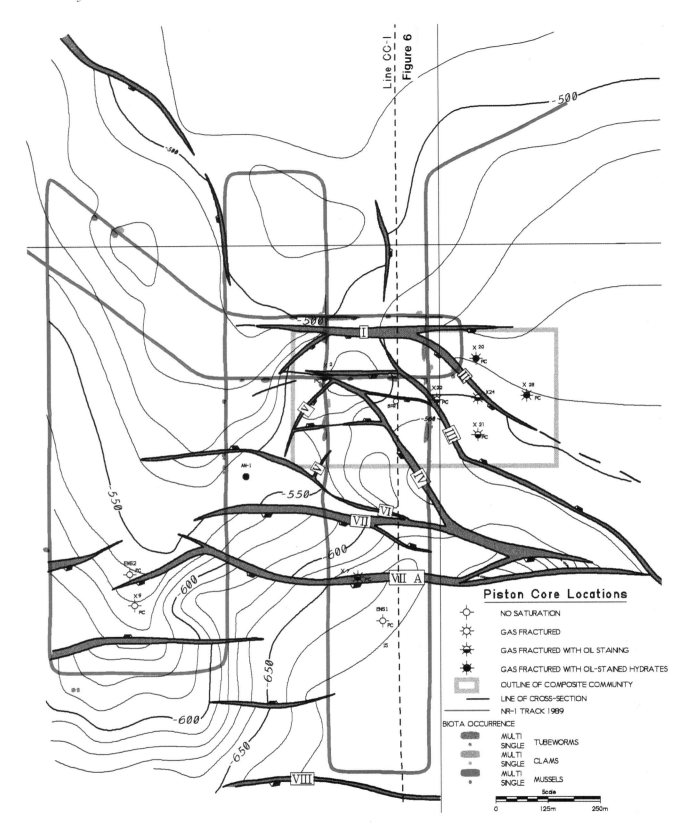

Figure 5—Sea floor structure map of the complex community site annotated with the 1989 *NR-1* survey observations and piston core locations.

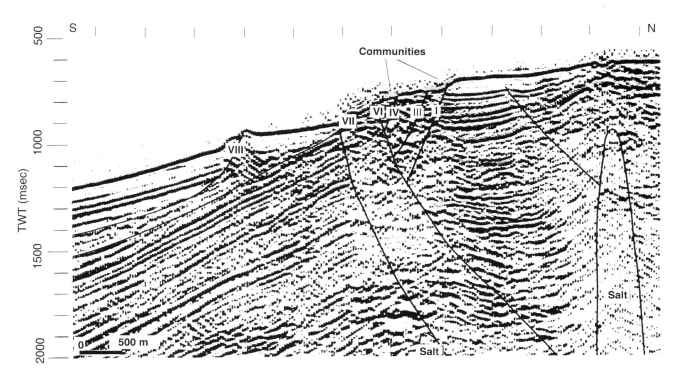

Figure 6—North-south 3-D CDP seismic profile CC-1 crossing the complex community site. See Figure 5 for location of numbered faults (I–VIII).

that mass failure postdates the formation of the normal faults. Although they exhibit extensive sea floor offset, the surface traces of these faults are barren of the tubeworm concentrations seen to the north. The lack of colonization near the headscarp suggests that the slump surfaces have little influence on the delivery of thermogenic seep fluids and may actually inhibit vertical migration at this site.

Salt is present north of the community as a shallow diapir that rises to within 300 m of the sea floor. Beneath the community, an intermediate-depth diapir rises to within 1000 m of the sea floor. These structures are part of a regional complex of diapirs that appear rooted to a common sill. Growth histories based on palinspastic reconstructions indicate that emplacement of the northern salt occurred recently, while the intermediate-depth structure demonstrates early growth followed by a series of reactivations to the current day. These periods of reactivation can be seen in the seismic section as stratigraphic thinning updip toward the salt. Interestingly, although the largest surface slicks are seen in satellite images in the vicinity of the shallow piercement north of the site and hydrocarbon saturation at the sea floor has been confirmed, there is no apparent colonization (N. L. Guinasso, personal communication, 1994).

In summary, the chemosynthetic communities at the GC 234 site are best developed along the surface expressions of normal faults that resulted primarily from sediment mass movement at intermediate depth with secondary structural development related to halokinesis. High concentrations of thermogenic hydrocarbons recovered at the sea floor confirm migration along these faults

of sufficient duration to generate conditions suitable for widespread colonization. Organisms are not present along the surface expression of the younger faults generated in response to slope failure in the southern part of the site nor at the northern shallow piercement exhibiting recent emplacement.

Simple Community

About 4 km west-southwest of the complex community site, the Mussel Beach community consists of dense unifaunal aggregates of the methanotrophic mussel (seep mytilid II n. sp.) which occupy shallow, linear depressions in the sea floor (Figure 4b). MacDonald et al. (1990a) described the community as a series of dense curvilinear clusters up to 5 m in length with living adults and shell remains in adjacent clusters. The presence of actively recruiting sites in close proximity to nonrecruiting or defunct clusters suggests a spatial variation in environmental conditions probably controlled by ephemeral seep fluid delivery.

The site as originally described covers an area of about 10 ha (Brooks et al., 1989; MacDonald et al., 1990a) at an average water depth of 650 m. The area of community coverage has since been enlarged as additional mytilid colonies along the 650-m isobath east of the community were noted during the 1989 NR-1 cruise. Within the community site, patches of the filamentous bacteria Beggiatoa are common, and free gas has been seen being discharged from the sea floor at several locations. Gas hydrates with a notable lack of high molecular weight hydrocarbons have been recovered in piston core samples and push

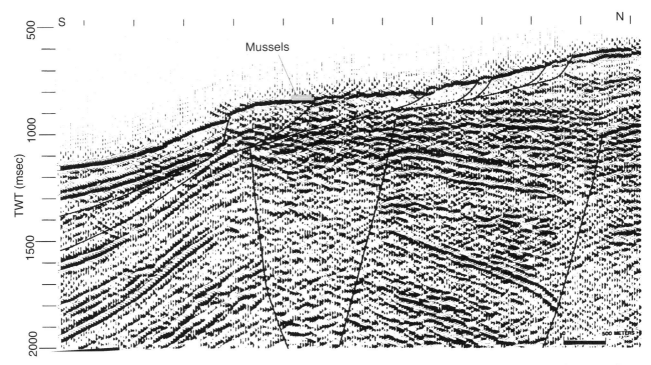

Figure 7—North-south 3-D CDP seismic profile MB-1 crossing the Mussel Beach community site illustrating organism occurrence and subsurface structure.

cores retrieved by submersible. Isotopic analyses of the recovered hydrocarbons, however, reflect a mixed thermogenic and biogenic source (Reilly, 1995). Brine discharges at the sea floor are likely based on the presence of darkly stained surficial sediments and elevated salinities in core samples (MacDonald et al., 1990a). Although some outcrops of authigenic carbonate were noted to the east during the 1989 cruise, there is a notable lack of carbonate present at the community site.

The community is located within the headscarp region of a recent slump at the head of the submarine canyon extending to the south-southeast of the site. The subsurface structure, illustrated on profile MB-1 (Figure 7), is dominated by the shallow slump above a north-dipping set of faults antithetic to a south-dipping listric normal fault. The north-dipping faults are equivalent to the north-dipping fault VI that forms the southern limit of the complex community. Structure is incidentally modified only by diapiric salt that rises to within 1000 m of the sea floor 1.5 km to the west of the site but is >2000 m below the community. Surficial faulting is developed in response to rotation and translation of the slump block, with little evidence of surface effects generated by fault development at depth.

Migration to the community is problematic because no direct pathway to deliver thermogenic components from depth to the site is recognized. The north-dipping faults that could form likely conduits have no surface expression and appear to be truncated by the failure surface of the slump block, which likely blocks vertical migration. However, a structure map of the detachment surface at

the base of the slump (Figure 8) show that the community location in the headscarp region of the slump is at a localized potentiometric high point on the surface. As a result, migrating hydrocarbons reaching the detachment surface downslope would be collected and delivered to the communities at the head of the slump. Collection of both thermogenic and microbial hydrocarbons would generate the mixed isotopic signature seen from the recovered samples.

NR-1 Brine Pool

The *NR-1* Brine Pool located at lat. 27° 43.4′ N, long. 91° 16.5′ W in 650 m water depth is an elliptical brine-filled depression surrounded by a ring of methanotrophic mytilids atop and inside a low-relief berm rising about 6 m above the surrounding sea floor. The pool was discovered during the 1989 *NR-1* reconnaissance cruise and is described by MacDonald et al. (1990b). It lies within a zone of subsurface acoustic signal attenuation and high-amplitude strength surface returns noted for the region that contains the complex community and Mussel Beach sites.

In the 25-kHz records taken during the survey (Figure 9), the brine exhibits a sharp, flat density interface with a second interface noted about 1.5–2 m below the first. A free stream of bubbles emanates from the center of the pool in a low-volume discharge. Compositional and isotopic analyses indicate that the gas is composed primarily of microbial methane ($\delta^{13}C = -63.8‰$) (MacDonald et al., 1990c) and is being released from decomposing

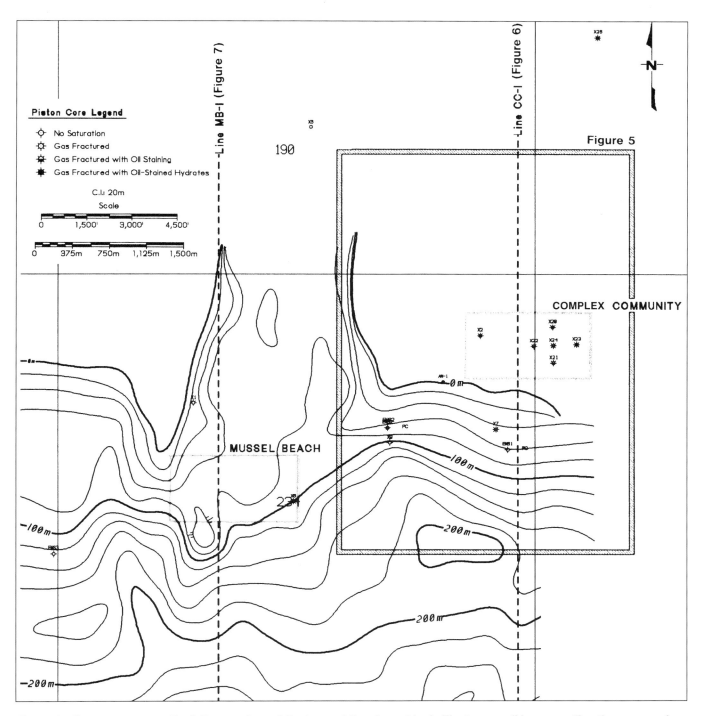

Figure 8—Structure map on the failure surface at the base of the slump block. The two small boxes outline the communities; the large box outlines the approximate boundaries of the sea floor structure map in Figure 5.

hydrates. Samples collected from the brine during 1993 show that the upper layer consists of clear, anoxic brine overlying a turbid, more concentrated brine. Temperatures in the upper layer are about 1.5–2°C warmer than the ambient seawater temperature, while the deeper layer is 10°C warmer. Thermal convection is therefore active in the lower layer, while the upper layer is relative-

ly stable. Gas discharge from the surface of the pool and the thermal convection cell in the lower brine infer that recharge is occurring from below where continuous agitation of the sediment, brine, and methane maintains a stable conduit.

Figure 10 (profile BP-1) shows the shallow subsurface dominated by the recent south-verging slump formed by

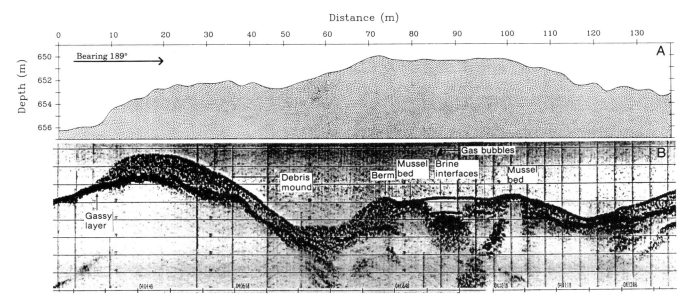

Figure 9—Subbottom seismic profile (at 25-kHz) crossing the *NR-1* Brine Pool community.

mass failure previously described for the Mussel Beach community. Beneath the slump block, a salt diapir rising to within 500 m of the sea floor forms the probable source of the brine. Normal faults developed above the diapir intersect the base of the slump and may displace the failure surface indicating postslump deformation. These faults apparently have little or no recent growth because they do not appear to offset the sea floor.

A model of pool formation assumes an initially stable sea floor with a stable hydrate layer formed near the sediment–seawater interface (Figure 11). Remobilization of the underlying salt diapir, generating increased pore pressures in the overlying sediments, helped to initiate mass failure. Following downslope movement of the slump block, upward fluid migration was initiated along the normal faults in response to the resultant reduction in overburden. With fluid movement, a thermal plume would likely develop, ultimately destabilizing the near-surface hydrate layer. As the hydrate decomposed from below, liberated gas built up under the remaining hydrate, eventually breaching the sea floor. The resulting rapid discharge excavated a deep, steep-sided crater. After the initial discharge, brine influx filled the crater, eventually forming a convection cell within the brine. Mytilid colonization occurred as the upper brine layer approached thermal equilibrium with the overlying water column, and a stable interface was formed.

Slow influx maintains the brine level near the spill point of the crater. Sulfide staining downslope from the pool indicates that occasional flow discharges do occur, although the lack of significant mortality within the community over which these discharges flow suggests that the pool is stable. The mechanism that maintains fluid level within the pool is unknown, but based on the small temperature differential between the upper brine layer

and the overlying water column, diffusive losses across the interface may be matching the influx volume into the lower layer. Deeper in the pool, a similar interaction between the layers must be occurring in a similar fashion to form the intermediate interface.

Although mytilid growth histories at this site are not completely understood, it is clear that colonization of the pool edge requires a stable settlement substratum. Comparisons with recent uncolonized pools suggest a sequence whereby the pool edge is initially steep but gradually erodes to the sediment's angle of repose. The mussel community at this site suggests that the formation of the pool is not a recent event. In any case, this site is unique in being a stable source of dissolved methane to a community in the form of a brine pool.

Garden Banks Blocks 388/386

The Garden Banks 388 and 386 sites are located about 140 km offshore Louisiana near lat. 27° 37′ N, long. 92° 13′ W in about 700 m and 600 m water depth, respectively (Figure 1). Regional water depths range from 400 to 850 m with locally steep bathymetric gradients north, northwest, and southwest of the GB 386 site. Widespread signal attenuation effects on high-frequency geophysical data similar to that seen at the GC 184 and GC 233/234 sites are recognized at both sites. A single piston core recovered in 1987 at the GB 388 site confirmed the presence of thermogenic hydrocarbons (Kennicutt et al., 1988a). Additional piston coring conducted in 1989, based on attributes recognized in 3-D CDP data, confirmed the presence of high molecular weight hydrocarbon and authigenic calcium carbonate at both sites. In addition, thermogenic hydrates were recovered at the GB 388 site.

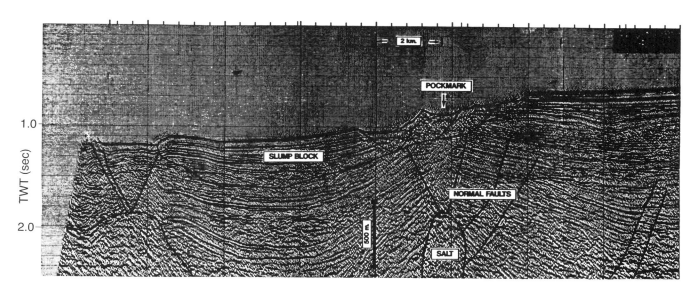

Figure 10—North-south 3-D CDP seismic profile BP-1 showing the location of the Brine Pool community above the shallow piercement and within the slump-altered sea floor. The top profile shows the corrected topography.

Sea floor and subsurface structure and seep attributes were mapped using commercially acquired 3-D CDP-processed seismic data acquired in 1988 and reprocessed in 1991. These data were migrated using a bin spacing of about 27 m × 27 m with high and low cut filters of 10 and 60 Hz, respectively. Depth control at the sea floor was provided from submersible data and engineering surveys conducted during drilling operations. Subsurface depth control was obtained from subsurface velocity surveys conducted in exploration wells drilled within 1–5 km of the sites. Geochemical and petrographic data were obtained on the piston cores acquired in 1988, 1989, and 1994.

Both sites are in areas where localized halokinetic structures are present. Diapirs present beneath the GB 388 and 386 sites rise to within 500 and 200 m of the sea floor, respectively. The diapir below the GB 388 site is a localized uplift apparently deeply rooted and independent of the salt massif to the northwest. The GB 386 dome, however, appears to be a parasite diapir developed on the southeastern margin of the massif across an intervening subbasin.

Garden Banks Block 388

The Garden Banks 388 site located near lat. 27° 36.75′ N, long. 92° 11.75′ W was one of 33 sites investigated as part of a 1987 study of potential chemosynthetic community sites (Kennicutt et al., 1988a; Brooks et al., 1989). A reconnaissance survey of the site was conducted with the *NR-1* during 1988. Additional site-specific surveys were conducted with the *Johnson Sea-Link I* submersible during 1992 and 1993. To document the biomass distribution at this site more accurately, a second *NR-1* survey consisting of seven north-south transits covering 17 line-km was carried out during 1993.

Sea floor topography at the site consists of a shallow depression within a low-relief break in an otherwise south-dipping slope (Figure 12). Visual observations of the sea floor in the depression reveal generally smooth, dark gray clay with little or no bioturbation. Carbonates, though present in isolated occurrences, do not form the massive outcrops as seen in the GC 234 and Bush Hill sites. The benthic community is limited to occasional scattered lucinid clamshells, one mytilid community about 3 m² in size, and several small (<0.3 m²), isolated, poorly developed vestimentiferan aggregations (Figure 13a). Lucinid occurrences are generally found in tightly concentrated groupings near or on local topographic highs (Figure 13b) thought to be sites where active discharge has excavated the sea floor. Several narrow, steep-walled craters in the area may represent recently active discharge sites. Overturned carbonate plates at the mouth of one of these craters suggest vigorous initial discharge. Seep discharge is evidently discontinuous, however, as active seepage observed at three locations during submersible operations in 1988 and ROV and sidescan sonar data acquisition during 1994 were not detected during submersible visits in 1989, 1991, and 1993. Extensive oil slicks were occasionally observed from ships at the sea surface. Also, the characteristic signature of a surface slick is not present on an *ERS-1* satellite image recorded on May 22, 1993.

On 3-D CDP geophysical data, a prominent high-amplitude strength, negative reflection coefficient return at the sea floor is seen in the depression. An rms amplitude strength extraction of an 80-msec envelope at the sea floor illustrates a distinct increase in relative amplitude strength at the observed sea floor discharge sites and where hydrocarbon saturated sediments were recovered in piston cores.

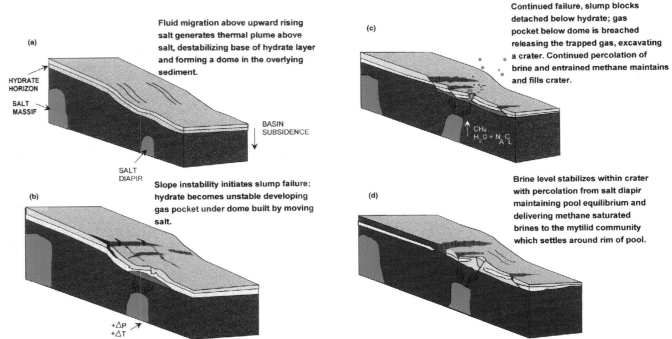

Figure 11—Conceptual model for the structural development of the *NR-1* Brine Pool community site, Green Canyon Block 233. (a)–(d) Block diagrams showing chronologic stages of development.

Headspace analyses of sediment samples recovered from piston cores across the site are plotted in Figure 14 as a north-south profile in two components: C_1–C_4 gases and C_{5+} gasoline-range liquids. An extracted CDP seismic profile closely paralleling this sampling profile illustrates the location of the core points relative to the near-surface structure and anomalous amplitude response (Figure 15).

Two observations can be made at this point. First, the presence of increased hydrocarbon content in the piston core samples is correlatable to increased reflection strength on the CDP data. Second, the increased C_{5+} concentrations are located in a graben formed between two south-dipping normal faults labeled A and A-1 in Figure 15 and a north-dipping antithetic fault labeled 1-A. Furthermore, virtually no free hydrocarbons are noted upthrown to fault A, while the C_{5+} range hydrocarbon saturations are concentrated between faults A-1 and 1-A.

Fault plane maps for faults A, A-1, and 1-A (Figure 16) show dip angles for each of the fault planes across the seep site and to the east and west away from the seep. Dip angles of about 60° at all sites suggest that faults A-1 and 1-A developed in response to Coulomb shear (σ_1 is vertical), possibly in response to recent movement of intermediate-depth salt. Fault A appears to be responding to Coulomb shear east and west of the site, but in the vicinity of the seep, the dip shallows to <40°, which suggests that the fault plane has been warped by the underlying salt.

Palinspastic reconstruction across the graben suggests that faults A-1 and 1-A are recent features. Fault A, however, appears to reflect early salt injection followed by diapiric stasis and subsequent reactivation (Reilly, 1995). Figure 17 shows a simplified model of this site. Early growth is represented as downslope mass failure generating a series of south-dipping normal faults that contain fault A. In the stage (Figure 17b), salt remobilization generates renewed uplift, which locks fault A and deforms the fault plane above the diapir. Resultant sealing of the fault traps migrating hydrocarbons in the shallow subsurface until renewed diapiric movement in the last stage generates extension above the diapir, reactivation of fault A, and development of faults A-1 and 1-A. Seep discharge is resumed as the shallow hydrocarbon traps are breached by the renewed fault movement.

Garden Banks Block 386

The Garden Banks block 386 site is located about 3 km west of the GB 388 site near lat. 27° 37' N, long. 92° 15 W in about 600 m water depth. The sea floor south of the site slopes 3–5° to the south and southeast, while to the north and northwest, the slope increases to 10–15° over a shallow piercement salt massif that rises to within 300 m of the sea floor. In the central part of the study area, a prominent plateau about 760 m in diameter rises about 20 m above the sea floor (Figure 18). An rms amplitude strength extraction exhibits enhanced signal strength on the plateau. Compared to the sea floor reflector at the GB 388 site, the sea floor return at this site displays a positive reflection coefficient on the CDP seismic data. Carbonate rubble recovered in piston core samples and submersible observations in 1989 and 1991 confirmed the presence of a carbonate veneer across the plateau.

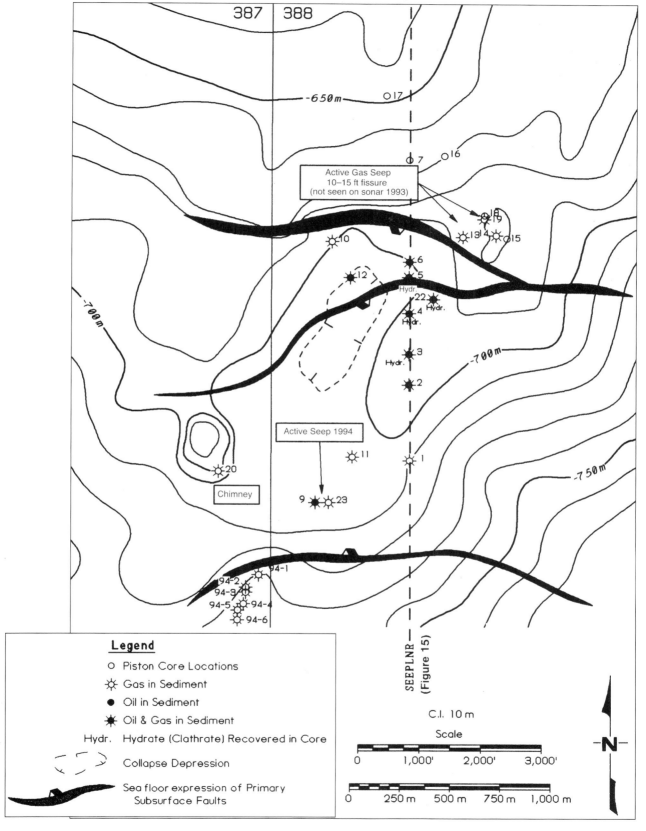

Figure 12—Sea floor structure map, Garden Banks 388 site. The locations of near-surface acoustic signal attenuation, active seeps observed during submersible surveys, and piston cores are noted.

Figure 13—(a) Photograph of the *Escarpia* sp. community at the Garden Banks 388 seep site; image is about 6 m across. **(b)** Photograph of a *Vesicomyid* community at the same site; image is about 5 m across.

To document the distribution of chemosynthetic organisms across the plateau, a series of seven submersible transects covering a total of 34.2 line-km were carried out with the *NR-1* during 1989 (Brooks et al., 1991). Organism types and relative concentrations mapped from video records and annotated against bathymetry displayed a distinctive segregation of faunal types across the plateau. Although locally dense concentrations of shells were recorded across the interior of the plateau, the communities appeared to be defunct, as no living bivalves were recovered during submarine surveys. At the margins of the plateau, isolated tubeworm bushes were noted. No concentrations of the seep mytilids were seen.

Two extracted seismic lines, profile CS-1 crossing the plateau from southwest to northeast and profile CS-2 crossing from northwest to southeast, illustrate the near-surface structure on the flat plateau (Figures 19a, b). To show organism distribution, both lines were oriented parallel to submersible transects where significant concentrations of clams were encountered. Surficial faulting is limited to the plateau area. The intermediate-depth fault complexes seen at the previous sites are not apparent here.

Sediment samples in piston cores recovered from the plateau contain tars in a dark, anoxic, sulfide-rich mud capped with carbonate nodules. Gas chromatography–flame ionization detector analyses of hydrocarbon extracts from the sediments indicate the presence of a highly degraded oil in the section. The lack of unaltered hydrocarbons suggests that migration to this site has ceased or has been blocked by the formation of the carbonate veneer.

An analogous structure, about 25 km to the west in GB Block 425, exhibits active discharge of fluidized mud, liquid, gaseous hydrocarbons, and possibly brine at two sites along the margins of a carbonate-capped plateau. Fluid expulsion is thought to occur when elevated pore pressures above a salt diapir are released at the sea floor generating a hydraulic gradient capable of supporting a mud

Figure 14—Histogram of sediment hydrocarbon saturations across the Garden Banks 388 seep site. See Figure 12 for core locations. White = low molecular weight gas concentrations; gray = gasoline-range concentrations.

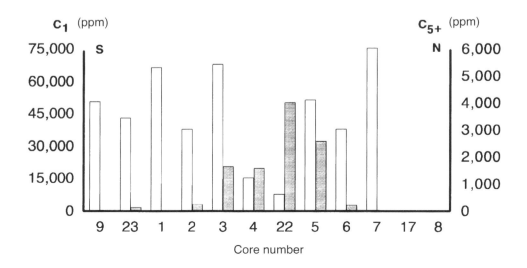

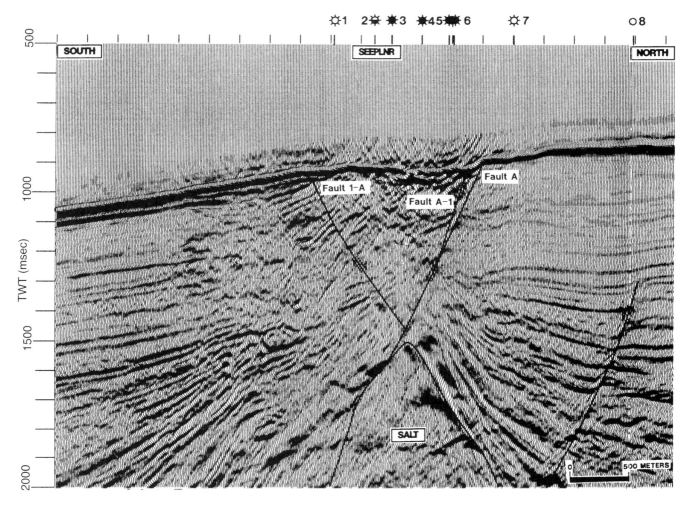

Figure 15—North-south 3-D CDP seismic profile (SEEPLNR) crossing the Garden Banks 388 seep site. See Figure 12 for location of seismic profile.

column about 20 m in height. As the weight of accreting sediment reaches hydraulic equilibrium, the active seep site migrates laterally. Hydrate formation and rapid carbonate lithification occurring across the surface of the plateau redirects fluid release to the plateau margins.

At the GB 386 site, overpressure release was apparently in response to mass failure above a salt diapir. This is evident from an incised, sediment-filled scour downdip from the plateau recognized in the 3-D CDP data. Mapping of this deposit revealed it to be a lobate wedge extending downslope to the southeast from the plateau, with maximum thickness of the deposit occurring at a break in slope below the plateau. Because the calculated volume of this wedge is greater than the volume of sediment modeled to restore the slope above the plateau, either auxiliary slump failure occurred laterally into the trough upslope from the plateau or fluidized mud eruptions from the plateau added material to the deposit.

A model of this mechanism at GB 386 (Figure 20) shows upward movement of the diapir oversteepening the overlying sediment and initiating slope failure. Fluid

expulsion initiated in response to overburden release vertically constructs a "mud volcano" at the site. When the lithostatic overburden reaches equilibrium with the hydraulic gradient, the eruption site shifts to the flank of the mud volcano. Site shifting continues laterally around the feature, forming a nearly circular plateau until fluid flow is restricted either laterally by the mass of the structure or vertically by pressure depletion or conduit blocking in the subsurface. Clams colonize the interior of the plateau where high concentrations of H_2S and low concentrations of unaltered hydrocarbons are present. Vestimentiferans, with their affinity for environments where seep flux is high, selectively colonize the rim of the plateau.

Green Canyon Blocks 286/287

The Green Canyon 286/287 site is located at lat. 27° 40.52' N, long. 90° 49.46' W with water depths ranging from 800 to 1250 m across the area. Sea floor topography is dominated by a single domal feature about 5500 m in

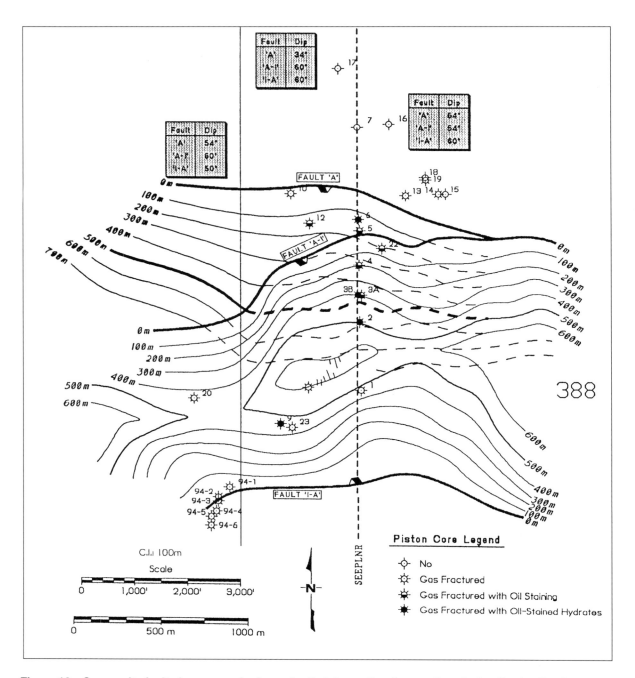

Figure 16—Composite fault plane map of primary faults intersecting the sea floor in the Garden Banks 388 seep site. Fault labels (A, A-1, and 1-A) are the same as on the seismic profile of Figure 15. Contours are in 100-m intervals below sea floor.

diameter separating an area to the northeast where the average water depth is about 950 m from a trough southwest of the dome where the average water depth is about 1200 m (Figure 21). A single trawl conducted in 1987 recovered calyptogenid and vesicomyid clams from a zone exhibiting shallow wipe-out on high-frequency geophysical data (Brooks et al., 1987a). Active seep discharge is evidenced by the presence of repeated slick develop-

ment in a time-series of orbital images (MacDonald et al., 1993; this volume). The position of the common source or head of these slicks (see Figure 21) is near the crest of the sea floor dome. The surface of the dome is highly irregular, as shown from a single survey conducted with the *Johnson Sea-Link I* during 1993. This survey, conducted near the crest of the dome, failed to encounter significant concentrations of chemoautotrophic organisms.

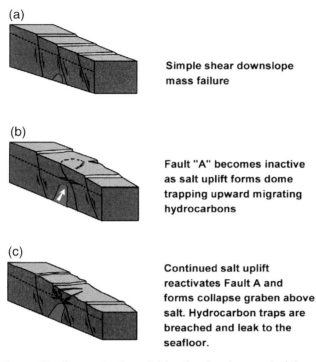

(a)

Simple shear downslope mass failure

(b)

Fault "A" becomes inactive as salt uplift forms dome trapping upward migrating hydrocarbons

(c)

Continued salt uplift reactivates Fault A and forms collapse graben above salt. Hydrocarbon traps are breached and leak to the seafloor.

Figure 17—Conceptual model for the development of the Garden Banks 388 seep site. (a)–(c) Block diagrams showing chronologic stages of development.

Two seismic lines extracted from the 3-D data set display the subsurface structure in the area of the seep (Figures 22a, b). Both lines are oriented perpendicular to the steep face of the dome, with profile SB-1 oriented east-west and profile SB-2 southwest-northeast. A near-surface salt diapir covered by a thin veneer of sediment dominates the near-surface structure on both lines. On profile SB-2 (Figure 22b), the salt appears to crop out on the sea floor along the steep southwest face of the sea floor dome. A subsea floor depth map constructed at top salt (Figure 23) illustrates how the sea floor dome mimics the underlying salt with the diapir rising to within 400 m of the sea floor throughout much of the area. At the slick source, the salt surface is mapped at <1000 m directly beneath the location of the slick confluence. A prominent fault scarp shown on profile SB-2 at the sea floor near the seep location is the likely discharge site.

Episodic movement of the salt is recognized on profile SB-2 with recent movement evidenced by the onlap of the sea floor reflector to the northeast. Although data quality is insufficient to delimit periods of active fault growth, activation and reactivation of the faults are expected to mimic salt movement. Visual observations show these faults to be a series of parallel swales with scattered bivalve shells and occasional communities present in the troughs.

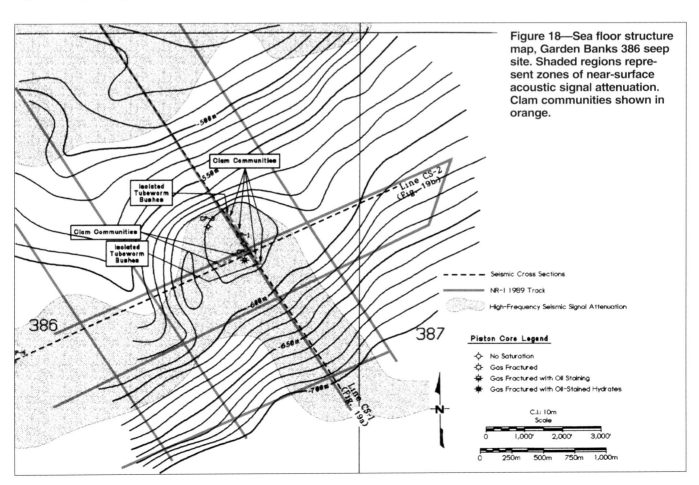

Figure 18—Sea floor structure map, Garden Banks 386 seep site. Shaded regions represent zones of near-surface acoustic signal attenuation. Clam communities shown in orange.

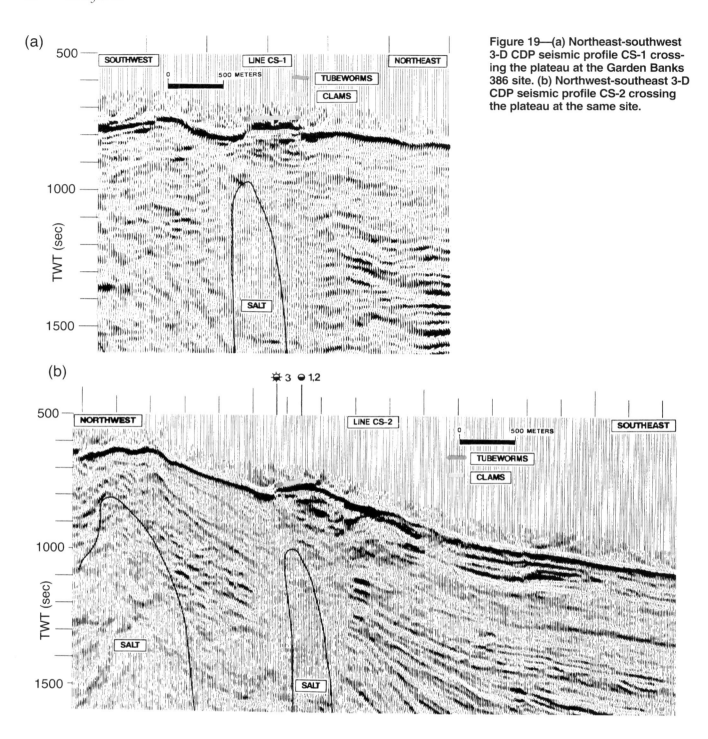

Figure 19—(a) Northeast-southwest 3-D CDP seismic profile CS-1 crossing the plateau at the Garden Banks 386 site. (b) Northwest-southeast 3-D CDP seismic profile CS-2 crossing the plateau at the same site.

DISCUSSION

The complex chemoautotrophic fauna in high-density communities require an environment where a narrow balance between marine oxic and sulfide-rich anoxic conditions is maintained (Fisher, 1990; MacDonald et al., 1990c). Maintenance of this environment relies on the delivery of thermogenic hydrocarbons to the sea floor in sufficient volumes and rates to preserve the oxic–anoxic interface at or very near the sediment–sea water interface

(Sassen et al., 1994). Vertical permeability of claystones forming the bulk of the sediments in the Gulf of Mexico is sufficiently low to discount significant interstratal flux rates at the sea floor (Bryant et al., 1991; Chiou et al., 1991), and bacterial degradation of high molecular weight hydrocarbons is generally rapid (Brakstad and Grahl-Nielsen, 1988). Thus, diffusion is unlikely to deliver at rates sufficient to maintain the anoxic horizon near the sea floor. Enhancements to vertical permeability require the formation of discrete conduits from source to surface

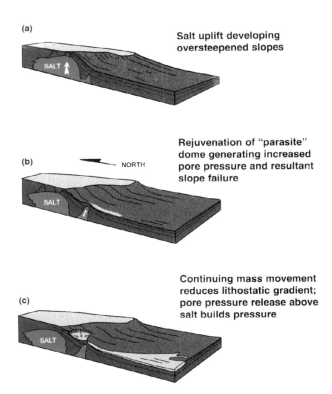

Salt uplift developing
oversteepened slopes

Rejuvenation of "parasite"
dome generating increased
pore pressure and resultant
slope failure

Continuing mass movement
reduces lithostatic gradient;
pore pressure release above
salt builds pressure

Figure 20—Conceptual model for the development of the Garden Banks 386 seep site. (a)–(c) Block diagrams showing chronologic stages of development.

(Abrams, 1992; Jones and Drozd, 1983). Furthermore, the vestimentiferan tubeworms forming the primary constituent to the complex communities appear to have a very slow growth rate (about 100 years to maturity; C. S. Fisher, personal communication, 1993). As a result, a temporal component requiring conduits of sufficient duration to exist may form an additional constraint to community development.

Xiao and Suppe (1989; 1992), Xiao et al. (1991), and Bischke and Suppe (1990) have described the mechanics and geometry of secondary fold and fault features in the Gulf of Mexico and have outlined the development of an antithetic system to a master normal fault. In their model, the antithetic system is generated in response to hanging wall collapse above the primary fault plane, with timing and orientation of the collapse dependent on fault shape, relative particle motion in hanging wall collapse, sediment type, and compaction. Dip angles on the antithetic faults are virtually always steeper than the master fault and, being a younger phase of shear deformation, preferentially maintain open conduits. With continued accommodation at depth in an active system, the antithetic conduits may be long-lived and relatively continuous sources of thermogenic hydrocarbons at the sea floor.

Therefore, presence of the fault surfaces intersecting the sea floor, especially those developed in response to

Coulomb shear, are likely to provide the primary means by which hydrocarbon fluids reach the sediment–seawater interface. Correlation of tubeworm communities with fault traces in the GC 234 site, as done in this study, and in the Bush Hill community (MacDonald et al., 1989) confirms a relationship between the complex communities and the location of the primary conduits for migrating thermogenic hydrocarbons required to sustain these communities.

The presence of high-volume conduits alone do not guarantee the presence of a complex community; therefore, a recent history of active seepage coupled with geochemical maturation is evidently required by the complex communities. For example, although the fault system on the sea floor at the GB 388 site is actively discharging thermogenic hydrocarbons to the water column and charging the near sea floor sediments, widespread colonization by complex chemoautotrophs has not occurred. Here, the formation of currently active fault conduits are recent events forming in response to remobilization of the underlying salt stock. Furthermore, on the basis of replicate observations where discharge to the water column is observed at locally active sites of relatively short duration, the delivery at any given site on the sea floor must be considered an ephemeral event. As a result of the discontinuous delivery, the ability of the vestimentiferans to colonize these areas may be restricted.

Vestimentiferan colonization is also apparently restricted at the GB 386 site where isolated occurrences are noted along the margins of the plateau but not across the interior. Again a temporal restriction on seep activity is thought to occur due to the lateral migration of the seep site as the hydraulic head is overcome by vertical sediment accretion.

On the basis of these observations and the apparent fluid specificity of the organisms, a first-order correlation of biological complexity with increasing molecular weight hydrocarbon content (Figure 24) may form a control for community type and development. An additional temporal constraint on community development appears to be active where maintenance of seep conduits to the sea floor is required to develop the chemical substrate sufficient to support the complex communities present in the Bush Hill and GC 234 sites.

CONCLUSIONS

Naturally occurring hydrocarbon seepage known to support chemoautotrophic communities is a widespread phenomenon in the Gulf of Mexico at scales visible to orbital platforms. Physical manifestations of these seeps at the sea floor are visible using commonly available geophysical techniques and can be recognized as acoustic turbidity in shallow penetration data and as high reflection strength returns (either positive or negative reflection coefficients) in lower frequency 3-D CDP-processed data. Comparison of geologic conditions in areas where high-density, complex communities are present with areas

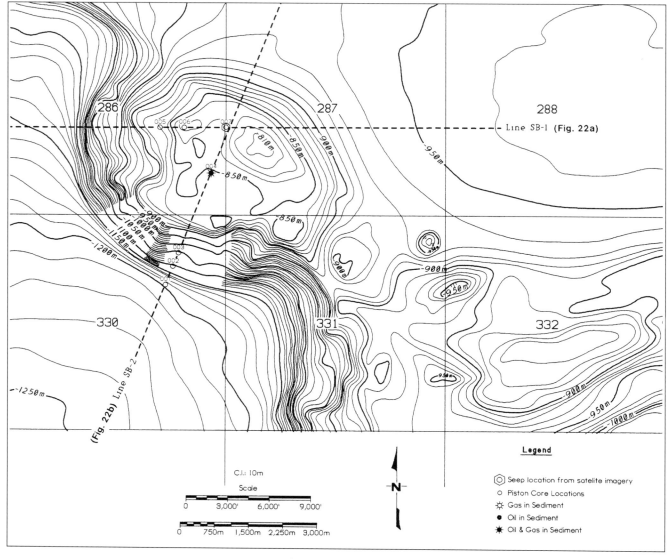

Figure 21—Sea floor structure map, Garden Banks 286/287 seep site. The observed origin of the surface slick is located at the intersection of seismic profiles SB-1 and SB-2.

where low-density, often unifaunal occurrences are present suggests that a structural forcing mechanism is required to sustain significant aggregations of the more complex chemoautotrophs. Based on the comparisons made in this paper, seep terrains can be broadly subdivided into either fault-dominated systems responding to simple shear or salt-related systems responding to pure shear stress generated by halokinesis.

In this study, the Bush Hill and GC 234 complex communities are fault-dominated, as the prime movers within the systems are listric-normal faults with antithetic faults developed in response to hanging-wall rotation. The Mussel Beach and NR-1 Brine Pool communities are also fault-dominated in a similar sense in that mass failure generating a shallow-dipping slump surface collects

the requisite hydrocarbons for the community. The NR-1 Brine Pool is unique in being co-dependent on salt movement to initiate formation of the feature.

The GB 386 and 388 and GC 287 sites are all salt-dominated features. The GB 386 and GC 287 sites exhibit near-surface salt injection as the prime mover to seep migration. Although the GB 388 site was initially formed in a fault-dominated system, migration here is apparently dependent on the formation of secondary faulting developed in response to recent salt movement. Discrete periods of structural reorganization of the fault blocks above the salt result in episodic seep discharge at these sites, thereby applying a temporal constraint on development of the environment required for widespread community growth.

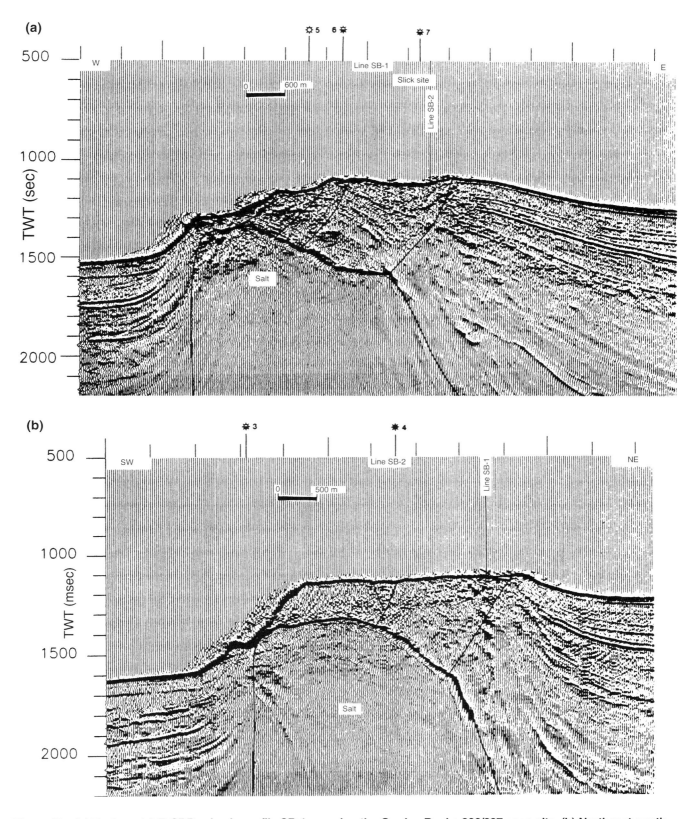

Figure 22—(a) East-west 3-D CDP seismic profile SB-1 crossing the Garden Banks 286/287 seep site. **(b)** Northeast-southwest 3-D CDP seismic profile SB-2 crossing the same site. Note the apparent salt outcrop on the sea floor at the base of the dome.

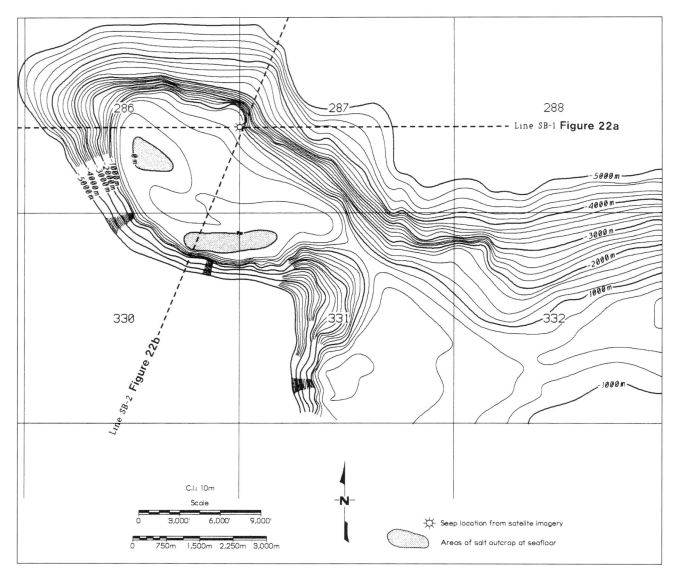

Figure 23—Structure map at top salt, Garden Banks 286/287 seep site. Shaded areas denote apparent outcrops on the sea floor. Contours are 10-m intervals below sea floor.

Acknowledgments—*This work would not have been possible without the efforts of the crews of the U.S. Navy's Submarine NR-1 and the Johnson-Sea Link II operated by Harbor Branch Oceanographic Institution. Kevin Gallatin and Gary Nix provided invaluable assistance in drafting illustrations. Michael Abrams and Dietmar Schumacher provided editorial comments and assistance that are greatly appreciated. Finally, the guidance and support of Richard Mitterer and Kent Nielsen of the Geosciences Department of The University of Texas at Dallas and the staff of the Geochemical and Environmental Research Group of Texas A&M University are greatly appreciated.*

REFERENCES CITED

Abrams, M. A., 1992, Geophysical and geochemical evidence for subsurface hydrocarbon leakage in the Bering Sea, Alaska: Marine and Petroleum Geology, v. 9, p. 208–221.

Anderson, R. K., R. S. Scanlon, P. L. Parker, and E. W. Behrens, 1983, Seep oil and gas in Gulf of Mexico sediment: Science, v. 222, p. 619–621.

Anderson, A. L., and W. R. Bryant, 1989, Acoustic properties of shallow gas: Proceedings of the Offshore Technology Conference, OTC 5955, p. 669–676.

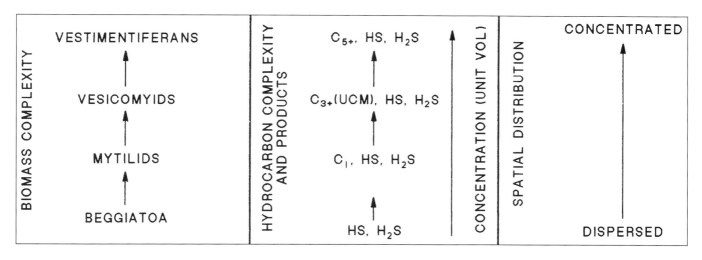

Figure 24—Matrix diagram showing first-order correlation of biomass and hydrocarbon complexity with an inverse relationship to spatial distribution.

Behrens, W. E., 1988, Geology of a continental slope oil seep, northern Gulf of Mexico: AAPG Bulletin, v. 72, p. 105–114.

Bischke, R. E., and J. Suppe, 1990, Geometry of rollover origin of complex arrays of small antithetic and synthetic faults (abs.): AAPG Bulletin, v. 74, p. 611.

Brakstad, F., and O. Grahl-Nielsen, 1988, Identification of weathered oils: Marine Pollution Bulletin, v. 19, p. 319–324.

Brooks, J. M., M. C. Kennicutt II, R. R. Fay, T. J. MacDonald, and R. Sassen, 1984, Thermogenic gas hydrates in the Gulf of Mexico: Science, v. 225, p. 409–411.

Brooks, J. M., M. C. Kennicutt II, R. R. Bidigare, and R. R. Fay, 1985, Hydrates, oil seepage, and chemosynthetic ecosystems on the Gulf of Mexico slope: EOS Transactions, American Geophysical Union, v. 66, p. 106.

Brooks, J. M., M. C. Kennicutt II, and R. R. Bidigare, 1987a, Chemosynthetic marine ecosystems in the Gulf of Mexico: Unpublished final cruise report to the Offshore Operators Committee, 102 p.

Brooks, J. M., M. C. Kennicutt II, C. R. Fisher, S. A. Macko, K. Cole, J. J. Childress, R. R. Bidigare, and R. D. Vetter, 1987b, Deep-sea hydrocarbon seep communities: evidence for energy and nutritional carbon sources: Science, v. 228, p. 1138–1142.

Brooks, J. M., M. C. Kennicutt II, I. R. MacDonald, D. L. Wilkinson, N. L. Guinasso, Jr. ,and R. R. Bidigare, 1989, Gulf of Mexico hydrocarbon seep communities, part IV—descriptions of known chemosynthetic communities: Proceedings of the Offshore Technology Conference, OTC 5954, p. 663–667.

Brooks, J. M., R. A. Burke, Jr., D. A. DeFreitas, S. R. Gittings, N. L. Guinasso, Jr., M. C. Kennicutt II, I. R. MacDonald, C. D. Perkins, and D. L. Wilkinson, 1990, Submarine *NR-1* cruise report, chemosynthetic community survey, Green Canyon Blocks 232/233/234: Unpublished technical report, Geochemical and Environmental Research Group, Texas A&M University, no. 89-050, 435 p.

Brooks, J. M., R. A. Burke, Jr., D. A. DeFreitas, S. R. Gittings, N. L. Guinasso, Jr., M. C. Kennicutt II, I. R. MacDonald, C. D. Perkins, and D. L. Wilkinson, 1991, Submarine *NR-1* cruise report, chemosynthetic community survey, Garden Banks Block 386: Unpublished technical report,

Geochemical and Environmental Research Group, Texas A&M University, no. 91-051, 96 p.

Bryant, W. R., R. H. Bennett, P. J. Burkett, and F. R. Rack, 1991, Microfabric and physical properties characteristics of a consolidated clay section: ODP Site 697, Weddell Sea, *in* R. H. Bennett, W. R. Bryant, and M. H. Hulbert, eds., Microstructure of fine-grained sediments: from mud to shale: New York, Springer-Verlag, p. 73–92.

Chiou, W. A., W. R. Bryant, and R. H. Bennett, 1991, Clay fabric of gassy submarine sediments, *in* R. H. Bennett, W. R. Bryant, and H. Hulbert, eds., Microstructure of fine-grained sediments: from mud to shale: New York, Springer-Verlag, p. 333–352.

Cook, D., and P. D'Onfro, 1991, Jolliet field thrust fault structure and stratigraphy, Green Canyon Block 184, offshore Louisiana: Transactions of the Gulf Coast Association of Geological Societies, v. XVI, p. 100–121.

Fisher, C. S., 1990, Chemoautotrophic and methanotrophic symbioses in marine invertebrates: Reviews in Aquatic Sciences, v. 2, p. 399–436.

Geyer, R. A., and C. P. Giammona, 1980, Naturally occurring hydrocarbons in the Gulf of Mexico and Caribbean Sea, *in* Marine environmental pollution, I. hydrocarbons: New York, Elsevier, p. 37–106.

Jones, V. T., and R. J. Drozd, 1983, Predictions of oil and gas potential by near-surface geochemistry: AAPG Bulletin, v. 67, p. 932–952.

Kennicutt II, M. C., J. M. Brooks, R. R. Bidigare, R. R. Fay, T. Wade, and T. J. MacDonald, 1985, Vent-type taxa in a hydrocarbon seep region on the Louisiana slope: Nature, v. 317, p. 351–353.

Kennicutt II, M. C., J. M. Brooks, R. R. Bidigare, and G. J. Denoux, 1988a, Gulf of Mexico hydrocarbon seep communities, I. Regional distribution of hydrocarbon seepage and associated fauna: Deep-Sea Research, v. 35, p. 1639–1651.

Kennicutt II, M. C., J. M. Brooks, and G. J. Denoux, 1988b, Leakage of deep reservoired petroleum to the near surface on the Gulf of Mexico continental slope: Marine Chemistry, v. 24, p. 39–59.

MacDonald, I. R., G. S. Boland, J. S. Baker, J. M. Brooks, M. C. Kennicutt II, and R. R. Bidigare, 1989, Gulf of Mexico hydrocarbon seep communities, II. Spatial distribution of seep organisms and hydrocarbons at Bush Hill: Marine Biology, v. 101, p. 235–247.

MacDonald, I. R., W. R. Callender, R. A. Burke, Jr., S. J. McDonald, and R. S. Carney, 1990a, Fine-scale distribution of methanotrophic mussels at a Louisiana cold seep: Progress in Oceanography, v. 24, p. 15–24.

MacDonald, I. R., J. F. Reilly, N. L. Guinasso, Jr., J. M. Brooks, and W. R. Bryant, 1990b, Chemosynthetic mussels at a brine-filled pockmark in the northern Gulf of Mexico: Science, v. 248, p. 1096–1099.

MacDonald, I. R., N. L. Guinasso, Jr., J. F. Reilly, J. M. Brooks, W. R. Callender, and S. G. Gabrielle, 1990c, Gulf of Mexico hydrocarbon seep communities, VI. Patterns in community structure and habitat: Geo-Marine Letters, v. 10, p. 244–252.

MacDonald, I. R., N. L. Guinasso, Jr., S. G. Ackelson, J. F. Amos, R. Duckworth, R. Sassen, and J. M. Brooks, 1993, Natural oil slicks in the Gulf of Mexico are visible from space: Journal of Geophysical Research, v. 98-C9, p. 16351–16364.

MacDonald, I. R., N. L. Guinasso, Jr., R. Sassen, J. M. Brooks, L. Lee, and K. T. Scott, 1994, Gas hydrate that breaches the sea floor on the continental slope of the Gulf of Mexico: Geology, v. 22, p. 699–702.

Neurauter, T. W., and W. R. Bryant, 1990, Seismic expression of sedimentary volcanism on the continental slope, northern Gulf of Mexico: Geo-Marine Letters, v. 10, p. 225–231.

Reilly II, J. F., 1995, Geological controls on the distribution of chemosynthetic communities in the Gulf of Mexico: Ph.D. dissertation, University of Texas at Dallas, Texas, 222 p.

Roberts, H. H., D. J. Cook, and M. K. Sheedlo, 1992, Hydrocarbon seeps of the Louisiana continental slope: Seismic amplitude signature and sea floor response: Transactions of the Gulf Coast Association of Geological Societies, v. XLII, p. 349–361.

Sassen, R., I. R. MacDonald, A. G. Requejo, N. L. Guinasso, Jr., M. C. Kennicutt II, S. T. Sweet, and J. M. Brooks, 1994, Organic geochemistry of cold hydrocarbon vents in the Gulf of Mexico, offshore Texas and Louisiana: Geo-Marine Letters, v. 14, p. 110–119.

Xiao, H., and J. Suppe, 1989, Role of compaction in the listric shape of growth normal faults: AAPG Bulletin, v. 73, p. 777–786.

Xiao, H., F. A. Dahlen, and J. Suppe, 1991, Mechanics of extensional wedges: Journal of Geophysical Research, v. 96, p. 10301–10318.

Xiao, H., and J. Suppe, 1992, Origin of rollover: AAPG Bulletin, v. 76, p. 509–529.

Guliev, I. S., and A. A. Feizullayev, 1996, Geochemistry of hydrocarbon seepages in
Azerbaijan, *in* D. Schumacher and M. A. Abrams, eds., Hydrocarbon migration
and its near-surface expression: AAPG Memoir 66, p. 63–70.

Chapter 5

Geochemistry of Hydrocarbon Seepages
in Azerbaijan

I. S. Guliev

A. A. Feizullayev

*Geology Institute of the Azerbaijan
Academy of Sciences
Baku, Republic of Azerbaijan*

Abstract

Oil and gas seeps on the western flank of the South Caspian Basin are associated with mud volcanoes, out-crops of oil-bearing strata, and mineral water springs. Most of the seepage occurs in faulted zones along the basin flanks. Gas seepage can be subdivided into three groups based on gas composition, isotopic composition, and the age of exposed rocks. Neogene–Quaternary seeps are predominantly methane and are derived from low-maturity organic matter and biogenic sources. Mesozoic–Paleogene seeps have more wet gas and nonhy-drocarbon components and were generated during late catagenesis. Paleozoic–Mesozoic seeps have the great-est amount of methane, nitrogen, and carbon dioxide and have undergone the highest degree of catagenesis. Oil seepage can be subdivided into isotopically light and heavy groups. The chemical and isotopic changes in the gas and oil seepages are regular and systematic, allowing the data to be used for correlation. Reservoir oils can also be subdivided into two groups: isotopically heavy Neogene oils and isotopically light Paleogene oils. In general, the surface hydrocarbon seepages correlate to the subsurface hydrocarbon distribution. A methodolo-gy was developed that predicts that the petroleum prospects of the deeply buried basin deposits are favorable.

INTRODUCTION

The South Caspian Basin (SCB) and surrounding mountain ranges are favorable for studying natural hydrocarbon seepages. Numerous gas and oil seepages of differing chemical composition, maturation, and type occur within a small region (Abikh, 1939; Veber, 1935; Kovalevski, 1940; Dadashev, 1963). This variety is related to the structural contrasts between the existing depressions of the SCB and the folded uplifts of the surrounding Great and Lesser Caucasus, Elburs, and Kopetdag mountains. The SCB is characterized by a high sedimentation rate (up to 1.3 km/m.y.) and an enormous thickness of sedimentary cover (up to 30 km). The Quaternary–Pliocene complex (up to 10 km thick) consists predominantly of shaly terrigeneous rocks. A low heat flow (25–50 mW/m²) and abnormally high pressures (with anomaly factors up to 1.8) also characterize this basin. Finally, there is considerable faulting and fracturing of the sedimentary cover and tectonic activity.

All of these factors create favorable conditions for the seepage of natural hydrocarbons to the surface, especial-ly in areas associated with mud volcanoes, groundwater, and faults and fractures. The most active gas seeps are within a thick series of Mesozoic and Cenozoic sedimen-tary rocks that have undergone intensive tectonism. The annual production of hydrocarbon gases by mud volca-noes is estimated to be 3×10^8 m³, about 2×10^7 m³ of which is produced during periods of little to no tectonic activity (Figure 1).

HYDROCARBON SEEPAGE

Gas Seeps

Comparable quantities of gas are discharged from the central Lesser and Great Caucasus mountain and fold range, where intensive volcanism occurred during the Mesozoic and Cenozoic. Near-surface gas seeps, which are 90% carbon dioxide, are confined mainly to the high-ly faulted and fractured areas in various volcanic, meta-morphic, and sedimentary terranes. Nitrogen gas seeps are confined to the southern and northern slopes of the Great Caucasus Mountains, which are less tectonically active (Dadashev and Guliev, 1981; Guliev, 1984).

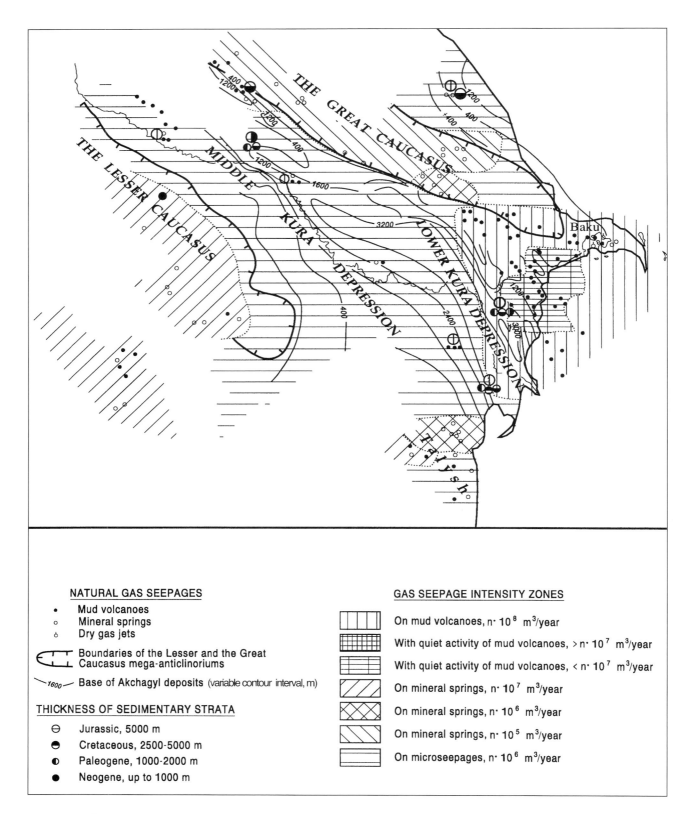

NATURAL GAS SEEPAGES

- •   Mud volcanoes
- ○   Mineral springs
- ⚬   Dry gas jets

⊏⊐   Boundaries of the Lesser and the Great Caucasus mega-anticlinoriums

⌐1600⌐   Base of Akchagyl deposits (variable contour interval, m)

THICKNESS OF SEDIMENTARY STRATA

- ⊖   Jurassic, 5000 m
- ⬤   Cretaceous, 2500-5000 m
- ◑   Paleogene, 1000-2000 m
- ●   Neogene, up to 1000 m

GAS SEEPAGE INTENSITY ZONES

On mud volcanoes, $n \cdot 10^8$ m^3/year

With quiet activity of mud volcanoes, $> n \cdot 10^7$ m^3/year

With quiet activity of mud volcanoes, $< n \cdot 10^7$ m^3/year

On mineral springs, $n \cdot 10^7$ m^3/year

On mineral springs, $n \cdot 10^6$ m^3/year

On mineral springs, $n \cdot 10^5$ m^3/year

On microseepages, $n \cdot 10^6$ m^3/year

Figure 1—Schematic map of the study area in the South Caspian Basin showing the nature and intensity of surface gas seepages.

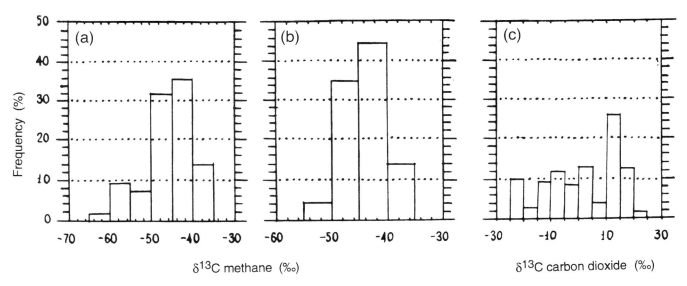

Figure 2—Distribution of (a) methane δ¹³C in mud volcanoes, (b) methane δ¹³C in petroleum fields, and (c) carbon dioxide δ¹³C in mud volcanoes of Azerbaijan.

Table 1—Isotopic and Geochemical Characterization of South Caspian Basin Seepages

Age	CH_4 (%)	$\delta^{13}C_{CH4}$ (‰)	CO_2 (%)	$\delta^{13}C_{CO2}$ (‰)	He (%)	$^3He/^4He$ (%)	N_2 (%)	He/Ar_r	He/Ar_a
Neogene–Quaternary	>90	−45 to −80	0.1–10	−7.6 to −15.8	$1–2 \times 10^{-3}$	$5–7 \times 10^{-8}$	1.5–5.0	0.1–0.5	<5
Mesozoic–Paleogene	75–90	−30 to −45	5–10	−5.0 to +23.4	$10^{-3}–10^{-1}$	$5–20 \times 10^{-8}$	5–10	0.5–1.0	5–10
Lower Mesozoic	0.1–90	−28 to −45	0.1–90	?	$10^{-2}–10^{-1}$	$5–20 \times 10^{-8}$	0.1–90	>1	>10

The carbon isotopic compositions ($\delta^{13}C$) of near-surface gas seeps, seeps confined to mineral water springs, and seeps from mud volcanoes are as follows (Figure 2):

- −80 to −45‰ for methane
- −29.6 to −23.3‰ for ethane
- 2.8×10^{-8} to 55.0×10^{-8}‰ for helium
- −236 to −150‰ for hydrogen
- −25 to +23‰ for carbon dioxide

These values show that the seepage gases come from a wide variety of sources and a wide depth interval (7–12 km) of generation and migration.

The analysis of natural gas seepages in the SCB and surrounding regions allows their subdivision into three groups based on gas composition, isotopic composition, and the age of exposed rocks at surface seepage sites (Table 1):

- **Neogene–Quaternary seepage** occurs in the shallower part of the intermontane depressions. These seeps are predominantly methane (>90%) with some nonhydrocarbon gas, with the $\delta^{13}C$ for

methane ranging from −45 to −80‰. The gases from these seeps are derived from low-maturity organic matter ($R_o = 0.8$–1.0) and biogenic sources that were not in contact with subsurface fluids.

- **Mesozoic–Paleogene (?) seepage** occurs in the deeper part of the intermontane depressions and was subjected to higher geothermal temperatures. The $\delta^{13}C$ for methane ranges from −30 to −45‰ for the Mesozoic–Paleogene seeps. These seeps have a greater amount of wet gases (e.g., ethane, propane, and butane) and nonhydrocarbon components than the Neogene–Quaternary seeps. The main source of gas is organic matter transformed to late catagenetic stages ($R_o = 1.5$–1.8).

- **Paleozoic–Mesozoic (?) seepage** is at the base of the intermontane depressions that are involved in the tectonics of the surrounding mountain system. The $\delta^{13}C$ for methane ranges from −28 to −45‰ for these seeps, which have the greatest concentrations of methane, nitrogen, and carbon dioxide. The dense rocks here are exposed to high temperatures and pressures typical of areas where there is a high degree of organic matter transformation.

Table 2—Rare Gas Quantities and Ratios for Distinctive Mud Volcano Groups in Azerbaijan

Age		Mud Volcano Name	He (%)	He/Ar$_r$	He/Ar$_a$
Neogene	I	Astrakhanka	0.0097	2.5	31
		Demirchi	0.0100	2.5	30.5
Paleogene	II	Cheildag	0.0011	0.5	6.3
		Shorbulag	0.0029	0.5	9.1
Mesozoic	III	Matrasa	0.0007	0.13	1.74
		Charagan	0.0006	0.17	1.6

Comparison of the three data sets presented in Table 1 shows that the changes in chemical and isotopic composition of these gases are regular and systematic, allowing the use of these data for gas–gas and gas–rock correlations. Table 2 illustrates some of these correlations for nonhydrocarbon gases (helium and argon). A study of the rare components of Azerbaijan mud volcano gases identified several characteristic groups (Ismet et al., 1972). The first group is associated with Mesozoic sediments, the second with Paleogene sediments, and the third with Neogene sediments.

Oil Seeps

Oil seepages are mainly confined to mud volcanoes and basin flanks where hydrocarbon-bearing strata outcrop. The oils are slightly biodegraded, having a naphtheno-paraffin base and less than 0.5% sulfur. The carbon isotopic composition of the crude oil ranges from –28.2 to –25.0‰ and the pristane/phytane (Pr/Ph) ratio ranges from 0.9 to 2.0.

The oil seeps from mud volcanoes were generated at relatively low temperatures and appear to be low-maturity products. We have classified the oil seeps into two types: isotopically light (Shorbulag, Demirchi, Cheildag, and Ayrantekyan) and isotopically heavy (Matrasa, Melikchobanly, and Gyrlykh). The gas chromatography–mass spectrometry (GC-MS) traces show low diasteranes and an almost equal distribution of C_{27-29} αα- and ββ-sterane compounds. C_{28} is often the dominate sterane (Figure 3).

RESERVOIRED OILS AND SOURCE ROCKS

Characteristics of Reservoired Oils

Reservoired oils from 20 oil- and gas-producing regions were studied. These hydrocarbons are associated with Upper Cretaceous volcanogenic sedimentary deposits and with Paleocene, Miocene, and Pliocene sediments. Two separate groups were defined based on physico-chemical and geochemical parameters, including

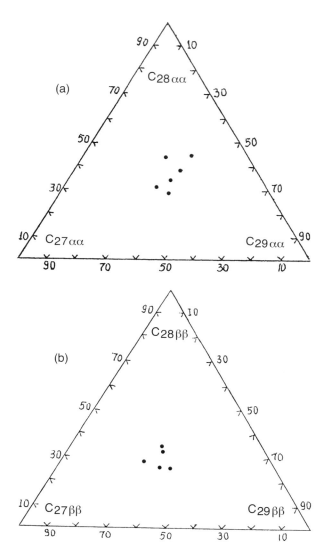

Figure 3—Distribution of C_{27} to C_{29} (a) αα-sterane and (b) ββ-sterane compounds.

density, group hydrocarbon composition, rare components, alkane distribution, Pr/Ph ratio, carbon preference index (CPI), and isotopic composition of the crude.

The first group includes oil from Neogene reservoirs (Miocene and Pliocene) and is characterized by heavy carbon isotopic composition (–25.5‰), a high percentage of aromatic hydrocarbons, a relatively low Pr/Ph ratio, and a CPI with the maximum *n*-alkane content ranging from C_{15}–C_{17} (Figure 4). The second group includes oils from Paleogene reservoirs that have a characteristically light carbon isotopic composition (–27.7‰) and a relatively high percentage of aromatic hydrocarbons. Some Upper Cretaceous oil seeps are also noteworthy. These oils have relatively low density, high asphaltene content, relatively high Pr/Ph ratios, high CPI values, and a reduced percentage of aromatic hydrocarbons.

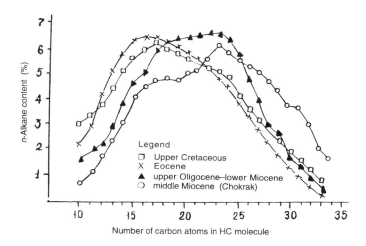

Figure 4—Comparison of oil *n*-alkanes from different age deposits in the South Caspian Basin.

Table 3—Mean TOC Values for Various South Caspian Basin Shales[a]

Stratigraphic Interval	Number of Samples	TOC (wt. %)
Middle Pliocene (productive series)	1693	0.16
Middle–upper Miocene (Diatomaceous suite)	639	0.56
Oligocene–Miocene (Maikopian)	1446	0.69
Eocene	500	0.43
Cretaceous	609	0.25
Middle Jurassic	158	0.60

[a]From Ali-zadeh et al. (1975).

Paleogeographic investigations have shown that natural distinctions exist between these three groups of hydrocarbon reservoirs. The Neogene reservoirs are generally offshore, shallow water, and deltaic. The Paleogene reservoirs are generally marine, deep water, and reducing, while the Cretaceous reservoirs accumulated in shallow-water environments (Guliev et al., 1987).

Potential Source Rocks

Outcrop, mud volcano ejecta, and core samples from the fine-grained rocks were examined for total organic carbon (TOC) and hydrocarbon potential. Table 3 lists the mean values of TOC for different stratigraphic units. Figure 5 shows the organic matter type and maturity determined by pyrolysis of mud volcanic ejectas. Selected rock samples from the Eocene, the Oligocene–Miocene Maikopian suite, and the middle–upper Miocene Diatomaceous suite do have sufficient organic richness and matter type to generate hydrocarbons similar to those found in the seeps.

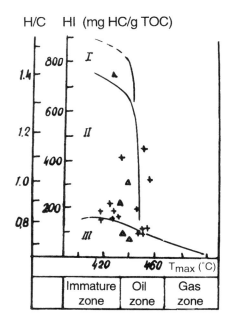

Figure 5—Pyrolysis type (I, II, or III) and degree of maturity of organic matter from mud volcano ejectas in Azerbaijan. Key: +, Neogene samples; ▲, Paleogene samples.

COMPARISON OF SURFACE SEEPAGE AND SUBSURFACE ACCUMULATIONS

Correlations based on $\delta^{13}C$ data show that the isotopically heavy oil seepages closely relate to Neogene reservoir oils and that the isotopically light seepages correlate to Paleogene reservoir oils (Dadashev et al., 1986). Many of the natural seepages are intermediate to these end members. Preliminary biomarker analyses indicate that the onshore Lower Kura depression oils from both subsurface accumulations and natural seepages were formed from mixed Neogene and Paleogene sources. In contrast, oils from fields offshore of the Apsheron Peninsula and the Lower Kura depression were formed primarily from Neogene sources, and the Shamakhy-Gobustan region

and Middle Kura depression from Paleogene sources.

With increasing stratigraphic age, a decreasing carbon isotopic weight of oils and an increasing carbon isotopic weight of gases were noted. An index D is proposed, with D defined as the difference between the oil and gas isotopic indicators:

$$D = \delta^{13}C_{oil} - \delta^{13}C_{gas}$$

As additional data are gathered, this index can be used to correlate and evaluate hydrocarbon maturation levels (Figure 6). With increasing maturation, isotopic characteristics (as well as other geochemical indicators) tend to merge and often approach graphite values (Galimov, 1973; Prosolov, 1990).

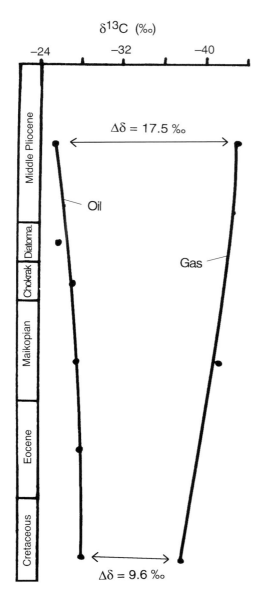

Figure 6—Plot of δ¹³C for methane gas and oil showing average data variation against stratigraphic depth.

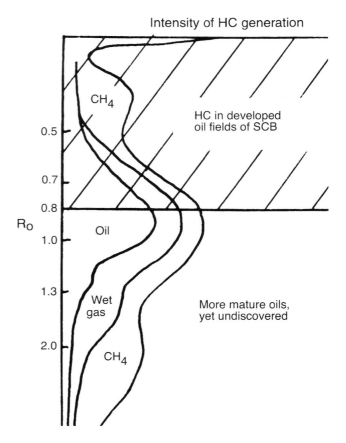

Figure 7—Schematic plot of maturity (R_o) against intensity of hydrocarbon generation, showing vertical zonality of hydrocarbon formation in sedimentary basins.

The maturity level of oil and gas from natural seepages and commercial fields is of great importance from both theoretical and practical viewpoints. Research conducted in the past decade has resulted in the development of a zonation scheme for hydrocarbon formation. This zonation is based on the oil and gas that is generated as a result of the varying stages of organic matter maturation. According to this zonation, the bulk of the oil is generated during the maturation level of $R_o = 0.6–1.3$, or early metacatagenic stages. Optimum gas formation occurs with higher organic matter maturity, at $R_o = 1.3–2.5$ (Figure 7). In the SCB, where the temperature in the deepest deposits is at or above 300°C, the whole hydrocarbon spectrum is generated. This includes biogenic gases ($R_o =$

0.1–0.2) and immature oils ($R_o = 0.5–0.7$), up to mature oils ($R_o = 0.8–1.3$), condensates ($R_o = 1.3–2.0$), and dry methane gas ($R_o = 2.0–3.0$).

Determining the maturation level of oil became possible with the development of isotope and biomarker geochemistry. This considerably increased the applicability of isotope and geochemical methods to practical oil and gas problems. In particular, by comparing the theoretical models of catagenesis zonation and the experimental determination of organic matter, oil, and gas maturation, it is possible to gather data on the generation, accumulation, and migration of oil and gas. We were the first to analyze the hydrocarbon maturation levels for the SCB, using the isotopic and geochemical characteristics of gases in petroleum fields and mud volcanoes.

The range of δ¹³C values for methane carbon in Azerbaijan mud volcano (Figure 8) gases is –61.2 to –35.9‰, with the mode ranging from –50 to –40‰. According to Galimov (1973), this corresponds to an intermediate stage of methane generation (the onset of liquid hydrocarbon generation). Isotopic composition values for δ¹³C of –40 to –30‰ correspond to the optimum phase of methane generation. These data agree with the results published by Prosolov (1990), who stud-

(a)

(b)

Figure 8—Mud volcanoes of Azerbajian. (a) General view of the Akhtarma Pahaly mud volcano. (b) Oil seepage from the Charagan mud volcano.

ied gas reserve distribution for 400 fields worldwide. Prosolov discovered that 45% of methane reserves range from –52 to –46‰ and that 40% range from –42 to –34‰. The 45% group was generated at temperatures of 120–160°C, and the 40% group at temperatures of 200–240°C (Prosolov, 1990). Thus, one can assume that considerable volumes of gas reserves (especially methane) with values for $\delta^{13}C$ within the range of –40 to –30‰ have yet to be discovered. Heavy methane, with carbon isotopic compositions reaching –36‰ in some mud volcanoes, lends support to this assumption.

The maturity level of oils was determined by the distribution of key biomarkers and maturation parameters. Specifically, maturation values for isosterane α-sterane in Apsheron oils are 1.7–1.9 (Petrov, 1984), which correspond to an intermediate degree of maturation (Naftalan oil is 6.0). Other biomarker data also indicate low oil maturation, not exceeding R_o = 0.8. The present-day view is that such low-maturity oils are generated either at early stages of catagenesis or at late stages of diagenesis. Therefore, it is reasonable to assume that the generation of higher maturity liquid and gaseous hydrocarbons is possible and may considerably exceed the volume of lower maturity hydrocarbons (Figure 7).

Higher maturity hydrocarbon gases may exist in the SCB subsurface in the form of "primary" droplike liquid oil, forming small, scattered concentrations. However, there is evidence for much larger accumulations. In particular, deep-seated deposits where the zones of hydrocarbon generation and accumulation coincide are of particular interest. Seismic data below 7 km suggest that there is good potential for hydrocarbon-bearing, reservoir-quality structures. Thermodynamic calculations at these depths indicate that mature oils (R_o = 0.8–1.3), condensates (R_o = 1.3–2.0), and dry gases (R_o = 2.0–3.0) should be generated. Core samples from deep wells and mud volcano ejectas show good reservoir quality, with up to 30% porosity and permeabilities of 200×10^{-15} m^2 (Guliev et al., 1991, 1992). These data indicate a strong potential for the generation and accumulation of commercial hydrocarbons in the deep SCB.

When a mud volcano erupts (Figure 8), a great quantity of gas is emitted. Equivalent volumes of liquid hydrocarbons are expected to have accumulated in reservoirs proximal to these mud volcanoes. Various tectonically, stratigraphically, and lithologically screened traps, as well as zones of contact with mud volcanoes, magmatic bodies, and shale diapirs, are present in the SCB.

FUTURE WORK

The thermodynamic conditions that exist in the SCB (relatively low temperatures and high pressures) are favorable for hydrocarbon preservation in both the gas and liquid phases. Our future investigations will concentrate on the following areas:

1. Construction of three-dimensional seismic models for the sedimentary complexes in which mud volcanoes occur
2. Isotopic and geochemical studies of petroleum and reservoir parameters of deeply buried rocks
3. Paleotectonic, paleogeographic, and paleoecologic studies of the conditions of hydrocarbon formation at different stages of basin development
4. Development of computer-aided basin modeling for hydrocarbon generation and accumulation.

Acknowledgments—This paper was reviewed by Michael Abrams and Deet Schumacher and was rewritten by Gail Bergan. Funding for the rewrite, as well as for the conference, was provided by Exxon Ventures (CIS) Inc.

REFERENCES CITED

Abikh, G. V., 1939, An island that appeared in the Caspian sea with mud volcano and ejecta materials, Caspian region (in Russian): Proceedings of Geological Institute of Azerbaijan, Branch of USSR Academy of Sciences, Baku, AzFAN, v. XII/63, p. 21–129.

Ali-zadeh, A. A, H. A. Ahmedov, H.-M. A. Aliyev, V. A. Pavlova, and N. I. Khatskevich, 1975, Oil generative properties assessment of Meso-Cenozoic deposits of Azerbaijan: Baku, Elm, 140 p.

Dadashev, F. G., 1963, Hydrocarbon gases of Azerbaijan mud volcanoes (in Russian): Baku, Azerneshr, 69 p.

Dadashev, F. G., and I. S. Guliev, 1981, Bildungsgesetzmabigkeiten der Gase in der Stratisphare der Sud-Kaspi-Senke und der angrenzenden Gebirgssysteme: Zeitschrift fur angewandte Geologie, Bd. 27, Heft 3, p. 145–148.

Dadashev, A. A., A. A. Feizullayev, and I. S. Guliev, 1986, On vertical zonality of oil and gas generation deduced from the data on isotopic composition of methane carbon from Azerbaijan mud volcanoes and petroleum fields (in Russian), *in* Oil and gas geology and geophysics: Express Information Series, Moscow, no. 6, p. 24–28.

Galimov, E. M., 1973, Carbon isotopes in petroleum geology (in Russian): Moscow, Nedra, 382 p.

Guliev, I. S., 1984, Quantitative evaluation and mapping of gas flows (in Russian): Izvestiya Akademii Nauk Azerb. SSR, Seriya Nauk o Zemle, no. 5, p. 52–60.

Guliev, I. S., N. I. Pavlencova, and M. M. Rajabov, 1987, Zone of regional deconsolidation in the sedimentary complex of South Caspian depression (in Russian): Izvestiya Akademii Nauk Azerb. SSR, Seriya Geologicheskaya, no. 6, p. 110–116.

Guliev, I. S., S. F. Suleymanova, and N. A. Klyatsko, 1991, Rocks reservoir properties prediction for sedimentary cover of the South Caspian depression (in Russian): Sovetskaya Geologiya, no. 7, p. 7–15.

Guliev, I. S., N. A. Klyatsko, S. A. Mamedova, and S. F. Suleymanova, 1992, Petroleum generative and reservoir properties of South Caspian depression deposits (in Russian): Litologiya i poleznye iskopaemye, no. 2, p. 110–119.

Ismet, A. R., R. S. Jafarova, and S. A. Jafarov, 1972, Argon and helium isotopes in mud volcanic gases (in Russian, abs.): All Union Seminar Meeting Materials, Baku, GIA, p. 74.

Kovalevski, S. A., 1940, Mud volcanoes of the South Caspian region: Azerbaijan and Turkmenistan: Baku, Azgostoptekhizdat, 200 p.

Petrov, A. A., 1984, Hydrocarbons of oil (in Russian): Moscow, Nauka, 264 p.

Prosolov, E. M., 1990, Isotopic geochemistry and origin of natural gases (in Russian): Leningrad, Nedra, 284 p.

Veber, V. V., 1935, Baku region (in Russian), *in* Natural gases of USSR: Leningrad–Moscow, ONTI NKTP SSSR, p. 345–365.

Schumacher, D., 1996, Hydrocarbon-induced alteration of soils and sediments, *in* D. Schumacher and M. A. Abrams, eds., Hydrocarbon migration and its near-surface expression: AAPG Memoir 66, p. 71–89.

Chapter 6

Hydrocarbon-Induced Alteration of Soils and Sediments

Dietmar Schumacher

Earth Sciences and Resources Institute
University of Utah
Salt Lake City, Utah

Abstract

The surface expression of hydrocarbon-induced alteration of soils and sediments can take many forms, including (1) microbiological anomalies and the formation of "paraffin dirt"; (2) mineralogic changes such as formation of calcite, pyrite, uranium, elemental sulfur, and certain magnetic iron oxides and sulfides; (3) bleaching of red beds; (4) clay mineral alteration; (5) electrochemical changes; (6) radiation anomalies; and (7) biogeochemical and geobotanical anomalies.

Bacteria and other microbes play a profound role in the oxidation of migrating hydrocarbons, and their activities are directly or indirectly responsible for many of the surface manifestations of hydrocarbon seepage. These activities, coupled with long-term migration of hydrocarbons, lead to the development of near-surface oxidation-reduction zones that favor the formation of a variety of hydrocarbon-induced chemical and mineralogic changes. This hydrocarbon-induced alteration is highly complex, and its varied surface expressions have led to the development of an equally varied number of surface exploration techniques, including soil carbonate methods, magnetic and electrical methods, radioactivity-based methods, and remote sensing methods.

Exploration methods based on what are assumed to be hydrocarbon-induced soil or sediment alterations have long been popular. Many claims of success have been made for these methods. However, well-documented studies are rare, and the claims are seldom substantiated by a scientifically rigorous program of sampling and analysis. The cause of these altered soils and sediments may well be hydrocarbon-related, but hydrocarbons are an indirect cause at best and not the most probable cause. Although the occurrence of hydrocarbon-induced geochemical alteration is well established, considerable research is needed before we understand the many factors affecting the formation of these anomalies in the near surface. Only then will we realize their full value for hydrocarbon exploration.

INTRODUCTION

The objective of this paper is to provide an overview of the major hydrocarbon-induced changes affecting soils and sediments and their implications for surface exploration methods and applications. Long-term leakage of hydrocarbons, either as macroseepage or microseepage, can set up near-surface oxidation-reduction zones that favor the development of a diverse array of chemical and mineralogic changes. The bacterial oxidation of light hydrocarbons can directly or indirectly bring about significant changes in the pH and Eh of the surrounding environment, thereby also changing the stability fields of the different mineral species present in that environment.

These changes result in the precipitation or solution and remobilization of various mineral species and elements, such that the rock column above a leaking petroleum accumulation becomes significantly and measurably different from laterally equivalent rocks (Pirson, 1969; Oehler and Sternberg, 1984; Price, 1986). This alteration chimney or plume has been documented empirically, and its surface expression can range from subtle biogeochemical anomalies, such as in Wyoming's Recluse field (Dalziel and Donovan, 1980), to the dramatic hydrocarbon-induced diagenetic aureoles (HIDAs) described from the Cement field area of Oklahoma (Donovan, 1974; Lilburn and Al Shaieb, 1983, 1984). Because such changes are measurable and mappable, they have formed the basis for many different surface exploration methods over the years. Unfortunately, our understanding of the complex physical, chemical, and biological processes responsible for these phenomena remains incomplete, with the result that these methods are viewed with skepticism and remain underutilized.

EARLY OBSERVATIONS

The association of mineralogic changes and hydrocarbon seepage has been recognized since the earliest days of petroleum exploration. Many of the early explorationists noted the correlation of productive areas with seeps, paraffin dirt, saline or sulfurous waters, surface mineralization, and topographic highs. Sawtelle (1936) reported that such features were instrumental in the discovery of about 70% of American Gulf Coast oil fields.

Harris (1908) was among the first to report the presence of pyrite and other sulfides in strata overlying oil fields associated with some Louisiana salt domes. Reeves (1922) observed the discoloration of surface red beds in the Cement field area of southwestern Oklahoma and noted the intense carbonate cementation over the crest of the Cement structure. Thompson (1933) observed that sulfur and pyrite are commonly associated with oil accumulations, and he reported that in the presence of hydrocarbon gases, gypsum and anhydrite are replaced by limestone in salt dome cap rocks in the Persian Gulf–Iraq oil belt. He also described another of the common alteration products found in Persia known as Gach-i-turush, an association of seepage petroleum, gypsum, jarosite, and sulfur caused by the interaction of petroleum with evaporites (Thompson, 1933).

McDermott (1940) and Rosaire (1940) reported the occurrence of secondary mineralization such as soil carbonates, caliche, and silicification in the vicinity of some Texas oil fields. Feely and Kulp (1957) demonstrated that the sulfur present in the cap rocks of Gulf Coast salt domes originated by bacterial action on the anhydrite and that calcite replaced anydrite as a result of bacterial oxidation of petroleum. More recent studies have documented these and other changes and discussed the processes involved (Oehler and Sternberg, 1984; Matthews, 1986; Price, 1986; Klusman, 1993; Thompson et al., 1994).

MICROBIAL EFFECTS

Bacteria and other microbes found in the soils and sediments above hydrocarbon accumulations play a profound role in the oxidation of migrating hydrocarbons. Their activities are directly or indirectly responsible for the varied and often confusing surface manifestations of hydrocarbon seepage. Their role is still largely unknown, or at least not fully recognized, by most of the investigators of hydrocarbon seepage and surface exploration technology.

Kartsev et al. (1959), Davis (1952, 1956, 1967), Krumbein (1983), and Atlas (1984) have discussed in great detail the oxidation of hydrocarbons by bacteria. In addition to the many varieties of aerobic bacteria that oxidize hydrocarbons, important anaerobes also exist (e.g., sulfate-reducing bacteria and denitrifying bacteria). Also, certain fungi and actinomycetes readily oxidize hydrocarbons in soil (McKenna and Kallio, 1965). Although bacterial activity is most pronounced in surface soils, it can occur at all depths above a leaking hydrocarbon accumulation. The most obvious result of hydrocarbon oxidation is a decrease in the concentration of free soil gas hydrocarbons (interstitial), hydrocarbons dissolved in pore fluids, occluded hydrocarbons, and adsorbed hydrocarbons. In addition, bacteria produce carbon dioxide and organic acids from hydrocarbon oxidation; the sulfate reducers produce hydrogen sulfide, and the denitrifiers produce free nitrogen and nitrous oxide.

One of the byproducts of bacterial oxidation of hydrocarbons is *paraffin dirt*, a yellow-brown waxy-appearing soil commonly associated with gas seepages in the onshore U.S. Gulf Coast, as well as in other areas with tropical to temperate climates such as Colombia, Romania, and Burma (Milner, 1925; Davis, 1967). Davis (1967) was able to create paraffin dirt in the laboratory that was indistinguishable from natural samples by encouraging bacterial growth by passing hydrocarbon gas through moist soils. He determined that paraffin dirt was an accumulation of dead cell walls of fungi and bacteria, consisting chiefly of carbohydrates (Davis, 1967).

In offshore areas, bacterial mats and suspensions are commonly associated with petroleum seepage, especially oil seepage. Such bacterial mats (*Beggiatoa*) have been reported from the North Sea (Gullfaks, Statfjord, and Tommeliten fields), the Santa Barbara Channel, and the Gulf of Mexico (Hovland and Judd, 1988; Sassen et al., 1993). *Beggiatoa* mats are also known from onshore seepages in Tunisia, Iraq, Papua New Guinea, the Congo, and Colombia. In addition to bacterial mats, offshore seep sites in the Gulf of Mexico also support a diverse chemosynthetic community including mussels, lucinid clams, and tube worms, all of which depend on methane and hydrogen sulfide seepage (Kennicutt et al., 1985; Childress et al., 1986; Brooks et al., 1987; MacDonald et al., 1989, 1990; Reilly et al., this volume).

Microbial hydrocarbon oxidation consumes either free oxygen or chemically bound oxygen (SO_4^{2-} or NO_3^{2-}) via one of two main metabolic pathways. First, aerobic bacteria oxidize hydrocarbons to form carbon dioxide or bicarbonate that eventually precipitates as carbonate. Second, once oxygen is depleted within the sediment or pore fluid, other bacteria reduce sulfate to produce hydrogen sulfide. These changes can significantly alter the oxidation-reduction potential (Eh) of the environment and can affect the pH of the system. Such pH/Eh changes can result in new mineral stability fields in which some minerals become unstable and are dissolved and mobilized, while others are precipitated from solution. In this setting, bacteria can produce minerals either through passive growth or as a result of metabolic activity.

Examples of passive microbial biomineralization include bacterial precipitation of amorphous silica in hot springs, as well as formation of some forms of authigenic iron oxides, phosphates, carbonates, and clays (Krumbein, 1983; Ferris et al., 1994; Ferris, 1995). Microbial mineral precipitation can also result directly from metabolic activity of microorganisms whereby bacterial activity simply triggers a change in the solution chemistry that leads to mineral precipitation. For example, an increase in pH can initiate the precipitation of calcium carbonate. Similarly, sulfide production by sulfate-reducing bacteria can bring about the precipitation of a number of iron sulfides and oxides, including pyrite, greigite, pyrrhotite, and maghemite (Reynolds et al., 1990; Ferris, 1995). Still other mineral phases precipitate directly from bacterial enzyme action, such as the formation of magnetite particles inside the cells of magnetotactic bacteria (Krumbein, 1983; Ferris, 1995).

Bacteria play a profound role in determining the nature and direction of physical, chemical, and biological changes in near-surface soils and sediments. Nowhere is their role more significant than in the presence of hydrocarbon seepage. Not only are bacteria responsible for the destruction of hydrocarbons at seeps, but they are also responsible for the formation of large volumes of authigenic minerals, including carbonate, elemental sulfur, and iron oxides and sulfides.

HYDROCARBON-INDUCED DIAGENETIC ALTERATION

Although the close association between surface mineralization or discoloration and oil accumulations has long been noted, detailed investigations of these changes and the processes that produce them have been conducted only since the 1970s. Donovan (1974) published the first of a series of studies describing the complex chemical and mineralogic changes observed in red beds over a number of Oklahoma oil fields. Lilburn and Al-Shaieb (1984) proposed the term *hydrocarbon-induced diagenetic aureole (HIDA)* for these near-surface alterations.

These diagenetic zonations are particularly well developed over Cement-Chickasha, Velma, Healdton, Eola, and Carter-Knox fields in southwestern Oklahoma. The geologic setting of these fields consists of tightly folded and faulted Pennsylvanian sedimentary rocks overlain by more than 600 m (2000 ft) of unfaulted Permian sandstones, red beds, and gypsum. Oil occurs in both the Pennsylvanian and Permian strata, with major production in the former. The geologic setting of these fields and their diagenetic relationships have been reviewed most recently by Al-Shaieb et al. (1994). Other studies of these near-surface diagenetic changes have been published by Olmstead (1975), Donovan and Dalziel (1977), Goldhaber et al. (1978), Ferguson (1979a, b), Donovan et al. (1981), Allen and Thomas (1984), and Reynolds et al. (1990).

Three major diagenetic facies are observed in the red beds and associated sediments overlying these oil fields (Figures 1–4): (1) an innermost zone of intense carbonate cementation and/or carbonate-replacing gypsum whose distribution coincides with the pre-Permian fault system, (2) a central zone of pyrite mineralization which may not be well exposed at the surface but is well developed at depth, and (3) a zone of bleached red beds surrounding the carbonate zone. Unaltered sediments occur beyond the zone of discolored red beds. The areal extent of these alteration zones approximates the productive limits of the reservoirs in the subsurface.

Figures 1 and 2 illustrate the observed diagenetic zonation at Velma field, southwestern Oklahoma. Identical diagenetic facies occur over nearby Cement-Chickasha field, as shown in Figures 3 and 4. Similar hydrocarbon-induced alterations have been reported in places as diverse as the Baku region of Azerbaijan (Kartsev et al., 1959), Lisbon field in southeastern Utah (Segal et al., 1984), Ashland field in southeastern Oklahoma (Oehler and Sternberg, 1984), Turkey Creek seep in Colorado (Reid et al., 1992), Mist gas field in Oregon and Brown-Bassett field in Texas (Campbell, 1994), and the Gulf of Mexico (Roberts et al., 1990; Sassen et al., 1994).

Carbonates

Diagenetic carbonates and carbonate cements are among the most common hydrocarbon-induced alterations associated with petroleum seepage. In offshore settings, the carbonates can form as slabs and rubble, large mounds and pillars, hard grounds, or pore-filling carbonate cement. On land, pore-filling calcite and replacement calcite is most common. These near-surface diagenetic carbonates are formed principally as a byproduct of petroleum oxidation, particularily of methane, using one of two reaction pathways, as summarized below:

1. Aerobic:

   $$CH_4 + 2\,O_2 + Ca^{2+} = CaCO_3 + H_2O + 2\,H^+$$

2. Anaerobic:

   $$CH_4 + SO_4^{2-} + Ca^{2+} = CaCO_3 + H_2S + H_2O$$

When these reactions occur, carbon dioxide evolves and reacts with water to produce bicarbonate. The bicarbonate bonds with calcium and magnesium in groundwater and precipitates as carbonate, or carbonate cement, that has an isotopic signature matching that of the parent hydrocarbon(s).

Normal calcite, whether its carbon is derived from the atmosphere, freshwater, or the marine environment, has a carbon isotopic value of about –10 to +5‰ relative to the PDB standard (Fairbridge, 1972; Anderson and Arthur, 1983). The carbon isotopic composition of most crude oils ranges from about –20 to –32‰, whereas that of methane can be range from –30 to –90‰. Calcite formed from oxidized petroleum incorporates carbon from the organic source which typically has an isotopic composition more negative than –20‰. Depending on

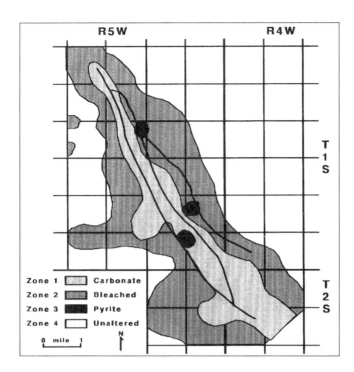

Figure 1—Surface diagenetic alteration zones and traces of pre-Permian faults over Velma field, Stephens County, Oklahoma. (From Al-Shaieb et al., 1994; reprinted by permission.)

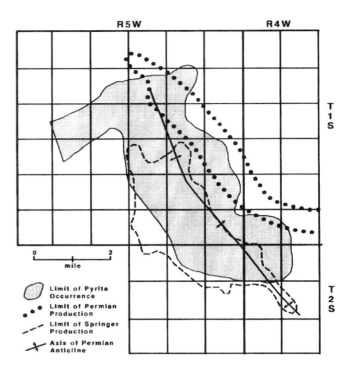

Figure 2—Subsurface limits of pyrite mineralization over Velma field, Stephens County, Oklahoma. The pyritization limits coincide with production and fault boundaries. (From Al-Shaieb et al., 1994; reprinted by permission.)

the proportion of oxidized hydrocarbon incorporated, the isotopic composition of the resultant carbonate can range from −10 to −60‰.

In Cement-Chikasha, Velma, and other southwestern Oklahoma oil fields, the Permian sandstones at the surface are highly cemented by secondary calcite and dolomite in the crestal areas of the field, but they contain little or no carbonate cement away from the field. A local gypsum-bearing formation grades from pure gypsum on the flanks of the field to gypsum entirely replaced by calcite along the structural axis of the field. The increased resistance to erosion of these carbonate-cemented units forms topographic highs along the anticlinal crest. Diagenetic carbonates present over these fields include calcite, ferroan calcite, high-Mg and high-Mn calcite, dolomite, ankerite, aragonite, siderite, and rhodochrosite (Donovan, 1974). The carbon isotopic composition of diagenetic carbonates at Cement-Chikasha field ranges from −2 to −35‰ (Donovan, 1974; Lilburn and Al-Shaieb, 1984), with the lightest (most negative) values occurring along the structural axis of the field (Figure 5). Donovan et al. (1981) report carbon isotope values for carbonate cements from Velma field of −7 to −36‰. This wide range in isotopic composition reflects more than one carbon source or a mixture of sources. For these fields, the data suggest a hybrid carbon source from both freshwater and hydrocarbons (oil and gas).

Diagenetic carbonates are also well developed over Ashland gas field in the Arkoma basin of southeastern

Oklahoma (Oehler and Sternberg, 1984). Geochemical analyses indicate that about 45% more total carbonate occurs (chiefly as calcite) in near-surface sandstones located over the field as compared to the same stratigraphic interval in off-field locations. The carbon isotopic composition for shallow calcites at Ashland ranges from −22 to −29‰, indicating that they have derived a significant portion of their carbon from oxidized hydrocarbons. These isotopically anomalous calcites are present in both on-field and off-field wells, although their concentration is greater in the on-field wells.

Isotopically light carbonates have been widely documented in salt dome cap rocks in the U.S. Gulf Coast and in modern hydrocarbon seep sites in the Gulf of Mexico. Roberts et al. (1990) report carbon isotope values of −16 to −48‰ for authigenic carbonates in the Green Canyon area, northern Gulf of Mexico. Sassen et al. (1994) report values of −22 to −29‰ for carbonate cements and −24 to −31‰ for carbonate cap rock in salt domes. Other values reported for calcite cap rocks of Gulf Coast salt domes range from −12 to −53‰ (Posey et al., 1987; Prikryl, 1990). Since the volume of diagenetic carbonate that can form at seafloor seep sites can be considerable, some workers have even suggested that some carbonate reefs might owe their origin to hydrocarbon seepage (Hovland, 1990).

Other areas with isotopically light carbonate include Recluse oil field in Wyoming (Dalziel and Donovan, 1980), Gulf of Alaska (Barnes et al., 1980), Davenport oil field in Oklahoma (Donovan et al., 1974), Ocho Juan field

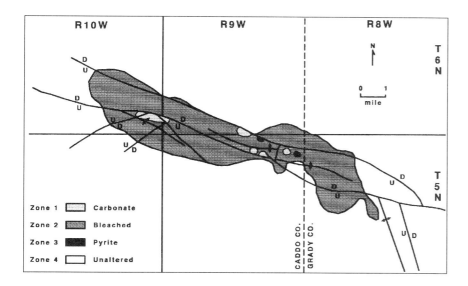

Figure 3—Surface diagenetic alteration zones and pre-Permian structural configuration of Cement-Chickasha field, Caddo County, Oklahoma. (From Al-Shaieb et al., 1994; reprinted by permission.)

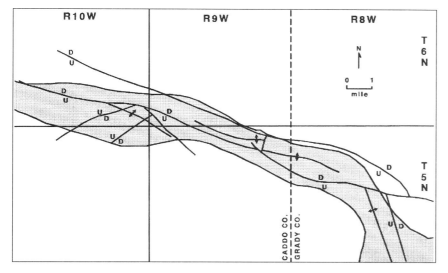

Figure 4—Subsurface limits of pyrite mineralization in Cement-Chickasha field, Caddo County, Oklahoma. The limits of the pyritized zone coincide with the pre-Permian structure. (From Al-Shaieb et al., 1994; reprinted by permission.)

in Texas, Fox-Graham field in Oklahoma (Duchscherer, 1984), and the carbonate hardgrounds formed around modern gas seeps on the Carolina continental rise (Paull et al., 1995) and near Fredrikshavn, Denmark (Dando et al., 1994).

Diagenetic carbonates related to hydrocarbon seepage appear to be widespread, although not all fields or hydrocarbon seep areas possess isotopically anomalous carbonates. Hydrocarbon microseepage is well documented for Patrick Draw field in Wyoming and Lisbon Valley field in Utah, but the carbon isotopic composition for their soil carbonates ranges from 0 to –9‰, indicating that little if any of the carbon is derived from the oxidation of hydrocarbons (Conel and Alley, 1985; Lang et al., 1985a). Similarly, calcite cements from outcropping sandstones in the Little Buffalo Basin oil field of Wyoming have isotopic values heavier than –10‰, suggesting little contribution from oxidized hydrocarbons, although the occurrence of the most negative values over the crest of the structure suggests that there may be some contribution (Bammel et al., 1994). Other fields, such as Coyanosa in west Texas,

show no evidence of hydrocarbon seepage, and carbonate cements from over the field have isotopic compositions of –1 to –9‰, results consistent with a nonpetroleum carbon source (Lang et al., 1985c).

It is tempting to relate near-surface carbonate diagenesis to leakage of reservoired hydrocarbons, but we should remember that geochemical anomalies caused by abnormal amounts of carbon dioxide are nonspecific for petroleum. Abnormal CO_2 concentrations in soils and sediments can result from processes other than microbial oxidation of hydrocarbons, such as hydrothermal activity, volcanic activity, catagenesis of organic matter, micropore filtration, and thermochemical sulfate reduction (Kartsev et al., 1959; Matthews, 1986, Sassen et al., 1994). Different mechanisms can yield similar end-products, and the formation of near-surface carbonates and carbonate cements may be more heavily dependent on the area's geology and groundwater chemistry. Unless each mechanism has a distinctive geochemical or isotopic "fingerprint," careful analysis is required to determine the nature and origin of shallow carbonates and carbonate cements.

Figure 5—Variation in the carbon isotopic composition of subsurface carbonate cements from Cement-Chickasha field, Caddo County, Oklahoma. (From Al-Shaieb et al., 1994; reprinted by permission.)

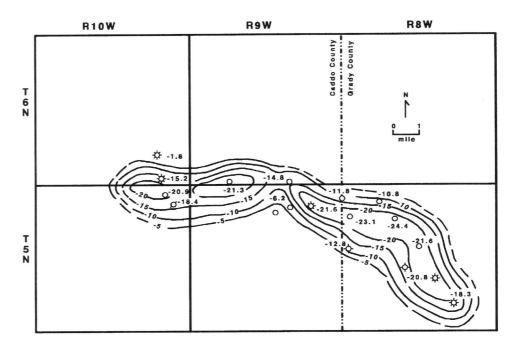

Sulfides

The formation of secondary pyrite and other sulfides has been documented for many petroleum fields, including the Cement-Chickasha, Velma, Eola, and Ashland fields (all in Oklahoma) by Ferguson (1979a, b), Lilburn and Al-Shaieb (1983, 1984), Donovan et al. (1981), Oehler and Sternberg (1984), and Hughes et al. (1986). Pyrite is the dominant sulfide mineral in these hydrocarbon-induced alteration zones, but pyrrhotite, marcasite, galena, sphalerite, and native sulfur are also found and may be locally abundant. The mechanisms responsible for the formation of sulfides in the hydrocarbon seep environment have been discussed by Sassen (1980), Oehler and Sternberg (1984), Sassen et al. (1988, 1989), Goldhaber and Reynolds (1991), and Reynolds et al. (1990, 1993).

Pyrite can be precipitated in a reducing environment, given a source of sulfur and iron. The major source of sulfur in a petroleum province is hydrogen sulfide gas from the petroleum itself, from anaerobic bacterial activity, or from the oxidation of petroleum in the near-surface. Sources of iron include iron oxide grain coatings in sandstone, pore-filling clays such as chlorite, rock fragment inclusions, and deeper meteoric waters. The reaction of hydrogen sulfide and iron (from hematite) to precipitate pyrite or marcasite can be summarized as follows (Oehler and Sternberg, 1984):

$$Fe_2O_3 + 4\ H_2S = 2\ FeS_2 + 3\ H_2O + 2\ H^+ + 2e^-$$

The development of a pyrite alteration zone depends on the sulfur content of the oils, the geology and groundwater geochemistry of the sedimentary sequence, and the nature of the bacterial degradation (Hughes et al., 1986). For example, if the sulfur content of oils is high and the groundwater is rich in iron, pyrite could be precipitated

at any depth within the migration plume that possesses sufficient porosity. If oils are free of sulfur, the sulfur required for the reaction must be derived from bacterial degradation, in which case pyrite precipitation would occur in the near-surface sediments due to environmental restrictions on bacterial activity. In each case, however, seepage-related pyrite would be concentrated in the reduced plume of hydrocarbon leakage (Hughes et al., 1986).

The pyrite zone is well-developed in Velma, Healdton, and Cement-Chickasha fields in Oklahoma (Figures 2, 4). Pyrite mineralization is concentrated in sandy intervals in the uppermost 100 m (330 ft) of section, and its distribution approximates the surface projection of the productive reservoirs. Pyrite is abundant in cuttings from on-field wells and generally absent in cuttings from off-field wells. The concentration of pyrite in the cuttings ranges from 5–20% over Cement field to 2–4% over Velma, Chickasha, and Eola fields (Campbell, 1994). The average value of the sulfur isotopic composition of the pyrite (–3.6‰) compares favorably with that of the sulfur in the associated oil (–4.7‰), strongly suggesting that the H_2S associated with the oil is the major source of the sulfur in the pyrite (Goldhaber et al., 1978; Lilburn and Al-Shaieb, 1983, 1984).

The presence of a pyrite zone has also been documented for Ashland gas field, a stratigraphic trap in the Arkoma basin, southeastern Oklahoma. A comparison of on-field and off-field pyrite and marcasite content shows that the on-field wells average nearly twice as much iron sulfide within the same sandy stratigraphic interval as the off-field wells do, 1.5–3.5% versus 0.5–1.5% (Oehler and Sternberg, 1984). Campbell (1994) has documented the development of pyrite zones in near-surface sediments overlying Mist gas field in Oregon, Hogback Ridge gas field in Utah, and Brown-Bassett gas field in west Texas.

For Brown-Bassett field, Campbell (1994) reports that the average concentration of pyrite in cuttings from on-field wells is more than three times as high as observed in off-field wells, 5.7% versus 1.7%.

Not all shallow pyrite anomalies result from hydrocarbon leakage. Oehler and Sternberg (1984) have described a "false" pyrite anomaly over a nonproductive petroleum prospect in Texas in which the pyrite mineralization was associated with fine-grained organic-rich mudstones and unrelated to petroleum accumulation or leakage. While high concentrations of pyrite have been observed over many oil and gas fields, such mineralization only occurs where the shallow stratigraphy and its geochemical environment are favorable.

Bleached Red Beds

The presence of bleached and discolored red sandstones at the surface above petroleum accumulations has been widely noted, but detailed studies are few. Bleaching of red beds occurs whenever acidic or reducing fluids are present to remove ferric oxide (hematite). Such conditions also favor the formation of pyrite and siderite from the iron released during the dissolution of hematite. The possible reducing agents responsible for bleaching red beds above petroleum accumulations include hydrocarbons, H_2S, and CO_2.

In the Cement field area of Oklahoma, Donovan (1974) reported that the color of the Permian Rush Springs Formation grades from reddish-brown for unaltered sandstone adjacent to the field, to pink, yellow, and white along the flanks of the Cement anticline, and to gray and white along the anticlinal axis, where maximum bleaching and iron loss occur. Similar changes are observed at nearby Velma, Eola, Healdton, and Chickasha fields (Ferguson, 1979a,b; Donovan et al., 1981).

The geology and geochemical alteration associated with Lisbon Valley field in southeastern Utah have been described in considerable detail by Segal et al. (1984, 1986) and Conel and Alley (1985). They report that the distribution of the bleached outcrops of the Triassic Wingate formation approximates the geographic limits of the oil and gas reservoirs at depth. The red color of the unbleached Wingate was found to result from a pervasive hematite-clay mixture coating virtually all sand grains, whereas the bleached Wingate appears white or gray due to the absence of these hematite grain coatings. Hematite is present in the bleached rocks at Lisbon Valley field, although it occurs only as pseudomorphs of pyrite and siderite rather than as hematite grain coatings (Segal et al., 1984, 1986; Conel and Alley, 1985).

Other petroleum fields associated with bleached red beds include those near Baku in Azerbaijan (Kartsev et al., 1959), several Wind River basin oil fields in Wyoming (Love, 1957), and Garza field in west Texas (Donovan et al., 1979). While the presence of bleached red beds over oil and gas accumulations is a highly visible manifestation of hydrocarbon-induced alteration, one must remember that the reducing fluids causing the observed discoloration are not limited to hydrocarbons and, even if hydrocarbons, might represent shallow biogenic methane rather than thermogenic oil or gas.

Clay Mineral Alteration

The production of CO_2, H_2S, and organic acids resulting from the microbial oxidation of hydrocarbons in near-surface soils and sediments can create reducing, slightly acidic conditions that promote the diagenetic weathering of feldspars to produce clays and may lead to the conversion of normally stable illitic clays to kaolinite. Clays thus formed remain chemically stable unless their environment is changed.

In Utah's Lisbon Valley field, Segal et al. (1984, 1986) report that bleached portions of the Wingate Sandstone directly overlying the field contain primarily kaolinite clays, whereas the unbleached areas of the sandstone located away from the field contain fresh plagioclase and muscovite. The bleached Wingate contains three to five times more kaolinite than the unbleached rock. The geographic distribution of the kaolinite is inversely related to that of the mixed-layer illite-smectite clays, suggesting that the enrichment of kaolinite is also related to the depletion of other clay minerals and not only to the alteration of feldspars (Conel and Alley, 1985).

Clay mineral diagenesis has also been documented at Cement-Chickasha field in Oklahoma, where kaolinite and mixed-layer illite-smectite clays of late origin have replaced detrital illite in red beds (Lilburn and Al-Shaieb, 1983, 1984). A third occurrence of hydrocarbon-induced formation of kaolinite has been described by Reid et al. (1992) from the Turkey Creek oil seep near Denver, Colorado. Turkey Creek is the site of an oxidation-reduction front that developed in the outcrop of the Cretaceous "J" sand due to active oil seepage. Coarse-grained authigenic kaolinite in concentrations up to 2.0 wt. % is present in the altered rocks of the outcrop.

Uranium

The occurrence of uranium has been linked to petroleum by many authors (Eargle and Weeks, 1961; Al-Shaieb, 1977; Goldhaber et al., 1978, 1983; Curiale et al., 1983). The association between petroleum and heavy metals such as uranium (as well as lead, zinc, and even gold) is due to the fact that the reducing environment created by migrating hydrocarbons and associated fluids favors the precipitation of uranium and other heavy metals. Oxidized uranium (UO_2^{2+}) is soluble in groundwater, although when reduced, it precipitates from solution as uraninite (UO_2), which is relatively insoluble.

A commercial uranium deposit occurs at Cement field, Oklahoma (Olmstead, 1975; Allen and Thomas, 1984). In Lisbon Valley field, Utah, the close spatial correspondence of uranium deposits in the Triassic Chinle Formation, outcrops of bleached Wingate Sandstone, and the geographic limits of the subsurface oil and gas accumulation suggests a genetic relationship among them (Conel

Figure 6—Controlled-source audiofrequency magnetotellurics (CSAMT) residual resistivity profile across Ashland gas field, Arkoma Basin, Oklahoma. Note the well-developed low-resistivity zone at depth that overlies the gas field and closely approximates the productive limits of the field. The shallow high-resistivity zone corresponds to an interval of carbonate-cemented sandstone. (After Oehler and Sternberg (1984) and Phoenix Geophysics.)

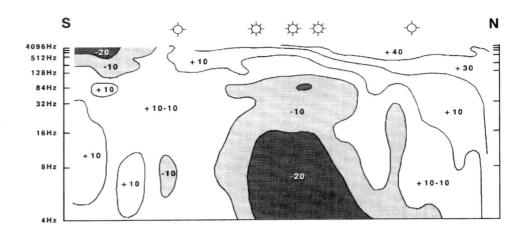

and Alley, 1985). The Turkey Creek oil seep near Denver, Colorado, has geologic and geochemical characteristics similar to uranium roll-front deposits in Texas and the Colorado Plateau, many of which are related spatially to petroleum accumulations (Reid et al., 1992). Eargle and Weeks (1961) reported an association between uranium roll-fronts in Karnes County, Texas, and oil and gas fields located down-dip; they speculated that H_2S seepage created the reducing environment responsible for deposition of uranium and the accompanying pyrite. A similar association was described from Live Oak County, Texas, where many uranium mines occur along the Oakville fault (Eargle and Weeks, 1973). Oil, gas, and H_2S leak up the fault and create the reducing environment that promotes pyrite and uranium deposition (Goldhaber et al., 1978, 1983).

Magnetic Minerals

The presence of magnetic anomalies over oil and gas fields has been noted for several decades, but it is only in relatively recent years that the phenomenon has been critically examined. The same hydrocarbon-induced reducing environment that promotes the formation of uranium and pyrite can lead to the precipitation of a variety of magnetic iron oxides and sulfides, including magnetite (Fe_3O_4), maghemite (γ-Fe_2O_3), pyrrhotite (Fe_7S_8), and greigite (Fe_3S_4).

Donovan et al. (1979) reported magnetite in altered Permian rocks overlying Cement field, Oklahoma, and speculated that hydrocarbon-related brines migrating from depth caused reduction of hematite to form magnetite. Reynolds et al. (1984, 1988, 1990) reexamined the occurrence of magnetite at Cement field and concluded that it represented drilling contamination. However, they did document the presence of ferrimagnetic pyrrhotite in the section and suggested that it precipitated as a result of hydrocarbon seepage and degradation. Although authigenic magnetite is absent at Cement field, its occurrence has been documented at many hydrocarbon seep sites by Elmore et al. (1987) and McCabe et al. (1987). The formation of pyrrhotite and other metals in hydrocarbon seep

environments at several U.S. Gulf Coast salt domes has been described by Sassen et al. (1988, 1989). Other studies have documented the presence of greigite and maghemite in near-surface sediments above petroleum accumulations and suggest that these minerals may be responsible for most of the magnetic anomalies associated with oil and gas fields (Foote, 1987; Foote and Long, 1988; Foote, this volume).

The presence, origin, and exploration significance of magnetic mineralization associated with petroleum accumulations remain controversial and have recently been addressed by Gay and Hawley (1991), Machel and Burton (1991a, b), Gay (1992), Reynolds et al. (1993), and Machel (this volume). The general agreement of elevated magnetic susceptibility of soils and sediments with light hydrocarbon soil gas anomalies supports the hypothesis that hydrocarbon microseepage may generate magnetic anomalies in near-surface soils and sediments (Henry, 1988; Saunders et al., 1991; Ellwood and Burkart, this volume). Gay (1992) agrees that recent measurements of soil magnetic susceptibility in oil fields constitutes evidence for anomalous near-surface magnetization associated with hydrocarbon leakage plumes. He urges caution, however, in ascribing the origin of all shallow magnetic anomalies to seep-induced alteration without considering possible syngenetic magnetic sources such as detrital magnetite, magnetic sedimentary formations, and burned coal seams. To further confuse the issue, it has been documented that increases in soil magnetic susceptibility can be due to pedogenic formation of magnetite and maghemite in soils and that such formation is closely tied to rainfall and climate, not hydrocarbon seepage (Maher and Thompson, 1991, 1992; Liu et al., 1994).

Electrochemical Changes

Considerable evidence has been cited to demonstrate that hydrocarbon-induced changes alter the mineralogy and chemical composition of the sediments and fluids overlying a petroleum accumulation. While these near-surface manifestations of hydrocarbon migration are varied and complex, only a few of these alterations have the

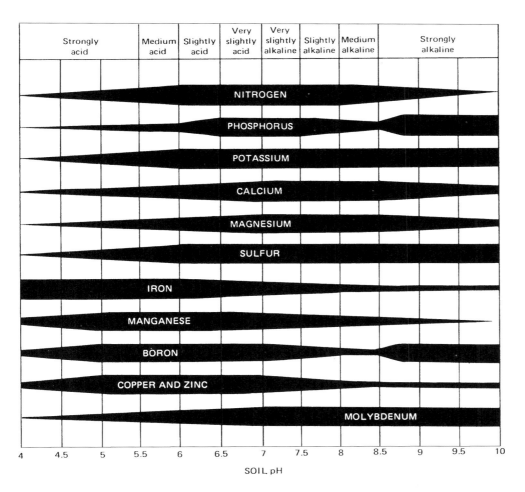

Figure 7—The effect of soil pH on the availability of nutrients to plants. The width of the horizontal band reflects relative solubility. (From Bidwell, R. G. S., Plant Physiology, ©1974, p. 252. Reprinted by permission of Prentice Hall, Upper Saddle River, New Jersey.)

potential to significantly affect the electrochemical properties of the rock column above an oil or gas accumulation. Hughes et al. (1986) described and discussed five such phenomena: pyrite mineralization, calcite cementation, clay alteration, brine effects, and redox potential cell. The first three effects have already been discussed in this paper and were shown to be important alteration products associated with petroleum migration and seepage. The existence of the two remaining phenomena is less well documented. The upward migration of brines has been postulated by Donovan et al. (1981) and others to explain the numerous conductive anomalies measured over known oil fields. There is, however, little published documentation for the existence of brine plumes, and that mechanism remains speculative (Hughes et al., 1986). Pirson (1969, 1976) proposed that the generation of a reducing plume or chimney above a hydrocarbon accumulation produces an excess of free electrons within the plume and that this produces the electrically conductive zone within the plume. However, there appear to be few if any quantitative studies to document the existence of redox potential cells.

Electrical measurements of oil and gas fields aim to detect one or more of the following hydrocarbon-induced alterations by their electrical response: (1) shallow pyrite and marcasite, which provide the source of induced

polarization anomalies; (2) pore-filling carbonate cements, which provide the source of shallow high-resistivity anomalies, and (3) the presence of a deeper low-resistivity zone representing the conductive plume or inferred hydrocarbon leakage chimney. The best documented study to date is probably that of Oehler and Sternberg (1984) for Ashland gas field, a stratigraphic trap in southeastern Oklahoma. Their results document the presence of a near-surface pyrite-marcasite zone and a shallow calcite-cemented zone. They show that these mineralized zones correspond to induced polarization and shallow high-resistivity anomalies, respectively, in surface electrical surveys. Furthermore, their results show excellent correlation between the lateral extent of the induced polarization and resistivity anomalies and the productive limits of the field. The resistivity anomaly for Ashland field is shown on Figure 6.

Results of other induced polarization and resistivity surveys have been reported for Red Oak and South Pine Hollow gas fields in Oklahoma (Sternberg, 1991), Masrab field in Sirte basin in Libya (Sternberg, 1991), Mist field in Oregon and Brown-Bassett field in Texas (Campbell, 1994), and the Turkey Creek seep in Colorado (Reid et al., 1992). All of these examples demonstrate a close correspondence between induced potential and resistivity anomalies and deeper petroleum production. As encour-

Figure 8—Diagram illustrating the empirical remote sensing exploration model developed for the Patrick Draw, Wyoming, NASA/Geosat test site. The stunted sage brush anomaly coincides with soils characterized by high concentrations of light hydrocarbons, zinc, and elevated pH. (From Lang and Nadeau, 1985; reprinted by permission.)

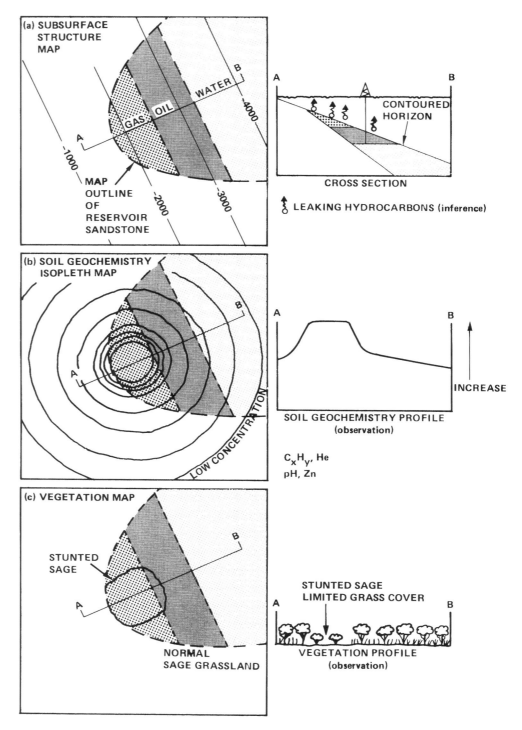

aging as such examples are, significant problems and uncertainties remain, as shown by the results of induced potential and resistivity surveys over Whitney Canyon and Ryckman Creek fields in Wyoming. Hughes et al. (1986) found a striking correlation between the conductive anomalies and the areal extent of the fields. However, they determined from modeling studies that these same anomalies could be explained as the result of cultural, geologic, and topographic factors unrelated to the underlying petroleum accumulations. The importance of minimizing cultural influences in electrical surveys has been most recently discussed by Carlson and Zonge (this volume).

Trace Elements and Biogeochemistry

Hydrocarbon microseepage creates a chemically reducing zone in the soil column at depths shallower than would be expected in the absence of seepage. Such leak-

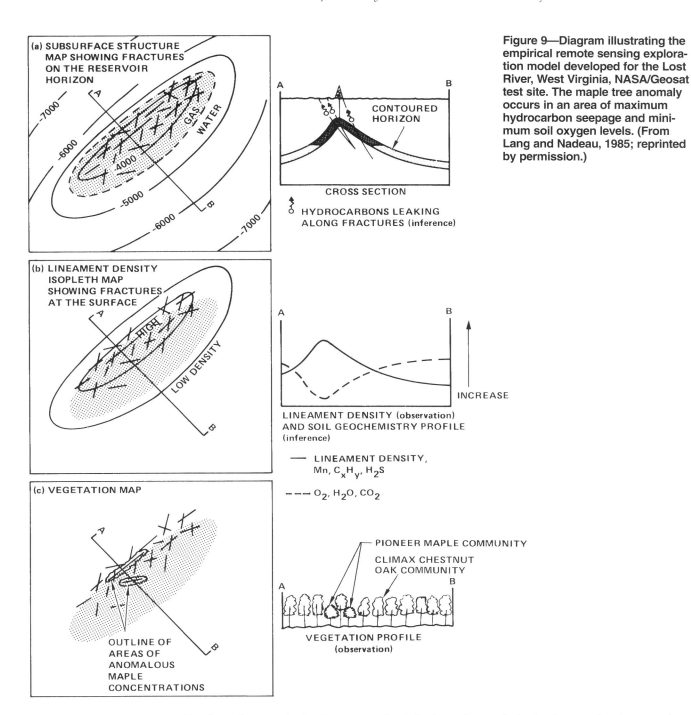

Figure 9—Diagram illustrating the empirical remote sensing exploration model developed for the Lost River, West Virginia, NASA/Geosat test site. The maple tree anomaly occurs in an area of maximum hydrocarbon seepage and minimum soil oxygen levels. (From Lang and Nadeau, 1985; reprinted by permission.)

age stimulates the activity of hydrocarbon-oxidizing bacteria, which decreases soil oxygen concentration while increasing the concentration of CO_2 and organic acids. These changes can affect pH and Eh in soils, which in turn affects the solubilities of the trace elements and consequently their availability to plants. Figure 7 illustrates the effect of soil pH on the relative solubility of common trace metals (Bidwell, 1979). This effect can be quite pronounced; for example, the solubility of iron at pH 6 is 10^5 times greater than at pH 8.5. The lack of essential nutrients such as iron, manganese, copper, and zinc—or their presence in excessively high concentrations—can lead to

physiologic and morphologic changes in plants and can alter their spectral reflectance.

Soil trace metal surveys have been used as indirect surface indicators of petroleum accumulations due to their ability to form organometallic complexes under the reducing conditions that can be found above petroleum accumulations. Duchscherer (1984) reports that V, Cr, Mn, Ni, Co, Cu, Mo, U, Fe, Zn, and Pb are among those trace metals found as geochemical halos over oil fields, and he cites examples of such anomalies from Ocho Juan field in Texas and Bell Creek field in Montana. Another element that has been cited as an indirect pathfinder for petroleum accumulations is iodine (Gallagher, 1984; Klusman, 1993; Tedesco, 1995). Iodine can be derived from clays or mineral assemblages in soil, from the breakdown of humic acids or humins, from the atmosphere, or from formation waters. Increases in soil iodine are thought to result from reactions between migrating hydrocarbons and iodine at the soil-air interface (Tedesco, 1995), but the mechanism is poorly understood and largely undocumented.

The most comprehensive investigation of the effects of hydrocarbon leakage on the chemistry of soils and vegetation was the joint NASA-Geosat study of Patrick Draw oil field in Wyoming, Lost River gas field in West Virginia, and Coyanosa field in west Texas (for summary, see Lang and Nadeau, 1985). The study documents a variety of hydrocarbon-induced effects on vegetation and soil over Patrick Draw and Lost River fields, but no apparent effect at Coyanosa. The most pronounced anomaly observed at Patrick Draw field was an area of stunted sagebrush and an associated tonal anomaly visible on Landsat imagery. The anomaly overlies the field's gas cap and occurs in a region of strong light hydrocarbon microseepage, as shown in Figure 8 (Lang et al., 1985a; Richers et al., 1982, 1986). The geology and production history of the field show that the sagebrush anomaly results from the upward migration of injected gases and waters used to maintain reservoir pressures in the field (Arp, 1992). These gases and waters produced anoxic, low-Eh (oxidation potential), high-pH, and high-salinity soils that are toxic to the overlying sagebrush (Lang et al., 1985a; Arp, 1992).

Hydrocarbon microseepage was also documented over the Lost River gas field in the Appalachian Mountains of West Virginia. The principal vegetation anomaly observed here was the presence of maple trees over the gas field at sites where more typical oak-hickory climax vegetation would be expected (Lang et al., 1985b). The results for the Lost River site, as summarized in Figure 9, show that the maple trees occur in an area of maximum hydrocarbon seepage (methane) and minimum soil oxygen content. The anomalous maple distribution may relate to anaerobic soil conditions that directly or indirectly influence the mycorrhizal fungi living on the trees' root hairs, favoring maple trees whose fungi appear to be better able to tolerate the anaerobic soils than their counterparts living on the roots of oaks (Lang et al., 1985b).

Applying biogeochemical techniques, Dalziel and Donovan (1980) measured reduced iron and manganese in the leaves of pine and sagebrush that grew over the Recluse oil field in Wyoming and found that the Mn:Fe ratio was highest over the field. Similar results were reported from Bell Creek oil field in Montana by Dalziel and Donovan (1980) and Roeming and Donovan (1985). At Bell Creek, as in most areas studied, soil and plant geochemical data are inversely related, with low concentrations of metals in soils from under plants with high metal concentrations in their leaves. McCoy and Wullstein (1988) analyzed leaves of sagebrush and greasewood from Blackburn oil field in Nevada and reported a halo

anomaly of high Mn:Fe ratios surrounding the productive part of the field. McCoy et al. (1989) revisited Blackburn field and determined that the spectral reflectance of sagebrush from the anomalous area was lower than that of sagebrush from background areas.

More recently, Klusman et al. (1992) compared the amounts of 20 trace elements in plants growing over and near two oil fields—Eagle Springs field in Nevada and Cave Canyon field in Utah. Klusman theorized that alkaline soil elements such as calcium, strontium, and barium are less available to plants growing in microseepage environments, whereas the transition trace elements such as iron, manganese, and vanadium increase in availability due to their increased solubility in the seep environment. Data from Eagle Springs field supported the expected relationship, but data from Cave Canyon did not.

There is no doubt that hydrocarbon microseepage can have a pronounced effect on soils and vegetation, but the specific response is not consistent for different species and sites. In addition, factors such as bedrock geology, soil type, slope, soil moisture, and climate can have a more pronounced effect than that due to the presence of hydrocarbons (Rock, 1984; Klusman et al., 1992).

MODEL FOR HYDROCARBON-INDUCED ALTERATION

Models and mechanisms to explain the diverse array of hydrocarbon-induced changes observed in soils and sediments have been widely proposed and discussed by Donovan (1974), Oehler and Sternberg (1984), Hughes et al. (1986), Price (1986), Klusman (1993), Al-Shaieb et al. (1994), and Thompson et al. (1994). A simplified summary of the basic reactions and processes is presented here and illustrated in Figure 10:

1. Hydrocarbons, chiefly methane through pentane, migrate upward from the reservoir to the surface.
2. When upward-migrating light hydrocarbons reach near-surface oxidizing conditions, aerobic hydrocarbon-oxidizing bacteria consume methane (and other light hydrocarbons) and decrease oxygen in pore waters:

$$CH_4 + O_2 = CO_2 + H_2O$$

3. With development of anaerobic conditions, the activity of sulfate-reducing bacteria results in sulfate ion reduction and oxidation of organic carbon to produce reduced sulfur species and bicarbonate ion:

$$2\,CH_2 + SO_4^{2-} = 2\,HCO_3^- + H_2S$$

$$2\,CH_2 + SO_4^{2-} = HCO_3^- + HS^- + H_2O + CO_2$$

4. Highly reactive reduced sulfur species then can

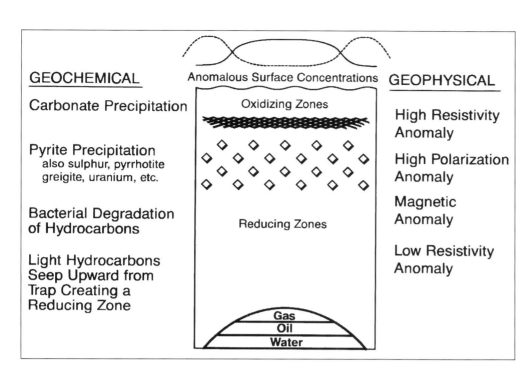

GEOCHEMICAL

Carbonate Precipitation

Pyrite Precipitation
also sulphur, pyrrhotite
greigite, uranium, etc.

Bacterial Degradation
of Hydrocarbons

Light Hydrocarbons
Seep Upward from
Trap Creating a
Reducing Zone

Anomalous Surface Concentrations

Oxidizing Zones

Reducing Zones

Gas
Oil
Water

GEOPHYSICAL

High Resistivity
Anomaly

High Polarization
Anomaly

Magnetic
Anomaly

Low Resistivity
Anomaly

Figure 10—Generalized model of hydrocarbon-induced geochemical and geophysical alteration of soils and sediments.

combine with available iron to form iron sulfides and oxides:

$$Fe_2O_3 + 4 H_2S = 2 FeS_2 + 3 H_2O + 2 H^+ + 2e^-$$

(Iron sulfide can be in the form of pyrite, marcasite, magnetite, pyrrhotite, greigite, or maghemite.)

$$Fe_2O_3 + 2 H_2S = FeS_2 + FeO + 2 H_2O$$

$$FeO + Fe_2O_3 = Fe_3O_4 \ \text{(magnetite)}$$

5. As a result of bacterial sulfate reduction, sulfate ion concentration is decreased. In addition, bicarbonate is added to pore waters, raising pH and thus promoting precipitation of isotopically light, pore-filling carbonate cements:

$$2 HCO_3^- + Ca^{2+} = H_2O + CO_2 + CaCO_3 \ \text{(calcite)}$$

This alteration model is not meant to represent all possible reactions and processes occurring in the chemically and biologically dynamic near-surface environment. Rather, it is intended to provide a general framework within which a wide range of reactions can occur.

IMPLICATIONS FOR EXPLORATION METHODS

The surface and near-surface expression of hydrocarbon seepage can take many forms, ranging from elevated hydrocarbon concentrations in soils and water to complex chemical, mineralogic, microbial, and botanical changes. These various surface manifestations have led to the development and marketing of an equally diverse number of surface exploration methods. Some are geochemical, some are geophysical, and others come under the category of remote sensing. Although this is not the place to discuss the advantages and limitations of each of the many commercially available exploration methods, it seems appropriate to comment on the major categories of exploration methods as they relate to the alteration phenomena on which they are based.

Soil Carbonate Methods

Near-surface diagenetic carbonates and carbonate cements are among the most common and widespread manifestations of hydrocarbon seepage, although geochemical anomalies caused by abnormal amounts of CO_2 are nonspecific for petroleum. High concentrations of CO_2 in soils and sediments can result from processes other than microbial oxidation of hydrocarbons, such as volcanic or geothermal activity, catagenesis of organic matter, micropore filtration, and thermochemical sulfate reduction (Kartsev et al., 1959; Price, 1986). Also, even if the CO_2 is a product of hydrocarbon oxidation, the hydrocarbon source could be shallow biogenic methane rather than thermogenic oil or gas from depth. If the origin is biogenic methane, the resulting alteration phenomena will have no relationship to deep subsurface exploration objectives.

Two of the more widely used surface geochemical exploration methods that depend on the presence of soil carbonate are Duchscherer's delta-carbonate (ΔC) method and Horvitz's "adsorbed" soil hydrocarbon method. The ΔC method measures the amount of CO_2

evolved from the thermal decomposition of soil carbonates at high temperature (Duchscherer, 1981, 1984, 1986). The Horvitz method utilizes an acid extraction technique to release hydrocarbons from the fine-grained fraction of the soil sample (Horvitz, 1945, 1985). Horvitz believed the hydrocarbons to be adsorbed onto clays or incorporated into soil calcites; the latter is more likely based on recent studies by Price (this volume). Price has found that the Horvitz technique is moderately dependent on soil calcite concentration, whereas Duchscherer's method is completely dependent on calcite concentration. Although soil calcite is nonspecific for thermogenic hydrocarbons, the Horvitz technique measures the concentration of ethane through pentane in the sample and is therefore specific for thermogenic hydrocarbons. Duchscherer's ΔC method cannot by itself discriminate thermogenic from nonthermogenic calcites.

Magnetic Methods

The presence of magnetic anomalies over oil and gas fields has been noted for several decades, but it is only in recent years that the phenomenon has been critically examined. Analysis of data from geologically and geographically diverse regions shows (1) that authigenic magnetic minerals may occur in near-surface sediments above petroleum accumulations; (2) that this hydrocarbon-induced mineralization is detectable with low-level, high-resolution aeromagnetic data; and (3) that magnetic susceptibility analysis of well cuttings (and sometimes soils) confirms the existence of the aeromagnetic anomalies.

While shallow sedimentary magnetic anomalies appear to be associated with many petroleum accumulations, hydrocarbon-induced mineralization is but one of several possible causes for such anomalies. Gay and Hawley (1991) and Gay (1992) urge caution in the interpretation of such anomalies and cite examples of many false anomalies caused by cultural contamination, geologic structure, and syngenetic magnetic sources such as detrital magnetite and burned coal seams. Not only is the interpretation of some shallow magnetic anomalies open to question but their very existence can be short-lived. Investigations by Ellwood and Burkart (this volume) document that nonmagnetic phases such as hematite, pyrite, and siderite can be oxidized to magnetite or maghemite, and more significantly for the explorationist, magnetic phases can revert to nonmagnetic pyrite or siderite under highly reducing conditions. The fluctuation in oxidizing and reducing conditions required to bring about these changes is most pronounced in soils and in the shallow subsurface.

Electrical Methods

Electrical geophysical methods have gained acceptance in recent years due to advances in both hardware and software technology. The electrical tools most appropriate for oil and gas exploration include induced polarization (IP), spectral IP, magnetotellurics (MT), and controlled-source audiofrequency magnetotellurics (CSAMT). Each of these methods is designed to detect electrochemically altered sediments, that is, the conductive plume or alteration chimney that may extend from the accumulation to the surface. Magnetotelluric data are the result of measuring natural fluctuations in magnetic and electrical fields at the earth's surface. CSAMT is a more recent electrical geophysical application that uses an artificial signal source, unlike MT, which uses naturally occurring signals. CSAMT now appears to have sufficient resolution to image the subsurface alteration plume across a wide range of geologic conditions. Despite encouraging reports, there are still relatively few published studies documenting electrochemical alteration over petroleum accumulations, and most have not fully addressed the possible contribution of geologic, topographic, and cultural effects.

Sternberg (1991) states that the IP-resistivity method for hydrocarbon exploration has significant limitations. Many areas do not appear to have the required geologic and geochemical conditions for the formation of IP or resistivity anomalies. IP and resistivity anomalies may also need to be tested with surface geochemistry and shallow drill holes to separate anomalies caused by hydrocarbon seepage from false anomalies due to other causes.

Radioactivity Methods

The existence of gamma radiation lows over oil and gas fields has long been known and forms the basis for radiometric surveys utilizing airborne or ground-based gamma ray spectrometers. Potassium-40 in clay is generally thought to be the major source of soil radioactivity, with lesser contributions from bismuth-214 and thallium-208. The low radiation values over petroleum accumulations have been attributed to either (1) precipitation of uranium salts at the oxidation-reduction boundary at the edge of the inferred hydrocarbon leakage plume, or (2) conversion of K-bearing clays and feldspars to kaolinite or other K-deficient clays (Pirson, 1969; Heemstra et al., 1979; Price, 1986; Saunders et al.,1993).

Despite numerous claims of success for surface radiation surveys by Weart and Heimberg (1981) and Curry (1984), among others, there have been few scientifically rigorous investigations of radiation surveys and the many factors that can influence their results. One such study by Heemstra et al. (1979) conducted in Kansas found no correlation between petroleum production and surface radiation. They did, however, find correlations among gamma radiation and topography, soil type and thickness, bedrock outcrops, and other factors. There seems little doubt that anomalous gamma radiation values are associated with some oil and gas fields, but the processes that produce them are poorly understood and even less well documented.

Remote Sensing Methods

Satellite-based remote sensing of hydrocarbon-induced alteration holds great promise as a rapid, cost-effective means of detecting anomalous diagenesis in surface soils and rocks. Research in the vicinity of Patrick Draw, Lost River, and Lisbon Valley fields during the NASA-Geosat test case project demonstrates that Landsat MSS and Thematic Mapper (TM) data can be used to detect three types of hydrocarbon-induced geochemical changes: (1) reduction of ferric iron (red bed bleaching), (2) conversion of mixed-layer clays and feldspars to kaolinite, and (3) anomalous spectral reflectance of vegetation. The potential for application of these techniques is greatest in areas of sparse vegetation and susceptible surface clays and red beds.

The outlook for trace element and biogeochemical surveys seems less encouraging. Anomalous distributions of iodine and trace metals have been reported in some oil and gas fields, but the mechanisms responsible for these anomalies are neither well known nor well documented. Hydrocarbon microseepage can have a pronounced effect on soils and vegetation, but the specific response appears to be inconsistent for different species and sites. Also, factors such as bedrock geology, soil type, soil moisture, topography, and climate can have a greater effect than that due to the presence of hydrocarbons.

CONCLUSIONS

Exploration methods based on what are assumed to be hydrocarbon-induced soil and sediment alterations have long been popular, but the processes that produce the observed effects are not well understood and even less well documented. The nature and extent of the alteration can vary significantly not only laterally and vertically but also temporally.

The cause of these altered soils and sediments may well be hydrocarbon related, but it is an indirect cause at best and may not be the most probable cause. We must evaluate seemingly "significant" alteration anomalies carefully to determine if they are related to hydrocarbon seepage. This requires answers to the following questions. Is the anomaly a function of geology or an artifact of culture? If geology, is the observed alteration syngenetic or authigenic? If authigenic, is the anomaly seep-related or of nonseep origin? If seep-related, does the anomaly result from an active hydrocarbon seep or a paleoseep? Finally, if the anomaly at the surface is to be related to a drilling objective at depth, does the anomaly result from mainly vertical migration or does the migration path follow a more complex route?

Numerous claims of success have been made for various exploration methods based on soil and sediment alteration anomalies. Well-documented case studies are rare, however, and the claims are seldom substantiated by a scientifically rigorous program of sampling and analysis. Although the occurrence of hydrocarbon-induced alteration is well established, considerable scientific research is needed before we understand the formation of these anomalies in the near-surface and realize their full value for hydrocarbon exploration.

REFERENCES CITED

Allen, R. F., and R. G. Thomas, 1984, The uranium potential of diagenetically altered sandstone of the Permian Rush Springs Formation, Cement district, southwest Oklahoma: Economic Geology, v. 79, p. 284–296.

Al-Shaieb, Z., 1977, Uranium potential of Permian and Pennsylvanian sandstones in Oklahoma: AAPG Bulletin, v. 61, p. 360–375.

Al-Shaieb, Z., J. Cairns, and J. Puckette, 1994, Hydrocarbon-induced diagenetic aureoles: Indicators of deeper leaky reservoirs: Association of Petroleum Geochemical Explorationists Bulletin, v. 10, p. 24–48.

Anderson, T. F., and M. A. Arthur, 1983, Stable isotopes of oxygen and carbon and their application to sedimentologic and paleoenvironmental problems, in M. A. Arthur, ed., Stable isotopes in sedimentary geology: SEPM Short Course #10, p.1-1–1-151.

Arp, G. K., 1992, An integrated interpretation of the origin of the Patrick Draw oil field sage anomaly: AAPG Bulletin, v. 76, p. 301–306.

Atlas, R. M., 1984, Petroleum microbiology: New York, Macmillan Company, 692 p.

Bammel, B. H., C. P. Chamberlain, and R. W. Birnie, 1994, Stable isotope evidence of vertical hydrocarbon microseepage, Little Buffalo Basin oil field: Association of Petroleum Geochemical Explorationists Bulletin, v. 10, p. 1–23.

Barnes, I., G. Plafker, L. D. White, and A. K. Armstrong, 1980, Potential natural gas in the Gulf of Alaska indicated by calcite depleted in carbon-13, in W. L. Coonrad, ed., The United States Geological Survey in Alaska—accomplishments during 1980: USGS Circular 844, p. 143–146.

Bidwell, R. G. S., 1974, Plant physiology: New York, Macmillan Company.

Brooks, J. M., M. C. Kennicutt, C. R. Fisher, S. A. Macko, K. Cole, J. J. Childress, R. R. Bidigare, and R. D. Vetter, 1987, Deep-sea hydrocarbon communities: evidence for energy and nutritional sources: Science, v. 238, p. 1138–1142.

Campbell, B. S., 1994, Induced polarization anomalies associated with hydrocarbon reservoirs: three case studies: Association of Petroleum Geochemical Explorationists Bulletin, v. 10, p. 49–60.

Childress, J. J., C. R. Fisher, J. M. Brooks, M. C. Kennicutt, R. Bidigare, and A. Anderson, 1986, A methanotrophic molluscan (Bivalvia, Mytilidea) symbiosis: mussels fueled by gas: Science, v. 233, p. 1306–1308.

Curiale, J. A., S. Bloch, J. Rafalska-Bloch, and W. E. Harrison, 1983, Petroleum-related origin for uraniferous organic-rich nodules of southeastern Oklahoma: AAPG Bulletin, v. 67, p. 588–608.

Curry, W. H., 1984, Evaluation of surface gamma radiation surveys for petroleum exploration in the deep Powder River basin, Wyoming, in M. J. Davidson and B. M. Gottlieb, eds., Unconventional methods in exploration for petroleum and natural gas, III: Dallas, Texas, Southern Methodist University Press, p. 25–39.

Conel, J. E., and R. E. Alley, 1985, Lisbon Valley, Utah uranium test site report, *in* M. J. Abrams, J. E. Conel, H. R. Lang, and H. N. Paley, eds., The Joint NASA/Geosat Test Case Project: final report: AAPG Special Publication, pt. 2, v. 1, p. 8-1–8-158.

Dando, P. P., S. C. M. O'Hara, U. Schuster, L. Taylor, C. J. Clayton, S. Bayliss, and T. Laier, 1994, Gas seepage from a carbonate-cemented sandstone reef on the Kattegat coast of Denmark: Marine and Petroleum Geology, v. 11, p. 182–189.

Dalziel, M. C., and T. J. Donovan, 1980, Biogeochemical evidence for subsurface hydrocarbon occurrence, Recluse oil field, Wyoming: preliminary results: USGS Circular 837, 11 p.

Dalziel, M. C., and T. J. Donovan, 1984, Correlations of suspected petroleum-generated biogeochemical and aeromagnetic anomalies, Bell Creek oil field, Montana, *in* M. J. Davidson and B. M. Gottlieb, eds., Unconventional methods in exploration for petroleum and natural gas, symposium III: Dallas, Texas, Southern Methodist University Press, p. 59–69.

Davis, J. B., 1952, Studies on soil samples from a "paraffin dirt" bed: AAPG Bulletin, v. 36, p. 2186–2188.

Davis, J. B., 1956, Microbial decomposition of hydrocarbons: Industrial and Engineering Chemistry, v. 48, no. 9, p. 1444–1448.

Davis, J. B., 1967, Petroleum Microbiology: New York, Elsevier, 604 p.

Donovan, T. J., 1974, Petroleum microseepage at Cement, Oklahoma—evidence and mechanisms: AAPG Bulletin, v. 58, p. 429–446.

Donovan, T. J., and M. C. Dalziel, 1977, Late diagenetic indicators of buried oil and gas: USGS Open File Report 77-817, 44 p.

Donovan, T. J., R. J. Forgey, and A. A. Roberts, 1979, Aeromagnetic detection of diagenetic magnetite over oil fields: AAPG Bulletin, v. 63, p. 245–248.

Donovan, T. J., I. Friedman, and J. D. Gleason, 1974, Recognition of petroleum-bearing traps by unusual isotopic composition of carbonate-cemented surface rocks: Geology, v. 2, p. 351–354.

Donovan, T. J., A. A. Roberts, and M. C. Dalziel, 1981, Epigenetic zoning in surface and near-surface rocks resulting from seepage-induced redox gradients, Velma oil field, Oklahoma: a synopsis: Oklahoma City Geological Society Shale Shaker, v. 32, no. 3, p. 1–7.

Donovan, T. J., P. A. Termain, and M. E. Henry, 1979, Late diagenetic indicators of buried oil and gas, II: direct detection experiments at Cement and Garza oil fields, Oklahoma and Texas, using enhanced Landsat I and II images: USGS Open File Report 79-243.

Duchscherer, W., 1981, Carbonates and isotope ratios from surface rocks, a geochemical guide to underlying petroleum accumulations, *in* B. M. Gottlieb, ed., Unconventional methods in exploration for petroleum and natural gas, symposium II: Dallas, Texas, Southern Methodist University Press, p. 201–218.

Duchscherer, W., 1984, Geochemical hydrocarbon prospecting, with case histories: Tulsa, Oklahoma, PennWell Publishing, 196 p.

Duchscherer, W., 1986, Delta carbonate hydrocarbon prospecting, *in* M. J. Davidson, ed., Unconventional methods in exploration for petroleum and natural gas, symposium IV: Dallas, Texas, Southern Methodist University Press, p. 173–182.

Eargle, D. H., and A. M. D. Weeks, 1961, Possible relationship between hydrogen sulphide-bearing hydrocarbons in fault line oil fields and uranium deposits in the southeast Texas coastal plain: USGS Professional Paper 424-D, p. D7–D9.

Eargle, D. H., and A. M. D. Weeks, 1973, Geologic relations among uranium deposits, south Texas coastal plain region, U.S.A., *in* G. C. Amstutz and A. J. Bernard, eds., Ores in sediments: New York, Springer-Verlag, p. 101–113.

Elmore, R. D., M. H. Engel, L. Crawford, K. Nick, S. Imbus, and Z. Sofer, 1987, Evidence for a relationship between hydrocarbons and authigenic magnetite: Nature, v. 325, p. 428–430.

Fairbridge, R. W., 1972, The encyclopedia of geochemistry and environmental sciences—encyclopedia of earth sciences series (IVA): Stroudsburg, Pennsylvania, Dowden, Hutchinson and Ross, p. 134.

Feely, R. W., and J. L. Kulp, 1957, Origin of Gulf Coast salt dome sulfur deposits: AAPG Bulletin, v. 41, p. 1802–1853.

Ferguson, T. D., 1979a, The subsurface alteration and mineralization of Permian red beds overlying several oil fields in southern Oklahoma, part I: Oklahoma City Geological Society Shale Shaker, v. 29, p. 172–178.

Ferguson, T. D., 1979b, The subsurface alteration and mineralization of Permian red beds overlying several oil fields in southern Oklahoma, part II: Oklahoma City Geological Society Shale Shaker, v. 29, p. 200–208.

Ferris, G. F., 1995, Microbes to minerals: Geotimes, v. 40, no. 9, p. 19–22.

Ferris, G. F., R. G. Wiese, and W. S. Fyfe, 1994, Precipitation of carbonate minerals by microorganisms: implications for silicate weathering and the global carbon dioxide budget: Geomicrobiology Journal, v. 12, p. 1–13.

Foote, R. S., 1987, Correlations of borehole rock magnetic properties with oil and gas producing areas: Association of Petroleum Geochemical Explorationists Bulletin, v. 3, p. 114–134.

Foote, R. S., and G. J. Long, 1988, Correlations of oil- and gas-producing areas with magnetic properties of the upper rock column, eastern Colorado: Association of Petroleum Geochemical Explorationists Bulletin, v. 4, p. 47–61.

Gallagher, A. V., 1984, Iodine: a pathfinder for petroleum deposits, *in* M. J. Davidson and B. M. Gottlieb, eds., Unconventional methods in exploration for petroleum and natural gas, symposium III: Dallas, Texas, Southern Methodist University Press, p. 148–159.

Gay, Jr., S. P., 1993, Epigenetic versus syngenetic magnetite as a cause of magnetic anomalies: Geophysics, v. 57, p. 60–68.

Gay, Jr., S. P., and B. W. Hawley, 1991, Syngenetic magnetic anomaly sources: three examples: Geophysics, v. 56, p. 902–913.

Goldhaber, M. B., and R. L. Reynolds, 1991, Relations among hydrocarbon reservoirs, epigenetic sulfidization, and rock magnetization: examples from south Texas coastal plain: Geophysics, v. 56, p. 748–757.

Goldhaber, M. B., R. L. Reynolds, and R. O. Rye, 1978, Origin of a South Texas roll-type uranium deposit, II: sulfide petrology and sulfur isotope studies: Economic Geology, v. 73, p. 1690–1703.

Goldhaber, M. B., R. L. Reynolds, and R. O. Rye, 1983, Role of fluid mixing and fault-related sulfide in the origin of the Ray Point uranium district, south Texas: Economic Geology, v. 78, p. 1043–1063.

Harris, G. D., 1908, Salt in Louisiana, with special reference to its geologic occurrence, part II—localities south of the

Oligocene: Louisiana Geological Survey Bulletin, v. 7, p. 18–27.

Heemstra, R. J., R. M. Ray, T. C. Wesson, J. R. Abrams, and G. A. Moore, 1979, A critical laboratory and field evaluation of selected surface prospecting techniques for locating oil and natural gas: Bartlesville, Oklahoma, Department of Energy, Bartlesville Energy Technology Center, BETC/RI-78/18, 84 p.

Henry, W. E., 1988, Magnetic detection of hydrocarbon microseepage in a frontier exploration region: Association of Petroleum Geochemical Explorationists Bulletin, v. 4, p. 18–29.

Horvitz, L., 1945, Recent developments in geochemical prospecting for petroleum: Geophysics, v. 10, p. 487–493.

Horvitz, L., 1985, Geochemical exploration for petroleum: Science, v. 229, p. 821–827.

Hovland, M. T., 1990, Do carbonate reefs form due to hydrocarbon seepage?: Terra Nova, v. 2, p. 8–18.

Hovland, M. T., and A. G. Judd, 1988, Seabed pockmarks and seepages: impact on geology, biology, and the marine environment: London, Graham and Trotman, 293 p.

Hughes, L. J., K. L. Zonge, and N. R. Carlson, 1986, The application of electrical techniques in mapping hydrocarbon-related alteration, *in* M. J. Davidson, ed., Unconventional methods in exploration for petroleum and natural gas, symposium IV: Dallas, Texas, Southern Methodist University Press, p. 5–26.

Kartsev, A. A., Z. A. Tabasaranskii, M. I. Subbota, and G. A. Mogilevskii, 1959, Geochemical methods of prospecting and exploration for petroleum and natural gas: Berkeley, Los Angeles, University of California Press, 349 p. (English translation by P. A. Witherspoon and W. D. Romey).

Kennicutt, M. C., J. M. Brooks, R. R. Bidigare, R. R. Fay, T. L. Wade, and T. J. McDonald, 1985, Vent-type taxa in a hydrocarbon seep region of the Louisiana slope: Nature, v. 317, p. 351–353.

Klusman, R. W., M. A. Saeed, and M. A. Abu-Ali, 1992, The potential use of biogeochemistry in the detection of petroleum microseepage: AAPG Bulletin, v. 76, p. 851–863.

Klusman, R. W., 1993, Soil gas and related methods for natural resource exploration: Chichester, John Wiley & Sons, 483 p.

Krumbein, W. E., 1983, Microbial geochemistry: Oxford, Alden Press, 330 p.

Lang, H. R., and P. H. Nadeau, 1985, Petroleum commodity report—summary of results obtained at the petroleum test sites, *in* M. J. Abrams, J. E. Conel, H. R. Lang, and H. N. Paley, eds, The Joint NASA/Geosat Test Case Project: final report: AAPG Special Publication, pt. 2, v. 2, p. 10-1–10-28.

Lang, H. R., W. H. Aldeman, and F. F. Sabins, Jr., 1985a, Patrick Draw, Wyoming—petroleum test case report, *in* M. J. Abrams, J. E. Conel, H. R. Lang, and H. N. Paley, eds., The Joint NASA/Geosat Test Case Project: final report: AAPG Special Publication, pt. 2, v. 2, p. 11-1–11-28.

Lang, H. R., J. B. Curtis, and J. C. Kovacs, 1985b, Lost River, West Virginia—Petroleum test site report, *in* M. J. Abrams, J. E. Conel, H. R. Lang, and H. N. Paley, eds., The Joint NASA/Geosat Test Case Project: final report: AAPG Special Publication, pt. 2, v. 2, p. 12-1–12-96.

Lang, H. R., S. M. Nicolais, and H. R. Hopkins, 1985c, Coyanosa, Texas, Petroleum test site report, *in* M. J. Abrams, J. E. Conel, H. R. Lang, and H. N. Paley, eds., The Joint NASA/Geosat Test Case Project: final report: AAPG Special Publication, pt. 2, v. 2, p. 13-1–13-81.

Lilburn, R. A., and Z. Al-Shaieb, 1983, Geochemistry and isotopic composition of hydrocarbon-induced diagenetic aureole (HIDA), Cement, Oklahoma: Oklahoma City Geological Society Shale Shaker, pt. I, v. 34, no. 4, p. 40–56.

Lilburn, R. A., and Z. Al-Shaieb, 1984, Geochemistry and isotopic composition of hydrocarbon-induced diagenetic aureole (HIDA), Cement, Oklahoma: Oklahoma City Geological Society Shale Shaker, pt. II, v. 34, no. 5, p. 57–67.

Liu, X.-M., J. Bloemendal, and T. Rolph, 1994, Pedogenesis and paleoclimate: interpretations of the magnetic susceptibility record of Chinese loess-paleosol sequences: comment: Geology, v. 22, p. 858–859.

Love, J. D., 1957, Stratigraphy and correlation of Triassic rocks in central Wyoming: 12th Annual Field Conference Guidebook, Wind River Basin, Wyoming Geological Association, p. 39–45.

MacDonald, I. R., G. S. Boland, J. S. Baker, J. M. Brooks, M. C. Kennicutt, and R. R. Bidigare, 1989, Gulf of Mexico hydrocarbon seep communities: Marine Geology, v. 101, p. 235–247.

MacDonald, I. R., J. F. Reilly, N. L. Guinasso, Jr., J. M. Brooks, R. S. Carney, W. A. Bryant, and T. J. Bright, 1990, Chemosynthetic mussels at a brine-filled pockmark in the northern Gulf of Mexico: Science, v. 248, p. 1096–1099.

Machel, H. G., and E. A. Burton, 1991a, Chemical and microbial processes causing anomalous magnetization in environments affected by hydrocarbon seepage: Geophysics, v. 56, p. 598–605.

Machel, H. G., and E. A. Burton, 1991b, Causes and spatial distribution of anomalous magnetization in hydrocarbon seepage environments: AAPG Bulletin, v. 75, p. 1864–1876.

Maher, B. A., and R. Thompson, 1991, Mineral magnetic record of the Chinese loess and paleosols: Geology, v. 19, p. 3–6.

Maher, B. A., and R. Thompson, 1992, Paleoclimatic significance of the mineral magnetic record of the Chinese loess and paleosols: Quaternary Research, v. 37, p. 155–170.

Matthews, M. D., 1986, The effects of hydrocarbon leakage on earth surface materials, *in* M. J. Davidson, ed., Unconventional methods in exploration for petroleum and natural gas, symposium IV: Dallas, Texas, Southern Methodist University Press, p. 27–44.

McCabe, C., R. Sassen, and B. Saffer, 1987, Occurrence of secondary magnetite within biodegraded oil: Geology, v. 15, p. 1351–1370.

McCoy, R. M., and L. H. Wullstein, 1988, Mapping iron and manganese concentrations in sagebrush over the Blackburn oil field, Nevada, *in* Proceedings of Sixth Thematic Conference on Remote Sensing for Exploration Geology: Ann Arbor, Michigan, Environmental Research Institute, p. 523–528.

McCoy, R. M., L. F. Scott, and P. Hardin, 1989, The spectral response of sagebrush in areas of hydrocarbon production, *in* Proceedings of Seventh Thematic Conference on Remote Sensing for Exploration Geology: Ann Arbor, Michigan, Environmental Research Institute, p. 751–756.

McDermott, E., 1940, Geochemical exploration (soil analysis), with speculation about the genesis of oil, gas, and other mineral accumulations: AAPG Bulletin, v. 24, p. 859–881.

McKenna, E. J., and R. E. Kallio, 1965, The biology of hydrocarbons: Annual Reviews of Microbiology, v. 19, p. 183–208.

Milner, H. B., 1925, "Paraffin dirt": its nature, origin, mode of occurrence, and significance as an indication of petroleum: Mining Magazine, v. 32, p. 73–85.

Oehler, D. Z., and B. K. Sternberg, 1984, Seepage-induced anomalies, "false" anomalies, and implications for electrical prospecting: AAPG Bulletin, v. 68, p. 1121–1145.

Olmstead, R. W., 1975, Geochemical studies of uranium in south-central Oklahoma: Master's thesis, Oklahoma State University, Stillwater, Oklahoma, 116 p.

Paull, C. K., F. N. Spiess, W. Ussler III, and W. A. Borowski, 1995, Methane-rich plumes on the Carolina continental rise: association with gas hydrates: Geology, v. 23, p. 89–92.

Pirson, S. J., 1969, Geological, geophysical, and geochemical modification of sediments in the environments of oil fields, *in* W. B. Heroy, ed., Unconventional methods in exploration for petroleum and natural gas, symposium 1: Dallas, Texas, Southern Methodist University Press, p. 159–186.

Pirson, S. J., 1976, Predictions of hydrocarbons in place by magneto-electrotelluric exploration: Oil and Gas Journal, May 31, p. 82–86.

Posey, H. H., P. E. Price, and J. R. Kyle, 1987, Mixed carbon sources for calcite cap rocks of Gulf Coast salt domes, *in* I. Lerche and J. O'Brien, eds., Dynamical geology of salt and related structures: London, Academic Press, p. 593–630.

Price, L. C., 1986, A critical overview and proposed working model of surface geochemical exploration, *in* M. J. Davidson, ed., Unconventional methods in exploration for petroleum and natural gas, symposium IV: Dallas, Texas, Southern Methodist University Press, p. 245–304.

Prikryl, J. D., 1990, Origin of limestone cap rock at the Damon Mound salt dome: a petrographic and geochemical model, *in* D. Schumacher and B. F. Perkins, eds., Gulf Coast oils and gases: Ninth Annual Research Conference, Proceedings, Arlington, Texas, GCS/SEPM Foundation, p. 325–336.

Reeves, F., 1922, Geology of the Cement oil field, Caddo county, Oklahoma: USGS Bulletin 726, p. 41–85.

Reid, J. C., B. S. Campbell, and S. D. Ulrich, 1992, Hydrocarbon-induced mineralogic and geochemical alteration, Turkey Creek, Colorado: Association of Petroleum Geochemical Explorationists Bulletin, v. 8, p. 64–95.

Reynolds, R. L., N. S. Fishman, and D. M. Sherman, 1988, Magnetite and maghemite from hydrocarbon wells: EOS Transactions, American Geophysical Union, v. 65, no. 44, p. 1156.

Reynolds, R. L., M. B. Goldhaber, and M. L. Tuttle, 1993, Sulfidization and magnetization above hydrocarbon reservoirs, *in* D. M. Aissaoui, D. S. McNeill, and N. F. Hurley, Application of paleomagnetism to sedimentary geology: SEPM Special Publication 49, p. 167–179.

Reynolds, R. L., N. S. Fishman, R. I. Grauch, and J. A. Karachewski, 1984, Thermomagnetic behavior and composition of pyrrhotite in Lower Permian strata, Cement oil field, Oklahoma: EOS Transactions, American Geophysical Union, v. 65, no. 45, p. 866.

Reynolds, R. L., N. S. Fishman, R. B. Wanty, and M. B. Goldhaber, 1990, Iron sulfide minerals at Cement oil field, Oklahoma: implications for magnetic detection of oil fields: GSA Bulletin, v. 102, p. 368–380.

Richers, D. M., V. T. Jones, M. D. Matthews, J. Maciolek, R. J. Pirkle, and W. C. Sidle, 1986, The 1983 Landsat soil-gas geochemical survey of Patrick Draw area, Sweetwater county, Wyoming: AAPG Bulletin, v. 70, p. 869–887.

Richers, D. M., M. J. Reed, K. C. Horstman, G. D. Michels, R. N. Baker, L. L. Lundell, and R. W. Marrs, 1982, Landsat and soil-gas geochemical study of Patrick Draw oil field, Sweetwater county, Wyoming: AAPG Bulletin, v. 66, p. 903–922.

Roberts, H. H., R. Sassen, R. Carney, and P. Aharon, 1990, The role of hydrocarbons in creating sediment and small-scale topography on the Louisiana continental slope, *in* D. Schumacher and B. F. Perkins, eds., Gulf Coast oils and gases: Ninth Annual Research Conference, Proceedings, Arlington, Texas, GCS/SEPM Foundation, p. 311–324.

Rock, B. N., 1984, Remote detection of geobotanical anomalies associated with hydrocarbon microseepage: Third Thematic Conference on Remote Sensing for Exploration Geology, Proceedings, v. 2, p. 183–195.

Roeming, S. S., and T. J. Donovan, 1985, Correlations among hydrocarbon microseepage, soil chemistry, and uptake of micronutrients by plants, Bell Creek oil field, Montana: Journal of Geochemical Exploration, v. 23, p. 139–162.

Rosaire, E. E., 1940, Geochemical prospecting for petroleum: AAPG Bulletin, v. 24, p. 1400–1423.

Sassen, R., 1980, Biodegradation of crude oil and mineral deposition in a shallow Gulf Coast salt dome: Organic Geochemistry, v. 2, p. 153–166.

Sassen, R., E. W. Chinn, and C. McCabe, 1988, Recent hydrocarbon alteration, sulfate reduction, and formation of elemental sulfur and metal sulfides in salt dome cap rock: Chemical Geology, v. 74, p. 57–66.

Sassen, R., C. M. McCabe, J. R. Kyle, and E. W. Chinn, 1989, Deposition of magnetic pyrrhotite during alteration of crude oil and reduction of sulfate: Organic Geochemistry, v. 14, p. 381–392.

Sassen, R., G. A. Cole, R. Drozd, and H. H. Roberts, 1994, Oligocene to Holocene hydrocarbon migration and salt dome carbonates, northern Gulf of Mexico: Marine and Petroleum Geology, v. 11, no. 1, p. 55–65.

Sassen, R., H. H. Roberts, P. Aharon, J. Larkin, E. W. Chinn, and R. Carney, 1993, Chemosynthetic bacterial mats at cold hydrocarbon seeps, Gulf of Mexico continental slope: Organic Geochemistry, v. 20, p.77–89.

Saunders, D. F., K. R. Burson, and C. K. Thompson, 1991, Observed relation of soil magnetic susceptibility and soil gas hydrocarbon analyses to subsurface petroleum accumulations: AAPG Bulletin, v. 75, p. 389–408.

Saunders, D. F., K. R. Burson, J. F. Branch, and C. K. Thompson, 1993, Relation of thorium-normalized surface and aerial radiometric data to subsurface petroleum accumulations: Geophysics, v. 58, p. 1417–1427.

Sawtelle, G., 1936, Salt dome statistics: AAPG Bulletin, v. 20, p. 726–735.

Scott, L. F., and R. M. McCoy, 1992, Near-surface soil and plant chemical properties over a deep, structurally complex reservoir (Lodgepole oil field, Summit county, Utah): Association of Petroleum Geochemical Explorationists Bulletin, v. 8, p. 18–42.

Segal, D. B., M. D. Ruth, and I. S. Merin, 1986, Remote detection of anomalous mineralogy associated with hydrocarbon production, Lisbon Valley, Utah: The Mountain Geologist, v. 23, no. 2, p. 51–62.

Segal, D. B., M. D. Ruth, I. S. Merin, H. Watanabe, K. Soda, O. Takano, and M. Sano, 1984, Correlation of remotely detected mineralogy with hydrocarbon production, Lisbon Valley, Utah: ERIM Proceedings of the International Symposium on Remote Sensing for Exploration Geology, p. 273–292.

Sternberg, B. K., 1991, A review of some experience with the induced polarization-resistivity method for hydrocarbon surveys: successes and limitations: Geophysics, v. 56, p. 1522–1532.

Tedesco, S. A., 1995, Surface geochemistry in petroleum exploration: New York, Chapman and Hall, 206 p.

Thompson, A. B., 1933, The economic value of surface petroleum manifestations: First World Petroleum Congress, Proceedings, London, p. 241–250.

Thompson, C. K., D. F. Saunders, and K. R. Burson, 1994, Model advanced for hydrocarbon microseepage, related alterations: Oil & Gas Journal, November 14, p. 95–99.

Weart, R. C., and G. Heimberg, 1981, Exploration radiometrics: post-survey drilling results, *in* M. J. Davidson and B. M. Gottlieb, eds., Unconventional methods in exploration for petroleum and natural gas, symposium II: Dallas, Texas, Southern Methodist University Press, p. 116–123.

Ellwood, B. B., and B. Burkart, 1996, Test of hydrocarbon-induced magnetic pat-
terns in soils: the sanitary landfill as laboratory, *in* D. Schumacher and M. A.
Abrams, eds., Hydrocarbon migration and its near-surface expression: AAPG
Memoir 66, p. 91–98.

Chapter 7

Test of Hydrocarbon-Induced Magnetic Patterns in Soils: The Sanitary Landfill as Laboratory

Brooks B. Ellwood

Burke Burkart

Department of Geology
University of Texas at Arlington
Arlington, Texas, U.S.A

Abstract

The magnetic susceptibility of soils has been studied at a sanitary landfill site, where upward-fluxing methane gas has caused changes in the magnetic mineralogy of the capping soils. Soil used as a cap on the Hillsboro, Texas, sanitary landfill was put into place 1, 10, and 20 years before sampling for this study. After 1 year in place, the susceptibility of the capping soil dropped below that of control samples not exposed to methane flux. Magnetic susceptibilities increased progressively from the control soils to the 10- and 20-year-old samples, with the highest values at depths of ~40 cm below the soil surface. New authigenic minerals accumulated in landfill caps, with longer exposure to infiltration during reducing conditions producing greater magnetic effects. Calcite along with maghemite, the principal authigenic magnetic mineral, accumulated below the 40-cm level, iron and calcium having dissolved from the upper soil of the landfill cap. Calcite also accumulated during times of soil desiccation, forming a barrier to fluid transfer. Landfill caps that have distinct zonation of Fe(II) minerals beneath those of Fe(III) are likely to have a well-established $CaCO_3$ barrier that separates redox environments.

Magnetic anomalies appear in capping soils exposed to high upward flux of methane and periodic infiltration of water, which produce a reducing environment favorable to the growth of magnetotactic bacteria. When the level of microbial catalysis is high, Fe(II) dissolved from the upper levels is transported deeper into the soil where it can reprecipitate as magnetic oxide or sulfide. Precipitation of nonmagnetic Fe(II) phases during wet winters followed by oxidation to magnetic phases during dry summers may take place, as observed in normal soils. Our study demonstrates that sanitary landfills can be used as convenient laboratories for studies of natural soil magnetism and are effective model systems for the study of magnetic effects in soils above areas of light hydrocarbon flux, such as petroleum reservoirs.

INTRODUCTION

Because unusual magnetic effects have been associated with oil and gas accumulations (Steenland, 1965), the magnetic characteristics of soils have been considered potentially important tools for petroleum exploration. Although a number of models have been proposed to account for observed patterns (e.g., Donovan et al., 1984; Fishman et al., 1989; Machel and Burton, 1991), the data on which these models are based are controversial. Many of these models invoke microseepage of hydrocarbon compounds upward from the oil or gas reservoir into surface rocks and soils where unusual alteration effects take place. Proponents of these microseepage models suggest that effects can be detected by airborne magnetic surveys and confirmed by direct magnetic and geochemical measurements of surface samples. Cause and effect relationships between hydrocarbon seepage and observed magnetic patterns are, however, poorly understood at this time, and quantitative studies are needed that focus on changes in soil component magnetization caused by differences in soil chemistry and in times and amounts of exposure to hydrocarbon flux. Models based on such quantitative data would provide a basis for understanding case studies of soil magnetism as a prospecting tool (Donovan et al., 1979a,b; Barton, 1990; Land, 1991; Saunders et al., 1993).

This is the first study of hydrocarbon-induced magnetic anomalies in landfill soils. This work is still in progress and does not include chemical data from the Hillsboro landfills that will help clarify the development of the observed magnetic effects. The importance of future studies of soil gas and interstitial water compositions, detailed control soil compositions, and other data are discussed here.

LANDFILLS AS LABORATORIES

The effects of hydrocarbon seepage on iron mineralogy are best understood when the seep is located where site-specific variables can be measured and to some degree controlled. We believe that landfills are ideal for such studies. Investigating soil that covers sanitary landfills, such as that found in Hillsboro, Texas, located 60 miles south of Dallas, allows one to measure most variables important in the development of magnetic signatures associated with seeping methane gas. We have found that the following characteristics of sanitary landfill systems make them useful as laboratories for the study of hydrocarbon-induced soil magnetism:

1. Such studies require a "long-term" flux of light hydrocarbons, and landfills are known to generate large amounts of methane. Indeed, at some sites, methane is exploited as fuel (Carolan, 1987). A typical soil gas composition of 55% CH_4, 42% CO_2, 2.5% N_2, and 0.5% O_2 is reported by Lu and Kunz (1981) from the Fresh Kills Landfill, Staten Island, New York. The Hillsboro site displays typical deleterious effects on grasses and other vegetation attempting to grow in anaerobic capping soils (Duell et al., 1986).
2. Landfills have soil caps in which changes in magnetic mineralogy can be observed. Capping soils vary from region to region, and sites can be selected on the basis of soil composition. A soil must have an iron content high enough that magnetic minerals can form at concentrations detectable by magnetic methods. The iron content at the Hillsboro sites is ample for the study undertaken. Thicknesses of landfill caps vary from site to site, and at the Hillsboro sites they are about 1 m thick.
3. Documentation of the ages of individual landfills is often available so that changes in magnetic effects over time can be studied. The Hillsboro landfills have overburden soils that were emplaced at three documented times: 1, 10, and 20 years before soil sampling for this study was accomplished.
4. "Control soils" may be available because the identity and place of origin of the capping soil is often known. The control can reveal magnetic properties and soil compositions that existed before emplacement on a landfill. At the Hillsboro landfills, the capping soil is weathered Eagle Ford Shale (Upper Cretaceous) from an adjacent excavation.
5. Landfill sites can be chosen to represent climates; thus, when selecting a site by geographic area, some of the variables, such as the temperature and the amount, frequency, and seasonality of rainfall, can be limited.

MAGNETIC SUSCEPTIBILITY

Magnetic susceptibility (MS) is a fundamental property of all materials and is generally considered a measure of iron mineral concentration when it is used in a geologic context (Nagata, 1961). All materials, especially those containing iron, are "susceptible" to becoming magnetized in the presence of small magnetic fields, with MS increasing as iron-containing grains increase. This is different from remanent magnetic measurements used to determine magnetic polarities. While MS measurements require that an inducing magnetic field is generated within the sample measuring space, remanent magnetic measurements are made in magnetic field–free spaces, where the spontaneous magnetization of samples is independent of external magnetic fields.

Magnetic susceptibility can be easily measured on small samples using a balanced coil induction system (susceptibility bridge). Here we report MS data in SI units relative to mass, given in m^3/kg. Conversion to m^3/kg can be made from volume cgs units, for an assumed 10 cm^3 sample, by multiplying by 4×10^{-3}. We measure MS using a susceptibility bridge calibrated with 50 samples made from standard salts (MnF_2 and MnO_2) and magnetite (Fe_3O_4). Calibration values were taken from the National Bureau of Standards (Swartzendruber, 1992) and calculated from tables of standards (CRC Handbook of Chemistry and Physics, 1983). The practical sensitivity of the University of Texas at Arlington (UTA) bridge is $3.0 \pm 0.2 \times 10^{-9} \ m^3/kg$.

SUSCEPTIBILITY OF MAGNETIC MINERALS

Iron Oxides

Magnetite (Fe_3O_4, or $FeO \bullet Fe_2O_3$) has the highest susceptibility of the iron oxide minerals (Table 1) and is found in many types of rocks and less commonly in soils. Authigenic magnetite has been documented by Karlin et al. (1987) from suboxic marine sediments, and its production in sulfide-rich anaerobic marine sediments by magnetotactic bacteria has been documented by Bazylinski et al. (1988). Lovley et al. (1987) described nonmagnetotactic bacteria that reduce ferric oxides to extracellular magnetite under anaerobic conditions and suggest that they may produce ultrafine-grained magnetite in anaerobic sediments. Authigenic magnetite in soils is rare, forming only under special conditions (Maher and Taylor, 1988; Schwertman and Taylor, 1989; Fassbender et al., 1990). With one of three iron atoms in valence state 2, magnetite

Table 1—Iron-Bearing Minerals in the Model of Figure 4 and Their Relative Magnetic Susceptibilities

Group	Mineral	Chemical Formula	Magnetic Susceptibility[a]
Oxides	Magnetite	Fe_3O_4	***
	Maghemite	γFe_2O_3	**
	Hematite	αFe_2O_3	*
Sulfides	Smythite (?)	Fe_9S_{11}	?
	Greigite	Fe_3S_4	***
	Mackinwite (?)	FeS_{1-x}, Fe_9S_8	?
	Pyrrhotite	$Fe_{1-x}S$	*/**
	Sulfided iron (?)	Fe_2S_3	?
	Marcasite	FeS_2	*
	Pyrite	FeS_2	*
	Chalcopyrite	$CuFeS_2$	*
Hydrated sulfates	Jarosite	$KFe_3(SO_4)_2(OH)_6$	?
	Coquimbite	$Fe_2(SO_4)_3 \cdot 9H_2O$	*** ?
	Melanterite	$FeSO_4 \cdot nH2O$	*** ?
	Rozenite	$FeSO_4 \cdot nH2O$	*** ?
	Siderotil	$FeSO_4 \cdot nH2O$	*** ?
	Szomolnokite	$FeSO_4 \cdot nH2O$	*** ?
	Halotrichite	$FeAl_2(SO_4)_4 \cdot 22H_2O$	*** ?
Oxyhydroxides	Geothite	$\alpha FeO(OH)$	*
	Lepidocrocite	$\gamma FeO(OH)$	*
	Ferrihydrate	$Fe_5O_7(OH) \cdot 4H_2O$	?
Carbonates	Ankerite	$Ca(Mg,Fe,Mn)(CO_3)_2$	*
	Siderite	$FeCO_3$	*
Clays	Chlorite (chamosite)	$(Mg,Fe,Al)_6(Al,Si)_4O_{10}(OH)_8$	*
	Berthierine	$(Fe_{1.7}Mg_{0.2}Al_{0.8})(Si_{1.2}Al_{0.8})_4O_5(OH)_4$	*
	Glauconite	$K(Fe,Mg,Al)_2(Si_4O_{10})(OH)_2$	*
Phosphates	Vivianite	$Fe_3(PO_4)_2 \cdot 8H_2O$	*

[a] Magnetic susceptibility: *** = high, ** = intermediate, * = low.

can alter to less magnetic minerals, such as maghemite (γ-Fe_2O_3), hematite (α-Fe_2O_3), and goethite [γ-$FeO(OH)$].

Maghemite, γ-Fe_2O_3, has a moderate MS, and as the dominant Fe(III) mineral in soils, it is the major producer of soil magnetism. Climate is clearly important in the development of magnetic signatures in sediments, due primarily to variations in magnetic mineral production during pedogenesis, as demonstrated by Tite and Linington (1975). They have shown that a two-stage process operates in some soils, which first alternate as reducing environments, where Fe(III) oxides such as hematite are reduced to Fe(II) and then oxidized to Fe(III) when the redox potential rises. Alternation of redox potential is typical of Mediterranean soils, where pore waters change from reducing during cool, wet winters to oxidizing during hot, dry summers. MS is generally higher in soils than in the sedimentary bedrock from which they were derived. This is due largely to *in situ* conversion of iron oxides from weakly magnetic phases such as hematite to maghemite, which exhibits much higher susceptibility (Le Borgne, 1955; Tite and Linington, 1975; Mullins, 1977; Singer and Fine, 1989).

Iron Sulfides

Pyrrhotite ($Fe_{1-x}S$) and greigite (Fe_3S_4) (Table 1) are known to have pronounced magnetic properties that contribute strongly to the MS of some rocks (Uyeda et al., 1963; Skinner et al., 1964). Not all pyrrhotite is magnetically important, as the mineral exhibits ferromagnetism only in the composition range where x, a measure of nonstoichiometry, is between 0.09 and 0.14. Pyrrhotite forms in strongly reducing environments in recent sulfide-rich sediments and sedimentary rocks exposed to hydrocarbon seepage (Kyle et al., 1987; Farina, 1990; Reynolds et al., 1990), but appears to be otherwise rare in marine sediments. When found in association with microseepage, pyrrhotite is generally magnetic.

Greigite, Fe_3S_4 (Skinner et al., 1964), is the sulfur analog of magnetite and exhibits a similar high MS. Although in some rocks it may contribute strongly to the observed magnetic properties, it appears to be chemically unstable in most sedimentary systems and is difficult to identify. Both pyrrhotite and greigite can be produced in sediments by sulfate-reducing bacteria (Farina et al., 1990;

Mann et al., 1990; Stanjek et al., 1994).

Pyrite (FeS_2) is a common, relatively stable paramagnetic (nonmagnetic) iron sulfide with an isometric crystal structure. Marcasite (FeS_2) is an orthorhombic polymorph that is considerably less stable in oxidizing sedimentary environments, readily hydrolyzing to hydrated iron sulfate phases such as melanterite, siderotil, and coquimbite, each of which can be highly magnetic. These minerals oxidize readily to hematite or an oxyhydroxide. Ellwood et al. (1989b) have argued that magnetic changes produced by alteration of marcasite to a hydrated iron sulfate in sandstone dikes located in Rockwall, Texas, are responsible for magnetic anomaly patterns observed above the dikes. Pyrite and marcasite can be generated by sulfate-reducing bacteria.

The metastable mineral mackinawite (FeS_{1-x}) can form in soils where sulfate is limited (Berner, 1984). However, its magnetic properties and those of two other iron sulfides, smythite and sulfided iron, are not well known. Some may be strongly magnetic within certain chemical composition ranges, and oxidation may free the iron to recombine as other magnetic phases. Sulfided iron can be produced, for example, by H_2S reduction of $Fe_2O_3 \cdot H_2O$.

Iron Oxyhydroxides

It is now apparent that the mineral goethite, γ-FeO(OH), is relatively nonmagnetic. This mineral often forms from alteration of iron-bearing precursors such as magnetite and pyrite.

Iron Carbonates

The iron carbonates, siderite, ankerite, and ferroan dolomite, are an important set of paramagnetic minerals that form in reducing environments. A series of experiments on rocks containing siderite has demonstrated their importance to paleomagnetic studies (Ellwood et al., 1986a, 1989a). It is shown that, while siderite itself is not magnetic, it readily oxidizes to magnetite or maghemite, which are very magnetic. These secondary minerals may dominate the magnetic properties of the rock, replacing or overprinting it with highly increased magnetization. The magnetization of rocks that have experienced such oxidation effects is high enough to be measured with portable magnetometers.

Other Paramagnetic Minerals

Magnetic properties of paramagnetic minerals are poorly known. In addition to the carbonates (discussed above), several iron-rich clay minerals may undergo oxidation that obscures the original magnetic properties of many rocks. For example, iron-bearing kaolinite can oxidize to iron-free kaolinite, with corresponding production of goethite and hematite (Rozenson et al., 1982). Oxidation of pyrite and glauconite can produce new, strongly magnetic phases that will dominate rock magnetism (Ellwood et al., 1986b).

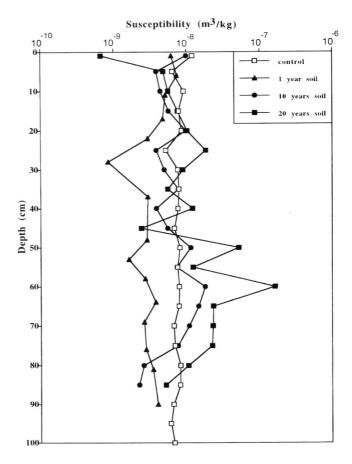

Figure 1—Graph of magnetic susceptibility (logarithmic scale) versus depth in Hillsboro sanitary landfill cores. Cores are through the control soil and the 1-, 10-, and 20-year-old capping soils.

MEASUREMENTS

Sampling

Cores 1 m in length were taken through the 1-, 10-, 20-year-old, and control soils at the Hillsboro, Texas, landfill. Laboratory measurements were made on samples taken from each core at 5-cm intervals. Other measurements of core samples included powder X-ray diffraction and calcium carbonate determinations by means of vacuum gasometric analysis.

Magnetic Results

Susceptibility was measured on a balanced coil induction system. Samples studied were those from the control soil and the 1-, 10-, and 20-year-old cores, with data presented in Figure 1. Control soil samples had relatively uniform susceptibilities, averaging about 5.8×10^{-8}

m^3/kg through the entire core and a maximum of 9.2×10^{-8} m^3/kg. The 1-year-old core susceptibilities were lower than the control, with a mean value of 3.6×10^{-8} m^3/kg and a maximum of 7.4×10^{-8} m^3/kg. The 10-year-old soil susceptibilities rose to a mean of 7.5×10^{-8} m^3/kg, with high MS values for the core beginning at about 40 cm in depth and a maximum of 18×10^{-8} m^3/kg at 60 cm depth. The 20-year-old samples had a mean susceptibility of 21.6×10^{-8} m^3/kg, with high MS values beginning at about 40 cm and rising to a maximum of 163×10^{-8} m^3/km at 60 cm depth. The following is a summary of susceptibility comparisons within and between cores:

1. Mean MS in the 1-year-old core is 62% of the control soil core MS. Except for the two samples nearest the surface, susceptibilities at equivalent depths are uniformly lower in the 1-year-old soil than those in the control.
2. Mean MS rises by a factor of 1.3 in the 10-year-old core and 3.7 in the 20-year-old core above that of the control soil.
3. The maximum MS of the 10-year-old soil is double that of the control soil after 10 years and higher by a factor of 17.7 in the 20-year-old soil.
4. The *core anomaly* is defined as the largest deviation from the mean MS of samples in a given core. The control, 1-, 10-, and 20-year-old soils have core anomalies of 0.34, –0.38, 1.1, and 14.1 m^3/kg, respectively. Normalized to the mean MS for each core (core anomaly ÷ mean MS value), the four soils have ratios of 0.59, –1.06, 1.40, and 6.53, respectively. The highest MS values are found in samples below the 40-cm level of the 10- and 20-year-old soil cores. The core anomaly appears to develop to a minor extent at the expense of MS values above the 40-cm level. This is based on the fact that six of the eight samples from the 10-year-old core and four of the eight samples from the 20-year-old core display MS values below those of control samples from the same depths.

Mineralogic Determinations

X-ray powder diffraction studies of core samples were performed using a Philips diffractometer with CuK-α radiation (45 kV, 40 Ma) and a graphite monochrometer. The fluorescence effects of iron were minimized by filtration with the monochrometer. X-ray diffraction revealed only two major mineral phases in samples from all cores, calcite, and montmorillonite clay. Low amounts of maghemite (γ-Fe_2O_3) were detected in the 10- and 20-year-old overburden soils.

The amount of calcium carbonate (in wt. %) in each core sample was determined by the vacuum-gasometric technique of Jones and Kaiteris (1983). Figure 2 indicates an increase in calcium carbonate to as much as 46 wt. % in the 20-year-old core across the 15–60-cm depth range, whereas the control soil maintains concentrations of approximately 25 wt. % across this range.

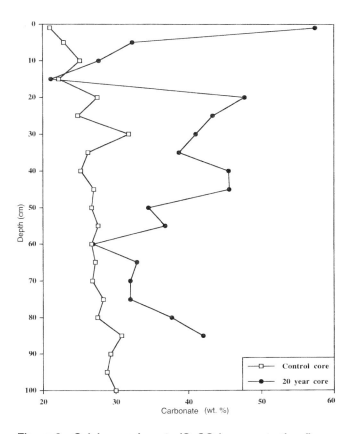

Figure 2—Calcium carbonate ($CaCO_3$) concentration (in weight percent) determined by vacuum gasometric analysis (from Jones and Kaiteris, 1983). Cores are through the control soil and the 20-year-old-old capping soil.

REDUCTION OF Fe(III) TO Fe(II)

According to our interpretation, the MS values indicate that iron dissolves from upper levels of the 10- and 20-year-old landfill soils and reprecipitates below the 40-cm level. Such movement of iron requires a reducing environment in the interstitial water so that Fe(III) is reduced to Fe(II), the soluble ionic species in sedimentary environments of moderate acidity (pH values above ~3).

Methane is the primary reductant in the conversion of Fe(III) to Fe(II) in landfill soils, whereas the organic component of normal soils can serve this function. Another possible reductant is hydrogen sulfide (H_2S) generated from SO_4 in landfills where gypsum board has been discarded. Because H_2S has a pronounced odor in trace amounts, it is easily detected, and it is not a significant gas at the Hillsboro landfills.

Lovley et al. (1987) point out that microbial catalysis is required in sedimentary environments for the reduction of Fe(III) to Fe(II) by most organic reductants, including hydrocarbons. Microbial reactions involving iron are catalytic, with the microbes having the capacity to drive oxidation or reduction only if the thermodynamic conditions are met. A low oxidation potential (Eh), low pH, or both

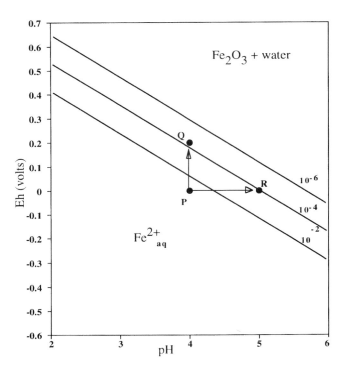

Figure 3—Eh–pH diagram showing the equilibrium between soluble Fe(II) ion (or Fe²⁺) and α-Fe₂O₃ (hematite) in water at 25°C (Faure, 1991). Equilibrium is a function of three variables: Eh (in volts, V), pH of water, and activity of Fe²⁺ in solution (in mol/L or *M*). Equilibrium is represented by a family of straight lines: $Eh = 0.65 - 0.177(pH) - 0.059 \log \alpha\text{-}Fe^{2+}$. Only three of the infinite number of isochemical contours, representing Fe²⁺ at activities of 10⁻², 10⁻⁴, and 10⁻⁶ *M*, are plotted. The effect of a rise in pH or Eh on iron solubility (as Fe²⁺) is illustrated for water with initial pH = 4 and Eh = 0 V, with Fe²⁺ activity of 10⁻¹ *M* (representing saturation at point P). If the water remains at pH = 4 but Eh increases to +0.2 V, Fe²⁺ activity will drop to 10⁻⁴·⁴ *M* (point Q) by oxidation of Fe²⁺ and precipitation of iron as Fe₂O₃. If the water remains at Eh = 0 V but pH increases to 5, its Fe²⁺ equilibrium (saturation) activity will drop from 10⁻¹ *M* to about 10⁻⁴ *M* (point R). Thus, oxidation leading to loss of Fe²⁺ can occur by an increase in Eh or pH or both, while reduction results in increased hematite solubility by a decrease in either or both.

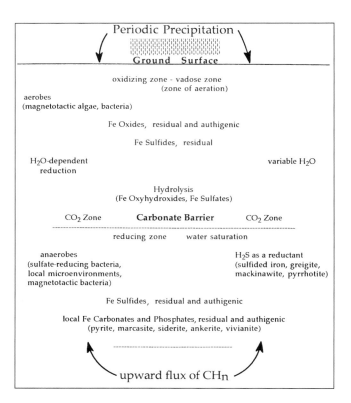

Figure 4—Schematic model of near-surface environments associated with hydrocarbon seepage and the iron minerals expected in those environments.

are required (Figure 3) to dissolve high amounts of Fe(II) (Garrels and Christ, 1965; Krauskopf, 1979; Faure, 1991).

The amount of water infiltration is a variable whose effect on transport and precipitation of iron in soils is apparent. Water is required for microbial metabolism. It excludes oxygen from soil pores to create a reducing environment (low Eh), and it dissolves iron, carrying it as Fe(II) during infiltration. Also, evaporation of water or an increase in its pH during infiltration promotes precipitation of magnetic minerals, primarily maghemite (Figure 3).

MODEL FOR HILLSBORO LANDFILL

We conclude that magnetism of the control soil results from the presence of maghemite, produced during pedogenesis of the upper soil. This conclusion is based on the existence of positive MS values below 40 cm in the 10- and 20-year-old cores. Data from the 1-year-old core indicate that magnetism is lost at all levels in the first year a soil is on a landfill. Secondary maghemite deposited in the 10- and 20-year-old soils replenishes this first year loss and progressively raises susceptibilities in the soils for 10 and 20 years. We suspect that the large MS values in these cores below 40 cm are caused primarily by authigenic maghemite that developed through oxidation to maghemite of otherwise low-MS minerals. Table 1 shows the range of possible candidate minerals. The distribution of these minerals in the subsurface is shown in Figure 4.

Redistribution of iron minerals in the soil comes about when Fe(III) is converted to soluble Fe(II) in reducing, mildly acidic pore water, when periodic infiltration of water is separated by periods of drying. Each infiltration event displaces oxygen and promotes growth of sulfate-reducing bacteria that feed off of landfill methane. Tite and Linington (1975) described alternating oxidizing and reducing conditions that produce maghemite in soils in which organic matter in the soil acts as the reductant (see above), whereas methane is the reductant in landfill soils.

Methane is metabolized by bacteria that generate CO_2, which reacts with water to form carbonic acid, H_2CO_3. The weak acid helps to maintain mildly acidic pore water that originated as rain with a pH of 4.9 (Hillsboro data from USGS, 1992). Although not measured in this study, we suspect that pH values during infiltration events compare with published values at other sites, characterized by those of a soil in New Jersey with a measured pH of 4.6 in the "aerobic" surface soil above a 2-year-old landfill (Duell et al., 1986). Fe(II) is the soluble form at these pH values if pore waters are sufficiently reducing (see above).

Fe(II) can be flushed from the soil into the landfill if infiltration periods are prolonged, or it can reprecipitate somewhere in the soil column if periods are short. During the short-term event, Fe(II) could be forced out of solution by evaporation in the partially air-filled pores, or it could precipitate by chemical oxidation caused by pH increase from hydrolysis of calcium carbonate (Figure 3). The latter would require catalysis by oxidizing bacteria for the reaction to proceed in a timely manner.

Dissolved calcite is also reprecipitated in the short-term event, accumulating at the 15–60 cm level in the 20-year-old soil (Figure 2). This is caused by a pH increase or by evaporation, the same factors that bring about Fe(III) precipitation. Calcite has accumulated in the 20-year-old soil at the 15–50 cm level, where it shows up as a caliche-like layer. In time, this barrier could isolate a zone of oxidized iron minerals (primarily maghemite) from an underlying zone of reduced minerals (pyrite, marcasite, makinawite, pyrrhotite, siderite, and magnetite) (Table 1). Methane would be the dominant gas beneath the carbonate barrier once it has been emplaced, but above this level, the redox potential would continue to fluctuate as oxygen and methane contents vary. Vertical migration of methane may diminish as the carbonate barrier grows.

DISCUSSION AND CONCLUSIONS

Magnetic susceptibility of capping soils in landfills changes with time of exposure to hydrocarbon flux. Magnetism has increased over time in the Hillsboro landfill soils after an initial drop during the first year. This decrease is attributed to wholesale leaching of iron through the loosely consolidated landfill cap. At 40 cm depth in each of the two older cores, MS highs indicate redistribution of some iron within the soil cap.

A model is proposed to explain how an initially homogeneous soil can develop magnetic anomalies that grow with time after placement on a landfill (Figure 4). Fluctuations in oxidation state in water of the upper soil zone occur as pores of soil are alternately occupied by upward-fluxing methane, downward-seeping meteoric water, atmospheric oxygen, and carbon dioxide. We propose that insoluble Fe(III) is reduced to soluble Fe(II), which is carried downward during infiltration. If pore water evaporates or the environment becomes oxidizing, the Fe(II) reprecipitates as $Fe(OH)_3$, the precursor to maghemite. Development of a caliche-like layer may

increase the likelihood of developing a lower zone of Fe(II)-bearing magnetic minerals, such as magnetite, greigite, or pyrrhotite. Such a layer may not develop in landfill soils in humid climates, where dissolved carbonate and iron can be carried out of the soil in a process similar to that which decreased the susceptibility of the 1–year-old-old soils of this study. Further investigations may elucidate the processes by which negative magnetic soil anomalies form above petroleum reservoirs.

Future studies should tie magnetic effects over time to a complete characterization of site-specific variables, including (1) landfill gas compositions and flow rates; (2) rates and times of infiltration; (3) soil pore water temperature, Eh, pH, and dissolved components; and (4) soil mineralogy. It is important to know how soil inhomogeneities evolve that might affect water or gas permeability, resulting in different magnetic effects at the same depths in a soil.

We have shown that changes in MS in landfill soils are rapid by geologic time standards, occurring in just a few decades. Studies of soil caps above sanitary landfills that have stopped generating methane could provide useful information on the persistence of this magnetism. Major declines in light hydrocarbons occur in soils above partially depleted petroleum reservoirs, a decline that appears to be related to a progressive drop in reservoir pressures (Horvitz, 1969). We are unaware of any observations suggesting that MS values in soils above depleted reservoirs experience a parallel drop.

Sanitary landfills appear to be viable laboratories for the study of hydrocarbon-induced soil magnetism because they can provide the experimenter with a limited number of fixed or narrowly defined variables, including knowledge of the time variable. Sanitary landfills are found in all climates and have a wide range of capping soils. Collectively, they may come close to being global analogs for study of soil magnetism above petroleum deposits.

Acknowledgments—We wish to thank James E. Cooper, who was instrumental in helping to plan this study and was a constant source of information as it progressed. We are grateful to William Balsam, who provided vacuum gasometric data of calcite concentrations.

REFERENCES CITED

Barton, R. H., 1990, Relationship of surface magnetic susceptibility variations to hydrocarbons and subsurface structure: Association of Petroleum Geochemical Explorationists Bulletin, v. 6, p. 1–11.

Bazylinski, D. A., R. B Frankel, and H. W. Jannasch, 1988, Anaerobic magnetite production by a marine, magnetotropic bacterium: Nature, v. 334, p. 518–519.

Berner, R. A., 1984, Sedimentary pyrite formation: an update: Geochemica et Cosmochimica Acta, v. 48, p. 605–615.

Carolan, M. J., 1987, The future of landfill gas projects: American City and County, v. 102, p. RR12.

Donovan, T. J., R. L. Forgey, and A. A. Roberts, 1979a, Aeromagnetic detection of diagenetic magnetite over oil fields: AAPG Bulletin, v. 63, p. 245–248.

Donovan, T. J., P. A. Termain, and M. E. Henry, 1979b, Late diagenetic indicators of buried oil and gas: II, direct detection experiment at Cement and Garza oil fields, Oklahoma and Texas, using enhanced Landsat I and II images: USGS Open File Report 79-243, 45 p.

Donovan, T. J., J. D. Hendricks, A. A. Roberts, and P. T. Eliason, 1984, Low-altitude aeromagnetic reconnaissance for petroleum in the Arctic National Wildlife Refuge, Alaska: Geophysics, v. 49, p. 1338–1353.

Duell, R. W., I. A. Leone, and F. B. Flower, 1986, Effect of landfill gases on soil and vegetation: Pollution Engineering, v. 18, p. 38–40.

Ellwood, B. B., W. Balsam, B. Burkart, G. L. Long, and M. L. Buhl, 1986a, Anomalous magnetic properties in rocks containing the mineral siderite: paleomagnetic implications: Journal of Geophysical Research, v. 91, p. 12779–12790.

Ellwood, B. B., J. G. McPherson, B. K. Sen Gupta, and M. Matthews, 1986b, The proposed Eocene–Oligocene stratotype, SW, Alabama: not ideal due to magnetostratigraphic inconsistencies: Palaios, v. 1, p. 417–419.

Ellwood, B. B., B. Burkart, K. Rajeshwar, R. L. Darwin, R. A. Neely, A. B. McAll, G. J. Long, M. L. Buhl, and C. W. Hickcox, 1989a, Are the iron carbonate minerals ankerite and ferroan dolomite, like siderite, important in paleomagnetism?: Journal of Geophysical Research, v. 94, p. 7321–7331.

Ellwood, B. B., J. Payne, and G. Long, 1989b, The Rockwall in Rockwall, Texas: a study of unusual natural magnetic effects in geoarcheological surveys produced by mineral oxidation: Geoarchaeology, v. 4, p. 103–118.

Farina, M. D., D. Motta, S. Esquivel, and H. G. P. Lins de Barros, 1990, Magnetic iron-sulfur crystals from a magnetotactic microorganism: Nature, v. 343, p. 256–258.

Fassbender, J. W. E., H. Stanjek, and H. Vali, 1990, Occurrence of magnetic bacteria in soil: Nature, v. 343, p. 161–163.

Faure, G., 1991, Inorganic Geochemistry: New York, MacMillan Publishing, 626 p.

Fishman, N. S., R. L. Reynolds, M. R. Hudson, and V. F. Nuccio, 1989, source of anomalous magnetization in areas of hydrocarbon potential: petrologic evidence from Jurassic Preuss Sandstone, Wyoming–Idaho thrust belt: AAPG Bulletin, v. 73, p. 182–194.

Garrels, R. M., and C. L. Christ, 1965, Solutions, minerals, and equilibria: New York, Harper and Row, 450 p.

Horvitz, L., 1969, Hydrocarbon geochemical prospecting after thirty years: *in* W. B. Heroy, ed., Unconventional methods in exploration for petroleum and natural gas: Symposium, Institute for the Study of Earth and Man, Southern Methodist University, Dallas, Texas, p. 205–218.

Jones, G. A., and P. Kaiteris, 1983, A vacuum-gasometric technique for rapid and precise analysis of calcium carbonate in sediments and soils: Journal of Sedimentary Petrology, v. 53, p. 655–660.

Karlin, R., L. Mitchell, and G. R. Heath, 1987, Authigenic magnetite formation in suboxic marine sediments: Nature, v. 326, p. 490–493.

Krauskopf, K. B., 1979, Introduction to geochemistry: New York, McGraw-Hill, 617 p.

Kyle, J. R., M. R. Ulrich, and W. A. Gose, 1987, Textural and paleomagnetic evidence for the mechanism and timing of anhydrite cap rock formation, Winnfield salt dome, Louisiana, *in* I. Lerche and J. J. O'Brien, eds., Dynamical geology of salt and related structures: Orlando, Florida, Academic Press, p. 497–542.

Land, J. P., Sr., 1991, A comparison of micromagnetic and surface geochemical survey results: Association of Petroleum Geochemical Explorationists Bulletin, v. 7, p. 12–35.

Le Borgne, E., 1955, Abnormal magnetic susceptibility of the topsoil: Annals of Geophysics, v. 11, p. 399–419.

Lovley, D. R., J. F. Stoltz, G. L. Nord, Jr., and E. J. P. Phillips, 1987, Anaerobic production of magnetite by a dissimilatory iron-reducing microorganism: Nature, v. 330, p. 252–254.

Lu, A. N., and C. O. Kunz, 1981, Gas flow model to determine methane production at sanitary landfills: Environmental Science and Technology, v. 15, p. 436–440.

Machel, H. G., and E. A. Burton, 1991, Chemical and microbial processes causing anomalous magnetization in environments affected by hydrocarbon seepage: Geophysics, v. 56, p. 598–605.

Maher, B. A., and R. M. Taylor, 1988, Formation of ultrafine-grained magnetite in soils: Nature, v. 336, p. 368–370.

Maher, S., N. H. C. Sparks, R. B. Frankel, D. A. Bazylinski, and H. W. Jannasch, 1990, Biomineralization of ferrimagnetic greigite (Fe_3S_4) and iron pyrite (FeS_2) in a magnetotactic bacterium: Nature, v. 343, p. 258–261.

Mann, S, N. H. C. Sparks, R. B., Frankel, D. A. Bazylinski, and H. W. Jannasch, 1990, Biomineralization of ferrimagnetic greigite (Fe_3S_4) and iron pyrite (FeS_2) in a magnetotactic bacterium: Nature, v. 343, p. 258–261.

Mullins, C. E., 1977, Magnetic susceptibility of the soil and its significance in soil science—a review: Journal of Soil Science, v. 28, p. 223–246.

Nagata, T., 1961, Rock magnetism: Tokyo, Maruzen, 350 p.

Reynolds, R. L., N. S. Fishman, R. B. Wanty, and M. B. Goldhaber, 1990, Iron sulfide minerals at Cement oil field, Oklahoma: implications for magnetic detection of oil fields: GSA Bulletin, v. 102, p. 368–380.

Rozenson, I., B. Spiro, and I. Zak, 1982, Transformation of iron-bearing kaolinite to iron-free kaolinite, goethite, and hematite: Clays and Clay Minerals, v. 30, p. 207–214.

Saunders, D. F., K. R. Burson, and J. F. Branch, 1993, integrated surface methods profiles two east Texas fields: Association of Petroleum Geochemical Explorationists Bulletin, v. 9, p. 32–50.

Schwertman, U., and R. M. Taylor, 1989, Iron oxides, *in* J. B. Dixon and S. B. Weed, eds., Minerals in soil environments, second edition: Madison, Wisconsin, Soil Science Society of America, p. 380–438.

Singer, M. J., and P. Fine, 1989, Pedogenic factors affecting magnetic susceptibility of northern California soils: Journal of the Soil Science Society of America, v. 53, p. 1119–1127.

Skinner, B. J., R. C. Erd, and F. S. Grimaldi, 1964: Greigite, the thio-spinel of iron; a new mineral: American Mineralogist, v. 49, p. 543–555.

Stanjek, H., J. W. E. Fassbinder, H. Vali, H. Wägele, and W. Graf, 1994, Evidence of biogenic greigite (ferrimagnetic Fe_3S_4) in soil: European Journal of Soil Science, v. 45, p. 97–103.

Steenland, N. C., 1965, Oil fields and aeromagnetic anomalies: Geophysics, v. 30, p. 706–739.

Swartzendruber, L. J., 1992, Properties, units and constants in magnetism: Journal of Magnetic Materials, v. 100, p. 573–575.

Tite, M. S., and R. E. Linington, 1975, Effect of climate on the magnetic susceptibility of soils: Nature, v. 256, p. 565–566.

U.S. Geological Survey, 1992, Precipitation chemistry in the United States: NADP/NTN Annual Data Survey, 480 p.

Uyeda, S., M. D. Fuller, J. C. Belshe, and R. W. Girdler, 1963, Anisotropy of magnetic susceptibility of rocks and minerals: Journal of Geophysical Research, v. 68, p. 279–291.

Machel, H. G., 1996, Magnetic contrasts as a result of hydrocarbon seepage and migration, *in* D. Schumacher and M. A. Abrams, eds., Hydrocarbon migration and its near-surface expression: AAPG Memoir 66, p. 99–109.

Chapter 8

Magnetic Contrasts as a Result of Hydrocarbon Seepage and Migration

Hans G. Machel

Department of Earth and Atmospheric Sciences
University of Alberta
Edmonton, Canada

Abstract

Seepage of hydrocarbons from traps and migration from source rocks result in hydrocarbon-contaminated plumes in ground and formation waters. Such plumes are characterized mainly by a marked reduction in the redox potential, causing the generation of magnetic ferrous iron oxides and sulfides and the destruction of ferric iron oxides. The resulting magnetic mineral assemblages can be predicted on the basis of thermodynamic criteria and microbiological processes. Moreover, these assemblages may result in positive, absent, or negative magnetic contrasts relative to the total magnetization prior to hydrocarbon invasion. Thermodynamic modeling further suggests that magnetic contrasts are more likely and tend to become more positive with depth and with closer proximity to the hydrocarbon source(s).

Magnetic mineral assemblages and the resulting magnetic contrasts, such as those predicted here, have been documented from several hydrocarbon seepage environments at or near the land surface. Such magnetic contrasts can be used for hydrocarbon exploration in association with other surface exploration methods. Magnetic exploration for hydrocarbons can also be conducted successfully on drill cores. Furthermore, migration pathways can be delineated by magnetic methods because migration from source to reservoir rocks may generate magnetic mineral assemblages similar to those in seepage from traps.

INTRODUCTION

Magnetic contrasts associated with hydrocarbon seepage have been reported by numerous investigators. Aeromagnetic surveys can detect such magnetic contrasts and have been proposed as a technique for hydrocarbon exploration (e.g., Donovan, et al., 1979, 1984; Foote, 1984; Saunders and Terry, 1985), based on the supposed genetic relationship between certain magnetic minerals (mainly magnetite) and hydrocarbons. In addition, similar magnetic contrasts may also be detectable in core (e.g., Reynolds et al., 1991), and magnetic studies may thus be a valuable addition to core analysis in assessing the presence and proximity of hydrocarbon traps. Furthermore, as shown in isolated cases, it also appears possible to detect or delineate migration pathways of hydrocarbons by paleomagnetic methods (e.g., Elmore et al., 1989).

The chief underlying cause for such magnetic contrasts is that the invasion of hydrocarbons leads to a marked reduction in the redox potential of the pore waters which, in turn, results in the formation of specific magnetic mineral assemblages. My collaborators and I have discussed these phenomena in seepage environments ranging from near-surface diagenetic settings to depths of about 6–7 km (Machel and Burton, 1991a,b; 1992a,b; Burton et al., 1993). It became obvious that in these diagenetic settings predictable magnetic mineral assemblages can form as a result of changes in the redox potential and of concomitant changes in temperature, pressure, pH, total dissolved inorganic carbon, total dissolved sulfur, and total dissolved iron. Our studies also permitted certain predictions about the resulting magnetic contrasts that could be used for hydrocarbon exploration.

The results of these studies also have ramifications for and applications to many other facets of paleomagnetism, including those that involve diagenetic remagnetization via the formation or destruction of magnetic minerals and associated nonmagnetic iron minerals, relative timing of hydrocarbon migration and accumulation, and refined timing of major diagenetic remagnetization events related to tectonism, regional hydrogeochemical regimes, and normal compaction and cementation reactions (Aïssaoui et al., 1993). Some of these applications are discussed in Machel (1995).

This paper focuses on seepage of hydrocarbons from subsurface traps and migration from source rocks. The objectives are to emphasize that (1) magnetic contrasts are normal by-products of hydrocarbon seepage and migration and that (2) such contrasts are to be expected in many places.

BOUNDARY CONDITIONS

For hydrocarbon "seepage" or "migration" (a purely semantic difference in the *present* context) to result in detectable magnetic contrasts, four general conditions must be met:

Condition 1—In the case of traps, magnetic contrasts within or below the source of hydrocarbons generally are undetectable at the surface. Hence, magnetic contrasts associated with hydrocarbon traps mainly reside in the overlying and adjacent strata. In the case of carrier beds for migration, magnetic contrasts depend on the (paleo-) magnetic signature of the carrier beds prior to migration (see below).

Condition 2—Detrital, authigenic (inorganic-geochemical), and bacteriogenic magnetic minerals in the sedimentary strata overlying or adjacent to hydrocarbon accumulations carry a total magnetization that is genetically unrelated to the hydrocarbon accumulation.

Condition 3—Hydrocarbon seepage from traps and/or migration from source rocks occurs or has occurred.

Condition 4—The escaping organic compounds, as well as the associated hydrogen sulfide, induce chemical reactions that result in the formation or destruction of magnetic minerals.

Condition 1 is often met, as shown by empirical evidence (e.g., Reynolds et al., 1991, and references therein). The magnetization of reservoir rocks generally is very weak and overpowered by the magnetization in juxtaposed seepage environments. Furthermore, as shown below, premigration magnetization may be overprinted by invading hydrocarbons.

Condition 2 is also commonly met, as shown by the fact that most sedimentary rocks carry magnetic signatures (remanent and induced) (e.g., Aïssaoui et al., 1993). This magnetization serves as the "background" against which hydrocarbon-induced contrasts are measured.

Condition 3, with reference to migration, is met regularly because source rocks and traps rarely coincide. Also, seepage from traps is very common, as few cap rocks are tight over geologically significant times (e.g., Ebanks et al., 1993). In the vicinity of traps, the area influenced by hydrocarbon leakage is known as a *seepage environment*, *geochemical halo*, or *geochemical chimney* (e.g., Davidson, 1982), but should be called a *plume* in accordance with the standard nomenclature of contaminant hydrogeology, which calls any three-dimensional zone containing con-

taminants a plume (e.g., Domenico and Schwartz, 1990).

Condition 4 is based on theoretical data (e.g., Garrels and Christ, 1965; Nordstrom and Munoz, 1986) as well as experimental and empirical evidence (e.g., Trudinger et al., 1985; Machel, 1987), which demonstrates that certain geochemical reactions forming magnetic minerals are likely to occur because of thermodynamic constraints. Preconditions are that iron must be available in minerals or dissolved as aqueous species, hydrocarbons must be reactive biologically or abiotically, and the rates of reactions involving hydrocarbons and iron minerals must be rapid relative to the rates of fluid transport. Under these conditions, the reactions involving hydrocarbons can alter the ambient water geochemistry sufficiently to cause magnetic minerals to precipitate or dissolve in quantities large enough to produce magnetic contrasts that are detectable in core or by aeromagnetic surveys.

MAGNETIC PROPERTIES

Magnetic contrasts, such as those discussed is this paper, can be expressed as contrasts in total magnetization (J), magnetic susceptibility (χ), or remanence (J_n), which are related according to

$$J = \chi H + J_n$$

where H = induced or applied field strength. Several relatively common types of remanent magnetization contribute to the natural remanent magnetization (NRM) of rocks in geochemical plumes: detrital remanent magnetization (DRM), postdepositional detrital remanent magnetization (pDRM), chemical remanent magnetization (CRM), bacterial/biological remanent magnetization (BRM) (in most documented cases of magnetosome deposition, this type would be considered pDRM), viscous remanent magnetization (VRM), and thermoviscous remanent magnetization (TVRM). NRM in rocks that are or were under the influence of hydrocarbons is usually dominated by DRM, CRM, BRM, or any combination of these. The other types of magnetization, albeit almost always present, are minor in such rocks. In the present context, it is principally irrelevant at what level magnetic contrasts are measured. Hence, when reference is made in the following discussion to magnetic contrasts, one may apply this to aeromagnetic, soil magnetic, or rock magnetic measurements on core or cuttings.

IMPORTANT MINERALS

Of all the magnetic and nonmagnetic minerals that are common in diagenetic settings, only a few are important in the present context—those that are magnetic, those that commonly form diagenetically at the expense of magnetic minerals, and those that liberate iron upon dissolution. Other minerals are excluded from the remaining discussion for simplicity.

Regarding magnetic minerals, the present discussion is limited mainly to magnetite (Fe_3O_4) (including such varieties as titanomagnetite), monoclinic pyrrhotite (Fe_7S_8), and hematite (Fe_2O_3). There are several reasons for this limitation. (1) The occurrence of these minerals is controlled by organic compounds, including hydrocarbons. (2) Magnetite and monoclinic pyrrhotite are the most abundant and widespread magnetic minerals in diagenetic environments. (3) These minerals have relatively strong magnetic remanence and susceptibility. (4) Hematite, although only weakly magnetic, is important because it is widespread and locally abundant, particularly in redbeds.

Magnetic contributions from other minerals are relatively small or negligible in most (but not all) diagenetic environments. This is because these minerals are relatively rare, metastable, restricted to few environments, or have low magnetic remanence and susceptibility. For example, greigite is fairly rare except in relatively young sediments (including "young" seepage environments) because it tends to recrystallize to nonmagnetic pyrite (e.g., Morse et al., 1987). Another example is maghemite which, albeit a common replacement of magnetite, usually retains the magnetization of its precursor and does not normally add to the formation of magnetic contrasts. However, these minerals may be important in some locations (e.g., Reynolds et al., 1991, 1993; Roberts and Turner, 1993; Thompson and Cameron, 1995).

The most common nonmagnetic minerals that form diagenetically at the expense of magnetic minerals are iron sulfides, particularly pyrite and siderite. They are included in the following discussion because their formation lowers the overall magnetic intensity of the rocks.

Minerals that liberate iron upon dissolution are principally all iron-bearing minerals, particularly those containing oxidized iron. This group includes goethite, limonite, hematite, ferric oxyhydroxides, various iron-bearing silicates, and nonmineralic compounds such as organic and inorganic Fe complexes. For simplicity in the following discussion, most of these minerals and compounds are lumped with hematite as representative of "the" oxidized iron-bearing mineral.

MICROBIAL FORMATION AND DESTRUCTION OF MAGNETIC MINERALS

Nearly all minerals of interest in the present context can be formed or destroyed by microbes. For example, several types of "iron bacteria" in oxygenated environments leave iron deposited on the cell surfaces, whereby several types of iron oxides and hydroxides are formed (Hirsch, 1974). Conversely, a variety of microbes can dissolve hematite via the excretion of substances that complex Fe^{3+}, whereby chelates are taken up by the organisms and the iron is used for intracellular metabolic processes (Hirsch, 1974).

Magnetite can be formed by aerobic as well as anaerobic bacteria. The most widely distributed group are the magnetotactic bacteria, which form intracellular chains of small (commonly a few nanometers), single-domain magnetite grains (magnetosomes). Magnetosomes can have a variety of shapes, ranging from cubic, via hexagonal, octahedral, teardrop, tooth-, arrowhead-, or bullet-shaped, to fibrous (Vali and Kirschvink, 1990). Upon death, the chains commonly disintegrate and the magnetosomes contribute to the NRM and induced magnetization. There are also nonmagnetotactic bacteria that form extracellular magnetite (e.g., Lovley et al., 1987). Such organisms may produce magnetite via the oxidation of certain organic compounds in anaerobic diagenetic settings, including hydrocarbon-contaminated plumes. The distribution and importance of these bacteria in natural environments is unknown.

Anaerobic magnetotactic bacteria that form pyrrhotite have been discovered living at depths from 20 cm to >1 m in marine and brackish water at the interface between water and sulfide-rich sediments (Farina et al., 1990). Bacteria that form intracellular greigite and pyrite also live in brackish, sulfide-rich water and sediments (Mann et al., 1990). One extraordinary bacterium was found to contain magnetosomes of an iron sulfide and an iron oxide, that is, greigite along with magnetite (Bazylinski et al., 1993). Like magnetite, pyrrhotite and greigite magnetosomes can potentially contribute to the NRM. The magnetic contribution of greigite, however, tends to be relatively short-lived on a geologic time scale (normally only a few million years) because it tends to recrystallize to pyrite over time (e.g., Morse et al., 1987).

Anaerobic sulfate-reducing bacteria are a factor in the formation of magnetic iron sulfides because they generate reduced sulfur (mainly as H_2S, HS^-, and polysulfides). Where dissolved Fe^{2+} is available, it reacts with the aqueous sulfur species to form iron sulfides, including greigite, pyrite, pyrrhotite, marcasite, and mackinawite (e.g., Morse et. al., 1987). Mineral precipitation generally is extracellular. The best known examples are clusters of framboidal pyrite that can be considered "pyritized bacteria" (note, however, that framboids are also known to form inorganically; e.g., Sawlowicz, 1993).

INORGANIC FORMATION AND DESTRUCTION OF MAGNETIC MINERALS

The relative stabilities of the previously mentioned minerals in diagenetic settings are governed by seven major geochemical parameters: temperature, pressure, dissolved sulfur, dissolved inorganic carbon, dissolved iron, pH, and Eh. These parameters vary over specific ranges in diagenetic settings: temperatures from $25°$ to $200°C$; pressures from 1 to 600 bars; activities of dissolved sulfur species (mainly HS^-, H_2S, and SO_4^{2-}) from 10^{-2} to 10^{-14}; activities of bicarbonate from 10^{-1} to 10^{-4}; activities

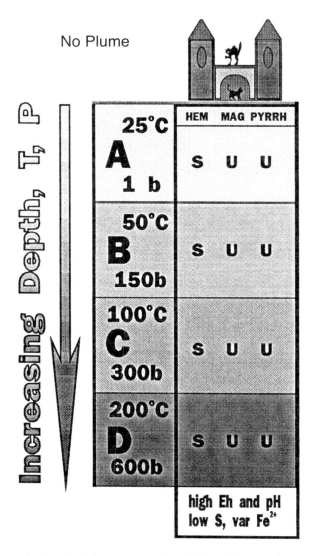

Figure 1—Simplified summary of modeling of magnetic mineral assemblages in geologic environments devoid of hydrocarbons (no plume). A, B, C, and D denote discrete depth intervals. The fates of the minerals are indicated as U (unstable) or S (stable). HEM = hematite, MAG = magnetite, PYRRH = pyrrhotite.

of Fe^{2+} from 10^{-4} to 10^{-6}; pH values from 5 to 9; and Eh values from −100 to −150 mV. These ranges were measured in ground and formation waters of near-surface to deep burial diagenetic settings (Baas-Becking et al., 1960; Billings et al., 1969; Hitchon et al., 1971; Plummer et al., 1976; Kharaka et al., 1977; Sverjensky, 1984; Connolly et al., 1990).

With these ranges as variables, magnetic mineral parageneses in the diagenetic settings of interest have been modeled and represented as a series of bivariate phase diagrams by Machel and Burton (1991a), Burton et al. (1993), and Machel (1995) (these papers also include extensive discussions of the calculation procedures and implications). Thus, diagenetic settings have been divid-

ed into four discrete layers (A, B, C, and D) that represent the following mean temperature and pressure values: 25°C at 1 bar, 50°C at 150 bars, 100°C at 300 bars, and 200°C at 600 bars. The essential results of this modeling are depicted here in a qualitative fashion (Figures 1, 2, 3).

Figure 1 illustrates diagenetic settings that are essentially devoid of hydrocarbons as reducing agents, with corresponding high Eh and pH values, low total dissolved S and variable Fe^{2+} activities, here designated as "no plume." Under such conditions, hematite is stable throughout the temperature and pressure ranges under consideration, but both magnetite and pyrrhotite (the other magnetic minerals of interest) are unstable (Figure 1). In this context, stable implies that a mineral either forms or does not dissolve. Unstable implies that a mineral cannot form and should dissolve without another mineral forming as a replacement at its expense. However, some minerals may survive under unstable conditions as metastable phases for a considerable time (see further discussed below).

Figure 2 represents diagenetic settings containing relatively low concentrations of hydrocarbons that correspond to medium to low Eh values, with pH > 7.5 and variable Fe^{2+} activities, here designated as "distant plume." (Low hydrocarbon concentrations can also occur close to a leaking source if the rate of leakage is low, but hydrocarbon concentrations are generally relatively high close to the source and decrease with distance.) Under these conditions, various possibilities exist for mineral stability, unstability, and replacement, depending mainly on variations in the activity of total dissolved sulfur (in this context only dissolved sulfur species are important). For example, in near-surface settings at medium total sulfur (MTS) activity, pyrite is stable relative to hematite and hence likely to replace hematite, barring metastable survival. Magnetite is rendered unstable, although no sulfide will form at its expense. Also, pyrrhotite is unstable because the sulfur activity is not high enough (Figure 2, layer A, line MTS).

At high total sulfur (HTS) activity in near-surface diagenetic settings, the fates of hematite and magnetite are unchanged, but pyrite is stable relative to pyrrhotite and therefore likely to replace it (Figure 2, layer A, line HTS). In some cases, more than one replacement is possible. For example, in layer B at low total sulfur (LTS) activity, magnetite or siderite can replace hematite (Figure 2). Which mineral is likely to replace hematite is determined by variations in pH and the bicarbonate activity of the pore water. Similarly, variations in pH can render one mineral either stable or unstable relative to another mineral. One such example is magnetite, which can be stable or unstable relative to siderite in deep and hot diagenetic settings under low and medium total sulfur activities (Figure 2, layer D, lines LTS, MTS). (For details on these phenomena, refer to the phase diagrams that form the basis of Figures 1, 2, and 3; see Machel and Burton, 1991a; Burton et al., 1993; Machel, 1995.)

Figure 3 illustrates diagenetic settings that have relatively high concentrations of hydrocarbons corresponding to medium to low Eh values, with pH < 7.5 and vari-

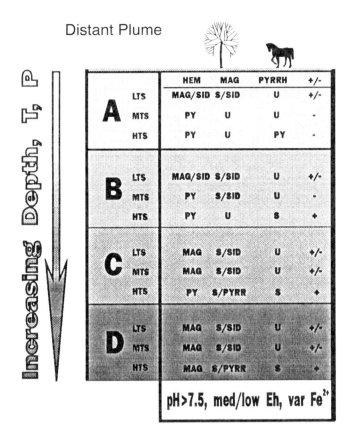

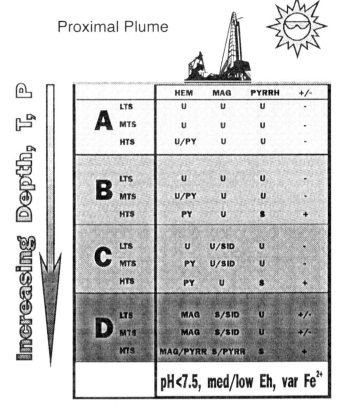

Figure 2—Simplified summary of modeling of magnetic mineral assemblages and resulting magnetic contrasts in geologic environments with moderate concentrations of hydrocarbons (distant plume). A, B, C, and D designate discrete depth intervals that correspond to those in Figure 1. LTS, MTS, and HTS denote low total sulfur, medium total sulfur, and high total sulfur activities, respectively. The fates of the minerals are indicated by U = unstable or S = stable, or replaced by MAG = magnetite, SID = siderite, or PY = pyrite. HEM = hematite, PYRRH = pyrrhotite. The expected magnetic contrasts are shown relative to the background levels as increased (+) or decreased (–).

Figure 3—Simplified summary of modeling of magnetic mineral assemblages and resulting magnetic contrasts in geologic environments with relatively high concentrations of hydrocarbons (proximal plume). Abbreviations and symbols as in Figure 2.

able Fe^{2+} activities, here designated as "proximal plume." The possibilities of stable, unstable, and mineral replacement are determined from Figure 3 analogously to Figures 1 and 2.

POSSIBLE MAGNETIC CONTRASTS

Figures 1, 2, and 3 can be used to qualitatively identify possible magnetic contrasts. Quantification of magnetic contrasts requires mass balance calculations of the mineral reactions involved, which is beyond the scope of this paper.

The conditions in Figure 1 can be considered "background" relative to environments under the influence of seeping or migrating hydrocarbons. The conditions in

Figures 2 and 3 result in *positive* or *negative contrasts* (corresponding to enhanced or reduced total magnetization or its component magnetizations, respectively, compared to the background), depending on which type of diagenetic reaction takes place and to what extent. For example, a negative contrast is expected where a magnetic mineral dissolves without replacement, such as magnetite at medium total sulfur activity in a near-surface setting (Figure 2, layer A, line MTS). Similarly, a negative contrast is expected where a nonmagnetic mineral forms at the expense of a magnetic mineral, such as pyrite at the expense of pyrrhotite at high total sulfur activity in a near-surface setting (Figure 2, layer A, line HTS), or where a weakly magnetic mineral forms at the expense of a strongly magnetic mineral. Conversely, a positive contrast is expected where a magnetic mineral forms where no magnetic mineral was stable before, as in the case of pyrrhotite at high total sulfur activity at intermediate burial depths (Figure 2, layers B, C, line HTS). A positive contrast is also likely where a strongly magnetic mineral, such as magnetite, forms at the expense of a weakly magnetic mineral, such as hematite, as is common in deep diagenetic settings irrespective of the total sulfur activity (Figure 2D).

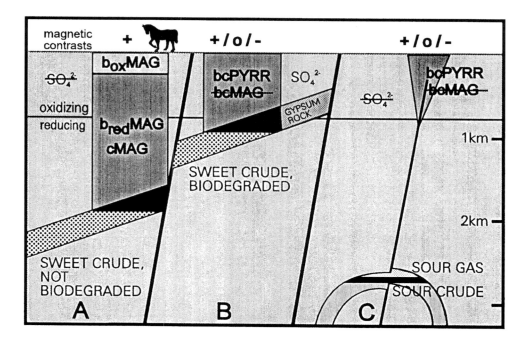

Figure 4—Schematic illustration of three common hydrocarbon plays with associated geochemical plumes. Detectable magnetic contrasts are possible in all cases. All rocks above petroleum traps are assumed to have no or insignificant natural remanent magnetization prior to hydrocarbon invasion. If hematite was present in the strata covering the petroleum traps, it would be destroyed (dissolved or replaced) in all plumes. SO_4^{2-} (crossed out) indicates that the system does not contain dissolved sulfate, and bcMAG (crossed out) implies that bacterial or chemical magnetite is dissolved or replaced.

This type of analysis yields several important results. First, compared to the original detrital or biogenic magnetization of the sediments and rocks, hydrocarbon-induced (re-)magnetization is likely with increasing depth (Figures 2, 3); very few minerals remain stable (S), but unstable and replacement conditions are common. Second, compared to mineral dissolution, there appear to be more possibilities for mineral replacement under relatively high pH settings compared to lower pH settings (compare Figures 2, 3). Third, in the vicinity of leaking hydrocarbon traps or source rocks, (re-) magnetization is likely to result in negative contrasts at relatively shallow depths and in increasingly positive contrasts at greater depths, whereby this pattern becomes more pronounced with greater proximity to plumes (compare +/– columns of Figures 2, 3).

THERMODYNAMIC DISEQUILIBRIUM ASSEMBLAGES

Kinetic and microbial effects may generate thermodynamic disequilibrium mineral assemblages, at least temporarily. Many of these effects are poorly understood and cannot be quantified at the present time, but four of them can be alluded to here.

First, diagenetic environments, particularly at temperatures below about 50°C, may not be in thermodynamic equilibrium. For example, it is possible for an environment to become increasingly reducing and alkaline (Eh < –0.25 V and pH > 7) without magnetite forming at the expense of hematite, as predicted by thermodynamic modeling. This happens if certain ions are absorbed to the hematite grain surfaces, thus inhibiting hematite dissolution (e.g., Berner, 1980). Such a condition is not represented by thermodynamic stability diagrams.

Another problem is that some minerals, such as pyrite, rarely form directly from aqueous solution. Rather, metastable intermediates form that recrystallize to pyrite over time (e.g., Morse et al., 1987).

A third problem is related to microbial processes. Microbes generally produce minerals that are thermodynamically stable under the prevailing conditions. However, in some cases, microbial metabolism can produce minerals under conditions at which they are thermodynamically unstable, or microbes can promote the dissolution of thermodynamically stable minerals. For example, aerobic magnetotactic bacteria live under conditions in which magnetite is thermodynamically unstable, that is, in the hematite fields of thermodynamic stability diagrams, or in oxygenated environments with more positive Eh. Hence, these bacteria may generate, at least temporarily, a thermodynamic disequilibrium assemblage.

Fourth, detrital minerals may not dissolve when they become thermodynamically unstable. For example, detrital magnetite is relatively abundant in a variety of shallow diagenetic environments (e.g., McCabe and Elmore, 1989) even though it is unstable in such settings (Figures 1, 2, 3).

In most cases, kinetics will tend to lead to the metastable survival of detrital minerals rather than to the formation of thermodynamically unstable minerals. Hence, the predictions advanced earlier in this paper are valid at least for authigenic minerals.

GEOLOGIC ENVIRONMENTS

The thermodynamic and microbial considerations previously discussed can be applied to geologic environments that include plumes caused by hydrocarbon invasion or regional migration. Several examples are shown

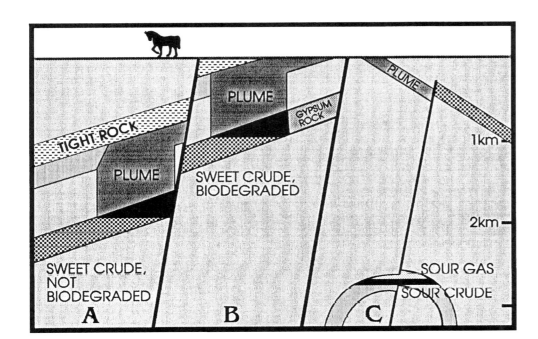

Figure 5—Schematic illustration of deflection of geochemical plumes along tight beds (domains A and B) and of funneling of plumes through zones with relatively high permeability, such as certain faults and porous sedimentary layers (domain C).

in Figures 4, 5, and 6. These examples are based on the necessary assumption that magnetite and pyrrhotite may form chemically (inorganically) or microbiologically in sufficient quantities to cause magnetic contrasts. Three additional assumptions have been made for simplicity: (1) none of the strata enclosing the hydrocarbon accumulations contain indigenous organic matter; (2) all shallow environments subjected to hydrocarbon seepage contain oxidized iron compounds that provide Fe^{2+} upon reduction; and (3) there is no significant lateral hydrologic flow, in other words, any active hydrodynamic forces are incapable of significantly deflecting or deforming the plumes.

In a first set of conditions, three major faults subdivide the cross section into domains A, B, and C (Figure 4). The faults between domains A and B and between B and C are sealing, and the fault within domain C is leaking. In domain A, a structural trap contains sweet crude at an intermediate depth. The trap is located well within the "reducing" part of the regional groundwater system, hence no aerobic biodegradation can take place. A plume extends from the top of the trap to the surface. All formation waters are assumed to be essentially devoid of sulfate and, thus no bacterial sulfate reduction occurs. The only magnetic minerals that can be formed are bacterial (bMAG) and chemical (cMAG) magnetite. Aerobic magnetotactic bacteria can form magnetite in the uppermost, partially oxidizing part of the plume (b_{ox}MAG). Anaerobic magnetotactic and nonmagnetotactic bacteria can form magnetite in the reducing part of the plume (b_{red}MAG). The rocks affected by hydrocarbon seepage are assumed to have insignificant total magnetization prior to hydrocarbon seepage. Therefore, a positive magnetic anomaly may be located above the trap.

In domain B (Figure 4), a stratigraphic trap of similar size and hydrocarbon volume is located within the oxidizing part of the regional groundwater system. Oil in

this trap is at least partially biodegraded aerobically. Also, the groundwater in domain B contains significant amounts of sulfate from the dissolution of gypsum. Therefore, bacterial sulfate reduction is taking place where hydrocarbons are supplied as nutrients (i.e., in the plume). Bacterial or chemical pyrrhotite (bcPYRR) may be formed, bacterial or chemical magnetite (bcMAG) is being replaced or dissolved, and hematite, if present, is being destroyed. The pyrrhotite may or may not be magnetic. Accordingly, magnetic contrasts in domain B are either positive, absent, or negative, depending on the amounts of magnetite lost and magnetic pyrrhotite generated.

The situation in domain C (Figure 4) differs from domians A and B in that an anticlinal trap is located in the deeper subsurface and contains sour gas with thermochemically generated hydrogen sulfide. The regional groundwater is (as in domain A) essentially devoid of sulfate. Hydrocarbons are leaking up along the fault that penetrates the apex of the structure, and a plume is formed near the surface. Bacterial and chemical pyrrhotite (bcPYRR) may form from the thermogenic sulfide, and bacterial/chemical magnetite (bcMAG) may be destroyed. Bacterial sulfate reduction cannot take place because there is no sulfate in the system. If sulfate were available, bacterial sulfate reduction would take place in the plume. Bacterial sulfide would then complement the thermochemical sulfide leaking up the fault. With or without bacterial sulfate reduction, the possibilities for magnetic anomalies are the same in domains C and B.

Variations in the conditions used to construct Figure 4 result in additional spatial possibilities for plumes and magnetic contrasts. The most important possibilities result from the nonvertical migration of organic compounds, which is likely if the geologic succession contains alternating units with high and low permeability and if

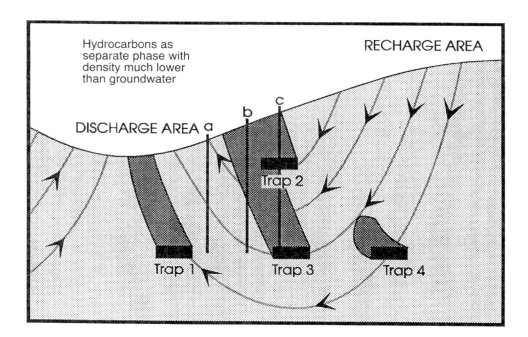

Figure 6—Schematic illustration of four hydrocarbon traps embedded in an active regional groundwater flow system of an isotropic, homogeneous medium, depicting the effect of advection on plume orientation. Separate-phase hydrocarbons have densities significantly lower than that of the groundwater. Drillholes at locations (b) and (c) would penetrate the combined plumes of traps 2 and 3, but only drillhole (c) would penetrate the traps. A drillhole at (a) would no longer penetrate a plume.

there is vigorous flow of the groundwater or formation water. For example, plumes can be deflected along relatively impermeable beds (Figure 5), thereby following more permeable beds. In such cases, it is possible that a plume may not reach the surface (Figure 5, domain A) or it may be laterally displaced, generating a magnetic contrast that is no longer located directly above the leaking trap (Figure 5, domains B, C).

Similar alternatives arise where there is an active regional groundwater flow system, as is common in relatively shallow (<2–3 km) settings with significant topography (topography-driven groundwater flow). Plumes tend to be deflected in the direction of regional groundwater flow. Thus, the degree of deflection (and dispersion) of the plumes depends mainly on the density contrast between the hydrocarbons and water and on whether the hydrocarbons are present as a dissolved or separate phase (Figure 6). In such cases, the shapes and sizes of the plumes depend on several factors, most importantly, flow direction and velocity, the size of the source, and dispersion. Dispersion becomes significant at low flow velocities, and the plumes tend to be shorter, broader, and have lower contaminant (hydrocarbon) concentrations within them (Domenico and Schwartz, 1990). Generally, the sizes and orientations of plumes depend on fluid potential gradients determined by topographic relief, buoyancy, anisotropy, bedding, structure, and capillary forces.

This discussion also implies that hydrocarbon exploration aided by magnetic measurements may lead to erroneous predictions of the locations of the leaking traps. However, it is possible that the migration paths of the hydrocarbons could be traced by following the plumes. The cases shown in Figures 6, for example, pose interesting challenges for hydrocarbon exploration. Trap 2 could still be found by surface exploration methods, including

magnetic exploration, if a wildcat is drilled at location c. Traps 3 and 4 would go either unnoticed or would be missed. For example, trap 3 would be missed because a wildcatter would normally not drill through the shallower trap 2 and because the bulk of the plume of trap 3 is deflected so much to the left that a well at location b would be misplaced. Trap 4 would go unnoticed because its plume would not reach close enough to the surface.

Finally, it is important to note that plumes may shift or vary in intensity through time and space. This can result in magnetic contrasts that indicate former hydrogeologic conditions or that are transient on a geologic time scale (as suggested by Henry and Thomlinson, 1991). Therefore, many variations on the examples shown here are possible, depending on the interplay of the previously discussed parameters.

CASE STUDIES

In this relatively new field of research, a number of studies have been published during the last 15 years that deal with magnetic contrasts in hydrocarbon-contaminated plumes. Donovan et al. (1979, 1984), Foote (1984), and Saunders and Terry (1985) were the first to advocate the use of aeromagnetic surveys as a complementary method to conventional exploration. Subsequently, Elmore et al. (1987), McCabe et al. (1987), Sassen et al. (1989), Kilgore and Elmore (1989), Elmore and Crawford (1990), and Reynolds et al. (1990a) documented authigenic magnetite and magnetic pyrrhotite as sources of magnetic contrasts from hydrocarbon-contaminated plumes. Henry (1989), Henry and Thomlinson (1991), and Saunders et al. (1991) combined soil and subsoil magnetic susceptibility measurements with soil gas hydrocarbon analyses and found magnetic contrasts associated with hydrocarbon seepage.

Saribudak (1992) reported high counts of certain microbes near magnetic anomalies in hydrocarbon-contaminated plumes. Reynolds et al. (1993) provided an account of the ongoing studies at the Cement oil field in Oklahoma, at the Simpson oil field in Alaska, and above deep Cretaceous oil and gas reservoirs in south Texas. Dating of hydrocarbon seepage by chemical remagnetization, combination of fluid inclusion and paleomagnetic data, and other paleomagnetic properties of hydrocarbon reservoirs are covered in several papers in Turner and Turner (1995).

In all these cases, magnetic contrasts were shown or inferred to be the result of the inorganic diagenetic formation and destruction of magnetic minerals. To date, there are no unequivocal cases of BRM in hydrocarbon seepage environments. Moreover, Reynolds et al. (1990b) and Land (1991) provided first attempts at forward modeling of anomalous magnetization in hydrocarbon seepage environments and compared them to aero- and ground magnetic data from specific study areas. Elmore et al. (1987, 1989) showed that the timing of hydrocarbon migration can be determined with paleomagnetic methods. Other aspects relevant to remagnetization in diagenetic environments, including thermoviscous remagnetization, implications for reconstruction of magnetic pole locations, and applications to magnetostratigraphy, core orientation, and susceptibility logging, are discussed in McCabe and Elmore (1989) and Turner and Turner (1995).

CONCLUSIONS

1. Magnetic mineral assemblages in diagenetic environments are governed by microbial and inorganic-chemical processes. Empirical evidence and thermodynamic modeling allows prediction of stable magnetic mineral assemblages in diagenetic environments.

2. Any significant change in the redox potential is likely to cause magnetic contrasts via formation or destruction of magnetic minerals, thus generating secondary (chemical) magnetization and resetting original detrital or bacterial magnetization.

3. As a result of significant changes in the diagenetic redox potential, positive, absent, or negative magnetic contrasts are generated relative to the total magnetization (or remanence, or susceptibility) prior to such changes, depending on the relative amounts of magnetic to nonmagnetic minerals formed and destroyed.

4. Magnetic contrasts measured from the air, at the surface, or on drill core can be used for hydrocarbon exploration in association with other exploration methods.

5. The predicted magnetic mineral assemblages can aid in the interpretation of paleomagnetic studies, particularly through the recognition and proper interpretation of remagnetization.

6. Considering that thermodynamic disequilibrium assemblages are quite common in diagenetic settings, the stable mineral assemblages discussed in this paper should only be taken as rough guidelines. Hence, predictions of magnetic contrasts associated with hydrocarbon seepage environments should be undertaken with caution.

Acknowledgments—E.A. Burton and J. Qi provided important contributions to the thermodynamic modeling. Constructive comments by M. A. Abrams, D. Schumacher, and one anonymous reviewer are much appreciated. This research was supported by the Natural Engineering and Research Council of Canada (NSERC) and by the Alexander von Humboldt Foundation.

REFERENCES CITED

Aïssaoui, D. M., D. F. McNeill, and N. F. Hurley, eds., 1993, Applications of paleomagnetism to sedimentary geology: SEPM Special Publication No. 49, 216 p.

Baas-Becking, L. G. M., I. R. Kaplan, and D. Moore, 1960, Limits of the natural environment in terms of pH and oxidation-reduction potentials: Journal of Geology, v. 68, p. 243–284.

Bazylinski, D. A., B. R. Heywood, S. Mann, and R. B. Frankel, 1993, Fe_3O_4 and Fe_3S_4 in a bacterium: Nature, v. 366, p. 218.

Berner, R. A., 1980, Early diagenesis: Princeton, New Jersey, Princeton University Press.

Billings, G. K., B. Hitchon, and D. R. Shaw, 1969, Geochemistry and origin of formation waters in the Western Canada Sedimentary Basin, 2. Alkali metals: Chemical Geology, v. 4, p. 211–223.

Burton, E. A, H. G. Machel, and J. Qi, 1993, Thermodynamic constraints on anomalous magnetization in shallow and deep hydrocarbon seepage environments, *in* D. M. Aïssaoui, D. F. McNeill, and N. F. Hurley, eds., Applications of paleomagnetism to sedimentary geology: SEPM Special Publication No. 49, p. 193–207.

Connolly, C. A., L. M. Walter, H. Baadsgaard, and F. J. Longstaffe, 1990, Origin and evolution of formation waters, Alberta Basin, Western Canada Sedimentary Basin, I. Chemistry: Applied Geochemistry, v. 5, p. 375–395.

Davidson, M. J., 1982, Toward a general theory of vertical migration: Oil and Gas Journal, v. 21, p. 288–300.

Domenico, P. A., and F. W. Schwartz, 1990, Physical and chemical hydrogeology: New York, John Wiley and Sons, 824 p.

Donovan, T. J., R. L. Forgey, and A. A. Roberts, 1979, Aeromagnetic detection of diagenetic magnetite over oil fields: AAPG Bulletin, v. 63, p. 245–248.

Donovan, T. J., J. D. Hendricks, A. A. Roberts, and P. T. Eliason, 1984, Low-altitude aeromagnetic reconnaissance for petroleum in the Arctic National Wildlife Refuge, Alaska: Geophysics, v. 49, p. 1338–1353.

Ebanks, J., J. Kaldi, and C. Vavra, eds., 1993, Seals and traps: a multidisciplinary approach (abs.): AAPG, Hedberg Research Conference Abstracts, June 21–23, Crested Butte, Colorado.

Elmore, R. D., and L. Crawford, 1990, Remanence in authigenic magnetite: testing the hydrocarbon-magnetite hypothesis: Journal of Geophysical Research, v. 95, no. B4, p. 4539–4549.

Elmore, R. D., M. H. Engel, L. Crawford, K. Nick, S. Imbus, and S. Sofer, 1987, Evidence for a relationship between hydrocarbons and authigenic magnetite: Nature, v. 325, p. 428–430.

Elmore, R. D., R. McCollum, and M. H. Engel, 1989, Evidence for a relationship between hydrocarbon migration and diagenetic magnetic minerals—implications for petroleum exploration: Association of Petroleum Geochemical Explorationists Bulletin, v. 5, no. 1, p. 1–17.

Farina, M., D. M. S. Esquivel, and H. G. P. Lins De Barros, 1990, Magnetic iron-sulphur crystals from a magnetotactic organism: Nature, v. 343, p. 256–258.

Foote, R. S., 1984, Significance of near-surface magnetic anomalies, in M. J. Davidson and B. M. Gottlieb, eds., Unconventional methods in exploration for petroleum and natural gas: Dallas, Southern Methodist University, Institute for Study of Earth and Man, p. 12–24.

Garrels, R. M., and C. L. Christ, 1965, Solutions, minerals and equilibria: New York, Harper and Row, 450 p.

Henry, W. E., 1989, Magnetic detection of hydrocarbon seepage in a frontier exploration region: Association of Petroleum Geochemical Explorationists Bulletin, v. 5, p. 18–29.

Henry, W. E., and W. D. Thomlinson, 1991, Evidence for near-surface magnetic anomalies formed in response to both recent and long-term hydrocarbon microseepage: AAPG Bulletin, v. 75, p. 593.

Hirsch, P., 1974, Budding bacteria: Annual Reviews of Microbiology, v. 28, p. 391–444.

Hitchon, B., G. K. Billings, and J. E. Klovan, 1971, Geochemistry and origin of formation waters in the Western Canada Sedimentary Basin, III. Factors controlling chemical composition: Geochimica et Cosmochimica Acta, v. 35, p. 567–598.

Kharaka, Y. K., E. Callender, and R. H. Wallace, 1977, Geochemistry of geopressured geothermal waters from the Frio clay in the Gulf Coast region of Texas: Geology, v. 5, p. 241–244.

Kligore, B., and R. D. Elmore, 1989, A study of the relationship between hydrocarbon migration and the precipitation of authigenic magnetic minerals in the Triassic Chugwater Formation, southern Montana: GSA Bulletin, v. 101, p. 1280–1288.

Land, J. P., 1991, A comparison of micro-magnetic and surface geochemical survey results: Association of Petroleum Geochemical Explorationists Bulletin, v. 7, p. 12–35.

Lovley, D. R., J. F. Stolz, G. L. Nord, and E. J. P. Philips, 1987, Anaerobic production of magnetite by a dissimilatory iron-reducing organism: Nature, v. 330, p. 252–254.

Machel, H. G., 1987, Some aspects of diagenetic sulfate-hydrocarbon redox-reactions, in J. D. Marshall, ed., Diagenesis of sedimentary sequences: Geological Society of London Special Publication 36, p. 15–28.

Machel, H. G., 1993, Magnetic contrasts caused by flow of petroliferous fluids—evidence for hydrocarbon seepage and/or regional fluid flow, in G. M. Ross, ed., Alberta Basement Transects Workshop, March 1–2, Lithoprobe Report #31: Vancouver, University of British Columbia, Lithoprobe Secretariat, p. 68–78.

Machel, H. G., 1995, Magnetic mineral assemblages and magnetic contrasts in diagenetic environments—with implica-

tions for studies of paleomagnetism, hydrocarbon migration, and exploration, in P. Turner and A. Turner, eds., Paleomagnetic applications in hydrocarbon exploration and production: Geological Society of London Special Publication 98, p. 9–29.

Machel, H. G., and E. A. Burton, 1991a, Chemical and microbial processes causing anomalous magnetization in environments affected by hydrocarbon seepage: Geophysics, v. 56, p. 598–605.

Machel, H. G., and E. A. Burton, 1991b, Causes and spatial distribution of anomalous magnetization in hydrocarbon seepage environments: AAPG Bulletin, v. 75, p. 1864–1876.

Machel, H. G., and E. A. Burton, 1992a, Magnetische Kontraste durch entweichende Kohlenwasserstoffe—mit Implikationen für Kohlenwasserstoffexploration: Zentralblatt für Geologie und Paläontologie, Teil I, v. 12, 2977–2994.

Machel, H. G., and E. A. Burton, 1992b, Comment on "Sediment magnetism: soil erosion, bushfires, or bacteria?": Geology, v. 20, p. 670–671.

Mann, S., N. H. C. Sparks, R. B. Frankel, D. A. Bazylinski, and H. W. Jannasch, 1990, Biomineralization of ferrimagnetic greigite (Fe_3S_4) and iron pyrite (FeS_2) in a magnetotactic bacteria: Nature, v. 343, p. 258–261.

McCabe, C., and R. D. Elmore, 1989, The occurrence and origin of late Paleozoic remagnetization in the sedimentary rocks of North America: Reviews of Geophysics, v. 27, p. 471–494.

McCabe, C., R. Sassen, and B. Saffer, 1987, Occurrence of secondary magnetite within biodegraded oil: Geology, v. 15, p. 7–10.

Morse, J. W., F. J. Millero, J. C. Cornell, and D. Rickard, 1987, The chemistry of the hydrogen sulfide and iron sulfide systems in natural waters: Earth Science Reviews, v. 24, p. 1–42.

Nordstrom, D. K., and J. L. Munoz, 1986, Geochemical thermodynamics: Palo Alto, California, Blackwell Scientific Publications, 477 p.

Plummer, L. N., Vacher, H. L., MacKenzie, F. T., Bricker, O. P. and Land, L. S., 1976, Hydrogeochemistry of Bermuda: A case history of ground–water diagenesis of biocalcarenites. GSA Bulletin, v. 87, p. 1301–1316.

Reynolds, R. L., N. S. Fishman, R. B. Wanty, and M. B. Goldhaber, 1990a, Iron sulfide minerals at Cement oil field, Oklahoma—implications for the magnetic detection of oil fields: GSA Bulletin, v. 102, p. 368–380.

Reynolds, R. L., M. Webring, V. S. Grauch, and M. L. Tuttle, 1990b, Magnetic forward models of Cement oil field, Oklahoma, based on rock magnetic, geochemical, and petrological constraints: Geophysics, v. 55, no. 3, p. 344–353.

Reynolds, R. L., N. S. Fishman, and M. R. Hudson, 1991, Sources of aeromagnetic anomalies over Cement oil field (Oklahoma), Simpson oil field (Alaska), and the Wyoming-Idaho-Utah thrust belt: Geophysics, v. 56, no. 5, p. 606–617.

Reynolds, R. L., M. B. Goldhaber, and M. L. Tuttle, 1993, Sulfidization and magnetization above hydrocarbon reservoirs, in D. M. Aïssaoui, D. F. McNeill, and N. F. Hurley, eds., Applications of paleomagnetism to sedimentary geology: SEPM Special Publication No. 49, p. 167–179.

Roberts, A. P., and G. M. Turner, 1993, Diagenetic formation of ferrimagnetic iron sulphide minerals in rapidly deposited marine sediments, South Island, New Zealand: Earth and Planetary Science Letters, v. 115, p. 257–273.

Saribudak, M., 1992, Significance of near-surface magnetic anomalies over Clarita prospect (Arkoma Basin, Oklahoma): Association of Petroleum Geochemical Explorationists Bulletin, v. 8, p. 43–54.

Sassen, R., C. McCabe, J. R. Kyle, and E. W. Chinn, 1989, Deposition of magnetic pyrrhotite during alteration of crude oil and reduction of sulfate: Organic Geochemistry, v. 14, p. 381–392.

Saunders, D. F., and S. A. Terry, 1985, Onshore exploration using the new geochemistry and geomorphology: Oil and Gas Journal, Sept. 16, p. 126–130.

Saunders, D. F., K. R. Burson, and C. K. Thompson, 1991, Observed relation of soil magnetic susceptibility and soil gas hydrocarbon analyses to subsurface hydrocarbon accumulations: AAPG Bulletin, v. 75, p. 389–408.

Sawlowicz, Z., 1993, Pyrite framboids and their development: a new conceptual mechanism: Geolgische Rundschau, v. 82, p. 148–156.

Sverjensky, D. A., 1984, Oil field brines as ore-forming solutions: Economic Geology, v. 79, p. 23–37.

Thompson, P., and T. D. F. Cameron, 1995, Paleomagnetic study of Cenozoic sediments in North Sea boreholes: an example of a magnetostratigraphic conundrum in a hydrocarbon-producing area, *in* P. Turner and A. Turner, eds., Paleomagnetic applications in hydrocarbon exploration and production: Geological Society of London Special Publication 98, p. 223–236.

Trudinger, P. A., L. A. Chambers, and J. W. Smith, 1985, Low-temperature sulphate reduction: biological versus abiological: Canadian Journal Earth Science, v. 22, p. 1910–1918.

Turner, P., and A. Turner, eds., 1995, Paleomagnetic applications in hydrocarbon exploration and production: Geological Society of London Special Publication 98, 301 p.

Vali, H., and J. L. Kirschvink, 1990, Observations of magnetosome organization, surface structure, and iron biomineralization of undescribed magnetic bacteria: evolutionary speculations, *in* R. B. Frankel and R. P. Blakemore, eds., Iron biominerals: New York, Plenum Press, p. 97–115.

Foote, R. S., 1996, Relationship of near-surface magnetic anomalies to oil- and gas-producing area, *in* D. Schumacher and M. A. Abrams, eds., Hydrocarbon migration and its near-surface expression: AAPG Memoir 66, p. 111–126.

Chapter 9

Relationship of Near-Surface Magnetic Anomalies to Oil- and Gas-Producing Areas

Robert S. Foote

Geoscience & Technology, Inc.
Irving, Texas, U.S.A.

Abstract

Observation of aeromagnetic patterns associated with the Cement oil field initiated a continuing program to investigate the association of shallow magnetic anomalies with the presence of oil and gas reservoirs. If local magnetic anomalies are detectable from an aircraft, the source of such anomalies will necessarily be present in the near-surface sedimentary strata as magnetic minerals sufficiently concentrated to allow airborne detection and identification by magnetic susceptibility measurement of drill cuttings. The results of analysis of cesium vapor magnetometer data are supported by exploration drill hole rock magnetic susceptibility data from four separated areas: (1) Caddo and Grady counties, Oklahoma; (2) Cheyenne County, Colorado; (3) San Juan County, Utah; and (4) southwestern Alabama. Results indicate a strong correlation of oil- and gas-producing areas with magnetic anomalies produced from the aeromagnetic data and with intervals of enriched magnetic rock strata in the shallow sedimentary environment. For the areas studied, 78–90% of oil and gas exploration drilling within the magnetic bright spot (MBS) anomalies is productive. Outside these anomalies, only 5–16% is productive. Analyses of minerals by X-ray diffraction and Mossbauer effect measurements have identified maghemite and greigite as the major iron-bearing minerals in the anomalous magnetic sedimentary intervals.

INTRODUCTION

The association between magnetic anomalies and petroleum accumulations has been observed for several decades (Steenland, 1965; Donovan et al., 1979). Although the precise relationship and processes remain uncertain, aeromagnetic surveys have been proposed as a cost-effective method for hydrocarbon exploration (Donovan et al., 1979; Foote, 1984, 1992; Saunders and Terry, 1985). Similar magnetic anomalies are reported detectable from the analysis of soils, sediments, and drill cuttings (Foote, 1987, 1992; Foote and Long, 1988; Henry, 1989; Reynolds et al., 1991; Saunders et al., 1991).

The mechanism whereby hydrocarbon microseepage can cause the formation of detectable magnetic anomalies in near-surface sediments and soils is not the subject of the study, but has been investigated by Elmore et al. (1989), Reynolds et al. (1990a,b, 1991, 1993), Machel and Burton (1991), Fruit and Elmore (1994), Machel (this volume), and Ellwood and Burkart (this volume). The objective of this study is to critically investigate the relation-

ship of aeromagnetic data, shallow sedimentary magnetic mineralization, and oil and gas fields.

Following observation of aeromagnetic patterns associated with the Cement oil field, Donovan et al. (1979) initiated a continuing investigative program on the association of local aeromagnetic anomalies with shallow magnetic minerals and oil and gas areas. The results of the Donovan et al. (1979) publication initiated a renewed interest in gathering improved higher sensitivity aeromagnetic data, improving data reduction techniques, and investigating the magnetic properties of altered drill cuttings from a large number of drill holes within and outside a variety of oil field areas. The studies described here began in 1984 to determine whether rocks in shallow sediments having higher concentrations of magnetic minerals can be correlated to the presence of oil and gas reservoirs. If such near-surface magnetic alterations are found and can be correlated with oil and gas fields, a valuable exploration method exists. This study examines four geologically diverse sites in Oklahoma, Colorado, Utah, and Alabama.

Table 1—Number of Exploration Test Wells Outside Cement Oil Field, Oklahoma

Status	In MBS	Not in MBS	In Cultural Contamination	Total
Oil and gas	39	17	21	77
Dry	9	109	37	155
Totals	48	126	58	232

Table 2—Wells and Dry Holes per MSRI Interval, Caddo and Grady Counties, Oklahoma

MSRI Interval	1	2	3	4	5	6	Total
In-field							
Oil and gas	2	8	10	15	11	3	49
Dry	0	1	1	0	0	0	2
Subtotal	2	9	11	15	11	3	51
Field-edge							
Oil and gas	1	5	2	4	2	4	18
Dry	3	0	0	2	1	0	6
Subtotal	4	5	2	6	3	4	24
Wildcat							
Oil and gas	0	1	0	1	0	0	2
Dry	13	2	0	0	1	2	18
Subtotal	13	3	0	1	1	2	20
Totals	19	17	13	22	15	9	95
% Oil & gas per interval	16%	82%	92%	91%	87%	78%	

AEROMAGNETIC AND MAGNETIC SUSCEPTIBILITY MEASUREMENTS

Aeromagnetic Data

Low terrain-clearance, high-sensitivity, cesium vapor aeromagnetic measurements of the earth's magnetic field give a composite of the effects on the earth's magnetic field from (1) magnetic basement rocks, (2) natural magnetic material in the sedimentary column between the basement and the surface, (3) cultural iron contamination on or near the surface, (4) volcanic and salt material, and (5) authigenic magnetic alterations in sedimentary rocks. The aircraft used to gather airborne data must not create false or noncorrectable variations in the total field as a result of yaw, pitch, or roll attitude changes during data gathering. The aircraft used must be magnetically "clean" or use magnetic compensation to correct for the effect of aircraft attitude changes on the measured magnetic field.

Data reduction techniques that remove the influence of magnetic basement rocks on the total magnetic field and, where possible, the effects of cultural iron contamination allow identification of *sedimentary residual magnetic* (SRM) anomalies as they exist along flight lines. When these SRM anomalies are positioned line-to-line above and below one another, they develop SRM anomaly clusters, the outline of which presents an area defined as a *magnetic bright spot* (MBS). Cultural contamination from oil field operations normally does not destroy the ability to define SRM anomalies where drilling density is 40 acres or greater. Oil and gas pipelines cause small to large oscillations in the total magnetic field created by the DC currents carried in the pipelines. In most cases, no data analysis is possible in close proximity to pipelines.

Magnetic Susceptibility

SRM anomalies positioned along a flight line profile exist over a short distance along the flight line and must be created by magnetic material in the near-surface environment, down to ~1000 m (~3000 ft) depth. Drilling and sampling of the drill cuttings allow measurement of rock magnetic properties as a function of surface depth. If drilling is made within the MBS, it is expected that the cause of any enclosed SRM anomalies will be identified by encountering rock intervals having substantially increased magnetic properties. Magnetic properties are measured as rock magnetic susceptibility with a sensitivity of about 10^{-7} (unitless) per gram of material using commercially available analytical instruments. *Magnetic susceptibility* the ratio of the intensity of magnetization produced in a substance to the intensity of the magnetic field applied. Cuttings have been obtained from various geologic sample libraries and from individual operators to provide the data presented.

CEMENT OIL FIELD AREA, OKLAHOMA

Cesium vapor aeromagnetic data were gathered over the Cement oil field, Caddo and Grady counties, Oklahoma, in 1984 at a surface elevation of ~100 m (300 ft). These data showed that the effect of oil field iron contamination on the measured magnetic field could not be removed accurately and that any airborne measurement of the earth's magnetic field over the Cement oil field would provide invalid SRM results. Areas outside the Cement field limits, however, do not have destructive cultural contamination, and data gathered give SRM results. Since SRM results are impossible to obtain over the Cement field, a study of the magnetic susceptibility (MS) of library rock drill cuttings is provided to show the association with oil wells and dry holes over the Cement field and beyond.

The combination SRM, MBS, and MS map in Figure 1 shows the location of the exploration drill holes on which magnetic susceptibility measurements are made and their degree of magnetic alteration plus the SRM, MBS, and MS anomalies outside the Cement oil field. Table 1 shows the number of oil and gas wells and dry holes in Caddo and Grady counties in the area of the MBSs, outside the MBSs, and in areas of cultural contamination for the 161 sq mi

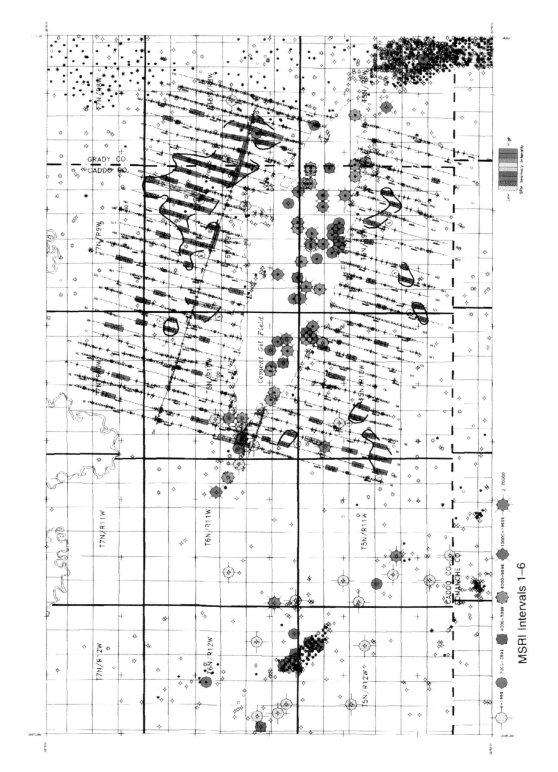

Figure 1—Map showing cesium vapor magnetometer sedimentary residual magnetic field intensity blocks, where magnetic field intensities (in nT [nanoteslas] × 1000) are as follows: gray, <40; brown, 40–<80; blue, 80–<120; green, 120–<160; yellow-green, 160–<200; yellow, 200–<240; orange, 240–<280; red, 280–<320; and violet, ≥320. The 69 wells, 26 dry holes, and 7 shallow core tests are classified according to MSRI intervals for the Caddo and Grady counties area, Oklahoma.

aeromagnetic survey, not including the Cement oil field area. The percentage of oil and gas wells in the MBS areas is 81.3%, with only 15.6% outside the MBSs.

The factor MSRI is defined as the product of the magnetic susceptibility (MS) alterations greater than MS background times the thickness of the altered rock interval (RI) $\times 10^7$, or

$$MSRI = (\text{Magnetic susceptibility}, 10^{-7}/\text{g})$$
$$\times (\text{Rock interval, ft}) \times 10^7$$

Six MSRI intervals are defined for the Caddo and Grady counties area (Figure 1) as follows: (1) 0–1999, gray; (2) 2000–3999, blue; (3) 4000–5999, green; (4) 6000–9999, orange; (5) 10,000–19,999, red; and (6) >20,000, violet. A total of 69 oil and gas wells, 26 dry holes, and 7 shallow (2500-ft) cores and tests were measured for their magnetic susceptibility in this area (Figure 1). The Cement oil field contains 65 of the 102 drill holes measured; cuttings were made available through the Oklahoma Geologic Survey and the Ardmore Sample Library.

Table 2 lists the number of exploration drill holes measured as a function of the MSRI intervals 1 through 6 calculated for each well and dry hole. Exploration hole cuttings measured are separated into in-field, field-edge, and wildcat categories in Table 2. Of the 95 exploration drill holes measured, 51 are in-field tests, 49 of which are oil and gas wells and 2 dry holes. There are 24 field-edge tests, 18 of which are oil and gas wells and 6 dry holes. The remaining 20 are wildcat tests, 2 of which are oil and gas wells and 18 dry holes. For MSRI interval 1, 16% of the wells are oil and gas producers, with 87% as oil and gas producers in intervals 2 through 6.

The first histogram (a) in Figures 2, 3, and 4 shows the number of oil and gas wells and dry holes for exploration in-field, field-edge, and wildcat drill holes as a function of MSRI intervals 1 through 6. For the Caddo and Grady counties area, the probability of oil and gas production for 76 exploration drill tests having MSRI values > 2000 is 86.8%. The probability of the 15 exploration tests for MSRI < 2000 producing a dry hole is 84%. The results of two MS measurements from the deep Anadarko gas and oil field north of the Cement field (SRM anomaly 9) show no MS alteration in cuttings from Sec. 14, T 6 N, R 9 W, while those from Sec. 29, T 6 N, R 9 W show a strong MSRI interval 5 alteration.

CHEYENNE COUNTY AREA, COLORADO

A cesium vapor, low terrain-clearance, aeromagnetic survey was flown in February 1986 which covered an area in Colorado from approximately T 41 W to T 50 W and from R 13 S to R 17 S and in Kansas from T 14 S to T 17 S and R 41 W to R 43 W. At the time of the survey, the eastern Cheyenne County Morrow Sand fields had not been discovered, thus there was no oil field iron to contaminate the aeromagnetic data. SRM analyses with drill

Table 3—Number of Exploration Tests in Cheyenne, Greeley, and Wallace Counties, Colorado

Status	In MBS	Not in MBS	In Cultural Contamination	Total
Oil and gas	212	51	133	396
Dry	0	0	77	501
Show	21	113	0	0
No show	50	240	0	0
Totals	283	404	210	897

Table 4—Number of Wells and Dry Holes per MSRI Interval in Cheyenne, Greeley, and Wallace Counties

MSRI	1	2	3	4	5	6	Total
In-field							
Oil and gas	0	6	17	11	9	9	52
Dry, show	0	1	0	0	0	0	1
Dry, no show	0	0	0	0	0	1	1
Subtotal	0	7	17	11	9	10	54
Field-edge							
Oil and gas	12	12	11	8	4	2	49
Dry, show	7	4	2	4	2	0	19
Dry, no show	14	6	2	2	5	2	31
Subtotal	33	22	15	14	11	4	99
Wildcat							
Oil and gas	0	1	2	0	0	1	4
Dry, show	33	12	3	3	1	2	54
Dry, no show	27	11	2	2	2	0	44
Subtotal	60	24	7	5	3	3	102
Totals	93	53	39	30	23	17	255
% Oil & gas per interval	13%	36%	77%	66%	57%	75%	

hole magnetic susceptibility studies for the Frontera and Second Wind fields have been previously reported (Foote, 1992). The results of the Cheyenne County, Colorado, and Greeley and Wallace counties, Kansas, survey area are given here, with the MBS results compared with oil and gas wells and dry holes for 897 exploration drill holes.

The aeromagnetic SRM data reduction has provided the many magnetic bright spots that lie in the surveyed region (Figures 5a, b, c). The total number of oil and gas wells and dry holes was determined within the MBSs, outside the MBSs, and within areas where cultural contamination has destroyed the ability to analyze the aeromagnetic data (Table 3). Of all the oil and gas wells drilled, 75% were within the MBSs and 12.6% were outside the MBSs.

A magnetic susceptibility cuttings evaluation was made for the extended region where 255 exploration drill hole cuttings sets were measured to provide the MSRI intervals. This evaluation extended into T 12 S and T 13 S, not covered by the aeromagnetic data. The results of MS

Text continues on p. 121

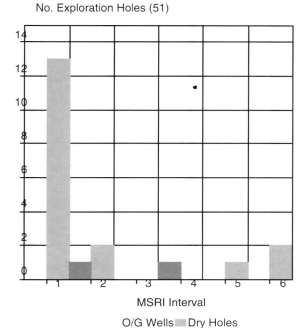

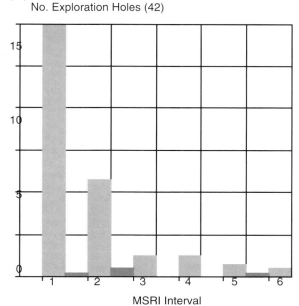

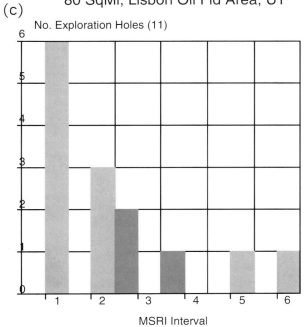

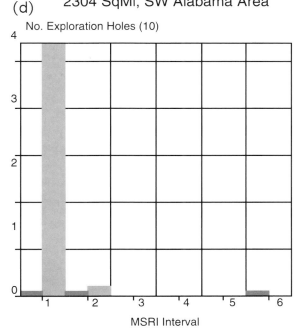

Figure 2—Histograms of the number of in-field exploration holes as a function of MSRI for the four areas covered in this study: (a) Caddo and Grady counties, Oklahoma; (b) Cheyenne County, Colorado; (c) Lisbon oil field area, San Juan County, Utah; (d) southwestern Alabama.

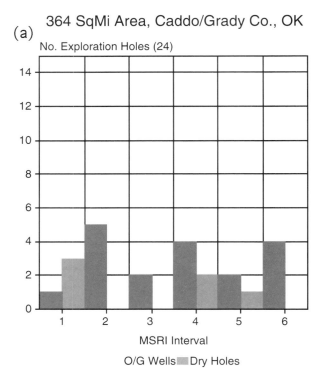

(a) 364 SqMi Area, Caddo/Grady Co., OK

No. Exploration Holes (24)

MSRI Interval

O/G Wells ▓ Dry Holes

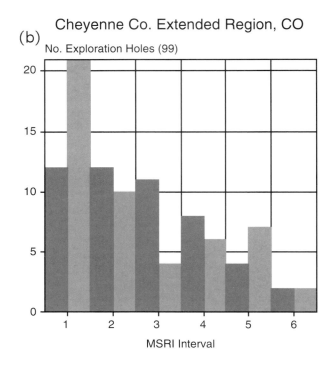

(b) Cheyenne Co. Extended Region, CO

No. Exploration Holes (99)

MSRI Interval

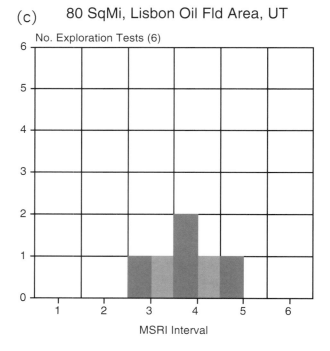

(c) 80 SqMi, Lisbon Oil Fld Area, UT

No. Exploration Tests (6)

MSRI Interval

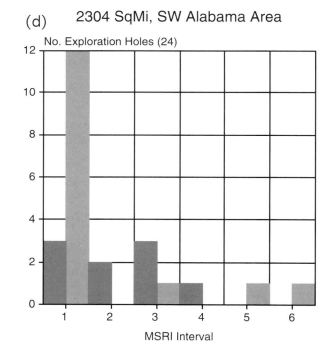

(d) 2304 SqMi, SW Alabama Area

No. Exploration Holes (24)

MSRI Interval

Figure 3—Histograms of the number of field-edge exploration holes as a function of MSRI for the four areas covered in this study: (a) Caddo and Grady counties, Oklahoma; (b) Cheyenne County, Colorado; (c) Lisbon oil field area, San Juan County, Utah; (d) southwestern Alabama.

(a) 364 SqMi Area, Caddo/Grady Co., OK

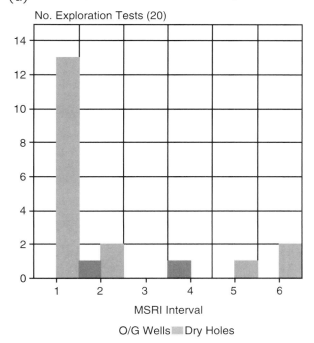

(b) Cheyenne Co. Extended Region, CO

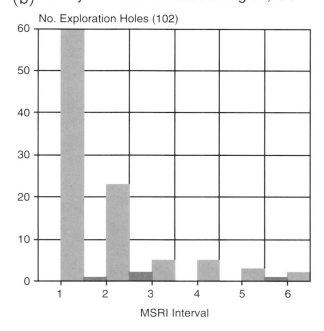

(c) 80 SqMi, Lisbon Oil Fld Area, UT

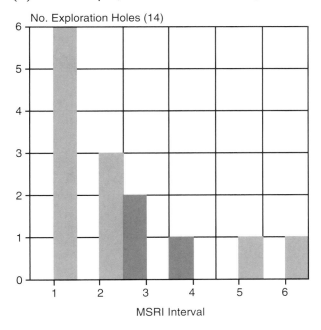

(d) 2304 SqMi, SW Alabama Area

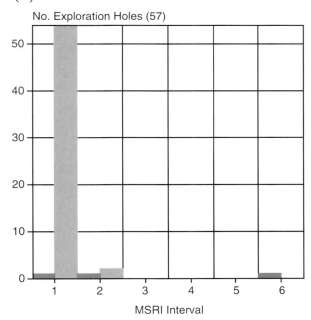

Figure 4—Histograms of the number of wildcat exploration holes as a function of MSRI for the four areas covered in this study: (a) Caddo and Grady counties, Oklahoma; (b) Cheyenne County, Colorado; (c) Lisbon oil field area, San Juan County, Utah; (d) southwestern Alabama.

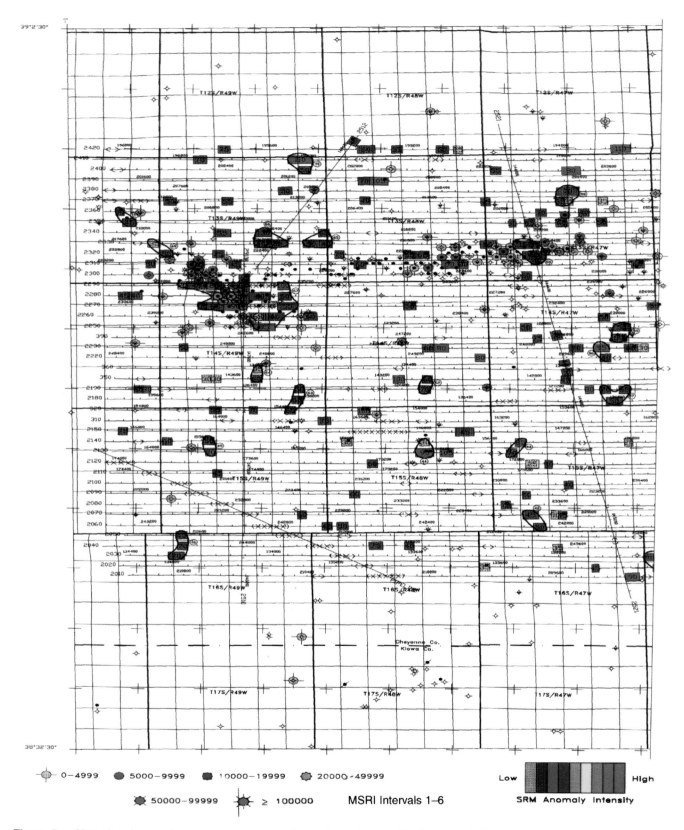

Figure 5a—Map showing cesium vapor magnetometer sedimentary residual magnetic field intensity values with wells and dry holes, which are classified according to MSRI intervals, for the western part of the three-part Cheyenne County area, Colorado. (See central and eastern parts in Figures 5b and 5c, which follow.)

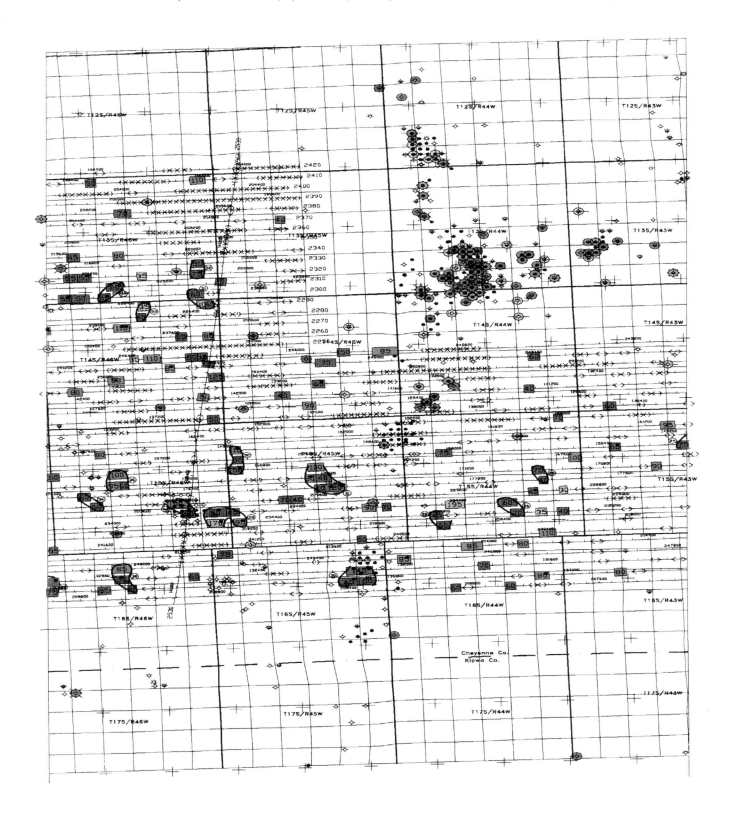

Figure 5b—Map showing cesium vapor magnetometer sedimentary residual magnetic field intensity values with wells and dry holes, which are classified according to MSRI intervals, for the central part of the three-part Cheyenne County area, Colorado. See Figure 5a for legends to symbols and colors.

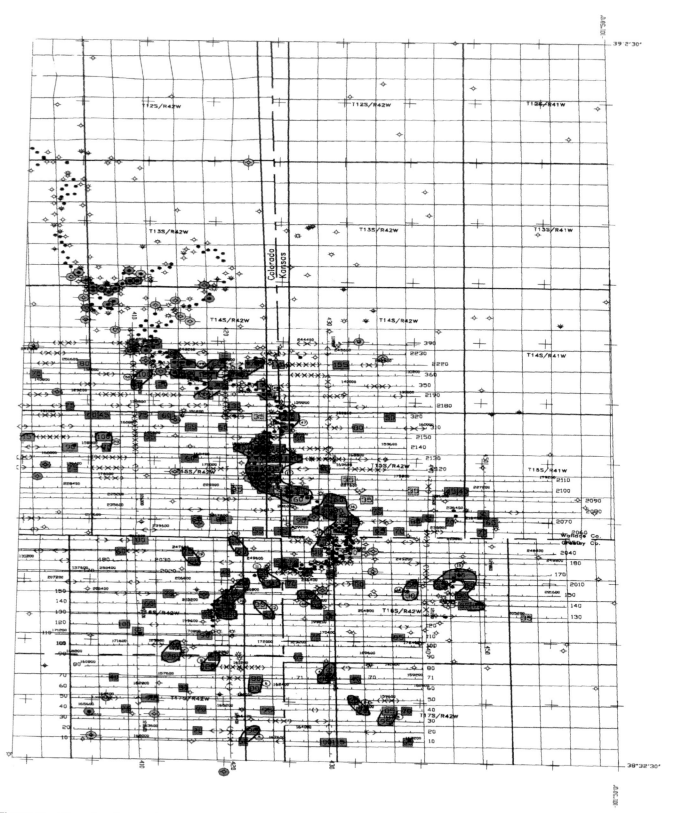

Figure 5c—Map showing cesium vapor magnetometer sedimentary residual magnetic field intensity values with wells and dry holes, which are classified according to MSRI intervals, for the eastern part of the three-part Cheyenne County area, Colorado. See Figure 5a for legends to symbols and colors.

Table 5—Number of Exploration Test Wells in San Juan County Area, Utah

Status	In MBS	Not in MBS	In Cultural Contamination	Total
Oil and gas	19	1	23	43
Dry	0	0	8	8
Show	2	9	0	11
No show	0	11	0	11
Totals	21	21	31	73

Table 6—Number of Wells and Dry Holes per MSRI Interval, San Juan County, Utah

MSRI	1	2	3	4	5	6	Total
In-field							
Oil and gas	0	0	4	2	2	3	11
Dry, show	0	0	0	0	0	0	0
Dry, no show	0	0	0	0	0	0	0
Subtotal	0	0	4	2	2	3	11
Field-edge							
Oil and gas	0	0	1	2	1	0	4
Dry, show	0	0	1	1	0	0	2
Dry, no show	0	0	0	0	0	0	0
Subtotal	0	0	2	3	1	0	6
Wildcat							
Oil and gas	0	0	2	1	0	0	3
Dry, show	4	1	0	0	1	1	7
Dry, no show	2	2	0	0	0	0	4
Subtotal	6	3	2	1	1	1	14
Totals	6	3	8	6	4	4	31
% Oil & gas per interval	0%	0%	88%	83%	75%	75%	

analyses of the samples from the U.S. Geological Survey in Denver and the Kansas Geological Survey in Wichita are tabulated in Table 4.

Magnetic alterations in the samples are more intense and extensive than those of Caddo County, Oklahoma, thus the MSRI intervals are larger. To distribute the data evenly, the six intervals were expanded for the Colorado and Kansas areas, as follows: (1) 1–4999, gray; (2) 5000–9999, blue; (3) 10,000–19,999, green; (4) 20,000–49,999, orange; (5) 50,000–99,999, red; and (6) >100,000, violet.

The results of Table 4 show a strong association with magnetic altered rocks for in-field exploration tests and a lack of magnetic rocks for the wildcat tests. The probability of oil and gas production for rock depths measured to ~600 m (2000 ft) for MSRI > 5000 within the area oil and gas fields (in-field) is 96%. Oil and gas wells and dry holes at the outer limits of fields are about equally likely to have small to large magnetic alterations. There are few wildcat oil and gas wells in the data set. Productive wildcat discoveries are normally followed by development wells and the area developed into a field, thus only the poorer wells are expected to remain. Table 4 shows the MSRI intervals as a function of show and no-show exploration dry holes. No strong correlation is seen for show and no-show dry holes and the MSRI intervals.

The second histogram (b) in Figures 2, 3, and 4 show the data of Table 4. Locations of exploration drill holes for which magnetic susceptibility was measured are shown in Figure 5. The SRM analyses are also shown in Figure 5, in which intensities of line anomalies are shown as gray, brown, blue, green, yellow-green, yellow, orange, red, and violet in order from low to high (0.20–3.2 nT, nanoteslas). Some original magnetic bright spots are not presented in the data of Figure 5 so as to retain the confidentiality of certain untested anomalies.

LISBON OIL FIELD AREA, UTAH

A large area was surveyed in 1988 using the cesium vapor magnetometer. An 80 sq mi region surrounding the Lisbon oil field is discussed here to show the relationship of oil and gas production with magnetic bright spot results of airborne magnetometer data analysis and with magnetic susceptibility measurement of exploration drill cuttings. Figure 6 presents all results from this area. This

80 sq mi area has 73 exploration drill tests, 43 of which are oil and gas wells and 30 dry holes. Table 5 shows the correlation of MBSs with drilled oil and gas wells and dry holes. The percentage of oil and gas is 90% inside the MBSs and 5% outside the MBSs, within the statistics of the number of exploration tests made. Cultural contamination destroyed the ability to evaluate 31 tests.

Magnetic susceptibility measurements have been made for 31 of the 73 exploration tests. Table 6 shows the number of tests for in-field, field-edge and wildcat wells as a function of MSRI interval; the results are also shown in the third histogram (c) in Figures 2, 3, and 4. Values for MSRI intervals 1 through 6 are the same as those used in the previous Cheyenne County, Colorado, area. MSRI intervals 1 and 2 together had 9 exploration drill holes, all of which were dry, giving a 0% correlation with oil and gas wells. MSRI intervals 3 through 6 together had 18 oil and gas wells, with 4 being dry holes, giving an 82% correlation with oil and gas wells.

SOUTHWESTERN ALABAMA AREA

An aeromagnetic cesium vapor survey and magnetic susceptibility study was done for areas in Escambia, Monroe, Conecuh, and Clarke counties, Alabama. Results of the correlation of oil and gas wells and dry holes within and outside magnetic bright spots are shown in Table 7. The percentage of oil and gas wells is 78% inside MBS anomalies and 5% outside the anomalies. Magnetic susceptibility studies have been done to depths of ~1200 m (4000 ft), which was necessary to penetrate through the

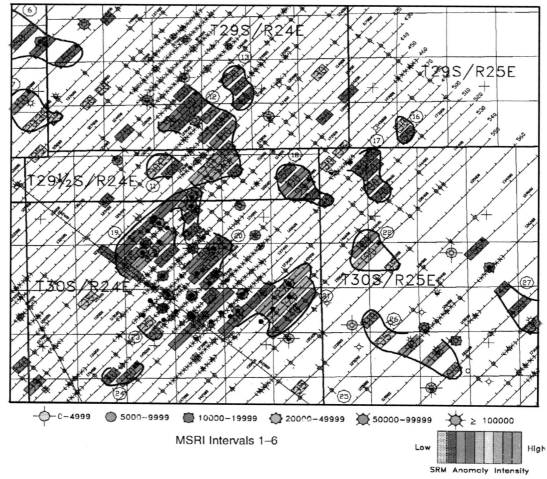

MSRI Intervals 1–6

Low ▓▓▓▓ High
SRM Anomaly Intensity

Figure 6—Map showing cesium vapor magnetometer sedimentary residual magnetic field intensities with 18 oil and gas wells and 13 dry holes, which are classified according to MSRI interval, for the Lisbon oil field area, San Juan County, Utah.

Selma Chalk. The majority of producing wells have shown strong magnetic alteration in the Selma Chalk, and when alterations do not exist, the probability of oil and gas production is low.

Table 8 presents the data in the final histogram (d) in Figures 2, 3, and 4 showing in-field, field-edge, and wildcat MSRI data. The MSRI intervals used for the areas in Alabama are (1) 1–99,999, (2) 100,000–199,999, (3) 200,000–299,999, (4) 300,000–399,999, (5) 400,000–499,999, and (6) ≥500,000.

MOSSBAUER EFFECT MEASUREMENTS

The *Mossbauer effect* has been used to identify iron minerals contained in drill cuttings found to have anomalous MS values (Foote and Long, 1988). The Mossbauer technique is superior to X-ray diffraction and Curie point methods originally used for mineral identification and has proven to be the only method capable of identifying magnetic minerals in small volumes and concentrations.

Figure 7 presents Mossbauer calibration standards for magnetite and maghemite at 295K and 78K (degrees Kelvin) (G. J. Long, personal communication, 1990).

Figure 8a shows the Mossbauer spectra obtained at 295K and 78K for a magnetic concentrate sample from a depth of 1035 ft in the Lisbon oil field. Analysis shows maghemite to be the magnetic mineral. Figure 8b shows the magnetic susceptibility intensity as a function of sample depth plot, the chi-log plot. Figure 9 gives examples of Mossbauer spectra showing maghemite to be the magnetic mineral in drill cuttings from one locality in central Kansas and two in eastern Colorado. Finally, Figure 10 shows the magnetic susceptibility chi-logs for a well in the Arnold SW oil field in Kansas, which has the most magnetically altered sample yet measured, and for the well from eastern Colorado shown in Figure 9c.

Mossbauer effect measurements of samples from the Cement oil field in Oklahoma (Figure 1) suggest the iron mineral to be an iron sulfide, Fe_xS_{x+1} (G. J. Long, personal communication, 1992; Reynolds et al., 1993), most likely greigite, Fe_3S_4. Figure 11 shows the Mossbauer effect

Table 7—Number of Exploration Test Wells, Escambia, Conecuh, Monroe, Clarke, and Choctaw Counties, Alabama

Status	In MBS	Not in MBS	In Cultural Contamination	Total
Oil and gas	52	15	67	134
Dry	15	297	52	364
Totals	67	312	119	498

Table 8—Number Wells and Dry Holes per MSRI Interval, Southwest Alabama

MSRI	1	2	3	4	5	6	Total
In-field							
Oil and gas	0	1	2	4	1	2	10
Dry	0	0	0	0	0	0	0
Subtotal	0	1	2	4	1	2	10
Field-edge							
Oil and gas	3	2	3	1	0	0	9
Dry	12	0	1	0	1	1	15
Subtotal	15	2	4	1	1	1	24
Wildcat							
Oil and gas	1	1	0	0	0	1	3
Dry	54	2	0	0	0	0	56
Subtotal	55	3	0	0	0	1	59
Totals	70	6	6	5	2	4	93
% Oil & gas per interval	4.3%	67%	83%	100%	50%	75%	

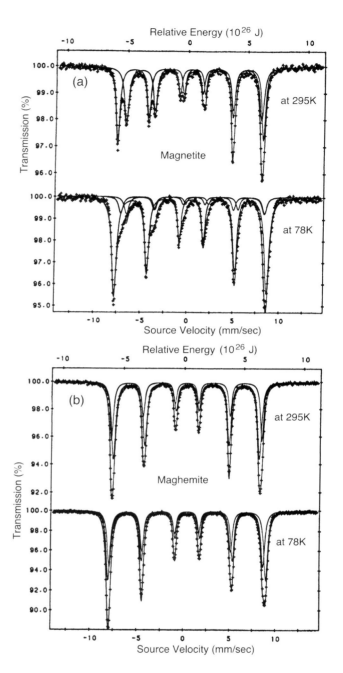

Figure 7—Mossbauer effect spectral standards for (a) magnetite at 295K and 78K and for (b) maghemite at the same temperatures.

measurements for two wells from the Cement oil field. Mossbauer measurements from the KWB oil field, Tom Green County, Texas, and the Lane-Chapel oil field, Smith County, Texas, have shown greigite to be the magnetic mineral. Figure 12 gives the chi-logs for the Cement oil field wells shown in Figure 11 and indicates the MS values for magnetic concentrates at specific depths.

The thicker altered magnetic intervals for the Alabama area cause the MSRI values there to be generally more intense than those from the Oklahoma, Colorado, and Utah regions. Mossbauer effect measurements in Figure 13 are from a sample from the top of the >300-m (1000-ft) thick Selma Chalk interval in the Vocation oil field, Monroe County, Alabama. The chi-log identifies maghemite to be the magnetic mineral.

CONCLUSIONS

The magnetic bright spot (MBS) method is shown to provide a high correlation between MBSs and the presence of oil and gas at depth in sedimentary basins and regions removed from exposed igneous mountains. Identification of altered magnetic rock strata can be made by MSRI rock cuttings measurements. Table 9 pre-

sents a summary of the MBS and MSRI measurements for the four areas covered in this study. The results of percentage of oil and gas correlations with MBSs and with MSRI intervals 1 and 2 through 6 combined are remarkably similar. For the areas studied, 78–90% of the wells within MBS areas were productive oil and gas wells, while outside the MBS areas only 5–16% of the wells were productive.

The use of the SRM method to develop MBSs can elim-

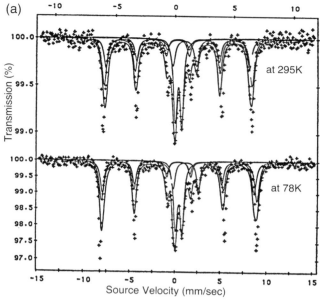

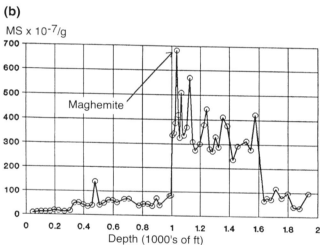

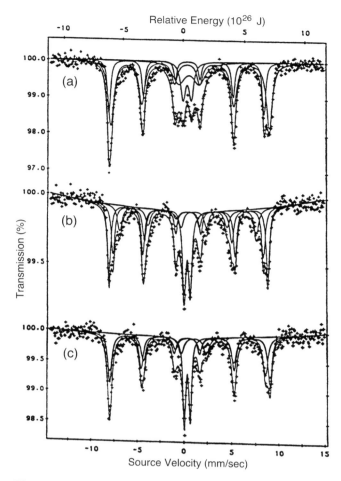

Figure 9—Mossbauer effect spectra at 78K for samples from (a) the Huber #1 Wildgren well, Arnold SW oil field, Kansas (oil); (b) the Pintail #1 Chamblin-Cobb well, Lisbon oil field, Colorado (oil); and (c) the Mull Drilling #1 Farms "A" well, eastern Colorado (dry hole, gas show). See Figure 10 for chi-logs of parts (a) and (c).

Figure 8—(a) Mossbauer effect spectra at 295K and 78K for a magnetic concentrate from 1035 ft depth in the Pure Oil #1 well in the Lisbon oil field, Utah. (b) A chi-log plot showing the magnetic susceptibility intensity as a function of sample depth.

inate less prospective and nonprospective areas that may represent up to 95% of any survey area. Seismic costs can be greatly reduced by concentrating on the remaining 5% of the area.

REFERENCES CITED

Donovan, T. J., R. J. Forgey, and A. A. Roberts, 1979, Aeromagnetic detection of diagenetic magnetite over oil fields: AAPG Bulletin, v. 63, p. 245–248.

Elmore, R. D., R. McCollum, and M. H. Engel, 1989, Evidence for a relationship between hydrocarbon migration and diagenetic magnetic minerals—implications for petroleum exploration: Association of Petroleum Geochemical Explorationists Bulletin, v. 5, p. 1–7.

Foote, R. S., 1984, Significance of near-surface magnetic anomalies, *in* M. J. Davidson and B. M. Gottlieb, eds., Unconventional methods in exploration for petroleum and natural gas: Institute for Study of Earth and Man, Southern Methodist University, Dallas, Texas, p. 12–24.

Foote, R. S., 1987, Correlations of borehole rock magnetic properties with oil and gas producing areas, Association of Petroleum Geochemical Explorationists Bulletin, v. 3, no. 1, p. 114–134.

(a)

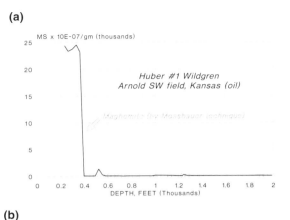

(b)

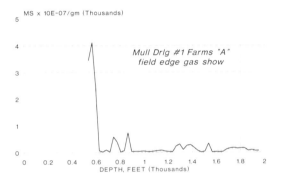

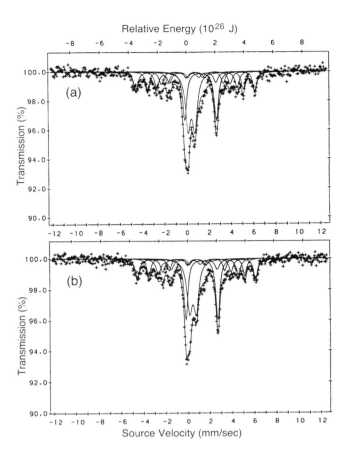

Figure 10—Chi-logs for (a) the Huber #1 Wildgren well, Arnold SW oil field, Kansas (oil); and (b) the Mull Drilling #1 Farms "A" field-edge gas show (dry hole), eastern Colorado.

Figure 11—Mossbauer effect spectra for drill cuttings from the Cement oil field, Oklahoma: (a) the Mobil Oil #19 McKenna well; (b) the Midstates #2 Gray well. See Figure 12 for the chi-logs of these samples.

Foote, R. S., 1992, Use of magnetic field aids oil search: Oil & Gas Journal, v. 90, no. 18, p. 137–142.

Foote, R. S., and G. J. Long, 1988,Correlations of oil and gas producing areas with magnetic properties of the upper rock column, eastern Colorado: Association of Petroleum Geochemical Explorationists Bulletin, v. 4, p. 47–61.

Fruit, D. J., and R. D. Elmore, 1994, Effects of hydrocarbon seepage in a frontier exploration region (abs.), *in* Near-surface expressions of hydrocarbon migration: AAPG Hedberg Research Conference Abstracts, April 24–28, Vancouver, British Columbia.

Henry, W. E., 1989, Magnetic detection of hydrocarbon seepage in a frontier exploration region: Association of Petroleum Geochemical Explorationists Bulletin, v. 5, p. 18–29.

Machel, H. G., and E. A. Burton, 1991, Chemical and microbial processes causing anomalous magnetization in environments affected by hydrocarbon seepage: Geophysics, v. 56, p. 598–605.

Reynolds, R. L., N. S. Fishman, R. B. Wanty, and M. B. Goldhaber, 1990a, Iron sulfide minerals at Cement oil field, Oklahoma—implications for the magnetic detection of oil fields: GSA Bulletin, v. 102, p. 368–380.

Reynolds, R. L., M. Webring, V. S. Grouch, and M. Tuttle, 1990b, Magnetic forward models of Cement oil field, Oklahoma, based on rock magnetic, geochemical, and petrological constraints: Geophysics, v. 55, p. 344–353.

Reynolds, R. L., N. S. Fishman, and M. R. Hudson, 1991, Sources of aeromagnetic anomalies over Cement oil field (Oklahoma), Simpson oil field (Alaska), and the Wyoming-Idaho-Utah thrust belt: Geophysics, v. 56, p. 606–617.

Reynolds, R. L., M. . Goldhaber, and M. L. Tuttle, 1993, Sulfidization and magnetization above hydrocarbon reservoirs, *in* D. M. Aissaouri, D. M. McNeil, and N. F. Hurley, eds., Applications of paleomagnetism to sedimentary geology: SEPM Special Publication 49, p. 167–179.

Saunders, D. F., and S. A. Terry, 1985, Onshore exploration using the new geochemistry and geomorphology: Oil & Gas Journal, Sept. 16, p. 126–130.

Saunders, D. F., K. R. Burson, and C. K. Thompson, 1991, Observed relation of soil susceptibility and soil gas hydrocarbon analyses to surface hydrocarbon accumulations: AAPG Bulletin, v. 75, p. 389–408.

Steenland, N. C., 1965, Oil fields and aeromagnetic anomalies: Geophysics, v. 30, p. 706–739.

(a)

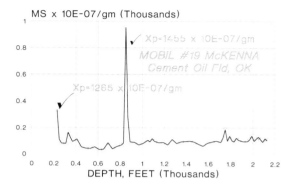

(b)

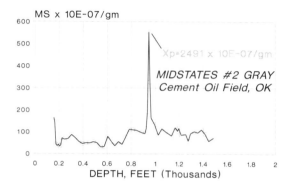

Figure 12—Chi-logs for drill cuttings from the Cement oil field, Oklahoma: (a) the Mobil Oil #19 McKenna well; (b) the Midstates #2 Gray well.

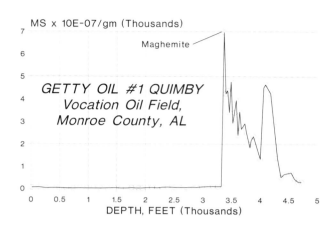

Figure 13—Chi-logs for drill cuttings from the Getty Oil #1 Quimby well, Vocation oil field, Monroe County, Alabama. Mossbauer effect measurements show maghemite as the magnetic mineral.

Table 9—Results of SRM and MSRI Over Four Widely Separated Areas

Area	SRM Results (% Oil and Gas)		MSRI Results (% Oil and Gas)	
	In MBS Anomaly	Outside MBS Anomaly	Interval	% in Interval
Caddo/Grady Co., Oklahoma	81.3%	15.6%	1	16%
			2–6	87%
Cheyenne Co., Colorado/Kansas	75%	12.6%	1 and 2	21%
			3–6	68%
San Juan Co., Utah	90%	5%	1 and 2	0%
			3–6	82%
SW Alabama, 2304 sq. mi. area	78%	5%	1	4.3%
			2–6	78%

Carlson, N., and K. L. Zonge, 1996, Induced polarization effects associated with
hydrocarbon accumulations: minimization and evaluation of cultural influences, *in* D. Schumacher and M. A. Abrams, eds., Hydrocarbon migration and
its near-surface expression: AAPG Memoir 66, p. 127–137.

Chapter 10

Induced Polarization Effects Associated With Hydrocarbon Accumulations: Minimization and Evaluation of Cultural Influences

Norman R. Carlson

Kenneth L. Zonge

Zonge Engineering and Research Organization, Inc.
Tucson, Arizona, U.S.A.

Abstract

The use of induced polarization (IP) methods in oil and gas exploration dates back to the 1930s, but the validity of anomalies has been difficult to establish. Although recent geochemical and downhole research has verified the source of IP anomalies in some geologic environments, the influence of cultural (anthropogenic) features
on the electrical data remains a serious stumbling block to the acceptance of electrical methods in oil exploration. Spurious effects from power lines, pipelines, fences, and well casings can be misinterpreted as anomalies
from hydrocarbon alteration or can mask true alteration anomalies.

The cultural problem is not insurmountable, however, and it is not valid to assume automatically that all IP
anomalies measured over oil fields are the result of culture. A case study of the development of an oil field near
Post, Texas, illustrates how proper survey design can be used to minimize and evaluate the effects of culture in
the interpretation of IP survey data. Evaluation of before-and-after IP data sets and two-dimensional finite
element modeling strongly support the interpretation that the observed IP anomaly results from hydrocarbon-
induced alteration and not from well casing or other cultural effects. Furthermore, the interpreted extent of the
IP anomaly as defined in 1982 agrees well with the productive limits of the field as it exists more than 12 years
later.

INTRODUCTION

In recent years, the sediments above some hydrocarbon reservoirs have been shown to be altered by
microseepage from the reservoirs (e.g., Donovan, 1974;
Ferguson, 1977; Donovan et al., 1981; Schumacher, this
volume). Some of the alterations in the sediments are
potential targets for surface-based electrical exploration
techniques, since these alterations can result in a change
in the ground resistivity, in the chargeability (induced
polarization or IP), and sometimes in both of these characteristics. Correlation of seepage-induced alteration with
electrical anomalies has been established in some environments (Oehler and Sternberg, 1984; Sternberg, 1991).
However, many case histories of electrical surveys over
known oil fields remain unconvincing when not corroborated by extensive core hole analysis from drill holes both
on and off the field. The electrical methods themselves
(measurements in various forms of ground resistivity and
IP effects) are well-established techniques and have been
used in mineral exploration for decades (e.g., Sumner,
1976; Telford et al., 1976). Their use over known oil fields,
however, is often compromised by the presence of electrically conductive cultural features such as fences, well casings, power lines, and pipelines.

Our discussion specifically addresses the cultural
problem and how it can be minimized and evaluated
when using electrical methods in oil and gas exploration.
If the cultural problems can be overcome, the use of electrical methods is attractive in several ways. An electrical
crew is typically made up of only three or four people and
is thus relatively low in cost, low in environmental impact
(no drilling or blasting and no large vehicles), and the
depth of investigation is variable, thus the target geochemical alterations need not extend to the surface to be
detectable.

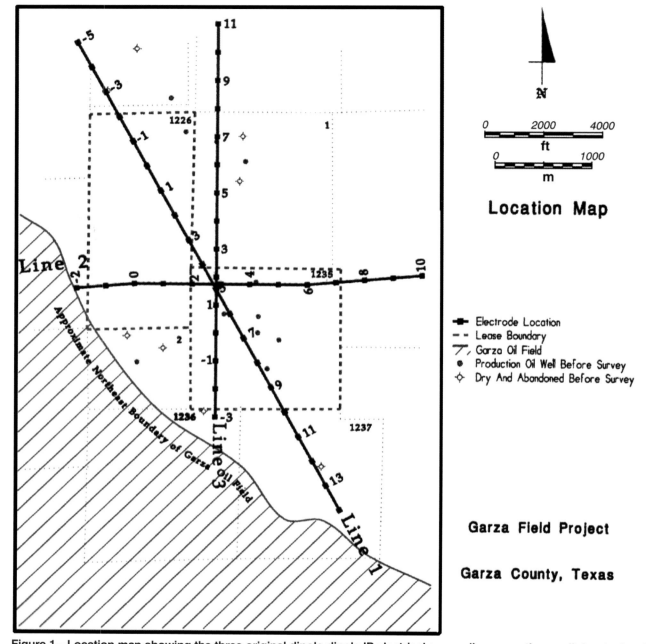

Figure 1—Location map showing the three original dipole-dipole IP electrical survey lines over the small developing field near Post, Texas (dipole length is 300 m, or 1000 ft).

BACKGROUND AND LOGISTICS

Background

The electrical survey performed in 1982 near Post, Texas, had two specific goals. Six production wells had been recently drilled in Lease Block 1235 (Figure 1), and leases in Block 1226 to the northwest were becoming available. Since oil production was only 10–50 bbl oil per day per well in Block 1235, it was important for the operator to minimize the number of dry wells drilled while

developing and extending the field. Within strict budget constraints, the goals of this project were

1. To determine whether production was likely in the leases becoming available in Block 1226 and, if so, how far to the northwest production extended
2. To determine the approximate extent of the oil field within Block 1235 since the low production rate of each well made minimizing the number of dry wells an important economic concern (a 10-acre well density was permitted)

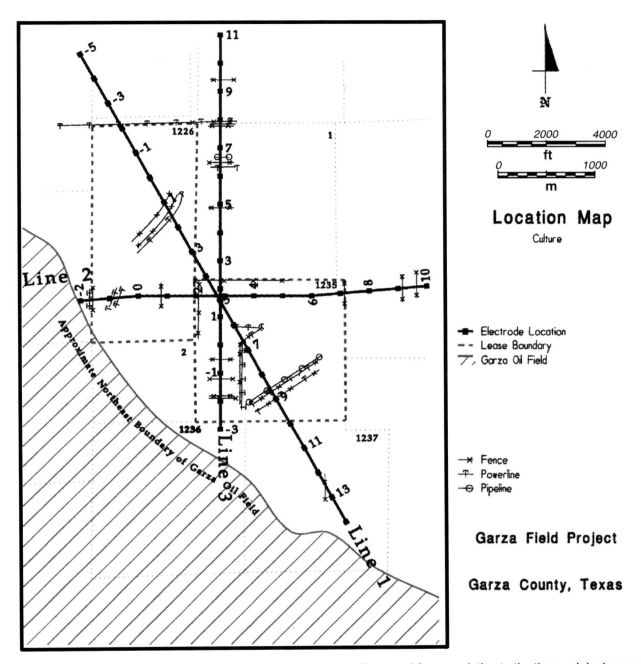

Figure 2—Culture map showing the location of pipelines, power lines, and fences relative to the three original survey lines (dipole length is 300 m, or 1000 ft).

Production in the Garza field itself (south and west of the lease blocks under study) is from the San Andres and Glorieta formations of the Lower Permian. The traps result from a loss of porosity on and flanking an anticline, and a permeability pinchout occurs to the east (Hild, 1986). There are several separate, thin producing zones in the lower two-thirds of the San Andres, stacked vertically and offset successively to the southwest. Production in the eastern part of the field is therefore stratigraphically deeper than in the west (Ward et al., 1986). The field pro-

duces 36–38° API oil from dolomites at depths of 887–994 m (2910–3260 ft) (Myres, 1977). Although there is no information on percentages of H_2S, most oils produced from the San Andres are described as borderline between sour and sweet. Production in Block 1235 and to the north in Block 1 is assumed to be physically separate from the main Garza field to the west and south on the basis of drilling results. Overlying the San Andres is the upper Guadalupian Yates Formation, composed primarily of sandstone and shale.

Figure 3—Induced polarization (IP) data in pseudo-section form for the three survey lines. All values are in milliradians; dipole length on all lines is 300 m (1000 ft).

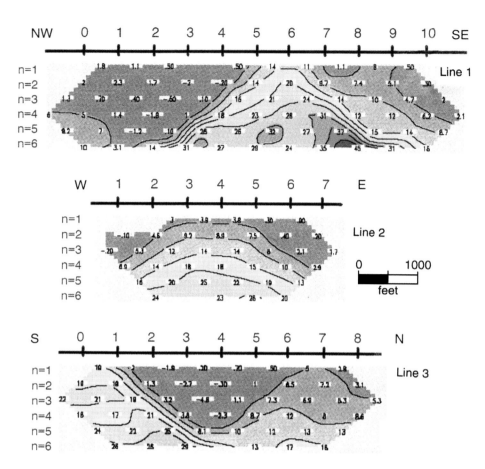

The three survey lines that were run intersected in the northwest quarter of Block 1235; the layout of the lines was chosen on the basis of culture and the need to evaluate Block 1226. Figure 2 shows the cultural features present at the time of the survey. Line 1 extended from past a dry well southeast of production, passed the producing wells, and proceeded northwest across Block 1226. The starting point for line 1 was chosen on the basis of the pipeline and power line combination along a road at about station 8.5. The power line was made up of two three-phase circuits capable of at least 20 kV. The combination of the power line and pipeline was expected to be the strongest cultural influence; thus, the field crew placed the line such that this culture would be in the middle of a dipole rather than close to an electrode position. The power line located along the northern boundary of Block 1226, however, turned out to be the strongest active noise source. Line 2 crossed the production west to east, providing a different orientation to the culture.

Line 3 crossed the production south to north and extended past a separate production field in the northern half of Block 1. The placement of line 3 was chosen to maximize the distance of the line from the two north-south fence sections that can be seen paralleling the line from approximately station 0 to 2 on the west and from station –2 to 0 on the east. Both fences were made of barbed wire strung on mostly wooden posts, with some

metal stakes. This placement put the line more than 150 m (500 ft, or one-half the dipole size) from the parallel culture. Budget constraints did not allow additional lines or extensions of the original lines.

Survey Logistics

The data were gathered using the dipole-dipole array, a common array in mineral exploration surveys. A square wave signal at 0.125-Hz was transmitted into a grounded dipole. The resulting magnitude (in millivolts) and phase shift (in milliradians) of the received signal was measured in a similar sized receiver dipole, collinear with the transmitter dipole. The dipoles were each 300 m (1000 ft) long, providing a depth of investigation of about 600 m (2000 ft) when the transmitter and receiver dipoles were six dipole lengths apart (n-spacing = 6). Matched quartz oscillators in the receiver and transmitter controller were synchronized and locked each morning prior to the field work to achieve absolute phase measurements. Data were stacked and averaged in the field until acceptable standard errors were achieved and were then stored on magnetic tape. Each stack consisted of at least 16 cycles, and a minimum of two stacks were made at each position to establish repeatability of the data. If the two stacks differed by 5% or more in resistivity, additional longer stacks were recorded until data blocks were repeatable within 5%. Transmitted currents ranged from 12 to 18 amps, depending on local contact resistance. One channel of the receiver monitored the transmitted wave form via a hard wire link between the transmitter and receiver.

The acquired data include calculated apparent resistivity (in ohm-meters) from the measured received magnitude, measured raw phase shift (in milliradians), and calculated three-point decoupled IP (in milliradians). The IP data (Figure 3) are displayed in standard pseudo-cross-section format, with stations along the top of each plot and increasing n-spacing or separation down the side of the plot, corresponding to increasing depth. Contours are in milliradians, and "warm" colors (red and orange) indicate high IP values, while "cool" colors (green and blue) indicate low IP values in these plots.

It is important to note that the depth of production in this field is substantially below the maximum depth of investigation of the survey. The intent of the survey was not direct detection of the oils but rather detection of alteration of sediments above the oils.

As noted, this discussion is concerned primarily with the IP data since this is the electrical property that showed the largest anomaly associated with the oil field. In the original interpretation and modeling in 1982, all electrical properties were considered.

Preliminary Results

Upon completion of line 1, a definite IP anomaly was evident; at that stage, however, it was not possible to state definitively whether the anomalous values were the result of alteration, culture, a combination of the two, or some unrelated geologic source. Lines 2 and 3 were then run at orientations designed to evaluate cultural influences and verify the correlation between the anomalies and the oil fields (regardless of the cause). All three lines of data showed a clear increase in IP values in the area of the six production wells (Figure 3). This was similar to IP anomalies evident over some other oil and gas fields we have studied, as well as some uncontaminated prospects which eventually proved to be productive. Line 3 also showed a weaker but definite IP anomaly associated with the production in the northern half of Block 1, with near zero background levels between the two production areas. Although the survey had been designed to minimize cultural effects, we felt it was necessary to further evaluate the effects to interpret the data set. As seen in these lines alone, the IP anomalies could be interpreted as the result of the surface culture (power lines, pipelines, well casings, and fences), alteration of sediments above the oil field, or a combination of both.

Note that there are small negative values seen adjacent to the IP anomalies. This is not unusual because a decrease in IP is often seen adjacent to polarizable bodies both in field measurements and in modeling of some horizontally layered environments in which the deepest layer is substantially more conductive than the overlying layers. Negative IP effects are also associated with culture

(Sumner, 1976). In this case, background IP values are near zero, and the slight decrease results in small negative values in the pseudo-section.

EVALUATION OF CULTURAL EFFECTS

Position Relative to Electrodes

The effects of grounded surface culture (such as power lines, pipelines, fences, and well casings) on electrical data are strongly dependent on the location of the culture relative to the survey line electrodes. Survey crews plan lines on this basis; spurious effects are greatly minimized when electrodes are placed symmetrically with respect to the culture. For example, if a power line crosses a survey line, the power line should cross in the middle of a dipole rather than near an electrode.

This strong variation in cultural effects with respect to location relative to electrodes is well documented in the literature by field data and in mathematical modeling (e.g., Nelson, 1977). An illustration of this dependence is shown in Figure 4, which shows the computer-generated model response of a small polarizable body placed at the surface midway between electrodes –1 and 0 (Figure 4a) and near electrode 0 (Figure 4b). These results are consistent with field experience. When the polarizable body is placed in the middle of the dipole, the resulting anomaly is weaker and more symmetrical than when the body is near an electrode. Note that in this simple example, the difference between placing the anomalous body midway between electrodes and near an electrode is almost 25 mrad in some parts of the anomaly. The anomaly that results from this surface polarizable body is shaped like an inverted V, called a "pants-leg" effect, which is also consistent with other models and field experience. The strongest IP effect extends diagonally downward in both directions from the source of the anomaly.

The size of the polarizable body used in the model for Figure 4 was 0.1 by 0.2 dipole lengths, which is much larger than normal cultural features such as pipelines and power lines. The size discrepancy is due to the limitations of the two-dimensional finite-element modeling program used to generate this illustration.

We should also note that based on our field experience, the strength of cultural anomalies is also dependent on grounding of the culture and on background resistivity of the ground itself. In high-resistivity environments, cultural effects are stronger than in low-resistivity environments. When culture is poorly grounded or background resistivities are low, we often see very weak or no cultural effects at all. Ground resistivities at this project site were low, ranging from 4 ohm-m in the shallow $n = 1$ data up to 20 ohm-m in the deeper $n = 6$ data.

It is possible to use this dependence of cultural effects on location (relative to electrodes) as a method of evaluating those cultural effects. This is done by repeating the suspect portion of the survey line with electrodes shifted along the line. Anomalous values that result from surface

Figure 4—2DIP IP modeling results for (a) a small surface polarizable body located between electrodes –1 and 0 and (b) the same body located at electrode 0. Note the strong difference between the anomaly created by the feature in the center of the dipole versus the anomaly created when the feature is at or near an electrode.

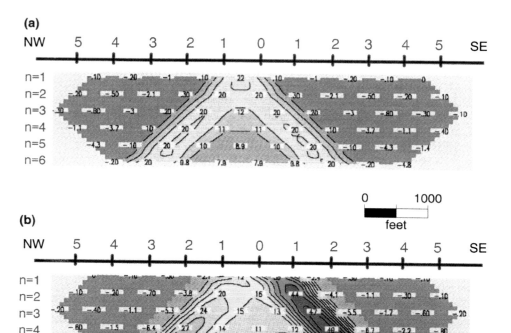

features or culture should show strong changes after shifting the electrode positions, similar to those seen in Figure 4. Deeper, larger features should show relatively little change with the shift in electrodes.

For this project, the portion of line 1 centered approximately at station 6 was repeated since this area shows the strongest IP values and line 1 data are critical in evaluating Lease Block 1226. After completing line 1 with electrodes located at integral station numbers (0, 1, 2, etc.), transmitting and receiving electrodes were then shifted 150 m (500 ft) along line. Thus, electrodes were then located at stations 0.5, 1.5, 2.5, etc. This shift is one-half the dipole size, thus a surface feature equidistant from electrodes in the original layout would be at an electrode in the shifted layout. Surface features causing an anomaly in the original layout should be minimized in the shifted layout, and features causing no anomaly in the original layout should show strong anomalies in the shifted layout (assuming the features are causing any anomalies at all). In Figure 5, the original IP data (Figure 5a) are compared with the IP data after shifting the electrodes (Figure 5b). Cultural features are plotted across the top of the line for reference.

Both the overall appearance and the strength of the anomaly are similar in the two data sets. The only significant change is along the diagonal of data extending downward to the left from dipole 8.5–9.5, where IP values have decreased relative to the original data set. This decrease is interpreted to be a cultural effect from the pipeline and power line combination located approximately at station 8.6. In the original data set, these cultural features were nearly centered between electrodes 8.0 and 9.0 to minimize their effect; after shifting electrodes,

the culture is close to the electrode at 8.5, maximizing the cultural effects. In this case, the cultural effect of the pipeline and power line appears to be a negative IP effect, reducing IP values along the affected diagonal.

Also note that a power line is located approximately at station 6.1, which is near the electrode in the original survey but midway between electrodes in the repeat survey. If this power line had caused a substantial portion of the anomaly in the original survey, we would expect a significant change between the two data sets in this area. There is little change in the anomaly after shifting the electrodes, particularly when compared to the change seen in the model results in Figure 4 discussed previously.

The absence of any significant change in either anomaly strength or anomaly appearance (except along the one diagonal extending from the pipeline and power line at station 8.6) strongly suggests that the anomalously high IP values seen in the original data set are not the result of surface or cultural features. The low-resistivity background (4–20 ohm-m) is probably the primary reason for the apparently weak cultural effects on the data. We frequently encounter pipelines, fences, and power lines that had little or no effect on the data, although in some other environments, these features make data interpretation impossible.

Before-and-After Comparisons

Well casing effects are a possible source of anomalous IP values and must be considered in the interpretation of electrical data over a producing field. While casing effects do not always appear similar to surface culture, they are

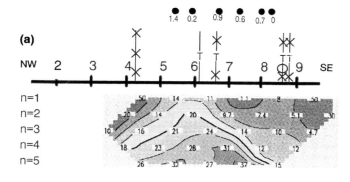

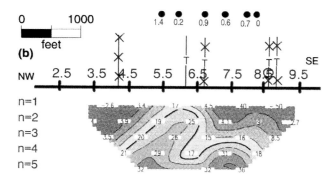

Figure 5—(a) IP data from the original line 1 measurements versus (b) measurements made after electrodes were shifted 150 m (500 ft) along line. Culture is shown along the top of the pseudo-section, with distances of the well casings off line (in dipole lengths). Note the overall similarity of the two anomalies, except along the diagonal extending downward to the left from the culture near electrode 8.5 in part (b). (See Figure 2 for key to culture symbols.)

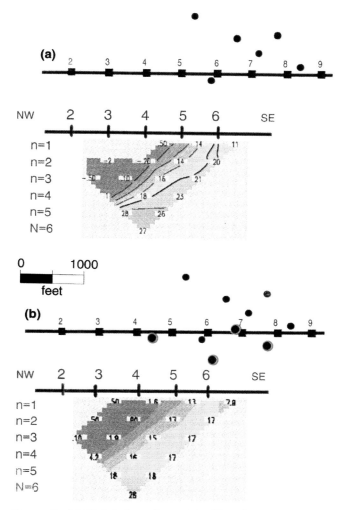

Figure 6—(a) IP data from the original line 1 measurements versus (b) repeated measurements after four new wells were drilled, cased, and in operation. There is a slight decrease in IP values, but otherwise little effect from the new wells, one of which is close to the electrode at station 7.

dependent on location and background resistivity, similar to surface culture. In this project, a total of six well casings were in place when the original three survey lines were acquired. Shortly after the survey lines were completed, four additional wells were drilled and casing was set. By returning to the site and repeating a portion of line 1 after the new wells were in place, a before-and-after comparison could be made on the effects of the four new wells.

Figure 6a shows part of the original IP data from line 1 with a plan view diagram of the oil wells relative to the electrodes. Figure 6b shows the part of line 1 that was repeated and a plan view with the location of the new wells added. This repeat of the line 1 data was done after the new oil wells had been producing for about 30 days, allowing the casings, cement, and pipelines to "settle" electrically.

The addition of the four new well casings had little effect on the data, suggesting that well casings are not a strong contributor to the IP anomaly in this environment. Based on proximity to the line and to electrodes, we might expect that the new well casings would have an even stronger effect on the data than the original casings,

but the shape of the anomaly is unchanged, and only one diagonal of data appears to be altered slightly by the new casings. In particular, note the new well casing on line and close to the electrode at station 7 and that two of the remaining three new wells are within 0.3 dipole lengths of the line. If well casings were affecting the data, these new casings should have been obvious in the data when the line was repeated. On the basis of this comparison, well casings are interpreted as having minimal effects on the data. Also, if the casings were affecting the data, it could be a negative IP effect, reducing the IP values, similar to the effect seen from the pipeline and power line combination discussed earlier.

In addition to the comparison of field data gathered before and after wells were drilled, the effect of well casings was also evaluated using a mathematical program

Figure 7—(a) IP data from the original line 1 measurements versus (b) IP results from the modeling program PIPE, with the complex impedance of the casings varied to force a best fit to the field data.

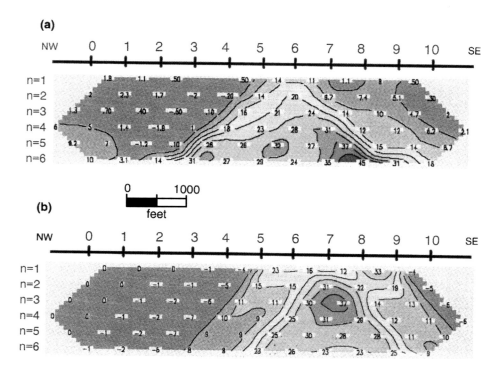

called PIPE written by Holladay and West (1984). The program models the effects of multiple well casings given information about the casing location, inner and outer diameters, longitudinal conductance, wave number, complex impedance, and background resistivity. Using the locations and casing diameters of the six wells in place at the time of the survey, plus conductance figures obtained from U.S. Steel, we varied the complex impedance in an attempt to achieve the best possible fit between the well casing model and the field data.

Figure 7a shows the IP data from line 1, and Figure 7b shows the closest fit obtained using the PIPE program. The strength of the PIPE anomaly is similar to the field data anomaly, although the PIPE anomaly is offset to the south and is shallower than the field data anomaly. The agreement between the field data and the PIPE results, however, could be considered a moderately good fit since it produces an anomaly of about the correct magnitude (although slightly offset from the field anomaly). Local variations in background (not taken into account in the program) and variations in contact impedance along the length of the casing could account for some of the discrepancy.

The four new wells discussed previously were then added to the input model, using the same casing characteristics that had provided the best fit in the model of the original wells. This provided a before-and-after comparison of the casing model effects. If the original results of the PIPE program (including the six original wells) are accepted as a good fit to the field data, then the PIPE program should predict little or no change in the data with the inclusion of the four new wells, since the repeat of line 1 field data after the four new wells showed little change.

Figure 8a shows the well casing model results for the original six wells, and Figure 8b shows the PIPE results for the original wells plus the four new wells (using the physical parameters that had provided the best fit to the original data). The data shown are for the segment of line 1 that was repeated in the field for comparison purposes.

The well casing model predicts a large change in IP values as a result of the new well casings, contrary to the field results. This suggests that even though a well casing model can be generated that fits the data moderately well, the IP anomaly observed in the field data is probably not the result of well casings. Based on the results of the before-and-after data sets, well casings are responsible for the anomaly only if the casing effects are very selective, that is, the original six well casings are different electrically from the four new well casings. This is considered unlikely because the geology is layered and relatively uniform, the casing sizes and materials are the same, and the new wells were allowed to settle 30 days before the field measurements were made.

ALTERATION MODELING

To verify that a deep, broad polarizable region similar to the alteration detected over some fields (Oehler and Sternberg, 1984) could be the cause of the IP anomaly measured at this site, an in-house 2-D finite-element mathematical program called 2DIP was used to model an oil field alteration anomaly. In this model, the polarizable area is broad, about 5 dipole lengths across (~1500 m, or 5000 ft), to represent an area of altered sediments above the oil field. The top of this polarizable region is ~230 m

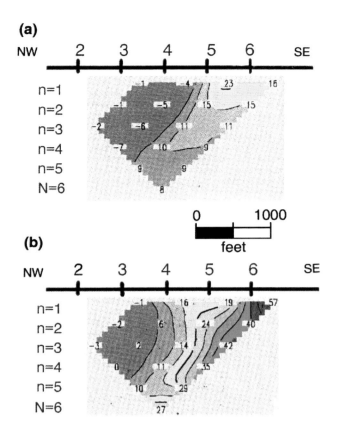

Figure 8—(a) IP data from the modeling program PIPE for the original six well casings versus (b) PIPE results after adding the four new well casings. Note the strong increase in the IP anomaly as compared to the actual field results shown in Figure 6.

(750 ft) beneath the surface. In this model, the polarizable region is relatively simple, with a higher IP response (80 mrad) assigned to the central ~1000 m (3000 ft), and a moderate IP response (20 mrad) assigned to the outer region. Background response is 0 mrad, and resistivities are represented by a layered 4 ohm-m overlying 10 ohm-m. More complex models are possible and may provide a better fit to the field data.

Figure 9a shows the IP field data for comparison with the 2-D model results (Figure 9b). The fit with the field data is moderately good and can be considered as good or better than the well casing model results. The model results produce weak negative values adjacent to the main anomaly. If a more complex model incorporating small localized variations were used, or if weak cultural effects could be added (as suggested here by the comparisons), a good fit with the field data could be generated. Of importance here is not that the model could be fine-tuned to a close fit, but that a broad, moderately deep polarizable region, about the size of the oil field, could generate the IP anomaly observed over this producing oil field.

FINAL INTERPRETATION

The final interpretation in 1982 of the original data set, the shifted-dipole test, and the before-and-after comparisons was that two areas appeared to exhibit electrical anomalies associated with the oil field at depth. These two areas are outlined in Figure 10, comprising the majority of Block 1235 and the northwestern quarter of Block 1. The interpretation of an alteration anomaly rather than a

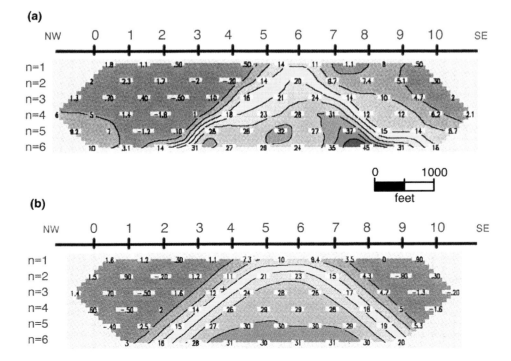

Figure 9—(a) IP data from the original line 1 measurements versus (b) IP results from the modeling program 2DIP simulating a seepage-induced polarizable region buried at ~230 m (750 ft), based on the interpreted size of the alteration anomaly.

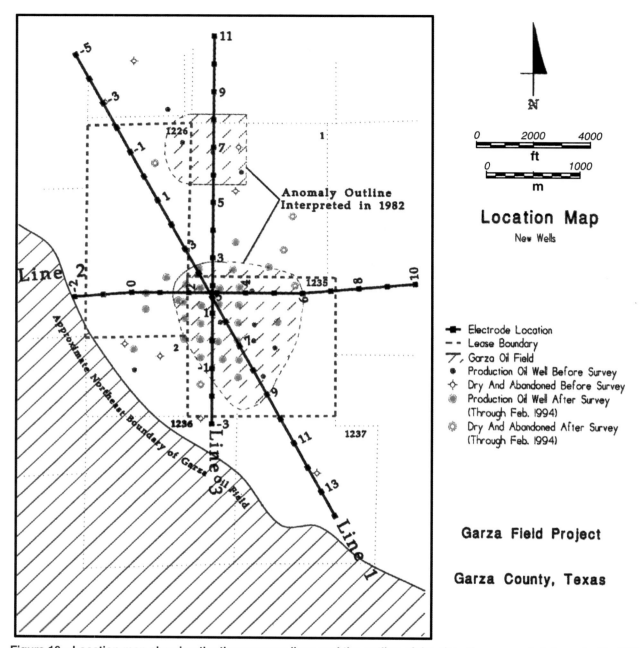

Figure 10—Location map showing the three survey lines and the outline of the alteration anomaly interpreted in 1982. Also shown are the new wells (producing and dry) drilled since the survey was completed.

surface culture or well casing anomaly is based on the following key elements:

1. The similarity in strength and appearance of the anomaly on all three lines, despite the different orientation and location of dipoles with respect to the culture. In the field data, the anomaly at the intersection of the lines is similar despite the difference in orientation of the receiver dipoles at the intersection and the completely different position of the associated transmitter dipoles (and therefore dif-

ferent cultural influences).

2. The similarity in strength and appearance of the anomaly before and after shifting electrodes on line 1 along line 500 ft. As discussed, this should have produced a significant change in any surface culture anomalies.

3. The similarity in strength and appearance of the line 1 segment after four new wells were drilled. As discussed, if well casing effects were significant, these new wells should have had a major effect on the repeated data.

As noted earlier, interpretation was also based on resistivity data and modeling and on prior experience over producing oil and gas fields.

There are no strong anomalies interpreted in Block 1226. The main alteration anomaly in Block 1235 does not appear to extend to the northwest into Block 1226, nor does the production in the northern half of Block 1 appear to extend very far to the west into Block 1226. In addition, because known production correlated well with the anomalies, there was no indication of lateral migration of the anomaly mechanism, and no lateral offset between the anomalies and production was suspected. On the basis of these interpretations, leases in Block 1226 were not acquired and development of the field in Block 1235 continued.

RESULTS OF THE SURVEY

Figure 10 shows the locations of all wells drilled through February 1994 and appears to confirm the 1982 interpretation of the data. Note in particular the dry well that was drilled (by a different operator) after the survey near station 0 of line 1 in Block 1226. It provides at least partial confirmation that the field in Block 1 does not extend to the west and that the main field in Block 1235 does not extend very far to the northwest. Dry wells near station 6 of line 2 and station −1.5 of line 3 seem to agree well with the interpreted boundary of the alteration anomaly. To date, a total of 23 production wells have been drilled within the outline of the interpreted anomaly, confirming the size of the production area (relative to the small size of the production area in the northern half of Block 1, for example). The excellent agreement between the alteration anomaly outline and the production limits, however, is at least in part a "lucky" interpretation because the lateral resolution of the electrical survey (using this particular dipole size) is probably no better than 150 m (500 ft). Thus, the outline of the actual alteration anomaly could be slightly smaller or larger. It is also important to note that the outline of the alteration anomaly does not necessarily define the exact limits of economic production.

Despite the presence of cultural influences, the final results of this project were of well-defined economic benefit to the operator. Budget money was not spent on leasing the northern half of Block 1226 (which so far has proven to be a relatively unproductive area), and only two dry wells were drilled in developing Block 1235. Had the budget existed, additional work could have added to the interpretation. Additional lines or data gathered with different arrays could have improved the lateral resolution, and shallow core holes could have verified the physical source of the electrical anomalies. Additional deep drilling would certainly be interesting and useful in verifying the correlation between the electrical anomalies and the outlines of production. The survey more than fulfilled its economic purpose, however, and serves as an excellent example of an effective evaluation of cultural effects on electrical geophysical data over a developing oil field.

Acknowledgments—We are grateful to the management of Matthews Oil Company of Lubbock, Texas, for permission to publish this data set with background information. Much of the data and discussion included in this paper appeared previously in Zonge Engineering's 1983 speculative "groupshoot" project called "Case histories of an electromagnetic method for petroleum exploration" authored by Larry Hughes of Environmental and Safety Engineering (EnSafe) of Memphis, Tennessee, whose efforts and contributions cannot be overstated.

REFERENCES CITED

Donovan, T. J., 1974, Petroleum microseepage at Cement, Oklahoma—evidence and mechanism: AAPG Bulletin, v. 58, p. 429–446.

Donovan, T. J., A. A. Roberts, and M. C. Dalziel, 1981, Epigenetic zoning in surface and near-surface rocks resulting from seepage-induced redox gradients, Velma oil field, Oklahoma (abs.): AAPG Bulletin, v. 65, p. 919.

Ferguson, J. D., 1977, The subsurface alteration and mineralization of Permian red beds overlying several oil fields in southern Oklahoma: Master's thesis, Oklahoma State University, Stillwater, Oklahoma, 95 p.

Hild, G. P., 1986, The relationship of San Andres facies to the distribution of porosity and permeability—Garza field, Garza County, Texas, *in* Hydrocarbon reservoir studies, San Andres/Grayburg formations, Permian Basin: Permian Basin Society–SEPM Publication No. 86-26, p. 429–446.

Holladay, J. S., and G. F. West, 1984, Effect of well casing on surface electrical surveys: Geophysics, v. 49, p. 177–188.

Myres, S. D., 1977, The Permian Basin, era of advancement: El Paso, Texas, Permian Press, 266 p.

Nelson, P. H., 1977, Induced polarization effects from grounded structures: Geophysics, v. 42, p. 1241–1253.

Oehler, D. Z., and B. K. Sternberg, 1984, Seepage-induced anomalies, "false" anomalies, and implications for electrical prospecting: AAPG Bulletin, v. 68, p. 1121–1145.

Sternberg, B. K., 1991, A review of some experience with the induced-polarization/resistivity method for hydrocarbon surveys: success and limitations: Geophysics, v. 56, p. 1522–1532.

Sumner, J. S., 1976, Principles of induced polarization for geophysical exploration: New York, Elsevier, 277 p.

Telford, W. M., L. P. Geldart, R. E. Sheriff, and D. A. Keys, 1976, Applied geophysics: New York, Cambridge University Press, 860 p.

Ward, R. F., C. G. St. C. Kendall, and P. M. Harris, 1986, Upper Permian (Guadalupian) facies and their association with hydrocarbons—Permian Basin, West Texas and New Mexico: AAPG Bulletin, v. 70, p. 239–262.

Matthews, M. D., 1996, Migration—a view from the top, *in* D. Schumacher and M. A. Abrams, eds., Hydrocarbon migration and its near-surface expression: AAPG Memoir 66, p. 139–155.

Chapter 11

Migration—A View from the Top

Martin D. Matthews

Texaco International Exploration Division
Bellaire, Texas, U.S.A.

Abstract

Many mechanisms have been proposed for hydrocarbon migration, and many processes have been described that modify the composition of migrating hydrocarbons. Examination of subsurface and surface data indicates that all the proposed mechanisms and processes are active. However, many play minor roles only recognizable in special situations. The dominant migration mechanism is as a free phase, rising under the forces of buoyancy within carrier and reservoir rocks, and capillary imbibition in the transition from sources and seal into carrier rocks. The migration pathway is determined by three-dimensional heterogeneity at all scales, from the individual pore systems to the interrelationships of facies. The dominant process modifying the composition of migrating hydrocarbons is phase partitioning, as evidenced by subsurface and surface data on hydrocarbon expulsion, migration, and accumulation.

In the near surface, many processes act to modify this seepage, particularly biogenic activity and diffusion (both chemical and mechanical). Free-gas surface hydrocarbon survey measurements, however, are dominated by this seepage mechanism. It explains (1) the spatial variability of the data; (2) the relationship of high-magnitude sites to high-permeability geologic features such as faults, fractures, unconformities, and outcropping reservoir units; (3) the compositional relationship of subsurface reservoired hydrocarbons to source rocks and the lack of relationship to ineffective source rocks; the variation of magnitude with time, both long and short; and current estimates of the rate of transport in the near surface.

INTRODUCTION

This paper is concerned with postexpulsion migration. With the exception of boundary conditions, there is little or no difference in the processes involved in secondary and tertiary migration. The knowledge gained by studying one directly constrains the behavior of the other. *Postexpulsion migration* is concerned with all aspects of migration, through mixed geologic sequences (reservoirs, conduits, seals, faults, and fractures) and ultimately to the surface. This mixed sequence of rocks is simplified into two classes: relatively large pore networks and relatively small pore networks. Because these terms are used in reference to adjacent systems, they may refer to a contact between sandstone and shale, fine and coarse sandstone, shale and evaporite, or even adjacent pore networks. Pore networks are chosen as the reference because they are a prime control on flow rate and volume.

Conventional conclusions on how hydrocarbons migrate are dominated by studies of flow in reservoirs by engineers, laboratory simulations, and theory of flow in porous media. These studies rarely encompass tens of years, while the time frame of reservoir charging is probably much longer. The transport of hydrocarbons through an entire stratigraphic section has been generally ignored. Notable exceptions are the work of hydrologists such as Tóth (1988 and this volume) and basin modelers such as Welte and Yükler (1980), Durand et al. (1983), and Nakayama and Lerche (1987).

Migration is like Einstein's watch. Observations concerning its operation can be made, but since opening the system is not permitted, only hypotheses about its operation, consistent with those observations, can be made. The movement of hydrocarbons in the deep and shallow subsurface is a complex balance of processes. We can draw conclusions based only on our understanding of the basic principles of science and our observations. We may never know if these conclusions are correct for any given situation.

First, I give an overview of the forces and processes responsible for hydrocarbon migration. This is followed by a description of constraints placed on migration by subsurface and near-surface studies. This information is then used to generate a model of migration, consistent with all these data.

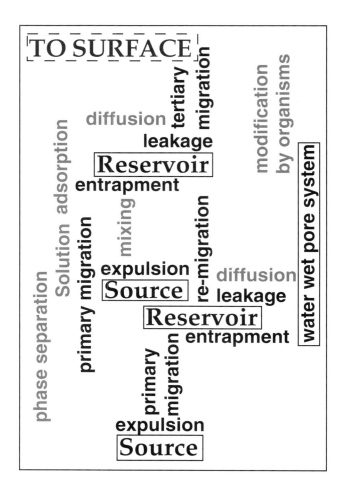

Figure 1—Schematic diagram of subsurface migration. Rock framework is shown in boxes, events in black lettering, and processes in gray.

MIGRATION OVERVIEW

Three principal questions about migration arise. (1) How is cross-formational flow of hydrocarbons accomplished? (2) Does the form change during migration and, if so, which form is dominant under what conditions? (3) How does the dominant form (or forms) control the volumes and compositions transported? To consider these questions, it is appropriate to review the constraints placed on our understanding of the subsurface migration process. The interrelationships of some of these constraints are shown in Figure 1.

Energy

The movement of hydrocarbons from one location to another requires energy. Many sources of energy are possible, and to one extent or another all may be active. Some arise directly as a result of the presence of hydrocarbons in given location, including buoyancy, chemical potential (related to concentration differences), and compaction of the sediment (squeezing the hydrocarbons from the collapsing pore space). Some act indirectly on the hydrocarbons as conditions around them change due to burial, such as thermal expansion and water motion due to compaction. Others come about due to decreases in pressure and temperature as a result of the upward motion of hydrocarbons through the rock, such as phase change and volume expansion.

Physical Form

Numerous models for the physical form of migrating hydrocarbons have been proposed. These can be grouped into three categories: dissolved in water, separate phase, and transport along a continuous kerogen network.

Hydrocarbons dissolved in water occur as true solution and micellar solution. Both of these forms enable the hydrocarbons to move one molecule at a time and thus present minimal restriction to movement. *True solution* is a function of pressure, temperature, salinity, molecular weight, and mixtures of components present. *Micellar solution* increases the capacity of water to carry molecular hydrocarbon species by the use of naturally occurring hydrocarbon solubilizers.

Separate phase migration of hydrocarbons occurs as discrete droplets (smaller than a pore throat) and as a larger continuous mass extending across several pores. Dispersions of separate phase hydrocarbons occur as discrete droplets, colloids, or emulsions, either gaseous or liquid. The taller the hydrocarbon mass, the greater the buoyancy force. Separate phase migration of hydrocarbons is subject to capillary forces at its contact with water. Continuous kerogen networks are only expected in rich source rocks and can be discounted as a dominant factor in post-expulsion migration.

Processes Causing Chemical Changes During Migration

During postexpulsion migration, many processes are capable of altering the chemical characteristics of the hydrocarbons expelled from the source rock. These include water washing, adsorption, phase partitioning, mixing, and biodegradation. *Water washing* is the selective removal of the more water-soluble components. Adsorption of migrating hydrocarbons onto the mineral substrate it passes through can lead to both selective removal of hydrocarbons and selective retardation of the migration rate of hydrocarbons, as in a chromatographic column. *Phase partitioning* is the concentration of different hydrocarbon species into gaseous and liquid phases with changes in pressure and temperature. Mixing alters composition through inclusion of hydrocarbons from other kerogen particles along the migration path, combining migration streams from two or more source rocks, and precipitation of asphaltenes and other high molecular weight compounds by the addition of methane. Biodegradation is the biologic alteration of hydrocarbons.

Concentration Mechanisms

Processes capable of concentrating hydrocarbons from the migrating stream into a reservoired accumulation include pressure minimums, critical orifices, exsolution, and capillary forces. Perfect seals, which do not leak at all, are the type of seals most petroleum geologists visualize, but they rarely occur. *Pressure minimums* are a perfect seal. When all the forces acting on a hydrocarbon mass are resolved, the hydrocarbons remain in the minimum as long as it exists. There is no migration out of that minimum. *Critical orifices* apply specifically to dissolved species. They act as molecular sieves, allowing particles smaller than the orifice to pass, while retaining particles larger than the orifice. If seals were uniformly composed of subcritical pore throats, this would also form a perfect seal. Nonporous unfractured rock, such as a salt, will form such a seal under certain circumstances. *Exsolution* of hydrocarbons is facilitated by increasing salinity, decreasing temperature, and decreasing pressure. Once a separate phase is formed, capillary forces become effective. *Capillary forces* arise at the interface between two immiscible mobile phases across a restricted opening. As the pressure difference across a capillary restriction increases, the interface deforms and the nonwetting phase eventually penetrates the restriction.

Near-Surface Boundary Conditions

Surface geochemical observations are made close to the end of the subsurface migration path. They offer the advantage of virtually complete spatial availability of sampling locations, but are subject to certain boundary conditions (Figure 2) that must be considered in interpretation of the data. Four interfaces are of interest: water-wet pores with free water, free water with the atmosphere, water-wet pores with soil air, and soil air with the atmosphere.

Water-Wet Pores with Free Water

The behavior of migrating hydrocarbons at the interface between water-wet pores and free water is particularly instructive because, in addition to sampling by coring and sniffers, it is subject to direct observation and measurement by submersibles and free diving. At shallow depths below the sediment–water interface, the concentration of hydrocarbons generally increases downward, as shown by Abrams (1992, and this volume). Detailed depth profiles (Brooks et al., 1978) show that this general increase is subject to considerable variation and local reversals of hydrocarbon concentration with depth. This variation is similar to that observed in high-sensitivity mud logs at deeper depths. Separate phase hydrocarbons escape through the sediment water interface as both bubbles (gas dominated) and globules (liquid dominated). The rate of escape varies greatly, from intermittent on a scale exceeding days to nearly continuous. The occurrence of separate phase hydrocarbons at the sea floor is direct proof that postexpulsion migration in a separate phase exists in the subsurface as well.

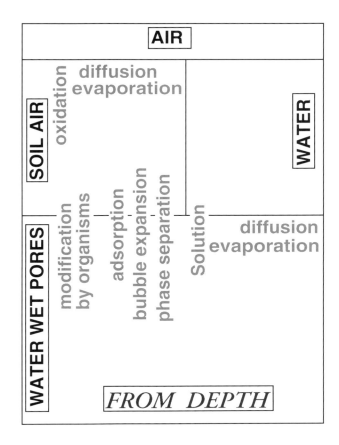

Figure 2—Schematic diagram of near-surface migration. Media through which hydrocarbons migrate are shown in boxes and processes are in gray lettering.

Sporadic escape suggests that the process often proceeds as small separate masses. Nearly continuous bubble trains support the occasional existence of longer continuous hydrocarbon stringers in the subsurface. As these stringers reach the free water surface, they break up due to surface tension. Gas bubbles rise rapidly due to their low density and expand as they move upward to lower pressure. If a bubble grows too large, it breaks up into smaller bubbles. Bubbles moving through the water column dissolve into the undersaturated water. If the bubble is small, it may dissolve before it reaches the water–atmosphere interface. If the bubble contains significant quantities of liquids, these liquids plate out at the bubble surface, preventing the light gases from dissolving. The rate of bubble transport in the water column is high, and under most conditions, the lateral transport is slight (<45°). The train of bubbles widens from a narrow point or line of entry at the sea floor to a cone that is most often narrower than 1 km at the surface. Surrounding this cone of bubbles is a zone of water with elevated dissolved hydrocarbons formed by solution of portions of the bubble train. This behavior is demonstrated by the occurrence of well-defined peaks on sniffer surveys surrounded by broad shoulders elevated above background.

Oil globules move more slowly and may be significantly displaced by water motion. Detection of dissolved hydrocarbons escaping through the sediment water interface is more difficult because this transport is not visible. Any dissolved hydrocarbons escaping through this interface are subject to rapid mixing and dilution by the overlying water mass. Broad regions of elevated hydrocarbon concentrations are detected by sniffer surveys. Where these occur near the sea floor they may indicate (1) escape of dissolved hydrocarbons over a broad area , (2) intermittent escape of bubbles over a large area, or (3) dispersive mixing and destruction of a discrete bubble train.

Free Water with the Atmosphere

Once a bubble reaches the water—atmospheric interface, it breaks and is rapidly mixed with the atmosphere. Globules form a slick, evaporating into the atmosphere and forming biodegraded masses or remaining as separate globules. Globules and biodegraded masses continue to lose volatiles to both the water and atmosphere and biodegrade until they sink or run aground. Hydrocarbons dissolved in the water column either escape to the atmosphere or are consumed by organisms. Williams et al. (1981) pointed out that the level of background hydrocarbons in sea water is higher in productive areas than in the open ocean. The degree to which solution of bubbles in the water column and transport of dissolved hydrocarbons through the sea floor each contribute to this phenomenon is unknown.

Water-Wet Pores with Soil Air

The behavior of hydrocarbons at the interface between water-wet pores and soil air is similar to that described above. Hydrocarbons in solution either come out of solution and join those already in the gas or liquid phase or are consumed by organisms. Hydrocarbons in a liquid phase spread out to form a free-floating mass subject to biological action, just like a slick at the sea–air interface. These hydrocarbons can be sampled from shallow water wells. Portions of this mass are slowly transported upward into the soil air by gaseous migration, through diffusion or effusion, either alone or with a carrier gas, such as methane. Hydrocarbons in a gas phase move upward due to buoyancy and diffusion, aided by barometric pumping.

The concentration of hydrocarbons decreases toward the soil air–free air interface due to dilution of the migrating gases with atmospheric gases. The concentration at any one point is therefore a function of the rate of hydrocarbon transport (controlling the volume of hydrocarbons delivered to that location), the rate of transport of atmospheric gases (controlling the volume of nonhydrocarbon gases delivered to that location), and the rate of destruction of hydrocarbons (and methane production) by organic activity. Statistically, the deeper the sample, the better, just like that observed offshore below the sediment–water interface. Studies at Gulf Research and Development Company suggest that for drilled holes, a depth of 3.6 m (12 ft) provides stable readings. Probe studies suggest that a depth of 1.2 m (4 ft) also provides stable data, but with reduced magnitudes. Depth profiles indicate that both increases and decreases in concentration can occur at any depth, with higher concentrations generally occurring below less permeable layers, similar to the behavior of mud logs but with reduced variation. This suggests that the concentration of hydrocarbons in the shallow surface (and in the deeper subsurface) is dependent on the three-dimensional permeability structure of the soil and rocks being sampled rather than a simple vertical mixing curve in a homogeneous medium. Despite this vertical variability, at any given depth, sites with high concentrations tend to be consistently high and sites with low values tend to be consistently low.

Soil Air with the Atmosphere

At the interface between soil air and the atmosphere, the concentration of hydrocarbons is fixed at or close to the average atmospheric concentration. Study of the concentration of hydrocarbons in tents sealed to the ground by Gulf Research and Development Company indicate that at low-level sites, the exchange occurs in both directions. The concentration of hydrocarbons in the air may increase above that in the atmosphere, presumably due to preferential expulsion of hydrocarbons from the soil air, or it may decrease, presumably due to preferential removal of hydrocarbons from the air by the soil. These exchanges are rapid, occurring within hours. At high-level sites, the concentrations also vary but remain above atmospheric concentrations. Antropov et al. (1981) found similar behavior in the atmosphere over petroleum fields using adsorption gas lasers.

LESSONS FROM SOURCE AND RESERVOIR STUDIES

I have made a wide variety of observations about the characteristics of rocks during burial, their ability to transmit fluids, and the static distribution of hydrocarbons in them. These are summarized in this section under the following topics: source–reservoir correlation, physical sizes, pore throat heterogeneity, fluid inclusions, phase behavior, hydrocarbon solution in water, diffusion, rates of hydrocarbon motion, mud log data, and seismic evidence.

Source–Reservoir Correlation

The fact that oils can be chemically correlated with a particular source rock places a fundamental constraint on any proposed migration process. It strongly indicates that the expulsion–migration process does not significantly effect the overall geochemistry of the migrated product. However, there are also some differences. Thompson (1988) points out that reservoired hydrocarbons in close proximity to one another and geochemically matched to the same family or source are frequently dissimilar in

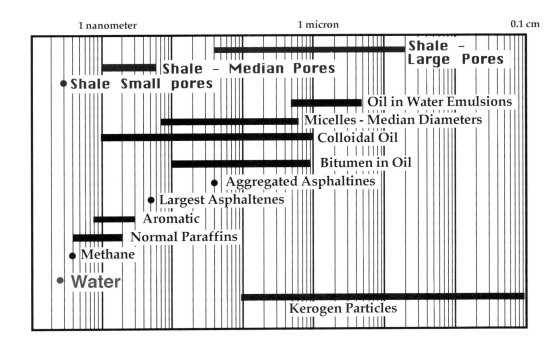

Figure 3—Comparative sizes of shale pores and molecules of hydrocarbon, water, and kerogen.

their light ends and other properties. This suggests that the migration and entrapment process has some control over the chemical characteristics of reservoired hydrocarbons, but in most cases this only affects the detailed chemistry, not the overall structure.

Physical Sizes

The majority of hydrocarbon molecules are larger than the size of small shale pores and approximately equal to the median pore size of shales (Figure 3). Methane molecules are slightly larger than the size of small shale pores, and water molecules are slightly smaller. The lower range of oil colloids is within the range of median shale pores, while larger colloids and other hydrocarbon particulates, such as micelles and bitumen, are in the range of the largest shale pores. This suggests that water, methane, and the smaller paraffins and aromatics are capable of moving through shales in solution, but that the larger molecules, aggregates, and separate phases undergo various degrees of molecular filtration and capillary blockage in attempting to pass through all but the largest shale pores or open fractures. The effect of burial on shale permeability (proportional to pore throat size) is shown in Figure 4. Note that shale compaction results in a logarithmic decrease in permeability down to a depth of about 3.5 km. Beyond this depth, two factors vie for dominance: continued collapse of the pore network and the formation of micro- and macrofractures as the shale becomes more brittle (Neglia, 1979).

This suggests that diffusion and aqueous solution transport may be practical for the smaller hydrocarbon molecules, but becomes increasingly less likely with increasing molecular size. Direct transport of larger separate phase particles, while possible through the large pores, becomes increasingly less likely as the traversed path lengthens due to the increased probability of a continuous large pore network terminating into a small pore throat. Indeed, even the flow of the comparatively small water molecule often requires significant pressure gradients to overcome the restrictions to flow common in shales.

Pore Throat Heterogeneity and Interstitial Continuity

The range of shale pore size is more than five orders of magnitudes (Figure 3). The effect of this heterogeneity is diagramatically represented in two dimensions in Figure 5, which shows the interconnectivity of 200 pore throats of five different sizes (1 to 5 units) randomly scattered in a regular pattern. Entry of hydrocarbons from a large pore network is assumed to occur at the bottom of the unit and exit into a similar unit at the top. There is no connection through this network that is composed exclusively of the largest (5-unit) pore throats. This is representative of a no-flow condition for hydrocarbons and is rarely if ever achieved in nature. It would represent a very fine grained, highly compacted, unfractured rock, such as an evaporite.

A more typical fine-grained condition is shown in Figure 5a. The continuous networks of 4- and 5-unit pore throats are highlighted. Note that only one is continuous through the unit (5% of the potential exit pore throats). Continuous pore throat networks of increasingly larger diameter are shown in Figure 5b (3, 4, and 5 units) and Figure 5c (2, 3, 4, and 5 units). Note that the percentage of exit pore throats increases to 50% and 80%, respectively. The connection of all pore throats (1 to 5 units) represents a total flow condition, such as is expected for a reservoir rock rather than a seal.

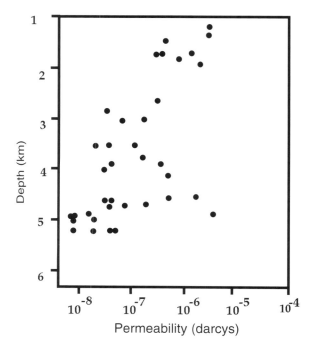

Figure 4—Graph of permeability in Tertiary shales with depth. (After Neglia, 1979.)

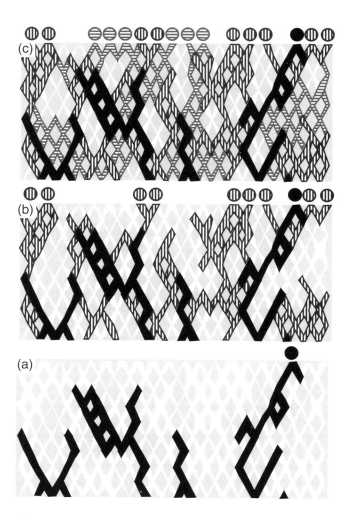

Figure 5—Diagrams of a pore network showing five possible pore throat sizes from 1 to 5 (smallest to largest, from bottom to top). (a) Pore throat sizes 4 and 5, with networks shown in black. (b) Pore throat sizes 3 and larger, with networks of sizes 4 and 5 shown in black and size 3 shown in vertical line pattern. (c) Pore throat sizes 2 and larger, with networks of sizes 3, 4 and 5 shown as before and size 2 shown in horizontal line pattern.

The relationship between net path permeability and frequency of occurrence of a given path is summarized in Figure 6. The net path permeability of a given pore network is determined by the smallest pore throat in that path. Net paths with small permeability are therefore more common than paths with large net permeability, and the longer any particular path is, the greater the chance it will include a pore throat smaller than it currently contains. A three-dimensional model would have similar behavior, but the potential connections to large pore throats would rise (connections to other small pore throats have no effect), increasing the probability of a large pore network. This pore network is a modification of Momper's (1978), Lindgreen (1985), and England's (1987) expulsion models.

The connectivity of these pore throats governs the extent to which hydrocarbons, either as separate phases or dissolved molecules, are retarded by the seal. The usual situation is the sealing of a separate phase because most shales have pores in the ranges shown in Figure 3. To examine the flow of separate phases, the pore throats in Figure 5 are considered as fixed and the diameter of the curvature at the hydrocarbon–water interface is allowed to vary among the three diagrams (Figure 5a, b, c). At a constant interfacial tension, capillary pressure depends on the curvature of the interface between the hydrocarbon phase and water. The radius of curvature is determined by the diameter of the pore throat through which it must pass and the difference in pressure across that pore throat. If that pressure is small and the pore throat is small, there cannot be any separate phase hydrocarbon

flow through the unit. The unit thus acts as a perfect seal, ignoring losses due to noncapillary processes.

As the pressure rises, however, the radius of curvature at the pore throat interface decreases until a point is reached when hydrocarbons move through the largest continuous pore throat network present (Figure 5a). If this flow balances the rate of pressure buildup, a steady state arises between pressure buildup below the seal and leakage through the seal. If, however, the rate of pressure buildup continues, the curvature of the interface at the pore throats will continue to decrease and flow will take place through increasingly small pore throats (Figures 5b, c). The heterogeneous nature of seals thus allows them to adjust their rates of hydrocarbon leakage dynamically.

One of the principal causes of pressure build up in reservoirs is the buoyancy of the hydrocarbon phase. As

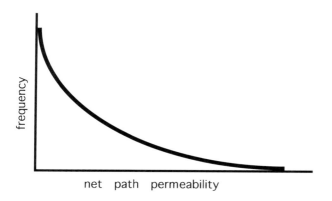

Figure 6—Graph of expected frequency of net path permeability.

the reservoired column increases in height, the pressure at the interface increases, decreasing the radius of curvature at the interface and increasing the leakage through the seal. Similarly, as the rate of reservoir charging slows down, the rate of leakage also slows down. This model is essentially in agreement with laboratory simulations of separate-phase flow in porous medium by Catalan et al. (1992) and Dembicki and Anderson (1989) and with Schowalter's (1979) observation that continuous filaments are formed only if the hydrocarbon concentration exceeds 4.5–17% of the pore volume. Examination of gaseous hydrocarbons in the cap rock of Snorre field in the North Sea (Lieth et al., 1993) indicates that, although diffusion plays a role in loss from the reservoir, bulk flow due to buoyancy is probably the dominant process.

On rare occasions, seals may be dominated by fine pore throats, in which case the rock acts like a molecular sieve. For dissolved phases, the size of the molecules is fixed. Therefore, the pore throat units are considered to increase from Figure 5a to 5c. A rock dominated by fine pore throats is diagramatically represented by Figure 5a. If we consider the pore throats of 4-unit width to represent 0.5 nm, the seal will slowly pass methane, and perhaps some ethane, due to the small percentage of available paths. The case in which the large pore (4-unit) throats are 0.2 nm is represented by Figure 5b. The larger normal paraffins and smaller aromatics can pass through the single large pore (4-unit) network, and methane and the smaller normal paraffins can pass through the more numerous 3-unit pore networks as well as the larger network. In Figure 5c, the large pore (4-unit) network is about 0.4 nm, the normal size of median pores. Under these conditions, the large pore network permits the passage of methane, the entire range of normal paraffins and aromatics, and all but the largest asphaltenes. The smaller pore networks pass the smaller hydrocarbons at a higher rate because of their greater frequency. Lindgreen (1987) documented molecular sieving in fine pore systems in source rocks and estimated that only one-fifth of the total pore volume was capable of transmitting normal paraffins.

Fluid Inclusions

Fluid inclusions in diagenetic cements are samples of paleopore fluids that were trapped during precipitation of that cement. By incorporating the composition and PVT properties of the inclusions with geologic history and the diagenetic sequence in the area, a series of snapshots of the pore fluid characteristics as a function of time can be obtained (Burruss et al., 1983). These studies show that reservoir charging occurs in a discrete time interval, either as a series of events or as a continuously evolving process (Karlsen et al., 1993). Jensenius and Burruss (1990) found that some oils in inclusions seem to be a mixture of a biodegraded hump of C_{12} to C_{30} compounds and water-soluble low carbon number compounds. They suggest that the low carbon number component may arise due to exsolution from a later warm aqueous phase rising from depth. McLimans (1987) suggests that hydrocarbon inclusions in reservoirs originate from a separate phase consisting of drops from 1–40 μm in diameter (the range of oil in water emulsions) (Figure 3), sufficiently dispersed to move freely within the reservoir pore space.

Phase Behavior

Figure 7 shows the phase behavior of hydrocarbons with depth. The exact configuration of this diagram is a function of the particular mixture of hydrocarbons present; for simplicity, a single composition is considered. As products are expelled and migrate vertically, phase partitioning occurs. Consider an expulsed oil at a depth of 5 km with 30% gas. This oil migrates vertically without any significant changes until it reaches its bubble point at about 4 km. At this point, both a gas and oil phase coexist. The single gas bubble would contain about 25% dissolved oil. If these two phases continued to maintain contact and migrate vertically, the proportion of gas in the oil phases would decrease and the proportion of oil in the gas phase would similarly decrease. At a depth of 2 km, the gas phase would contain about 5% oil and the oil phase would contain about 15% gas, but the total proportion of gas would still be 0.3. Silverman (1965) combined phase behavior with selective trapping to produce a "separation-migration" mechanism.

If the two phases separated at 3 km and continued to migrate vertically, at a depth of 2 km, the migrated gas phase would have 85% gas and the migrated oil phase would have 20% gas. The migrated gas and oil phases would both have a gas cap with 10% dissolved oil and oil legs with 5% gas, identical to the original unseparated product. The amounts of gas caps and their chemical compositions would, however, be different from one another and from the gas cap of the unseparated product. Separation-migration can significantly alter the gross composition of the migrated and trapped intervals. The general trends with migration from depth are as follows: hydrocarbon liquids lose low molecular weight compounds to a gaseous phase, becoming more dense, and hydrocarbon gases lose high molecular weight com-

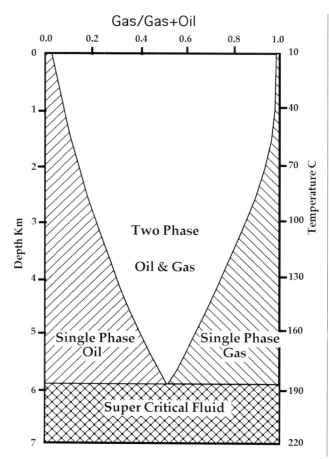

Figure 7—Graph of depth and temperature versus gas to gas + oil ratio showing hydrocarbon phase behavior. (After Pepper, 1991.)

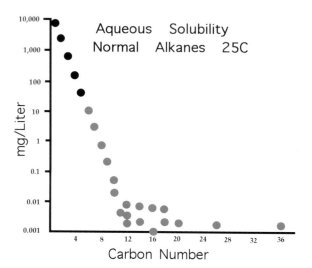

Figure 8—Graph of aqueous solubility versus carbon number for hydrocarbons at 25°C. (After Tissot and Welte, 1984.)

approach to a solubility threshold or the microdroplets in the laboratory studies (Tissot and Welte, 1984). This compositional relationship differs significantly from that found in most reservoired oils. There are, however, a few light oils that show this relationship (Tissot and Welte 1984; Quanxing and Qiming, 1991). The occurrence of these light oils as a separate phase demonstrates that solution transport does occur and that it can reach conditions causing it to separate into a separate phase in considerable quantity. The infrequent occurrence of oils with this signature suggests that this process of migration is an exception rather than the rule.

Diffusion

Studies of source and reservoir contacts show that diffusion of dissolved hydrocarbons occurs in the subsurface. Connan and Cassou (1980) and Leythaeuser et al (1984) have studied the relationship of sandstone and shale extracts in immature terrestrial organic sequences. They have shown that small quantities of hydrocarbons have been selectively depleted in the shales and accumulated in the sandstones. The greatest depletion in the shales occurred near the sandstone contacts while the greatest accumulation in the sandstone was associated with fractures. Paraffins, particularly the lighter compounds, were preferentially migrated. The sandstone extracts resemble a condensate or light oil while the depleted shale extract looked less mature and slightly biodegraded compared to the center of thick shales.

The gradients shown in Figure 9 are explained by diffusion (Leythaeuser et al., 1983). However, benzene and toluene (water-soluble aromatics) had anomalously high transfer rates from the siltstone into the sandstone. This suggests that some transport of hydrocarbons may have been assisted by transport of water from the siltstone into the sandstone.

pounds to a liquid phase, becoming less dense (England et al., 1987).

Price et al. (1983) showed experimentally that sufficient oil can be transported in gas solution to charge reservoirs. Thompson (1988) documented that migration fractionation is a viable mechanism to explain the geochemistry of the Gulf Coast and more than half of all other U.S. oils. Migration fractionation involves reservoired oil being partially dissolved in an excess of methane, subsequent loss of the gaseous phase, leaving behind a residual oil with elevated light aromatics. The leaked gaseous phase condenses as pressure and temperature are reduced during vertical migration, forming retrograde condensates.

Hydrocarbon Solution in Water

Figure 8 shows the aqueous solubility of normal alkanes at 25°C. The solubility decreases linearly with increasing carbon number, up to C_{12}. Normal alkanes with carbon numbers greater than 12 show a much higher solubility than would be expected by extrapolation of the lower carbon number data. This may represent an

Fraction of *n*-Alkanes

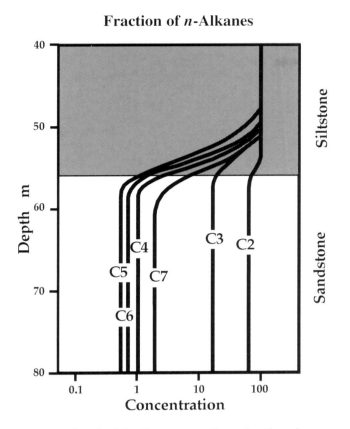

Figure 9—Graph of depth versus *n*-alkane fraction showing the concentration of hydrocarbons at the transition from a source to a sandstone.

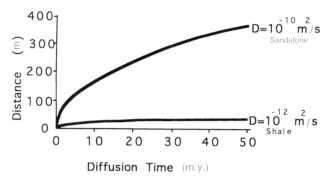

Figure 10—Graph showing the time needed for hydrocarbons to diffuse a given distance and reach a concentration of one-half the concentration of a nondepleted source.

Rates of Hydrocarbon Motion

Estimation of rates of hydrocarbon migration from source to reservoir are difficult at best. Estimates of rates of diffusion can be calculated theoretically, measured experimentally, and compared to concentration gradients away from concentrated sources. Active transport systems are much more difficult to constrain due to the uncertainties in the time of expulsion and entrapment.

Diffusion of light hydrocarbons in a water-filled porous system is extremely slow. Figure 10 shows the time it takes a light hydrocarbon to diffuse a given distance and reach a concentration level equal to one-half the concentration in an irreducible source. The larger diffusion coefficient is more typical of sandstone and the smaller one is more typical of shale. This is consistent with the findings of Zarrella et al. (1967) who showed that dissolved benzene and toluene followed a diffusion gradient horizontally for miles in sandstone but was vertically absent in the over- and underlying sections. This suggests that diffusion is not a practical transport mechanism in shale, except for short distances or at its boundary with sandstone. England et al. (1987) calculated that for distances greater than 10–100 m, diffusion is insignifi-

cant relative to bulk flow. An exception to this is methane and ethane. Diffusion would be expected to transport light hydrocarbons only tens of meters into a shale capping a reservoir. Despite this, significant quantities of light hydrocarbons can diffuse into a cap rock (Krooss et al., 1992). If the shale was fractured, the diffusion coefficient would only increase slightly , because the fractures would occupy a small percentage of the shale. Diffusion of larger hydrocarbons is less likely as an appreciable transport mechanism because the diffusion of hydrocarbons is related to the size of the molecule (Figure 9).

Cathles and Smith (1983) estimated that periodic fluid migration events greater than 1000 times the normal rates were necessary to form Mississippi Valley type lead-zinc deposits by expulsion of basin brines. They suggest that overpressure conditions build to a point where the section fractures, releasing large volumes of fluid over short periods of time. As the fluids escape, the pressure is released and the fractures close. The pressures then build up and the cycle repeats at intervals of about 1 m.y. at depths of more than ~3000–4500 m (~10,000–15,000 ft).

The flux rate of hydrocarbon transport in the subsurface is viewed as both parallel and series processes. *Parallel processes* are diffusive transport, aqueous transport in solution and as micelles, and separate phase transport. *Series transport* is the set of processes that are dominant sequentially along the most effective migration path. Any compositional constraint placed by any process (such as solution in water) can impact the characteristics of the migrated product from then on. From a linear rate standpoint, the least efficient process along this path is the rate-limiting step, controlling the overall rate of the process. In a sequence of sandstone and shale, the rate delimiter would be the least permeable shale. However, flux rates of a nonwetting separate phase, particularly through shales, are self-adjusting. This is accomplished by changing the area through which the linear process operates. Small dips in the section focus water and hydrocarbon flow laterally to an area where vertical flow becomes dominant (Roberts, 1980). Free-phase hydrocarbons accumulate in large pore systems underlying the restrictive shale, spreading out into a commercial or non-commercial accumulation. This increases the hydrocarbon flux rate through the overlying shale by increasing its surface area, making up for the low rate per unit area.

Mud Log Data

Mud log data are the most abundant subsurface geochemical data available. In general, they are an adequate representation of vertical variations in light gases and fluorescing compounds. Vertical trends in light gases are highly variable, but generally show an increase as the source-generative section is approached. They also show orders of magnitude changes in a few feet to tens of feet and rapid compositional shift of light gases and heavier compounds. High values are generally related to the occurrence of mature source rocks, top of overpressured zones, faults, and reservoirs.

Detailed geochemical studies, including isotopic analysis, often show a smoothly increasing gradient, consistent with increasing maturity. Superimposed on this gradient are isolated spikes of hydrocarbons with maturities characteristic of much deeper conditions. This suggests migration from a deeper source, through an intervening section, with little mixing or other effects from the intervening section.

Seismic Evidence

Vertical chains of seismic bright spots have been seen in the Gulf of Mexico, Nigeria, and other offshore areas. In many cases, the bright spots are associated with sandstones adjacent to faults and are occasionally recognized within fault zones. The most dramatic chain of bright spots is at Ekofisk in the North Sea where they are associated with a gas chimney of poor data and a low-velocity sag at the reservoir level (Van den Bark and Thomas, 1980).

LESSONS FROM NEAR-SURFACE STUDIES

Surface geochemical surveys represent the end of the migration pathway. As such, they offer the best opportunity for understanding the spatial patterns of migration and cross-formational flow. The compositional and magnitude information is, however, subject to the boundary conditions previously discussed. Therefore, caution must be exercised in comparing this information with deeper studies. Of particular importance is the volume change of free gas associated with decreasing pressure as the hydrocarbons migrate toward the surface. This volume increase makes the gas more visible to seismic profiling, makes it more measurable through collection, and increases its rate of motion due to buoyancy. Topics discussed in this section include seepage patterns, compositional relationships to the subsurface, variation with time, and the rate of transport.

Seepage Patterns

Surface macroseepage occurs in areally restricted locations and is direct evidence of hydrocarbon migration in a separate phase. Link (1952) studied worldwide occurrence of macroseeps and concluded that they occur dominantly along high-permeability pathways such as faults, fractures, unconformities, and pore networks in outcropping reservoirs. The linearity and spatial variability of macroseeps is particularly well illustrated by Preston (1980).

Microseep studies by Jones and Drozd (1983), Zorkin et al. (1977), and others often demonstrate the preferential pathway concept, particularly the role of faults and fractures. Preferential seepage up outcropping reservoir rock and through the fractured crest of an anticline is demonstrated by Matthews et al. (1984). The role of faults in focusing seepage is demonstrated by Jones and Thune (1982), Matthews et al. (1984), Burtell et al. (1986), and Abrams (1992). Wakita et al. (1978) has shown that gas microseepage up faults occurs in spatially distinct areas ("spots"), similar to that shown for gas macroseepage by Preston (1980). In macroseepage, there is a clear, visible distinction between where seepage is occurring and where it is not occurring. The replacement of visual observation with sensitive instrumentation removes the ability to conveniently distinguish areas of no seepage from seepage. Instead, there is a seepage continuum from the smallest level the machine can detect to visible seepage.

The dominant mechanism of seepage at large-magnitude sampling locations is the same—bulk flow of a separate phase. Microseepage magnitudes of light hydrocarbons in soil air vary from atmospheric levels, to elevated moderate levels, to percent levels. Atmospheric levels in soil air undoubtedly represent as close to no seepage as possible. Percent levels represent massive active seepage as a gas phase. Moderate levels also exist and are ambiguous in their mode of origin and their significance. They may represent the end of a dissolved phase migration path or diffusive mixing of a moderately active gas phase with atmospheric air in the soil profile.

Experience shows, however, that when closely spaced samples are taken, the boundaries between high- and low-magnitude sites are usually less than 400 m and often just a few meters. In one instance, soil gas magnitudes varied several orders of magnitude within 100 m. The boundary was recognized to occur over a distance of centimeters by the presence of bubble formation at the soil surface as rain began to saturate the soil and by the drying pattern when the rain had ceased (V. T. Jones, personal communication, 1978). The spatial proximity of surface hydrocarbon concentrations that differ by orders of magnitude, in both land and marine surveys, requires effusive transport of a separate phase, not solution transport.

High-resolution marine seismic surveys reveal the presence of shallow gas in the upper 600 m of sediment and up into the water column. The existence of considerable quantities of shallow gas has been proved many times by the collapse of offshore platforms. Analysis of hazard surveys by Salisbury (1990) suggests two pathways, fault planes, and localized pervasive vertical leakage (gas chimneys). Where the vertical seepage is interrupted by a competent seal, horizontal migration along siltstone and sandstone layers is observed.

This pattern of migration is consistent with the onshore pattern of gas migration over a leaking gas storage reservoir at Mont Belview, Texas, studied by Gulf Research. Bubble trails in the water column, revealed by echo sounders and side-scan sonar, indicate that in many cases the transport of gases is an active process (Judd, 1990). Side-scan sonar shows that these gas plumes in the water column often originate from pockmarks or mounds on the sea floor, arranged along what appear to be the surface trace of faults. McQuillin and Fannin (1979) report occasional plumes of sediment in the water column up to heights of 10 m from the sea bed. Numerous sniffer surveys by Gulf Research and others have documented the association of these plumes with both thermogenic and biogenic hydrocarbons. In a few cases, unidentified nonhydrocarbon gases were also observed. Gravity core studies (Kennicutt et al., 1988) and analysis of collected bubbles reveals that oils reach the sea floor as a separate phase in considerable quantities (Clarke and Cleverly, 1991).

Compositional Relationships with Subsurface

Jones and Drozd (1983) and Williams et al. (1981) demonstrated that light gas compositions in the surface and subsurface are similar in broad compositional classes for both soils and marine waters. The majority of oil seeps resemble reservoired hydrocarbons (Kennicutt et al., 1988). These correlations indicate bulk flow with little or no modification by other processes along the migration pathway to the surface. Numerous studies by Gulf and Texaco Research on land and in the oceans demonstrate the spatial proximity of zones of high-magnitude compositionally dissimilar samples that are correlatable to the composition of reservoired hydrocarbons in the subsurface. These relationships indicate that multiple pathways from depth exist and that often little or no mixing or alteration occurs along these paths. However, Illich et al. (1984) reported an oil seep that is compositionally similar to that expected from solution transport. The occurrence of an oil phase of this composition at the surface may mean that it leaked from a subsurface accumulation formed by solution transport or that it evolved from solution transport near the surface.

Sweeney (1988) found that in the organically lean shallow sediments of the North Sea, both diffusion and free-phase transport of thermogenic hydrocarbons occurs. The diffusion transport of these hydrocarbons to the near surface was effectively eliminated by bacterial oxidation in the sediment. The free-phase transport, however, has limited contact area with the bacterial zone, resulting in leakage to the sediment–water interface. Where the

process is efficient, a continuous stream of bubbles moves from the sediment into the water column. Where the process is inefficient, a patchy distribution of shallow gas pockets is formed. These pockets increase in size until they overcome the capillary restrictions of the sediment or, more likely, catastrophically break through to the water column by rupturing the sediment.

Variation with Time

Surface seepage at any particular location is time variable. Attempts to reoccupy a previous sampling site usually result in merely being close to the original location and confounding spatial changes with temporal changes. Broad patterns hold over time, provided the subsurface conditions do not change significantly within that time interval. The survey of Dickinson and Matthews (1993) was performed over two field seasons, but the pattern of anomalies held between the two surveys. In subsequent years, other surveys were undertaken enlarging the survey area. These surveys extended edge anomalies of the previous surveys seamlessly. There is, however, a change in surface magnitudes that is recognized in a longer time frame. Significant reservoir pressure decreases related to production lower the magnitudes of surface gases (Horvitz, 1969).

Permanent stations remove the uncertainty of spatial positioning in temporal sampling. These stations show remarkable repeatability, usually less than a factor of ten, if human activity is minimal. The near surface, however, is not a static system. Variations in surface hydrocarbon concentrations are linked to atmospheric pressure changes and earthquakes. Barometric pressure changes propagate downward in excess of 150 m (Figure 11) (Meents, 1958). This periodic pressure change is capable of causing minute expansions and contractions of gaseous free phases within water-wet systems. Seismic events also cause short-term variations (in hours) in hydrocarbon concentration (Zorkin et al., 1977; Antropov, 1981; Burtell, 1989).

Rate of Transport

Linear rate measurements at the surface are seldom made because of the uncertainty associated with depth of origin and cross-sectional area. Linear rate estimates in the upper 200 m are on the order of tens of meters per day, based on known times of injection of gas into storage reservoirs and subsurface coal burns (Jones and Thune, 1982). Arp (1992) summarized estimates of vertical seepage velocities as 75–300 m per year. Volume rate estimates have been made in the marine environment by collecting bubbles. The average oil seep flow rate compiled by Clarke and Cleverly (1991) is 50 m^3 (300 bbl) per year. These rates clearly indicate separate phase migration along narrow migration pathways. Diffusion rates and solution transport are much slower but operate over much larger areas.

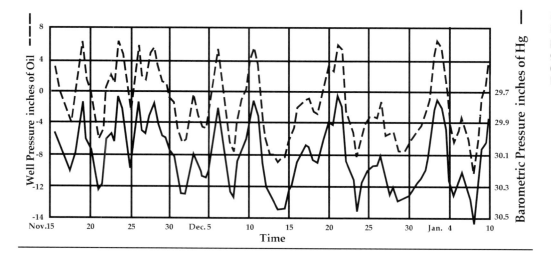

Figure 11—Graph showing atmospheric-driven pressure changes in a gas sandstone at ~150 m (500 ft) depth. (After Meents, 1958.)

CONCLUSIONS

Much variation exists in the characteristics of subsurface and surface hydrocarbons. Since it is clear that a single mechanism is incapable of explaining these variations, we must conclude that hydrocarbon migration occurs by many mechanisms and each case may or may not have significant local effects. Thus, our task is to discover the dominant pattern of migration and the nature of the exceptions. It is my opinion that the dominant migration mechanism is as a free phase interacting with a heterogeneous rock framework.

Except in rare instances, separate phase migration occurs along the entire preferential migration path, from source to surface. The only significant process responsible for modifying the composition of the migrating hydrocarbons is phase separation, aided by the characteristics of the seals along the migration path. The success of geochemical source–reservoir correlation techniques is the strongest of many arguments indicating that the dominant migration mechanism is as a free phase and not as solution. Solution transport, both by diffusion and active water transport, does occur and probably represents the largest reservoir of light gas in the subsurface. It is dominantly a dispersive mechanism, selectively removing the more soluble compounds from hydrocarbon masses, both stationary and in transport, leaving what have been called water-washed accumulations. The kilometer-long diffusion gradients of benzene and toluene within reservoirs demonstrate the effectiveness of this process.

The lack of significant transport of benzene and toluene across adjacent shales argues against both diffusion and active aqueous solution transport as a dominant mechanism of accumulation. The rapid variation with depth of mud log composition and magnitude indicates that diffusion generally does not extend smoothly to the surface or great distances vertically. The ability to concentrate hydrocarbons by solution, either through diffusion or advection, is also limited by the constraints of low solubility and the limited supply of compaction water at maturation depths. The saturation of water by hydrocar-

bons and subsequent migration is, however, supported by a few instances of surface measurements characteristic of solution transport. The existence of a few oil fields with compositions similar to that expected from solution demonstrates that accumulations may form by this process, but they are the exception, not the rule. Most reservoired hydrocarbons have compositions that are controlled by the characteristics of their source rock, not the processes of their transportation. Diffusion of light hydrocarbons into cap rocks explains the abundance of reservoirs undersaturated with light gases.

Micellar transport of hydrocarbons is not considered a dominant mechanism. The amount of solubilizer in the micelle is many times the amount of solubilized hydrocarbon. These solubilizers are seldom detected in reservoirs and source rocks in abundant quantities. Transport of hydrocarbons along a kerogen network is limited to rocks with TOC contents over about 1%. These constitute a small percentage of most migration paths. It may be an important expulsion mechanism within source rocks, but it would seem to render these source rocks relatively ineffective as a top seal. Since many source rocks do act as top seals, the mechanism is thought to be an exception.

Selective adsorption of hydrocarbons has been proposed to occur along the migration path. This is not thought to be a significant factor. The movement of a free phase through the pore network minimizes contact area and time, reducing the opportunity for adsorption. The pore network does not act as a chromatographic column, except as one that is saturated. Only in the first and last stages of a migration event would any chromatographic separation be noticeable.

General Model

Separate phase migration is significantly effected by the heterogeneous nature of rocks, particularly those classified as seals. The heterogeneity of shales is such that, over a large area, capillary restrictions are believed to rarely be 100% effective. In most cases, there is at least one pathway that will leak, albeit slowly. As the accumulation grows, the number of leakage pathways also grows, both because of increased buoyancy pressure opening up previously restricted pathways and because the increased areal extent of the growing accumulation intersects other leakage pathways. Figure 5, previously discussed from a pore system viewpoint, also serves as a model for macroscopic leakage behavior. Depositional facies are not internally random; there is structure to their deposition and spatial characteristics. They are fractile in nature. Thus, there are local concentrations of large pore throat networks and more extensive regions dominated by small pore networks. This translates into areally restricted regions of high flux rate and broad regions of almost no flow due to capillary heterogeneity. The rate-limiting step in postexpulsion migration of hydrocarbons is the transition from coarse to fine pore throat systems. This system is self-adjusting. A consequence of this process is the dynamic formation, maintenance, and destruction of oil and gas fields.

Postexpulsion Model

The motion of hydrocarbons during postexpulsion migration involves four repetitive conditions. First, as capillary imbibition causes hydrocarbons to move into a carrier from discrete pore networks in the source or a seal, they thin and separate into bubbles or droplets. Second, within the carrier bed, buoyancy (assisted by gas expansion) is a major control on migration. Water velocities generated by compaction are generally insignificant compared to buoyancy-driven velocities and as a first approximation can be ignored. In these cases, lateral transport of hydrocarbons is caused by the dip of the contact between the seal and the carrier bed. Water velocities due to topographic driven flow are, however, significantly higher and can locally dominate buoyancy. This can result in significant lateral transport due to water flow, both updip under the seal and downdip (Hubbert, 1953). Once a trapping configuration is reached, the hydrocarbons accumulate under a seal. At this point, the hydrocarbons become a coherent slug and cease to act as discrete droplets. Buoyancy pressure increases as the mass grows. Third, the flow of hydrocarbons from a reservoir into a seal is dominated by the penetration of a separate phase. This is accomplished by buoyant pressure overcoming capillary restriction.

Finally, within the seal, the hydrocarbons move through discrete pathways of one or more continuous threads until they encounter a carrier bed. At this point, capillary imbibition once again becomes dominant and the first condition (above) is repeated. This breaks the thread at each carrier unit, constraining the buoyant force to reservoir–seal couplets. As the hydrocarbons move vertically through the section, they continuously reequilibrate into gas and oil phases. These phases may stay in contact, but are often partitioned into different accumulation sites due to differences in migration pathways (spillage of oil and capillary leakage of gas or oil) and the vector differences in gas and oil when water flow is significant.

The rates of leakage through the sequence of seals between the source rock and the earth's surface control the pattern of hydrocarbon accumulations along the postexpulsion migration route and both the magnitudes and spatial properties of surface seepage. This complex system is idealized into three successive seals for purposes of discussion (Figure 12).

Shortly after hydrocarbons begin to be expelled from the source rock (time 0–2, Figure 12) they begin to appear at the surface in low-magnitude, spatially dispersed sites. This results in an elevated background population with occasional spikes (abnormally high magnitudes) caused by preferential migration pathways extending from the source rock to the surface. Some of the spikes are located near the crest of structures, while others are scattered along the surface projection of migration pathways to these structures. Once the hydrocarbons reach the upper boundary of the first carrier unit above their source (seal A), they migrate to a trapping configuration (time 0–1, Figure 12). The hydrocarbons accumulate there, building up both buoyancy pressure (because of the increased height of the hydrocarbon mass) and water pressure (because the hydrocarbons restrict the area the water has to flow through). If seal A has many large pore networks, only a small pressure gradient is required to carry hydrocarbons through it, leaving a small amount behind. If the volume rate of charge into the carrier is greater than can be accommodated by the large pore networks at the top of the trapping configuration, both hydrocarbon height and areal extent increase until an accommodation of reservoir charge and leakage is accomplished, and the reservoir enters a steady-state condition (time 1–6, Figure 12). Leakage through seal A increases steadily as pressure opens new pathways and the lateral expansion of the hydrocarbon mass encounters new pore networks capable of leaking. The accumulation below seal A disappears (time 6–8, Figure 12) as the rate of charging falls below the rate of leakage.

The hydrocarbons leaking through seal A migrate in the new carrier unit until they reach a trapping configuration under seal B (for convenience, the finest seal in the system). A large mass of hydrocarbons can be accommodated under seal B. As this reservoir begins to grow (time 1–4, Figure 12), both in size and thickness, the rate of leakage also grows due to increased pressure and increased opportunity. The reservoir may even fill to the spill point (time 4, Figure 12). As leakage through seal B increases, regional surface seepage increases in magnitude and number (times 2–4, Figure 12). Surface locations at the end of preferential migration pathways connected to the trapping configuration under seal B markedly increase in spike spatial density due to the subsurface concentration of hydrocarbons that are beginning to form coherent clusters of higher magnitudes rather than isolated spikes. When seal B reaches maximum capacity (time 4, Figure 12), the rate of leakage is controlled by the rate of expulsion from the source rock.

The hydrocarbons leaking through seal B (time 2, Figure 12) progress through a carrier system until they

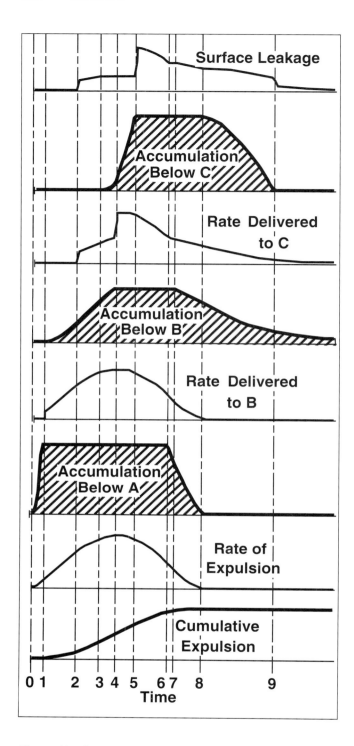

Figure 12—Schematic diagram of the postexpulsion model showing the evolution of subsurface accumulations and seepage from a source through three seals (A, B, and C) to the surface. See text for details.

encounter seal C. For illustrative purposes, seal C only accumulates hydrocarbons under it when the leakage rate from B is near a maximum (time 4, Figure 12). At lower leakage rates from B, seal C is essentially transparent to migrating hydrocarbons. Surface leakage grows in magnitude, extent, and density as long as the reservoir under seal B is growing or being charged beyond capacity (time 2–5, Figure 12) and will reach a maximum when reservoir C is fully charged (time 5). From time 5 to 6 in Figure 12, the reservoirs below all three seals are charged to maximum capacity and surface seepage is also at a maximum, being controlled by the rate of expulsion from the source rock. During these times, surface seepage magnitudes are at a maximum and the clustering of these spikes is best developed, in response to migration up the maximum number of preferential pathways connecting the surface to the accumulations. When the rate of expulsion from the source rock falls below the rate needed to maintain the accumulation below seal A (after time 6, Figure 12), the accumulation below seal B is at first maintained by a high rate of leakage of hydrocarbons stored below seal A (time 6–7, Figure 12).

Surface seepage remains constant during this time with high magnitudes and a well-developed clustering of spikes, supported by equilibrium leakage from the accumulations below seals B and C. The accumulation below seal B decreases in thickness and areal extent as charging from below decreases and becomes insignificant (after time 7, Figure 12). The rate of leakage through seal B simultaneously decreases as the reservoir diminishes. Surface seepage decreases slightly at his time (time 7–8, Figure 12), with the magnitudes diminishing and the clustering of spikes beginning to degrade as more and more preferential pathways are abandoned due to capillary forces overcoming the upward pressure of the hydrocarbons. Seepage is maintained by the excess rate of charging below seal C. Once the accumulation below seal C begins to decrease, the rate of surface seepage is supported by the steady-state seepage through seal C (time 8–9, Figure 12). The surface pattern of spike clustering begins to degrade and the magnitudes decrease slightly. Once the accumulation below seal C is depleted, the surface seepage falls sharply (time 9, Figure 12) and is supported entirely by the low rate of seepage through the best seal of the sequence (seal B). Seal C is totally transparent at these seepage rates. As the accumulation below seal B winds down, the surface seepage does also, decreasing in magnitude and both spatial coherence and frequency of spikes, once again returning to low levels with scattered low-magnitude spikes, indicating a general lack of focused migration of hydrocarbons.

As expulsion stops, the system winds down. As each successive reservoir ceases to leak, the residual hydrocarbons in the reservoirs and along the migration pathways are purged by diffusion and active transport in solution by migrating waters. Where temperature and other biological requirements permit, biodegradation also assists these processes. Indeed, these processes are active from the moment maturation starts, but are not dominant during active expulsion, when separate phase migration overwhelms their abilities. As the hydrocarbons move into lower temperature and pressure conditions, phase separation proceeds, partitioning compounds into gaseous and liquid phases. During postexpulsion migration, these two phases may be physically separated by a combination of trapping configurations and seal behavior.

These processes, combined with the potential for multiple source rocks, at different levels of maturity, and at different migration routes to different reservoirs and mixing within the same reservoir, can lead to a very complex system with spatially distinct and mixed surface hydrocarbon compositions. The preferential occurrence of hydrocarbon-filled inclusions along fractures and at specific diagenetic stages supports the hypothesis that hydrocarbon migration occurs in a relatively narrow time interval. The combination of a relatively narrow migration route with a short time interval minimizes contact of the migrating hydrocarbons with water during migration. Thus, hydrocarbon migration can be thought of as "slug" flow, an idea supported by the observations of Zarrella et al. (1967). If migrating oil were exposed to water over a long time with a large area-to-volume ratio, one would expect that much of the benzene and toluene produced in the source rock would have been removed. This does not seem to be the case, supporting some sort of solution-protected migration.

As a slug of hydrocarbons migrates through shallower hydrocarbons, the small contact area and time minimizes compositional changes in the migrating phase. If the upward-driving force is greater than the driving force of the shallow accumulations, the amount of mixing is slight. This is similar to an open hole completion: the formations with higher excess pressure generally dominate flow into the borehole. Thus, the general background trend of increased concentration and increased gas wetness observed in many mud logs is dominantly a product of the static local generation of unmigrated hydrocarbons with no drive. The higher concentrations observed in sandstones, with compositions similar to the overall trend of compositional change in the well, represent locally generated hydrocarbons that have accumulated in a sandstone near the source. The higher concentrations, with compositions dissimilar to the nearby background composition and isotopic values suggestive of higher maturation conditions, represent migrated hydrocarbons that were generated from a deeper source. These actively migrating hydrocarbons are the ones that make it to the surface, retaining their compositional similarity to the reservoired hydrocarbons, despite the presence of intervening minor accumulations.

This theory predicts that at every coarse–fine interface the rate of transport slows to adjust itself with time to balance input and output. Siddiqui and Lake (1992) point out that one result of capillary trapping theory is that a different hydrocarbon column height should exist at every coarse–fine grain transition and that this is generally not noted. Wells are usually drilled overbalanced. Despite this, mud logs show a wide range of concentra-

tion variation that may be evidence of this hydrocarbon concentration. A well is essentially a one-dimensional sampling of a three-dimensional field. It should not be expected that all sandstone–shale transitions should have hydrocarbon concentration at them. Well-executed mud logs show, however, many small variations in concentration occur and that the higher concentrations are generally correlated with sandstones below shales.

Surface seepage is the end of the migration process. The narrowness of the migration pathways, from both reservoirs and source rocks, to the surface and the speed of this migration maintain the chemical signature of the migrating hydrocarbons. Chemical and mechanical diffusion broadens and decreases the magnitude of seepage as the separate hydrocarbon phase reaches the surface. Biogenic activity uses part of the diffused hydrocarbon seepage as a feedstock, altering its composition and adding biogenic methane. The presence of mature hydrocarbons at the surface of the earth is indisputable proof that an active source rock exists in the subsurface. The composition of these hydrocarbons is similar to those generated by the source or modified by phase separation during migration and entrapment. The spatial pattern of seepage is controlled by the complexities of the migration pathway to the surface. Thus, the challenge in joining surface geochemistry with subsurface information is defining this subsurface migration pathway and improving our ability to predict subsurface accumulations along this pathway.

Acknowledgments—I wish to thank Hollis Hedberg, Ted Weismann, Ed Driver, Bill Glezen, Vic Jones, Dick Mousseau, Bob Pirkle, and numerous other workers at Gulf Research for the many discussions that helped generate my interest in this topic and formulate many of the opinions presented here. I particularly wish to thank Grover Schrayer, Bill Roberts, and Vaughn Robison for discussion and reviewing the manuscript.

REFERENCES CITED

Abrams, M. A., 1992, Geophysical and geochemical evidence for subsurface hydrocarbon leakage in the Bering sea, Alaska: Marine and Petroleum Geology, v. 9, p. 208–221.

Antropov, P. Y., 1981, Laser gas analysis in solving geologic and production problems: International Geological Review, v. 23, no. 3, p. 314–318.

Arp, G. K., 1992, Effusive microseepage: a first approximation model for light hydrocarbons movement in the subsurface: Association of Petroleum Geochemical Explorationists Bulletin, v. 8, p. 1–17.

Barker, C., 1980, Distribution of organic matter in a shale clast: Geochimica et Cosmochimica Acta, v. 44, p. 1483–1492.

Brooks, J. M., B. B. Bernard, and W. M. Sackett, 1978, Characterization of gases in marine waters and sediments, *in* J. R. Waterson and P. K. Theobald, eds., 7th International Geochemical Symposium: Association of Exploration Geochemistry, Golden, Colorado, p. 337–345.

Burruss, R. C., K. R. Cercone, and P. M. Harris, 1983, Fluid inclusion petrography and tectonic history of the Al Ali No. 2 well: evidence for the timing of diagenesis and oil migration, northern Oman foredeep: Geology, v. 11, p. 567–570.

Burtell, S. G., 1989, Geochemical investigations at Arrowhead springs, San Bernardino, and along the San Andreas fault in southern California: Master's thesis, University of Pittsburgh, Pittsburgh, Pennsylvania.

Burtell, S. G., V. T. Jones, R. A. Hodgson, K. Okasa, M. Kuniyasu, and T. Ando, 1986, Remote sensing and surface geochemical study of Railroad Valley, Nye County, Nevada, detailed grid study: Fifth Thematic Mapper Conference, Remote Sensing for Exploration Geology, Reno, Nevada, Sept. 29–Oct. 2.

Catalan, L., F. Xiaowen, I. Chatzis, and F. A. L. Dullien, 1992, An experimental study of secondary oil migration: AAPG Bulletin, v. 76, p. 638–650.

Cathles, L. M., and A. T. Smith, 1983, Thermal constraints on the formation of Mississippi Valley-type lead-zinc deposits and their implications on episodic basin dewatering and deposit genesis: Economic Geology, v. 78, p. 983–1002.

Clarke, R. H., and R. W. Cleverly, 1991, Petroleum seepage and post-accumulation migration, *in* W. A. England and E. J. Fleet, eds., Petroleum migration: Geological Society of London Special Publication No. 59, p. 265–271.

Connan, J., and A. M. Cassou, 1980, Properties of gases and petroleum liquids derived from terrestrial kerogen at various maturation levels: Geochimica et Cosmochimica Acta, v. 44, p. 1–23.

Dembicki, Jr., H., and M. J. Anderson, 1989, Secondary migration of oil: experiments supporting efficient movement of separate, buoyant oil phase along limited conduits: AAPG Bulletin, v. 73, p. 1018–1021.

Dickinson, R. G., and M. D. Matthews, 1993, Regional microseep survey of part of the productive Wyoming-Utah thrust belt: AAPG Bulletin, v. 77, p. 1710–1722.

Durand, B., P. Ungerer, A. Chiarelli, and J. L. Oudin, 1983, Modelisation de la migration de L'Huile: application à deux exemples de bassins sédimentaires: Eleventh World Petroleum Congress, London, Chichester, v. 1, no. 3, p. 3–11.

England, W. A., A. S. Mackenzie, D. M. Mann, and T. M. Quigley, 1987, The movement and entrapment of petroleum fluids in the subsurface: Journal of the Geological Society of London, v. 144, p. 327–347.

Horvitz, L., 1969, Hydrocarbon geochemical prospecting after thirty years, *in* W. B. Heroy, ed., Unconventional methods in exploration for petroleum and natural gas: Dallas, Southern Methodist University Press, p. 925–940.

Hubbert, M. K., 1953, Entrapment of petroleum under hydrodynamic conditions: AAPG Bulletin, v. 37, p. 1954–2026.

Hunt, J. M., 1984, Primary and secondary migration of oil, *in* R. F. Meyer, ed., Exploration for heavy crude oil and natural bitumen: AAPG Studies in Geology 25, p. 345–349.

Jensenius, J., and R. C. Burruss, 1990, Hydrocarbon–water interactions during brine migration: evidence from hydrocarbon inclusions in calcite cements from Danish North Sea oil Ffields: Geochemica et Cosmochimica Acta, v. 54, p. 705–713.

Jones, V. T., and R. J. Drozd, 1983, Prediction of oil or gas potential by near-surface geochemistry: AAPG Bulletin, v. 67, p. 932–952.

Jones, V. T., and H. W. Thune, 1982, Surface detection of retort gases from an underground coal gassification reactor in steeply dipping beds near Rawlings, Wyoming: Society of Petroleum Explorationists, SPE 11050, 24 p.

Judd, A. G., 1990, Shallow gas and gas seepages: a dynamic process?: Safety in Offshore Drilling, v. 25, p. 27–50.

Karlsen, D. A., Nedkvitne, T., Larter, S. R., Bjørlykke, K., 1993, Hydrocarbon composition of authigenic inclusions: application to elucidation of petroleum reservoir filling history, Geochimica et Cosmochimica Acta, v. 57, p. 3641–3658.

Kennicutt II, M. C., J. M. Brooks, and G. J. Denoux, 1993, Leakage of deep, reservoired petroleum to the near surface of the Gulf of Mexico continental slope: Marine Geochemistry, v. 24, p. 39–59.

Krooss, B. M., D. Leythaeuser, and R. G. Schaefer, 1988, Light hydrocarbon diffusion in a caprock: Chemical Geology, v. 71, p. 65–76.

Leith, T. L., I. Kaarstad, J. Connan, J. Pierron, and G. Caillet, 1993, Recognition of caprock leakage in the Snorre field, Norwegian North Sea: Marine and Petroleum Geology, v. 10, p. 29–41.

Leythaeuser, D., A. Mackenzie, R. G. Schaefer, and M. Bjørøy, 1984, A novel approach for recognition and quantification of hydrocarbon migration effects in shale-sandstone sequences: AAPG Bulletin, v. 68, p. 196–217.

Leythaeuser, D., R. G. Schaefer, and H. Pooch, 1983, Diffusion of light hydrocarbons in subsurface sedimentary rocks: AAPG Bulletin, v. 67, p. 889–895.

Lindgreen, H., 1985, Diagenesis and primary migration in Upper Jurassic claystone source rocks in North Sea: AAPG Bulletin, v. 69, p. 525–536.

Lindgreen, H., 1987, Experiments on adsorption and molecular sieving and inferences on primary migration in Upper Jurassic claystone source rocks, North Sea: AAPG Bulletin, v. 71, p. 308–321.

Link, W. K., 1952, Significance of oil and gas seeps in world oil exploration: AAPG Bulletin, v. 36, p. 1505–1541.

McLimans, R. K., 1987, The application of fluid inclusions to migration of oil and diagenesis in petroleum reservoirs: Applied Geochemistry, v. 2, p. 589–603.

McQuillin, R., and N. G. T. Fannin, 1979, Explaining the North Sea's lunar floor: New Scientist, v. 83, p. 90–92.

Matthews, M. D., V. T. Jones, and D. M. Richers, 1984, Remote sensing and surface hydrocarbon leakage: Proceedings of the International Symposium on Remote Sensing and the Environment, Third Thematic Conference, Colorado Springs, April 16–19, p. 663–670.

Meents, W. F., 1958, Tiskilwa drift–gas area, Bureau and Putnam counties, Illinois: Illinois State Geological Survey Circular 253, 15 p.

Momper, J. A., 1978, Oil migration limitations suggested by geological and geochemical considerations, *in* C. Barker, W. H. Roberts, and R. J. Cordell, eds., Physical and chemical constraints on petroleum migration: AAPG Continuing Education Course Notes 8, p. B1–B60.

Nakayama, K., and I. Lerche, 1987, Two-dimensional basin analysis, *in* B. Doligez, ed., Migration of hydrocarbons in sedimentary basins: Second IFP Exploration Research Conference, Technip, Paris, p. 597–611.

Neglia, S., 1979, Migration of fluids in sedimentary basins: AAPG Bulletin, v. 63, p. 573–597.

Pepper, A. S., 1991, Estimating the petroleum expulsion behavior of source rocks: a novel quantitative approach, *in* W. A. England and E. J. Fleet, eds., Petroleum migration:

Geological Society of London Special Publication 59, p. 9–31.

Preston, D., 1980, Gas eruptions taper off in northwest Oklahoma: Geotimes, October, p. 18–20.

Price. L. C., L. M. Wagner, T. Ging, and C. W. Blount, 1983, Solubility of crude oil in methane as a function of pressure and temperature: Organic Geochemistry, v. 4, p. 210–221.

Quanxing, Z., and Z. Qiming, 1991, Evidence of primary migration of condensate by molecular solution in aqueous phase in Yacheng field, offshore South China: Journal of Southeast Asian Earth Sciences, v. 5, p. 101–106.

Roberts, W. H., 1980, Design and functionof oil and gas traps, *in* W. H. Roberts and R. K. Cordell, eds., Problems in petroleum migration: AAPG Studies in Geology 10, p. 217–240.

Sajgo, C., J. R. Maxwell, and A. S. Mackenzie, 1983, Evaluation of fractionation effects during the early stages of primary migration: Organic Geochemistry, v. 5, p. 65–73.

Salisbury, R. S. K., 1990, shallow gas reservoirs and migration paths over a central North Sea diapir: Safety in Offshore Drilling, v. 25, p. 167–180.

Schowalter, T. T., 1979, Mechanics of secondary hydrocarbon migration and entrapment: AAPG Bulletin, v. 63, p. 723–760.

Siddiqui, F. I., and L. W. Lake, 1992, A dynamic theory of hydrocarbon migration: Mathematical Geology, v. 24, p. 305–327.

Silverman, S. R., 1965, Migration and segregation of oil and gas, *in* A. Young and J. E. Balley, eds., Fluids in subsurface environments: AAPG Memoir 4, p. 53–65.

Sweeney, R. E., 1988, Petroleum-related hydrocarbon seepage in a recent North Sea sediment: Chemical Geology, v. 71, p. 53–64.

Thompson, K. F. M., 1988, Gas-condensate migration and oil fractionation in deltaic systems: Marine and Petroleum Geology, v. 5, p. 237–246.

Tissot, B. P., and D. H. Welte, 1984, Petroleum formation and occurence: Berlin, Springer-Verlag, 699 p.

Tóth, J., 1988, Groundwater and hydrocarbon migration, *in* W. Back, J. S. Rosenshein, and P. R. Seaber, eds., Hydrogeology: Geology of North America, v. O-2, p. 485–502.

Van den Bark, E., and O. D. Thomas, 1980, Ekofisk: first of the giant oil fields in western Europe, *in* M. T. Halbouty, ed., Giant oil and gas fields of the decade 1968–1978: AAPG Memoir 30, p. 195–224.

Wakita, H., 1978, Helium spots: caused by diapiric magma from the upper mantle: Science, v. 200, April 28, p. 430–432.

Welte, D. H., and A. Yükler, 1980, Evolution of sedimentary basins from the standpoint of petroleum origin and accumulation, an approach for a quantitative basin study: Organic Geochemistry, v. 2, p. 1–8.

Williams, J. C., R. J. Mousseau, and T. J. Weismann, 1981, Correlation of well gas analysis with hydrocarbon seeps: Proceedings of the American Chemical Society National Meeting, Atlanta, Georgia, March 29–April 3.

Zarrella, W. M., R. J. Mousseau, N. D. Coggeshall, M. S. Norris, and G. J. Schrayer, 1967, Analysis and significance of hydrocarbons in subsurface brines: Geochimica et Cosmochimica Acta, v. 31, p. 1155–1166.

Zorkin, L. M., S. L. Zabairaevi, E. V. Karus, and K. K., Kilmetov,1977, Experience of geochemical prospecting in petroleum and gas deposits in the seismically active zone of the Sakhalin Island: Izvestiya Vysshikh Uchebnykh Zavedeniy. Geologiya i Razvedka, v. 20, p. 52–62.

Klusman, R. W., and M. A. Saeed, 1996, Comparison of light hydrocarbon microseepage mechanisms, *in* D. Schumacher and M. A. Abrams, eds., Hydrocarbon migration and its near-surface expression: AAPG Memoir 66, p. 157–168.

Chapter 12

Comparison of Light Hydrocarbon Microseepage Mechanisms

Ronald W. Klusman

Department of Chemistry and Geochemistry
Colorado School of Mines
Golden, Colorado, U.S.A.

Mahyoub A. Saeed

Faculty of Science
University of Sana'a
Sana'a, Republic of Yemen

Abstract

Surface geochemistry applied to hydrocarbon exploration has gained little acceptance due to the lack of a satisfactory mechanism for vertical migration that is plausible and explains all observations. Acceptance of surface geochemistry is also hampered by a widely held belief that reservoirs do not leak. A satisfactory mechanism must apply in a wide variety of geologic environments and must be verified by laboratory and field observations. Three mechanisms are proposed for vertical migration of light hydrocarbons: diffusion, transport in aqueous solution, and buoyancy of microbubbles.

Diffusion fails to explain the rapid disappearance of surface anomalies after production from a reservoir begins. Diffusion is sufficiently rapid to dissipate gas reservoirs quickly in the geologic sense. As a vertical migration mechanism, it also cannot account for the resolution observed in surface anomalies. Diffusion as a mechanism for primary migration of hydrocarbons from source rocks and as a transport mechanism in the near-surface unsaturated zone have been demonstrated. Solubilities of light hydrocarbons in water are low at ambient temperatures, but increase dramatically with increasing temperatures at depth in basins. Transport with water, either in solution or as a separate hydrocarbon phase, is important in secondary migration. Computer modeling of the process using finite-difference techniques fails to explain the observed resolution and rapid disappearance of surface anomalies.

We favor the vertical migration mechanism of displacing water by ascending gas bubbles, that is, the "buoyancy of microbubbles." Computer modeling of this mechanism does explain surface observations. The close correspondence of surface anomalies with surface projections of a reservoir and the rapidly disappearing surface anomalies after the start of production are predicted by this model.

INTRODUCTION

The acceptance and use of surface geochemistry in the past 60 years has been limited by two major problems. The first is a two-part problem: direct observation of the microseepage process and development of a defensible theory about how microseepage occurs. The second problem is the complexity of the secondary reactions occurring in light hydrocarbons that are undergoing microseepage. These reactions, dominated by oxidation in the upper part of the sedimentary section or in the soil column, can result in many changes and alterations in the near surface. These manifestations are the basis of indirect surface exploration techniques (Klusman, 1993).

We focus on the microseepage process itself and the more prominent mechanisms for the transport of light hydrocarbons from the level of the reservoir to the surface. Each proposed mechanism is discussed along with the evidence for and against it. To be favored, a mechanism must apply in most sedimentary environments. The favored mechanism must also explain most of the empirical observations that have accumulated during the 60+ years of surface exploration.

The existence of microseepage is supported by a large amount of empirical evidence. Foremost are (1) an increase in nonmethane light hydrocarbons as a reservoir is approached during the mud logging of many wells; (2) an increase in C_2/C_1, C_3/C_1, and C_4/C_1 ratios in soil gas over hydrocarbon reservoirs; (3) sharp lateral changes in these ratios at the edges of the surface projections of hydrocarbon reservoirs; (4) the similarity of stable carbon isotopic ratios for methane in soil gases to those found in underlying hydrocarbon reservoirs; and (5) undetectable amounts of ^{14}C in soil gas hydrocarbons and secondary oxidation products, suggesting ancient sources of carbon.

The rate of transport of light hydrocarbons from the reservoir to the surface is hypothesized to be rapid, on the order of months or a few years, and not requiring significant geologic time (MacElvain, 1969). Evidence for rapid transport is primarily from the common observation that a light hydrocarbon soil gas anomaly disappears after initiation of production from a reservoir (Horovitz, 1969). Light hydrocarbons are also observed to increase in a soil gas over pressurized demand gas storage reservoirs (Coleman et al., 1977; Araktingi et al., 1982). Hydrocarbons observed in soil gas over storage reservoirs have stable carbon isotope compositions similar to that of the reservoired gas (Coleman et al., 1977). This frequent observation of rapid transport may be the most critical for a hypothesized mechanism to predict adequately.

We consider three mechanisms for the migration of light hydrocarbons from reservoirs to the surface: (1) diffusion, (2) transport in solution in ascending water, and (3) buoyancy of microbubbles rising through a water-saturated sedimentary column.

MECHANISMS FOR LIGHT HYDROCARBON MICROSEEPAGE

Transport of Hydrocarbons by Diffusion

Diffusion was one of the original mechanisms cited as being responsible for microseepage (Stegena, 1961). Diffusion rates through water-saturated pore space have been calculated but not directly measured in the environment of a hydrocarbon reservoir. Buckley et al. (1958) measured the concentration of dissolved light hydrocarbons in formation waters and found that high concentrations were localized adjacent to known petroleum and gas reservoirs. This suggests that diffusion controlled transport has not moved dissolved gases appreciable distances. Diffusion has been directly measured, and a satisfactory theory developed for gas transport in soil columns, above the saturated zone.

Stegena (1961) applied Fick's second law to the diffusion of hydrocarbons in a sedimentary column. The change in concentration with respect to time is

$$\frac{\partial C}{\partial t} = D \nabla C \tag{1}$$

where D is the coefficient of diffusion and C is the concentration. ∇ is the "del operator," a short-hand symbol for the partial derivative of the variable that follows, in this case concentration C. This partial derivative is with respect to the change in concentration in three dimensions, x, y, z. Only diffusion in the vertical direction is of primary concern in this discussion. We assume that reservoirs fill instantly, and thereafter gases lost by diffusion

are continually replaced from source rocks. The one-dimensional equation for vertical migration through a plane at the top of the reservoir is

$$\frac{\partial C}{\partial t} = D \frac{\partial^2 C}{\partial z^2} \tag{2}$$

Integration of equation 2 as shown by Stegena (1961) is

$$C_{tz} = C_0 \left\{ \frac{z}{h} + \frac{2}{\pi} \sum_{n=1}^{\infty} \frac{(-1)^n}{n} \right.$$
$$\left. \exp\left[-\left(n^2 \pi^2 / h^2\right)Dt\right] \sin\left(\frac{n\pi z}{h}\right) \right\} \tag{3}$$

Equation 3 allows calculation of the concentration of the gas dissolved in the pore water, C_{tz}, at a depth z, at time t, between $z = 0$ and $z = h$. The concentration of the gas in the reservoir must remain constant at C_0.

Stegena (1961) estimated diffusion coefficients for methane to be 10^{-10} to $10^{-12} m^2/sec$. Stegena's calculation indicates that a steady-state concentration of methane would be achieved at the surface only after hundreds of millions of years, even for a relatively shallow reservoir at 1000 m. Calculated flux rates for methane to the surface are about 2 $g/m^2/100$ m.y. This exceedingly slow transport rate contrasts with measured methane flux rates at the surface in sedimentary basins on the order of 0.0–0.002 $g/m^2/day$ (Jakel and Klusman, 1995).

Leythaeuser et al. (1980) measured diffusion rates for light hydrocarbons in shale cores from a thermally mature, hydrocarbon-rich shale in Greenland. The original weathered surface was planed off by glaciation, and the section is presently in permafrost. This suggests that microbial consumption of hydrocarbons was minimal. Ice retreat began about 6000 years ago, allowing hydrocarbon diffusion from the shale to the atmosphere.

Hydrocarbon concentrations in pore ice were constant below 8 m depth, which Leythaeuser et al. (1980) defined as C_0. The concentration at the surface was assumed to be zero, even though this is not equal to the atmospheric concentration (Warneck, 1988). Figure 1 shows the distribution of ethane and *n*-butane concentrations in the rock matrix of the core. Initial estimated concentrations C_0 from Figure 1 were 3.1 ng ethane/g rock and 165 ng *n*-butane/g rock. The diffusion coefficients were calculated by Leythaeuser et al. (1980) at 6.4×10^{-11} m^2/sec for ethane and 3.2×10^{-11} m^2/sec for *n*-butane. Calculated flux rates were about 1 $g/m^2/year$ for ethane and 0.1 $g/m^2/year$ for pentane. The diffusion rates for light hydrocarbons found by Leythaeuser et al. (1980) are similar in magnitude for the C_2–C_5 hydrocarbons. Extrapolation to methane suggests a flux for methane similar to that of ethane. This flux rate is comparable to that found by Jakel and Klusman (1995) for methane in sedimentary basins of temperate climates.

Leythaeuser et al. (1982) estimated the time required to fill a gas reservoir and the time required to empty a reservoir by diffusive loss. They focused on the Harlingen gas field in The Netherlands. They estimated the half-life of this gas field to be 4.5 m.y., by diffusive loss through a cap rock, if the gas were not replenished from a source.

Krooss et al. (1992a) reevaluated the calculations of Leythaeuser et al. (1982) and suggested that the model was too simple and did not account for the equilibration between methane in the gas phase and dissolved methane in the water-saturated cap rock. Krooss et al. modified equation 3 to calculate only vertical total flux, Q, into a cap rock:

$$\frac{Q(t)}{hC_0} = \frac{Dt}{h^2} + \frac{2}{\pi^2} \sum_{n=1}^{\infty} \frac{1}{n^2} \left[1 - \exp\left(-Dn^2\pi^2t/h^2\right) \right] \quad (4)$$

where h is now the thickness of the cap rock. For $t \rightarrow \infty$, equation 4 becomes

$$\frac{Q(t)}{hC_0} = \frac{Dt}{h^2} + \frac{1}{3} \quad (5)$$

To calculate loss of one-half of a reservoired gas, equation 5 was modified by Krooss et al. (1992a) as follows:

$$\frac{Q_{1/2}}{hC_0} \approx \frac{Dt_{1/2}}{h^2} + \frac{1}{3} \quad (6)$$

which is rearranged to solve for $t_{1/2}$:

$$t_{1/2} \approx \left(\frac{Q_{1/2}}{hC_0} - \frac{1}{3} \right) h^2 / D \quad (7)$$

Krooss et al. (1992a) gave the cap rock thickness, h, as 390 m, with D as 2.12×10^{-10} m^2/sec, $Q_{1/2}$ as 173.9 kg CH$_4$/m^2 in the cap rock, and $t_{1/2}$ as 70 m.y.

This is complicated by two factors. Once the methane passes the cap rock, it must traverse the overlying sedimentary column, which will presumably have a higher permeability than the cap rock. Therefore, the traverse of this part of the section will be more rapid. Also, if the adsorptive sites in the overlying section are not saturated, migrating gas will be removed until adsorbing sites are saturated. Applied to surface exploration, these factors can be ignored, if it is assumed the reservoir has existed for sufficient time for methane and light hydrocarbons to have traversed the cap rock and the overlying sedimentary column and to have saturated the sites capable of adsorbing methane and light hydrocarbons.

Nelson and Simmons (1992) criticized the Krooss et al. (1992a) calculations, and Krooss et al. (1992b) replied. Nelson and Simmons suggested that the value for D must be an effective D that incorporates porosity, permeability, and tortuosity terms. They calculated the value for $t_{1/2}$ to be 11.5 m.y. for the Harlingen gas reservoir. Nelson and Simmons (1992) also considered the effect of temperature on the diffusion coefficient and of salinity on the viscosi-

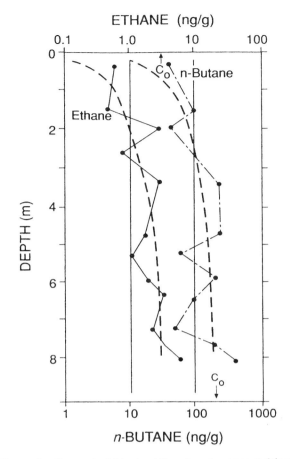

Figure 1—Computed (dashed lines) and measured (solid lines) concentration versus depth trends for ethane and *n*-butane from corehole E in Campanian–Maastrichtian shale, Niaqorssuaq, West Greenland. Initial concentrations, C_0, used for calculation of diffusion coefficients for ethane and *n*-butane were 3.1 and 165.0 ng/g, respectively (indicated by arrows). (From Leythaeuser et al., 1980. Reprinted with permission from Nature, v. 284. Copyright © 1980 Macmillan Magazines Limited).

ty of water. With a cap rock temperature of 70°C and 3.6 wt. % salinity, Nelson and Simmons calculated $t_{1/2}$ to be 4.4 m.y., which was essentially the same as the original 4.5 m.y. given by Leythaeuser et al. (1982). Krooss et al. (1992b) reiterated that the original assumptions about porosity effects on the value for the diffusion coefficient were correct and stated that the effect of temperature on D would be addressed in a future publication.

Nelson and Simmons (1995) estimated diffusive losses of methane and ethane for the McClave gas field on the Las Animas Arch in southeastern Colorado. They included the effect of temperature on the diffusion coefficient and the effect of salinity on water viscosity. The cap rock is shale, with an estimated porosity of 5%, and the entire methane volume in the reservoir must be replaced from a source (by catagenesis) every 2.2 m.y. and ethane every 5.3 m.y. For a simulated porosity of 10% in the overlying shale, the replacement time in the McClave field was calculated by Nelson and Simmons (1995) as 0.48 m.y. for methane and 1.2 m.y. for ethane.

Krooss et al. (1992a) calculated a total of 347.8 kg CH_4/m^2 under the cap rock of the Harlingen gas reservoir. Using the 4.4-m.y. value of Nelson and Simmons (1992) as the half-life of the reservoir and assuming a linear rate of loss, Nelson and Simmons calculated the average flux through the cap rock as 0.11 mg CH_4/m^2/day. Nelson and Simmons (1995) calculated fluxes from the McClave field for methane and ethane. Using the 5% porosity value for the cap rock, the flux for methane was 0.007 mg/m^2/day and at 10% porosity the flux was 0.03 mg/m^2/day. The calculated ethane flux was about 0.1 that of methane. It should be emphasized that the calculated losses would be a first-order decay process proportional to the amount present, and the loss would not be linear with respect to time unless the reservoir were maintained by secondary migration or catagenesis. The fluxes of other light hydrocarbons would be less, but when the adsorptive capacity of the overlying sedimentary column was exceeded, measurable concentrations could be observed at the surface.

The changes in concentration in a flux chamber at the surface for methane are within the realm of detection, allowing calculation of flux rates. This requires that the methane reach the surface without being consumed by methanotrophic bacteria. A single flux determination would not be adequate to determine low rates, just as a single soil gas measurement would not reliably detect a subsurface reservoir. Measured fluxes of 0–2 mg CH_4/m^2/day given by Jakel and Klusman (1995) are in this order of magnitude. We must emphasize that methanotrophic bacteria in the soil column do consume methane and light hydrocarbons. The result is apparent negative fluxes. That is, methane is removed from the atmosphere as the soil acts as a sink for methane. Negative methane fluxes are commonly determined for soils of temperate climates, particularly in areas not underlain by sedimentary basins with hydrocarbon potential or basins not actively generating hydrocarbons in the subsurface. There is currently not enough known about light hydrocarbon fluxes to effectively use such measurements in surface prospecting. Significant positive fluxes of methane do occur in wet environments, independent of underlying geology, due to methanogenesis.

Thomas and Clouse (1990a,b,c) determined diffusion coefficients for heavier hydrocarbons. Using C_{15} as a model compound, a diffusion coefficient, D, of 5.7×10^{-11} m^2/sec was determined for migration through a kerogen layer on micritic sediments. Whelan et al. (1984) determined diffusion coefficients for black shales as 1.7×10^{-12}, 2×10^{-12}, 2.6×10^{-12} m^2/sec for *n*-pentane, *n*-hexane, and *n*-heptane, respectively. With these lower diffusion coefficients and greater adsorption capacities, we expect that heavier hydrocarbons will not move at appreciable rates from reservoirs because gasoline-range hydrocarbons are retained for much longer times than are methane and other gaseous hydrocarbons.

The previous discussion suggests that if diffusion were a primary mechanism of vertical migration, it should be possible to detect elevated concentrations of methane and light hydrocarbons in soil gases above reservoirs that are undergoing microseepage. That is, diffusion as a transport mechanism would not preclude the use of surface geochemistry for exploration. However, the response at the surface to a pressure decrease in a reservoir associated with production would not be detected in the short time frame often seen in surface geochemical data (Horovitz, 1969). The effects of nonvertical dispersion must also be considered, which may not allow an adequate explanation of the sharp definition observed for many surface anomalies (Rice, 1986).

Light hydrocarbon transport calculated by equations 3 and 4 needs to be incorporated into a computer model of the diffusion process in order to assess the interactions of the many parameters. Krooss et al. (1992a, b) and Nelson and Simmons (1992, 1995) calculated the loss of gas through the cap rock. Since this gas is no longer recoverable, it is not of economic interest. To test a hypothesis of diffusion as a vertical migration mechanism and to apply it to surface geochemical exploration, diffusion through the entire sedimentary column must be considered. Layered sedimentary columns must be incorporated for each modeled layer, with each stratum having a specific value for the diffusion coefficient. Also, the change in the diffusion coefficient with temperature in response to the geothermal gradient and the change in salinity as it affects the viscosity of pore water must be considered. Furthermore, the equations should be extended to three dimensions, rather than one dimension, to account for dispersion in the horizontal direction and lateral migration of connate water. Only then can a more thorough understanding of the role of diffusion in vertical migration be determined.

Transport of Hydrocarbons by Water

Transport of hydrocarbons by water dissolved in ascending water is a second possible mechanism for the transport of hydrocarbons. The solubilities of hydrocarbons and of various petroleum fractions have been measured by Price (1976) and McAuliffe (1979). The aqueous solubility is relatively constant for ethane through *n*-butane, then decreases rapidly with increasing carbon number from *i*-butane through C_{12}; then the rate of decrease in solubility is much less for heavier hydrocarbons (McAuliffe, 1979). Table 1 lists the solubility of the light hydrocarbons from methane through *n*-octane at 25°C. Water solubilities of individual hydrocarbons and petroleum fractions increase rapidly with increasing temperature (Price, 1976). Figure 2 shows that steeply increasing solubilities are particularly evident for the lighter alkanes and lighter distillation fractions.

The removal of hydrocarbons from the ascending water must be considered. In the formation where a hydrocarbon is reservoired, the decreased solubility with increased salinity can be invoked as a removal mecha-

Table 1—Solubility of Light Hydrocarbons in Water at 25°C[a]

Hydrocarbon	Solubility (ppm by weight)
Methane	24.4
Ethane	60.4
Propane	62.4
n-Butane	61.4
i-Butane	48.9
n-Pentane	38.5
n-Hexane	9.5
n-Heptane	2.93
n-Octane	0.66

[a]From McAuliffe (1979).

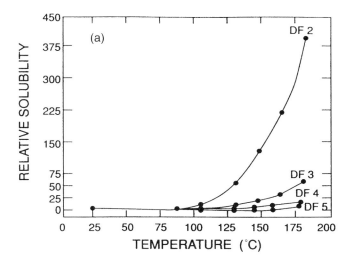

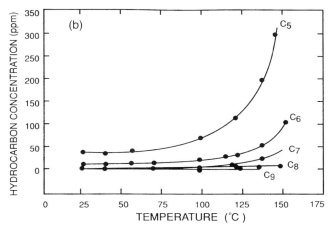

Figure 2—(a) Aqueous solubilities of second (DF2, 132–193°C), third (DF3, 193–232°C), fourth (DF4, 232–316°C), and fifth (DF5, 316–371°C) distillation fractions of Ghawar Arabian crude oil as a function of temperature at the pressure of the system. (b) Aqueous solubility of pentane (C_5), hexane (C_6), heptane (C_7), octane (C_8), and nonane (C_9) as a function of temperature at the pressure of the system. (From Price, 1976.)

nism. The well-known "salting-out" of dissolved molecular species estimated using the Setchenow relationship may be a suitable mechanism for separating out a hydrocarbon phase. Price (1976) demonstrated decreasing hydrocarbon solubilities with increasing salinity. Jones (1984) demonstrated secondary migration of hydrocarbons by a combination of solution and two-phase transport. Solubilization of liquid hydrocarbons in a gas phase was described by Neglia (1979) and would facilitate secondary migration.

For vertical migration and its application in surface geochemistry, hydrodynamics must be considered (Tóth, 1988; Rostron and Tóth, this volume). Connate water in the sediments, meteoric water added at the margins of the basin, and structural water derived from the conversion of smectite clays to illite all contribute to the source of ascending fluids for two-phase transport. With increasing depth of burial, temperature and pressure increases, resulting in fluid volume increases (Magara, 1978). Spencer (1987) suggested that generation of hydrocarbons with increasing depth and temperature adds hydrocarbons to the fluid phase, further increasing pressure.

Bethke (1985) showed a schematic of water flow in a subsiding basin (Figure 3). In the absence of faults, the flow is primarily lateral through permeable layers. Vertical flow becomes more important if the ratio of vertical to horizontal path lengths becomes less than the ratio of vertical to horizontal permeabilities. Figure 3 shows horizontal flow velocities at about two orders of magnitude greater than vertical velocities. If faulting has occurred or if high potential gradients are present (bottom, Figure 3), there may be significant vertical flow.

Advection and hydrodynamic dispersion must be considered. In the transport of hydrocarbons by water, advective or bulk vertical transport of hydrocarbons by water is required to be effective for surface exploration. Evaporative loss may occur as the fluid approaches the surface, and light hydrocarbons would partition into the gas phase according to Henry's law. The dispersion term would decrease maximum concentrations and decrease resolution at the boundaries of anomalies. Tortuosity of the sediment and transport by gaseous diffusion above

the water table would also reduce resolution in a surface anomaly (Devitt et al., 1987; Krooss et al., 1992b; Nelson and Simmons, 1992, 1995). Transport of light hydrocarbons in the gas phase above the water table is primarily by diffusion, modified by complex meteorologic and seasonal variables (Klusman, 1993).

The transport of a single hydrocarbon molecule due to advection and dispersion can be written as follows (Bear, 1979):

$$Q = \varepsilon\left[C(V - D_t)\nabla C\right] \qquad (8)$$

where Q is the hydrocarbon flux (per liter, or L^{-1}), V the velocity of water (L/sec), ε the water content (L), D_t the

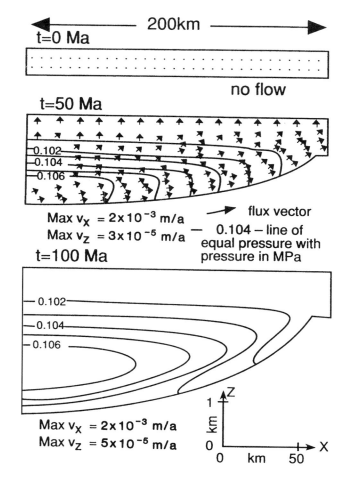

Max $v_x = 2 \times 10^{-3}$ m/a   ➤ flux vector
Max $v_z = 3 \times 10^{-5}$ m/a   — 0.104 — line of
                                    equal pressure with
                                    pressure in MPa

Figure 3—Calculated compaction-driven flow within a basin cross section. Equipotentials are at intervals of 0.001 MPa (0.01 atm). The flow velocities v_x and v_z are relative to the subsiding medium. (From Bethke, 1985.)

diffusion coefficient, and C the concentration of dissolved hydrocarbon. If there are no chemical reactions resulting in production or consumption of the light hydrocarbons and no adsorption, or if the adsorption sites are already saturated, the mass balance for a dissolved hydrocarbon becomes

$$\frac{\partial(\varepsilon C)}{\partial t} = -\nabla QC \qquad (9)$$

Substitution of equation 8 into equation 9 gives a hydrocarbon transport equation in water of

$$\frac{\partial C}{\partial t} = \nabla \varepsilon D_t (\nabla_C - \nabla_\varepsilon)VC \qquad (10)$$

Thomas (1982) solved equation 10 by a finite-difference method, and Saeed (1991) applied it to vertical migration. The finite-difference method divides the

region of interest into small blocks. Time is divided into small steps, which are solved in sequence. The partial differential equations are replaced by their finite-difference equivalents, which results in algebraic equations that are approximations of the original partial differential equations. We determined the concentration of the hydrocarbon at the center of the blocks by solving the algebraic equations using the "strongly implicit procedure" of Thomas (1982).

Equation 10 was rewritten for each grid point of volume v within the pore solutions as

$$\int_v \frac{\partial(\varepsilon C)}{\partial t} dv = \int_v \nabla \varepsilon D_t \nabla C dv - \int_v \nabla \varepsilon \bar{v} C dv \qquad (11)$$

Thomas (1982) transformed the two volume integrals of equation 11 to surface integrals by using the Gauss divergence theorem. If the volume of each block is small, its porosity and density are assumed to be constant, as is the concentration C of the hydrocarbon. The transformation to surface integrals gives

$$v \frac{\partial(\varepsilon C)}{\partial t} = \int_s -\varepsilon D_t \nabla C \bar{n} d\bar{s} - \int_s \bar{v} \varepsilon C \bar{n} d\bar{s} \qquad (12)$$

We approximated the dispersive integral in equation 12 for two-dimensional flow by considering only four faces of the finite difference block:

$$\int_s \varepsilon D_t \nabla C \bar{n} d\bar{s} = \sum_{i=1}^{4} \int_{s_i} \varepsilon D_t \nabla C \bar{n} d\bar{s}_i \qquad (13)$$

To solve the equations, both initial and boundary conditions must be defined. An initial hydrocarbon gas concentration is entered as a constant value at the level of the reservoir, C_0. The flux is the sum of the dispersive and advective fluxes. The dispersive term decreases the flux, and the concentration changes vertically. The computer program GASINWAT was modified by Saeed (1991) from one documented by Healy (1990) for modeling of solute flow in water. Modifications by Saeed allow varying porosities and permeabilities in the overlying sedimentary column, the compressibility of water, and the solubility change of a hydrocarbon with pressure.

Figure 4 shows the calculated dissolved concentration of light hydrocarbons dissolved in water ascending at 10^{-12} m/sec for a period of 10,000 years; other input parameters are given in the figure caption. Little dissolved hydrocarbons reach the surface in this short period of time. By increasing the permeabilities of the layers and using much longer times, appreciable light hydrocarbons are shown to reach the surface (Figure 5).

The exsolution of the dissolved hydrocarbon into the gaseous phase near the surface is controlled to some extent by Henry's law, although this is a heterogeneous process that may result in some degree of gas supersaturation in the aqueous phase. Slow evaporation of ground-

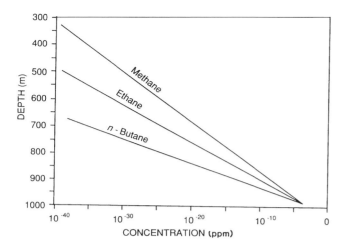

Figure 4—Calculated concentration of methane, ethane, and *n*-butane (in ppm by weight) versus depth after 10,000 years of migration from a reservoir at 1000 m depth. The reservoir is overlain by a single lithology with a permeability of 10^3 md, a diffusion coefficient of 10^{-10} m^2/sec, a porosity of 20%, and a water flux rate of 10^{-12} m/sec. (Modified from Saeed, 1991; Klusman, 1993.)

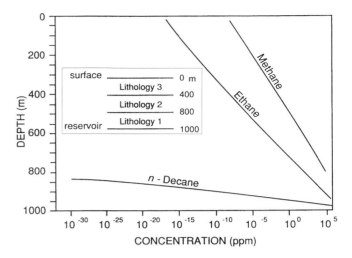

Figure 5—Calculated concentration of methane, ethane, and *n*-decane (in ppm by weight) versus depth after 500,000 years of migration from a reservoir at 1000 m depth. The reservoir is overlain by three lithologies: lithology 1 is 200 m thick and has a permeability of 10^5 md; lithology 2 is 400 m thick with a permeability of 10^3 md; and lithology 3 is 400 m thick with a permeability of 0.1 md. All lithologies have a porosity of 20%, a diffusion coefficient of 10^{-10} m^2/sec, and a water flux rate of 10^{-9} m/sec. (Modified from Saeed, 1991; Klusman, 1993.)

water would increase the concentration observed in the soil gas, although this cannot be quantitatively predicted. If the process operated for a long period of time, adsorption sites in the unsaturated sediment and soil would become saturated. The free soil gas concentration of hydrocarbons would reach dynamic equilibrium, but vary over short time periods due to meteorologic and seasonal parameters (Klusman, 1993). The adsorbed soil gas concentrations would likely remain more constant than free soil gas, particularly over the time frame of meteorologic variations. Seasonal variations would likely remain significant in samples taken in the top few meters (Klusman, 1993).

The upward migration of connate waters associated with hydrocarbon reservoirs is supported by heat flow measurements. Meyer and McGee (1985) measured higher temperature gradients over 15 of 22 oil and gas fields sampled in the Rocky Mountain province. Bodner and Sharp (1988) found elevated subsurface temperatures along the Wilcox growth fault in the Gulf Coast. They proposed upward migration of fluids along the fault, increasing the temperature gradient. Hanor (1987) demonstrated convective transport of connate waters associated with the Iberia salt dome in Louisiana.

The transport of hydrocarbons by water should allow detection at the surface and not preclude water transport as a mechanism for vertical migration. The locations of anomalous free or adsorbed soil gas samples relative to a subsurface reservoir would be determined largely by hydrodynamic flow. This mechanism is not sufficiently rapid, however, to account for the rapid change in the character of the surface anomaly with initiation of production. The data of Hanor (1987) suggest flows of a few meters per year.

Transport of Hydrocarbons by Buoyancy of Microbubbles

Transport of hydrocarbons by buoyancy of microbubbles occurs when capillary pressure of a microbubble of a gaseous hydrocarbon exceeds the water displacement pressure of the largest interconnected pores in a cap rock. Watts (1987) classified these as *membrane seals*. Berg (1975) described membrane seals in stratigraphic traps, and Schowalter (1979) compared the effectiveness of seals against the migration of gas compared to penetration by oil. The threshold displacement pressure required to displace water increases as the permeability of the cap rock decreases (Katz and Coats, 1968). Another variation in the cap rock seal is the *hydraulic seal* (Watts, 1987). This seal functions until there is some degree of overpressure in the reservoir, when the seal opens as a fracture. The scale of fracturing is important in the manifestations of the leakage (Watts, 1987). Microfractures are sufficient to allow seal penetration by gaseous hydrocarbons. Larger fractures and faults may result in upward streaming of two-phase hydrocarbon and water mixtures (Watts, 1987).

MacElvain (1969) was apparently the first to describe the buoyancy of microbubbles in the context of surface exploration. He described rates of vertical migration that was possibly on the order of several millimeters per sec. The overlying sedimentary column has some capacity for adsorption of light hydrocarbons, as indicated by thermal stripping of well cuttings, but when this capacity is exceeded, the microbubbles move unimpeded in an approximately vertical direction. MacElvain (1969) proposed bubbles of a size that would generate significant Brownian motion, which would prevent adsorption. This seems unlikely, and does not have to be prevented if the adsorptive capacity of the overlying sediments has been satisfied by earlier microseepage (Klusman, 1993). Differential adsorption of alkanes by the sedimentary matrix results in the occasional occurrence of a chromatographic effect. If the system is in steady-state equilibrium with a reservoir of constant composition, the adsorptive sites will not have any additional chromatographic capacity and additional fractionation will not occur. Also, the solubility of heavier hydrocarbons in the gas phase decreases with temperature and pressure, resulting in the lighter hydrocarbons dominating in a soil gas (Price et al., 1983; Price, 1986).

The basic equations describing hydrocarbon gas and water flow in porous media are the continuity equations for each phase. For both gas and water, Darcy's law is proportional to the potential gradient of each phase (Tek et al., 1966). The velocity of transport for water and gas can be written as follows (Saeed, 1991; Klusman, 1993):

$$\overline{V}_w = -K \frac{K_w}{\mu_w} \nabla \phi_w \quad (14)$$

and

$$\overline{V}_g = -K \frac{K_g}{\mu_g} \nabla \phi_g \quad (15)$$

where V_w and V_g are the velocities of water and gas, K is the rock permeability, K_w and K_g are the relative permeabilities of water and gas, μ_w and μ_g are the viscosities of water and gas, and ϕ_w and ϕ_g are the water and gas potentials, both defined as $P + \rho g h$. Gas and water relative permeabilities are related to the degree of gas or water saturation. The pressure in the gas phase exceeds that in the water by the capillary pressure, P_c, and

$$P_c = P_g - P_w = \left(\phi_g - \phi_w\right) + \left(\rho_w - \rho_g\right)gh \quad (16)$$

where P_g and P_w are the pressures of water and gas, g is the acceleration due to gravity, ρ_g and ρ_w are the densities of water and gas, and h is the column height. The relationship between capillary pressure and water saturation, s, with respect to time is

$$\frac{\partial s}{\partial t} = \frac{\partial s}{\partial P_C} \times \frac{\partial P_C}{\partial t} \quad (17)$$

This can be combined with equation 16 to give the relationship between capillary pressure and time:

$$\frac{\partial P_C}{\partial t} = \frac{\partial\left(\phi_g - \phi_w\right)}{\partial t} \quad (18)$$

The following simultaneous non-linear partial differential equations for both gas and water potentials as dependent variables are obtained:

$$\nabla K \frac{K_g}{\mu_g} \nabla \phi_g + Q_g = -\varepsilon \frac{\partial s}{\partial P_C}\left(\frac{\partial \phi_g}{\partial t} - \frac{\partial \phi_w}{\partial t}\right) \quad (19)$$

and

$$\nabla K \frac{K_w}{\mu_w} \nabla \phi_w + Q_w = -\varepsilon \frac{\partial s}{\partial P_C}\left(\frac{\partial \phi_g}{\partial t} - \frac{\partial \phi_w}{\partial t}\right) \quad (20)$$

where Q_g and Q_w are gaseous hydrocarbon and water fluxes and ε is the water content.

Gas displaces water, so the sum of the fluxes is one, and migration is only considered in the upward direction because of buoyancy. Equations 19 and 20 modified by Crichlow (1977) allow solution by finite-difference methods. The original computer code written by Tek et al. (1966) was modified by Saeed (1991) to model hydrocarbon gas migration by buoyancy of microbubbles.

Figure 6 shows hydrocarbon gas potential with depth over short time periods as calculated by the program GASWATER. The simulated reservoir was at a depth of 1000 m and had a single lithology as overburden, a permeability of 0.001 md, and a porosity of 10%. The reservoir was overpressured at a total pressure of 3000 psi (20,700 kPa). There were rapid changes in gas potential above the reservoir, but not readily discernible at the surface.

Figure 7 shows the hydrocarbon gas potential over short time periods at three different reservoir pressures. As in Figure 6, the simulated reservoir was 1000 m deep and had a single lithology as overburden, a permeability of 0.001 md, and a porosity of 10%. The modeled reservoir pressures were 1500 psi (10,300 kPa), 2000 psi (13,800 kPa), and 3000 psi (20,700 kPa). These pressures are slightly below hydrostatic pressure, just above hydrostatic pressure, and well above hydrostatic pressure, respectively. Curves A, B, and C on Figure 7 are for a location 10 m below the surface. Curves D, E, and F are for a location 100 m above the reservoir. Again, only small differences are noted at the near-surface locations relative to the location just above the reservoir.

Figure 8 is an expanded view of the upper left corner of Figure 7, showing only the data for the location 10 m below the surface. The greater the magnitude of the reservoir pressure, the greater the change in gas potential at the surface in short time intervals. If production were to occur from the simulated reservoir, rapid changes would be observed both in the reservoir and at the near-surface location. This is consistent with observations of enhanced leakage discussed in the introduction on pressure changes in demand gas storage reservoirs.

The plot in Figure 9 shows the calculated change in gas potential pressure 10 m below the surface for different simulated reservoir pressures, again at a depth of 1000 m. If the reservoir pressure were abruptly changed from

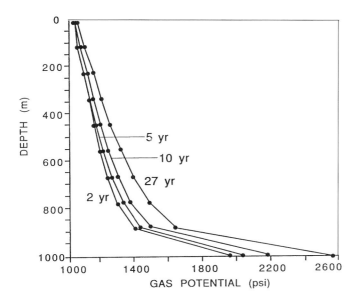

Figure 6—Calculated hydrocarbon gas potential versus depth after 2, 5, 10, and 27 years. A single lithology with a permeability of 0.001 md and a porosity of 10% overlies the reservoir at 1000 m depth. Reservoir pressure is 3000 psi (20,700 kPa). (Modified from Saeed, 1991; Klusman, 1993.)

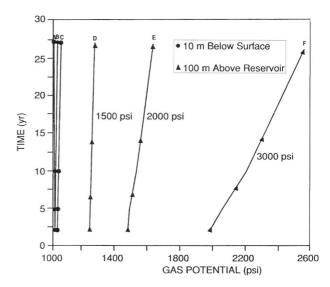

Figure 7—Calculated hydrocarbon gas potential versus time as a function of reservoir gas potential. A single lithology with a permeability of 0.001 md and a porosity of 10% overlies the reservoir at 1000 m depth. Curves A, B, and C (far left) are 10 m below the surface with reservoir pressures of 1500 psi (10,300 kPa), 2000 psi (13,800 kPa), and 3000 psi (20,700 kPa), respectively. Curves D, E, and F are 100 m above the reservoir with reservoir pressures of 1500 psi (10,300 kPa), 2000 psi (13,800 kPa), and 3000 psi (20,700 kPa), respectively. (Modified from Saeed, 1991; Klusman, 1993.)

2000 psi (13,800 kPa) to 1500 psi (10,300 kPa), a change in gas potential of 41 kPa would be observed in the near surface in 10 years. A change in reservoir pressure from 3000 psi (20,700 kPa) to 1500 psi (10,300 kPa) would change the gas potential by 170 kPa in the same time period. The simulated change in gas potential declines relatively rapidly after 10 years for a low-pressure reservoir and after 40 years for a high-pressure reservoir.

Modeling of the buoyancy of microbubbles of gas by Saeed (1991) supports this mechanism as responsible for vertical migration. It satisfies our difficult requirement that migration be close to vertical, so that anomalies remain sharply defined at the surface. The simulation of this mechanism also predicts the rapid response in gas potential and in observed soil gas concentration changes at the surface.

Araktingi et al. (1982) described the failure of a fault seal in the Leroy gas storage reservoir in southwestern Wyoming. The fault seal failure was intermittent, in direct response to the reservoir pressure changes. Copious leakage was noted at the surface a few years after overpressuring the reservoir. They described this phenomenon as acting like a "check valve," although a more appropriate term might be "pressure relief valve" such as on a water heater. With overpressuring, the fault (valve) opened, allowing copious leakage until the pressure dropped below the capillary entry pressure. Then the fault (valve) closed again. The transport rate of gas to the surface was on the order of 1000 ft/year (~300 m/year). In this case, the system was acting as a hydraulic seal, using Watt's (1987) terminology.

Arp (1992) described pressure-induced leakage from the Patrick Draw field in southwestern Wyoming. Produced gas, as well as water, were reinjected into the field to maintain pressure. Initial gas injection is thought to have begun about 1960–1962; there were reports of surface vegetation die-off by 1980 due to saline water and because of anaerobic soils as hydrocarbons displaced oxygen. This translates to a transport rate 250 ft/year (~75 m/year). Large petroleum production increases began in the 1973–1975 period. If the die-off that started in 1980 is related to the production increase, the transport rate for hydrocarbons and water exceeds 1000 ft/year (~300 m/year).

We must be careful not to interpret these examples strictly as buoyancy of microbubbles. The Leroy storage reservoir described by Araktingi et al. (1982) and the Patrick Draw oil field leakage described by Arp (1992) represent atypical flow rates. The leakage from Patrick Draw was certainly two-phase, as evidenced by observations of saline water and extreme concentrations of hydrocarbons in the soil gas.

CONCLUSIONS

Three mechanisms proposed for vertical migration and microseepage have been evaluated: diffusion, transport by ascending water, and buoyancy of microbubbles. All three mechanisms can produce surface anomalies.

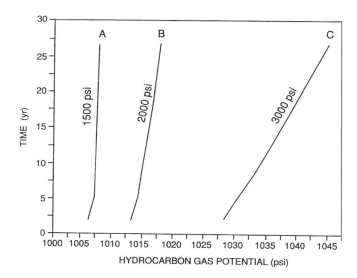

Figure 8—An expanded view of the upper left corner of Figure 7 showing only the data for the location 10 m below the surface. See Figure 7 for parameters. A significant change in gas potential occurs in the near-surface with change in reservoir pressure over a short period of time. (Modified from Saeed, 1991; Klusman, 1993.)

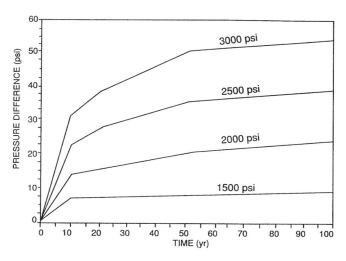

Figure 9—Calculated change in potential pressure 10 m below the surface versus time at different reservoir pressures. The significant change in potential pressure with time results in rapid changes in soil gas anomalies with production. (Modified from Saeed, 1991; Klusman, 1993.)

Diffusion may be the dominant transport mechanism in the short interval above the water table, but it cannot be the dominant mechanism through the entire saturated zone. One major problem is the speed of transport, even though Leythaeuser et al. (1982), Krooss et al. (1992a, b), and Nelson and Simmons (1992, 1995) proposed that gas reservoirs are somewhat ephemeral during geologic time. Other observations not satisfied by a diffusion mechanism are (1) close correspondence of a projection of the subsurface reservoir with the anomaly on the surface, (2) sharp lateral variations that would be blurred if diffusion were the principal mechanism (Rice, 1986), and most seriously, (3) the disappearance of anomalies after production starts (Horvitz, 1969). The role of diffusion in primary migration remains viable (Thomas and Clouse, 1990a,b,c).

The mechanism of vertical transport by dissolution in water suffers many of the same problems as transport by diffusion. Significant lateral offset of surface anomalies would be much more common, and gas anomalies directly above the reservoir would be rare. Dispersion, resulting in blurred anomalies, would be the norm, and halo anomalies would be rare. The rapid disappearance of a surface anomaly with initiation of production also would not occur with the water transport mechanism. The role of water transport in secondary migration is likely a major factor (Tóth, 1988).

Ascending buoyant microbubbles of gaseous hydrocarbons is the third mechanism that was discussed. Modeling by Saeed (1991) supports this as the dominant mechanism for vertical migration and as responsible for surface anomalies. This was also the conclusion of Price

(1986). This favored mechanism can explain near-vertical migration in the absence of faulting and normal slow rates of lateral water flow in a basin. It also explains anomalies that are directly above the surface projection of reservoirs. Indirect effects such as secondary cementation of part of the sedimentary column result in halo anomalies due to diversion of microseepage around the cemented "chimney" produced by oxidation-reduction processes (Klusman, 1993). Most importantly, the computer simulations of Saeed (1991) predict rapid changes in surface anomalies once production from the reservoir starts.

Acknowledgments—Support was provided to M. A. S. by the U.S. Agency for International Development during part of his tenure as a student at the Colorado School of Mines. The manuscript was improved by the reviews of M. A. Abrams, D. Schumacher, and particularly W. A. Young.

REFERENCES CITED

Araktingi, R. E., M. E. Benefield, Z. Bessenyei, K. H. Coats, and M. R. Tek, 1982, Leroy Storage, Uinta County, Wyoming: a case study of attempted control of gas migration: 57th Annual Society of Petroleum Engineers of AIME, Fall Technical Conference, SPE–11180, 11 p.

Arp, G. K., 1992, An integrated interpretation for the origin of the Patrick Draw oil field sage anomaly: AAPG Bulletin, v. 76, p. 301–306.

Bear, J., 1979, Hydraulics of groundwater: New York, McGraw-Hill, 567 p.

Berg, R. R., 1975, Capillary pressures in stratigraphic traps: AAPG Bulletin, v. 59, 939–956.

Bethke, C. M., 1985, A numerical model of compaction-driven groundwater flow and heat transfer and its application to the paleo-hydrology of intracratonic sedimentary basins: Journal Geophysical Research, v. 90, p. 6817–6828.

Bodner, D. P., and J. M. Sharp, Jr., 1988, Temperature variations in south Texas subsurface: AAPG Bulletin, v. 72, p. 21–32.

Buckley, S. E., C. R. Hocott, and M. S. Taggart, Jr., 1958, Distribution of dissolved hydrocarbons in subsurface waters, *in* L. G. Weeks, ed., Habitat of oil: Tulsa, Oklahoma, AAPG, p. 850–882.

Coleman, D. D., W. F. Meents, C.-L. Liu, and R. A. Keogh, 1977, Isotopic identification of leakage gas from underground storage reservoirs—a progress report: Illinois State Geological Survey, Report 111, 10 p.

Crichlow, H. B., 1977, Modern reservoir engineering—a simulation approach: Englewood Cliffs, New Jersey, Prentice-Hall, 354 p.

Devitt, D. A., R. B. Evans, W. A. Jury, T. H. Starks, B. Eklund, A. Gnolson, and J. J. van Ee, 1987, Soil gas sensing for detection and mapping of volatile organics: Dublin, Ohio, National Water Well Association, 270 p.

Hanor, J. S., 1987, Kilometre-scale thermohaline overturn of pore water in the Louisiana Gulf Coast: Nature, v. 327, p. 501–503.

Healy, R. W., 1990, Simulation of solute transport in variably saturated porous media with supplemental information on modification to the U.S. Geological Survey computer program VS2D: USGS Water Resources Investigation Report 90-4025, 125 p.

Horvitz, L., 1969, Hydrocarbon geochemical prospecting after thirty years, *in* W. B. Heroy, ed., Unconventional methods in exploration for petroleum and natural gas: Dallas, Texas, Southern Methodist University Press, p. 205–218.

Jakel, M. E., and R. W. Klusman, 1995, Methane fluxes to the atmosphere due to natural microseepage from sedimentary basins (abs.): American Geophysical Union, EOS, v. 76, no. 46, p. F126.

Jones, P. H., 1984, Deep water discharge: a mechanism for the vertical migration of oil and gas, *in* M. J. Davidson and B. M. Gottlieb, eds., Unconventional methods in exploration for petroleum and natural gas, III: Dallas, Texas, Southern Methodist University Press, p. 254–271.

Katz, D. L., and K. H. Coats, 1968, Underground storage of fluids: Ann Arbor, Michigan, Ulrich's Books, Inc., 575 p.

Klusman, R. W., 1993, Soil gas and related methods for natural resource exploration: Chichester, U.K., John Wiley & Sons, 483 p.

Krooss, B. M., D. Leythaeuser, and R. G. Schaefer, 1992a, The quantification of diffusive hydrocarbon losses through cap rocks of natural gas reservoirs—a reevaluation: AAPG Bulletin, v. 76, p. 403–406.

Krooss, B. M., D. Leythaeuser, and R. G. Schaefer, 1992b, The quantification of diffusive hydrocarbon losses through cap rocks of natural gas reservoirs—a reevaluation: Reply: AAPG Bulletin, v. 76, p. 1842–1846.

Leythaeuser, D., R. G. Schaefer, and A. Yukler, 1980, Diffusion of light hydrocarbons through near-surface rocks: Nature, v. 284, p. 522–525.

Leythaeuser, D., R. G. Schaefer, and A. Yukler, 1982, Role of diffusion in primary migration of hydrocarbons: AAPG Bulletin, v. 66, p. 408–429.

McAuliffe, C. D., 1979, Oil and gas migration—chemical and physical constraints: AAPG Bulletin, v. 63, p. 761–781.

MacElvain, R., 1969, Mechanics of gaseous ascension through a sedimentary column, *in* W. B. Heroy, ed., Unconventional methods in exploration for petroleum and natural gas: Dallas, Texas, Southern Methodist University Press, p. 15–28.

Magara, K., 1978, Compaction and fluid migration; practical petroleum geology: Amsterdam, Elsevier, 319 p.

Meyer, H. J., and H. W. McGee, 1985, Oil and gas fields accompanied by geothermal anomalies in Rocky Mountain region: AAPG Bulletin, v. 69, p. 933–945.

Neglia, S., 1979, Migration of fluids in sedimentary basins: AAPG Bulletin, v. 63, p. 573–597.

Nelson, J. S., and E. C. Simmons, 1992, The quantification of diffusive hydrocarbon losses through cap rocks of natural gas reservoirs—a reevaluation: Discussion: AAPG Bulletin, v. 76, p. 1839–1841.

Nelson, J. S., and E. C. Simmons, 1995, Diffusion of methane and ethane through the reservoir cap rock, and its implications for the timing and duration of catagenesis: AAPG Bulletin, v. 79, p. 1064–1074.

Price, L. C., 1976, Aqueous solubility of petroleum as applied to its origin and primary migration: AAPG Bulletin, v. 60, p. 213–244.

Price, L. C., 1986, A critical overview and proposed working model of surface geochemical exploration, *in* M. J. Davidson, ed., Unconventional methods in exploration for petroleum and natural gas, IV: Dallas, Texas, Southern Methodist University Press, p. 245–309.

Price, L. C., L. M. Wenger, T. Ging, and C. W. Blount, 1983, Solubility of crude oil in methane as a function of pressure and temperature: Organic Geochemistry, v. 4, p. 201–221.

Rice, G. K., 1986, Near-surface hydrocarbon gas measurement of vertical migration, *in* M. J. Davidson, ed., Unconventional methods in exploration for petroleum and natural gas, IV: Dallas, Texas, Southern Methodist University Press, p. 183–220.

Saeed, M. A., 1991, Light hydrocarbon microseepage mechanisms: theoretical considerations: Ph.D. dissertation, Colorado School of Mines, Golden, Colorado, 128 p.

Schowalter, T. T., 1979, Mechanics of secondary hydrocarbon migration and entrapment: AAPG Bulletin, v. 63, p. 723–760.

Spencer, C. W., 1987, Hydrocarbon generation as a mechanism for overpressuring in Rocky Mountain region: AAPG Bulletin, v. 71, p. 368–388.

Stegena, L., 1961, On the principles of geochemical oil prospecting: Geophysics, v. 26, p. 447–451.

Tek, M., J. O. Wilkes, and D. Katz, 1966, New concepts in underground storage of natural gas: New York, American Gas Association, 342 p.

Thomas, C. W., 1982, Principles of hydrocarbon reservoir simulation: Boston, IHRDC, 207 p.

Thomas, M. M., and J. A. Clouse, 1990a, Primary migration by diffusion through kerogen: I. model experiments with organic-coated rocks: Geochimica et Cosmochimica Acta, v. 54, p. 2775–2779.

Thomas, M. M., and J. A. Clouse, 1990b, Primary migration by diffusion through kerogen: II. hydrocarbon diffusivities in kerogen: Geochimica et Cosmochimica Acta, v. 54, p. 2781–2792.

Thomas, M. M., and J. A. Clouse, 1990c, Primary migration by diffusion through kerogen: III. calculation of geologic fluxes: Geochimica et Cosmochimica Acta, v. 54, p. 2793–2797.

Tóth, J., 1988, Ground water and hydrocarbon migration, *in* W. Back, J. S. Rosenshein, and P. R. Seaber, eds., The geology of North America—hydrogeology, GSA, v. O-2, p. 485–502.

Warneck, P., 1988, Chemistry of the natural atmosphere: San Diego, Academic Press, 757 p.

Watts, N. L., 1987, Theoretical aspects of cap-rock and fault seals for single- and two-phase hydrocarbon columns: Marine and Petroleum Geology, v. 4, p. 274–307.

Whelan, J. K., J. M. Hunt, J. Jasper, and A. Huc, 1984, Migration of C_1–C_8 hydrocarbons in marine sediments: Organic Geochemistry, v. 6, p. 683–694.

Clayton, C. J., and P. R. Dando, 1996, Comparison of seepage and seal leakage rates, *in* D. Schumacher and M. A. Abrams, eds., Hydrocarbon migration and its near-surface expression: AAPG Memoir 66, p. 169–171.

Chapter 13

Comparison of Seepage and Seal Leakage Rates

Chris J. Clayton

School of Geological Sciences
Kingston University
Kingston upon Thames, Surrey, U.K.

Paul R. Dando

School of Ocean Sciences
University of Wales
Bangor, Gwynedd, U.K.

Abstract

Rates of petroleum seepage must ultimately be related to the rate of supply from below. This could be either from a leaking accumulation or directly from the source rock itself. However, many things can happen during migration to the surface, such that the relationship between trap leakage rates and surface seep rates remains obscure. We calculate the potential flux rates of gas leakage across seals and compare these with measurements of fluxes for three seepage sites on the European continental shelf. We conclude that seepage flow rates can be modeled effectively by assuming Poiseuille flow through the matrix of mudstone seals. Flow rates calculated in this way are about 0.4–1.0 m^3 gas/m^2/year, consistent with field observations.

INTRODUCTION

Petroleum flow in the subsurface is limited ultimately by mudrocks. It is these deposits that form the seals to accumulations, dictate migration directions in the enclosed carrier beds, and limit the ultimate flow rate through mixed lithology sections. Flow of petroleum through mudrocks can be either via the narrow pore throats between mineral grains (capillary flow) or via fractures. However, Clayton and Hay (1994) demonstrate that capillary failure will always occur in preference to fracturing in mudrocks if buried deeper than 1000–1500 m (Figure 1). Furthermore, such capillary failure will be extremely widespread. This is because even moderate depositional rates lead to sufficient overpressure in up-dip structural and stratigraphic traps to ensure that such leakage usually occurs (Figure 2). In most sedimentary basins, we should therefore observe widespread seepage, although at low rates because capillary flow through mudrocks is slow. Fracture flow occurs preferentially only in near-surface environments (<1000 m depth and usually only in the upper 200 m), but even near the surface, the absolute flow rate is likely to be limited by the supply of petroleum by capillary leakage from below.

CALCULATION OF FLUX RATES

To estimate fluxes across seals (i.e., the supply rate), we have used the Darcy equation modified to take account of Poiseuille flow through the network of pores in the shale. This equation relates the rate of fluid flow to the driving force or "potential gradient" ($\nabla\Phi$), the permeability of the rock (k), and the fluid viscosity (μ).

$$q = (k/\mu)\,\nabla\Phi$$

where q is the rate of advance of the seeping petroleum front following initial failure. The Darcy equation is converted to a flux (Q) with dimensions of m^3/m^2/sec by multiplying by the cross-sectional area. The permeability can be calculated from modeled porosity for any depth from a knowledge of how the pore throat sizes relate to porosity (Clayton, 1993):

$$k \approx S\,(\phi r^2/8\Theta^2)$$

where r is the pore throat radius, Θ is the tortuosity of the pathway (generally taken to be $\sqrt{3}$), and S is the petroleum saturation (typically about 0.1). Taking the potential

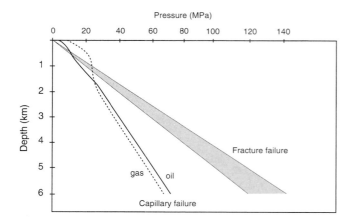

Figure 1—Minimum absolute pressure required to cause seal failure by fracturing (shaded area) and to cause capillary failure. The fracture failure envelope is based on normal uncemented mudstone seals. The gas curve is for pure methane, and the oil curve is for 35° API oil with a gas–oil ratio of 600 scf/bbl. (Modified after Clayton and Hay, 1994.)

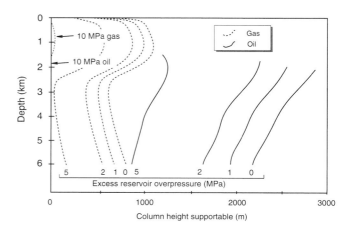

Figure 2—Influence of excess reservoir overpressure on mudstone seal capacity for oil and gas. When reservoir pressure exceeds seal pressure by 10 MPa (1500 psi), seal capacity is effectively zero, resulting in ubiquitous and pervasive seal failure. (After Clayton and Hay, 1994.)

gradient term ($\nabla\Phi$) to be the sum of the normal overpressure gradient in North Sea-type shales (based on the model of Mann and Mackenzie, 1990) plus the gas buoyancy potential, we obtain the superficial flow velocities shown in Figure 3.

FIELD STUDIES

In order to compare the calculated seal leakage rates with seeps, we consider three active gas seepage sites. Because local focusing and ponding in the shallow surface cause local temporal and spatial variations in seep rate, we use the average rate of gas release over the total area of seepage for each site.

In UKCS block 15/25, gas is escaping from an active pockmark at about 170 m water depth. The gas appears to be of biogenic origin ($\delta^{13}C \approx -73$ to -79 ‰), sourced in carbonaceous sediments of the Tertiary Hordaland and Nordland groups. Submersible measurements of seepage rates on individual gas vents range from 0.14 to 0.6 L/hr at ambient conditions, equivalent to a mean flow rate of 5.74 L/hr at standard temperature and pressure. Taking into account the total number of seep vents and the area of the pockmark, we estimate a net flux of about 440 L/m^2/year. The long-term net flux, however, may be higher than this since the pockmark morphology implies periodic eruptive gas activity. Such eruptive behavior has not yet been observed.

Hovland and Sommerville (1985) reported gas seepage rates in 75 m of water in the Ekofisk area in the North Sea. Extrapolating their estimate of 1000 L/hr (derived from bubble counts using an ROV) to the total area of gas seepage of about 100,000 m^2, containing about 140 seeps, we obtain a net flux of 890 L/m^2/year.

Finally, in the intertidal zone on the Kattegat coast of Denmark, biogenic gas is leaking from a presumed source in Eemian–early Weichelian deposits. The source depth of the gas is uncertain but is clearly shallower than 1000 m. For the calculations given below, we have assumed a maximum possible depth, thus the calculated fluxes are minimum estimates. Seepage here covers an area of about 1700 m^2, and individual seep vents vary in flow rate from 0.15 to 21.8 L/hr. Although individual gas vents vary in flow rate over time, the net gas flux is approximately constant through time at 520 L/m^2/year (Dando et al., 1994). In practice, isolated high flow rate seepage vents also occur in this area, so this estimate must also be considered a minimum.

Thus, all three seepage sites are characterized by similar net flow rates on the order of 400–1000 L/m^2/year, despite the varied geologic settings and differences in source depth of the leaking gas.

COMPARISON OF CALCULATED AND MEASURED FLOW RATES

Measured flux rates are compared with calculated fluxes across mudstone "seals" at the appropriate source or leakage depth in Figure 4. All three seeps can be modeled to within about 15% using the Poiseuille equation. We assume that migration rate is limited by the gas supply across the seal in the Ekofisk example and by escape from the source mudstones in the other two cases.

The similarity between modeled and observed methane fluxes implies that surface seepage is in approximate equilibrium with capillary leakage across the seal at depth. This is surprising because it implies that gas ponding, local focusing, and flow rate variations at shallower depths are not influencing the net flux, even though we

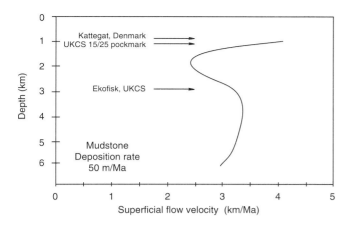

Figure 3—Calculated superficial flow velocity of pure methane through mudstone seal assuming Poiseuille flow. (Pressure and porosity modeling based on Mann and Mackenzie, 1990.) Geothermal gradient assumed to be 30°C/km. Source depths of case studies are indicated by arrows.

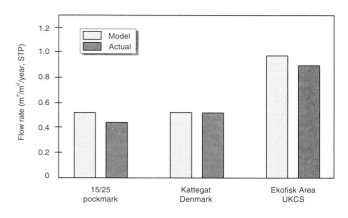

Figure 4—Comparison of modeled flux rates (assuming rate is controlled by source depth) and measured surface flux rates for three examples studied.

would expect highly variable behavior in these zones. This can be understood by considering the two extremes of behavior. If the migrating gas meets a comparatively "tight" zone, it should slow down. However, this leads to higher petroleum saturation at this level as the following, faster moving petroleum catches up, akin to a traffic jam on a freeway. This in turn leads to higher petroleum flow rates since more of the pore network is now available for fluid flow, and the system stays in dynamic equilibrium. In the opposite case of the petroleum encountering a "fast" migration pathway, the reverse is true. Petroleum can only pass this zone as fast as it is supplied from below. Thus, migrating petroleum is held up, waiting for more to be supplied from below.

Finally, it would appear from these data that all three of the seeps studied result from dominantly vertical leakage of gas. In other geologic situations, considerable lateral diversion of flow would be expected. These, too, are amenable to such modeling, providing that the geometry of the migration pathway can be constrained. Such treatment potentially provides a method of interpreting seeps at prospect level.

CONCLUSIONS

1. Observed seepage rates are consistent with dominantly vertical migration in which the flow rate is in approximate equilibrium with flow rates through seals at depth.

2. Seepage flow rates can be modeled effectively by assuming Poiseuille flow through the matrix of mudstone seals. It is not necessary to invoke microfractures to account for observed seepage rates.

3. Flow rates calculated in this way are about 0.4–1.0 m^3 gas/m^2/year. Significantly higher flow rates must be attributed to fracture leakage of near-surface secondary accumulations.

REFERENCES CITED

Clayton, C. J., 1993, Differential flow rates of oil and water in fine-grained sediments (abs.): AAPG, International Hedberg Research Conference Abstracts, October 17–20, The Hague, The Netherlands, p. 41.

Clayton, C. J., and S. J. Hay, 1994, Gas migration mechanisms from accumulation to surface: Bulletin, Geological Society of Denmark, v. 41, p. 12–23.

Dando, P. R, S. C. M. O'Hara, U. Schuster, L. Taylor, C. J. Clayton, S. Baylis, and T. Laier, 1994, Gas seepage from a carbonate-cemented sandstone reef on the Kattegat coast of Denmark: Marine Petroleum Geology, v. 11, p. 182–189.

Hovland, M., and J. H. Sommerville, 1985, Characteristics of two natural gas seepages in the North Sea: Marine Petroleum Geology, v. 2, p. 319–326.

Mann, D. M., and A. S. Mackenzie, 1990, Prediction of pore fluid pressures in sedimentary basins: Marine Petroleum Geology, v. 7, p. 55–65.

Krooss, B. M., and D. Leythaeuser, 1996, Molecular diffusion of light hydrocarbons in sedimentary rocks and its role in migration and dissipation of natural gas, *in* D. Schumacher and M. A. Abrams, eds., Hydrocarbon migration and its near-surface expression: AAPG Memoir 66, p. 173–183.

Chapter 14

Molecular Diffusion of Light Hydrocarbons in Sedimentary Rocks and Its Role in Migration and Dissipation of Natural Gas

B. M. Krooss

Institute of Petroleum and Organic Geochemistry Forschungszentrum Jülich GmbH (KFA) Jülich, Germany

D. Leythaeuser

Department of Geology University of Cologne Köln, Germany

Abstract

The role of molecular diffusion during different stages of hydrocarbon migration has been an issue of recurrent interest during the past 40 years. Controversial views on the importance of this transport mechanism can be partially attributed to inconsistent definitions and computational errors. Considerable amounts of light hydrocarbons can be released by diffusion, even from thick source rock sequences. At higher generation rates, primary migration of gas is, however, dominated by volume flow (Darcy flow). Molecular transport of C_5–C_9 hydrocarbons in source rocks deviates strongly from ideal diffusion behavior.

Secondary migration is dominated by volume flow, and diffusion plays only a subordinate role. Diffusion may be important in the dismigration of natural gas if seal leakage by volume flow does not occur over extended periods of time. Average diffusive fluxes for methane through seals reported in various studies range from 0.16 to 89 $m^3/km^3/year$. For comparison, the compressible volume flow of gas through a shale 50–450 m thick with a permeability of 1 nanodarcy (10^{-21} m^2) was calculated between 100 and 1000 $m^3/km^3/year$. Numerical simulation of diffusion in the context of integrated two-dimensional basin modeling improves the quantification of molecular transport of hydrocarbon gas in petroleum systems.

INTRODUCTION

Over the past 40 years, various attempts have been made to assess the potential relevance of molecular diffusion during certain phases of natural gas and oil migration. Disagreement about the importance of this transport mechanism—partially due to inconsistent definitions of the relevant parameters—have resulted in substantial controversy even in recent years (e.g., Krooss et al., 1992a,b; Nelson and Simmons, 1992). The main prerequisites for a quantification of diffusive processes in petroleum and natural gas systems over geologic time are the correct and consistent mathematical description of molecular transport in porous media and the availability of reliable transport parameters (effective diffusion coefficients).

The mathematical basis for the treatment of molecular diffusion processes in geologic systems was created by P. Antonov (1954, 1958, 1964, 1968, 1970a,b), who adapted the well-established algorithms of heat flow (Carslaw and Jaeger, 1959) and diffusion theory (Crank, 1975) to sedimentary rocks. In addition, Antonov conducted the first comprehensive study to measure effective diffusion coefficients experimentally for hydrocarbons in different types of sedimentary rocks. Starting with the work of Leythaeuser et al. (1980, 1982) in the late 1970s, the investigation of hydrocarbon diffusion in sedimentary rocks became one of the research issues of the Institute of Petroleum and Organic Geochemistry at KFA. Diffusion work has since been conducted experimentally, theoretically, and in the context of case histories, with the goal of reaching a comprehensive understanding of the processes and arriving at an improved quantification of molecular transport in petroleum migration.

In this paper, we give a brief overview of the state of the art and current research issues in hydrocarbon diffusion research. With respect to the topic of this volume, our contribution emphasizes the potential role of diffusion in the dismigration of hydrocarbons through seals above natural gas reservoirs and its relevance to surface expressions of hydrocarbon migration.

DEFINITIONS

Diffusion denotes the transport process resulting from random thermal motion (Brownian motion) of individual molecules. This motion causes a permanent redistribution of molecules in fluids (gases and liquids) and, at a much smaller rate, in solids. In the absence of external forces or fields, diffusion results in the equilibration of concentration gradients. The rate of diffusive transport in fluids is related to the concentration gradient of the compound under consideration by Fick's first law of diffusion:

$$J = -D \text{ grad } C \qquad (1)$$

where J = diffusive flux (mass/area/time), D = diffusion coefficient (length^2/time), and C = volume concentration (mass/volume). This diffusion law, which strictly holds only for low concentrations and small concentration gradients, represents the basis for most mathematical descriptions of diffusion problems (Crank, 1975). The material balance equation for diffusive transport processes,

$$\frac{\partial C}{\partial t} = D \frac{\partial^2 C}{\partial x^2} \qquad (2)$$

is commonly termed *Fick's second law of diffusion.*

For the treatment of diffusion processes in porous media, which in this context is sedimentary rocks, the diffusion law is formulated in the analogous form:

$$J = -D_{\text{eff}} \text{ grad } C_{\text{bulk}} \qquad (3)$$

where D_{eff} = effective diffusion coefficient (length^2/time) and C_{bulk} = bulk volume concentration (mass/volume of porous medium). Here the fluid phase diffusion coefficient (D) is replaced by the *effective diffusion coefficient* (D_{eff}) and the fluid phase concentration (C) by the bulk volume concentration (C_{bulk}) of the diffusing compound. This convention was introduced explicitly in the work of Antonov (1954, 1958). The effective diffusion coefficient reflects the influence of the porous medium on the mobility of the diffusing substances. The mobility is affected mainly by the geometry of the pore system and by sorption-desorption processes on the pore surfaces or within sedimentary organic matter (kerogen). The bulk volume concentration (C_{bulk}) is controlled by the porosity of the sedimentary rock, the solubility of the diffusing species in the pore fluid, and the sorption capacity of pore surfaces or organic matter for the diffusing species.

For the simple case of a water-saturated porous medium with porosity ϕ and negligible sorption of the diffusing hydrocarbon (e.g., methane), the bulk volume concentration is given by the product of the aqueous phase concentration and the porosity:

$$C_{\text{bulk}} = C_{\text{fluid}}\phi \qquad (4)$$

where ϕ = porosity. It can be shown (Krooss et al., 1992b)

that under these conditions, the effective diffusion coefficient is equal to

$$D_{\text{eff}} = D_{\text{fluid}}/\tau \qquad (5)$$

where τ is a geometric factor representing the decrease in mobility of a diffusing molecule due to the tortuosity of the pore channels as compared to its mobility in the pure fluid.

As stated above, the definition of the diffusion law assumes that external fields (gravity and magnetic) and temperature and pressure gradients are absent. These conditions can be realized with good approximation in laboratory experiments. On the size scale of geologic systems and the time scale of petroleum generation and migration, however, pressure gradients resulting from the earth's gravity field and geothermal gradients can be expected to bear direct or indirect influence on the molecular transport processes.

In tight overpressured sedimentary sequences, pressure must be taken into consideration as an additional and possibly predominant driving force for molecular migration. At the extremely low permeabilities of certain rock types (<<1 nanodarcy) a differentiation between volume flow and molecular transport becomes impossible, that is, mass flow can only occur by the movement of individual molecules. For the flow of gas in porous media, this transport regime is known as *Knudsen diffusion*. It occurs when the pore diameter is smaller than the mean free path length of the gas molecules.

Krooss et al. (1992b) have demonstrated that the mass balance equation for diffusion in homogeneous sedimentary rocks can be expressed both on the basis of bulk rock (mass per unit volume of rock) and fluid phase (mass per unit volume of pore fluid) concentrations and that in both cases the effective diffusion coefficient, as defined in equation (5), must be used.

DIFFUSION IN PRIMARY AND SECONDARY MIGRATION

Light Hydrocarbons

Leythaeuser et al. (1980, 1982) observed characteristic concentration gradients of light hydrocarbons (C_2–C_7) in outcrops of source rock type shales (Figure 1), which they interpreted as a result of diffusive gas losses from the lithosphere toward the atmosphere (Figure 2). These observed effects can be considered as surface expressions of (primary) hydrocarbon migration in a broader sense. The evaluation of the concentration profiles using the mathematical model of diffusion from a semi-infinite medium, in combination with an estimate of the time of onset of gas diffusion, resulted in the determination of effective diffusion coefficients for ethane through *n*-heptane in source rock type shales.

The objectives of the work of Leythaeuser et al. (1982) were (1) to estimate, on the basis of these diffusion coeffi-

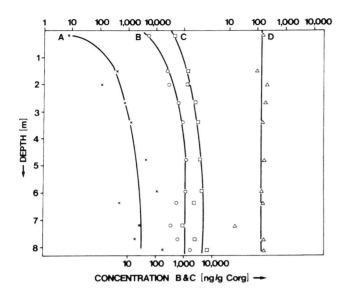

Figure 1—Depth trends in light hydrocarbon concentrations in a near-surface interval of an outcropping source rock of Campanian–Maestrichtian age in western Greenland. Graphs: (A) propane, (B) isobutane, (C) *n*-butane, and (D) *n*-heptane. (From Leythaeuser et al., 1982; reprinted by permission.)

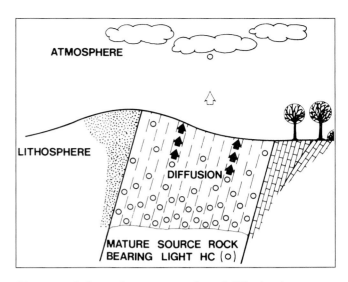

Figure 2—Schematic representation of diffusion in a near-surface interval of an outcropping source rock. (From Leythaeuser et al., 1982; reprinted by permission.)

cients, the amount of light hydrocarbons that can leave a source rock interval of given thickness by diffusion over periods of several million years and (2) to analyze the effects of the diffusive release from source rocks on the composition of the associated reservoir gases. It was concluded that gas quantities released by diffusion from an interval of mature source rock 200 m thick and 1000 km² in area within a period of less than 1 m.y. (1.5×10^9 m³ or 5.3×10^{10} scf in 0.54 m.y.) correspond to amounts found in commercial-sized accumulations. Furthermore, it was shown that a diffusion-controlled supply of gas into the carrier beds can be expected to result, for an initial period of several million years, in the accumulation of gas enriched in compounds with high diffusivity (e.g., methane). These calculations were performed under the assumption that gas generation had ceased in the source rocks. However, the rate of gas production in actively generating source rocks can readily exceed the rate of diffusive transport, particularly in thick layers. This leads to the formation of a free gas phase in the pore space, followed by phase flow of the gas toward adjacent carrier beds. As shown by Leythaeuser and Poelchau (1991), gas expelled in this way can act as a solvent for higher molecular weight compounds, resulting in compositional fractionation of the expelled petroleum products.

Higher Molecular Weight Hydrocarbons

Today it is widely held that diffusive transport is only of subordinate importance during the primary migration and expulsion of petroleum from source rocks. Stainforth and Reinders (1990) reconsidered diffusion as a primary

migration mechanism and concluded that activated diffusion in an organic network is a rate-controlling step for petroleum expulsion. Their reasoning was based on the fact that concentration profiles in natural source rock carrier systems and in laboratory experiments can be described by a diffusion model.

Experimental work at our institute (Hanebeck, 1994) has shown that C_5–C_9 hydrocarbons in source rocks indeed have relatively high molecular mobilities. Under the comparatively large concentration gradients and at experimental temperatures of 150°C, breakthrough velocities ranged from 10 to 100 μm/hr. Molecular mobilities were found to depend on compound class, (saturated, naphthenic, or aromatic), molecular weight, and source rock type. The connectivity of both the pore network and the kerogen network plays an essential role in this context. Aromatic hydrocarbons diffuse more rapidly if an interconnected water-saturated pore network exists, whereas in cases where the pore network is discontinuous and diffusion is restricted to the kerogen network, saturated hydrocarbons migrate preferentially. During this experimental study, it emerged that strong deviations from ideal diffusion behavior occur as a consequence of solvent swelling of the organic matter. Thus, the experimental breakthrough curves for C_5–C_9 hydrocarbons could not be described by simple diffusion equations.

Although the molecular mobility of light hydrocarbons in the organic matter network of many source rocks appears to be high, a simple diffusion model with constant boundary conditions falls short of describing the entire complex process of expulsion from source rocks. Concentration gradients are only part of the driving forces acting on hydrocarbon molecules in source rocks. Pressure gradients resulting from the redistribution of lithostatic stress from load-bearing kerogen (Palciauskas, 1991) to the fluid phase generated during thermal conversion of this kerogen are the main driving force for petroleum expulsion, as shown by both volume balance calculations and experiments. Thus, pressure is the most important component in primary petroleum migration resulting in material transport and deformation of the system. As outlined above, in tight lithologies, the bulk flow regime can be expected ultimately to turn over into a molecular migration regime.

The issue of molecular transport during primary migration is still associated with a number of unresolved problems. These include the transition from volume flow to molecular transport; the timing and mechanism of the formation of a free hydrocarbon phase at the source rock–carrier boundary; and the complex interrelationship of kerogen conversion, pressure build-up, source rock deformation, and hydrocarbon expulsion.

Diffusion in Secondary Migration

In secondary migration, which takes place in highly porous and permeable carrier and reservoir rocks, diffusion plays a role in the homogenization of compositional gradients (England and Mackenzie, 1989). Otherwise, secondary migration is dominated by multiphase volume flow.

DIFFUSION IN DISMIGRATION

Dismigration is the loss of hydrocarbons from a trap through overlying seals or roof rocks (Tissot and Welte, 1984). The term *cap rock,* which has been widely used as a general name for roof rocks or seal rocks above hydrocarbon reservoirs, should be restricted to the secondary sheath around the tops of salt domes (see North, 1985) and is therefore avoided in this paper.

The quantification of diffusive losses, particularly hydrocarbon gases, from reservoirs into the overburden has been an issue of recurrent interest but also a source of controversy. Predictions of reservoir half-life based on identical assumptions have resulted in differences of more than one order of magnitude (Krooss et al., 1992a,b; Nelson and Simmons, 1992).

Figure 3 depicts the typical idealized diffusion situation involving a membrane seal (Watts, 1987). Gas is trapped in a reservoir underneath a fine-grained roof rock due to the high capillary entry pressure (Berg, 1975, Schowalter, 1979; Watts, 1987). The free gas phase in the reservoir is in equilibrium with the pore water of the overlying seal. The concentration of gas in the seal is determined by the aqueous solubility as a function of pressure, temperature, and salinity. If additional effects such as the sorption on the pore surfaces or in the organic matter of the roof rock can be neglected, the bulk concentration (mass per volume of rock) of methane within it can be calculated from a solubility function (e.g., Haas, 1978) and its porosity.

The seal diffusion problem has been addressed by several workers over the past 40 years. Antonov (1958, 1964)

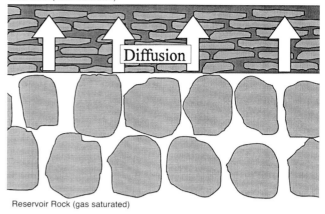

Roof Rock (water saturated)

Diffusion

Reservoir Rock (gas saturated)

Figure 3—Schematic illustration of roof rock–reservoir boundary.

carried out pioneering work in this field by formulating and partially answering the questions related to diffusive transport of hydrocarbons from reservoirs into the overburden:

1. How far does a diffusion front advance within a given period of time?
2. How long does it take until the hydrocarbon concentration exceeds the detection limit at a given distance above the seal–reservoir interface?
3. How much time is required to reach an approximately steady-state of diffusion?
4. What amounts of gas are finely dispersed in the overburden during the various phases of diffusion from natural gas reservoirs?

Krooss (1992) compared and recalculated some scenarios and quantified diffusive losses from gas reservoirs for specific situations. Published seal diffusion scenarios can be subdivided into two classes. The calculations based on *diffusion in a plane sheet* assume a roof rock of finite thickness, which implies that during the time period considered, a diffusion front passes through the seal and a stationary diffusive flux is ultimately established across the layer. However, when the thickness of the roof rock sequence is such that the diffusion front cannot reach the upper boundary within a reasonable time, a model for *diffusion in a semi-infinite medium* can be used. Figure 4 schematically shows the four principal scenarios for gas diffusion in seals.

Diffusion in a Plane Sheet

Stationary State

The simplest approach for the quantification of diffusive flux in a barrier rock is the calculation of the stationary state flux based on Fick's first law of diffusion (see above) as the product of a concentration gradient and an

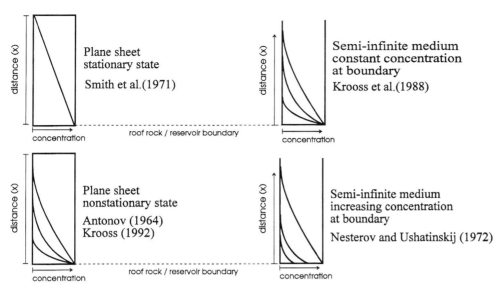

Figure 4—Overview of roof rock diffusion models used by various workers.

effective diffusion coefficient. This method was used by Smith et al. (1971) to calculate diffusive fluxes and residence times of hydrocarbon gases for two reservoir conditions derived from statistical data on U.S. gas reservoirs: scenario (1) was the most common U.S. reservoir size and scenario (2) was an exceptionally large reservoir. For their computation, Smith et al. (1971) used aqueous phase diffusion coefficients (2.7–1.4×10^{-9} m^2/sec for methane through butane in the temperature range of average reservoirs) reduced by a "geometric factor" of 39.3 ($D_i = D_{aq}/39.3$). Concentration gradients were calculated from estimated hydrocarbon solubilities in brines and an assumed overburden thickness of 1737 m. The porosity of the overburden was taken as 15.7% and reservoir conditions were characterized by a temperature of 70°C and a pressure of 142 atm.

Detailed examination of the calculation reveals that the usage of the diffusion coefficients by Smith et al. (1971) is not strictly consistent with the definitions for the effective diffusion coefficient given above and in Krooss et al. (1992b) and that their geometric factor contains both a tortuosity and a porosity contribution. If this porosity term is removed, the geometric factor takes on a value of 6.2, which is in fairly good agreement with experimental results for the geometric factor τ in equation (5).

Smith et al. (1971) computed residence times of methane through butane in the reservoirs of 74–99 m.y. for scenario 1 and 190–253 m.y. for scenario 2. These residence times correspond to the time in which all hydrocarbons are lost from the reservoir by diffusion. Smith and co-workers point out that the assumption of a diffusive transport through the entire overburden is probably unrealistic and that the upward migration of hydrocarbons is affected by water-filled joints and faults and intervening aquifers. They therefore consider the case of a 150-m-thick reservoir seal immediately above the reservoir with a substantially lower diffusion coefficient (geometric factor of 500 compared to 39.3) and a porosity of 4.5%. The computed rate of gas loss in this scenario is decreased

by 50%, thus doubling the residence times in the reservoir to values of 150 and 380 m.y. for methane.

The assumption of steady-state diffusion in the assessment of gas losses from reservoirs is a simplification because the rate of gas loss is greatest during nonsteady-state diffusion due to the higher concentration gradients (see also Smith et al., 1971). As shown by Krooss (1992), the time required for the establishment of a nearly stationary diffusive flux starting from the first influx of gas into the reservoir is more than 400 m.y. for the scenarios considered by Smith et al. (1971). This renders the assumption of stationary diffusive flux unrealistic. One important result of the calculations of Smith et al. (1971) is that the amount of gas dissolved in the overburden above the reservoir in the stationary state exceeds the gas content of the reservoir itself. This finding indicates the potential importance of dissolved gas in the gas balance of sedimentary basins.

Nonstationary State

The establishment of a stationary state of gas diffusion through the roof rock of a gas accumulation may require excessive time, depending on the thickness of the seal and the effective diffusion coefficient. Assuming zero initial concentration, the diffusive flux (J) in a plane sheet (roof rock) of finite thickness l at position x as a function of time (t) is given by

$$(x,t) = -\frac{D}{l}\left[(C_2 - C_1) + 2\sum_{n=1}^{\infty}(C_2 \cos n\pi - C_1)\right.$$
$$\left. \cos\frac{nx\pi}{l}\exp\left(-\frac{n^2\pi^2 Dt}{l^2}\right)\right] \quad (6)$$

where D = diffusion coefficient, C_1 = concentration in the plane sheet at boundary $x = 0$ (roof rock–reservoir interface), and C_2 = corresponding concentration at boundary $x = l$ (top of roof rock). Both C_1 and C_2 are constant with time. This equation can be used to calculate both the

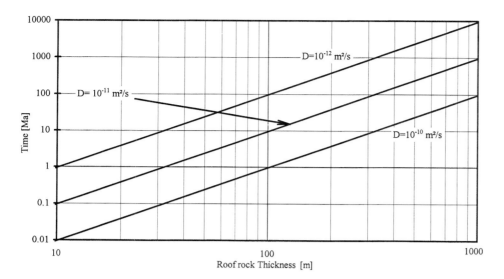

Figure 5—Nomogram for estimating the time required to reach a 90% stationary-state diffusive flux through a water-saturated roof rock.

influx of diffusing substance into the roof rock and the flux out of the roof rock at its upper boundary. It can be shown (Krooss, 1992) that, due to the high concentration gradients, the diffusive loss from a gas accumulation into the roof rock is highest during the initial phase of the non-stationary diffusion process and then decreases gradually to approach the stationary-state value. The onset of diffusive flux out of the roof rock occurs with a certain delay (time lag). During this period, the concentration of the diffusing substance in the seal gradually approaches the stationary-state concentration profile.

For an appraisal of the potential importance of diffusive transport in a given geologic situation, it is helpful to obtain a rough estimate of the time required to establish a stationary diffusive flux through a given seal. Figure 5 presents plots of three different diffusion coefficients (high, intermediate, and low) to estimate the times required to establish 90% of stationary diffusive flux through roof rocks 10–1000 m thick. Evidently, even with intermediate diffusion coefficients (10^{-11} m²/sec), the establishment of nearly stationary diffusive flux through sedimentary sequences of 1000 m requires periods of time exceeding the age of most geologic structures relevant to petroleum exploration. However, note that a diffusion front with threshold concentrations above the detection limit will penetrate the roof rock sequence in a significantly shorter time. Estimates on the migration distance of a "frontal concentration" can be performed, for example, using the formula for the concentration in a plane sheet as a function of distance and time (Crank, 1975):

$$C = C_1 + \left(C_2 - C_1\right)\frac{x}{l} +$$
$$\frac{2}{\pi}\sum_{n=1}^{\infty}\frac{C_2 \cos n\pi - C_1}{n}\sin\frac{n\pi x}{l}\exp\left(-\frac{Dn^2\pi^2 t}{l^2}\right) \quad (7)$$

Assuming diffusion coefficients between 5×10^{-9} and

5×10^{-13} m²/sec, Antonov (1964) calculated the times required to establish frontal concentrations at distances between 1000 and 3000 m from the roof rock reservoir boundary. These times range from 0.18 to 16,200 m.y.

Semi-Infinite Medium

Constant Boundary Concentration

If the roof rock thickness exceeds the expected penetration depth of molecular diffusion in the time frame under consideration, the diffusion process can be treated as *diffusion into a semi-infinite medium* (Figure 4). The formula for the concentration of a diffusing substance as a function of distance and time is

$$\frac{\left(C - C_1\right)}{\left(C_0 - C_1\right)} = \mathrm{erf}\left(\frac{x}{2\sqrt{Dt}}\right) \quad (8)$$

where C_0 = initial concentration throughout the medium, C_1 = constant concentration maintained at the surface, and erf is the error function.

This model in its reverse form (diffusion *out of* a semi-infinite medium) was used by Leythaeuser et al. (1980) to quantify diffusive losses from a near-surface source rock to the atmosphere. Krooss et al. (1989) used this approach to analyze light hydrocarbon concentration profiles in a shale interval above a gas condensate accumulation in North America. Light hydrocarbons had diffused into this sequence to various extents. The penetration distances ranged up to 140 m and depended on the molecular mobility of the hydrocarbons. In combination with diffusion coefficients estimated from laboratory measurements on comparable lithologies, the evaluation of the observed concentration profiles yielded an estimate for the time of onset of the diffusion process, which represents the time of filling of the underlying reservoir.

Table 1—Comparison of Average Diffusive Fluxes of Gas Through Roof Rocks for Published Case Histories

Scenario	Roof Rock Type	Seal/Overburden Thickness (m)	Effective Diffusion Coefficient (m²/sec)	Average Flux (m³/km²/year)
Antonov et al. (1958)	Rock salt (Devonian)	N/A	2.8×10^{-10}	88.5
Smith et al. (1971)	Shale (U.S. average)	1737 (overburden) 150 (seal) 1590 (overburden)	6.9×10^{-11} [a] 5.4×10^{-12} [a] $\Big\}$ 6.9×10^{-11} [a]	1.9 1.0
Nesterov et al. (1972)	Shale (Class I)	Semi-infinite	3.0×10^{-12}	0.16
Harlingen gas field (NL), Leythaeuser et al. (1982), Krooss et al. (1992a,b), and Nelson and Simmons (1992)	Shale	390	2.1×10^{-10}	3.7
Montel et al. (1993)	Claystone, siltstone, limestone (multilayered)	550	*D(T, t)*	10.0

[a] Do not conform to the definition of *effective diffusion coefficient* used in this paper; see text.

Variable Boundary Concentration

The model of molecular diffusion into a semi-infinite medium can be refined by taking into account that, with increasing gas pressure during subsidence and continuous filling of the reservoir, the gas pressure and thus the boundary concentration at the roof rock–reservoir interface increases. This approach (Figure 4) was used by Nesterov and Ushatinskij (1972) to estimate the diffusive gas loss from the Urengoy field in western Siberia.

Taking the boundary concentration as a linearly increasing function of time,

$$C_{(x=0)} = kt \qquad (9)$$

the concentration within the semi-infinite medium at position x and time t is given by (Crank, 1975) as

$$C = 4kti^2 \mathrm{erfc}\, \frac{x}{2\sqrt{Dt}} \qquad (10)$$

The function $i^2 \mathrm{erfc}$ is derived from the error-function complement (erfc) (Crank, 1975). The formula for the total amount M_t of diffusing substance (e.g., methane) per unit area that has entered the semi-infinite medium (seal rock) until time t takes the simple form

$$M_t = \frac{4}{3} kt \left(\frac{Dt}{\pi} \right)^{1/2} \qquad (11)$$

Using a very small effective diffusion coefficient of 3×10^{-12} m²/sec for the shale sequences overlying the Urengoy gas reservoir, Nesterov and Ushatinskij (1972) computed a total diffusive gas loss of 17×10^9 m³ since the

presumed start of reservoir filling at the Paleogene–Neogene boundary (25 Ma). This quantity amounts to less than 0.3% of the present-day reserves of this field (6500×10^9 m²). Diffusive losses from this field are therefore considered negligible.

Assessment of Diffusive Fluxes in the Subsurface

Because of the great variety of different geologic situations involving diffusive hydrocarbon losses into and through seal rocks of gas reservoirs, a general statement about the importance or relevance of molecular diffusion is not possible. To outline the range of potential diffusive flux rates of gas in geologic systems, the relevant data have been compiled from the literature. The resulting values, including case history information, are listed in Table 1.

The first case study by Antonov et al. (1958) investigated the diffusion of hydrocarbons through salt, which is commonly considered to possess ideal sealing properties. The high diffusion coefficients reported in their study (2.8×10^{-10} m²/sec), as well as observations of fracture patterns and the occurrence of oil and gas shows in rock salts, render this assumption questionable. Antonov et al. (1958) performed a simple calculation to assess the stationary diffusive flux through a salt sequence under a gas concentration gradient corresponding to a hydrostatic pressure gradient. The calculated flux of 88.5 m³/km²/year is the highest value in Table 1 and indicates the necessity of more detailed investigations on the sealing efficiency of evaporites.

Smith et al. (1971), in their scenario (1) based on an average U.S. gas reservoir situation (see above), arrived at a stationary state diffusive flux of methane of 1.9 m³/km²/year. In the semi-infinite medium model by Nesterov et al. (1972), the average flux from the reservoir was calculated by dividing the overall flux of 17×10^9 m^3 for methane by the field area (4200 km^3) and the diffusion time (25 m.y.). The very low flux of 0.16 m^3/km^2/year is essentially due to the extremely low diffusion coefficient attributed to the shale sequences in this calculation. These values mark the lower boundary of the experimentally accessible range of diffusion coefficients in sedimentary rocks.

The case study involving the assessment of diffusive losses from the Harlingen gas field in The Netherlands has been a source of substantial controversy. The original calculation by Leythaeuser et al. (1982) arrived at a diffusive loss of half the present-day reservoir content within a period of 4.5 m.y. This calculation was, however, based on an erroneous assumption concerning the boundary conditions at the roof rock–reservoir interface. Reevaluation by Kross et al. (1992a) with the correct boundary conditions but otherwise identical parameters showed that a substantially longer time (70 m.y.) would be required to lose these gas quantities.

Nelson and Simmons (1992) questioned this new computation and asserted that relatively large losses of natural gas even through thick seal rocks may occur in only a few million years. In their reply, Kross et al. (1992b) showed that use of the diffusion coefficients in the computations by Nelson and Simmons (1992) was not consistent with the definitions used in their assessment (both experimentally and by evaluation of concentration gradients in natural systems as performed by Leythaeuser et al., 1980). This common pitfall is partially a consequence of two different and confusing definitions of the *effective diffusion coefficient* in the literature. Diffusive gas losses of the magnitude considered for the Harlingen field require times on the order of several tens of millions of years. The diffusive flux ranges about 4 m^3/km^2/year (Table 1).

Montel et al. (1993) used a one-dimensional numerical diffusion model to evaluate hydrocarbon gas leakage through a seal rock overlying a reservoir in the North Sea. The advantage of such a model is that it can take into account layered roof rocks with different diffusion coefficients and porosities as well as the temperature dependence of gas solubility and diffusion coefficients. Furthermore, variations in the scenario and their influence on the diffusive loss can readily be modeled. For their case study in the North Sea, Montel et al. (1993) concluded that the diffusive loss was 15.4×10^9 m^2 over the past 20 m.y., which is not negligible in comparison to the present-day gas cap of 70×10^9 m^2.

In conclusion, the data in Table 1 show that the values computed for diffusive flux of gas from reservoirs vary by almost two orders of magnitude and therefore range from negligible (Nesterov et al., 1972) to considerable (Montel et al., 1993). They depend strongly on various parameters such as effective diffusion coefficients, seal thickness, pressure, and gas solubility. This rules out a general statement on the importance of diffusive transport as a gas loss mechanism and makes an evaluation of each individual case history indispensable.

Table 2—Dismigration of Gas by Compressible Single-Phase Darcy Flow in Kross' (1992) Scenario

Parameter	Value
Roof rock type	Shale
Roof rock thickness	50–450 m
Permeability	1 nanodarcy
Average flux	100–1000 m^3/km^2/year

Molecular Diffusion Versus Darcy Flow

The transport capacity of phase flow (Darcy flow) on the length scales relevant for petroleum and gas migration is indisputably much higher than the transport capacity of molecular diffusion. A comparison of diffusive flow and the flux rate of compressible Darcy flow of gas has been performed by Kross (1992). The results of the Darcy flow calculation are presented in Table 2. The compressible volume flux through an overburden 50–450 m thick with a permeability of 1 nanodarcy (10^{-21} m^2) is 100–1000 m^3/km^2/year, in other words, it is several orders of magnitude larger than the diffusive fluxes reported in Table 1. Thus, once the capillary entry pressure of the roof rock is exceeded and seal failure occurs, the flow rate of compressible gas flow will be orders of magnitude higher than the diffusive leakage. It must be pointed out that these Darcy flow calculations presume single-phase gas flow through a completely gas-saturated roof rock. The actual roof rocks in natural systems, however, are largely water saturated, and therefore a relative permeability relationship for the gas–water system in the roof rock lithology is required to assess gas permeability values as a function of gas saturation. To our knowledge, relative permeability relationships for tight lithologies are practically nonexistent, so that estimates on hydrocarbon leakage by Darcy flow remain highly speculative.

Gas Generation Scenario

It is illustrative to compare both the flux rates of diffusive transport and Darcy flow to the generation rate obtained from a simple source rock scenario. The scenario chosen for this computation is outlined in Table 3. The generation rate amounts to a value of 200 m^3/km^2/year. If we assume exclusively vertical migration, this value lies between the flux rates of diffusive transport and compressible Darcy flow. Thus, accumulation of gas can occur as long as the Darcy flow condition through the roof rock is not met.

Two-Dimensional Modeling of Gas Diffusion

The crude comparisons given above show that diffusive losses do not affect hydrocarbon accumulation dramatically if roof rock thicknesses are reasonable and gas supply is occurring. They make clear, however, that it is

Table 3—Scenario for Gas Generation from Type III Source Rock

Parameter	Value
Source rock (type III) thickness	100 m
TOC content	5%
Generation potential	110 mg CH_4/g C_{org}
Generation time	10 m.y.
Generation rate	200 m^3/km^2/year

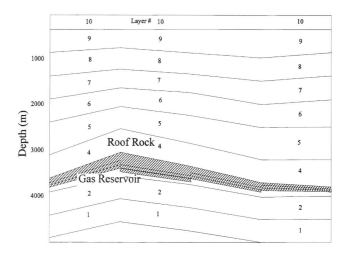

Figure 6—Assessment of diffusive gas losses for a hypothetical reservoir situation. This roof rock diffusion scenario was used for a test run of the PetroMod diffusion module.

desirable to have better control of the various parameters that influence hydrocarbon accumulation and dismigration in order to reach an improved quantitative understanding of these processes. The one-dimensional numerical diffusion model by Montel et al. (1993) provides higher flexibility in the assessment of molecular transport of gases through rock sequences in natural systems. But, as these authors state in their concluding remarks, an extension to a two-dimensional or even three-dimensional model is desirable.

To treat diffusion in the context of a comprehensive analysis of hydrocarbon generation and accumulation, a diffusion module has been developed as an integrated part of two-dimensional basin modeling software (PetroMod®, Integrated Exploration Systems). This new modeling tool makes use of our data base of experimental diffusion coefficients established during recent years. It provides the opportunity to compute concentration profiles at different grid points in a reservoir structure and to follow the evolution of diffusive transport through time. Furthermore, it makes possible a methane balance over carrier rocks and roof rock sequences along a geologic cross section composed of different lithologies. This diffusion module is now being used for the assessment of diffusive losses in complex geologic structures and scenarios.

Figure 6 shows a simple scenario for a roof rock overlying a gas reservoir. Methane diffusion into the roof rock sequence began with the initial filling of the reservoir structure. The present-day concentration profiles for methane (in units of mass per volume of pore water) at the grid points of this cross section are shown in Figure 7. In the source and reservoir sequences where a free gas phase is present or has passed through during secondary migration, the concentration profiles represent the maximum aqueous solubility of methane as a function of pressure and temperature. Above the roof rock–reservoir interface the concentration profiles result from molecular diffusion of methane into the completely water-saturated seal. In this example, the concentration front has moved into the overburden over a distance of more than 500 m.

CONCLUSIONS

Although molecular transport has a much lower transport capacity than volume flow (Darcy flow), the fact that diffusion occurs ubiquitously, permanently, and uncondi-

tionally can make it a relevant transport process in geologic systems under certain conditions. The present contribution focused on the role of diffusive loss from reservoirs into and through overlying roof rocks, which presently appears to be one of the major concerns in hydrocarbon exploration and has caused some controversy. No general statement is possible concerning the relevance of diffusive transport, and each particular case history must be analyzed individually. Diffusion should, however, be considered as a relevant dismigration mechanism under the following conditions:

1. absence of volume flow (leakage) over extended periods of geologic time,
2. tectonically stable areas, and
3. no further hydrocarbon supply from source rocks (e.g., inverted basins).

It can be shown that, due to high concentration gradients, diffusive loss from reservoirs are most effective during the nonstationary state, which implies that substantial amounts of gas are required to establish a steady-state concentration profile in the roof rock sequence. Assumption of stationary flux from the time of first filling of the reservoir will therefore underestimate diffusive losses. When performing a gas balance of an entire sedimentary basin, particular attention must be paid to the quantities of dispersed gas in the overburden, which may substantially exceed the reservoir contents. It is important to note that the time scale of significant diffusive losses from gas reservoirs is in the range of 10 to 100 m.y., rather than just a few million years (Krooss et al., 1992a,b; Montel et al., 1993).

As demonstrated by Montel et al. (1993) for a North Sea oil and gas field, one-dimensional diffusion models can yield relatively good estimates of diffusive gas losses

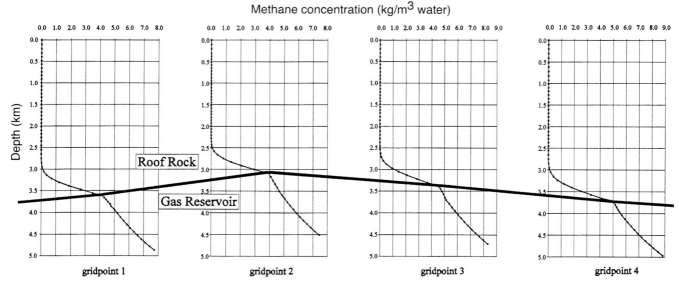

Figure 7—Methane diffusion into the roof rock of a gas reservoir. Shown here are the concentration profiles for methane in the aqueous phase of the carrier bed and the roof rock sequence.

for layered sequences, taking into account aqueous solubility of hydrocarbons and the temperature dependence of the diffusion coefficients. The required data base of diffusion coefficients for different lithologies is presently being extended by laboratory measurements at 150°C at our institute.

The next step in numerical modeling of hydrocarbon diffusion in sedimentary rocks has been taken by developing a two-dimensional diffusion module as an integral part of the IES PetroMod basin modeling software. This new tool will enhance the quantification of diffusive gas losses over geologic time by taking into account such factors as drainage area, filling time of the reservoir, and reservoir geometry.

The relevance of diffusive transport for the surface exploration of hydrocarbon accumulations appears to be limited considering the low propagation rate of diffusion fronts as compared to the reportedly rapid response of surface exploration methods to reservoir conditions. With the tools now available, however, it is possible to make fairly straightforward estimates for specific case histories.

Acknowledgments—We thank T. Hantschel of IES GmbH, Jülich, for his support in the development and testing of the numerical model for hydrocarbon diffusion. We also thank D. H. Welte for his encouragement and continuous interest in this work. Furthermore, we wish to thank D. Hanebeck and R. G. Schaefer for fruitful discussions. The paper benefited from a review by E. C. Simmons with helpful comments and suggestions.

REFERENCES CITED

Antonov, P. L., 1954, On the diffusion permeability of some claystones (in Russian): Trudy NIIGGR, Geokhim. Met. Poisk. Nefti i Gaza, Gostoptekhizdat, Moscow, v. 2, p. 39–55.

Antonov, P. L., 1958, Investigation of the laws of gas saturation of rocks with depth (in Russian): Trudy VNIGNI, Gostoptekhizdat, Moscow, v. 10, p. 241–256.

Antonov, P. L., 1964, On the extent of diffusive permeability of rocks: direct methods of oil and gas exploration (in Russian), *in* Direct methods of oil and gas exploration: Moscow, Nedra, p. 5–13.

Antonov, P. L., 1968, Some results of the research on molecular migration of hydrocarbon gases in rocks (in Russian): Trudy VNIIYaGG, Izd. Nedra, Moscow, v. 4, p. 132–154.

Antonov, P. L., 1970a, On the gas saturation of rock above reservoirs of finite thickness: depletion of reservoirs during upward gas migration (in Russian): Trudy VNIIYaGG, Izd. Geochimii, Moscow, v. 8, p. 38–50.

Antonov, P. L., 1970b, Results of the investigation of diffusion permeability of sedimentary rocks for hydrocarbon gases (in Russian): Trudy VNIIYaGG, Izd. Geochimii, Moscow, v. 8., p. 51–59.

Antonov, P. L., G. A. Gladyshewa, and W. P. Koslow, 1958, The diffusion of hydrocarbons through rock salt (in German): Zeitschrift für angewandte Geologie, no. 8, p. 387–388.

Berg, R. R., 1975, Capillary pressures in stratigraphic traps: AAPG Bulletin, v. 59, no. 6, p. 939–956.

Carslaw, H. S., and J.C. Jaeger, 1959, Conduction of heat in solids: Oxford, Clarendon Press, 510 p.

Crank, J., 1975, The mathematics of diffusion: Oxford, Clarendon Press, 414 p.

England, W. A., and A. S. Mackenzie, 1989, Some aspects of the organic geochemistry of petroleum fluids: Geologische Rundschau, v. 78, no. 1, p. 291–303.

Haas, J. L., 1978, An empirical equation with tables of smoothed solubilities of methane in water and aqueous sodium chloride solutions up to 25 weight percent, 360°C, and 138 MPa: USGS Open File Report 78-1004, 41 p.

Hanebeck D., 1994, Experimentelle Simulation und Untersuchung der Genese und Expulsion von Erdölen aus Muttergesteinen: Ph.D. dissertation, Rhenish-Westphalian Technical University (RWTH), Aachen, Germany, 326 p.

Krooss, B. M., 1992, Diffusive loss of hydrocarbons through cap rock: Erdöl und Kohle–Erdgas–Petrochemie/ Hydrocarbon Technology, v. 45, p. 387–396.

Krooss, B. M., D. Leythaeuser, and R. G. Schaefer, 1988, Light hydrocarbon diffusion in a cap rock: Chemical Geology, v. 71, p. 65–76.

Krooss, B. M., D. Leythaeuser, and R. G. Schaefer, 1992a, The quantification of diffusive hydrocarbon losses through cap rocks of natural gas reservoirs—a reevaluation: AAPG Bulletin, v. 76, p. 403–406.

Krooss, B. M., D. Leythaeuser, and R. G. Schaefer, 1992b, The quantification of diffusive hydrocarbon losses through cap rocks of natural gas reservoirs—a reevaluation: Reply: AAPG Bulletin, v. 76, p. 1842–1846.

Leythaeuser, D., and H. S. Poelchau, 1991, Expulsion of petroleum from type III kerogen source rocks in gaseous solution: modelling of solubility fractionation, *in* W. A. England and A. J. Fleet, eds., Petroleum migration: Geological Society of London Special Publication 59, p. 33–46.

Leythaeuser D., R. G. Schaefer, and A. Yükler, 1980, Diffusion of light hydrocarbons through near-surface rocks: Nature, v. 284, p. 522–525.

Leythaeuser, D., R. G. Schaefer, and A. Yükler, 1982, Role of diffusion in primary migration of hydrocarbons: AAPG Bulletin, v. 66, p. 408–429.

Montel, F., G. Caillet, A. Pucheau, and J. P. Caltagirone, 1993, Diffusion model for predicting reservoir gas losses: Marine and Petroleum Geology, v. 10, p. 51–57.

Nelson, J. S and E. C. Simmons, 1992, The quantification of diffusive hydrocarbon losses through cap rock of natural gas reservoirs—a reevaluation: Discussion: AAPG Bulletin, v. 76, p. 1839–1841.

Nesterov, I. I., and I. N. Ushatinskij, 1972, Sealing properties of shales above petroleum and natural gas accumulations in Mesozoic sediments of the West Siberian plain (in German): Zeitschrift für angewandte Geologie, v. 18, no. 12, p. 548–555.

North, F. K., 1985, Petroleum geology: Winchester, Massachusetts, Allen & Unwin, 607 p.

Palciauskas, V. V., 1991, Primary migration of petroleum, *in* R. K. Merrill, ed., Source and migration processes and evaluation techniques: AAPG Treatise of Petroleum Geology, Handbook of Petroleum Geology, p. 13–22.

Schowalter, T. T., 1979, Mechanics of secondary hydrocarbon migration and entrapment: AAPG Bulletin, v. 63, no. 5, p. 723–760.

Smith, J. E., J. G. Erdman, and D. A. Morris, 1971, Migration, accumulation, and retention of petroleum in the earth: Proceedings of the 8th World Petroleum Congress, Moscow, v. 2, p. 13–26.

Stainforth, J. G., and J. E. A. Reinders, 1990, Primary migration of hydrocarbons by diffusion through organic matter networks, and its effect on oil and gas generation: Organic Geochemistry, v. 16, nos. 1–3, p. 61–74.

Tissot, B., and D. H. Welte, 1984, Petroleum formation and occurrence: Berlin, Springer-Verlag, 699 p.

Watts, N. L., 1987, Theoretical aspects of cap rock and fault seals for single- and two-phase hydrocarbon columns: Marine and Petroleum Geology, v. 4, p. 274–307.

Rostron, B. J., and J. Tóth, 1996, Ascending fluid plumes above Devonian pinnacle
reefs: numerical modeling and field example from west-central Alberta,
Canada, *in* D. Schumacher and M. A. Abrams, eds., Hydrocarbon migration
and its near-surface expression: AAPG Memoir 66, p. 185–201.

Chapter 15

Ascending Fluid Plumes Above Devonian Pinnacle Reefs: Numerical Modeling and Field Example from West-Central Alberta, Canada

B. J. Rostron

Department of Earth and Atmospheric Sciences
University of Alberta
Edmonton, Alberta, Canada

Present address:

Department of Geological Sciences
University of Saskatchewan
Saskatoon, Saskatchewan, Canada

J. Tóth

Department of Earth and Atmospheric Sciences
University of Alberta
Edmonton, Alberta, Canada

Abstract

Hydrocarbon plumes emanating from breached reservoirs alter the subsurface environment through which they pass. The detection of these plumes and their alteration effects form the basis for most surface geochemical exploration programs. The mechanics of plume generation and migration, however, remain poorly understood in a quantitative sense, which can lead to reduced exploration success. A two-part study incorporating numerical simulations and field mapping was conducted to better understand the generation and subsequent migration of hydrocarbon plumes in the subsurface.

Numerical simulations of oil and water flow show that plume generation is controlled by the hydraulic properties of the system: the driving or "leaking factors" (oil/water density contrast and regional hydraulic gradient) and the resisting or "sealing factors" (entry capillary pressure and intrinsic permeability). Field mapping delineated a plume of saline water (>100 g/L dissolved solids) in the Mannville Group aquifer that appears to result from the mixing of vertically migrating saline Devonian waters with the more dilute Mannville waters. Saline water and oil leak upward out of Leduc pinnacle reefs, through the overlying Ireton aquitard, and into the Nisku aquifer. Numerous smaller oil plumes coalesce in the Nisku aquifer and continue to migrate vertically up into the Mannville Group aquifer.

The results have four implications for hydrocarbon exploration: (1) the formation of saline plumes above pinnacle reefs is controlled by the hydraulic properties of the flow domain; (2) geochemical exploration for Devonian pinnacle reefs has to be conducted at the Mannville Group level; (3) there will be little, if any, surface expression of Devonian hydrocarbon plumes in west-central Alberta; and (4) the saline plume may reach the surface in other areas of the basin where ascending fluid flow occurs and the Mannville Group is closer to the surface.

INTRODUCTION

Most geochemical exploration techniques are based on the principle that a hydrocarbon plume moving through the subsurface alters the physical, chemical, electrical, magnetic, and thermal properties of the rock framework and fluids through which it passes. If the resultant alteration effects, such as a geochemical chimney or a resistivity anomaly, are strong enough to be detected, then it may be possible to locate the plume and trace it back to its source at a breached reservoir. In the past, the controls on plume generation and migration have not been clearly understood or have been ignored, to the detriment of geochemical exploration techniques (Davidson, 1994). The processes that control both the origins of the plume and its subsequent migration path must be clearly understood in order to predict the location of the reservoir accurately. The purpose of this paper is to demonstrate the importance of hydrogeology and the hydraulic properties of the rock framework in the creation and migration of a hydrocarbon plume. Geochemical exploration programs can benefit from a better understanding of plume generation and migration in the subsurface.

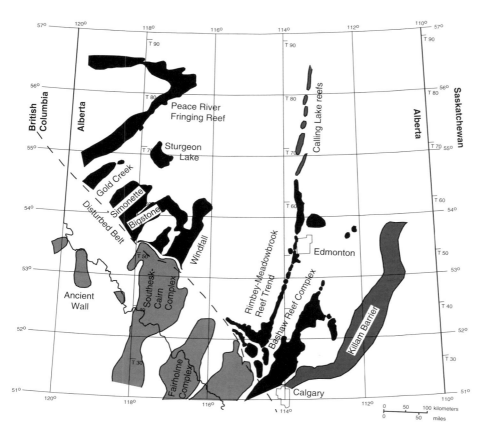

Figure 1—Map of Woodbend Group equivalent (Upper Devonian) reef trends in the Western Canada Sedimentary Basin. Lighter shading to the west of the disturbed belt indicates where reefs have been palinspastically restored. (Modified after Switzer et al., 1994, and others.)

From a hydrogeologic perspective, two key aspects of plume generation and migration remain problematic. First, the effects of formation-fluid flow on plume migration are commonly disregarded. In static environments, the basic mechanisms of plume generation are well understood. Seal failure, or plume generation, is controlled either by capillary failure (Hubbert, 1940; Berg, 1975; Schowalter, 1979; Watts, 1987) or by fracturing of the seal unit (Watts, 1987; Sales, 1993). In hydrodynamic systems, the treatment of plume generation has not received as much attention. Hydrodynamic effects are either ignored or limited to simple correction factors applied in the calculation of maximum hydrocarbon heights (Schowalter, 1979; Watts, 1987). Vertical migration of oil in dynamic subsurface environments following seal failure has been dealt with only in a qualitative manner (e.g., Sales, 1993), and the implications of vertical migration between the trap and the surface remain unstudied. There is a similar lack of quantitative studies on the effects of lateral fluid flow on the surface expressions of vertical hydrocarbon migration (Davidson, 1994), with some notable exceptions (Holysh, 1989; Machel and Burton, 1991; Holysh and Tóth, 1994).

The second problem is the lack of hydraulic data on the low-permeability rock units that comprise the flow systems. This lack of data limits quantitative analysis of plume generation and migration. At the present time, we are unaware of any published examples of a comprehensive hydraulic data set (intrinsic permeability, relative per-

meability, and capillary pressure curves) from a low-permeability hydrocarbon "seal." This lack of data appears to result from the high cost of obtaining such data, the perceived futility of measuring permeabilities on so-called impermeable rocks, and the lack of use of these data in reservoir trap studies (as mentioned previously). The paucity of actual hydraulic data has lead to many misconceptions about the role of low-permeability rocks in fluid migration in basins (Neuzil, 1986; Tóth, 1991).

This paper presents the results of a two-part study conducted to address some of these concerns. First, results of a numerical modeling study of plume generation and migration from breached pinnacle reef reservoirs are presented. Simulations were conducted using a flow domain based on the Devonian pinnacle reefs that form two large-scale reef trends in the Western Canada Sedimentary Basin: Rimbey–Meadowbrook and Bashaw (Figure 1). These reefs are suitable candidates for study for several reasons: they are prolific hydrocarbon producers; there are abundant available geologic data; and it has been postulated that they leak hydrocarbons (Davis, 1972; Creaney and Allan, 1990; Rostron, 1993). Second, a field hydrogeologic and hydrogeochemical study of the Bashaw Reef Complex is presented. A primary goal of the field study was to locate hydrocarbon plumes above leaky pinnacle reef reservoirs. The implications of this study for surface geochemical exploration in general, and for Devonian pinnacle reefs in the Western Canada Sedimentary Basin in particular, are discussed.

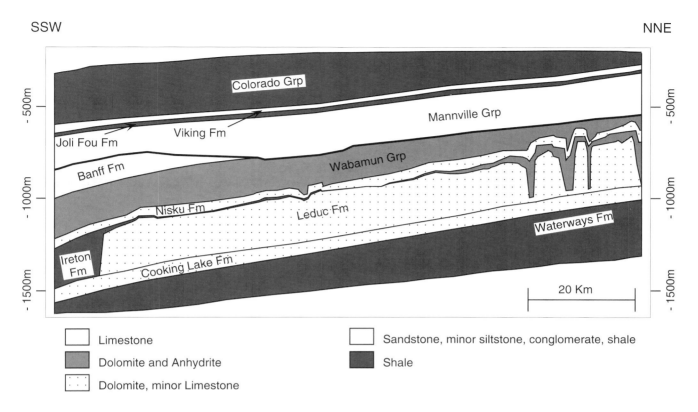

SSW

NNE

Figure 2—Structural cross section trending SSW-NNE through the Bashaw Reef Complex showing structural and stratigraphic relationships.

NUMERICAL MODELING OF PLUME GENERATION AND MIGRATION

Numerical simulations were performed with the transient two-dimensional multiphase finite-difference flow model SWANFLOW-2D (Faust, 1985). SWANFLOW-2D was coupled to the IMAGE output analysis program (National Center for Supercomputing Applications, 1991) to produce contour plots of fluid pressure and saturation at specified time-steps for each run. Simulations were conducted using Macintosh microcomputers and Silicon Graphics workstations, with execution times ranging from 1.5 to 9 hours per case. More than 40 different simulations were completed while systematically varying the rock permeabilities, capillary pressure curves, oil densities, and regional formation-fluid gradients of the system. Only the results of one run are presented. This run was chosen as a typical result of a case where a hydrocarbon plume is generated above a leaky reef. Selected end-time saturation distributions for other modeled cases are presented to show the relationship between the hydraulic parameters of the flow system and plume generation and migration. The importance of hydraulic properties on controlling the size and shape of hydrocarbon plumes is illustrated. Surface geochemical techniques would benefit from an understanding of these same factors.

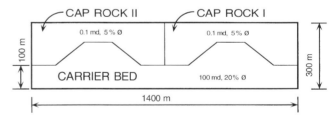

Figure 3—Schematic illustration of the modeled flow domain.

The simulation results presented here have been presented previously in more detail (Rostron, 1993). The model results are summarized briefly, and the implications to plume migration are explored.

Simulation Data

Numerical simulations were conducted on a flow system patterned after a typical cross section through one of the Devonian reef trends that cut across the Western Canada Sedimentary Basin (Figure 2). Reservoirs in these reef trends belong to the Woodbend and Winterburn groups, which are Late Devonian (Frasnian–Famennian) in age (Switzer et al., 1994). The Woodbend Group consists of three main units. The lowermost unit is a shallow-water platform carbonate (Cooking Lake Formation) that forms a common basal aquifer for the reef trend. On this unit are situated numerous platform margin reef buildups or "pinnacle reefs" (Leduc Formation) forming the main reservoirs. These reefs are encased in a layer of basin-filling shales and limestones (Ireton and Duvernay formations). Overlying the Woodbend Group are carbonates, primarily dolomites, of the Nisku Formation (Winterburn Group). Flow is confined to the Woodbend and Winterburn groups by the extremely low permeability shales and carbonates of the Waterways Formation (Beaverhill Lake Group) and Wabamun Group, situated below and above the modeled strata, respectively.

Geologic structure data were used to construct a hypothetical model through two typical reefs in the reef trend (Figure 3). A 60×15 block (horizontal × vertical) finite-difference grid was used to discretize the flow domain. Two main rock types were assigned to the grid: a reservoir rock and a cap rock. Reservoir rock units included the basal aquifer and pinnacle reefs, with cap rock comprising the remainder.

Two different boundary conditions were assigned along the edges of the grid. Along the top and bottom, no-flow boundaries were assigned, confining flow to the carrier bed and cap rock. Along the upstream and downstream ends of the domain (right and left ends, respectively), no flow boundaries were specified in the cap rock and constant heads were specified in the carrier bed. This configuration enabled the establishment of a horizontal flow field for water and allowed for vertical flow inside the reefs and cap rock.

The input parameters used in this simulation and in others in the suite are summarized in Table 1. These data, although derived from actual rock samples, were selected arbitrarily. At the time of modeling, capillary pressure curves and relative permeability curves for the cap rocks were unavailable, as was the bulk oil source rate. Parameters were selected such that each simulation could be completed in a reasonable amount of computer time. This arbitrary selection was justified because of the lack of any real data on the system and because the purpose of the simulations was to study the process of plume generation and migration in a conceptual sense. Results of model studies presented here cannot be used as a quantitative measures of migration in the real system of reefs.

Simulations were normally conducted in two stages: an initial stage of no oil input that was done to establish a steady-state water flow field across the grid and a subsequent transient (oil input) migration phase. Runs were halted once the migrating oil either had filled and bypassed both reefs or had leaked out of the reef and accumulated along the top of the grid. In specific cases, a third stage was added in which, upon filling of the reservoir, the oil source was discontinued and the system allowed to return to steady-state conditions.

In this study, only water and oil flow were considered because of a complete lack of three-phase hydraulic data

Table 1—Simulation Data

Characteristics	Values
Flow domain	1400×300 m (horizontal × vertical)
Fluid viscosity	2.0×10^{-3} Pa-sec (oil) and 1.0×10^{-3} Pa-sec (water)
Water density	1000 kg/m³
Porosity	20% for carrier bed and 5% for cap rock
Carrier bed permeability	100 md
Formation compressibility	1.0×10^{-10} kPa⁻¹
Injected fluid composition	90% water, 10% oil
Oil injection rate	3.71×10^{-1} m³/year
Irreducible oil saturation	20%
Irreducible water saturation	45%
Time-step length	50 year
Oil densities	700, 850, and 1000 kg/m³
Regional hydraulic gradients	0.005, 0.02, 0.08, and 0.2 m/m
Cap rock permeabilities	0.0001, 0.001, 0.01, 0.1, and 100 md
Critical oil column of cap rock	25.8, 75, 100, 150, and 285 m

on low-permeability rocks and computer limitations on simulating three-phase flow.

Simulation Results: Leaky Cap Rocks

Oil leakage and the creation of a hydrocarbon plume from the top of a pinnacle reef begins when the combined driving forces on oil, which are provided by buoyancy and hydrodynamic flow, exceed the resistive forces of the cap rock. Initial resistance to leakage through the cap rock is provided by the difference in entry pressures of the reservoir and cap rock. The resistive force can be expressed in terms of a *critical oil column (COC)* for a given set of reservoir and cap rock capillary pressure curves and fluid densities under static conditions. The COC can be expressed mathematically as

$$COC = \frac{Pe_{cr} - Pe_{cb}}{(\rho_w - \rho_o)g} \qquad (1)$$

where COC is the critical oil column for a specified pair of oil and water densities and a given pair of carrier bed and cap rocks (m), Pe_{cr} and Pe_{cb} are the entry capillary pressures of the cap rock and carrier bed (Pa), ρ_w is the water density (kg/m³), ρ_o is the oil density (kg/m³), and g is the gravitational constant (m/sec²).

The COC (with its units of length) provides a useful measure of sealing capacity of a cap rock. To initiate leakage in a hydrostatic system, the vertical component of the hydrocarbon column must exceed the COC. If the maximum attainable column height for a reservoir (defined as the vertical distance between the crest of the trap and the

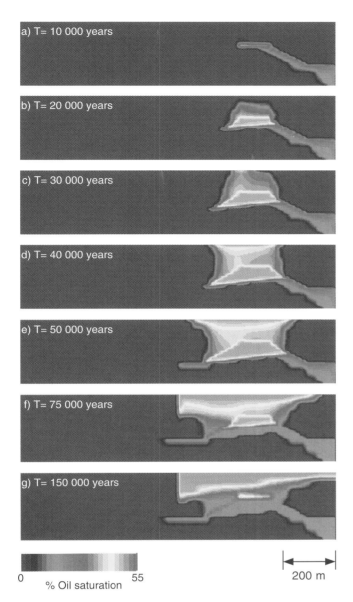

a) T= 10 000 years

b) T= 20 000 years

c) T= 30 000 years

d) T= 40 000 years

e) T= 50 000 years

f) T= 75 000 years

g) T= 150 000 years

0   % Oil saturation   55

200 m

Figure 4—Oil saturation distributions through time (10,000–150,000 years) for a typical leaky pinnacle reef (COC = 25.8 m). No vertical exaggeration.

horizontal water gradient and a vertical buoyancy driving force. Once introduced into the flow domain, the oil forms a stringer that moves to the top of the carrier bed. Inside the stringer, oil saturations remain below 21%. This occurs because as oil saturations in the grid blocks at the saturation front increase above the irreducible saturation of 20%, oil becomes mobile in those grid blocks. Once mobile, the front moves downstream into the next grid block(s). Oil saturations within the stringer never exceed 21% unless the stringer encounters some type of boundary. In these simulations, the stringer encounters a boundary when it reaches the crest of the reef structure.

At T = 10,000 years (Figure 4a), oil reaches the crest of the reef. At the crest, oil saturations quickly reach 55%, the maximum permitted with the specified irreducible water saturation of 45%. Once the pore space reaches 55% saturation, no further increase in saturation is possible, so the area of accumulation begins to expand downward. In this case, the downward movement of the oil/water contact begins at about T = 11,000 years.

As the oil/water contact descends, it generates an increasingly taller column of oil pushing up on the cap rock. When the oil column equals the height of the COC (25.8 m), oil penetrates into the cap rock (Figure 4b), producing a hydrocarbon plume in the cap rock above.

In the cap rock, the hydrocarbon plume moves vertically upward, reaching the top of the grid at about 24,000 years. Oil begins to accumulate along the top boundary of the flow domain (Figures 4d–e) solely as a result of the imposed no-flow boundary along the top of the grid. In reality, the plume would freely migrate vertically until reaching the ground surface.

Inside the reef, the oil/water contact moves downward with continued oil input through time (Figures 4b–d). Once the oil/water contact reaches the spill point, a second stringer with 21% oil saturation forms in the carrier bed and continues to move downstream (Figure 4e). Once the stringer reaches the second reef structure, the filling and subsequent plume generation stage is repeated.

The final phase of the simulation was conducted to study the effects of dissipation of the plume and entrapped oil after the discontinuation of the upstream source of oil. At T = 50,000 years, the upstream source of oil was shut off. With no new input from below, the oil/water contact moves upward displacing previously emplaced oil in the reef into the cap rock (Figure 4f). Oil that leaks out of the reef is added to an ever-expanding oil plume in the cap rock.

By T = 110,000 years, the oil/water contact has risen about 75 m, leaving about a 25-m column of oil inside the reef. At this time, the oil column has shrunk down to the COC. The oil column inside the reef becomes stable once the column height decreases to the COC because a column of oil larger than the COC is necessary for leakage to occur (Rostron, 1993). From this time on, no changes in saturation occur within the reef.

Outside the reef, the plume continues to spread above the reef (Figure 4g). The final saturation profile at T = 150,000 years (Figure 4g) illustrates the 25-m oil column

spill point) is less than the COC, then the trap will not leak and no hydrocarbon plume will form. On the contrary, if the COC is less than the maximum attainable column height, then it is highly probable that the trap will leak.

In this paper, we present the result of a potentially leaky trap model in which the COC of the cap rock and reservoir pair (25.8 m) is less than the maximum attainable hydrocarbon column (100 m).

Oil saturation distributions through time for a typical "leaky" reef are shown in Figure 4. At time T = 0 years, oil is introduced into the upstream end of the carrier bed. Oil moves downstream under the combined influence of a

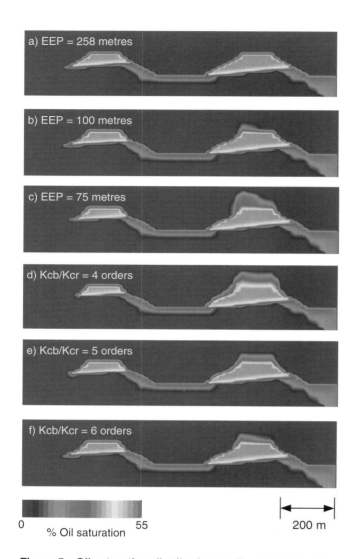

a) EEP = 258 metres

b) EEP = 100 metres

c) EEP = 75 metres

d) Kcb/Kcr = 4 orders

e) Kcb/Kcr = 5 orders

f) Kcb/Kcr = 6 orders

0   % Oil saturation   55   200 m

Figure 5—Oil saturation distributions at T = 50,000 years with variations in the equivalent entry pressure (COC) and intrinsic permeability of the first cap rock. No vertical exaggeration.

inflow from upstream was greater than the rate leaking into the cap rock. In other modeled cases, the reef was unable to fill to the spill point because the bulk of the oil entering the reservoir leaked into the cap rock. This leads to the conclusion that the larger the plume above the reef, the smaller the accumulation within the reef itself. However, this conclusion cannot be universally applied without examining the hydraulics of the flow system. A knowledge of the hydraulics of the flow system leads to a better understanding of flow directions, flow rates, and event timing. Only after the hydraulics of the flow system are known can the distinction be made between a large plume that has resulted from catastrophic draining of a reservoir (such as in Figure 4) and slow leakage from a relatively full reservoir. Formation of oil plumes and migration and entrapment in pinnacle reefs depends on the interaction between the forces driving migration ("leaking factors") and those resisting migration ("sealing factors"). The role of these interactions is discussed next.

Role of Sealing Factors in Plume Generation and Migration

Sealing factors are defined as those physical properties or processes that act to retain hydrocarbons within reservoirs. Simulation results show the two sealing factors are the entry capillary pressure and the intrinsic permeability of the cap rock. Initial formation of a hydrocarbon plume is governed by the magnitude of the critical oil column of the cap rock. If the vertical component of the oil column equals the COC, oil will begin to leak from the trap (Figure 4). If sufficient oil column cannot be built up to equal the COC, then a plume will not form.

Varying the COC component of the sealing factors produces vastly different final time oil distributions (Figure 5). With the COC greater than the trap height (Figure 5a), no plume forms above the reef. When the COC equals the trap height (Figure 5b), trace amounts of oil still leak from the trap because the COC is only a static measure of trapping capacity (Rostron, 1993). In hydrodynamic environments such as this, an additional driving force on the oil particles is provided by water flow. In this case, the vertical component of water flow combined with the buoyant force generated by the column of oil exceeds the static COC. Although the regional water flow field has been set up for horizontal flow, a small component of vertical flow arises when water flows from the carrier bed into the reservoir. In this case, the small size of the plume above the trap reflects the relatively small component of localized vertical water flow.

For the intermediate case with an COC equal to 75 m, the plume at T = 50,000 years (Figure 5c) is significantly smaller than in the original case (Figure 4e). The difference in plume size above the reef in the two figures is directly related to the COC. In the first case (COC = 25.8 m), the oil column reaches the COC value sooner than in the second case (COC = 75 m). With a longer period of time for oil to leak into the cap rock, a larger plume is generated in the first case.

retained within the reef and a large hydrocarbon plume above. The saturation distribution at T = 150,000 years reflects steady-state conditions for this reef and cap rock system. Past this time, negligible changes occur in the saturation distribution because of the boundaries imposed on the flow domain.

In this simulation, the plume does not penetrate the cap rock above the second reef because the COC of the second cap rock (Table 1, Figure 3) is equal to 258 m. Thus, the vertical discontinuity in the saturation distribution at the mid-point of the grid is solely an artifact of the imposed COC for the second cap rock.

In this case, the trap filled to its maximum capacity, that is, the oil/water contact reached the spill point, even though a plume formed above the reef. Complete filling and plume formation were possible because the rate of

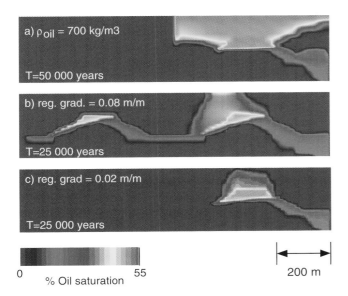

Figure 6—Oil saturation distributions at a constant COC of 25.8 m with variations in the oil density (ρ_{oil}) and the regional hydraulic gradient. No vertical exaggeration.

If a plume does form, the second sealing factor controls its subsequent migration. Oil movement within the cap rock is governed by the intrinsic permeability of the cap rock, subject to the overall hydrodynamic conditions of the flow system. If the intrinsic permeability of the cap rock is reduced, the plume sizes above the leaky reef are likewise smaller (Figures 5d–f). End-time saturation profiles for step-wise magnitude decreases in intrinsic permeability show that when the intrinsic permeability of the cap rock has been reduced to 1×10^{-4} md (Figure 5f), there is almost no plume in the cap rock above. This sequence of saturation distribution profiles demonstrates how reducing the intrinsic permeability of the cap rock decreases the size of the plume and has the same effect as reducing the COC. Also, with a low enough intrinsic permeability, the saturation distribution appears as though the cap rock has not failed, even though it has.

Role of Leaking Factors in Plume Generation and Migration

Leaking factors are defined as those physical properties or processes that act to drive hydrocarbons from reservoirs. Simulation results show that the two main leaking factors are buoyancy caused by the oil/water density contrast and the regional hydraulic gradient of water.

Buoyancy forces are generated by density differences among the immiscible phases, with the magnitude of the forces proportional to the density contrast. In a static environment, buoyancy forces act vertically. In hydrodynamic environments, a component of the regional water hydraulic gradient provides an additional vertical driving force. Changing the fluid densities or water hydraulic gradient can affect the formation and subsequent migration of a hydrocarbon plume. Results show how varying these leaking factors influences the end-time saturation distributions (Figure 6).

When the oil density is reduced from 850 to 700 kg/m^3, the net effect is to double the density contrast and to double the leaking factor. With increased buoyant forces, the size of the hydrocarbon plume in the cap rock (Figure 6a) is increased dramatically from the originally modeled case. Reducing the density from 850 to 700 kg/m^3 for a given cap rock entry pressure effectively reduces the COC of the cap rock by about 50% of the original value (Eq. 1). The resulting end-time saturation distribution (Figure 6a) illustrates the case in which a trap cannot fill to its spill point, as most of the migrating oil passes through the trap into the cap rock. Lighter oils generate larger driving forces and therefore larger plumes.

Plume size and shape also depend on the regional water hydraulic gradient. Increasing the gradient increases both the horizontal and vertical driving forces on the oil. This results in faster downstream migration through both the carrier bed and the cap rock. If the hydraulic gradient is increased by a factor of four (from 0.02 to 0.08 m/m), then the oil reaches the downstream boundary of the flow domain in less than 25,000 years (Figure 6b). With increased horizontal hydraulic gradients, the plume emanating from the breached reservoir is offset from the crest of the trap approximately 100 m downstream. When the saturation distribution under the increased hydraulic gradient (Figure 6b) is compared to the saturation distribution at T = 25,000 years for the original case (Figure 6c), it is clear that the increased lateral driving force offsets the plume and increases its size within the cap rock. These results demonstrate the importance of lateral fluid flow in controlling the size and shape of hydrocarbon plumes and their position above the hydrocarbon source.

Results of modeling leaky pinnacle reefs can be summarized as follows: (1) the critical oil column of the cap rock controls whether oil will begin to leak from the trap and whether or not a hydrocarbon plume will form; (2) when the vertical component of the hydrocarbon column meets the COC, a plume will begin to form in the cap rock; (3) once the plume begins to leak from the cap rock, the hydraulic properties of the cap rock and the flow domain control its migration path to surface; (4) after the source of hydrocarbons is terminated, leakage from the trap (plume sourcing) will continue until equilibrium conditions are reached between the oil column height and the COC of the cap rock; and (5) interaction between forces driving migration (leaking factors) and factors resisting migration (sealing factors) determines whether or not a plume will form and what its migration path will be.

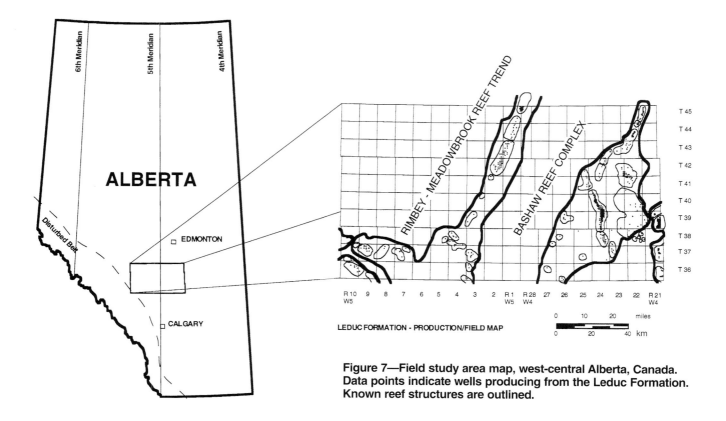

Figure 7—Field study area map, west-central Alberta, Canada. Data points indicate wells producing from the Leduc Formation. Known reef structures are outlined.

HYDROCARBON AND SALINE WATER PLUMES IN THE SUBSURFACE

The preceding modeling results illustrate three points: (1) that reefs can leak and create plumes in the cap rocks above them, (2) that leakage is controlled by the rock properties of the reservoir and cap rock, and (3) that knowledge of the subsurface hydrogeology is required to understand plume generation and subsequent migration. To corroborate the numerical modeling, a field study was conducted to search for evidence of plumes emanating from breached pinnacle reefs.

Field Study Area and Geology

The field study examined the hydrogeology and hydrochemistry of the Devonian and Cretaceous formations in an 18,000-km² area of west-central Alberta, Canada (Figure 7). The study area encompasses the Bashaw Reef Complex and the southern part of the Rimbey-Meadowbrook Reef Trend. In addition to the Cooking Lake, Leduc, and Ireton formations discussed previously, the subsurface mapping included (in ascending order): the Nisku Formation carbonates overlying the Ireton shales; the relatively impermeable Wabamun Group (Devonian) dolomites and anhydrites; the highly permeable sandstones, siltstones, shales, and carbonates of the Mannville Group (Lower Cretaceous); the thin

shales of the Joli Fou Formation; and finally, the highly permeable sandstones, siltstones, and conglomerates of the Viking Formation (Lower Cretaceous–Albian). Structural relationships among these units are illustrated in a SSW-NNE cross section through the Bashaw Reef Complex (Figure 2).

Abundant geologic, formation pressure, fluid chemistry, and hydrocarbon production data were available from the more than 11,000 oil and gas wells drilled in the study area. Formation pressure data used in the study consisted of more than 5500 drill-stem test (DST) pressures obtained from the Energy Resources Conservation Board. Extrapolated DST data were processed to remove pressures influenced by production-induced drawdown, using the method described by Rostron (1994). About 7500 water chemistry analyses were obtained and screened to remove nonrepresentative samples, such as drilling fluids and spent acid-frac fluids, using standard culling techniques (Hitchon and Brulotte, 1994).

Geologic, hydrogeologic, hydrochemical, and petroleum production data were used to synthesize the three-dimensional flow field in the study area. For each aquifer, pressures were converted to hydraulic heads using

$$h = z + (p/\rho_w g) \qquad (2)$$

where h is the hydraulic head at a point (m), z is the elevation of the point of measurement with respect to sea level (m), p is the formation pressure (Pa), ρ_w is the den-

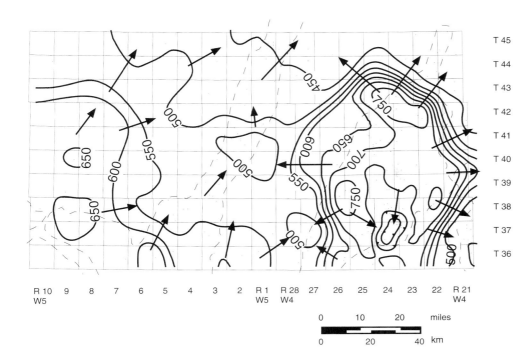

sity of freshwater (kg/m³), and g is the gravitational constant (m/sec²). Potentiometric surfaces were created by contouring the distribution of hydraulic heads in each aquifer. Pressure data were also used to construct pressure versus depth and pressure versus elevation plots for different aquifers and locations. Chemical data were converted to maps of major ion chemistry and total dissolved solids (TDS) for each aquifer. Oil and gas field production data were used to delineate known areas of hydrocarbon production. Results from this mapping identified a large-scale saline water plume in the study area.

Hydrogeology

Hydrogeologic mapping delineated three major flow regimes related to the Devonian reefs in the study area. They are the Upper Devonian hydrogeologic group (UDHG), the Mannville Group aquifer (MGA), and of lesser importance, the Viking aquifer (VA). The structural relationship of these three hydrostratigraphic units is shown in Figure 2. Although there appears to be evidence that Devonian fluids have or are affecting the Viking aquifer, the visible effects are small in comparison to those in the Mannville Group aquifer. The details of possible interaction between Devonian reefs and the Viking aquifer go beyond the scope of this paper, which focuses on the relationship between the Upper Devonian hydrogeologic group and the Mannville Group aquifer.

Upper Devonian Hydrogeologic Group

The Upper Devonian hydrogeologic group (UDHG) consists of two main aquifers: the Cooking Lake–Leduc aquifer and the Nisku aquifer. Fluid flow in the UDHG is exemplified by the potentiometric surface of the Nisku

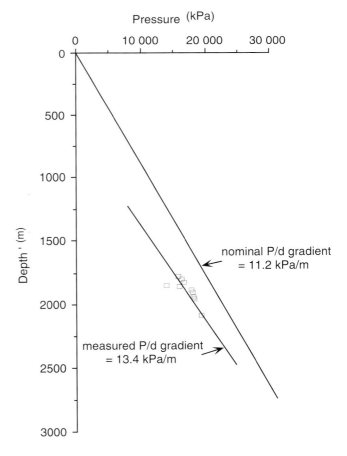

Figure 9—Pressure versus depth plot from the Leduc aquifer in the Bashaw Reef Complex. (Modified after Paul, 1994.)

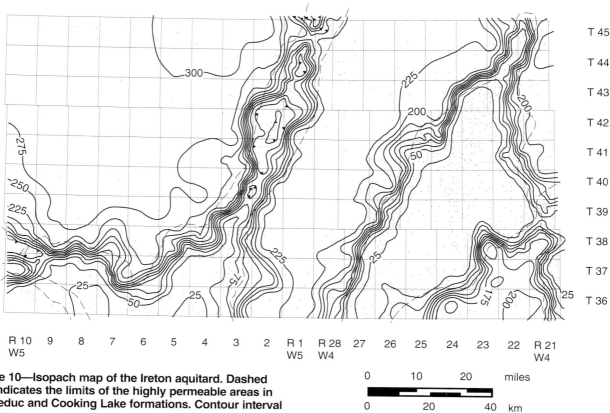

Figure 10—Isopach map of the Ireton aquitard. Dashed line indicates the limits of the highly permeable areas in the Leduc and Cooking Lake formations. Contour interval is 25 m.

aquifer (Figure 8). Hydraulic heads in the Nisku aquifer range from more than 750 m over the Bashaw Reef Complex to less than 450 m in the northeast corner of the study area. There are two distinct areas of fluid flow in the Nisku aquifer, delineated by the outline of the underlying Bashaw Reef Complex in the Leduc Formation. Where the Bashaw Reef Complex underlies the Nisku aquifer, there is a mounding of hydraulic heads in the potentiometric surface of the Nisku aquifer. Horizontal flow directions inferred from the potentiometric surface generally radiate outward from the reef complex.

Pressure versus depth plots constructed for the Bashaw Reef Complex indicate vertical fluid flow from the Leduc aquifer upward into the Nisku aquifer (Paul, 1994). A typical plot from the Bashaw Reef Complex (Figure 9) illustrates how the measured vertical pressure gradient in the UDHG (13.4 kPa/m) exceeds the density-corrected nominal value (11.2 kPa/m), indicating upward flow (Tóth, 1978). In the core of the complex, there are several localized spikes in the surface, indicating excellent communication with the higher potential Leduc aquifer beneath.

The second flow system in the Nisku aquifer is a regionally moving system that exists outside the area of the Bashaw complex. The fluid flows up dip, with fluid potentials ranging from more than 600 m in the southwest to less than 450 m in the northeast corner of the study area.

Flow directions in the Nisku aquifer are controlled by

the connection with the underlying Cooking Lake–Leduc aquifer. Although not shown here, fluid potentials in the Cooking Lake–Leduc aquifer are elevated 50–100 m above potentials at similar locations in the Nisku aquifer. Where the intervening aquitard (the Ireton Formation) is thinner over the Leduc Formation pinnacle reefs (Figure 10), a preferential pathway is provided for fluids to move upward from the Cooking Lake–Leduc aquifer into the Nisku aquifer. Over the Bashaw complex, the Ireton aquifer is generally less than 25 m thick (Figure 10), and in several places it is absent. Where the Ireton aquitard is thicker, such as on top of the Rimbey–Meadowbrook Reef Trend, the higher potentials in the Cooking Lake–Leduc aquifer do not increase hydraulic heads on the Nisku aquifer. Where not affected from below by the Cooking Lake–Leduc aquifer, flow in the Nisku aquifer is laterally up dip.

Mannville Group Aquifer

Fluid flow patterns inferred from the potentiometric surface of the Mannville Group aquifer (MGA) (Figure 11) are much more complicated than flow patterns in the Nisku aquifer. Both the MGA and UDHG exhibit regional-scale lateral up-dip flow systems, and both aquifers are intersected from below by a vertically moving flow system. However, three major differences exist between fluid flow in the MGA and the UDHG.

The first difference between the aquifers is that, in the MGA, a boundary exists with the Deep Basin (Masters,

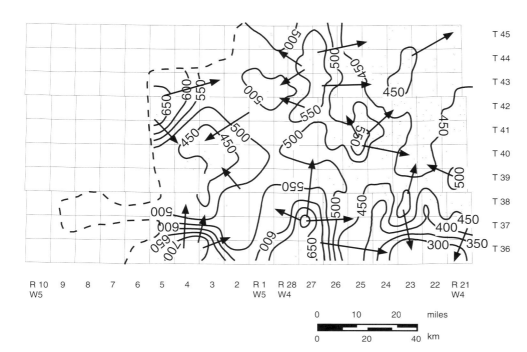

Figure 11—Map of hydraulic head distribution in the Mannville Group aquifer. Contour interval is 50 m of equivalent freshwater hydraulic head. Dashed line indicates the boundary with the Deep Basin.

1979) in the study area (Figure 11). To the west of the boundary line, the pore space is hydrocarbon saturated, mostly with gas. Wells producing from the Deep Basin produce negligible amounts of formation water, leading to speculation that the pore water is in a discontinuous state (Masters, 1979). To the east of the boundary line, conventional hydrocarbon pools are found where wells produce hydrocarbons with varying amounts of water. In the study area, the position of the Deep Basin boundary was delineated using well production data and DST recoveries. Since little or no mobile formation water is present west of the boundary line, pressure data from the Deep Basin were not included in the construction of the potentiometric surface used to infer water flow patterns.

The second major difference between the MGA and the UDHG is found in the much more subtle nature of the lateral flow system. Examination of the potentiometric surface (Figure 11) reveals that hydraulic heads in the MGA range from less than 300 m to more than 700 m, but with no systematic decrease in any one direction. Numerous closed areas of high and low values of hydraulic head are shown on the potentiometric surface. To discern the regional flow system, one must examine basin-scale potentiometric surfaces of the Mannville Group (Hitchon, 1969; Abercrombie and Fullmer, 1992) that illustrate lateral flow from the southwest to northeast across the Western Canada Sedimentary Basin. Fluid flow in this study area can be tied to the regional scale system by careful examination of the potentiometric surface (Figure 11). Generally, values of higher hydraulic heads (>650 m) are found in the southwest corner, to the west, and south of the study area. Intermediate values (500 m) are found along a band running generally north-south through Range 26, and values decrease systematically to 400 m toward the northeast. With this in mind, a subtle decrease in hydraulic heads from southwest to northeast across the study area can be

observed. This, however, disregards the values under 300 m in the southeast corner, the cause of which is unknown at the present time. It is possible that the region of hydraulic heads less than 300 m is at the northern edge of a regional-scale underpressured region, noted previously (Hitchon, 1969).

The third major difference between the MGA and the UDHG is the lack of uniform decrease in hydraulic heads across the study area. Numerous closed areas of low and high fluid potentials are superimposed on the large scale up-dip flow system. Examples of closed low areas are observed in T 43, R 1 and T 40, R 3–4. Closed high areas are found in T 41, R 25 and T 43–44, R 23. Closed features on potentiometric surfaces are indicative of significant components of vertical flow in heterogeneous aquifer systems. Since the potentiometric surface provides only a measure of the horizontal component of flow, upward or downward flow directions plot as closed areas of high and low potential on a potentiometric surface. The perturbations in the potentiometric surface of the MGA appear to be caused by the presence of subaquifers within the MGA. These subaquifers are relatively thin (up to 50 m thick) compared to the overall thickness of the MGA in the study area, which is 175–225 m. Although they are relatively discontinuous, the sand bodies that compose the Glauconite and Ellerslie subaquifers are reservoirs for many of the hydrocarbon pools in the study area.

As mentioned previously, the final component of the flow system in the MGA is the reflection in the potentiometric surface of the intersection with the vertically ascending UDHG. This intersection occurs in the northeast corner of the study area, where there is relatively little variation in hydraulic heads for 450 m. The noticeable flattening of the potentiometric surface in this area reflects a decrease in the lateral hydraulic gradient because a large component of the fluid flow field is directed in a

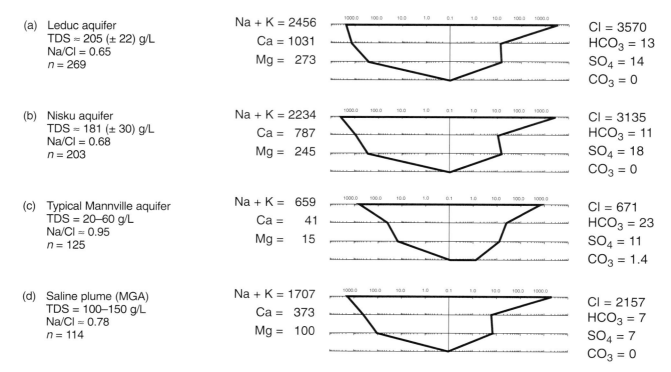

(a) Leduc aquifer
TDS ≈ 205 (± 22) g/L
Na/Cl = 0.65
n = 269

Na + K = 2456
Ca = 1031
Mg = 273

Cl = 3570
HCO$_3$ = 13
SO$_4$ = 14
CO$_3$ = 0

(b) Nisku aquifer
TDS ≈ 181 (± 30) g/L
Na/Cl = 0.68
n = 203

Na + K = 2234
Ca = 787
Mg = 245

Cl = 3135
HCO$_3$ = 11
SO$_4$ = 18
CO$_3$ = 0

(c) Typical Mannville aquifer
TDS = 20–60 g/L
Na/Cl ≈ 0.95
n = 125

Na + K = 659
Ca = 41
Mg = 15

Cl = 671
HCO$_3$ = 23
SO$_4$ = 11
CO$_3$ = 1.4

(d) Saline plume (MGA)
TDS = 100–150 g/L
Na/Cl ≈ 0.78
n = 114

Na + K = 1707
Ca = 373
Mg = 100

Cl = 2157
HCO$_3$ = 7
SO$_4$ = 7
CO$_3$ = 0

Figure 12—Stiff diagrams of averaged formation water composition from the Upper Devonian hydrogeologic group and Mannville Group aquifer. All chemical data are expressed in units of milliequivalents per liter.

vertical direction. Pressure versus depth plots constructed for T 43–45, R 21–24 (not shown here) indicate measured vertical pressure-gradients up to 13.4 kPa/m, well above the nominal gradient for the MGA of 10.6 kPa/m. As will be shown later, the vertical flow system in this area produces a plume of saline water in the MGA.

Hydrogeochemistry

The chemical characteristics of formation waters were examined by plotting distributions of major ions and total dissolved solids (TDS) for the various aquifers in the two main hydrogeologic groups. Differences between formation waters are best illustrated on Stiff diagrams of averaged chemical composition for each aquifer (Figure 12). According to the classification scheme of Hem (1985), all of the formation waters in the study area are brines.

Upper Devonian Hydrogeologic Group

Formation waters from the Cooking Lake–Leduc aquifer are Na-Ca-Cl brines (Connolly et al., 1990). They are the most concentrated waters found in the study area, with average TDS values of 205 g/L (Figure 12a). There is little variation in the dissolved solids content across the study area, with all 269 samples in the range of 183–227 g/L TDS. This tight range does not warrant a separate figure of TDS distribution. Chloride is the dominant anion in the samples. Sodium and potassium are the dominant cations, but as is typical for Cooking Lake–Leduc waters

in the basin, a significant proportion of the cations are made up of calcium. Average ratios of the reacting value of sodium to the reacting value of chloride (rNa/rCl) are 0.65, indicating an origin for these waters as a concentrated brine that has been subsequently diluted with meteoric water (Connolly et al., 1990).

Waters of the Nisku aquifer, although similar to the underlying Cooking Lake–Leduc waters, are typically less concentrated. Total dissolved solids for the Nisku aquifer average 181 g/L (Figure 12b). There is a greater range in dissolved solids throughout the study area, with values falling between 151 and 211 g/L, but again this variation does not warrant the inclusion of a separate map of TDS distribution. Proportionally, these waters contain slightly less calcium, which likely indicates a slightly different origin and dilution history (Connolly et al., 1990).

Mannville Group Aquifer

Formation waters in the MGA have a more variable chemical composition than waters in the UDHG. Total dissolved solids range from 20 to 150 g/L in the study area. TDS values generally increase from southwest to northeast (Figure 13), with the highest values clustered in the northeast corner of the study area.

There appears to be a mixture of two types of formation waters in the MGA. The first type is referred to as typical Mannville aquifer water and is characterized by a rNa/rCl value of about 0.95 (Figure 12c), with TDS values of 20–60 g/L. This chemical composition is similar to

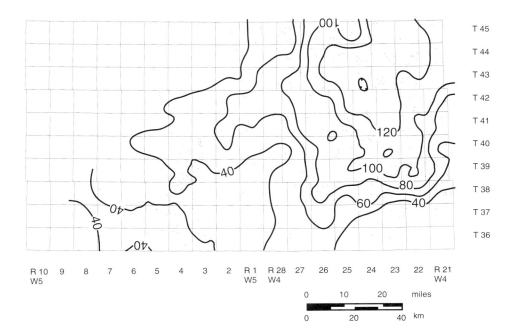

Figure 13—Map of total dissolved solids (TSD) distribution for the Mannville Group aquifer. Contour interval is 20 g/L.

waters found elsewhere in the Mannville Group in the Western Canada Sedimentary Basin (Connolly et al., 1990; Abercrombie et al., 1994; Cody and Hutcheon, 1994). Ambient Mannville formation waters owe their origin to a mixture of meteoric water recharging in southern and western Alberta (Cody and Hutcheon, 1994) and a brine of concentrated seawater (Connolly et al., 1990).

The second type of water in the MGA, referred to here as the saline plume, is characterized by higher TDS values and lower rNa/rCl values. Total dissolved solids in the saline plume generally exceed 100 g/L, with some values up to 140 g/L. These are some of the highest values of TDS found in this aquifer throughout the entire basin (Abercrombie et al., 1994). Average rNa/rCl values for the 114 samples in the plume are 0.78 (Figure 12d). The origin of the saline plume of formation waters in the MGA is explained in the next section.

Water samples with TDS values between 60 and 100 g/L are interpreted as mixtures of typical Mannville waters and the saline plume. These waters are found in an approximately 20-km-wide band bounded by the 60 g/L contour on the west and south and the 100 g/L contour on the northeast (Figure 13). Chemical characteristics of these samples are highly variable depending on the relative proportions of the two end-members present.

Origin of the Saline Plume of Formation Waters in the Mannville Group

The geology, hydrogeology, and hydrochemistry of the formations and fluids at this particular location in the Western Canada Sedimentary Basin all contribute to the production of a saline plume in the Mannville Group aquifer. Three factors are responsible for the creation of the saline plume here. First, saline water and hydrocarbons leak through the tops of the pinnacle reefs. This

leakage occurs where the cap rock (Ireton aquitard) is thin or absent (Figure 10). Second, the ascending nature of the formation fluid flow system in the Bashaw area (vertical flow from the plot in Figure 9) both assists the leakage of fluids out of the reefs and drives fluids upward out of the Nisku aquifer.

The third key factor in the creation of the saline plume in this particular location is the nature of the subcrop of the UDHG beneath the Mannville Group (Figure 2). The relatively impermeable dolomite, anhydrite, and minor halite that comprise the Wabamun Group aquitard separate the UDHG from the MGA. In the northeast corner of the study area, the Wabamun Group aquitard subcrops the Mannville Group (Figure 14). Outside the area of subcrop, the aquitard is in excess of 200 m thick. Northeast of the line of subcrop, the aquitard thins to under 60 m at the boundary of the study area (Figure 14). The thinning of the Wabamun aquitard in the northeast corner of the study area provides the pathway for the vertically ascending Devonian waters to pass upward into the Mannville Group, creating the saline plume. Comparison of the map of total dissolved solids for the Mannville Group (Figure 13) with the isopach map of the Wabamun Group (Figure 14) shows that the plume is found in the Mannville Group aquifer where the aquitard begins to thin along the subcrop line.

The chemical composition of the formation fluids supports the argument that the saline plume originates from the Upper Devonian hydrogeologic group. Formation waters that make up the UDHG are markedly different in composition from the waters in the MGA (Figure 12). With their characteristically high TDS values, high proportion of calcium cations, and low rNa/rCl ratios, they are recognizable, even when mixed with more dilute waters. The Stiff diagrams reveal that the overall shape of a typical saline plume water is similar to the shape of the water from the Cooking Lake–Leduc aquifer. The saline plume

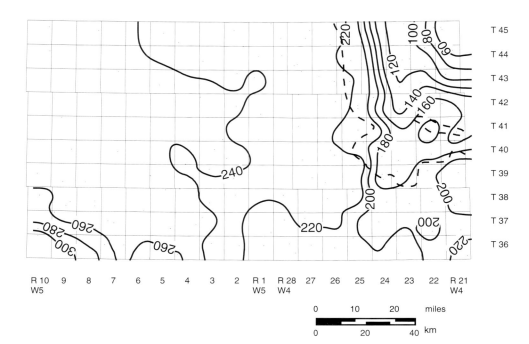

Figure 14—Isopach map of the Wabamun Group aquitard. Contour interval is 20 m. Dashed line indicates where the Wabamun Group aquitard subcrops the Mannville Group aquifer.

and Cooking Lake–Leduc waters only differ in the magnitude of their respective TDS contents. The characteristically high calcium content of waters from the UDHG that are also found in the saline plume is also diagnostic. Typical analyses for other formations in the study area (not shown here), do not match the Stiff diagram pattern of the saline plume. Furthermore, none of the other formation fluids exhibit the elevated amounts of calcium found in the UDHG and in the saline plume waters.

It could be argued that the saline plume is simply caused by laterally moving typical Mannville Group waters dissolving the carbonates of the Wabamun Group. It is possible that the low TDS Mannville waters are undersaturated with respect to calcium (or other ions), and when they came into contact with the carbonates and evaporites in the subcrop area, there is a rapid increase in TDS. This explanation is possible, but it is not supported by the flow system in the MGA because pressure versus density plots illustrate the vertical nature of fluid flow in the area of the saline plume. The fact that the fresher typical Mannville waters come into contact with the carbonates of the Wabamun Group probably plays only a minor role in the formation of the saline plume.

Organic geochemical analyses of produced oils from Mannville Group reservoirs east of the study area further support the input of UDHG fluids into the MGA. Riediger et al. (1994) used gas chromatography–mass spectrometer analysis of hydrocarbons and source rocks from the Provost field to demonstrate that Mannville Group oils contain a significant component of Devonian-sourced hydrocarbons. It is logical to assume that if oil can be shown to be migrating from the Devonian, then other formation fluids can also follow similar flow pathways, although results presented here do not include any oil to source rock correlations.

Based on these geologic, hydraulic, and geochemical data, it is reasonable to conclude that the saline plume in the Mannville Group aquifer is formed by formation fluids emanating from pinnacle reefs in the Leduc Formation. Implications of the dynamic nature of saline plumes on geochemical exploration in general, and on Leduc Formation pinnacle reefs in particular, are discussed next.

IMPLICATIONS FOR SURFACE GEOCHEMICAL EXPLORATION

A better understanding of the mechanics of plume formation above a breached pinnacle reef can improve the exploration efficiency of surface geochemical techniques.

Surface Expression Versus Hydrocarbon Retention

Creation of a plume or initiation of leakage from a reservoir can have both positive and negative aspects, depending on whether one is interested in the surface expression of that leakage or the amount of trapped hydrocarbons in the reservoir (Table 2). If there is no leakage from the trap and thus no formation of a hydrocarbon plume, there will be no surface expression, hence no surface indication of a subsurface reservoir. This would be positive from a trapping perspective because it means the cap rock did not fail and the trap could be filled to its spill point height. These situations occur when the COC of the cap rock exceeds leaking factors (column height and hydraulic gradient), allowing no penetration of oil into the cap rock (Rostron, 1993).

Table 2—Implications of the Sealing–Leaking Factor Interaction on the Subsurface Expression of Hydrocarbon Migration

	Possibility of Surface Expression	Possibility of Retention
No leakage	None	Excellent
Partial leakage	Some	Some
Complete leakage	Excellent	None

The complete leakage case is the extreme opposite to the no leakage case. If the cap rock cannot impede sufficient hydrocarbons to allow for trap filling (e.g., Figure 6a), then the reservoir cannot fill to capacity. In this case, there is minimal volume of retained hydrocarbon because the bulk of the migrating hydrocarbons have been lost to leakage. However, from the exploration perspective of obtaining the strongest possible surface expression of the presence of a trap, this is the best case. In other words, the best surface expression of the subsurface presence of a pinnacle reef is obtained when the least amount of hydrocarbons are retained in the subsurface trap.

The optimum case for both trapping and surface expression of migration lies between the complete and no leakage end-members (Table 2). Ideally, traps would leak enough to form a detectable plume while retaining as much hydrocarbons as possible. The cap rock above the reef must fail to be detectable by surface geochemical methods. To maximize the retained hydrocarbons once failure has occurred, the trap must either be found while being actively sourced (as in Figures 4a–e), or if not being actively sourced, then before the emplaced hydrocarbons escape (as in Figure 4f–g).

The applicability of surface geochemical techniques to a given trap depends on the degree of leakage from the trap. A trap's position in a leakage spectrum depends on the sealing and leaking factor interaction that occurs in each trap. Numerical results and field mapping show that sealing–leaking factor interactions depend on rock properties such as COC and intrinsic permeability. Furthermore, the subsurface flow fields affect both initiation of plume generation and subsequent migration toward the surface. Thus, measured values of rock parameters and a good understanding of the subsurface flow system are critical. Hydrogeologic studies must be completed as part of any surface geochemical exploration program.

Implications for Geochemical Exploration of Devonian Pinnacle Reefs

Results of field mapping presented here have four implications for the surface expression of hydrocarbon migration from Devonian pinnacle reefs. First, pinnacle reefs of the Leduc Formation in the Bashaw Reef Complex are leaking or have leaked hydrocarbons and saline water upward into overlying formations. This leakage is occurring where the Ireton aquitard is very thin or absent over individual pinnacle reefs. Formation fluids have crossed the Ireton aquifer into the Nisku Formation under the influence of buoyancy and the generally ascending UDHG flow system. Within the Nisku Formation, these individual plumes coalesce and continue to migrate up dip. Where the Wabamun aquitard subcrops and thins, formation fluids in the Nisku aquifer rise up into the Mannville Group. This generates the saline plume in the Mannville Group aquifer.

Second, there is a definite mappable signature of Devonian hydrocarbons and saline waters occurring in the Mannville Group aquifer. Within the saline water plume, the TDS of the formation waters are two to five times higher than the ambient Mannville formation waters. This plume is easily recognizable over large parts of the study area and is a direct indicator of upward-moving Devonian formation fluids. Mapping the TDS of formation waters in the Mannville Group could be used to find other areas where Devonian oils are entering the Mannville Group. Mapping TDS distributions elsewhere in the basin could lead to the discovery of as yet unknown pinnacle reefs or unknown areas where flow systems bring Devonian hydrocarbons closer to the surface. Saline plumes are useful because they can be detected easily using water salinities derived from geophysical well logs.

Third, the signature of the Leduc reefs is best mapped at the Mannville Group level because the flow system at the Mannville level changes from a dominantly vertical one to a more lateral one. Once in the MGA, formation fluids are redirected into a laterally moving flow system, thus offsetting any hydrocarbon plume above leaky pinnacle reefs. The fact that hydrocarbons from the UDHG are offset laterally from the leakage conduit in this study area is shown in two ways: by Devonian sourced hydrocarbons produced from Mannville pools up dip of the study area (Riediger et al., 1994) and by the extension of the saline plume up dip beyond the limits of the study area (Abercrombie et al., 1994).

Fourth, it is unlikely that the saline plume in the MGA and any associated hydrocarbons have any surface expression in this study area because the plume and hydrocarbons would have to pass through at least two major aquitards. These include the regionally extensive Joli Fou aquitard, which is known to be a restriction for hydrocarbons (Creaney and Allan, 1990), and the Colorado Group aquitard, which comprises more than 300 m of shale. This is in agreement with the findings of the surface geochemical study conducted on an area overlapping our study area to the south (McCrossan et al., 1972). The surface geochemical study of McCrossan et al. (1972) found little correlation between soil gas anomalies and underlying Leduc hydrocarbon pools. Thus, surface geochemical methods would not be useful in searching for Devonian reservoirs in our study area. However, this does not preclude saline plumes from reaching the surface in other areas of the basin where the overlying shales are thinner, vertical flow is stronger, the Mannville Group is shallower, or some combination of these.

CONCLUSIONS

Numerical modeling and field mapping of saline plumes emanating from Devonian pinnacle reefs has shown the following:

1. The formation of saline plumes above pinnacle reefs is controlled by the sealing–leaking factor interaction of the flow system. The sealing factors (capillary pressure curves and intrinsic permeabilities) and leaking factors (fluid densities and regional hydraulic gradients) are measurable quantities and can be predicted. Measurements of these properties, especially in the cap rock, must be conducted to quantify the formation of hydrocarbon plumes and their subsequent migration to surface. Exploration efficiency can be improved by an understanding of the hydraulics of the flow system.
2. Devonian pinnacle reefs in the Bashaw Reef Complex of west-central Alberta are leaking saline water and hydrocarbons. These fluids are carried upward by an ascending flow system in the Paleozoic aquifers. When these saline brines intersect the Mannville Group, they create a mappable plume of saline waters in the study area. Mapping such plumes in the Mannville Group can indicate areas of upwelling Devonian oils and indicate the presence of as yet undiscovered pinnacle reefs beneath the Mannville Group.
3. Geochemical exploration for Devonian pinnacle reefs and hydrocarbon plumes has to be conducted at the Mannville Group level because the Mannville intercepts the vertically ascending fluid flow and redirects it laterally up dip.
4. There will be little, if any, surface expression of Devonian hydrocarbon plumes in west-central Alberta because of the lateral interception of the plume by the Mannville Group and the isolation of the Devonian and Mannville aquifer systems from the surface by the Colorado Group aquitard. There may be surface expression of hydrocarbon migration in other areas of the basin where the Mannville Group is closer to the surface or where the ascending fluid flow reaches the surface.
5. Regional fluid flow and cross-formational migration play an important role in migration and entrapment in the subsurface and how these effects relate to surface expression of hydrocarbon migration. An understanding of subsurface hydrogeology and hydrochemistry, as shown in this paper, can increase the exploration efficiency of surface exploration techniques.

Acknowledgments—This project was funded in part by a Natural Sciences and Engineering Research Council operating grant to J. Tóth. Data used in the study were obtained from the Canadian Institute of Formation Evaluation, Rakhit Petroleum Consulting Limited, CDPubco Limited, and the Alberta Energy Resources Conservation Board. Well logs were made available at the Alberta Research Council. Computer time was made available by ARCO Exploration and Production Technology Company. Portions of this work were completed using software from the National Center for Supercomputing Research. Scott Juskiw of the Apple Research Partnership Program coupled the numerical model to the graphics routines. Geotrans Incorporated provided the original version of the numerical model. The paper benefitted from the reviews of CJMR and one other anonymous reviewer.

REFERENCES CITED

Abercrombie, H. J., and E. G. Fullmer, 1992, Regional hydrogeology and fluid geochemistry of the Mannville Group, Western Canada Sedimentary Basin: synthesis and reinterpretation, *in* Y. K. Kharaka and A. N. Maest, eds., Water–rock interaction: Rotterdam, The Netherlands, Balkema, p. 1101–1104.

Abercrombie, H. J., J. D. Cody, I. E. Hutcheon, and T. R. Myers, 1994, Fluid geochemistry of the Mannville Group, Alberta: physical and chemical processes, implications for basin evolution: CSEG and CSPG Annual Conference, Calgary, Alberta, Program with Expanded Abstracts, p. 307–308.

Berg, R. R., 1975, Capillary pressures in stratigraphic traps, AAPG Bulletin, v. 59, p. 939–956.

Cody, J. D., and I. E. Hutcheon, 1994, Regional water and gas geochemistry of the Mannville Group and associated horizons, southern Alberta: Bulletin of Canadian Petroleum Geology, v. 42, p. 449–464.

Connolly, C. A., L. M. Walter, H. Baadsgaard, and F. J. Longstaffe, 1990, Origin and evolution of formation waters, Alberta Basin, Western Canada Sedimentary Basin, I. chemistry: Applied Geochemistry, v. 5, p. 375–395.

Creaney, S., and J. Allan, 1990, Hydrocarbon generation and migration in the Western Canada Sedimentary Basin, *in* J. Brooks, ed., Classic petroleum provinces: Geological Society of London, Special Publication No. 50, p. 189–202.

Davidson, M. J., 1994, On the acceptance and rejection of surface geochemical exploration: Oil & Gas Journal, v. 92, p. 70–76.

Davis, T. L., 1972, Velocity variations around Leduc reefs, Alberta: Geophysics, v. 37, p. 584–604.

Faust, C. R., 1985, Transport of immiscible fluids within and below the unsaturated zone: Water Resources Research, v. 21, p. 587–596.

Hitchon, B., 1969, Fluid flow in the Western Canada Sedimentary Basin, 2. effect of geology: Water Resources Research, v. 5, p. 460–469.

Hitchon, B., and M. Brulotte, 1994, Culling criteria for standard formation water analyses: Applied Geochemistry, v. 9, p. 637–645.

Hem, J. D., 1985, Study and interpretation of the chemical characteristics of natural water: USGS Water Supply Paper #2254, 253 p.

Holysh, S., 1989, Petroleum related geochemical signatures and regional groundwater flow, Chauvin area, east-central Alberta: M.Sc. thesis, Department of Geology, University of Alberta, Edmonton, Alberta, 208 p.

Holysh, S., and J. Tóth, 1994, Flow of formation waters—a likely cause for poor definition of soil-gas anomalies over oil fields in east-central Alberta, Canada (abs.), *in* Near-surface expressions of hydrocarbon migration: AAPG Hedberg Research Conference Abstracts, April 24–28, Vancouver, British Columbia.

Hubbert, M. K., 1940, The theory of ground-water motion: Journal of Geology, v. 48, p. 785–944.

Machel, H. G., and E. A. Burton, 1991, Causes and spatial distribution of anomalous magnetization in hydrocarbon seepage environments: AAPG Bulletin, v. 75, p. 1864–1876.

Masters, J. A., 1979, Deep basin gas trap, western Canada: AAPG Bulletin, v. 63, p. 152–181.

McCrossan, R. G., N. L. Ball, and L. R. Snowdon, 1972, An evaluation of surface geochemical prospecting for petroleum, Olds-Caroline area, Alberta: Geological Survey of Canada, Paper 71-31, 101 p.

National Center for Supercomputing Applications, 1991, IMAGE 3.0 User's Guide: Champaign, University of Illinois.

Neuzil, C. E., 1986, Groundwater flow in low-permeability environments: Water Resources Research, v. 22, p. 1163–1195.

Paul, D., 1994, Hydrogeology of the Devonian Rimbey-Meadowbrook reef trend of central Alberta: M.Sc. thesis, Department of Geology, University of Alberta, Edmonton, Alberta, 152 p.

Riediger, C. L., M. G. Fowler, and L. R. Snowdon, 1994, Organic matter characteristics and biomarker analysis of the Lower Cretaceous Ostracode Zone, a source for some Mannville oils in Alberta: CSEG and CSPG Annual Conference, Calgary, Alberta, Program with Expanded Abstracts, p. 311.

Rostron, B., 1993, Numerical simulations of how cap rock properties can control differential entrapment of oil, *in* Formation evaluation and reservoir geology: Proceedings, SPE Annual Technical Conference and Exhibition, SPE Paper 26442, p. 263–275.

Rostron, B., 1994, A new method of culling pressure data used in hydrodynamic studies (abs.): AAPG Annual Convention Program with Abstracts, Denver, Colorado, p. 247.

Sales, J. K., 1993, Closure vs. seal capacity—a fundamental control on the distribution of oil and gas, *in* A. G. Doré, ed., Basin modeling: advances and applications: NPF Special Publication 3, Amsterdam, Elsevier, p. 399–414.

Schowalter, T. T., 1979, Mechanics of secondary migration and entrapment: AAPG Bulletin, v. 63, p. 723–760.

Switzer, S. B., W. G. Holland, D. S. Christie, G. C. Graf, A. S. Hedinger, R. J. McAuley, R. A. Wierzbicki, and J. J. Packard, 1994, Devonian Woodbend-Winterburn strata of the Western Canada Sedimentary Basin, *in* G. D. Mossop and I. Shetsen, comps., Geologic atlas of the Western Canada Sedimentary Basin: Canadian Society of Petroleum Geologists and Alberta Research Council, Calgary, p. 165–202.

Tóth, J., 1978, Gravity-induced cross-formational flow of formation fluids, Red Earth region, Alberta, Canada: analysis, patterns, and evolution: Water Resources Research, v. 14, p. 805–843.

Tóth, J., 1991, Hydraulic continuity in large sedimentary basins, *in* Proceedings of the International Conference on Groundwater in Large Sedimentary Basins: Canberra, Australian Government Publishing Service, Australian Water Resources Council, Conference Series 20, p. 2–14.

Watts, N. L., 1987, Theoretical aspects of cap rock and fault seals for single- and two-phase hydrocarbon columns: Marine and Petroleum Geology, v. 4, p. 274–307.

Jones, V. T., and S. G. Burtell, 1996, Hydrocarbon flux variations in natural and
anthropogenic seeps, *in* D. Schumacher and M. A. Abrams, eds., Hydrocarbon
migration and its near-surface expression: AAPG Memoir 66, p. 203–221.

Chapter 16

Hydrocarbon Flux Variations in Natural and Anthropogenic Seeps

Victor T. Jones III

Exploration Technologies, Inc.
Houston, Texas, U.S.A.

Stephen G. Burtell

Exploration Technologies, Inc.
Houston, Texas, U.S.A.

Abstract

Methodologies for conducting surface geochemical surveys and measuring the hydrocarbon flux rates of hydrocarbons migrating to the surface are addressed with examples from natural seeps and from anthropogenic seepage from underground gas storage reservoirs, leaky well casings, and underground coal gasification reactors. Natural gas flux was monitored for 1 year at Arrowhead Hot Springs, San Bernardino County, California, as part of an earthquake prediction program. The hot spring is on a splay of the San Andreas fault and releases 40 mL/min of free gases containing helium, hydrogen, light hydrocarbon gases, and radon. The volume of released gases varied by a factor of two within 7 months. Changes in gas flux could be a precursory signal of earthquake activity on the locked southern section of the fault and demonstrated that rapid changes were related to tectonic activity along this major basement fault.

Gas flux associated with pressure changes in underground storage reservoirs confirms the rapid variations observed for natural seeps. Other changes in gas concentrations over a propane storage cavern are related to barometric and meteorologic variations. The rapidity with which natural gas can migrate through the earth was also demonstrated by measuring the gas flux in 122 boreholes over an underground coal gasification reactor. Baseline gas concentrations were established one month before the 180-m- (600-ft-) deep retort was pressured and fired. Leaked gases were detected at the surface in 2 to 15 days, depending on the location of the boreholes with respect to the retort at depth.

The underground coal gasification (UCG) reactor provided an outstanding vehicle for migration flux measurements because of the unique gases generated in the reactor. Also, pressure and compositional changes in the reactor occur at known times in direct response to operational procedures. Individual gas pulses exhibited chromatographic effects as the gases migrated away from the source at depth. These chromatographic changes existed for only a few hours at the onset of a pressure pulse in the subsurface reactor and quickly returned to steady-state conditions in which the composition of the reactor gases matched those escaping at the surface.

INTRODUCTION

During the early development of soil gas prospecting at Gulf Research, it became clear that the magnitude of variations in soil gas and dissolved gas sniffer data required an explanation that appeared to be strongly affected by fault and fracture systems (Jones et al., 1995). Migration of oil and gas from the subsurface follows a complex pathway that is rarely imaged, as illustrated by the "bright spots" in Figure 1. Migration of deep gases are expected to form shallow micro-accumulations along the pathway to the shallow subsurface. Thus, the sniffer anomaly in Figure 1 must be projected to depth to find the field associated with the surface anomaly. This example illustrates the expected relationship of sniffer anomalies to their subsurface sources and was an actual Gulf

Oil Company discovery in the East Cameron area, offshore Louisiana. The accumulation is located downdip to the southwest, as shown by the bright spots marching down the fault (Weismann, 1980).

The Gulf of Mexico sedimentation and water compaction rates were also evaluated and compared to calculated diffusion rates for migrating light hydrocarbon gases. The conclusion was that neither macro- nor microseeps could occur by diffusion since the downward flux of water and sediments exceeds the upward diffusion of gases (R. J. Mousseau and G. Glezen, personal communication, 1974). The presence of both macro- and microseepages in the Gulf of Mexico suggests that diffusion is not the dominant migration mechanism. The comparison of seep compositional information with known production proves that the microseeps are real since seeps

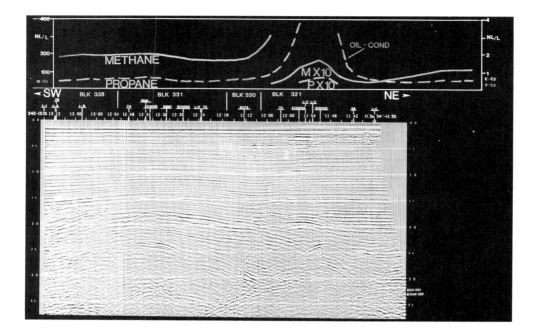

Figure 1—(Top) Graph of dissolved methane and propane data and (bottom) seismic profile showing the migration of oil and gas from the subsurface following complex pathways that can only rarely be imaged by geophysical data. The bright spots in the profile show the direct relationship between sniffer anomalies and their subsurface sources.

can be correlated to their respective source rocks at depth by comparing the compositional information of near-surface gases directly to their underlying sources (Williams et al., 1981; Jones and Drozd, 1983).

Some typical reservoir gas analysis data that were instrumental in this comparison were published by Nikonov (1971), who compiled gas data from 3500 different reservoirs in the United States, Europe, and the former Soviet Union and grouped them into useful subpopulations. Gases from basins containing only dry gas were shown to have less than 5% heavy homologs, whereas gases dissolved in oil pools have an average of 12.5–15% heavy homologs. The heavy homologs plotted by Nikonov included ethane, propane, butane, and pentane.

Compositional information from near-surface soil gas data sets collected in large gridded surveys of more than 600 sites each over the dry gas Sacramento basin, the gas-condensate deposits in the Alberta foothills, and two oil fields (Abo Reef and Sprayberry) in the Permian basin showed that the concept put forth by Nikonov could be used to establish similar relationships between surface seepage data and their respective sources in these areas (Jones and Drozd, 1983; Drozd et al., 1981).

Given this encouragement from compositional data, continued data gathering and generation of additional examples further demonstrated the strong structural control on microseep magnitudes. This is shown by examples from the Pineview and Ryckman Creek fields located in the Utah-Wyoming overthrust belt (Jones and Drozd, 1983). In both fields, many of the major faults mapped at the surface verified a direct relationship between the faults and the magnitudes of the anomalies. Additional examples from east Texas and the offshore Green Canyon area in the Gulf of Mexico (Figure 2) demonstrate the deflection and control of seepage pathways by growth faults (Pirkle, 1985). The Green Canyon

example was collected in 1983 directly over the Jolliet field before the field was discovered.

These examples helped in the reevaluation of basic concepts and proved that the conclusions reached about the association of macroseeps with production (Link, 1952) must also apply to microseeps. The Infantas field in Colombia had hissing gas seeps when first discovered. As the field was produced, the seeps disappeared. Later, during water flood operations, the hissing seeps reappeared upon repressuring, demonstrating a strong correlation with subsurface reservoir pressure.

Chekalin and Timofeev (1983) published examples in which reservoir pressure could be directly correlated with seepage magnitudes and seismic activity, further demonstrating the relationship of seep magnitudes to tectonic activity and subsurface pressure. Hunt (1981) stated that over 70% of the reserves in the world are associated with visible macroseeps. According to Link (1952), the mechanisms by which macroseeps migrate to the surface are fractures, joints, fault planes, unconformities, bedding planes, and diffusion through porous beds. In the previously cited examples, diffusion is only one of six migration mechanisms and does not appear to be the main process for migration of hydrocarbons to the surface.

MIGRATION MODEL

Early surveys conducted in west Texas revealed that larger magnitude CO_2 seeps occur with a higher frequency directly over deep-seated fault zones along the west flank of the Puckett field (M. D. Matthews and V. T. Jones, unpublished results, 1976). This observation was made in spite of the fact that the faults do not come to the surface and are covered with more than 4200 m (14,000 ft) of supposedly unfaulted sediments. A migration model pro-

(a)

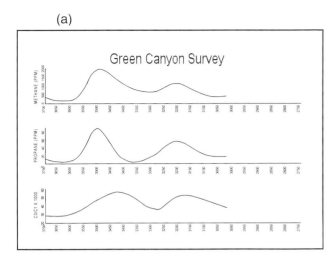

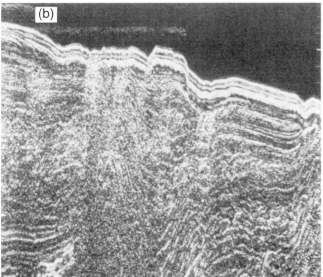

Figure 2—(a) Gas geochemical and fluorescence results for C$_3$/C$_1$ × 1000, propane, and methane analyzed from gravity cores collected in 1983 over the Jolliet field in the Gulf of Mexico. (b) Seismic profile of the Jolliet field. Note the control of seepage pathways by near-surface faults.

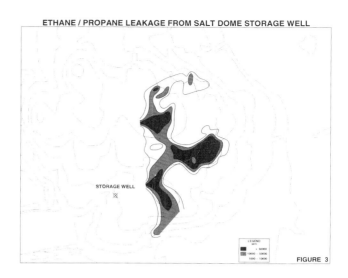

Figure 3—Contour map of gas concentrations (a 70/30% ethane/propane mixture) measured from about 500 permanent monitoring stations at 9 m (30 ft) depth located over a salt dome gas storage reservoir. The mixture has migrated more than 900 m (3000 ft) laterally from the leaking storage well. Legend: light gray = 1000–10,000 ppm, medium gray = 10,000–50,000 ppm, and dark gray = >50,000 ppm.

EXAMPLES OF UNDERGROUND STORAGE RESERVOIR LEAKAGE

Soil gas surveys have been used to detect leaks from a variety of underground storage complexes. One example outlined the leakage area affected by an ethane and propane mixture discharged into the cap rock through a hole in the casing at a depth of ~170 m (~570 ft) from an underground salt dome storage well. Deep relief wells were drilled to determine the vertical extent of the contamination and to relieve the pressure at depth. Finally, a nitrogen flood was conducted in the shallowest 9-m (30-ft) aquifer to drive the contamination back toward the source area for final surface clean up. In addition, compositional data from gas chromatographic analysis indicated whether the soil gas was derived from the product well or from a previous spill or pipeline.

Leakage from an Underground Salt Dome

Topography (Figure 3) defines the surface expression of a Gulf Coast salt; the salt dome storage well that developed the casing leak is also shown, along with a contour map that defines the horizontal extent of the migration, as shown by the isoconcentration contours of the ethane and propane mix measured in 9-m- (30-ft-) deep permanent monitoring stations. The hook-shaped seep located in the northern part of this anomaly consisted mainly of propy-

posed at that time was further demonstrated by a data set collected over the Patrick Draw field as part of the joint industry GEOSAT program (Matthews et al, 1984; Richers et al., 1986). Numerous studies have continued to suggest the validity of this model (Jones et al., in press).

An outstanding visual example of this model was demonstrated by macroseeps in Oklahoma when a shallow gas well was overpressured (Preston, 1980). Many of the vents and bubble trains were clearly aligned parallel to fracture orientations, even in unconsolidated river deposits. Attempts to plug the surface vents would be expected to merely divert the migration path to the surface to an alternate fracture system. Many of the initial vents were apparently abandoned by natural plugging, and new vents formed during the later states of activity.

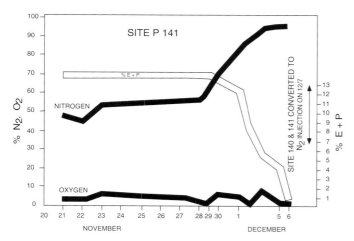

Figure 4—A typical response curve illustrating the time required for lateral migration of a nitrogen flood past an ethane-propane charged monitor well.

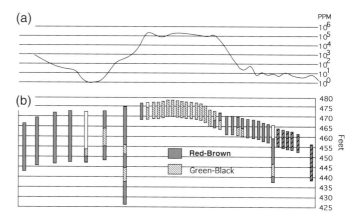

Figure 5—(a) Propane concentrations (from samples taken at 3 m, or 10 ft, depth) and (b) a corresponding lithologic cross section demonstrating soil color alteration caused by propane seepage over a mined underground storage cavern. Horizontal scale is 1 inch = 150 ft (1 cm = 18 m).

lene and was traced to a earlier reported spill, which was previously thought to have no known near-surface expression.

The ethane and propane concentrations shown in Figure 3 were based on the analysis of gases collected from about 500 monitor wells, 9 m (30 ft) deep, installed on 15–30-m (50–100-ft) centers over and adjacent to the affected area. The relationship between topography and high gas concentrations suggests that the location of some of the surface drainages may be related to subsurface control that directs and focuses the migration of the lost gases. Based on the analysis of these near-surface gases, 32 relief wells, some as deep as 150 m (500 ft), were drilled and logged to determine the vertical extent of the contamination. Combining recent drilling data with all available well logs from the older production wells indicated that the gas had charged a sandstone 60 m (200 ft) below the surface. Two deeper and one shallower sandstone that had not been charged were also found above the cap rock. Several high-magnitude gas anomalies lay close to old, possibly uncased wells, suggesting that the main avenue from the cap rock to the sandstone at 200 ft depth was along the uncased wells.

Clean-up was facilitated by turning the 30-ft-deep geochemical monitor wells into eductor sites by installing a venturi tube on the top of the well casing. Nitrogen was run through the venturi tube to produce a small vacuum on the hole and then injected into the ground through a monitor well located near the edge of the plume. The nitrogen migrated through the ground toward the eductor sites, where it escaped to the atmosphere. A typical response curve showing the advance of the nitrogen front and clean-up of the ethane and propane product mix are shown in Figure 4.

The rate of clean-up response over the entire area illustrated lateral variations in lithology and permeability. Although visual inspection of the sands within the 30-ft aquifer did not exhibit any obvious stratigraphic differ-

ences, the variability in N_2 flow clearly demonstrated a significant influence related to lateral variability. In some cases, the lateral permeability was so low that 15 lbs of nitrogen pressure at 9 m (30 ft) were not sufficient to push the gas 15 m (50 ft) laterally to the next soil gas station, even though water poured on the ground near the injection site would froth and bubble from the nitrogen escaping vertically through 30 ft of clay. Most of the recalcitrant hot spots also appeared to have a better association with deeper vertical migration pathways that charged the 30 ft sandstone rather than from lateral migration.

Following clean-up operations, an advance warning system was installed to detect leaks from storage wells before they could become a problem, and two permanent monitoring stations were installed at the casing of every storage well. This type of monitoring system has been in operation since 1980 and, if properly sampled on a regular basis, allows early detection of well leakage before it appears at the adjacent boundary of the property. Application of such a system is beneficial for preventive maintenance and is highly recommended.

Leakage from a Propane Storage Cavern

Another excellent example of gas leakage from an underground storage reservoir is provided by a study conducted over a 60-m- (200-ft-) deep propane storage cavern. The immediate objective was to determine the leakage rates and to test whether ongoing remedial efforts to repair the leaks were successful. This particular case also provided an opportunity to determine the product leakage distribution and to conduct pressure pulse tests by injecting helium into the cavern as a tracer. Although this was a small facility, 455 geochemical measuring stations on 3-m (10-ft) centers were installed to depths of 6 m (20 ft) and lined with PVC pipe. Figure 5 shows the propane collected over the cavern plotted on a

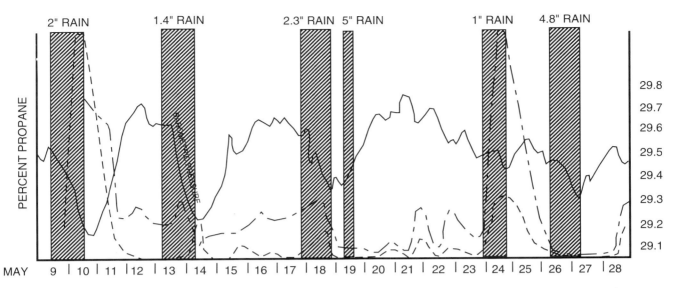

Figure 6—Graph of propane percentages by volume (two dashed lines) with respect to rainfall showing barometric pumping of propane flux under a ground sheet. Rainfall (vertical bars) expelled a significant amount of propane from under the ground sheets when first installed. Solid line shows barometric pressure.

log scale. In addition to the propane anomalies detected, the soil changed color from red-brown to green-black over the top of the cavern, coincident with the largest gas anomalies. The color changes appear to be related to hydrocarbon seepage and confirm areas where the gas leakage has occurred over a long time period above this cavern.

Relationship Between Barometric Pressure and Gas Flux

An underground storage cavern is also a good place to observe gas flux related to atmospheric phenomena. Plastic ground sheets about 5 ft square were installed directly over known leaks to measure propane gas flux related to meteorologic and barometric changes. Rainfall (shown as vertical bars in Figure 6) produced a significant change when the ground sheets were first installed. The 2 in. of rain that occurred on the evening of May 9 caused the sheets to balloon up from soil gas being forced up under the sheets by rainwater infiltration adjacent to and around the edges of the sheets. The rain probably displaced the gas in the ground and caused it to come up underneath the ground sheet. Since this large gas flux did not occur again after the next rainfall events, it was assumed that this initial gas flux was caused by buildup of longer term gas leakage that was trapped directly under the sheet and forced out by the first influx of rainwater. Except for perhaps the 1-in. event on May 24, subsequent rain events did not again show such a dramatic change once equilibrium was established.

Continued barometric monitoring (shown by the solid line on Figure 6) produced several small barometric changes on May 19 through 22. These barometric lows had clearly expressed positive gas fluxes (shown by the dashed lines). Every time the barometer took a dip, some

gas flux popped up under the ground sheet. Gas flux was measured as the percentage of propane measured under the sheet. Data from two separate sheets are shown in Figure 6 for comparison.

Migration Timing Related to Pressure Pulse

Additional opportunities to observe pressure-related gas fluxes were encountered during this remedial sampling conducted over the mined propane cavern. The cavern pressure was decreased to ambient levels for repairs and then repressured to 80 psi, allowing the recharge leakage rate from the cavern to be measured. Upon recharging the cavern, we determined the time it took for the propane gas to reestablish its previously measured maximum leakage values at the surface. Following the recharge of the cavern with propane to its original pressure, a value of over 90% of the original soil gas propane concentration was observed in the observation test hole within 15 days. As shown by Figure 7a, most of this leakage appeared to come from around the central shaft. However, because the storage site had been known to leak for over 20 years, there was a large propane background in the soil that made it impossible to determine whether the product reappearing at the surface came directly out of the reservoir during the sampling period.

Injection of Helium Tracer into Cavern

As a second, more definitive test, helium was injected into the cavern to create a concentration in the cavern of about 600 ppm. Results showed that in 15 days, not only had the propane moved to the surface, but as shown in Figure 7b, the helium concentration detected at the surface was over 75% of that in the cavern. Helium injection not only showed the leakage around the central cavern,

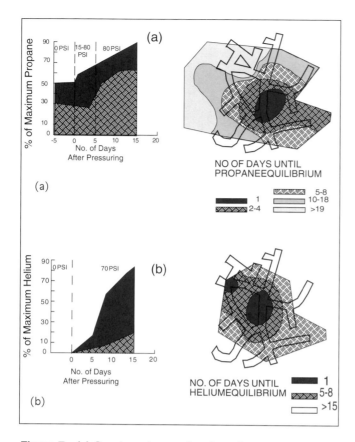

Figure 7—(a) Graph and map showing oil gas propane recharge rates measured in permanent stations in direct response to a subsurface pressure pulse in a propane storage cavern and (b) in a propane storage cavern spiked with 600 ppm helium. The outline of the storage cavern is shown as a series of tunnels in the maps.

but also found a leak at the end of one of the drifts that would have been missed looking only at propane. The amount of helium used for this test was not enough to damage the product for sale and yet still gave more than adequate sample for analysis. These two examples suggested that migration was rapid and controlled by faults, fractures, and other permeable migration pathways and not by diffusion. A similar case involving an underground coal gasification reactor is discussed next.

Leakage from an Underground Coal Gasification Reactor

Another educational example of gas leakage was taken from an underground coal gasification reactor near the Rock Springs uplift near Rawlins, Wyoming (Jones and Thune, 1982; Jones, 1983). The North Knobs underground coal gasification (UCG) facility is about 8 mi west of Rawlins, Wyoming. It is situated on the southwestern flank of the asymmetric Rawlins uplift adjacent to the Washakie Basin. Throughout the area, the exposed resis-

tant sandstones have well-developed rectilinear joint sets striking N 14° W and N 49° W. The nearly vertical beds dip too steeply to do anything except gasify the coal in-place. The gasification reactor is about 180 m (600 ft) below the surface. A total of 122 permanent monitoring wells, 5.4 m (18 ft) deep, were installed over the general area of the retort to facilitate measurement of the soil gas concentrations.

One objective of this soil gas survey was to determine if gases generated during the burn leaked into the near surface. We also wanted to know the rates of leakage, migration paths, and composition of the gases in order to evaluate the economic importance of such leakage and to assess their hazard potential to human safety and the environment.

All geochemical monitoring wells (here referred to as sample sites) were drilled with a 3-in. diameter auger to a nominal depth of 18 ft and established as permanent observation sites. This was accomplished by installing a 20-ft length of 1 in. ID PVC pipe, perforated with about 30 1/4-in.-diameter holes in the lower 2 ft of the pipe. During installation, sufficient pea gravel was installed to provide a permeable zone for collection of soil gases leaking from the adjacent formations.

Plots of methane and propane magnitudes with time are shown in Figures 8 and 9, respectively, for several sample sites selected to represent the typical changes noted in the response time of gas migration. Note that sites 22 and 27 exhibit an almost identical quick response to the retort pressuring even though they are about 15 m (50 ft) apart. Site pairs 1 and 2, 5 and 6, and 13 and 20 also show similar responses within close pairs, which are several hundred feet apart. The leakage patterns that emerge over time are clearly not random, but are systematically changing in relation to subsurface controls.

It took about 3–5 days after the beginning of the system air pressure test and ignition of the coal before any significant increases in the magnitude of the hydrocarbon gases were recognized in the near surface. Propane leakage was related to migration from the retort of the initial products in ignition because propane was used to achieve ignition and was not a major retort gas generated during the burn. Figure 9 illustrates this because site 22 was directly updip from the retort and clearly showed a sharp rise in propane upon pressuring and a fairly rapid decrease during the initial phases of the burn. The maximum pressure in the retort at 180 m (600 ft) was 700 psi. This excess pressure, used to link the slant and vertical wells, caused a rapid change in the surface signature that occurred within 2 days.

An examination of the change in methane flux from various sites suggested that the soil gas vapor data could initially be divided into at least four discrete time periods for mapping, as shown in Figures 10 through 14. These selected periods are defined by vertical lines in Figures 8 and 9 and are as follows: (1) Figure 10, preburn, prepressure, July 22 to August 16, 1981; (2) Figure 11, pressure, August 17 to 23, 1981; (3) Figure 12, burn, August 24 to November 10, 1981; and (4) Figure 13, postburn, Nov-

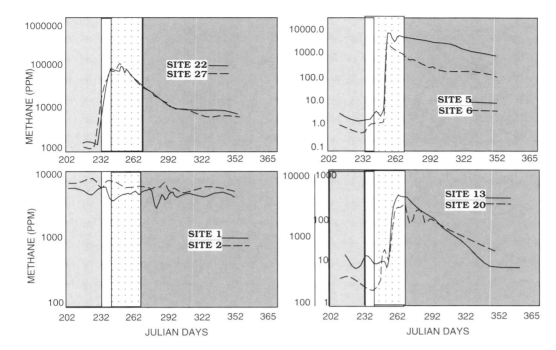

Figure 8—Methane concentrations measured at four permanent stations installed over a subsurface coal gasification reactor plotted versus Julian days of the year. Shaded panels represent changes in operational activities: gray on left = preburn/prepressure; white = pressure; dotted = burn; gray on right = postburn.

ember 11 to December 12, 1981 (end of field survey). A final measurement was made 6 months later, from July 21 to 30, 1982, essentially 1 year after the initial measurements were made. Figure 14 shows this final data set.

By selectively averaging the gas flux measurements from each of the five time windows for each site, we were able to construct a series of contour maps illustrating the most significant changes that occurred over the five time periods. As shown in detail in Figures 10–14, the leakage patterns changed with time in direct response to pressure variations in the subsurface retort (Jones and Thune, 1982; Jones, 1983). Evaluation of these contour maps allowed us to make estimates of how long gas remains in the surface sediments and on what magnitude of fluxes might occur under various pressure conditions. For example, Figures 13 and 14 were both measured after the retort had been depressured and filled with water. Within 6 months, the leakage directly updip at the outcrop had decreased an order of magnitude from about 10,000 to 1000 ppm.

It was suggested to the Department of Energy that this site should be maintained for future research on soil gas analysis because we could conduct various depth probe measurements to check the influence of joints and soil types on both vertical and lateral gas migration and further investigate dissipation of the retort leakage gases with time. This site would also be particularly useful for a research study because of the uniqueness of the gases generated, including carbon dioxide, carbon monoxide, hydrogen, and methane. Because these gases are unique to the coal gasification process, they could have come only from the subsurface retort. The maximum concentration of gas in the bedding plane at the outcrop was about 50,000 to 100,000 ppm (5–10%) at the peak generation of the retort. As shown by Figures 13 and 14, this falls off as the reactor pressure was reduced and finally filled with water at the conclusion of operations. The raw data

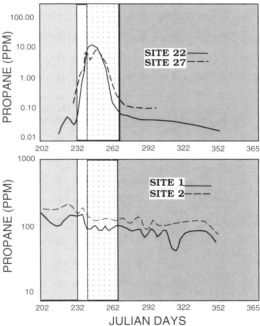

Figure 9—Propane concentrations measured at two permanent stations installed over a subsurface coal gasification reactor plotted versus Julian days of the year. Shaded panels represent changes in operational activities: gray on left = preburn, prepressure; white = pressure; dotted = burn; gray on right = postburn.

produced adequate flux information for modeling, thus providing an excellent resource for further study (Department of Energy, 1982).

Maximum gas values were observed in monitor wells 22 and 27, directly updip from the outcrop of the coals

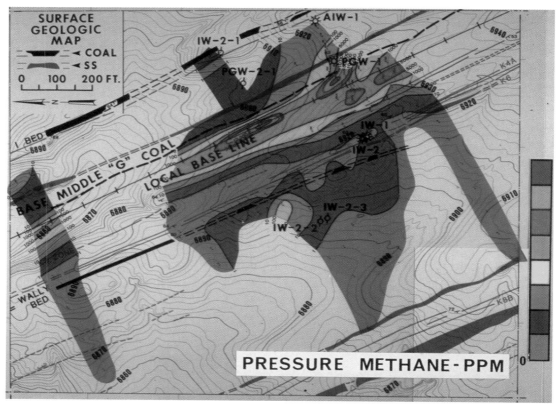

Figure 10—Color contour map of methane concentrations measured at permanent stations installed over an underground coal bed retort. Data were averaged over preburn, prepressure time window, July 22 to August 16, 1981.

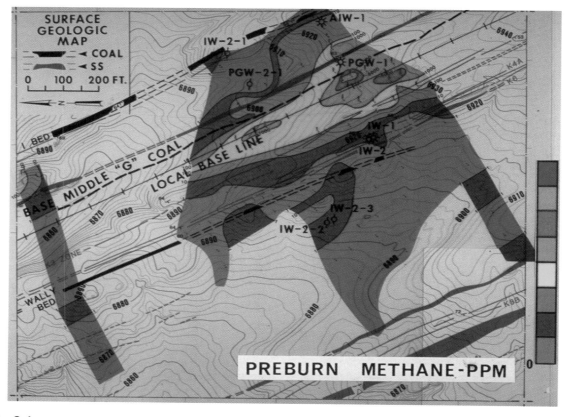

Figure 11—Color contour map of methane concentrations measured at permanent stations installed over an underground coal bed retort. Data were averaged over pressure time window, August 17 to 23, 1981.

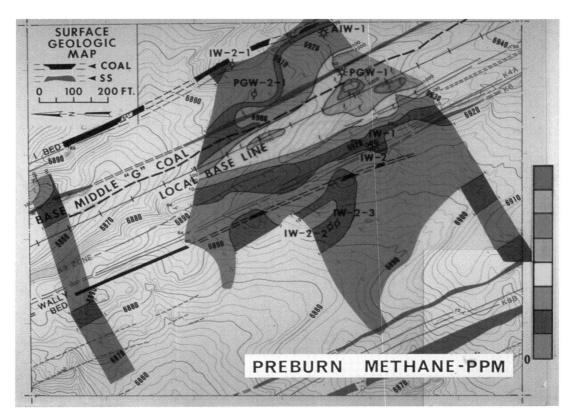

Figure 12—Color contour map of methane concentrations measured at permanent stations installed over an underground coal bed retort. Data were averaged over burn (burn A) time window, August 24 to November 10, 1981.

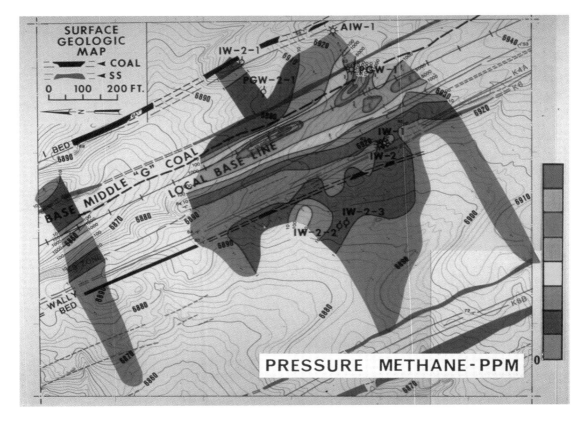

Figure 13—Color contour map of methane concentrations measured at permanent stations installed over an underground coal bed retort. Data were averaged over postburn (burn B) time window, November 11 to December 12, 1981.

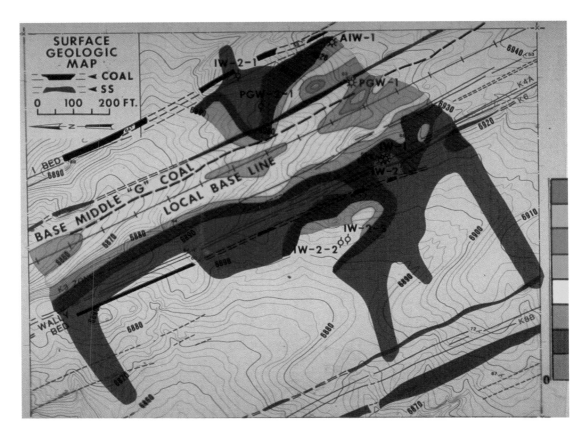

Figure 14—Color contour map of methane concentrations measured at permanent stations installed over an underground coal bed retort. Data were collected on July 21, 1982, about 6 months (200 days) after shutdown of subsurface reactor.

being gasified. This maximum value occurred in the sandstone directly over the coal with 100,000 ppm methane showing up at the surface. Figure 8 illustrates the location and change in shape of the methane magnitudes observed at the surface at four pairs of sites over the time windows selected for mapping. In contrast to sites 22 and 27, sites 85 and 86 had about 1200 ppm natural levels of hydrocarbon that had nothing to do with the active retort. They appeared to represent a natural seep that had existed previously in the area. The residual gases were pumped out of these two sites by the sampling process and apparently did not further recharge during the retort operation. The seep located to the northeast at sites 1 and 2 and along the baseline provided another anomaly that was not influenced by the gasification process. These latter two sites were charged by a previous UCG reactor (Department of Energy, 1982) that lies directly downdip from these sites. Both the vertical and slant hole product wells for both the 1979 and 1981 retorts are plotted on Figures 10–14 for reference.

Injection of Helium Tracer into UCG Reactor

Helium was occasionally injected during the stable part of the burn in order to estimate the transit time for gases passing through the reactor. This helium appeared in the surface seep gases at sites 22 and 27 within 2 days of its injection. It disappeared just as quickly.

Hydrocarbon Spots

Three major methane anomalies were observed along the strike of the bedding plane (Figures 13–17). These indicated that the leakage gases did not just migrate updip along the bedding plane and then laterally along the strike of the beds to fill the surface sediments with gas. Instead, the leakage gases came up almost simultaneously within three localized areas, or "hydrocarbon spots." This concept suggests a migration pattern with gases moving along a complex mixture of bedding planes and fracture avenues at depth. The location of the hydrocarbon spots at the surface are then controlled by these complex pathways and are thus somewhat predictable from geologic and geochemical mapping. Once these hydrocarbon spots are determined by geochemical sampling at a site, they provide the pathways for all future pressure relief from depth and can be used to establish a permanent monitoring system. In this example, the vertical migration zones were apparent and formed the main hydrocarbon spots during both the charging and discharging periods, as the surface seepage gases depleted over time.

Table 1—Gas Concentrations in Near-Surface Rocks Before (Above Line) and After (Below Line) Earthquake Activity at Mukhto Oil Field, Sakhalin Island [a]

Date	Strength of Shock	Distance from Epicenter to Deposit (km)	Well No.	Time of Sample (days)	CH_4 (10^{-4} vol. %)	HC (10^{-4} vol. %)	H_2 (vol. %)	CH_4 (% of HC fraction of gas)
91 X 1974	K = 9	100	8	6	135.4	1.9	0	98.56
				3	283.7	4.2	0	98.50
81 V 1975	M = 4[b]	12	8	2	73.6	0.81	0.53	98.35
				4	213.5	2.34	1.22	98.96
24 V 1975	K = 6.2	25	8	2	188.5	3.18	3.10	98.50
				1	525.0	2.52	7.10	99.50
4 VI 1975	K = 7.2	9	8	1	152.0	7.36	17.80	98.50
				2	852.0	8.85	15.70	98.97
8 VI 1975	K = 7.5	25	11	5	935000.0	11908.7	0.09	98.70
				2	954000.0	12465.0	0.26	98.70
5 X 1975	K = 9.5	100	8	6	58.8	3.8	26.8	98.90
				1	369.0	5.8	33.6	98.60
		100	61	5	256000.0	1273.0	3.6	99.54
				1	273000.0	1399.0	4.2	99.54

[a]From Zorkin et al. (1977).
[b]Intensity of 2 to 3.

RELATIONSHIP OF DEEP MOBILE GASES TO EARTHQUAKES

Large volumes of diverse gases continually escape from the earth's crust into the atmosphere (Jones et al., in press). Areas of especially high activity appear to be related to zones of deep tectonic fracturing and the accompanying jointing in which mineralization is sometimes found. Typical gases derived from depth are CO_2, N_2, CH_4, H_2, He, Ar, Rn, Hg, SO_2, COS, and H_2S. The major components are CO_2, N_2, CH_4, and H_2, with the remainder of this list generally found as minor or trace components. The isotope ratios of hydrogen, carbon, oxygen and uranium have considerable potential for helping to define the sources of these gases. The magnitudes of deep gas anomalies are strongly governed by tectonic and magmatic activity, thus stronger patterns are encountered in seismically active areas of late orogenic activity. Accordingly, weaker patterns are observed in platform and shield areas (consolidated blocks of the crust) that are relatively quiescent. Numerous published examples of gas flux related to earthquakes have been reported, including Kartsev et al. (1959), Fursov et al. (1968), Elinson et al. (1970), Sokolov (1971), Eremeev et al. (1972), Ovchinnikov et al. (1972), Zorkin et al. (1977), Wakita (1978, 1980), Melvin et al. (1978, 1981), Barsukov (1979), Borodzich (1979), Mamyrin (1979), King (1980), Reimer (1980), Shapiro et al. (1981, 1982), Mooney (1982), and Pirkle and Jones (1983).

That earthquakes are possibly preceded or accompanied by the escape of deep mobile gases was apparently first observed in the former Soviet Union in 1966. In a study of the Tashkent earthquake zone, Fursov et al. (1968) found air aspirating from boreholes over faults contained as much as 15 times more mercury than air not located over fault zones. This work points out that faults can be the channel ways through which mercury vapor migrates, but it also indicates that tectonic activity can release mercury not necessarily related to economic mineral deposits. Studies of this type were also undertaken in China at about the same time and in Japan in 1973.

In other areas, the Soviets showed that soil gas values increase dramatically at faults shortly after earthquakes in which fault movement was involved (Zorkin et al., 1977). An extensive study involving 105 observation wells 3–5 m (~9–15 ft) deep was done over the Mukhto oil field in northeastern Sakhalin Island. A total of 3700 samples were collected and analyzed over a 4-month period, with the most active wells sampled daily. The range of hydrocarbon seepage gases varied from 0.2 to 271,000 ppm (27.1%) for methane and from 0.3 to 13,000 ppm (1.3%) for methane homologs. Hydrogen and carbon dioxide ranged as high as 90% and 30%, respectively. The largest anomalies occurred on thrust faults, and the concentrations increased in direct response to seismic shocks. As shown by the data in Table 1, the composition of the gases changed toward a gassier (higher methane relative to heavier hydrocarbons) signature immediately after an earthquake. Relative magnitude changes were greatest in anomalous wells, whereas background areas showed little or no change. Table 1 (from Zorkin et al., 1977) provides impressive evidence for the tectonic relationship of this leakage gas flux. His study also left no doubt that faults and fractures are the main control on the effusion of gases from the subsurface.

Chemical monitoring of earthquake activity has not been widely practiced in the United States, where most efforts were geophysical until about 1975, when limited studies were initiated using radon. Gulf Research also

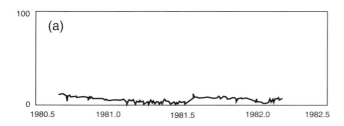

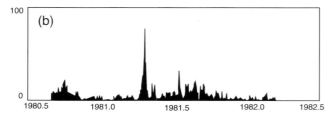

Figure 15— (a) Helium and (b) hydrogen concentrations (in ppm) sparged from monitor well at Pacioma Dam gas geochemical earthquake monitoring station. The concentrations are plotted versus time (three measurements per day) from June 1980 to June 1982. Note the hydrogen anomaly of about 75 ppm that occurred just before a 5.6 magnitude earthquake hit Westmoreland, California.

conducted their first measurements of helium and hydrogen in 1975 (Jones and Drozd, 1983).

Helium Spots

A landmark publication in *Science* by Wakita (1978) has shown that helium was observed in anomalous quantities along faults. In 1978, helium concentration was observed to be as high as 350 ppm in a nitrogen vent on the Matsushiro fault swarm. Wakita proposed that these anomalous areas be called "helium spots" because the helium leakage was not homogeneous throughout the fault zone. These unevenly distributed helium spots were reported to occupy areas of about 30 × 50 m. Extensive experience in soil gas prospecting indicates that soil gas anomalies generally occur in irregularly shaped and spaced spots. This is because the migration of gases are dominated by faults and fractures, on either a macro- or microscale.

A second paper by Wakita (1980) reported 70 measurements for hydrogen in the Yamasaki fault zone. These measurements, made in 0.5–1-m-deep holes, gave hydrogen anomalies ranging from 2 to 30,000 ppm in the fault zone, with ambient background values of 0.5 ppm measured outside the influence of the fault. Wakita postulated that hydrogen was formed by the reaction between groundwater and fresh rock surfaces created by fault movement.

Limited programs using radon as an earthquake-sensitive gas began in 1975 at about the same time that Gulf Research and Development Company first made measurements of light hydrocarbons, helium, and hydrogen

along the San Andreas fault in the Cholame Valley, California (Jones and Drozd, 1983). These data confirmed helium as a deep basement or tectonic indicator that is commonly independent of oil and gas deposits. At Cholame, where these data were gathered, the fault moved in 1857, 1906, and 1922 and most recently in 1966 (Iacopi, 1976). At the time these measurements were made in 1975, it was reported by the foreman of the Hurst Ranch that nearby doors and gates changed their overnight fit on a daily basis, indicating that the San Andreas fault remained active in this area at the time of the survey.

Earthquake Predictions Based on Gas Flux

After studying both microseeps and anthropogenic macroseeps, it became apparent that the next step was to acquire gas flux data from an active fault, such as the San Andreas. The level of seismic activity associated with this geologic feature makes it an obvious choice. To accomplish our objective, Gulf Research approved a corporate level research project called "Gas Flux Related to Earth Motions." Evaluation of the known earthquake prediction programs at the U.S. Geological Survey and various universities revealed that the Kellogg Radiation Laboratory at California Institute of Technology in Pasadena had the only computerized system that could provide automatic data collection of a series of geochemical variables.

A network of automated radon-thoron monitors operated by a microcomputer collected data every 8 hr and stored it on-site in the computer memory and then transmitted the data back to a central laboratory over regular telephone lines on command from a remote computer (Shapiro et al., 1981). Since Gulf's research objective was to map short-term flux changes, the computer link was essential. Operating stations were located at Fort Tejon, Lake Hughes, Pasadena, Santa Anita, Stone Canyon Reservoir, Big Dalton Canyon north of Glendora, Lyle Creek, Sky Forest in the San Bernardino Mountains, and Pacoima Dam.

Pacoima Dam was chosen for a site because Caltech scientists thought microseismic activity might be generated by mass loading and unloading within this steep and fractured valley. Initial measurements at Pacoima Dam indicated that hydrocarbon concentrations were low, so a helium-hydrogen gas chromatograph (GC) was set up at this station. A computer and GC automatically sampled the dissolved gases sparged from the well three times a day. The data were archived on tape and transmitted daily by modem to the VAX computer at CalTech and ultimately plotted on a time axis, as shown in Figure 15. A strong hydrogen peak of 75 ppm was measured in April 1981, just before a 5.6-magnitude earthquake hit Westmoreland, California. This hydrogen anomaly lasted nearly 3 weeks and peaked sharply at about 75 ppm. Whether a coincidence or not, this classic response provided considerable encouragement to the joint Gulf Research–CalTech program.

Previous experience in using carbon dioxide to map soil gas anomalies encouraged Gulf to introduce instruments for measuring carbon dioxide at several established CalTech stations. Within a short time, results from the Lake Hughes station showed the presence of correlated radon and carbon dioxide anomalies. It appeared that carbon dioxide reached saturation levels in the water and then served as a carrier for the radon (Shapiro et al., 1982). Although not an earth-shaking result (no pun intended), each improvement in measuring and relating the natural gases emanating from these stations increased the possibility of producing interpretable data.

ARROWHEAD HOT SPRINGS

In 1975, Scripps Institute of Oceanography began earth gas monitoring studies for possible fluid phase precursors to earthquakes with sampling at Arrowhead Hot Springs. Grab samples of spring gases were collected at 1-month intervals at the concreted hot spring and analyzed for dissolved radon, helium, and nitrogen, along with temperature and conductivity. Beginning in 1977, methane was also measured in each sample. Results of these compiled data reflected a variety of short-term variations in measured gas content for comparison with seismic events along the San Andreas fault in southern California. The most significant correlation identified was a large increase in measured gases (radon, helium, nitrogen, and methane) in 1979 before the Big Bear earthquake of magnitude 4.8 (Craig et al., 1980). This significant increase was interpreted as the result of an increase in the deep gas component dissolving into hot springs waters. The success of the Scripps' grab sampling program suggested that this location would provide even more valuable data for earthquake prediction studies with on-site computer-controlled continuous monitoring of gases.

Preliminary gas monitoring at Arrowhead Hot Springs began in early 1981 by collecting gas bubbles with a funnel and gas cylinder. Samples were analyzed by GC for methane, ethane, propane, *i*-butane, *n*-butane, ethylene, propylene, helium, and hydrogen. The initial results shown in Table 2 indicated that the hot springs gases contained 3918 ppm methane, 17.5 ppm ethane, and 1780 ppm helium. The overall high magnitude of measured gases observed with continued sampling suggested that the location was ideal for continuous gas emission monitoring and inclusion in the Gulf Research–CalTech earth gas research programs. A variety of sample collection and analysis methods were used at Arrowhead Hot Springs, including a cross-check analysis by J. Whelan of Scripps on a sample collected in May 1982. The Scripps data showed 6900 ppm (0.690%) methane, 96.68% nitrogen, 1.50% argon, 1.12% oxygen, and a trace of hydrogen. Methane homologs were not analyzed. On the basis of these early results, it was clear that Arrowhead was the type of active seepage site Gulf scientists wanted to sample and to include in the CalTech earthquake prediction program.

Geology of Arrowhead Hot Springs

Arrowhead Hot Springs is located at the base of the San Bernardino Mountains in the Transverse Range Province of southern California (Hadley and Kanamori, 1977; Miller, 1979). The Transverse Ranges are a unique east-west structural and geomorphic belt that crosses the San Andreas fault. The province is bound by the Coast Ranges to the north, the Peninsular Ranges to the south, and the Mojave Desert to the east and northeast. Despite apparent strike-slip movement along the San Andreas fault, the Transverse Ranges seem to be continuous across the bend of the fault in southern California. The San Bernardino Mountains are located northeast of and are bound by the San Andreas fault zone.

Hot springs are located in two canyons in the foothills of the San Bernardino Mountains on a splay of the San Andreas fault. The splay seems to be related to the bifurcation of the fault into northern and southern segments that continue southeast toward the Salton Trough. It is apparent that spreading and subsidence in the trough is moving northward along the San Andreas fault. Basement rocks of the San Bernardino Mountains, which were uplifted in late Quaternary time, can be divided into two structural blocks separated by the north branch of the San Andreas.

Outcrops in the vicinity of the field area range from Precambrian granite to Paleozoic metasedimentary limestones and schists and Mesozoic intrusive rocks. This suite of formations reflects the complex history of the San Bernardino structural block. The Precambrian formations seem to be related to similar units in both the San Gabriel Mountains and the Mojave Desert region.

The San Andreas fault cuts through the base of the San Bernardino Mountains as two distinct branches that continue to the southeast and a third branch that is truncated in the vicinity of Arrowhead Hot Springs. The San Andreas in this area shows extensive vertical displacement that has been active through recent times, as can be seen from terraces on alluvial fans in the two canyons of our study. Fault splays that cross the area do not show obvious recent movement.

The faults crossing the field area were identified on the surface and from low-level areal photographs where not covered by recent alluvial sequences in the two canyons. The faults are most easily identified by linear drainage and contacts of metamorphosed Paleozoic carbonates with Precambrian gneiss formations. Small amounts of vertical displacement of ~0.3–1.2 m (1–4 ft) can be seen in the outcrop that continues until covered by alluvium. The three mapped faults are inferred for a short distance past the hot springs area and are not easily identified in the rough topography of this area.

The hot springs in the field area are divided into two distinct groups and are located in two canyons about 0.5 mi apart on the Campus Crusade for Christ International property. The main group of springs to the east are located in and around Penyugal Canyon and can be divided into three groups by location as east of,

Table 2—Summary Results of Gas Concentrations (in ppm) Measured at Arrowhead Hot Springs from December 1981 to December 1982[a]

Obs. No.	Date	C_1	C_2	C_3	$i\text{-}C_4$	$n\text{-}C_4$	He	H_2	C_3/C_1
1	12/08/81	3918.08	17.500	2.697	0.420	0.785	1784	46	0.69
2	12/19/81	3800.36	17.410	2.875	0.468	0.912	1772	236	0.76
3	12/27/81	4374.00	19.458	3.201	0.546	1.001	2022	311	0.73
4	01/03/82	4469.61	19.663	3.185	0.507	1.887	2082	0	0.71
5	01/09/82	4313.75	20.238	3.300	0.533	1.970	1978	120	0.76
6	01/17/82	4506.54	18.923	3.135	0.518	1.983	1893	16	0.70
7	01/22/82	4721.68	23.275	4.124	0.659	1.472	2161	1253	0.87
8	01/31/82	4463.72	22.263	3.822	0.643	1.732	2069	60	0.86
9	02/06/82	4495.50	21.250	3.888	0.667	1.451	2049	92	0.86
10	02/15/82	4264.65	20.188	3.726	0.627	1.319	1970	0	0.87
11	02/21/82	4721.63	21.679	3.560	0.575	1.453	1811	4426	0.75
12	02/28/82	4965.41	22.022	3.632	0.518	1.259	1849	5701	0.73
13	04/20/82	5016.73	22.250	3.768	0.604	1.404	2015	0	0.75
14	04/04/82	5132.20	22.136	3.789	0.575	1.404	2008	61	0.74
15	04/16/82	6030.57	25.190	4.248	0.719	1.624	2212	0	0.70
16	05/11/82	5507.68	27.634	4.525	0.722	1.618	2228	127	0.82
17	05/21/82	5240.56	27.509	4.459	0.733	1.635	1745	66	0.85
18	05/28/82	2358.00	28.101	4.468	0.642	1.380	1760	74	1.89
19	06/06/82	5431.39	28.457	4.595	0.770	1.678	1783	177	0.85
20	06/11/82	5313.22	28.084	4.534	0.752	1.705	1853	154	0.85
21	06/20/82	5299.27	27.970	4.407	0.645	1.398	1912	58	0.83
22	06/28/82	5263.21	27.530	4.570	0.702	1.589	1883	60	0.87
23	07/06/82	4959.05	25.783	4.324	0.622	1.335	1835	54	0.87
24	07/12/82	5197.85	27.137	4.425	0.688	1.517	1931	61	0.85
25	07/18/82	5210.53	27.932	4.515	0.696	1.493	1924	93	0.87
26	07/27/82	5834.13	30.055	4.646	0.666	1.562	1874	60	0.80
27	08/01/82	5767.66	27.745	4.639	0.630	1.504	1857	82	0.80
28	08/02/82	5795.78	30.111	4.744	0.663	1.574	1791	92	0.82
29	08/03/82	5827.74	30.499	4.828	0.684	1.621	1794	96	0.83
30	08/11/82	5783.00	30.330	4.819	0.669	1.590	1793	97	0.83
31	08/12/82	6193.13	31.350	4.663	0.681	1.559	1789	100	0.75
32	08/13/82	6052.19	30.938	4.653	0.687	1.552	1739	117	0.77
33	08/16/82	5958.46	30.398	4.660	0.683	1.572	1739	65	0.78
34	08/17/82	5915.84	30.498	4.677	0.659	1.504	1706	66	0.79
35	08/19/82	5976.75	29.598	4.969	0.749	1.617	1827	70	0.79
36	08/27/82	5949.13	29.499	4.614	0.743	1.637	1806	63	0.79
37	09/11/82	5765.50	28.377	4.200	0.708	1.547	1858	76	0.78
38	10/19/82	5701.33	28.297	4.196	0.713	1.586	1994	89	0.77
39	10/20/82	5587.96	25.446	4.273	0.702	1.598	2067	89	0.74
40	10/21/82	5203.66	27.984	4.179	0.752	1.526	2148	96	0.77
41	10/22/82	5527.42	29.164	4.375	0.569	1.628	2110	89	0.80
42	10/25/82	5455.69	28.688	4.348	0.724	1.542	2356	82	0.79
43	10/26/82	5642.88	30.053	4.691	0.684	1.505	2066	84	0.80
44	10/28/82	5452.60	29.200	4.298	0.197	1.533	2117	83	0.83
45	10/29/82	5478.15	28.497	4.338	0.705	1.517	2274	89	0.79
46	11/02/82	5615.54	28.820	4.375	0.706	1.558	2102	66	0.79
47	11/03/82	5417.59	28.589	4.281	0.736	1.588	2102	113	0.78
48	11/04/82	5431.95	28.556	4.362	0.744	1.613	2204	85	0.79
49	11/05/82	5396.20	28.537	4.197	0.721	1.591	2239	83	0.80
50	11/08/82	5423.18	28.710	4.303	0.690	1.524	2044	92	0.78
51	11/09/82	5628.28	29.718	4.308	0.698	1.523	2077	78	0.79
52	11/10/82	5615.49	29.689	4.232	0.662	1.497	2219	88	0.76
53	11/11/82	5544.38	29.434	4.276	0.677	1.518	2150	84	0.75
54	11/12/82	5539.42	29.379	4.252	0.689	1.530	2082	93	0.77
55	11/15/82	5431.11	28.727	4.209	0.675	1.552	2124	99	0.77
56	11/17/82	5450.10	29.952	4.210	0.668	1.504	2156	84	0.77
57	11/18/82	5760.84	31.103	4.389	0.710	1.526	2188	89	0.76
58	11/19/82	5736.69	29.443	4.370	0.691	1.514	2292	80	0.76
59	11/22/82	5789.43	30.601	4.346	0.702	1.531	2180	108	0.75
60	11/24/82	5786.98	30.173	4.365	0.697	1.540	2139	101	0.75
61	11/29/82	5798.82	29.705	4.271	0.708	1.564	2180	90	0.73
62	12/01/82	5785.36	29.338	4.355	0.677	1.462	2036	76	0.75

[a]From Burtell (1989). Observations were selected from a much larger data base in order to show the longer term variations observed over a year.

west of, and in the canyon. The area has been developed as a bathing spa and recreational area since the late 1800s, and most of the original hot springs have been altered from their original character and location by the construction of baths and collection pools.

In the case of the springs to the west in Waterman Canyon, four caves were dug into the side of the alluvial fill covering the fractured bedrock, from which steam and hot water flow. The caves are presently bulldozed over to keep out trespassers and still show signs of warm ground and surface steam. The best descriptions of the area, before substantial development obliterated many surface geologic features associated with the springs, are provided by an assessment of the geothermal potential of hot spring areas adjacent to Southern Sierra Power Company transmission lines (Southern Sierra Power Company, 1925). The temperatures of the steam caves were described as being dependent on their locations with respect to the fault, which lies directly to the east. The temperatures were observed to decrease regularly as the distance from the fault increased.

Geochemical Sampling and Analysis

Field sample collection progressed in stages from simply filling an evacuated 1-L sample cylinder through an inverted funnel, to collecting and analyzing the gases continuously. To aid in this process, a Plexiglas sample collection bucket was installed above a stainless steel collection pan placed at the bottom of the spring. Samples collected from the collection bucket had a natural flow rate of 40–50 mL/min of free gas bubbles and provided an integrated sample over time. The water flows at a rate of ~53 L/min (14 gals/min) at a temperature of 87.8°C.

Gas chromatographic analysis of gas samples were completed using a dual GC with two columns and two detectors designed by Gulf Research. Light methane through butane hydrocarbons were analyzed using a 3-ft alumina column and a Gulf-designed flame ionization detector. Helium and hydrogen were analyzed using an 11-ft mol sieve 5A column coupled to a thermal conductivity detector. Both columns were set up with a timer controlled backflush that prevented hydrocarbons heavier than butane from entering the alumina column and any components heavier than helium from entering the hydrogen-helium mol sieve column. This feature reduced contamination of the GC and increased the sensitivity and reliability of the C_1-C_4/He-H_2 analysis. Samples could be analyzed either by flow-through or hand-injecting methods.

Continuous Gas Monitoring

As a preview to continuous monitoring, an on-site laboratory trailer was parked at the concreted hot spring and the monitoring instruments were run continuously during daytime hours. The continuous analyses, at 6-min intervals, provided a large body of data available for interpreting short-term fluctuations in the gas emission magnitude and compositions with time. Real-time sample analysis was completed on March 16, April 22, and October 19 to December 2, 1982. These numerous measurements, along with multiple analyses from single days, have been compiled into graphs for evaluation (Burtell, 1989).

To conduct these continuous monitoring tests, the Gulf Research–CalTech team had to design and build the first computer-controlled GC system. The system was controlled by an LSI 1123+ Digital computer with TU-58 tape drives for remote nonvolatile storage of data (Melvin et al., 1981).

Methane

As shown by the initial grab samples (Table 2) collected in 1-L cylinders from December 1981 to January 1982, there was a steady rise in methane from 3918 to 4507 ppm during the first month. More continuous measurements, which overlap these data, continued to show a steady but variable increase in methane content in the hot spring gases, ranging as high as 6000 ppm. As shown in Figure 16a, these data indicated gases migrating into the spring have a proportionately larger methane content throughout the sampling period.

Propane

Propane magnitude data from the concreted hot spring have also been plotted to examine propane content versus time and to compare these variations with the methane results (Figure 16b). Although propane exhibits a much narrower range of values and varies from 2.697 to 4.828 ppm, the propane data clearly mimicked the gross changes recorded by methane. However, the net propane increase with time was almost twice the initial magnitude, as observed for methane. Individual high-magnitude events identified for methane correlate well over the entire time period with propane magnitudes. However, on a sample-to-sample basis, propane does exhibit some independence from methane. Since ethane, *i*-butane and *n*-butane results show a close relationship with one another and parallel propane results, they were not plotted.

Propane/Methane Compositional Ratio

As a further means of evaluating light hydrocarbon seepage at the concreted hot spring, the composition of the migrating gases was closely examined. Compositional indicators included ratios of one gas to another and percentages of individual hydrocarbon gases as compared to the entire hydrocarbon content of the migrated gas. The propane/methane × 1000 light hydrocarbon ratio is commonly used in soil gas geochemical exploration activities to identify whether a seep originates from an oil, oil and gas, or natural gas type source in the subsurface (Jones and Drozd, 1983). This ratio (Figure 16c) ranges from 0.69 to 0.87, indicating a mature dry gas source.

Although both methane and propane generally correlate with each other, their ratio clearly exhibits variations

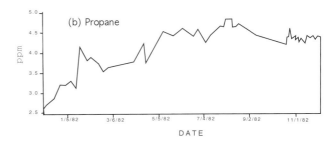

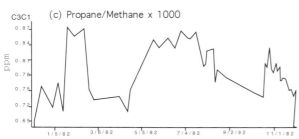

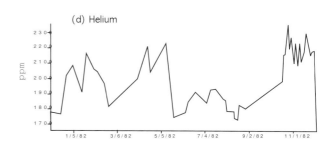

Figure 16—Gas concentrations in free gases evolving naturally from a concreted well at Arrowhead Hot Springs, near San Bernardino, California, measured during 1 year from December 1981 to December 1982. (a) Methane nearly doubled in that year. (b) Propane generally followed methane and also nearly doubled. (c) Propane/methane ratio ($\times$ 1000) varied significantly within variable time windows, suggesting a relationship to tectonic activity in the San Andreas fault zone. (d) Helium exhibited significant variations that do not follow those of the hydrocarbon gases.

similar to those noted for the Mukhto oil field on the seismically active Sakhalin Island. These well-established changes are probably a result of changes in the earth's stresses and its influence on gas emission. Ethane through butane also follow this trend and suggest that over this time period, a larger proportion of mature hydrocarbons was emitted from this spring.

Helium

Helium magnitudes fluctuated independently of the light hydrocarbon gases, suggesting some independence of the helium (Figure 16d). Overall helium magnitudes ranged between 1700 and 2350 ppm as free gases from the spring, indicative of a concentrated source of helium (more than 350 times atmospheric levels). Although the initial data from Scripps suggested a positive correlation between methane and helium for the 1979 Big Bear earthquake, this more detailed analysis showed that this correlation was much more complex and that helium may be affected by different geologic and tectonic events than methane.

Stable Carbon Isotopes

As noted earlier, the composition of the measured hydrocarbon gases reflects a dry gas signature. The presence of the methane homologs of heavier ethane through butane suggests a deeply buried sedimentary source for the hydrocarbons. Stable carbon isotope measurements made on two methane samples collected on January 9

and 17, 1982, had consistent results of –23.7‰, confirming the presence of a very mature source.

The methane and higher hydrocarbon gases measured at the site suggest a migrated product derived by normal sedimentary processes, typical of very mature oil and gas accumulations. Small blocks of sediment could have been squeezed across the fault plane. A small sedimentary block could be thrust below the San Bernardino Mountains in the vicinity of Arrowhead Hot Springs, where the additional heat could produce the measured hydrocarbons.

Helium Isotopes

Isotopic measurements of He_3/He_4 in waters from Arrowhead Hot Springs were found to have an average of 0.431 ± 1.2‰ for five measurements (Craig et al., 1980). This value is not nearly as thermally mature as measurements made over hot spot thermal reservoirs in Iceland and in Yellowstone National Park, Wyoming, and over helium spots in the Matsushiro area of Japan. Although these measurements are in the range of 7 to 18 times atmospheric concentrations, their low ratios relative to other known thermal reservoirs suggest a dilution of normal granitic helium generated by radioactive decay within the crystalline rocks of the San Bernardino Mountains, mixed with limited mantle helium from depth. The presence of only a slight amount of mantle-derived helium suggests that the thermal waters are a result of either frictional heat and resultant fractures or possibly a small intrusive body.

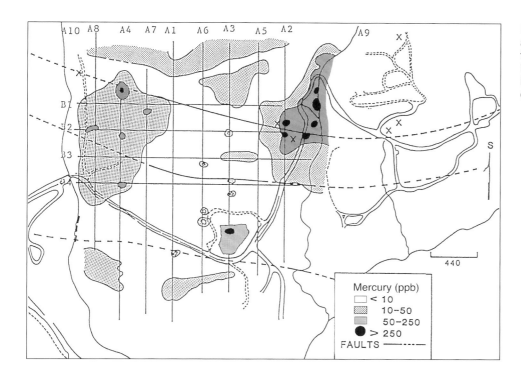

Figure 17—Contour map of adsorbed mercury extracted from surface soils at Arrowhead Hot Springs, California, illustrating its relationship to the concreted hot well.

Mercury Associations

To identify the size and shape of the hot springs system and its relationship to other geologic features, a low-temperature soil mercury mapping program was completed. A total of 525 soil samples were collected at 15-m (50-ft) intervals on a grid of ten north-south and four east-west survey lines 75 m (250 ft) apart and analyzed for adsorbed mercury content (Jones and Maciolek, 1984). Contoured mercury magnitude data (Figure 17) show two distinct and well-controlled zones of anomalous values that clearly define the area of thermal springs and wells. Anomalous zones are centered around the surface spring outlets in each area and on the northern edge of the study area where Precambrian rocks crop out. Anomalous zones have sharp boundaries and tend to be located within and north of mapped east-west trending fault zones that cross the property.

The highest magnitude mercury anomalies are located in Penyugal Canyon where several samples had concentrations in excess of 250 ppb. The distribution of mapped mercury values conclusively located and confirmed the distribution of thermal springs and wells on the Arrowhead Hot Springs property. Areas of anomalous mercury do not correlate with any particular mapped geologic unit, suggesting that the measured mercury migrates from depth and accumulates in the soils, rather than forming as a residuum of weathering of surficial rock units. The overall distribution of mercury anomalies correlates well with surface thermal spring outlets, indicating that the subsurface mercury source is controlled more by thermal systems and less by the east-west trending faults that splay from the San Andreas fault.

High concentrations of mercury in the soil are coincident with surface thermal outlets, suggesting that mer-cury sampling can be used to identify potential thermal areas that are not otherwise evident. In the hot spring, mercury is more highly concentrated in the gas bubbles than in the spring water. Therefore, mercury may be enriched in soil as a result of vapor phase migration. The mapped mercury anomalies record the areal extent of the mercury rich vapor which probably migrated to the surface from a subsurface thermal reservoir.

Origin of Arrowhead Hot Springs Gases

Light hydrocarbons, helium, and mercury have been monitored over time at Arrowhead Hot Springs and found to exhibit significant changes in magnitude and composition. The origin of the measured helium appears to be primarily from crustal radioactive decay, with minor input from mantle sources. Methane and other light hydrocarbons have a very mature signature. This mature organic hydrocarbon gas strongly suggests that sedimentary rocks are present below crystalline and metamorphic rocks of the San Bernardino Mountains. The possibility is strongly supported by recent geologic research indicating that low-angle thrust faults of Laramide–Tertiary age occur in the Mojave Desert to the east. Tectonic activity may have thrust the San Bernardino block over sedimentary rocks that now lie buried below the San Bernardino Mountains. Subsequent burial and maturation of sedimentary source rocks may have produced significant quantities of hydrocarbons. If this possibility can be substantiated by additional data, exploration for an untested petroleum province with commercial potential below the San Bernardino Mountain and other parts of the Transverse Range Province may be warranted.

Discussion

This project began as a spin-off of a joint Gulf Research–CalTech fault-migrated gas monitoring program to help CalTech develop vapor phase earthquake prediction techniques and to improve Gulf's understanding of gas migration mechanisms. The resultant investigations have tested various assumptions and theories about deep fault and fracture system gas emanations and their expression at near-surface sampling sites. Each phase of the program was completed to help develop geochemical sampling techniques, to start a database for future programs, and to aid in the evaluation of potential earthquake prediction sites.

This program included only the study of a single hot spring site in an attempt to relate gas emissions to earthquakes. Light hydrocarbons, helium, radon, hydrogen, carbon dioxide, and carbon monoxide were monitored for magnitude and compositional changes for correlation with earthquake occurrences. Although no clearly earthquake-related events were identified, significant gas magnitude and compositional changes were recorded. Regional stress variations are suggested as the most probable cause for the recorded changes in gas flux at this site. Plans for additional sites and longer term monitoring were made by Gulf and CalTech scientists and are highly recommended for future research. Industry downsizing and takeovers prevented the final implementation of these research plans. No comment needs to be made about the value these data might have had to California and to the nation if this program had been allowed to continue until today.

CONCLUSIONS

This paper presents a variety of near-surface geochemical measurements made over both natural and anthropogenic seeps, selected more by opportunity than design. The main advantage of these selected examples is that they generally represent opportunities to measure actual changes in flux rates of gases that are unique in chemical composition, and as such can be directly related to their respective sources. Understanding and mapping the actual migration pathway remains the difficult part of the process. However, in spite of these difficulties, these measurements do demonstrate that the migration model must be both pressure and permeability dependent. Magnitude changes related to pressure events at depth clearly occur in these examples on a time scale measurable in hours to days rather than in years. Suggestions that vertical migration from a field such as Hartzog Draw in the Powder River Basin might take a million years is no longer reasonable. In fact, measurements made in 1976 (before the field was discovered) and again in 1978 (after the wells were put on pump) indicated that the initial soil gas measurements made in 1976 had decreased by nearly an order of magnitude by 1978.

These studies confirm that surface geochemical data reliably reflect subsurface hydrocarbon sources and their compositions. The magnitude of a microseep from a reservoir is related to the permeability of the migration pathway. Since we do not know whether the surface signal comes from an economic or an uneconomic reservoir, it is risky to speculate whether a particular seep necessarily represents an economic accumulation of hydrocarbons. A surface geochemical survey is not a stand-alone prospect tool. However with judicious use, this technology can provide information on the maturity of source beds in a basin and the composition of subsurface hydrocarbons.

REFERENCES CITED

Barsukov, V. L., 1979, Geochemical methods of predicting earthquakes: Geochemistry International, v. 16, no. 2, p. 1–13.

Borodzich, E. V., 1979, Preliminary results on helium pattern variations in seismically active zones: Geochemistry International, v. 16, no. 2, p. 37–41.

Burtell, S. G., 1989, Geochemical investigations at Arrowhead Springs, San Bernardino, and along the San Andreas fault in Southern California: Master's thesis, University of Pittsburgh, Pittsburg, Pennsylvania.

Craig, H., Y. Chung, J. E. Lupton, S. Damasceno, and R. Poreda, 1980, Investigation of radon and helium as possible fluid phase precursors to earthquakes: USGS Technical Report 13, contract #14-08-0001-18348, p. 36.

Chekalin, L. M., and G. I. Timofeev, 1983, Methodology of geochemical methods in exploration, *in* Geochemical exploration for oil and gas: Academy of Science of USSR, Moscow, Nauka.

Department of Energy, 1981, Phase II report, results of Rawlins test no. 1, July 1981: Prepared for the Division of Oil, Gas, and In Situ Technology, DOE, #LETC 13108-70UC090c.

Department of Energy, 1982, Phase III report, results of Rawlins test no. 2, July 1982: Prepared for the Division of Oil, Gas, and In Situ Technology, DOE, #DE-AC20-77-ETI13108.

Drozd, R. J., G. J. Pazdersky, V. T. Jones, and T. J. Weismann, 1981, Use of compositional indicators in prediction of petroleum production potential (abs.): American Chemical Society Meeting, Atlanta, Georgia, March 29–April 3.

Elinson, M. M., Y. N. Pashkov, G. M. Agababov, and I. B. Ignatiev, 1970, Measurements of gases distributed above Cu-Mo ore bodies, *in* The geochemistry of Hg, Mo, and S in hydrothermal processes: Moscow, Nauka.

Eremeev, A. N., V. A. Sokolov, A. P. Solovov, and I. N. Yanitskii, 1972, Applications of helium surveying to structural mapping and ore deposit forecasting, *in* M. J. Jones, ed., Geochemical exploration 1972: London, Institution of Mining and Metallurgy, p. 183–192.

Fursov, V. Z., N. B. Vol'fson, and A. G. Khvalorski, 1968, Results of a study of mercury vapor in the Tashkent earthquake zone: Dolk Akademiya Navk USSR, v. 179, no. 5, p. 1213–1215.

Hadley, D., and H. Kanamori, 1977, Seismic structures of the transverse ranges, California: GSA Bulletin, v. 88, p. 1469–1478.

Hunt, J. H., 1981, Surface geochemical prospecting pro or con: AAPG Bulletin, v. 65, p. 939.

Iacopi, R., 1976, Earthquake country: Menlo Park, California, Lane Books, 87 p.

Jones, V. T., 1983, Surface monitoring of retort gases, from an underground coal gasification reactor: time dynamics (abs.): 1983 American Chemical Society Annual Meeting, September, Washington, D.C.

Jones, V. T., and R. J. Drozd, 1983, Predictions of oil and gas potential by near-surface geochemistry: AAPG Bulletin v. 67, no. 6, p. 932–952.

Jones, V. T., and J. B. Maciolek, 1984, Mercury as a pathfinder for deep-seated ore bodies: American Chemical Society Meeting, April 10, St. Louis, Missouri.

Jones, V. T., and H. W. Thune, 1982, Surface detection of retort gases from an underground coal gasification reactor in steeply dipping beds near Rawlins, Wyoming: Society of Petroleum Engineers, SPE 10050, p. 24.

Jones, V. T., M. D. Matthews, and D. Richers, in press, Light hydrocarbons in petroleum and natural gas exploration: handbook of exploration geochemistry, *in* Gas geochemistry: New York, Elsevier, v. 7.

Kartsev, A. A., Z. A. Tabarsaranskii, M. I. Subbota, and G. A. Mogilevskii, 1959, Geochemical methods of prospecting and exploration for petroleum and natural gas: Berkeley, California, University of California Press, 349 p.

King, Chi-Yu, 1980, Geochemical measurements pertinent to earthquake prediction: Journal of Geophysical Research, v. 85, no. 136, 3051 p.

Link, W. K., 1952, Significance of oil and gas seeps in world oil exploration: AAPG Bulletin, v. 36, p. 1505–1541.

Mamyrin, B. S., 1979, ^{3}He/^{4}He ratios in earthquake forecasting: Geochemistry International, v. 16, no. 2, p. 42–44.

Matthews, M. D., V. T. Jones, and D. M. Richers, 1984, Remote sensing and surface hydrocarbon leakage: International Symposium on Remote Sensing for Exploration Geology, Colorado Springs, Colorado.

Melvin, J. D., M. H. Shapiro, and N. A. Copping, 1978, An automated radon-thoron monitor for earthquake prediction research: Nuclear Instrumentation and Methods, v. 153, p. 239–251.

Melvin, J. D., M. H. Mendenhall, R. McKneown, and T. A. Tombrella, 1981, VAX/LSI-11/CAMAC nuclear data acquisition system under development at the W. K. Kellogg Radiation Laboratory, CalTech: IEEE Transactions on Nuclear Science, v. NS-28, p. 3738.

Miller, K. F., 1979, Geologic map of the San Bernardino North Quadrangle, California: USGS Open File Report 79-770, scale 1:24,000.

Mooney, J. R., 1982, Sniffing for oil and earthquakes: The Orange Disk, v. 25, no. 9, p. 26–29.

Nikonov, V. F., 1971, Distribution of methane homologs in gas and oil fields: Akademiya Nauk SSSR Doklady, v. 206, p. 234–246.

Ovchinnikov, L. N., V. A. Sokolov, A. I. Fridman, and I. N. Yanitskii, 1972, Gaseous geochemical methods in structural mapping and prospecting for ore deposits, *in* M. J. Jones, ed., Geochemical exploration 1972: London, Institution of Mining and Metallurgy, p. 177–183.

Pirkle, R. J., 1985, Hydrocarbon seeps in the Gulf of Mexico: Proceedings, 189th American Chemical Society National Meeting, Miami Beach, Florida, April 28–May 3, paper no. 78.

Pirkle, R. J., and V. T. Jones, 1983, Helium and hydrogen anomalies associated with deep or active faults: Proceedings, 181st American Chemical Society National Meeting, Atlanta, Georgia, March 29–April 3.

Preston, D., 1980, Gas eruptions taper off in northwest Oklahoma: Geotimes, October, p. 18–26.

Reimer, D. M., 1980, Use of soil gas helium concentrations for earthquake prediction, limitations imposed by diagonal variation: Journal of Geophysical Research, v. 85, no. 136, p. 3107–3114.

Richers, D. M., V. T. Jones, M. D. Matthews, J. Maciolek, R. J. Pirkle, and W. C. Sidle, 1986, The 1983 Landsat soil gas geochemical survey of the Patrick Draw area, Sweetwater County, Wyoming: AAPG Bulletin, v. 70, no. 7, p. 869–887.

Shapiro, M. H., J. D. Melvin, T. A. Tombrello, M. H. Mendenhall, P. B. Larson, and J. H. Whitcomb, 1981, Relationship of the 1979 southern California radon anomaly to a possible regional strain event: Journal of Geophysical Research, v. 86, p. 1725.

Shapiro, M. H., J. D. Melvin, T. A. Tombrello, J. Fong–liang, L. Gui-ru, M. H. Mendenhall, and A. Rice, 1982, Correlated radon and CO_2 variations near the San Andreas fault: Geophysical Research Letters, v. 9, no. 5, p. 503–506.

Sokolov, V. A., 1971, Geochemistry of natural gases: Moscow, Niedra.

Southern Sierra Power Company, 1925, Geothermal areas adjacent to the transmission lines of the Southern Sierra Power Company, *in* Geothermal exploration in the first quarter century: Geothermal Resource Council Special Report #3, p. 145–161.

Wakita, H., 1978, Helium spots caused by diapiric magma from the upper mantle: Science, v. 200, April 28, p. 430–432.

Wakita, H., 1980, Hydrogen release: new indicator of fault activity: Science, v. 210, October 10, p. 188–190.

Weismann, T. J., 1980, Developments in geochemistry and their contribution to hydrocarbon exploration: Tenth World Petroleum Congress, Bucharest, Romania, v. 2, p. 369–386.

Williams, J. C., R. J. Mousseau, and T. J. Weismann, 1981, Correlation of well gas analyses with hydrocarbon seep data: Proceedings, American Chemical Society National Meeting, March, Atlanta, Georgia, p. 3.

Zorkin, L. M., S. L. Zabairaevi, E. V. Karus, and K. Kh. Kilmetov, 1977, Experience of geochemical prospecting in petroleum and gas deposits in the seismically active zone of Sakhalin Island, IZU (Russia): UZSSh Uchebn. Zaved, Geol. Razved, v. 20, p. 52–62.

Thrasher, J., A. J. Fleet, S. J. Hay, M. Hovland, and S. Düppenbecker, 1996a,
Understanding geology as the key to using seepage in exploration: spectrum
of seepage styles, in D. Schumacher and M. A. Abrams, eds., Hydrocarbon
migration and its near-surface expression: AAPG Memoir 66, p. 223–241.

Chapter 17

Understanding Geology as the Key to Using Seepage in Exploration: The Spectrum of Seepage Styles

Jane Thrasher

*BP Exploration
Research and Engineering Centre
Sunbury-on-Thames, U.K.*

Present address:
*Sir Alexander Gibb & Partners
Reading, U.K.*

Andrew J. Fleet

*BP Exploration
Research and Engineering Centre
Sunbury-on-Thames, U.K.*

Stephen J. Hay

*BP–Statoil R & D Alliance
Trondheim, Norway*

Present address:
*Statoil
Stavanger, Norway*

Martin Hovland

*BP–Statoil R & D Alliance
Trondheim, Norway*

Present address:

*Statoil GASS T & T
Stavanger, Norway*

Stephan Düppenbecker

*BP Exploration
Research and Engineering Centre
Sunbury-on-Thames, U.K.*

Abstract

In most basins, lateral subsurface petroleum migration occurs over tens or even hundreds of kilometers between source rock and trap and between accumulations and the surface. In general, this means, first, that seepage can only provide information for risking petroleum charge at the basin scale, and second, that there is no direct spatial relationship between filled prospects and surface seepage. Understanding the geology, and hence petroleum dynamics, of a basin is the key to understanding and using seepage in exploration. A spectrum of seepage styles can be used to focus exploration thinking. The spectrum ranges from prolific seepage (e.g., offshore California), through focused point-source seepage (e.g., offshore Colombia), to basins where long-distance lateral migration concentrates seepage on basin margins (e.g., Western Canada Sedimentary Basin). Related controls on fluid flow and seepage range from active tectonism, through high fluid potential gradients resulting from rapid muddy deposition, to fault and salt structures and basinwide carrier bed systems. Case studies of offshore oil seepage from the Gulf of Mexico, Central North Sea, Haltenbanken (offshore mid-Norway), and North Viking Graben are used to illustrate the spectrum of seepage styles and the factors that control different styles. Understanding seepage in terms of basin geology and petroleum dynamics is not only necessary for interpreting seepage for exploration but is also critical for planning seep collection, particularly in offshore areas. In all but basins with prolific seepage, likely seepage sites, which occur at the surface end of migration pathways, need to be targeted if seeps are to be sampled for analysis.

INTRODUCTION

Explorers have used surface seepage as a pathfinder for oil occurrence throughout the history of petroleum exploration (e.g., Hunt, 1979). Ideally, they would like to use seepage to locate individual prospects prior to drilling (e.g., Horvitz, 1980). Unfortunately, unequivocal petroleum seepage does not usually occur directly above prospects, but is found at the end of migration pathways often tens or even hundreds of kilometers away (on the seabed or land surface) (e.g., Link, 1952; Hunt, 1979). This means that both lateral petroleum migration, which is responsible for this displacement, and vertical migration must be assessed if identified seepage is to be linked to possible prospects. The overall migration pathway is determined by the interrelationship of sediment fill, sedimentation rate, tectonics, and fluid flow, that is, the geology of the basin. Wilson et al. (1974) were among the first to stress this concept. They related seepage predominantly to tectonic style in estimating the amount of seepage into the marine environment. More recently, Macgregor (1993) discussed the relationship between visible seepage and tectonics based on a study of southeast Asian basins. He concluded that, at a subbasinal scale, seeps are concentrated over tectonic features such as active diapirs, active faults, and uplifted basin margins. He also observed that it is relatively rare for visible seeps to directly overlie major petroleum fields.

In this paper, we set out to illustrate this causal relationship between seepage and geology by considering the spectrum of seepage styles. The relationship implies constraints for using seeps to understand the location of subsurface accumulations. Our examples are solely from offshore, but the same principles apply onshore. In both offshore and onshore cases, near-surface lateral displacement of seepage must be taken into account: offshore water currents can have displaced seeps detected as slicks, and onshore displacement may occur at the water table (Clarke and Cleverly, 1991).

The evidence we cite is both from visible seepage and "microseepage" (detected by geochemical techniques) and from indirect indicators such as seismic gas blanking. We are not, however, concerned here with detection techniques themselves. Offshore seepage is best used in integrated studies involving searches for slicks and shallow coring surveys planned and interpreted on the basis of probable petroleum migration pathways in the basin. Onshore seepage detection requires ground surveys planned in the light of a similar prognosis of migration pathways and their probable near-surface displacements.

Detected oil is easier to correlate back to possible source successions (via its molecular and isotopic composition) than is gas, which by its nature is compositionally simpler. Because of the broad-scale conclusions and ambiguities often associated with interpreting gas compositions, we concentrate on oil in this paper while recognizing that numerous claims are made for gas surveys (e.g., Horvitz, 1980), particularly onshore, whether identified by direct measurement or through microbiological activity (e.g., Hitzman et al., 1994a).

SPECTRUM OF SEEPAGE STYLES

The spectrum of seepage styles ranges from spectacular (e.g., California and Trinidad), through prolific (e.g., Gulf of Mexico), to weak but highly focused (e.g., Haltenbanken), and finally to absent (e.g., North Viking Graben). In reality, the spectrum of seepage styles is continuous, not artificially divisible, but for the sake of this discussion, we recognize crude groupings of seepage styles that we can use to outline the key factors controlling seepage.

Spectacular seepage appears to result from high rates of sedimentation and from active tectonism. Rapid sedimentation leads to overpressures, steep gradients in fluid potential, and mud diapirism. Active tectonism causes petroleum to be focused structurally. Transcurrent tectonism, such as that in California, the southern Caribbean, and mid-Caspian, appears to be the key trigger in promoting spectacular seepage by providing both structural focus and sites for rapid sedimentation. An example of this is the intense localized seepage that occurs in the Santa Barbara Channel, offshore California, above faulted crests of anticlines in areas of compression and transpression (Fischer and Stevenson, 1973; references in Hovland and Judd, 1988). Other examples of spectacular seepage include offshore South Trinidad (Hedberg, 1980; Nash and Wood, 1994; Hitzman et al., 1994b) and the mid-Caspian area, Azerbaijan (Guliev and Feizullayev, 1994). The seeps of Iran and Iraq (Macgregor, 1993), which brought exploration to the Middle East at the beginning of this century, lie somewhere near this end of the spectrum.

Second only to areas of spectacular seepage are those where seepage is prolific. In the Gulf of Mexico, rapid sedimentation and petroleum migration, controlled largely by halokinesis, results in widespread focused seepage. Other rapidly deposited Tertiary deltas with active petroleum systems and mud diapirism (rather than salt diapirism) would be expected to promote similar prolific seepage, although none are documented.

More localized, smaller scale seepage occurs where petroleum migration is channeled by lateral carrier bed systems. Mud or salt diapirs, basement highs, and bounding faults can act as foci for vertical seepage from such systems. Mud diapirs in offshore Colombia and salt diapirs in the Central North Sea seem to focus seepage to the surface. In offshore Colombia, thrust faults may play a similar role, but it is difficult from the available to differentiate between the effects of faults and the effects of buried mud diapirs (Thrasher, Strait et al., this volume). Baffin Bay (MacLean et al., 1981) is an area where basement appears to control seepage to the surface. Elsewhere, migration along lateral carrier beds can take petroleum to offshore seabed outcrops, such as in the Haltenbanken area, offshore mid-Norway, or can carry it far away from the area of exploration targets, as probably occurs in the North Viking Graben sector of the North Sea. The Western Canada Sedimentary Basin and its enormous tar sand accumulation represent the supreme example of such a "gathering" system (e.g., Piggott and Lines, 1991).

In the following sections, we consider four case studies to illustrate in detail how seepage styles vary and make up the spectrum outlined above. These case studies are based on surface geochemical surveys carried out for exploration or research by BP over the past few years. Recent interest in seabed coring and surface geochemistry as exploration tools grew from results in the deep-water Gulf of Mexico, where cores containing mature thermogenic petroleum were regularly being retrieved from seabed sediments in the vicinity of oil discoveries. We set out to assess the value of such seabed coring and allied techniques as an exploration tool in other geologic settings with different seepage styles. Our offshore studies over the past few years have certainly not covered all the possible styles of seepage that might be encountered. Further styles can be recognized from literature descriptions, such as onshore Iran and Iraq (Macgregor, 1993, and references therein).

PROLIFIC OIL AND GAS SEEPAGE: NORTHERN GULF OF MEXICO CONTINENTAL SLOPE

Oil and gas seepage is prolific in the deep-water Gulf of Mexico and is readily detectable at many stages along both the subsurface migration pathway and the seepage pathway through the seabed to the sea surface. Seepage rates are enhanced by regional overpressuring and active salt movement. Most seepage is focused along good migration pathways up active faults and salt margins, although in the absence of such focused migration pathways, some vertical migration probably occurs.

Background

Our aim in studying the Gulf of Mexico was to understand what seepage could tell us about prospectivity on a subregional and target-specific scale. Our work built on the burgeoning interest in seabed coring and surface geochemistry as an exploration tool which grew from results in the deep-water Gulf of Mexico, where cores containing mature thermogenic petroleum were regularly being retrieved from seabed sediments in the vicinity of oil discoveries.

Geologic Setting

The northern Gulf of Mexico continental slope is an emerging exploration area with some very large discoveries in fairly deep-water. The main features of the geologic environment are the high Plio-Pleistocene sedimentation rate and the presence of allochthonous salt.

The high sedimentation rate has resulted from the large volumes of sediment delivered to the northern Gulf of Mexico by the Mississippi River. Loading of the thick sediment pile onto the Jurassic Louann Salt has caused the salt to extrude out and up into the Tertiary sequence, with the emplacement of an allochthonous salt nappe. There is extensive ongoing salt movement and diapirism through the Tertiary across the continental slope, creating significant sea floor topography and partially controlling sedimentation. The main petroleum plays are turbidite sand reservoirs structurally bounded by salt or faults with, in some cases, a stratigraphic component. Reservoired oils indicate the presence of deep subsalt Mesozoic carbonate source rocks (e.g., Kennicutt et al., 1992). These source rocks have not been encountered by drilling, but their prognosis implies extensive vertical migration. This migration must occur through holes in the salt; where subbasins are floored by unwelded salt, structures are dry. Thermal modeling shows that oil expulsion began in the Miocene and is still continuing to the present day in some areas. Thus, oil seepage may be direct from the source rock (or minor accumulations along a migration pathway), as well as from leaking accumulations.

Manifestations of Seepage

Oil and gas seeps are common in the deep-water Gulf of Mexico (Figure 1). Leakage and seepage can be detected and sampled at many stages along the migration pathway. Leaking hydrocarbons can be identified as anomalies on seismic (e.g., Lovely and Ruggiero, 1994). Their presence can be confirmed from well data (as high wet-gas concentrations in mudlogs) or from high abundances of extractable petroleum in cuttings, in cap rocks above accumulations, or in salt-associated migration conduits (BP data). On the Gulf of Mexico shelf, a recent well cored a salt-associated growth fault and sampled oil, gas, and naturally fractured silty shales (Anderson et al., 1994).

Oil and gas at the seabed are detected as anomalies on high-resolution and site survey seismic profiles (e.g., Tinkle, 1973) and are often sampled by seabed coring (e.g., Brooks et al., 1986) or by manned submersible (e.g., Sassen et al., 1993). Gas seepage is detectable in the water column as plumes on echo sound records (e.g., Williams et al., 1995) or as dissolved carbon anomalies (Aharon et al., 1992). Oil slicks from natural seepage are common at the sea surface in some areas and can be detected by airborne surveys (Williams et al., 1995) or from space using satellite images (Landsat thematic mapper and Space Shuttle sunglint photographs; e.g., Macdonald et al., 1993). Oil slicks can be effectively sampled for geochemical analysis using a nylon swab (Kennicutt et al., 1988; Williams et al., 1995). Macdonald et al. (1993) estimate the total natural oil seepage in the northern Gulf of Mexico to be about 120,000 bbl/year.

Controls on Seepage

Many thousands of seabed cores have been collected in the deep-water Gulf of Mexico, and analysis of these data, together with subsurface information, allows a better understanding of the controls on seepage in this environment.

(a)

(b)

(c)

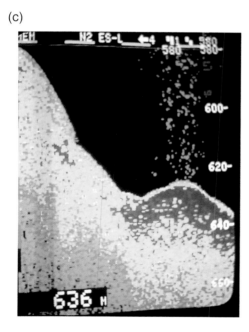

(d)

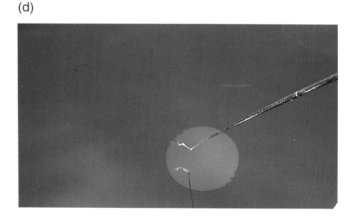

Figure 1—Manifestations of prolific seepage in the Gulf of Mexico. (a) Natural oil seepage slick viewed from the air. (b) Shipboard view of a natural oil seepage slick. (c) The source of a slick–water column gas plume recorded on fish-finding sonar. (d) Seeping oil droplets rise and form "pancakes" ~1 m in diameter on the sea surface; these can be sampled with a nylon swab. (Photographs by Roger Duckworth.)

Most macroseepages, which show visible oil staining in cores and may have related oil slicks at the sea surface, are located at the seabed above active salt bodies or close to salt-related faults (Figure 2a) (see Behrens et al., 1988; Lovely and Ruggiero, 1994; Williams et al., 1995). The most famous oil seepage site, the Bush Hill chemosynthetic community and hydrate mound (Macdonald et al., 1989), is situated over a salt-related fault above the Jolliet producing oil field (Figure 2b). Cores in these areas contain high concentrations of extractable hydrocarbons, which when extracted and analyzed by gas chromatography–mass spectrometry (GC-MS) show a very good match with the subsurface oils (e.g., Kennicutt et al.,

1988). Cores collected away from salt or active faults usually contain much lower concentrations of extractable hydrocarbons. Leakage and seepage is focused up the good migration pathways to the surface, so that seepage appears to be concentrated around the margins of intrasalt basins.

There is good regional correlation between the locations of macroseepage and of cores containing GC-MS "truthable" oil and oil discoveries along the continental slope in the Gulf of Mexico (Figure 3). On a local scale, however, the seeps cannot be considered prospect specific and are not necessarily directly related to accumulations (Figure 4). Seepage detected by seabed coring or by

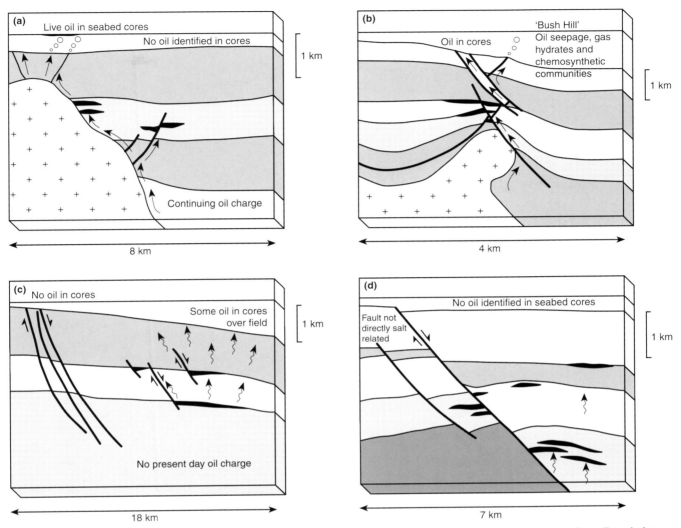

Figure 2—Seepage associated with some deep-water Gulf of Mexico discoveries. (a) Antares discovery: live oil and abundant gas found in seabed cores directly above the salt wall. (b) Jolliet field: abundant evidence for seepage, including the Bush Hill hydrates and chemosynthetic community, concentrated around salt-associated faults (structure from Cook and d'Onfro, 1991). (c) Trap not associated with salt: weak seepage signal seen at surface immediately above discovery, apparently direct vertical migration in absence of good migration pathways. (d) Snapper field: little evidence for thermogenic seepage to surface, even associated with active growth faulting.

other means can give valuable information on whether there has been an effective charge to an intrasalt subbasin but not whether specific traps are charged or in communication with effective migration pathways.

Only in special circumstances, where there are no clear migration pathways to the surface, can seepage be considered prospect specific. In such circumstances, traps are not directly related to active salt. In one such case, relatively high concentrations of extractable hydrocarbons have been detected at the seabed immediately above an accumulation (Figure 2c). Direct vertical migration may occur via stacked channel sands and subseismic faults and may be enhanced by overpressuring from rapid sedimentation. A discrete seepage signal such as this, directly above an accumulation, would not be detectable if a sourcing system were still active in the area. An active

sourcing system would be expected to produce seepage over a wider area, swamping out the discrete signal, assuming the area was able to leak uniformly.

Faults as Migration Pathways

The question of when faults act as pathways to the surface for petroleum is perennial, but intuition and observations provide hints as to which circumstances are favorable. Faults in mudstones, for instance, are most likely to be effective as migration pathways when the faults are active and a network of open fractures exists. Faults related to salt (Figure 2b) are more likely to be continually active, on a small scale, as the salt moves. The distribution of detected seeps above salt structures is good evidence that salt-related faults and salt margins make the best

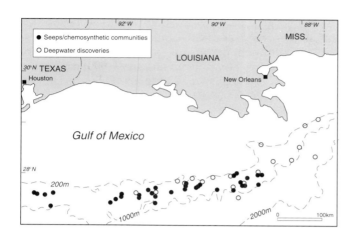

Figure 3—General association of seeps and discoveries in the deep-water Gulf of Mexico. (Adapted from Sassen et al., 1993.)

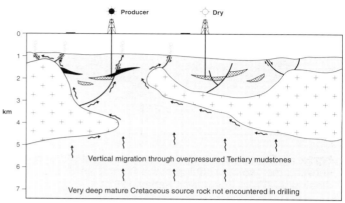

Figure 4—Schematic model of hydrocarbon migration and seepage in the deep-water Gulf of Mexico. Seepage can indicate where there is a charge to the intrasalt subbasin, but it cannot predict the presence of an effective migration pathway to a specific trap or the presence of an effective trap.

migration conduits. Growth faults resulting from slumping at high sedimentation rates are less likely to be continually active than salt-associated faults and are not expected to be as effective migration conduits to the surface. Data from seabed coring in the western Gulf of Mexico suggest that this is in fact the case. Cores collected above trap-associated growth faults do not necessarily contain any more petroleum than cores collected away from faults (Figure 2d). It is possible that an inactive muddy fault may focus seepage if it acts as a permeability barrier, displacing small-scale vertical migration laterally.

Summary

The following points summarize the main characteristics of petroleum seepage in the deep Gulf of Mexico:

- Prolific seepage of both oil and gas occurs.
- Seepage rates are enhanced by high sedimentation rates and allochthonous salt.
- Seepage is focused up good migration pathways—salt margins and active faults.
- Seepage can be detected and sampled at many stages along the migration and seepage pathways.
- Seepage gives information about the charge to intrasalt subbasins, but not about individual prospects.

SEEPAGE FOCUSED AROUND SALT DIAPIRS: CENTRAL NORTH SEA

Abundant seismic evidence exists for gas leakage and seepage associated with salt diapir structures in the Central North Sea (CNS), and good evidence for migrated thermogenic oil was identified in a seabed core collected above the diapir-related Machar oil field. However, although most diapir structures have gas chimneys, they

do not all appear to have valid traps or significant accumulations. The diapirs appear to act as structural highs focusing regional hydrocarbon migration in the Tertiary sediments, with faulting and fracturing associated with diapirism enhancing vertical migration.

Background

Seismic gas chimneys and other evidence for seepage above salt diapirs in the CNS have been documented (e.g., Cayley, 1987; Salisbury, 1990). Bubbling thermogenic gas has been collected by a remotely operated vehicle (ROV) from seepage sites above the diapir-related Tommeliten discovery in the Norwegian sector (Hovland and Judd, 1988). For the present study, a small number of seabed cores were collected over several diapirs to look for geochemical evidence of seepage to the surface. The coring data were integrated with existing seismic and well log data to improve understanding of the leakage on a regional scale and to evaluate the potential for using seepage to rank the prospectivity of individual diapirs.

Geologic Setting

Key geologic features of the salt diapirs of the CNS are illustrated in Figure 5. Extension and rifting started in the Early Permian, and up to 1800 m of predominantly halite evaporites were deposited in the rapidly subsiding basin that had restricted marine influx. Subsequent loading on this salt, combined with extensional tectonics, resulted in the formation of salt pillows, particularly along graben-bounding faults (Foster and Rattey, 1993). Throughout the Mesozoic, these salt pillows appear to have continued to rise, driven by sediment loading, as Triassic and Jurassic synrift sedimentation continued preferentially in the grabens. The Jurassic deposits include deltaic and shoreline clastics that provide important reservoirs. The

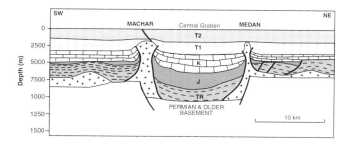

Figure 5—Cross section of the North Sea Central Graben, showing the geologic setting of the Central North Sea diapir play. (From Foster and Rattey, 1993.)

main source rock, the Kimmeridge Clay Formation, was deposited in restricted marine anoxic conditions toward the end of Jurassic time. Subsequent Cimmerian erosion denuded some horst blocks; Jurassic reservoirs exposed by this event were later sealed by onlapping Cretaceous marls.

The salt seems to have remained in buoyant equilibrium with the sediments throughout the Cretaceous and Paleocene, staying at or near the seabed. This resulted in condensed Chalk deposition immediately above the salt during the Cretaceous and possibly in ponding of deepwater siliciclastic turbidites around sea floor highs during the Paleocene (Foster and Rattey, 1993). From Eocene to Recent time, the basin filled with as much as 2750 m of undercompacted mudstones. Regional tectonic inversion in the middle Miocene caused active salt diapirism from the existing salt highs and cap rock fracturing, which in turn produced good reservoir quality in the Chalk above the diapirs. Rapid regional subsidence since the middle Miocene has caused diapirs such as Machar to be buried more deeply than at any previous time in their history. Structural relief has resulted in significant petroleum column heights and, thus, high pressures at the seal interface.

Leakage and seepage associated with the salt diapir structures would be expected to occur, not only because of reservoir overpressuring and overburden fracturing but also because the structures resulted in regional topographic highs in the Tertiary. These highs are likely to have acted as foci for lateral migration in sandy carrier beds.

Evidence for Leakage and Seepage over Machar Oil Field

Leakage and seepage associated with the Machar oil field have been studied in detail. Machar comprises a fractured chalk and Paleocene sand reservoir over the diapir crest and flank. The field is estimated to contain about 91 million bbl recoverable oil equivalent, with an oil column of more than 1100 m. The structural evolution of the field has been described by Foster and Rattey (1993).

A seismic chimney attributable to gas seepage is clear-

ly seen over the Machar structure (Figure 6). High-resolution site survey seismic data also clearly show the presence of shallow gas pockets and sea floor disturbance typical of gas seepage (Figure 7). Salisbury (1990) has shown that most of the higher amplitude anomalies on the site survey data are close to either small faults or gas chimneys. There also appears to be small-scale lateral displacement of seepage in shallow channel sands. Reservoir leakage can also be recognized on well logs, particularly the formation evaluation gas chromatograph logs, which show thermogenic gases (ethane, propane, and butane) to be present in higher concentrations in the drilling mud from sediments above the crest of the Machar diapir compared to sediments in the flank wells (Figure 8). The Machar reservoir is approximately 1200 psi overpressured, well below the pressure required for leakage via fluid induced fracturing (e.g., Clayton and Hay, 1994), suggesting that the primary mechanism for leakage is via capillary failure, with reservoir pressures exceeding top seal capacity.

Twenty seabed cores were collected in the vicinity of the Machar diapir, many of which were targeted at seabed disturbances identified on the site survey data. One core, which retrieved sandy substrate, had a soluble extract with a recognizable thermogenic hydrocarbon signature (Figure 9a). This extract is in some ways significantly different from reservoired Machar oil (Figure 9b), notably in the lower diasterane and C_{30} and C_{31} content. It is not known whether this results from alteration during seepage or whether the oil and extract are from different sources, although the latter does not match any known oil from the Central Graben.

Leakage and Seepage over Other Central North Sea Diapirs

Gas chimneys and high gas concentrations in mud logs have been recognized over many CNS diapirs and salt structures, and a number of these structures contain significant hydrocarbon reserves, including the Mungo, Medan, Scoter, and Tommeliten fields and the UKCS (U.K. continental shelf) 30/1a well (Figure 10). Other wells, such as NOCS (Norway offshore continental shelf) 1/6-5 and UKCS 22/25-T8, contain only shows and residual oil, suggesting that no valid trap exists. These diapirs show good evidence for hydrocarbon leakage, with seismic gas chimneys, shallow gas, and high wet hydrocarbon concentrations in the cuttings recorded on the mud logs. A diapir structure tested by NOCS 1/9-5 in the Ekofisk area did not even have shows in the Cretaceous, but shallow sands containing wet (thermogenic) gas were encountered in the Pliocene above the diapir, which has a marked seismic gas chimney over it.

A small number of seabed cores were collected over diapirs in Scoter, UKCS 30/1a, UKCS 22/25-T8, and NOCS 1/6-5, but no good evidence for higher thermogenic hydrocarbons was identified in any of these cores. The small number of cores (10 over each structure) may

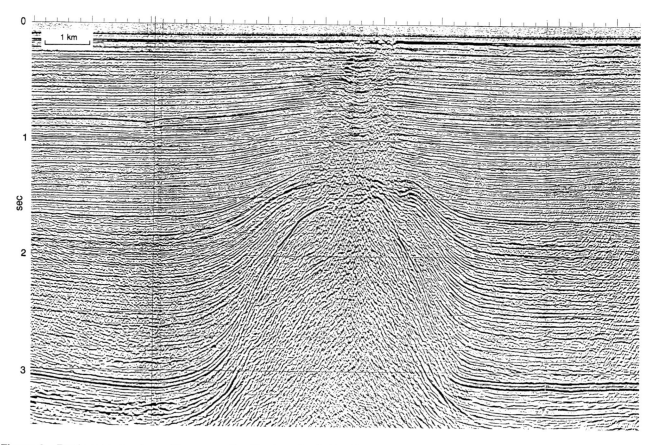

Figure 6—Regional seismic profile across the Machar diapir (A–A'), showing the clear seismic chimney over the salt high. (See Figure 10 for location.)

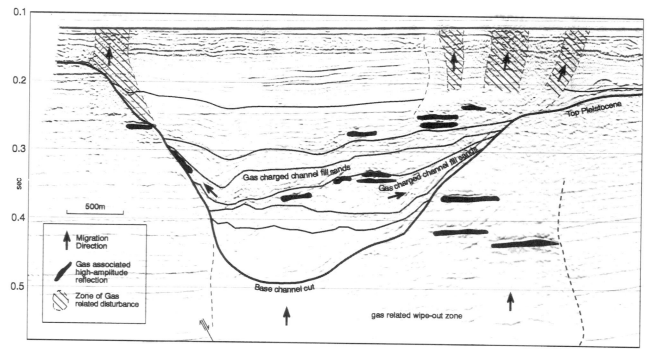

Figure 7—Interpreted high-resolution site survey seismic section showing abundant bright spots and wipeout zones indicating shallow gas in the Plio-Pleistocene sediments. Some small-scale lateral displacement of seepage by channel structures is also observed.

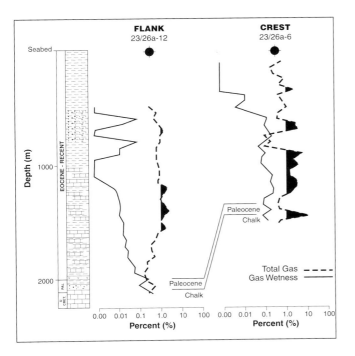

Figure 8—Gas distribution in the Machar structure, determined from formation evaluation gas chromatograph logs, showing higher concentrations of thermogenic C^{2+} gases above the crest as compared to the flank.

have played a part in the poor success rate where there was seismic evidence for seepage.

Regional Focusing of Seepage

Mud logs from wells drilled over these diapirs and elsewhere in the Central North Sea show that thermogenic gases in the shallow sections are generally found in the Tertiary sands. Laterally extensive Tertiary sand bodies could act as regional "gathering systems" for vertical leakage from diapir structures or from deeper source kitchens or leaky structures. Tertiary structures such as the giant Forties Field on the graben flank appear to be charged from such a gathering system. Within the graben, the diapirs cause regional highs in the Tertiary deposits and could thus be focusing regional leakage and allowing seepage to the surface through diapir-associated faults, regardless of whether there has been migration to deeper prospects in the structure. Trap relief, resulting petroleum column heights, reservoir pressures, and shallow seals all doubtlessly play parts in this leakage. Seeps above a CNS diapir should not be used as evidence of effective charge and migration pathway to potential reservoirs in the diapir structure, and cannot provide evidence for valid trap and commercial accumulation. Sampling the seepage should, however, give some information about the regional source system.

Summary

The following points summarize the main characteristics of petroleum seepage in the Central North Sea, as discussed here:

- Excellent evidence for seepage occurs over most Central North Sea salt diapirs.
- Seepage manifests as seismic gas chimneys, shallow gas, and high cuttings gas concentrations.
- Thermogenic oil seepage was identified in one seabed core over Machar diapir.
- Seepage was observed over dry structures and structures with no valid traps.
- The relationship of seepage to diapirs may be the result of regional Tertiary highs, not direct leakage from Cretaceous or Paleocene reservoirs.
- Seeps over diapirs do not reduce the exploration risk of finding effective migration pathways to reservoirs and cannot identify valid diapir-associated traps.

SIGNIFICANT LATERAL DISPLACEMENT UP DIPPING CARRIER BEDS: HALTENBANKEN

Although many of the petroleum accumulations in the Haltenbanken area of Norway appear to leak, oil seepage has not been detected at the seabed in the areas of the discoveries or the source kitchens. Instead, seepage is found 50 km up dip where carrier beds crop out along the uplifted Norwegian margin.

Background

Prior to the present study, it was known that some Haltenbanken accumulations leaked, but only gas with biogenic signatures had been detected at the surface or in the shallow subsurface here (Vik et al., 1992). A number of possible seabed seepage indicators had been identified along the Tertiary subcrop edge by Statoil pipeline surveys (M. Hovland, personal communication, 1993). An integrated research project was initiated to generate a case study of leakage and seepage in an uplifted margin and to ground truth the features identified in pipeline surveys. The project included a short research cruise using an ROV (remotely operated vehicle), seabed coring, and shallow seismic profiling. Exploration data for Haltenbanken was reevaluated in terms of leakage and seepage.

Geologic Setting

A generalized map of the Haltenbanken region in Norway is shown in Figure 11. The relatively stable Trondelag platform is separated from the more faulted Halten terrace by the Halten fault zone; most of the fault

Figure 9—GC-MS *m/z* 191 and *m/z* 217 mass fragmentograms of (a) Machar oil and (b) an extract from a seabed core collected above the Machar field. The core extract is clearly thermogenic, suggesting oil seepage, but it is significantly different from the Machar oil in some respects, such as the relatively low diasterane (D) content.

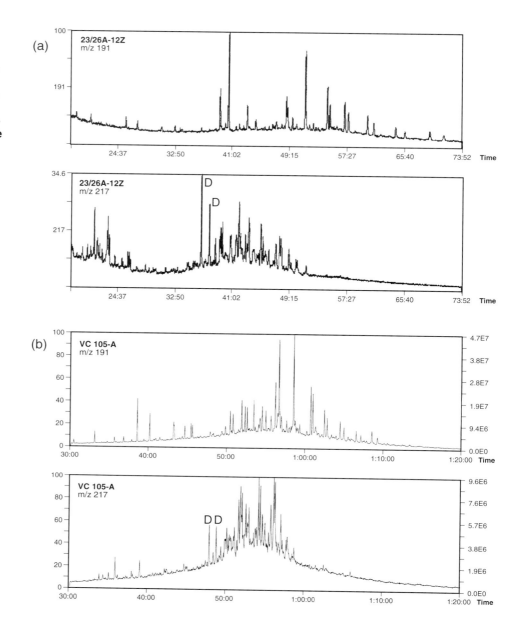

Manifestations of Leakage and Seepage

activity took place during the Late Jurassic–Early Cretaceous. Most of the oil and gas accumulations have been found in clastic reservoirs in prerift tilted fault block structures. Synrift deposits include the major oil source rock, the Spekk Formation, which is equivalent to the Kimmeridge Clay Formation, and minor reservoir formations. The postrift sequence comprises mainly passive infill, predominantly outer shelf mudstones, except for the coarser Paleocene tuffaceous Tare Formation, associated with the opening of the North Atlantic. In the past 3 m.y., uplift of the Scandinavian hinterland and Norwegian margin has resulted in the progradation of a Pliocene clastic wedge over the Halten terrace and has produced a gentle westward dip in the regional postrift stratigraphy, exposing Jurassic–Pliocene stratigraphy at the seabed (Bugge et al., 1984; Ellenor and Mozetic, 1986).

Seabed features identified in pipeline surveys and thought to be related to seepage include a number of plume-like features situated on Tertiary outcrop scarps (Figure 12). These were initially interpreted as gas plumes, with the high-amplitude reflectors in the substrate interpreted as shallow gas. Close examination of one of these features using side-scan sonar and ROV showed it to be a cold-water *Lophelia* coral reef. Seabed coring in the vicinity of the reef did not retrieve gassy cores, but one core was found to contain thermogenic petroleum, identified by GC and GC-MS data (Figure 13). This oil showed a reasonable match with Haltenbanken reservoir oils (e.g., Cohen and Dunn, 1987). The thermogenic petroleum is unlikely to be derived either from

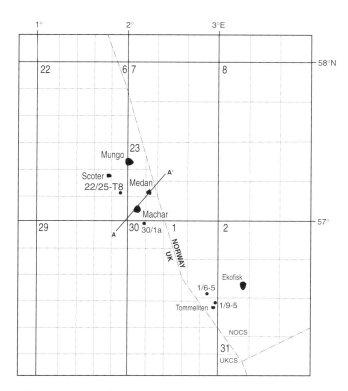

Figure 10—Locations of fields and wells in the Central North Sea discussed in this paper. (Section A–A' is shown in Figure 6.)

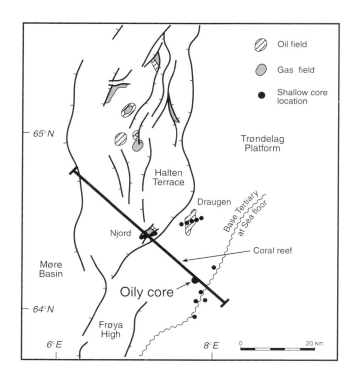

Figure 11—Map of Haltenbanken area showing the location of fields, wells, seabed coring sites, and geologic cross section (in Figure 14).

source rocks reworked into the sediments or from local subcrops of the Spekk Formation, because both are known to be immature (Continental Shelf Institute data). Thus, although the seabed plume feature is not a flowing gas seep, there is good evidence for small-scale oil seepage at this site and the high-amplitude reflectors are likely to be related to thermogenic gas.

Seabed cores were also collected directly above the Draugen and Njord discoveries and at ROV dive sites on the Halten terrace, where possible seepage-related features including pockmarks and side-scan sonar "dark patches" had been identified. No evidence of seeping thermogenic petroleum was identified in these seabed cores at sites away from the Tertiary subcrop edge.

Evidence for Leakage in Halten Terrace

Although no seepage of thermogenic petroleum to the seabed was found in cores from the Halten terrace, there is good evidence for leakage and vertical migration from deep structures into the overlying sediments.

Previous studies have indicated that reservoir overpressuring is the primary factor in promoting top seal failure (e.g., Clayton and Hay, 1994). A study of formation pressures in the Jurassic reservoirs of Haltenbanken has shown that several pressure-independent aquifer systems exist in the area, with formation pressures ranging from near hydrostatic to highly overpressured. In general, the highest overpressures exist in the western part of

the region where the Jurassic section is deeper and more compartmentalized (e.g., Mann and Mackenzie, 1990).

Direct evidence of leakage from reservoirs comes from formation evaluation GC logs, which detect the presence of thermogenic gases (ethane, propane, and butane) in the drilling mud from the overlying sediments. Thermogenic gases are recorded in the mud logs from the cap rocks of most of the overpressured accumulations, indicating that leakage is occurring. Some of the most highly overpressured structures are dry, with abundant traces of thermogenic hydrocarbons in the cap rocks, suggesting that the charge to the traps has leaked off. Significant volumes of thermogenic gases are not observed in drilling mud from above the tuffaceous Paleocene Tare Formation, suggesting that the overlying mudrocks act as a regional seal on migration.

Evidence of leakage from overpressured traps can also be observed on regional seismic sections, with numerous well-defined seismic chimneys interpreted as gas leakages (e.g., Vik et al., 1992). The seismic chimneys stop near the top of the Cretaceous at a similar level to the shallowest thermogenic gases in the drilling mud.

Interpretation and Implications for Future Exploration

Both mud gas and seismic data suggest that the Paleocene forms an effective barrier to vertical migration. The tuffaceous Tare Formation is the sandiest horizon in

Figure 12—Shallow sparker seismic section across Paleocene subcrop on the Trondelag platform. The plume-like feature on the crest of the scarp is a 35-m-high *Lophelia* cold-water coral reef. High-amplitude reflectors in the Paleocene probably indicate shallow gas. Geochemical data from a seabed core collected at the foot of this reef are shown in Figure 13.

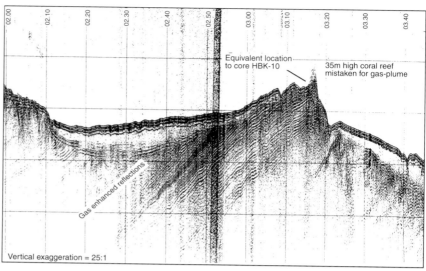

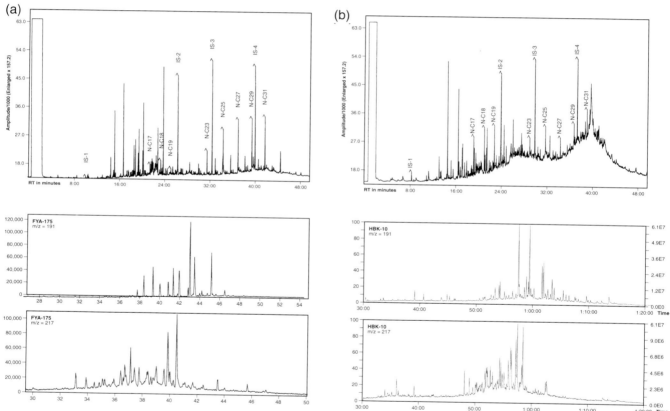

Figure 13—Whole-extract GC (top) and GC-MS *m/z* 191 and *m/z* 217 mass fragmentograms (bottom) of extracts from (a) a seabed core from Paleocene outcrops on the Trondelag platform and (b) a seabed core collected immediately above the Draugen oil field. The Trondelag platform core extract is clearly thermogenic and resembles oils from the Haltenbanken fields. In contrast, there is no evidence for thermogenic petroleum in the core from above the Draugen field.

a predominantly muddy section and appears to act as a regional carrier bed, the gentle dip of which focuses migration laterally upward (Figure 14). The discovery of oil seepage in shallow cores at the seabed subcrop of this formation on the Trondelag platform adds credence to this interpretation. In retrospect, using the subsurface

data, it is possible to follow the migration route from the overpressured reservoir, through the caprock, and up the dipping carrier bed to the surface, about 50 km from the source kitchen and traps. The Haltenbanken case study shows that, in the presence of good lateral migration pathways that can carry leakage away from a trap, a seep-

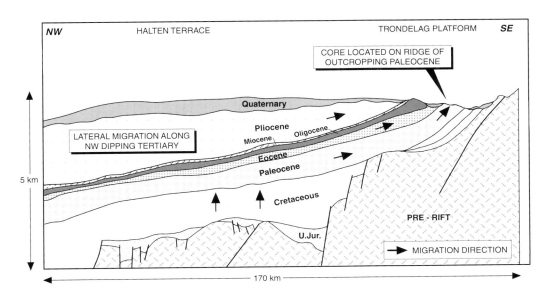

Figure 14—Geologic cross section across the Halten terrace and the Trondelag platform (after Bugge et al., 1984). (See Figure 11 for location.) Vertical leakage from Jurassic reservoirs and source rocks is displaced laterally up dipping sandier beds in Paleocene deposits and is found at the surface displaced as far as 50 km from the fields.

age signal should not be expected directly above an accumulation. If Haltenbanken were undrilled, the seeps along the Trondelag platform could be used as evidence for a working Upper Jurassic source system in the deeper part of the Halten terrace and as proof of an effective migration pathway from that source into the overlying section. They cannot be expected to give block-specific or prospect-specific information. More detailed mapping of likely carrier beds and the distribution of seismic chimneys, along with conventional exploration methods, would be needed to rank blocks or prospects.

Summary

The following points summarize the main characteristics of petroleum seepage in the Haltenbanken area of Norway:

- Leakage from Jurassic accumulations was identified on well logs and as gas chimneys in Cretaceous deposits.
- No vertical migration was observed above the dipping Paleocene sandy tuffs.
- Shallow gas and oil occur at the seabed where the Paleocene crops out along the Norwegian margin.
- Oil seepage is displaced up to 50 km from accumulations and source kitchens.

LITTLE VERTICAL MIGRATION TO SURFACE: NORTH VIKING GRABEN

Seabed coring surveys in the North Viking Graben have failed to identify clear evidence for thermogenic petroleum seepage that may be related to the underlying multibillion barrel oil and gas province. Evidence from well logs suggests that leakage does occur from overpressured accumulations, but thermogenic hydrocarbons

are seldom found above the Paleocene interval. Integrated two-dimensional basin modeling predicts the vertical distribution of thermogenic hydrocarbons observed in the well logs and suggests that some leakage has also been displaced laterally up carrier beds to the basin margins.

Background

Although the North Viking Graben (NVG) is well established as a multibillion barrel oil and gas province, a number of new blocks were offered in the 1992 Norwegian 14th Licensing Round, resulting in significant new data acquisition. This data including a consortium seabed coring survey that covered the main areas of interest as well as existing discoveries.

Geologic Setting

The NVG is the most northerly of three half-graben systems that trend en echelon north-northeast from the Moray Firth and Central Graben systems of the North Sea (Pegrum and Spencer, 1990). Rifting commenced in this area of the northern North Sea in Middle Jurassic time and continued through the Late Jurassic (Rattey and Hayward, 1993). The major regional carrier beds for petroleum migration occur in the Lower–Middle Jurassic sequence (Statfjord Formation and Brent Group). Thin coals occur in these carrier beds which have considerable potential for gas generation. The oil of the basin is sourced from overlying Middle–Upper Jurassic mudrocks (Heather and Draupne formations) (Düppenbecker and Dodd, 1993). Plays in the NVG occur in Triassic, Jurassic, and Paleogene deposits, and there are minor reservoir intervals in the Lower Cretaceous (Pegrum and Spencer, 1990). Thick Cretaceous shales blanket the graben and constrain petroleum migration above the Draupne and older Jurassic seals.

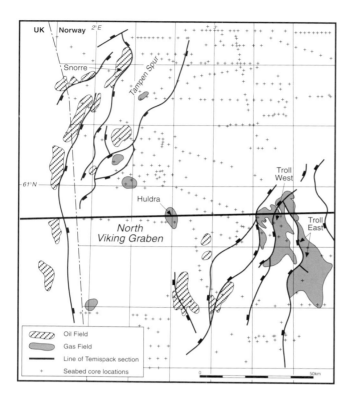

Figure 15—Map showing locations of seabed cores, cross section, and oil and gas fields in the North Viking Graben.

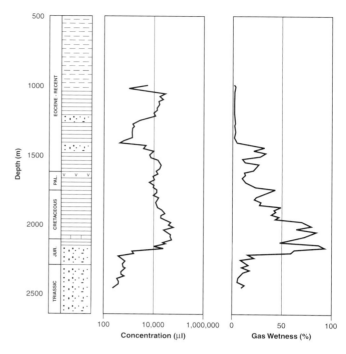

Figure 16—Typical gas distribution in a well from the Snorre field (after Leith, 1993). Wet thermogenic gas (with significant C^{2+} components) is common in the deeper sections of all wells studied, but the shallow gas is predominantly methane and probably biogenic in origin. There is little evidence for vertical migration of thermogenic petroleum above the lower Tertiary.

Results of Surface Geochemical Surveys

The consortium coring survey collected 564 seabed cores, averaging about 3 m long, across the area considered most prospective and potentially available for license. Locations were picked in traverses across discoveries and prospects, but relatively few cores were collected over any one accumulation (Figure 15).

Many of the cores contained high concentrations of headspace gas (up to 32%), which consists predominantly of isotopically light methane (–60 to –100‰ PDB), implying a shallow biogenic origin. The distribution of high gas concentrations shows good correlation with the location of shallow sands.

No good evidence for migrated mature thermogenic hydrocarbons was identified in extracts from any of the cores, which were analyzed by total scanning fluorescence (TSF) or gas chromatography (GC). Gas chromatography–mass spectrometry (GC-MS) analysis of 10 selected core extracts show that the *m/z* 191, *m/z* 217, *m/z* 231, and *m/z* 253 traces (expected to show thermogenic biomarkers if present) contain only low concentrations of unsaturated and functionalized biomarkers, typical of recent organic matter.

Evidence for Trap Leakage in North Viking Graben

As in Haltenbanken, there is good evidence for some leakage from overpressured traps in the NVG, such as in the Snorre field (Leith et al., 1993). Leith and co-workers found oil shows as high as 400 m above the reservoir, and petroleum extracted from cuttings in this interval showed a good match with reservoir fluids. The main top seal of the Snorre reservoir is Cretaceous Shetland Group mudstones. These are of poorer seal quality than the Jurassic Kimmeridge Clay Formation, which is the more common top seal for the Jurassic reservoirs, but has been eroded off the top of the Snorre structural high. The poor seal quality of the Shetland Group, combined with overpressuring, appears to be a major control on leakage. Mud gas concentrations decrease above the reservoir, and few wet thermogenic gases are found above the Paleocene Balder Tuff (Figure 16). Gas concentrations in many NVG wells increase in the Tertiary Hordaland and Nordland groups, but this gas is invariably dry (only methane recorded) and was probably derived as biogenic gas from the Tertiary lignites that are widespread in the Northern North Sea (e.g., Knight et al., 1994).

Use of Integrated Two-Dimensional Basin Modeling in Understanding Migration

Integrated two-dimensional basin modeling, focused on Brent targets, was used for acreage assessment by BP Norway in the fourteenth licensing round. Petroleum migration was modeled as separate phase flow controlled by buoyancy, hydrodynamics, and capillary pressure. The primary purpose of the modeling was to test the various conceptual models of basin geologic evolution, taking into account the effects on timing of petroleum generation, expulsion, and migration. It could also be used to aid seep interpretation by modeling petroleum flow to the present-day sediment surface and by comparing the model predictions with observed seepage phenomena.

In the case of the NVG, two-dimensional modeling suggests that present-day vertical migration does not extend beyond the Paleocene and that it only reaches the Paleocene above basin center structural highs such as the Huldra fault block (Figure 17). The major source of vertically migrating petroleum is predicted to be the Upper Jurassic Draupne Formation, while the main charge to the reservoirs is derived from the Middle Jurassic Heather Formation. These results compare well with the subsurface fluid geochemistry (BP data) and the mud log evidence for wet gas seepage. Migration up dipping carrier beds displaces petroleum toward the basin margins. We have not looked for seepage where these carrier beds approach the seabed at the graben margins, but if seepage occurs, we would expect to identify it there. Alternatively, the petroleum may be effectively contained in the broad subregional "structure" of the basin margins and no significant seepage may occur.

Surface Geochemistry as Exploration Tool in North Viking Graben

For a number of reasons, the geology of the NVG is not ideally suited for the application of surface geochemistry to petroleum charge risk reduction. As in Haltenbanken, dipping carrier beds overlying the prospective structures may displace seepage laterally. An additional complication arises from the presence of multiple effective source rocks, which means that any seeping fluids are more likely to be representative of the shallowest effective source system, which has not necessarily provided the main reservoir charge. Integrated two-dimensional basin modeling improves the understanding of the complex migration processes in this environment and allows the results of the seabed coring survey to be put into perspective.

Summary

The following points summarize the main characteristics of petroleum seepage in the North Viking Graben:

- The NVG is a multibillion barrel oil and gas province, but only gas seepage has been identified at the surface.
- Near-surface gas is predominantly methane and almost certainly biogenic.
- Well logs show that leakage does occur from overpressured reservoirs, but wet gases are seldom observed above the Paleocene.
- Integrated two-dimensional basin model predictions compare well with observations from well logs.
- Some leakage may be displaced laterally up carrier beds but is not identified by coring surveys

IMPLICATIONS FOR EXPLORATION

Petroleum seeps provide samples of the petroleum charge to a basin. They thus provide information for risking petroleum charge at the basin scale. Because most migration to the surface involves lateral migration of tens or even hundreds of kilometers, it is generally unrealistic to risk prospect charge from seep data.

Understanding the geology of a basin is the key to relating seepage to basin charge and petroleum migration. Varying geology dictates a spectrum of seepage styles (Figure 18). Active tectonism can result in minimal lateral migration and prolific seepage (e.g., offshore California, Figure 18a). High depositional rates and muddy sedimentation also tend to produce near-vertical migration and prolific seepage, although salt structures and faults can focus migrating petroleum to spectacular point-source seeps (e.g., Gulf of Mexico, Figure 18b).

In areas of less dramatic sedimentation, structural features related to active tectonism or halokinesis focus laterally migrating petroleum upward to produce seeps at only specific points over the basin (e.g., offshore Colombia, Figure 18c, and Central North Sea, Figure 18d). In tectonically quiescent basins, petroleum migration to the surface is governed by relatively permeable and sandy carrier beds that take petroleum laterally to near-surface subcrops or sea floor outcrops (e.g., Haltenbanken, Figure 18e, and North Viking Graben, Figure 18f). To use seeps in exploration, it is necessary to understand where the basin fits into the spectrum of seepage styles. Figure 18 provides a schematic template of the spectrum. In any actual study, attention must be given to determining present and past migration within the basin in terms of timing of petroleum generation, migration pathways, reservoir leakage, and seismic indicators.

Understanding basin geology and petroleum migration is the key not only to interpreting seeps but also to finding them in the first place. In most areas where migration is focused to give point seeps at the surface, hunting for seeps needs to be specifically targeted on features where seeps are likely to occur if samples are to be retrieved. This is particularly true in offshore areas where sample collection depends on such techniques as shallow coring, which requires multiple samples at any one locality to give a reasonable chance of seep collection in all areas but those of prolific seepage.

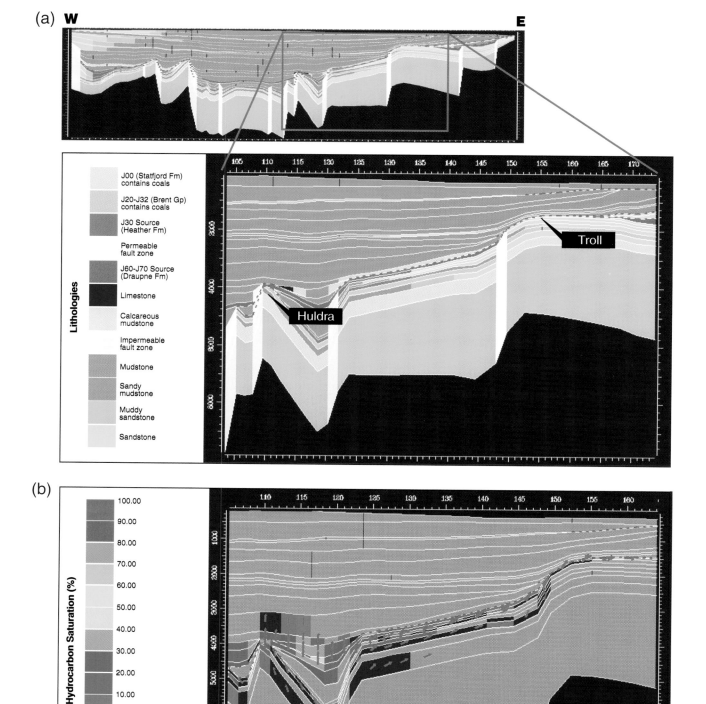

Figure 17—Integrated two-dimensional basin model created using IFP Temispak software. (a) Lithologic template of the modeled section across the Huldra and Troll fields. (b) Results of modeling showing simulated present-day hydrocarbon saturation and flow, suggesting little vertical migration above the Paleocene, and lateral displacement of leakage up dipping carrier beds to the shallow subsurface at the basin margins.

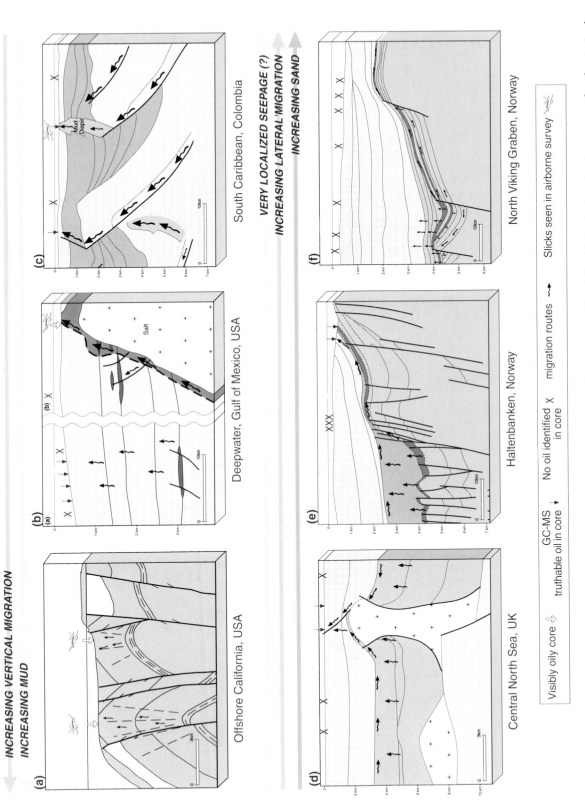

Figure 18—Spectrum of seepage styles dictated by various geologic settings. (a) Prolific seepage with minimal lateral migration in an area of active tectonism—offshore California (no vertical exaggeration) (adapted from Hovland and Judd, 1988). (b) High depositional rate and muddy sedimentation leading to near-vertical migration, with focusing of petroleum up salt walls and active faults leading to major point-source seeps—deep-water Gulf of Mexico (3× vertical exaggeration). (c) More localized seepage associated with mud diapirism—offshore Colombia (3.5× vertical exaggeration) (see Thrasher, Strait et al., this volume). (d) Laterally migrating petroleum focused upward over salt diapirs—Central North Sea (no vertical exaggeration). (e) Lateral displacement of seepage up dipping carrier beds—Haltenbanken (5.5× vertical exaggeration). (f) No leakage to surface detected but significant lateral migration seen in tectonically quiescent basin—North Viking Graben (7× vertical exaggeration).

Acknowledgments—Our sincere thanks go to numerous colleagues at BP and Statoil who worked on the "Understanding Seepage" project and the exploration areas studied, particularly Tony Barwise, Chris Clayton, Tim Dodd, Roger Duckworth, Richard Miller, Neil Piggott, Roger Sassen, Alan Williams, and the BP Norge explorers. We also thank the BP Exploration drawing office at Sunbury for drafting the figures, BP and Statoil managements for permission to publish the paper, and NEPCON AS for permission to show the survey locations given in Figure 15.

REFERENCES CITED

Aharon, P., E. R. Graber, and H. H. Roberts, 1992, Dissolved carbon and δ13C anomalies in the water column caused by hydrocarbon seeps on the northwestern Gulf of Mexico slope: Geo-Marine Letters, v. 12, p. 33–40.

Anderson, R. N., P. Flemings, S. Losh, J. Austin, and R. Woodhams, 1994, Gulf of Mexico growth fault drilled, seen as oil, gas migration pathway: Oil and Gas Journal, June 6, p. 97–104.

Behrens, E. W., 1988, Geology of a continental slope oil seep, northern Gulf of Mexico: AAPG Bulletin, v. 72, p. 105–114.

Brooks, J. M., M. C. Kennicutt, and B. D. Carey, 1986, Offshore surface geochemical exploration: Oil and Gas Journal, October 20, p. 66–72.

Bugge, T., R. Knarud, and A. Mørk, 1984, Bedrock geology on the mid-Norwegian continental shelf, *in* A. M. Spencer et al., eds., Petroleum geology of the north European margin: Norwegian Petroleum Society, London, Graham and Trotman, p. 271–283.

Cayley, G. T., 1987, Hydrocarbon migration in the Central North Sea, *in* J. Brooks and K. Glennie, eds., Petroleum geology of northwest Europe: London, Graham and Trotman, p. 549–555.

Clarke, R. H., and R. W. Cleverly, 1991, Petroleum seepage and post-accumulation migration, *in* W. A. England and A. J. Fleet, eds., Petroleum Migration: Geologic Society Special Publication 59, p. 265–271.

Clayton, C. J., and S. J. Hay, 1994, Gas migration mechanisms from accumulation to surface, in Proceedings of the 2nd Conference on Gas in Marine Sediments, Bulletin of the Geologic Society of Denmark, v. 41, p. 12–23.

Cohen, M. J., and M. E. Dunn, 1987, The hydrocarbon habitat of the Haltenbank-Trænabank area offshore mid-Norway, *in* J. Brooks and K. Glennie, eds., Petroleum geology of northwest Europe: London, Graham and Trotman, p. 1091–1104.

Cook D., and P. D'Onfro, 1991, Jolliet field thrust fault structure and stratigraphy, Green Canyon Block 184, offshore Louisiana: Gulf Coast Association of Geologic Societies Transactions, v. 41, p. 100–121.

Düppenbecker, S. J., and T. Dodd, 1993, Petroleum charge model for Brent accumulations—application of integrated basin modeling (abs.): Fifth Conference, European Association of Petroleum Geoscientists and Engineers Extended Abstracts, Stavanger, Norway.

Ellenor, D. W., and A. Mozetic, 1986, The Draugen oil discovery, *in* A. M. Spencer et al., eds., Habitat of hydrocarbons on the Norwegian continental shelf: Norwegian Petroleum Society, London, Graham and Trotman, p. 313–316.

Fischer, P. J., and A. J. Stevenson, 1973, Natural hydrocarbon seeps along the northern shelf of the Santa Barbara Basin, California: Offshore Technology Conference 1973 Preprints, v. 1, p. 159–168.

Foster, P. T., and P. R. Rattey, 1993, Evolution of a fractured chalk reservoir: Machar oilfield, UK North Sea, in J. R. Parker, ed., Petroleum geology of northwest Europe: Proceedings of the Fourth Conference, Geological Society of London, p. 1445–1452.

Guliev, I. S., and A. A. Feizullayev, 1994, Natural hydrocarbon seepages in Azerbaijan (abs.), in Near-surface expressions of hydrocarbon migration: AAPG Hedberg Research Conference Abstracts, April 24–28, Vancouver, British Columbia.

Hedberg, H. D., 1980, Methane generation and petroleum migration, in W. H. Roberts III and R. J. Cordell, eds., Problems of petroleum migration: AAPG Studies in Geology No. 10, p. 179–206.

Hitzman, D. C., J. D. Tucker, and J. P. Lopez, 1994a, Hydrocarbon microseepage signatures of seismic structures identified by microbial surveys, sub-Andean region, Bolivia (abs), in Near-surface expressions of hydrocarbon migration: AAPG Hedberg Research Conference Abstracts, April 24–28, Vancouver, British Columbia.

Hitzman, D. C., J. D. Tucker, and P. D. Heppard, 1994b, Offshore Trinidad survey identifies hydrocarbon microseepage: Offshore Technology Conference, May, Houston, Texas, OTC 7378.

Horvitz, L., 1980, Near-surface evidence of hydrocarbon movement from depth, in W. H. Roberts III and R. J. Cordell, eds., Problems of petroleum migration: AAPG Studies in Geology No. 10, p. 241–269.

Hovland, M., and A. G. Judd, 1988, Seabed pockmarks and seepages: impact on geology, biology and the marine environment: London, Graham and Trotman, 293 p.

Hunt, J. M., 1979, Petroleum geochemistry and geology: San Francisco, W. H. Freeman and Company, 617 p.

Kennicutt, M. C., J. M. Brooks, and G. J. Denoux, 1988, Leakage of deep reservoired petroleum to the near surface on the Gulf of Mexico continental slope: Marine Chemistry, v. 24, p. 39–59.

Kennicutt, M. C., T. J. McDonald, P. A. Comet, G. J. Denoux, and J. M. Brooks, 1992, The origins of petroleum in the northern Gulf of Mexico: Geochimica et Cosmochimica Acta, v. 56, p. 1259–1280.

Knight, J. L., P. Dolan, and D. C. Edgar, 1994, Coals on the United Kingdom continental shelf: Geoscientist v. 4, no. 5, p. 13–16.

Leith, T. L., I. Kaarstad, J. Connan, J. Pierron, and G. Caillet, 1993, Cap rock leakage in the Snorre field: Marine and Petroleum Geology, v. 10, p. 29–50.

Link, W. K., 1952, Significance of oil and gas seeps in world oil exploration: AAPG Bulletin, v. 36, p. 1505–1539.

Lovely, D. A., and R. W. Ruggiero, 1994, Seabed analysis of deep-water Gulf of Mexico 3-D seismic data in an effort to better understand hydrocarbon migration (abs.), in Near-surface expressions of hydrocarbon migration: AAPG Hedberg Research Conference Abstracts, April 24–28, Vancouver, British Columbia.

MacDonald, I. R., G. S. Boland, J. S. Baker, J. M. Brooks, M. C. Kennicutt II, and R. R. Bidigare, 1989, Gulf of Mexico hydrocarbon seep communities, II. Spatial distribution of seep organisms and hydrocarbons at Bush Hill: Marine Biology, v. 101, p. 235–247.

MacDonald, I. R., N. L. Guinasso, Jr., S. G. Ackleson, J. F. Amos, R. Duckworth, R. Sassen, and J. M. Brooks, 1993, Natural oil slicks in the Gulf of Mexico visible from space: Journal of Geophysical Research, v. 98, p. 16,351–16,364.

Macgregor, D. S., 1993, Relationships between seepage, tectonics, and subsurface petroleum reserves: Marine and Petroleum Geology, v. 10, p. 606–619.

MacLean, B., R. K. Falconer, and E. M. Levy, 1981, Geologic, geophysical, and chemical evidence for natural seepage of petroleum off the northeast coast of Baffin Island: Canadian Petroleum Geology, v. 29, p. 75–95.

Mann, D. M., and A. S. Mackenzie, 1990, Prediction of pore fluid pressures in sedimentary basins: Marine and Petroleum Geology, v. 7, p. 55–66.

Nash, G. D., and L. Wood, 1994, High-resolution geophysical imaging of hydrocarbon migration patterns and sea floor seepage offshore southeastern Trinidad (abs.), *in* Near-surface expressions of hydrocarbon migration: AAPG Hedberg Research Conference Abstracts, April 24–28, Vancouver, British Columbia.

Pegrum, R. M., and A. M. Spencer, 1990. Hydrocarbon plays in the Northern North Sea, *in* J. Brooks, ed., Classic petroleum provinces: Geological Society of London Special, Publication No. 50, p. 441–470.

Piggott, N., and M. D. Lines, 1991, A case study of migration from the West Canada Basin, *in* W. A. England and A. J. Fleet, eds., Petroleum migration: Geological Society of London, Special Publication No. 59, p. 207–225.

Rattey, R. P., and A. P. Hayward, 1993, Sequence stratigraphy of a failed rift system: the Middle Jurassic to Early Cretaceous basin evolution of the Central and Northern North Sea, *in* J. R. Parker, ed., Petroleum geology of northwest Europe: Proceedings of the Fourth Conference, Geological Society of London, p. 215–249.

Salisbury, R. S. K., 1990, Shallow gas reservoirs and migration paths over a Central North Sea diapir, *in* D. A. Ardus and C. D. Green, eds., Safety in offshore drilling: the role of shallow gas surveys: Underwater Technology, Ocean Science and Offshore Engineering, v. 25, p. 167–180.

Sassen, R., J. M. Brooks, M. C. Kennicutt, I. R. MacDonald, and N. L. Guinasso, 1993, How oil seeps, discoveries relate in deep-water Gulf of Mexico: Oil and Gas Journal, v. 91, no. 16, p. 64–69.

Tinkle, A. R., 1973, Natural gas seeps in the northern Gulf of Mexico: a geological investigation: Master's thesis, Texas A&M University, College Station, Texas.

Vik, E., O. R. Heim, and K. G. Amaliksen, 1992, Leakage from deep reservoirs: possible mechanisms and relationships to shallow gas in the Haltenbanken area, mid-Norwegian shelf (abs.), *in* W. A. England and A. J. Fleet, eds., Petroleum migration: Geologic Society of London, Special Publication No. 59, p. 273.

Williams, A. K., A. Kloster, R. Duckworth, and N. Piggott, 1995, The role of the airborne laser fluorosensor (ALF) and other seepage detection techniques in exploring frontier basins, *in* S. Hanslein, ed., Petroleum exploration and exploitation in Norway: Norwegian Petroleum Society Special Publication 4, p. 421–431.

Wilson, R. D., P. H. Monaghan, A. Osanik, L. C. Price, and M. A. Rogers, 1974, Natural marine oil seepage: Science, v. 184, p. 857–865.

Matthews, M. D., 1996, Importance of sampling design and density in target recognition, *in* D. Schumacher and M. A. Abrams, eds., Hydrocarbon migration and its near-surface expression: AAPG Memoir 66, p. 243–253.

Chapter 18

Importance of Sampling Design and Density in Target Recognition

Martin D. Matthews

Texaco International Exploration Division
Bellaire, Texas, U.S.A.

Abstract

The design and density of surface geochemical sampling programs can significantly influence the interpretability of the survey. Testing hypotheses by purposeful sampling is the most straightforward application of surface geochemistry and the easiest to interpret but requires *a priori* geologic knowledge and provides limited information. Designing a spatial sampling program to produce a map is a much more difficult problem. Two common techniques are line profiles and areal surveys. The interpretation of these techniques, as a function of sampling density, is simulated by examining the relationships of artificial surveys to a known subsurface target. A deterministic model of hydrocarbon migration is used to constrain the randomized mixing of two nonoverlapping populations (anomalous and background). A high-resolution grid of simulated surface measurements is decimated to create lower resolution grids and line profiles. For regional high-grading, a grid with a minimum of two samples across the narrowest expected surface signal (minimum subsurface target width plus a dispersion zone) appears to be adequate. A higher density of four samples is suggested for prospect high-grading. In addition, the sampled area must include sufficient "background" measurements to recognize the existence of an anomaly. About 80% of the samples should be obtained outside the expected area of interest. The most cost-effective sampling program consist of two stages: a low-density regional survey to high-grade the area, followed by a higher density survey within the high-graded area. Undersampling is probably the major cause of ambiguity and interpretation failures involving surface geochemical studies.

INTRODUCTION

Soil gas data, like other natural data, is typically noisy. There is no clear cut spatial or magnitude boundary between anomalously high magnitude sites (above a defined threshold and within a homogeneous compositional class) and lower level (below the threshold) background sites. Methodology to separate background and anomalous sites are covered in Klusman (1993) and Jones et al. (in press). To optimize the recognition of an anomaly (spatial cluster of anomalous sites), while minimizing survey costs, several factors must be considered when planning a survey:

1. The goal of the survey—release of acreage, comparison of two or more separate areas, regional identification of areas of interest, and prospect scale evaluation

2. The size and shape of the expected anomaly—typically assumed to be proportional to the size and shape of the subsurface target (*target* is defined as the subsurface feature that gives rise to a surface anomaly)—and the geologically expected relationship between the two (directly over the target or displaced by dipping faults or stratigraphy)

3. The expected natural variation in surface measurements—both random and those from known geologic factors, such as identified for macroseepage by Link, 1952)

4. The magnitude of the expected signal-to-background ratio.

It is important to recognize that defining background values adequately is required before anomalies can be properly identified. Estimates of background values and signal-to-background ratios are best achieved by pilot studies over nearby producing fields or by a combination of experience and small preliminary surveys in the study area if known producing fields are unavailable. Both pilot and preliminary surveys should include estimates of surface hydrocarbon magnitudes, effect of sample depth, and optimal sample spacing. In practice, these studies are seldom done, except for occasional calibration studies over existing production, performed at the same time as the survey to minimize costs.

It is beyond the scope of this paper to discuss the statistical methods of selecting optimal sample spacing and depth, methods of choosing a threshold between background and anomalously high magnitude measurements, or statistical tests. The reader is referred to several texts covering these techniques, such as Dixon and Massey (1957) or Krumbein and Graybill (1965). Discussions of the use of variograms to estimate the effects of sample spacing by Davis (1973) or Journel and Huijbregts (1978) are recommended. Restricting sampling programs to the vicinity of the expected anomaly is a false economy that is common to the design of most surface sampling programs. It is difficult to characterize an elephant if you only sample its side or leg and never find its limits or other characteristics. Experience at Gulf Research and Development Company suggests that, for each sample taken within an expected anomalous area, more than five times as many samples should be taken in the surrounding area.

The purpose of this paper is to examine the consequences of interpretions made on the basis of sample design and density. Several categories of sampling designs are briefly reviewed, followed by a simulation of the effects of sampling density on line profile and grid designs. This is accomplished through the use of a known subsurface target, defined anomalous and background populations, and a defined relationship between target and seepage measurements at the surface. The method of obtaining these relationships is explained, and the consequences of both line and grid sampling designs are illustrated through simulations of sampling results obtained at different sampling densities along line profiles and regular grids. The effect of both noise and variation in sampling density are considered and related to the ability to define a "known" surface anomaly. The results of one interpretation scheme are then summarized and a cost-benefit analysis presented. Readers are encouraged to examine these simulations using their own style of interpretation. This should be done with respect to the technique's ability to predict the known subsurface target, remembering that it is easier to find something when you know what it looks like than when its shape, size, and location are unknown.

SAMPLING DESIGNS

The goal of the survey determines the nature of the sampling plan. Purposeful selection of sampling sites is a traditional choice of geologists; it maximizes information gained per unit cost and has great use in surface geochemical surveys in circumstances that I group under the umbrella of hypothesis testing. Marine bottom sampling surveys often use purposeful selection of sampling sites due to the high cost of keeping the ship on station during sampling. Typical sample sites include faults, mud volcanoes, sea bottom pockmarks, shallow seismic bright spots, seismically detected gas chimneys, and bubble trains in the water column. Under these circumstances, one is generally trying to determine if hydrocarbon seepage is associated with this feature and what the compositions of any subsurface hydrocarbons are. Prior knowledge of the area is required to properly select the sample locations. Extra samples are generally needed because of the uncertainty between the sampling location determined on a map and the actual position obtained in the field.

It cannot, however, be emphasized enough that the information gained by purposeful sampling pertains only to that particular sample site and should not be used to infer anything about areas not sampled. For instance, if a fault is sampled by taking a sample at three widely separated locations in the vicinity of the fault trace, and no seepage was found, it would be incorrect to say that the fault does not leak. It is possible that at some sample locations the fault was missed and at the unsampled locations seepage was occurring. Seepage is known to occur in spots along fault traces (Wakita, 1978; Preston, 1980) rather than uniformly along their length. A better approach would be to take several samples in a cluster (to minimize the risk of missing the fault's surface trace) at several locations along a fault where reservoir leak points would be expected to occur, such as near the crest and expected spill points of the trap. Then, if no seepage was found, it would be more reasonable to infer that the fault was not leaking hydrocarbons, although nothing could be inferred about other unsampled areas.

Random sampling of an area is particularly useful if one wishes only to detect the presence or absence of thermogenic hydrocarbons or to obtain compositional information. Random sampling does not ensure uniform coverage of an area. Indeed, it usually results in spatial clustering (Krumbein and Graybill, 1965). This type of sampling is appropriate when spatial information is unimportant to answering the questions posed and the area is considered as a single homogeneous population. Random sampling is also useful in obtaining unbiased estimates of two areas for comparison purposes. This is particularly useful in comparing two basins, two parts of a basin, or two prospective areas. The error of estimating the population mean is directly proportional to the variation of the population and inversely proportional to the square root of the number of samples taken (Dixon and Massey, 1957). Thus, a larger number of samples are needed for an area with a greater range of magnitudes than for one where all the values are similar.

Most land hydrocarbon surveys and marine sniffer or airborne surveys are generally concerned with high-grading selected areas within a larger region. This requires that spatial variation of hydrocarbon seepage be mapped in adequate detail to unambiguously resolve features that are large enough to be considered interesting (targets) from the natural background variation and smaller features. Two sampling techniques are commonly used: *line profiles* (one dimensional) and *grids* (two dimensional).

Designing a line profile sampling program is relatively straightforward. Once the number, orientation, and location of the lines are selected, the remaining parameters to be chosen are the method of sampling, the depth or depths of sampling, and the sample spacing along the line. Information gathered along a line is, strictly speaking, only useful for inferences in behavior along that line because it does not provide an unbiased sample of the area. This sampling design is typical of offshore surveys by sniffers or onshore in seismic shot holes. An array of several line surveys is transitional between one- and two-dimensional information and is often presented as a map. If this is done, care should be exercised in forming trends between lines because of the bias introduced by the line locations.

When two-dimensional information is needed, some type of gridded or stratified survey is preferred to ensure an approximately equal distribution of points. For many people, the word *grid* brings up the image of a regular, square grid, such as the squares on a checkerboard. There are, however, many types of grid designs (Yates, 1960; Krumbein and Graybill, 1965). Within each grid element, a fixed number of samples is taken. These can be chosen by a variety of methods, such as regular positions (corners or centers), random positions (using a random number generator), haphazard (chosen for convenience), or even purposeful selection.

In practice, I prefer a combination of sampling methods. Within each grid element, two or more sample locations are chosen, with the first one purposefully located (if there is a reason to do so). The others are chosen randomly or to achieve a combination of (1) no large internal unsampled areas, (2) ease of access, and (3) close proximity to other samples to enable estimation of short-distance variability and the potential of aliasing. It should be remembered, however, that as the sampling program departs from unbiased selection criteria, statistical inferences become weaker. Included within each survey should be a few replicate samples to estimate sampling repeatability. I believe the majority of surface seepage studies that have failed or yielded ambiguous results have done so because of undersampling. This results in aliasing the targets and the inability to resolve a target from natural variations in background noise and stray signals.

Surface seep detection is considered to be an inexpensive technique, but in our zeal to keep it inexpensive, it has been made "cheap" and does not always provide the data necessary to evaluate its usefulness properly. By analogy, the cost of acquiring and processing seismic data can be significantly reduced if only single-fold data are gathered. The redundancy in multifold data reduces noise, justifying the additional cost.

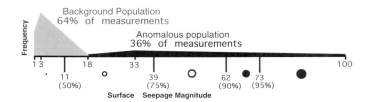

Figure 1—Background and anomalous populations used in Figures 5 through 8. The anomalous population is bounded between 18 and 100, with a mode at 33 and comprises about 36% of the total population. The background population comprises the remaining 64% and is bounded between 1 and 18 with a mode at 3. Symbols for magnitude classes are used in Figures 6a–d.

I believe similar results can be achieved in surface geochemical surveys by increasing the spatial density of sampling and looking for spatial clustering of high-magnitude sites of homogeneous composition. This can be achieved by visual inspection of the data and application of a variety of spatial statistical techniques, such as the one used by Dickinson and Matthews (1993) or those summarized in Davis (1973). The increased ability to interpret geochemical survey data resulting from increased sampling density is shown by Burtell et al. (1986). To quantify the effect of both sample design and density, the following simulation technique was used.

DESIGN OF SIMULATION

Traditionally, surface geochemical measurements from a survey are treated as either a log normal distribution (with part of the high-magnitude tail of the population defined as anomalous) or as two or more normal populations with the lowest magnitude population considered as background and the highest as anomalous. Whether there are, in reality, multiple populations or a continuous multicompositional population is an academic question. In practice, a threshold value is selected for each compositional class and magnitudes above this threshold are considered to be anomalous and values below are considered to be background. Dividing a distribution of seepage magnitudes into these two populations is commonly achieved by the selection of a cut-off value, often coincident with a natural break or frequency minimum in the measurements as determined by inspection of a histogram, plotting on probability paper (Krumbein and Pettijohn, 1938; Harding, 1949), or other statistical tests (Court, 1949; Dixon and Massey, 1957).

For simplicity in these simulations, I have defined two nonoverlapping populations: (1) an anomalous population representative of the target and (2) a background population representative of the area in general. These two populations are shown in Figure 1. In these examples, the anomalous population is bound between measured values of 18 and 100 (with a mode at 33) and com-

Table 1—Division of Population into Five Classes

Class	Range of Values	Percentage of Population
1	1–11	Lower 50%
2	11–39	50–75%
3	39–62	75–90%
4	62–73	90–95%
5	73–100	Upper 5%

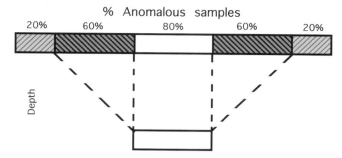

Target at one target width depth

Figure 2—Relationship of anomalous surface data to the subsurface target used in Figures 5 through 8. In the heart (directly over the target), 80% of the data is defined as belonging to the anomalous population. In the fringe (extending outward from the heart a distance equal to the vertical distance from the target to the surface), 60% is defined as belonging to the anomalous population. In the background region (at the greatest distance from the heart), 20% belongs to the anomalous population.

prises about 36% of the total population. The background population comprises the remaining 64% of the total population and is bounded between values of 1 and 18 (with a mode at 3).

Two techniques have been used for purposes of display. In the line profile examples, the data have been presented as a continuously varying bar. In the grid examples, for ease of recognition, the continuous population has been divided into five classes, shown in Table 1. The lowest class is entirely within the background class, while the upper three classes are entirely within the anomalous population. The class with values between 11 and 39 comprises 25% of the population and contains samples from both background and anomalous populations. This class thus represents an uncertainty in defining the separation point of the two anomalous populations. This is commonly encountered in the real world, where the boundary between the two populations is unknown and subject to finite and biased sampling.

The spatial distribution of these two populations, and therefore the five classes, is defined by artificially dividing the sampled area into three zones (Figure 2). Those samples in the heart (directly over the subsurface target) are defined to have an anomalous population percentage of 80%. Those samples in the fringe (extending outward from the heart a distance equal to the vertical distance from the target to the surface) are defined to have an anomalous population percentage of 60%. Those samples in the background region (at the greatest distance from the heart) are defined to have an anomalous population percentage of 20%.

Surface geochemical data are notoriously noisy, with large magnitude changes in concentration over short distances. I believe much of this variation is due to undersampling. The nature of the migration process (Matthews, this volume) is also responsible for some of the noise, as illustrated in Figure 3 (Matthews et al., 1984). In these simulations, I have assumed homogeneous strata from the subsurface target to the surface with randomly scattered fractures that act as high-permeability pathways. The frequency of occurrence of anomalous or background populations in each of the three zones is therefore controlled by the connection of these fractures to a subsurface occurrence of hydrocarbons and the location of a sampling point relative to these fractures. The relationship in the heart zone is based on the assumption that migration of hydrocarbons is dominantly vertical. The greatest concentration of subsurface hydrocarbons is

within the target and therefore the greatest concentration of anomalous seepage is directly over the target.

The relationship in the fringe zone is based on the assumption that migration of hydrocarbons from the target also moves outward in a zone no wider than 45° off vertical from the edge of the target to the surface. The dispersion of signal and decrease in frequency of occurrence is caused by a mechanical dispersion process as the hydrocarbons move through the pore network as a separate phase (Matthews, this volume). This angle is chosen because it is halfway between dominantly vertical and dominantly horizontal migration. The background zone is dominated by hydrocarbons in low concentrations derived from biogenic activity and solution-diffusion transport in the water column (Matthews, this volume). The occurrences of samples from the anomalous population are assumed to have originated directly from the source rock or from subeconomic accumulations between the source and the surface.

Under these assumptions, single points with anomalously high concentrations of surface hydrocarbons, such as those that occur in the background zone, are not of spatial interest. Although they do support the existence of a subsurface hydrocarbon source, there is a significant chance that they are not associated with the tail end of a major migration route. However, a large number of adjacent sites with anomalously large values are of interest. The real question is "How are numbers of adjacent sites between these two extremes classified?" Unfortunately, there are no fixed rules to aid in this decision process. In the interpretation section, an oversimplified rule-based decision process is used for illustrative purposes. In an interpretation of a real survey, the relationships that are defined for this simulation are unknown. They are, however, inferred from an understanding of the subsurface

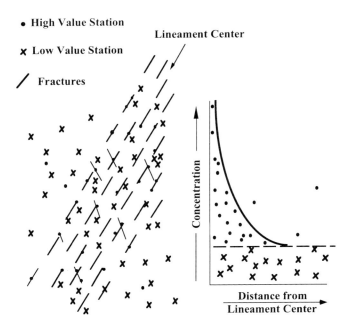

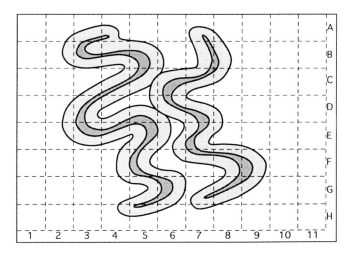

Figure 4—A "mask" showing the location of two hypothetical meandering river deposits superimposed on the sampling design. The idealized heart (dark gray) and fringe (light gray) areas were used to generate Figures 5 through 8.

Figure 3—In the seepage model, seepage is focused along preferential pathways, shown here as fractures associated with a lineament zone. The high-magnitude sites are preferentially associated with the lineament zone. Note that even inside the lineament zone, the samples do not always intersect fractures.

geology of the area, compositional information, and appropriate analog surveys over known accumulations.

It should be emphasized that this is only one of several relationships between the target and the surface pattern. It is a simple one, constructed to allow the interpreter to focus on the effect of sample spacing. The real world is much more complex and variable. One of the significant challenges in interpretation of surface geochemical data is understanding the relationship of its pattern to the subsurface.

THE MODEL

In the simulations that follow, I have defined the existence of hydrocarbon accumulations in two meandering river channels. Therefore, the patterns of anomalous samples are expected to be longer than they are wide, to have a preferred trend with some variation about this trend, and to have variable width. The sands are defined to be as deep as their average width, causing the fringe zone around the main trend to be approximately as wide as the target itself. There is no implied scale to the simulations. The relationships are entirely dependent on the sampling density per minimum anomaly width and the continuity of the anomaly along is greatest length.

A regular grid of simulated sampling locations was chosen to be 81 samples wide and 61 samples tall (giving

a total of 4941 locations). A mask showing the location of the two idealized river channels and their representative fringe areas was superimposed on the sampling design to determine which stations would fall into background, fringe, and heart areas (Figure 4). Within each area, stations were chosen randomly and assigned to either anomalous or nonanomalous populations in the appropriate proportions (20%, 60%, and 80% anomalous, respectively) and to which of the five magnitude classes they belonged. Once this gridded data set had been created, it was decimated to create the various line and grid sampling patterns.

Reference cells (A–H, 1–11) have been provided to facilitate comparison between the target locations and simulations of the various surface expressions. Note that the widest heart area is one-half the width of a reference cell and the width of the fringe plus heart is about 1–1.5 cells. Note also that even longer adjacent line segments encompass both heart and fringe areas and are dependent on the angular relationship of the sampling program with the orientation of the target. Since interpretation is a personal task, the reader is encouraged to interpret the simulations shown in Figures 5 and 6 and compare their interpretation to Figure 4.

Line Profiles

Line profiles, which have been used for many years, offer a compromise between cost and density of information. This can be a real advantage in regional investigations or if the target's shape and orientation are known. This sampling design is particularly cost-effective for any technique that is approximately continuous, such as seismic profiles or marine sniffers. Figures 5a–d have a constant line separation of one cell and increase from a den-

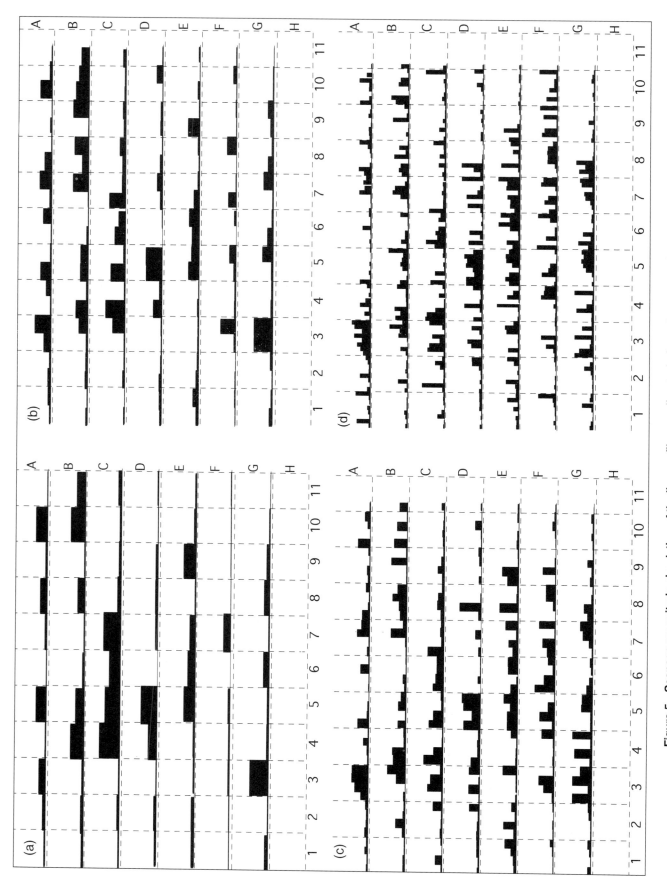

Figure 5—Seepage magnitude simulation of the line profile sampling technique shown at various sampling densities: (a) one sample per cell, (b) two per cell, (c) four per cell, and (d) eight per cell.

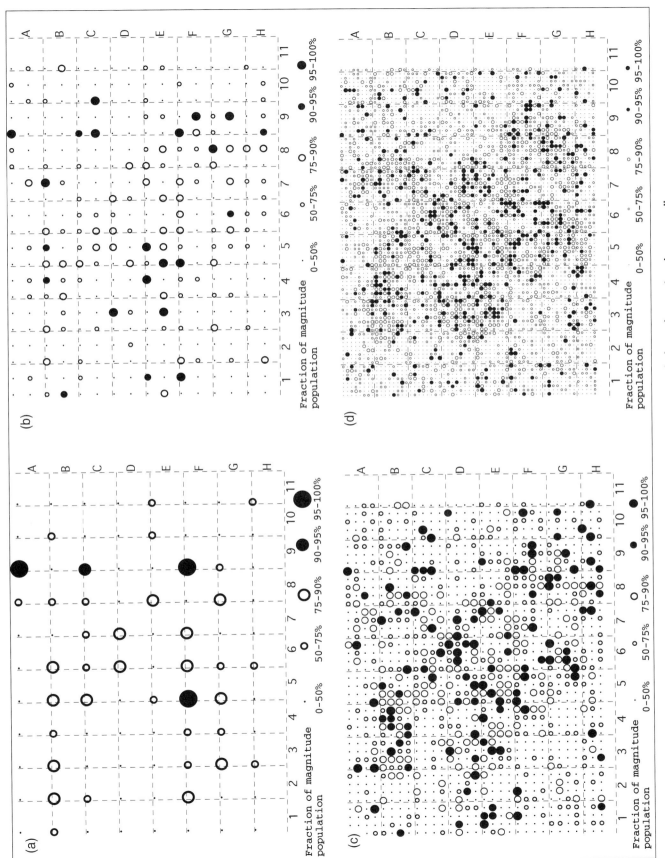

Figure 6—Seepage magnitude simulation of the grid sampling technique shown at various sampling densities: (a) one sample per cell, (b) four per cell, (c) sixteen per cell, and (d) sixty-four per cell.

Figure 7—Four-point running average of the seepage magnitude simulation in Figure 5c.

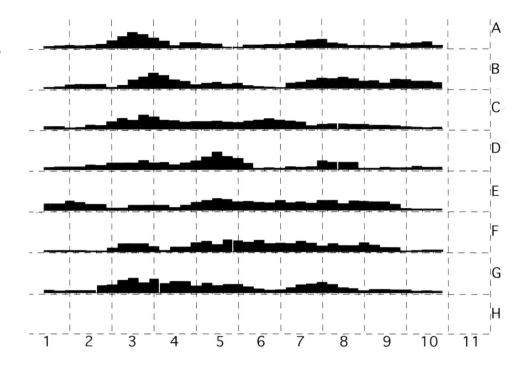

sity along the line of one station per cell to a maximum of eight per cell. In Figure 5a, it is relatively easy to pick out major trends by emphasizing clusters of high-magnitude values. There are few samples, and most of the statistical chance of identifying the background area works for you. Hence, it should be easy to divide the region into areas of potential interest and no interest.

However, there is a high degree of aliasing of information, as can be seen by a comparison with Figure 5d. Figure 5a uses the first of the eight samples in each cell shown in Figure 5d. If the second or any other of the eight samples had been chosen, the results would have been quite different. Because there is a high chance of not identifying some of the potential target area, some opportunities may not be recognized. If individual high values are identified as being of interest, the risk of selecting background locations as areas of interest is increased. As the density along the lines increases (Figures 5b–d), decisions become more important. The chance of spurious high values increases, while the clustering tends to break into smaller areas. The chance of aliasing along the line decreases, but the aliasing between lines remains constant because the line spacing has not changed. It becomes increasingly difficult to correlate the smaller areas of interest detectable along the lines across the much greater distances between lines. It is difficult to identify the trend because of this spatial bias, except as a broad generality.

Grid Surveys

Grid designs keep the spatial density of sampling approximately constant. This is a significant advantage over line profiles in preparing maps. In Figures 6a–d, sample density increases from a spacing of one sample

per cell to sixty-four samples per cell. Sample density in Figure 6a is identical to that of Figure 5a. The only difference is that the first row of data in each cell was used instead of the last, and the magnitude information has been grouped into the five classes. All discussion of Figure 5a applies equally well to Figure 6a. The effect of aliasing can be seen by comparison with Figure 6d. As the sampling density increases (Figures 6b–d), decisions again become more important because as the aliasing decreases, the visual appearance of the sampling noise increases. The tendency to focus on the individual rather than the larger pattern can be overcome by defocusing your eyes while looking at Figure 6d or by looking at it from a distance. A good relationship can be seen with the heart and fringe area of Figure 4. Simulation of higher density grids does not increase the relationship but simply reaffirms the assumed statistical chances that anomalous values correspond to the heart, fringe, and background areas.

Comparing Line and Grid Surveys

Because Figures 5c and 6b both have the same sampling density, they can be directly compared to determine the information gained for similar costs. The effect of smoothing the data by calculating a four-station running average is shown in Figures 7 and 8. This was accomplished in discrete steps for the line profiles that are shown at the same scale as Figure 5c. The gridded data were averaged by a running box and rescaled into five classes with the same percentage values previously discussed. Note that this spatial stacking suppresses the magnitude of isolated stray sites (see cell D3 in Figures 6b and 8), reinforces clustered high values (cell E5), and

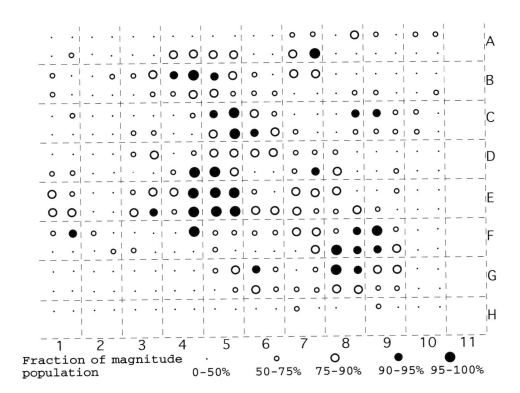

Figure 8—Four-point running average of the seepage magnitude simulation in Figure 6b.

Fraction of magnitude population

·	○	O	●	●
0-50%	50-75%	75-90%	90-95%	95-100%

diminishes but spreads out smaller isolated clusters of high values (cell C9). The readers are encouraged to draw their own conclusions about the relative values of the information derived from Figures 5–8. My personal preference is to examine a combination of raw and smoothed gridded data and to use an objective statistical test such as the technique described by Dickinson and Matthews (1993).

REGIONAL AND PROSPECT INTERPRETATION

Although the readers are encouraged to interpret the figures for themselves, the following observations are offered as a guide to a numerical comparison of interpreting the different sampling densities. The interpretation style chosen for this comparison is a simple rule-based technique. Sampled areas are considered to be homogeneous squares extending from a single data point located in the upper left-hand corner to the next data point to the right and down. For the purpose of regional high-grading, anomalous areas are those that contain a sample in the upper 50% of the measurements and are adjacent to at least four other such samples. The measure of efficiency in this paper is the percentage of total target retained in the high-graded region. For the purpose of prospect high-grading, anomalous areas are those that contain a sample in the upper 25% of the measurements and are adjacent to at least four other such samples. The percentage of the anomalous areas occupied by target is

calculated and used as an efficiency measure. Different interpretation schemes will generate different but often not too dissimilar results.

The results of applying this technique to regional high-grading of the grid data are shown in Figure 9. The percentage of the total target area included in the high-graded region is maximized at slightly above four samples per cell. These values can be compared to an expected 4.5% value achieved by a random selection of 50% of the total area (about one-half the 8.9% of the target area). Actual high-graded areas were 40–43% of the total area for a sampling density of greater than four per cell and 48% for a sampling density of one per cell.

The results of applying this technique to high-grade prospect areas using the grid survey data are shown in Figure 10. The efficiency percentages on the graph are equivalent to the probabilities of being vertically over a portion of the target. They can be compared to the probability of hitting a portion of the target by random drilling in the entire area (8.9%), and the probability of hitting the target by confining random drilling to only the heart plus fringe areas (24.8%). Note that the efficiency of the survey, as defined by this measure, increases rapidly to a density of sixteen samples per cell and then increases much more slowly as the sampling density increases. This density ensures that four or more adjacent samples will be taken through the heart and fringe areas in any grid direction. The continuity along the length of the heart and fringe areas builds spatial sample correlation, while the requirement that four or more adjacent values must occur in the upper 25% of the population minimizes the selection of areas in the background region.

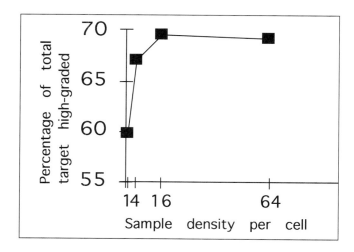

Figure 9—Percentage of the target that is high-graded by regional sampling using a grid survey.

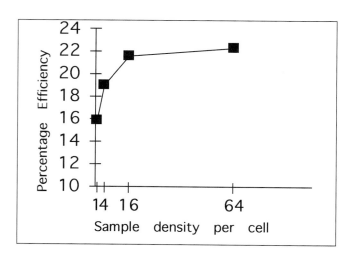

Figure 10—Efficiency (in %) of the target definition using a grid survey.

COST–BENEFIT ANALYSIS

One of the premises of this study is that taking too few samples is a false economy. To judge this, it is necessary to compare the expected direct cost of a survey to an estimate of the total exploration costs that could be saved by increasing the sampling. The costs used are only an example of one way to evaluate this decision. The comparison requires a measure of scale. Assuming the target occurs at a depth of 1.6 km (1 mi), the width of the fringe area (about one-half a cell width) is 1.6 km on each side of the heart of the anomaly. Thus, each cell is about 3.2 km (2 mi) wide. For purposes of discussion, I use a cost of $100 per station as the cost of a surface geochemical sample.

As an example of the cost versus benefit for a regional survey, consider the cost of increasing the sampling density from one to four samples per cell. For the surveys shown in Figures 6a and b, this would increase the number of samples from 88 to 336 and increase costs by $24,800. The cost of seismic data and processing is estimated at $3750/km ($6000/mi). A conservative regional program for the surveyed area (25.7 km north-south and 32 km east-west) would be three east-west regional lines, each 32 km long (ignoring the extra data needed to attain full stack over the entire length of the line). The total cost of this survey would thus be $360,000. The elimination of 6.7 km (4.2 mi) from this regional program (7%) would pay for the survey. If the regional seismic lines can be optimally placed by a surface geochemical survey to provide needed information, the total cost of the exploration program can be reduced.

As an example of the cost versus benefit for a program of several wildcat wells, consider the cost of increasing the sampling density on each prospect from four to sixteen samples per cell. For the surveys shown in Figures 6b and c, this would increase the number of samples from 336 to 1312 and increase costs by $97,600. The dry hole cost of drilling each well to 1.6 km is estimated at

$250,000. Thus the break-even point for a wildcat program would be the elimination of 0.4 dry hole.

The greatest improvement in cost effectiveness would be achieved by a two-stage surface geochemical sampling program. For the 336 mi² area described above, this would entail an initial low-density regional survey at four samples per cell (one per square mile) at a cost of $33,600, followed by a prospect-scale survey of sixteen samples per cell over the 168-mi² high-graded area at a cost of $67,200. The total cost of this two-stage survey would be $100,800. The break-even point would be the cost equivalent of either 0.4 dry hole or 17 mi of seismic, or some combination of these.

SUMMARY

True comparisons of the effectiveness of particular sampling designs or densities can only be achieved if the answer is known without error. In the real world, there are too many uncontrolled variables, such as the permeability structure of the earth, the location of the reservoir, the action of bacteria, the location of the source rock, the effects of groundwater, and other factors. In this paper, the comparison is achieved by creating a known, simulated target and its constructed relationship to a survey whose results are determined by fixed probabilities of results. It should be emphasized that the measurement patterns shown here are not only a function of target shape and sampling design. They are also dependent upon the assumed model of seepage (vertical with a 45° fringe area), the probability of anomalous and background sample occurrences (80%, 60%, and 20% in the heart, fringe, and background, respectively), the range of values within the two populations, and the break points chosen for plotting purposes. Real-world measurements also include compositional information that can be helpful in identifying populations.

The effect of assumptions about the existence of a fringe zone on the interpretation of these simulations deserves special consideration. The proportion of anomalies assigned to the fringe zone is much closer to that assigned to the heart than it is to the background regions. This effectively enlarges the areal extent of the interpreted anomalous region, increasing the chance of the target being included within the regional interpretation scheme but simultaneously reducing the percentage of prospect-scale anomaly that is underlain by target. The larger the fringe zone, the lower the sample density needed for a regional survey, but the more difficult it becomes to locate the target within this zone. Conversely, a smaller fringe zone requires a denser sample spacing for the regional survey but results in a higher probability of locating the target within it.

Sampling designs for surface geochemical surveys must be designed to meet the goal of the survey. Is the goal to determine (1) if there is an active hydrocarbon system, (2) the type of hydrocarbons present, if any, (3) regions of general interest for further work, or (4) prospect definition? Because interpretation of data involves many decisions and choices, the readers will ultimately be required to draw their own conclusions and weigh the cost versus benefit of interpretation derived from the two designs and the different densities.

The most straightforward application of surface geochemistry, and the easiest to interpret, involves hypothesis testing using purposeful sampling. Spatial high-grading of areas within a larger region is more difficult. On the basis of simulations presented here, I believe the sampling method that is most interpretable with respect to defining the original target is the highest density grid (sixty-four samples per cell, Figure 6d). It seems, however, that a gridded sampling density of four samples per cell could provide sufficient regional resolution. A density of sixteen samples per cell would provide sufficient prospect resolution. Interpretation of a regional survey, followed by detailed sampling in the high-graded region, is the most cost-effective technique. If a different model was constructed, with a different fringe zone, proportions of anomalous and background populations, and continuity of target, the details of the results would certainly be different. However, the following basic principles should still hold: (1) at least two samples per line within the heart and fringe zone for regional high-grading, (2) a minimum of four samples per expected target width for prospect-scale surveys, and (3) sufficient samples in the background region to define the areas of interest (about five times as many as within the area of interest).

REFERENCES CITED

Burtell, S. G., V. T. Jones, M. D. Matthews, R. A. Hodgson, O. K. Okada, T. Ohhashi, M. Kuniyasu, T. Ando, and J. Komai, 1986, Remote sensing and surface geochemical study of Railroad Valley, Nye County, Nevada—detailed grid study: Proceedings of the Fifth Thematic Conference on Remote Sensing for Exploration Geology, ERIM, v. 2, p. 745–759.

Court, A., 1949, Separating frequency distributions into two normal populations: Science, v. 110, p. 500–501.

Davis, J. C., 1973, Statistics and data analysis in geology: New York, John Wiley and Sons, 550 p.

Dickinson, R. G., and M. D. Matthews, 1993, Regional microseep survey of part of the productive Wyoming–Utah thrust belt: AAPG Bulletin, v. 77, p. 1710–1722.

Dixon, W. J., and F. J. Massey, Jr., 1957, Introduction to statistical analysis: New York, McGraw-Hill, 488 p.

Harding, J. P., 1949, The use of probability paper for the graphical analysis of polymodal frequency distributions: Journal of the Marine Biological Association, v. 28, p. 141–153.

Jones, V. T., M. D. Matthews, and D. M. Richers, in press, Light hydrocarbons in petroleum and natural gas exploration, *in* M. Hale, ed., Handbook of exploration geochemistry: Gas Geochemistry, Amsterdam, Elsevier.

Journel, A. G., and C. H. Huijbregts, 1978, Mining geostatistics: New York, Academic Press, 600 p.

Krumbein, W. C., and F. A. Graybill, 1965, An introduction to statistical models in geology: New York, McGraw-Hill, 475 p.

Krumbein, W. C., and F. J. Pettijohn, 1938, Manual of sedimentary petrology: New York, D. Appelton-Century Co., 549 p.

Link, W. K., 1952, Significance of oil and gas seeps in world oil exploration: AAPG Bulletin, v. 36, p. 1505–1541.

Matthews, M. D., R. G. Jones, and D. M. Richers, 1984, Remote sensing and surface hydrocarbon leakage: Proceedings of the Third Thematic Conference on Remote Sensing and the Environment, Colorado Springs, April 16–19, p. 663–670.

Preston, D., 1980, Gas eruptions taper off in northwest Oklahoma: Geotimes, October, p. 18–20.

Wakita, H., 1978, Helium spots: caused by diapiric magma from the upper mantle: Science, v. 200, April 28, p. 430–432.

Yates, F., 1960, Sampling methods for censuses and surveys: London, Charles Griffin and Co., 440 p.

Holysh, S., and J. Tóth, 1996, Flow of formation waters: likely cause of poor definition of soil gas anomalies over oil fields in east-central Alberta, in D. Schumacher
and M. A. Abrams, eds., Hydrocarbon migration and its near-surface expression:
AAPG Memoir 66, p. 255–277.

Chapter 19

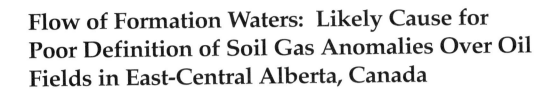

Flow of Formation Waters: Likely Cause for Poor Definition of Soil Gas Anomalies Over Oil Fields in East-Central Alberta, Canada

Stephan Holysh

Regional Municipality of Halton
Oakville, Ontario, Canada

József Tóth

Department of Earth and Atmospheric Sciences
University of Alberta
Edmonton, Alberta, Canada

Abstract

Anomalous amounts of hydrocarbon gases in soil pores have long been used in search of commercial accumulations of petroleum by surface geochemical explorationists. Sufficient new field discoveries are made on the
basis of soil gas surveys to maintain interest, but the results are too inconsistent to warrant routine use of the
method. We hypothesize that the main reason for the unconvincing track record of soil gas surveys and of other
surface geochemical exploration techniques is a poor understanding of the mechanisms and pathways of hydrocarbon migration from petroleum deposits to the surface. Thus, these techniques are often used when their
application is inappropriate.

The intensity and pathway geometry of the hydrocarbon vertical leakage is strongly affected by groundwater
flow between a deposit and the surface. To test this hypothesis, we conducted a comprehensive study of the
dynamics, major ion chemistry, halogen ratios, $\delta^{18}O$ and δ^2H, and ^{14}C ages of groundwaters; the distribution of
groundwater recharge and discharge areas; and a survey and MS analysis of hydrocarbon components in soil
gas in an area with known oil fields in the Lower Cretaceous Mannville Group in east-central Alberta. We found
that the groundwater flow pattern is regionally downward and into the underpressured Mannville reservoirs.
Shallow flow systems are associated with local topographic highs and depressions. Soil gas anomalies, shown
best by benzene and toluene, were observed but are ill defined. Their geographic association seems to be with
local groundwater discharge areas rather than with subsurface petroleum accumulations.

Our interpretation is that descending groundwater flow (underpressuring of the reservoirs) prevents the formation of well-defined soil gas anomalies, regardless of the mechanism of migration from pool to surface. A
slight increase in soil gas signatures in local discharge areas is probably due to lateral transport of gas by shallow groundwater flow. Our findings provide an explanation for the unsuccessful attempts to apply surface geochemical exploration in Alberta, where most of the near-surface petroleum fields are strongly underpressured,
thus inducing downward flow of groundwaters and obliterating near-surface geochemical signatures.

INTRODUCTION

Anomalous amounts of hydrocarbon gases in pores of
the soil have long been considered as possible indicators
of deep-seated petroleum deposits and, consequently,
have been used as a prospector's guide for more than 60
years (Laubmeyer, 1933; Sokolov, 1933, 1936). The rationale for this approach to exploration is based on the
hypothesis of *vertical migration* or *vertical leakage* of hydrocarbons and is considered to be a logical extension of our
knowledge of macroseepages of petroleum hydrocarbons. Indeed, direct observation of visible gas and oil

seeps has resulted in the discovery of a significant proportion of all worldwide oil and gas fields (Debnam,
1969). Thus, acceptance of the possibility of microseepages of petroleum hydrocarbons should readily follow.

According to the hypothesis, gaseous, but also liquid,
petroleum may escape from accumulations through
leaky cap rocks and, driven by diffusion, buoyancy, or
pore pressure gradients, ascend to the land surface as a
column of finely dispersed hydrocarbons. The possibility
of lateral channeling by dipping faults and fracture zones
is also occasionally recognized (Jones and Drozd, 1983;
Price, 1980, 1986). In its path, the rising organic matter

causes chemical reduction of the subsurface environment. In turn, the reduced conditions result in a wide array of physical, chemical, mineralogical, botanical, microbiological, and other phenomena which collectively form the *geochemical chimney*. This is a vertical columnar space in the subsurface characterized by anomalous physical, chemical, and other observable properties (Pirson, 1969; Duchscherer, 1981). The detection, analysis, and interpretation of these anomalies constitute the basis of a theoretically powerful branch of petroleum prospecting, namely, surface geochemical exploration.

Notwithstanding the conceptual plausibility of the vertical migration hypothesis, the advanced levels of detection and analytical techniques, and the millions of dollars spent annually worldwide on geochemical exploration and on soil gas surveys in particular, the results are inconsistent and therefore unconvincing, in terms of new field discoveries. Many dry holes are drilled on anomalies and many fields are found without associated anomalies. This unsettling state of affairs is clearly reflected by statements made in most discussions on the topic. For example, suffice only to quote two prominent specialists of surface geochemical exploration writing in the *American Petroleum Geochemical Explorationists Newsletter*. Referring to the numerous prospecting techniques on the market and their competing proponents, Schumacher (1990, p. 1) asks, "Is it any wonder explorationists are confused or at least skeptical?" The President of APGE, A. H. Wadsworth, Jr. (1991, p. 1) summarizes his impressions from geologists and geophysicists in the United States by saying, "I am struck by their lack of understanding and, therefore, their lack of acceptance of the concept of vertical migration. In my opinion, this is a primary cause for the dismay we all feel about the acceptance of surface prospecting by the exploration community."

In our view, the exploration results of the surface geochemical approach are inconsistent because the effects of groundwater flow on the migration trajectories of hydrocarbons between the deposits and the land surface are ignored. Groundwater flow is the most and, in most cases, the only effective transport mechanism in the subsurface. Its direction is vertical and upward only in relatively rare special situations. In most other cases, it tends to deflect, suppress, or negate the fluxes of hydrocarbons even if they should be driven by buoyancy or diffusion toward the land surface. In such cases, anomalies are either displaced laterally, possibly over sufficiently large distances to lose their value as indicator signatures for the deposit that is their source, or they do not develop at all.

Warnings of possibly deleterious effects of groundwater flow on the development of geochemical signatures, including soil gas anomalies, and on the predictability of their positions with respect to petroleum accumulations have been voiced by several authors, starting with Pirson (1946). Later, in their comprehensive review of the topic, Philp and Crisp (1982, p. 1) put the cause of the problem succinctly: "Failures in prospecting to date are attributable to the simplistic manner in which

data have been interpreted; insufficient attention has been paid to the hydrological and geologic factors which modify the upward migration of indicator species to the surface." Their proposed solution is also stated explicitly (p. 4): "Any application of geochemical prospecting methods to the search for oil or gas therefore needs to study the effects of groundwater movement upon the migration of gases."

The purpose of the present study was to test our working hypothesis, which postulates that the results of exploration for petroleum by surface geochemical techniques, in general, and by soil gas surveys, in particular, are inconsistent because the effects of groundwater flow are ignored. To this end, a field-based study was initiated in a petroliferous area of the Western Canada Sedimentary Basin. The study sought to discern the functional interrelationships of four aspects of the region: (1) geologic setting, (2) oil field locations, (3) groundwater flow, and (4) soil gas anomalies. It was conducted during 1986–1989 as a Master of Science research project at the Geology Department of the University of Alberta, Canada.

STUDY AREA

The area selected is near the town of Chauvin in east-central Alberta, Canada (Figure 1). It appeared ideal for the study because it contains several distinct oil pools within a restricted geographic region. The pools are located in the Lower Cretaceous Mannville Group. They are positioned such that some of them coincide vertically with topographic uplands, presumed to be areas of descending groundwaters or recharge, while others are beneath topographic depressions, suspected to be areas of groundwater discharge.

The Chauvin area also contains sizeable tracts of solonetzic soils which are indicative of groundwater discharge. According to our working hypothesis, discharge areas (areas of upward flow) produce the strongest petroleum geochemical signatures. Based on the work of Christopher (1980), who completed a regional petroleum geology study of the Mannville Group in Saskatchewan, the Chauvin area was initially interpreted as an area of regional groundwater discharge.

The study was undertaken on two different scales. On a small scale, the potentiometric surface of the Mannville Group was mapped to determine the regional distribution of groundwater flow. The soil gas survey and the water chemistry analyses were conducted on a larger scale, over an area of about 840 km^2, in the immediate vicinity of Chauvin. Both the regional and local study areas are shown in Figure 1.

General Physiography

The region was covered by glacial till during the Wisconsin glaciation, which left the surface with a gently undulating topography (Figure 2). Topographic eleva-

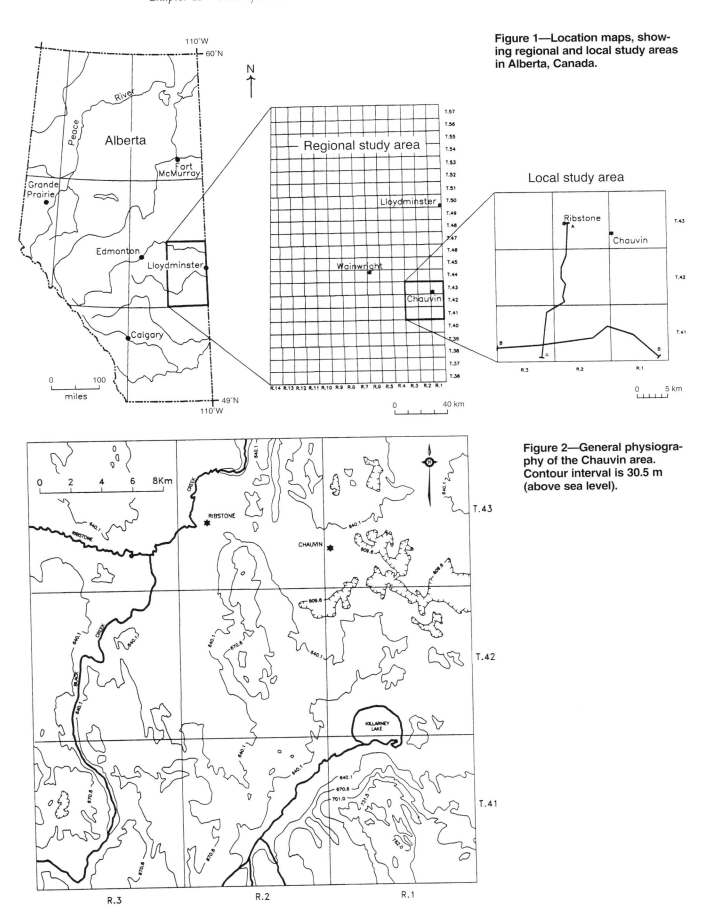

Figure 1—Location maps, showing regional and local study areas in Alberta, Canada.

Figure 2—General physiography of the Chauvin area. Contour interval is 30.5 m (above sea level).

tions in the general area range from about 600 m in low-lying wet regions to over 750 m in the southeastern corner of the local study area. The dominant physiographic features include a topographically high ridge running north-south through the center of the study area in T 42, R 2 (W4M); a topographically low wet region along the eastern border in T 42–43, R 1; hills in T 41, R 1; and Killarney Lake, a low-lying lake with no surface water outlet, in T 41–42, R 1 (Figure 2).

The eastern part of the study area is internally drained. Drainage in the western part, where waters drain northeastward via Black Creek into Ribstone Creek, is, however, more open. The climate of east-central Alberta is humid continental (LeBreton, 1963) with temperatures ranging from about 32°C in the summer to –30°C in the winter.

Geologic Setting

The Chauvin area was glaciated by the Keewatin ice sheet of Pleistocene age. The soils consist largely of tills directly overlying the bedrock. Locally, the till is overlain by glaciolacustrine and glaciofluvial deposits. The overburden thickness can vary from 20 m in lower topographic areas to about 125 m in the hills in T 41, R 1.

Figure 3 shows a stratigraphic section of the subsurface geologic units in the study area. The uppermost bedrock unit is the Upper Cretaceous Belly River Formation. The formation dips gently to the southwest and can be differentiated into alternating marine shale and continental sandstone sequences (LeBreton, 1963). Of importance from a hydrogeologic perspective are two dominant nearshore sandstone members of the Belly River—the Ribstone Creek Member and underlying Victoria Member. The Ribstone Creek aquifer is used both by farmers for domestic and agricultural water supplies and by the petroleum industry for water-flooding projects. The Victoria sandstones do not appear to be a water-producing aquifer over the whole region.

Beneath the Belly River Formation lies the Lea Park Formation, a dominantly shale unit with localized dirty sandstone beds and lenses. It is about 200 m thick and is underlain by shales of the Colorado Group. Both of these shale units were deposited in a broad, slowly subsiding epeiric sea. The Viking Formation, a relatively thin (20–30 m) gas-producing sandstone, lies near the bottom of the Colorado Group. The Joli Fou Formation is composed of aerially extensive dark gray marine shales that extend beneath the Viking Formation 25–30 m to the Mannville Group, which is the major oil-producing unit in the Chauvin area. There are few wells in the area that penetrate into the Lower Devonian carbonate strata beneath the Mannville.

The shales of the Lea Park Formation and Colorado Group constitute a significant hindrance to groundwater flow, while the Joli Fou shale is a second, less extensive aquitard. The Ribstone Creek and Victoria members of the Belly River Formation, the Viking Formation, and the Mannville Group are the major aquifers in the area.

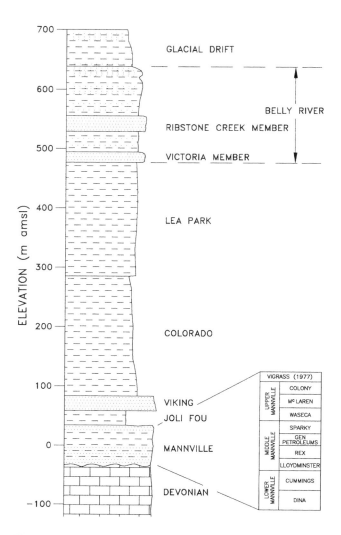

Figure 3—Stratigraphic section of the Chauvin area.

Petroleum Geology

The Chauvin area is part of the Lloydminster heavy oil region of Alberta, in which petroleum is found throughout the Lower Cretaceous Mannville sandstones. The Mannville Group is a complex assemblage of sandstones, siltstones, shales, and coals of both marine and continental origin. The lack of extensive chronostratigraphic units, combined with frequent drastic changes in facies composition, has impeded the development of a rigorous stratigraphic framework. The Mannville has been informally divided into lower, middle, and upper subgroups and more specifically into nine informal members: Colony, McLaren, Waseca, Sparky, General Petroleums (G.P.), Rex, Lloydminster, Cummings, and Dina (Vigrass, 1977). The Colony, McLaren, and Waseca are considered the upper Mannville; the Sparky, G.P., Rex, and Lloydminster are the middle Mannville; and the Cummings and Dina are the lower Mannville (after Vigrass, 1977) (Figure 3).

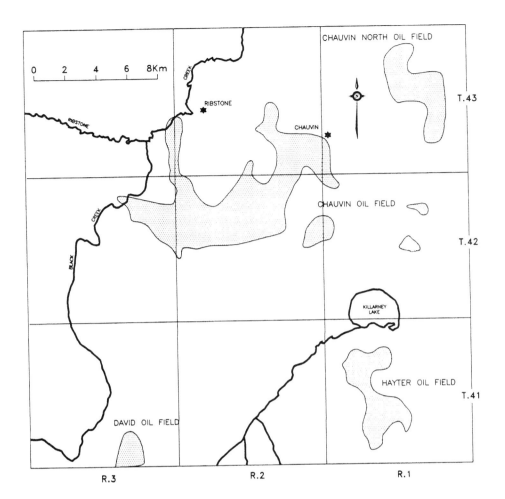

Figure 4—Location map of oil fields (shaded) in study area.

Early Mannville sedimentation was influenced by the topography on the Paleozoic erosional surface. With the emergence of land that produced the unconformity, the Upper Devonian units were eroded and karstified (Tóth, 1978). Nonmarine, dominantly fluvial, clastic deposition of the Dina sands occurred in the topographic lows, creating tabular sand bodies.

The Cummings, Lloydminster, Rex, G.P., and Sparky units were subsequently laid down in marine to nearshore marine environments. The cyclic nature of the deposition reflects transgressive and regressive phases of the Boreal and Gulfian Early Cretaceous seaways. Sand bodies deposited during this time were dominantly sheet sands formed by the progradation of beach sand facies. Rapid changes in sea level during this time shifted the focus of sand deposition seaward or landward such that sheet sand deposition was interrupted by low-permeability lagoonal or deeper water sediments. In addition to these lateral facies changes, the sheet sands were also broken by ribbon-shaped sand bodies deposited in estuarine channels, tidal inlet channels, tidal creek channels, and distributary channels (Putnam, 1982).

Sandstone units of the upper Mannville Group (the Colony, McLaren, and Waseca) are dominated by similar sheet sandstones and ribbon-like channel sand bodies. Sheet sandstones are not as regionally extensive in this interval, and the channel sandstones are more abundant. The sands were deposited in a continental fluvial environment with deltaic and tidal influences. There is a distinct northwest trend to the channel sandstones.

Petroleum accumulations have been found throughout the entire thickness of Mannville sandstones. In the Chauvin area, the Mannville is found at a depth of about 650 m and dips to the southwest at about 2 m/km (Dunning et al., 1980). There are four oil fields within the local Chauvin area: the Hayter, David, Chauvin, and Chauvin North oil fields (Figure 4).

On a regional scale, oil in the Lloydminster heavy oil belt has been trapped through a variety of mechanisms. The dissolution of the underlying Devonian Prairie Evaporite Formation has resulted in draping of the overlying Mannville sediment and subsequent accumulation of petroleum in the resultant folds. In addition, oil is often trapped in sandstones that onlap lower permeability Devonian strata, that abut against shale-filled channels, or that pinch out laterally into more shale-rich facies.

GROUNDWATER FLOW

The objective of our project was to determine if identifiable soil gas anomalies exist in the study area and, if so, what effects the flow of groundwater might have on their intensity and their positions with respect to the oil fields from which, presumably, they originate. One of the two principal parameters needed to answer these questions is the distribution of groundwater flow. The other parameter—the nature and distribution of soil gas anomalies—is discussed in the section on soil gas survey.

Because of its importance in the present context, groundwater flow in the area was evaluated by various methods. The methods of studying regional groundwater flow distributions can be divided into two main groups: (1) those of direct evaluation based on fluid dynamic parameters and (2) methods of indirect evaluation based on flow-related effects and phenomena.

All the methods and techniques used in this study rely on data that are available from, or can be easily generated through, the conventional activities of petroleum exploration. Six approaches were used: (1) field patterns of hydraulic heads and pressure–depth gradients; (2) surface manifestations of groundwater flow; (3) major ionic species of dissolved chemical constituents; (4) concentration ratios of the halogen elements chloride, bromide, and iodide; (5) concentration of the stable isotopes ^{2}H and ^{18}O; and (6) ^{14}C age dating of groundwater.

Evaluation of Groundwater Flow Patterns Based on Fluid Dynamic Parameters

Principles and Definitions

The basic physical principle that governs the movement of groundwater is that flow takes place from regions of high mechanical energy contained by the water toward regions of low mechanical energy. The energy content of the water is expressed by the fluid potential, ϕ, which is defined (Hubbert, 1940) as

$$\phi = gh = gz + p/\rho \qquad (1a)$$

where ϕ = mechanical energy per unit mass of fluid (L^2/T^2); g = acceleration due to gravity (L/T^2); z = vertical coordinate of the point of measurement, or *elevation head*, which is positive upward with respect to a datum plane at which $z = 0$ (L); h = *hydraulic head*, or the elevation to which water rises with respect to the datum plane in an open vertical tube from the point of measurement (L); p = pore fluid pressure at point z (M/T^2L); and ρ = fluid density (M/L^3).

The hydraulic head h (also called the *potentiometric elevation*) is a practically useful and convenient measure of the fluid potential. It can be determined from water level or pore pressure measurements in wells by the following relation:

$$h = z + p/\rho g \qquad (1b)$$

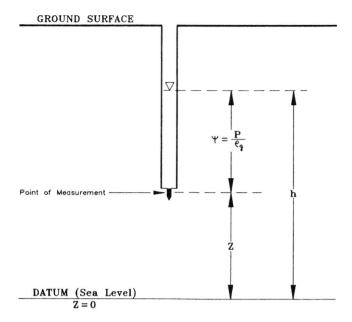

Figure 5—Schematic drawing of groundwater monitoring, showing the relationship of the hydraulic head (h), pressure head (Ψ), elevation head (Z), and vertical pore pressure gradient (ρ_g).

where $\rho g = \gamma$, the rate of vertical change in pore pressure or the vertical pore pressure gradient (M/L^2T^2).

The intensity and direction of fluid flow are related to the hydraulic head by Darcy's law:

$$q = -K \,\text{grad}h = -K \,dh/dl \qquad (2)$$

where q = specific volume discharge or flux (L/T), K = hydraulic conductivity of the rock-fluid system (L/T), and $dh/dl = \text{grad}h$ is the hydraulic gradient, which is taken positive in the direction of increasing hydraulic head, h. The negative sign in Darcy's law indicates that the fluid impelling force, and thus the flow, is oriented opposite to the hydraulic gradient, that is, in the direction of decreasing hydraulic head.

In the case of hydrostatic equilibrium, there is no flow and thus $q = 0$. Consequently, the condition of no flow in the vertical direction is that the vertical component of the hydraulic head gradient must also be zero. This condition can be expressed from equation 2 as

$$q_z = \partial q/\partial z = -K \,\partial h/\partial z = 0 \qquad (3a)$$

from which the vertical head gradient $\partial h/\partial z$, obtained from equation 1b, must also be zero because the value of K can only be finite:

$$\partial h/\partial z = (1 + 1/\rho g)(\partial p/\partial z) = 0 \qquad (3b)$$

The vertical pressure gradient in a static (st) body of water is therefore

$$\gamma_{st} = \rho g = -(\partial p/\partial z)_{st} = (\partial p/\partial d)_{st} \qquad (4)$$

Equation 4 explicitly states that the condition for no flow in the vertical direction is that the vertical pressure gradient must be equal to the product of the fluid's density, ρ, and the acceleration of gravity, g. The negative and positive signs indicate that the pressure decreases with increasing elevation, z, and increases with increasing depth, d, respectively.

However, flow upward is reflected by a dynamic vertical pressure gradient, γ_{dyn}, which is greater (while flow downward by one that is less) than the hydrostatic or nominal pressure gradient, $\rho g = \gamma_{st}$ (Figure 6):

$$\gamma_{dyn,up} = (\partial p / \partial d)_{up} > \rho g = \gamma_{st} = (\partial p / \partial d)_{st} \qquad (5a)$$

and

$$\gamma_{dyn,dn} = (\partial p / \partial d)_{dn} < \rho g = \gamma_{st} = (\partial p / \partial d)_{st} \qquad (5b)$$

By means of calculated and observed values of these fluid dynamic parameters, namely, the hydraulic head, h, and the vertical pressure gradient, the distribution of groundwater flow can be characterized in three different ways: the potentiometric surface, hydraulic cross sections, and pressure–elevation or pressure–depth profiles.

1. The *potentiometric surface* shows the lateral distribution of fluid potential in a horizontal confined aquifer by means of a map of contours of equal hydraulic heads, $h(x, y) =$ constant. Such a map can be used to infer the lateral (x, y) directions of groundwater flow in the aquifer; flow is normal to the contours and takes place toward decreasing head values. The intensity of the flow can also be evaluated by equation 2 if the hydraulic conductivity K is known.
2. *Hydraulic cross sections* portray flow patterns by hydraulic head contours, $h(l, z) =$ constant, in vertical planes that can be oriented in any arbitrary direction, l. However, because the hydraulic head values depend on the fluid's density (equation 1b), the magnitude and even the sense of the vertical component of the driving force, and thus of the flow, may not be correctly reflected by the contours if the flow regime contains fluids of significantly different densities (Lusczynski, 1961; Davis, 1987).
3. *Pressure–elevation* or *pressure–depth profiles* show the values of pore pressures along a vertical line, such as in a borehole, usually presented in two-dimensional coordinate systems. Pressures, p, are shown on the horizontal axis, while the vertical positions of the points of measurement are on the vertical axis with respect to elevation, z, or depth, d.

The rate of change in pressures shown on these profiles is the vertical pressure gradient, as expressed in equations 4 and 5. The above-mentioned problem caused by variable fluid densities can be reduced by comparing profiles of *actual*, possibly dynamic, vertical pressure gradients (γ_{dyn}) determined from measurements, to profiles

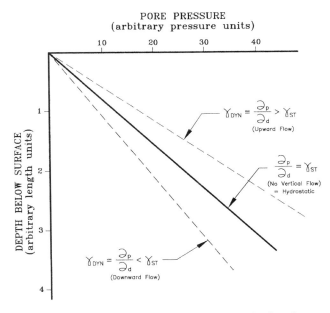

Figure 6—Generalized pressure versus depth plot showing conditions for vertical fluid movement.

of *nominal* static gradients (γ_{st}) calculated by equation 5 with the fluid densities prevailing at the various depth ranges (equations 5a, b; Figure 6).

Field Observations of Fluid Dynamic Parameters

Potentiometric Surface Maps—Potentiometric surface maps were drawn (Holysh, 1989) from well water levels and from formation fluid pressures at a detailed scale for the glacial drift (Figure 7) and for the shallow bedrock (Figure 8) in the local study area covering T 41–43, R 1–3. Potentiometric surface maps were also constructed at a smaller scale for four 100-m-thick horizontal zones in the Lower Cretaceous, including the Viking and Mannville aquifers, for the region covering T 36–57, R 1–14. From the latter set, only a representative map for the depth range of +100–0 m (above sea level) is reproduced here (Figure 9).

The similarity between the potentiometric surfaces in the glacial drift (Figure 7) and shallow bedrock (Figure 8) and the topography of the Chauvin area (Figure 2) shows that the hydraulic head distribution (and thus the groundwater flow) in these two uppermost units is closely adjusted to the configuration of the land surface. In both units, the areas of high hydraulic heads coincide with topographic mounds or ridges, such as in T 41, R 1; T 42, R 2; and T 43, R 1–2. Conversely, low topographic elevations cause corresponding depressions in both potentiometric surfaces. Furthermore, hydraulic heads decline generally under the hills as well depths increase from the drift to the underlying bedrock, indicating descending water flow. In the depressions, however, water levels tend to rise with increasing well depth,

Figure 7—Potentiometric surface map of shallow overburden wells.

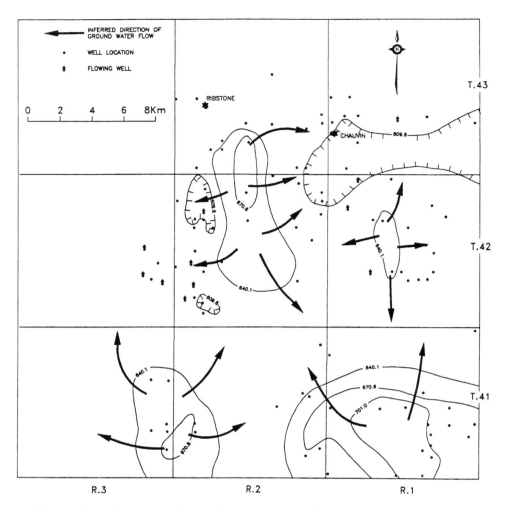

resulting in ascending flow. This condition is also reflected by flowing artesian wells at various low-lying locations (Figures 7, 8).

The groundwater flow in the Lower Cretaceous strata, including the petroliferous Viking and Mannville aquifers, was inferred from potentiometric maps prepared for four 100-m-thick, highly permeable horizontal zones (Holysh, 1989, his figures 3.6–3.9). These zones underlie the 300–400-m-thick, low permeability marine shales of the Lea Park and Colorado formations (Figure 3). The principal features of the potential distributions are similar in all four zones and are exemplified here by the potentiometric surface map for the arbitrarily selected zone of the +100–0 m elevation range in Figure 9.

Regional flow is toward the northeast, although local convergences and channeling are indicated by several deep reentrants of the hydraulic head contours. These features can be attributed to variations in permeability caused by lithologic changes. Some closed potentiometric mounds and depressions may instead reflect topographic effects that are apparently transmitted through the Lea Park and Colorado aquitards. Such effects may be exemplified by the potentiometric mounds in the glacial drift in T 41–42, R1 (Figure 7), which seem to extend downward and coalesce in the depth range of +100–0 m (Figure

9), and by the topographic and potentiometric depressions in T 43, R 1 (Figures 2, 7), reflected by the low-potential reentrant in T 42–43, R 1 (Figure 9).

In general, the hydraulic heads decline with increasing depth through the Mannville Group, as evidenced by the four mapped rock zones of different elevation ranges. This head decline clearly indicates that water flow is downward in the oil-bearing basal Cretaceous units.

Pressure–Depth Profiles—Pressure versus depth profiles have been prepared (Figure 10) for several regions of the study area in which sufficient drill-stem test data were available for the Mannville Group and over- and underlying units (Holysh, 1989). Results from four of these regions are discussed in this paper. On average, each of the regions covers nine townships (29 km²). Individual pressure–depth profiles were generated from field data collected from all parts of these regions. They thus present a generalized distribution of pore pressures with respect to depth. Each plot is identified by the legal description of the central township of the sampled region that it represents (Figure 10).

To compare the actual formation fluid pressures and the actual vertical pressure gradients to pressures and gradients that would occur in the absence of a vertical

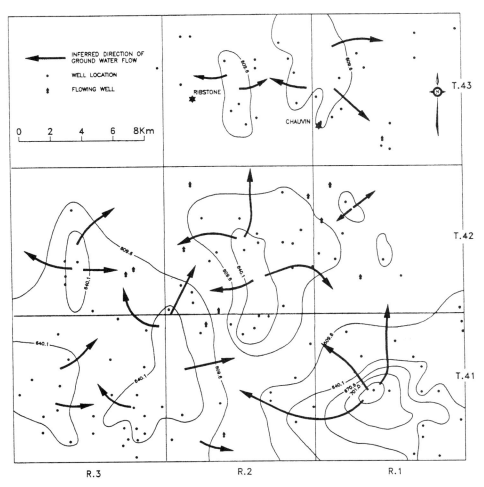

Figure 8—Potentiometric surface map of shallow bedrock wells.

flow component, a nominal hydrostatic pressure–depth profile is shown in Figure 10 for each set of field data curves. Two important observations can be made:

1. In general, formation fluid pressures are less than hydrostatic (i.e., less than normal). The ratio of actual to nominal (i.e., dynamic to hydrostatic) pressures averages 0.7. The actual formation pressures are from 2/3 to 3/4 of the normal hydrostatic values. The greatest pressure deficiencies occur in the lower Mannville Group.

2. The vertical pressure gradients are subhydrostatic ($\partial p/\partial d < \gamma_{st}$) above and superhydrostatic ($\partial p/\partial d > \gamma_{st}$) below the Mannville reservoirs. According to equations 5a and b and Figure 6, these gradients indicate downward-oriented fluid driving forces above the Mannville Group and upward-directed forces below it, on other words, from the Devonian strata.

When considered together with the northeastward decline of the potentiometric surface (Figure 9), the vertical distribution of pore pressures (Figure 10) shows the Mannville Group to draw groundwater from the land surface downward, as well as deeper formation fluids from the underlying Devonian formations upward. It

then conducts these fluids laterally toward subcrop or outcrop regions of low elevations to the northeast, mostly in the neighboring province of Saskatchewan.

Hydraulic Cross Sections—Vertical distribution patterns of groundwater flow are shown in hydraulic cross sections A–A' and B–B' (Figures 11a, b). The patterns have been inferred from water levels measured in shallow-water supply wells, from surface manifestations of groundwater discharge, and from hydraulic heads calculated from virgin formation pressures determined by drill-stem tests in oil wells. The flow distribution interpreted from the vertical sections confirms the distribution that is inferred from the potentiometric surfaces and vertical pressure gradients.

Clearly, groundwater flow is oriented uniformly downward below the Ribstone Creek Member along both cross sections. Ascending flow occurs only in local discharge areas in the glacial drift or in the shallow parts of the uppermost bedrock unit, namely, the Belly River shales and sandstones. The hydraulic head differences across a depth interval of about 400 m between the Belly River and Mannville formations are on the order of 150–200 m. The resulting vertical gradients of about 0.3–0.5 m/m represent a strong downward-oriented driving force across the Lea Park and Colorado shale aquitards.

Interpretation of Fluid Dynamic Parameters

Based on the observed distributions of fluid dynamic parameters just presented, groundwater flow conditions in the study area can be summarized as follows:

1. Groundwater flow is adjusted to the relief of the water table in the glacial drift and shallow bedrock, that is, to less than the depth of the Ribstone Creek Member of the Belly River Formation. The discharge areas receive water from recharge areas in adjacent local highlands.
2. Groundwater flow is descending regionally through the Lea Park Formation and Colorado Group shales into the Mannville Group.
3. The Mannville receives deep formation waters from the underlying Devonian strata.
4. Descending meteoric waters and waters rising from the Devonian strata mix in the Mannville Group and move northeastward into subcrop or outcrop regions in Saskatchewan.

Evaluation of Groundwater Flow Patterns Based on Flow-Related Effects and Phenomena

Flow-Related Effects and Phenomena

It is generally recognized that, at a regional scale, formation water flow is distributed into systems of varying lengths, depths, shapes, and intensities (Freeze and Cherry, 1979; Domenico and Schwartz, 1990; Fetter, 1994). It is also known that moving formation water interacts with the rock framework chemically, mechanically, thermally, and in several other ways at and beneath the land surface. The combination of a geometrically distributed flow field configuration with the effects of rock–water interaction results in areal patterns of natural processes and phenomena. Because of their functional relationship to the flow systems, these processes and phenomena can be interpreted in terms of flow directions.

Five types of flow-related effects and phenomena were used in this study to infer information on directions of groundwater flow: (1) surface manifestations of groundwater discharge; (2) major ionic species in formation waters; (3) concentration ratios of the halogen elements chloride, bromide, and iodide; (4) concentration of the stable isotopes ^{18}O and ^{2}H; and (5) ^{14}C age dating of groundwater. The basic principles of the origin and application of flow-related effects and phenomena, the field observations, and their interpretation in terms of flow directions are briefly reviewed here.

Surface Manifestations of Groundwater Discharge

Discharge, or ascending flow, of groundwater to the land surface can give rise to various observable natural phenomena (Tóth, 1971, 1980, 1988). The type and nature of these flow manifestations depend on the hydraulic characteristics of the discharge itself (e.g., areal extent,

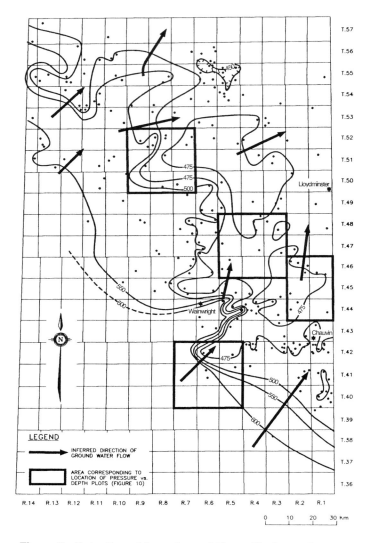

Figure 9—Potentiometric surface of Mannville Group from 0 to 100 m above sea level. Rectangles indicate areas for which pressure versus depth plots are given in Figure 10.

intensity, temporal variations, and orifice geometry) and on the geologic and physiographic environment (e.g., rock type, topography, climate, and season). These phenomena include flowing wells, springs, seeps, marshes, saline soils, playas, phreatophytic and halophytic plants, quick sands, slumping slopes, positive geothermal anomalies, and anomalous formation water chemistry.

Hydrogeologic field mapping and a water well survey in the Chauvin area have identified a number of groundwater discharge features (flowing wells, saline soils, marshes, phreatophytes, and halophytes) and enabled the delineation of several distinct discharge areas (Figure 12). A comparison of the mapping results (Figure 12) with the topographic map (Figure 2) clearly shows that depressions in the land surface serve as groundwater discharge areas. No groundwater discharge phenomena were observed in the higher elevations of the study area.

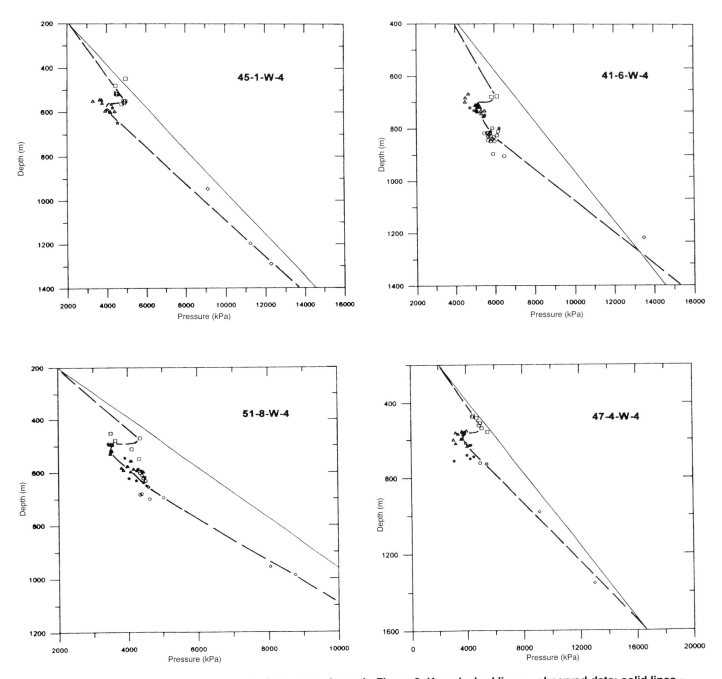

Figure 10—Pressure versus depth plots for four areas shown in Figure 9. Key: dashed lines = observed data; solid lines = hydrostatic gradient; square = Viking Formation; triangle = upper Mannville Group; star = middle Mannville; circle = lower Mannville; diamond = Devonian strata.

Major Ionic Species in Formation Waters

The chemical constituents found most commonly in formation waters are the four cations Na^+, K^+, Ca^{2+}, and Mg^{2+} and the three anions HCO_3^-, SO_4^{2-}, and Cl^-. Their relative and absolute amounts are functions of two principal mechanisms: (1) the sequential chemical reactions between the water and the rock framework along the flow path and (2) the mixing of chemically different waters originating from different localities. Since the water's chemical evolution resulting from these mechanisms is dependent on its movement, the spatial distribution patterns of the ions can be interpreted in terms of flow directions.

In the course of the field mapping and well survey, 77 samples were collected from shallow domestic water wells and 32 samples from oil wells in the Mannville Group (Figure 13) for chemical analysis. Based on the

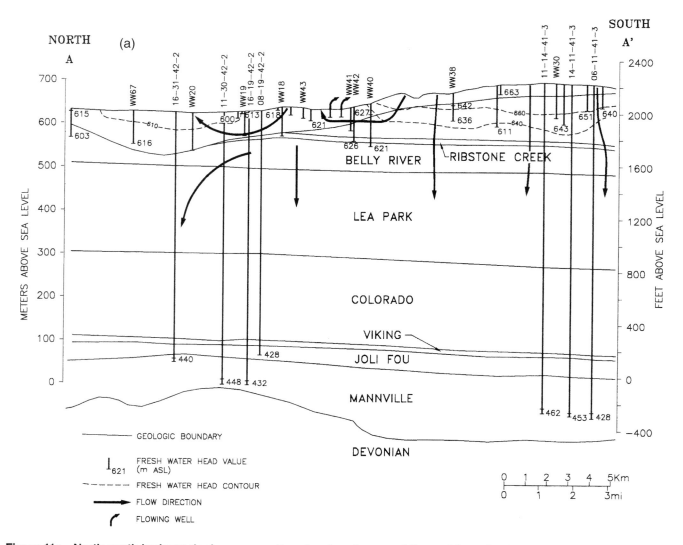

Figure 11a—North-south hydrogeologic cross section showing downward flow to Mannville Group. [Part (b) on next page.]

analytical results, the water samples can be grouped into four distinctly different chemical types (Figure 14):

- **Type 1** waters are rich in Ca^{2+} and HCO_3^- and is typical of fresh meteoric water.
- **Type 2** waters plot toward the bottom left of the Piper's diamond plot (Figure 14). The shift can be explained by replacement of the water's calcium with sodium from the clays in the soil and glacial till. The enrichment in sulfates is attributable to dissolution of gypsum.
- **Type 3** waters are from the sandstones of the Ribstone Creek Member. They are characterized by low sulfate content, highly variable chloride concentrations (360–1800 mg/L), and sodium as the principal cation. The variability in the chloride concentration can be related to the subsurface geology and to the position of the sampling point within the flow system. Under conditions of thin

overburden and high topographic altitude (i.e., where the Ribstone Creek is close to the land surface and in a potential groundwater recharge area), the concentration of chlorides is relatively low. Conversely, high chloride concentrations occur in the Ribstone Creek Member where it lies at greater depths and farther away from highlands. Also, adjacent to relatively transmissive buried bedrock channels, waters in this unit are elevated in bicarbonate, while farther away they contain more chloride.

- **Type 4** waters characterize the Mannville Group. Most of the major ions are represented in these waters, but chloride and sodium dominate. Total mineralization (total dissolved solids, TDS) varies within a wide range from 54 to 106 g/L.

In terms of the direction of formation water flow, interpretation of these observations can be summarized as fol-

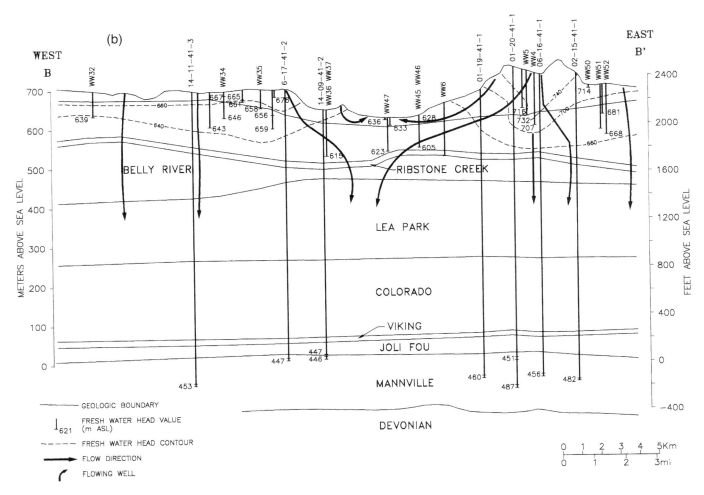

Figure 11b—East-west hydrogeologic cross section showing downward flow to Mannville Group. [Part (a) on previous page.]

lows. Meteoric waters infiltrating into the subsurface have low TDS, with calcium and bicarbonate being the dominant ions (type 1). With increasing depth and residence time in the subsurface, cation exchange reactions increase the sodium concentration (type 2). Further evolution of the water takes place by the addition of residual chloride from the slightly permeable silts and clays of the Belly River Formation (type 3). Continued descending flow into the Mannville Group results in further increase in the water's major ion concentrations and TDS (Type 4).

Within the Mannville Group, downward-moving waters appear to mix with waters moving upward from the Lower Devonian units. The addition of Devonian waters is indicated by much higher TDS concentrations (90–120 g/L) in Mannville water samples from the Chauvin area than in more typical Mannville samples (15–30 g/L) to the south and west (Jardine, 1974; Abercrombie and Fullmer, 1992). This interpretation is entirely consistent with that based on the pressure–depth profiles, which also indicate upward flow of Devonian groundwaters to the Cretaceous Mannville Group.

Concentration Ratios of the Halogen Elements Chloride, Bromide, and Iodide

The halogen ions of chloride, bromide, and iodide can be used to study the historical evolution of groundwater because they are conservative, and in a halite-free subsurface environment, few processes other than physical mixing (such as dissolution, precipitation, and biological uptake) affect their concentration. High concentrations of bromide and iodide in subsurface brines can be attributed to bioconcentration in marine organisms. Seaweeds have been found to concentrate iodide to 8000 ppm and bromide up to 6800 ppm, whereas some corals have iodide and bromide concentrations as high as 69,200 and 19,800 ppm, respectively (Chave, 1960; Collins et al., 1967)

High concentrations of halogens can also result from evaporation of sea water in restricted marine basins. As beds of calcium carbonate and gypsum initially form, only a minimal loss of halogens from solution occurs; their concentrations thus increase. Even with the loss of chloride at the onset of halite precipitation, both bromide and iodide concentrations continue to increase in the

Figure 12—Map of surface discharge features in study area.

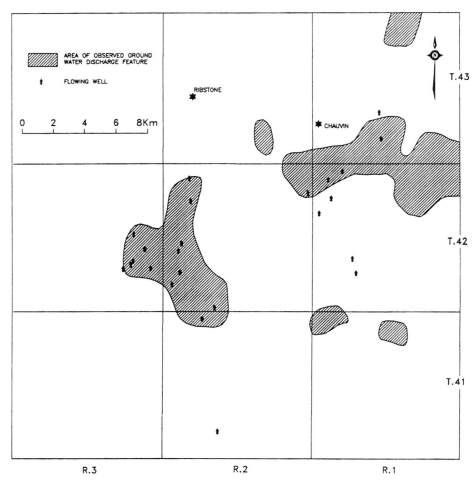

remaining solution (Rittenhouse, 1967; Lloyd et al., 1982). Elevated halogen concentrations are therefore expected in groundwaters associated with marine evaporites or with extensive accumulations of organic materials (e.g., shales).

All shallow waters and oil field brines in the Chauvin area were analyzed for chloride, bromide, and iodide. From Figure 15a, it appears that type 1 waters have a low bromide concentration and a low ratio of Na/Cl of about 1.0. Type 2 waters are also low in bromide, but their Na/Cl ratio is higher than that of type 1 waters. The trend toward increasing bromide concentrations and Na/Cl ratios continues for type 3 waters in the Ribstone Creek sandstones, whereas type 4 waters of the Mannville Group show still higher bromide concentrations but slightly lower Na/Cl ratios than those of the Ribstone Creek waters.

Figure 15b shows the concentration of bromide versus iodide in both the shallow Chauvin area waters and the deeper Mannville brines. The plot uses linear concentration scales so that conservative mixing trends appear as a straight line (Hanor, 1987). The plot shows two distinct, generally straight lines, indicating the conservative nature of the halogens within each group of waters. Within the Mannville brines, varying degrees of mixing between already altered meteoric waters with Devonian

brines produce a straight line. At depths above the Mannville Group, however, the various stages of evolution and mixing of meteoric waters are represented by a second straight line. The distinctness of these lines indicates that there is no direct mixing of unaltered meteoric waters with the Mannville brines.

Directions of formation water flow can be inferred from the previous observations. Type 1 waters are meteoric waters. These waters evolve into type 2 by descending through near-surface argillaceous sediments to greater depths, thereby undergoing exchange of their original Ca^{2+} ions for Na^+ ions from the clays, without significantly affecting the halogen ratios. The halogen concentrations increase in the brackish deposits of the deeper Belly River Formation. Type 3 waters of the Ribstone Creek Member reflect both an increase in halogens and continued exchange of cations. Halogen concentrations in type 4 waters of the Mannville Group are significantly higher than in the shallower groundwater samples. These brines are interpreted to reflect the brackish to marine depositional environment of the Mannville sediments and possible upward flow of groundwater from the underlying Devonian carbonates.

The halogen analyses indicate that there is no direct mixing of shallow waters with Mannville brines, either in the shallow subsurface or within the deeper Mannville

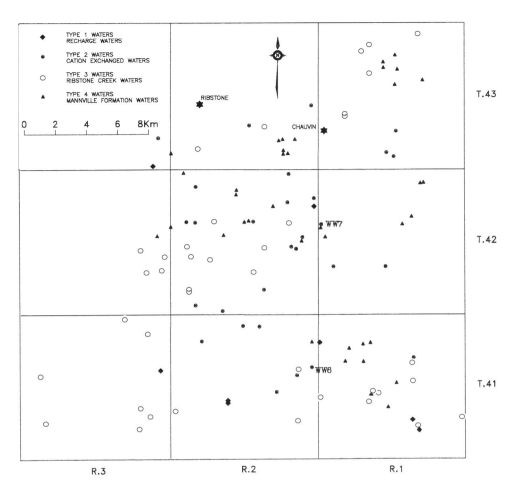

Figure 13—Map of ground-water sampling locations in study area showing chemical grouping of water types. (See Figure 14 for Piper plot of these water types.)

Group. It is inferred, therefore, that the Mannville waters represent a mix of gradually evolving meteoric waters moving downward from the surface with waters ascending from the deeper Devonian strata, but that Mannville waters cannot rise to mix with the near-surface groundwater.

Concentration of Stable Isotopes

Because of their conservative behavior, the stable isotopes ^{18}O and ^{2}H have become standard tools in tracing the sources of waters and studying their evolution in the hydrologic cycle (Fontes, 1986). Stable isotopes are reported as delta (δ) values in parts per thousand (‰) relative to standard mean ocean water (SMOW) for oxygen and hydrogen. (The relative homogeneity of oxygen and hydrogen in the oceans makes them suitable as a reference standard.)

As water evaporates from ocean basins, a higher proportion of lighter isotopes are incorporated into the atmosphere, and the ratio of light to heavy isotopes is thus greater in atmospheric waters than that in the oceans. As air masses move inland, heavy isotopes precipitate preferentially, thereby increasing the ratio of light isotopes remaining in the atmosphere even more. Each subsequent rainfall farther inland becomes more depleted in heavy

isotopes. This rain-out effect is enhanced by high altitudes and cooler climates. Craig (1961) has shown a systematic worldwide relationship between $\delta^{2}H$ and $\delta^{18}O$ that has become known as the *meteoric water line.*

In certain continental areas where precipitation results more from local evaporation than from oceanic air masses moving inland, slight departures from the meteoric line are observed. The prairie region of North America is one such area, and a *prairie meteoric line* has been defined (Figure 16). Variations in the isotopic composition of meteoric waters entering the groundwater system can also arise from seasonal differences in precipitation, temporal climatic shifts, and evaporation of waters before infiltration into the ground.

Selected shallow groundwater samples and selected Mannville brines were analyzed for the stable isotopic composition of hydrogen and oxygen. Shallow waters (types 1 and 2) are isotopically light, lying on the prairie meteoric water line (Figure 16). This suggests a recent meteoric origin for these waters. The deeper shallow waters, namely, those from the Ribstone Creek Member (type 3), are the isotopically heaviest waters, which indicates some degree of equilibration with the rock framework. The deeper Mannville brines are positioned near the prairie meteoric water line but are enriched in heavy

Figure 14—Piper plot showing different water types (located in Figure 13) and the chemical evolution of the groundwater with depth.

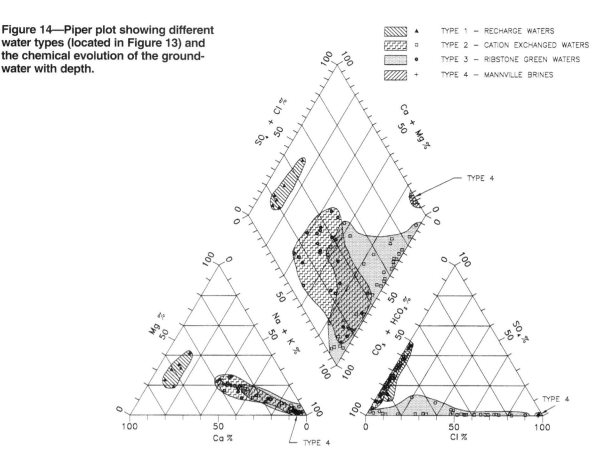

oxygen. A possible interpretation is that there is no mixing of Mannville brines with the meteoric waters in the near-surface units, that is, no upward flow of these brines. Shallow meteoric waters, however, may have moved to depth and influenced the isotopic composition of brines in the Mannville Group.

Thus, based on the stable isotope analyses, the shallow waters in the Chauvin area are interpreted to be of recent meteoric origin, showing no indication of mixing with (ascending) Mannville brines. Isotopically, the Mannville brines are similar to other brines analyzed in the Alberta Basin (Hitchon and Friedman, 1969). They are heavier than the shallow waters, possibly indicating an isotopic equilibration or exchange with the surrounding rock framework. The isotopic composition of these brines may, however, be influenced to some extent by shallow meteoric waters descending to depth.

14C Age of Groundwaters

To investigate the question of vertical fluid migration between Mannville reservoirs and the surface in as many different ways as possible, the ^{14}C ages of waters in the shallow surface and the Mannville Group were evaluated. The radioactive isotope of carbon, ^{14}C, is continuously produced in the earth's upper atmosphere through nuclear reactions. The ^{14}C atoms oxidize to form $^{14}CO_2$ molecules that mix with nonradioactive CO_2 in the atmosphere and, through a series of exchange and assimilation reactions, enter the biosphere and hydrosphere. The ^{14}C decays at a known rate, and by measuring the remaining ^{14}C in a carbonaceous material, its age can be determined.

The rate of radioactive decay is described by the decay equation

$$A_0 = A e^{\lambda t} \qquad (6)$$

where A_0 and A = original and measured specific radioactivities of the material, respectively, e = natural log function, t = age, and λ = decay constant.

For our purposes, four Mannville brine samples and two shallow-water samples were analyzed. The samples were obtained and preserved using the procedures outlined by Geyh and Wagner (1979). Table 1 shows the results of the ^{14}C analyses, as well as the corrected ages using the Ingerson–Pearson equation (Fontes and Garnier, 1979). This equation uses the ^{13}C value to correct for subsurface carbon sources that can alter the carbon composition of water once it enters the subsurface. From the ^{14}C analyses, some difference in the ages of the shallow waters versus the Mannville brines is apparent. Although they both vary in age, in general, the Mannville brines are considerably older than the shallow waters.

The two shallow-water samples (Table 1, WW6 and WW7) were obtained from the Ribstone Creek sand-

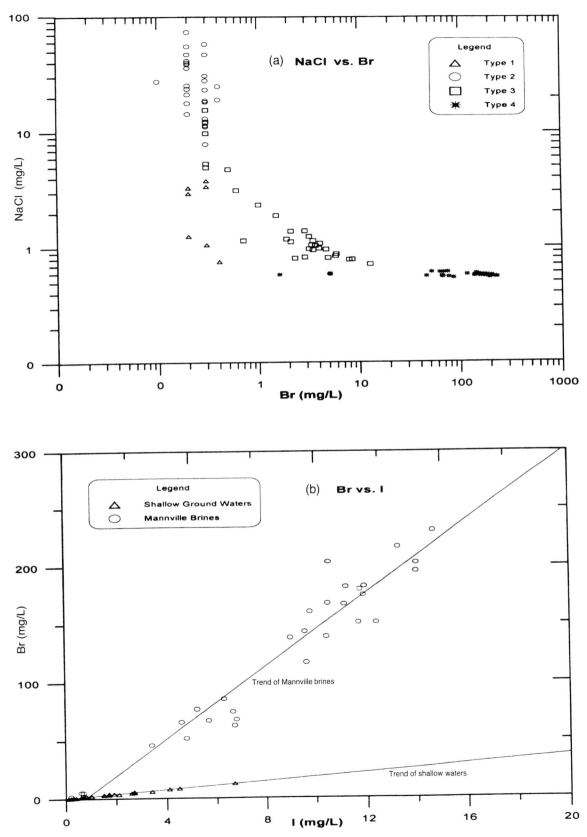

Figure 15—Halogen plots showing evidence of mixture of shallow waters with Mannville brines. (a) Sodium chloride versus bromide. (b) Bromide versus iodide.

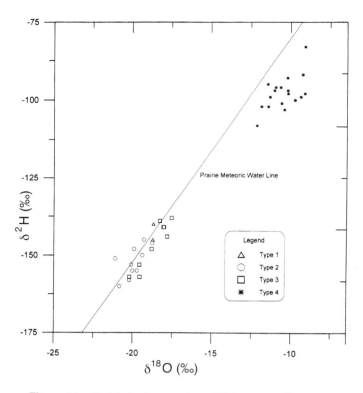

Figure 16—Stable isotope results (δ²H versus δ¹⁸O) showing no evidence of mixing of shallow waters with Mannville brines.

Table 1—¹⁴C Age Dating Results

Sample No.[a]	¹³C (‰)	Original ¹⁴C Age (Years)	Corrected Age[b] (Years)
WW6	−7.0	>40,190	>10,523
WW7	−18.1	37,600	2,670
10-29-41-1	−5.0	>40,020	>13,305
10-20-41-1	−3.5	>40,750	>16,253
7-28-43-1	−8.5	>40,500	>8,918
11-26-42-2	−1.3	34,750	24,440

[a]WW6 and WW7 are shallow-water wells; the remaining samples are Mannville brines.

[b]Ages corrected using the Ingerson–Pearson equation.

stones and have been classified as type 3 waters. The calculated (corrected) age of sample WW7 was less than 3000 years, whereas WW6 was greater than 10,000 years. In the Mannville brines, one determination suggested an age of over 24,000 years, whereas the remainder of the samples contained no live ¹⁴C, so no accurate age could be determined.

The range in the age obtained for the two shallow Chauvin waters can be explained by the location of the samples in the groundwater flow system. WW6 was sampled from beside the high hills in the southeastern corner of the study area (Figure 13). Water recharging the Ribstone Creek Member in this area has to travel a much longer distance, through thicker overburden sediments, to reach WW6, than the water recharging the Ribstone Creek from the nearby local topographic ridge in the vicinity of sample WW7.

One sample from the Mannville Group contained live ¹⁴C, thus suggesting that relatively recent meteoric waters may migrate to depth in the Chauvin area. This supports our interpretation that the regional flow of groundwater is directed downward.

Summary of Groundwater Flow Distribution

All methods of evaluating groundwater flow directions in this study showed that groundwater flow is

downward from the surface to the Mannville Group. Groundwater flow is adjusted to the water table and is controlled by the local relief of the land surface in the glacial overburden and shallow bedrock. Shallow-flow systems are recharged in the topographic highs and terminate in local depressions where various surficial discharge phenomena occur. The chemical composition of the shallow groundwater samples reflects a recent meteoric calcium bicarbonate character nearer to the surface and a simple evolution with depth to water dominated more by sodium chloride.

Below the depth of the Ribstone Creek Member, the pressure–depth profiles indicate that, on a regional scale, groundwater moves downward through the Lea Park, Colorado, and Joli Fou shales toward the Mannville Group. The underpressured Mannville also draws deeper formation waters from the Devonian carbonates. This influences the chemistry of the Mannville brines such that simple mixing between shallow waters and deeper Mannville waters apparently does not occur. The potentiometric maps of the Mannville Group suggest that lateral flow is in a northeastward direction toward the subcrop and outcrop regions of northern Saskatchewan.

All fluid dynamic, major ion, halogen ratio, stable isotope, and ¹⁴C evidence indicates that there is no upward groundwater flow from the Mannville Group to the near surface. Rather, all indications show unequivocally that groundwater moves downward from the surface to the Mannville Group.

This conclusion disagrees with the interpretation of Creaney et al. (1994) in which the Joli Fou shale acts as a regional hydraulic barrier. We reject that interpretation on the following grounds:

1. The conclusion of Creaney et al. (1994) is based on circumstantial chemical evidence that there is no geochemical correspondence between the oils above and below the Joli Fou shales; it is not based on hydraulic evidence.
2. The Joli Fou shale is a "dark marine shale with small proportions of interbedded fine- to medium-grained sandstone" (Leckie et al., 1994. p. 341), which is also known to be fractured in cores (B. J. Rostron, personal communication, 1993) and vary-

ing from 0 to 25 m in thickness. Such a rock unit cannot act as a regional hydraulic barrier over tens of thousands of square kilometers and for long periods of geologic time.

3. In our hydrogeologic experience, including nuclear fuel waste problems, no perfect hydraulic seal exists in nature, and Darcy's law is valid. The combination of these two circumstances inherently results in flow in the direction of the driving force, regionally downward in our case.

4. Most importantly, we believe that our hydraulic and chemical evidence is sufficiently strong to prove flow downward through the Colorado Formation, including the Joli Fou shale.

SOIL GAS SURVEY

Introduction

Anomalous geochemical constituents in the shallow subsurface have been considered as possible indicators of deep-seated petroleum accumulations since the early 1930s (Laubmeyer, 1933; Sokolov, 1936). The rationale for surface geochemical exploration of petroleum is rooted in the assumption that petroleum cap rocks are not totally impermeable and that some hydrocarbons can escape from the reservoir and reach the shallow subsurface. The migration of gases to the surface is driven by diffusion, buoyancy, and pore pressure gradients. The rising hydrocarbon gases generate a chemically reducing environment that results in a wide variety of physical, chemical, mineralogical, botanical, and microbiological anomalies collectively referred to as the *geochemical chimney* (Pirson, 1969; Duchscherer, 1981).

Geochemical surveys rely mostly on methods that directly detect hydrocarbon gases in the shallow subsurface. Other techniques use indicators that are indirectly related to migrating petroleum, such as carbon isotopes, microorganisms, and trace metals, or indicators that are not related to migrating hydrocarbons, such as helium anomalies.

An absence of soil gas anomalies over a petroleum reservoir is rarely, if ever, explained by groundwater flow and its potential to divert plumes of ascending hydrocarbon gas. Pirson's (1946) study appears to be the only investigation in which a correlation was seen between shallow groundwater flow systems and soil gas anomalies. During more than 5 years of petroleum geochemical studies, Pirson (1946) found that areas of anomalously high ethane gas emanations were located over regions of groundwater discharge. Conversely, over topographically high groundwater recharge areas, the ethane gas concentrations were anomalously low.

The Chauvin area soil gas study was designed to evaluate the potential of groundwater flow to influence the surficial geochemical expression of the petroleum accumulations that are located in the underlying Mannville Group. As a working hypothesis, it was assumed that oil fields located beneath recharge areas of local flow systems would display either weakened geochemical signatures or no signatures at all. This would be due to downward-flowing groundwater hindering the rise of upward-moving hydrocarbon gases or causing these gases to shift laterally from the recharge zone. Over discharge areas, however, upward flow of groundwater would coincide with upward migration of hydrocarbon gases and would thereby enhance the development of stronger geochemical signatures.

Chauvin Area Soil Gas Survey

Hydrocarbon gases in the soil can occur in three different forms: as free gas in the soil pores, as dissolved gases in shallow groundwater, and as gases adsorbed on soil particles. McRossan et al. (1972), in a survey of adsorbed soil gas in the Caroline, Alberta, area, concluded that the quantities of hydrocarbon gases released upon acid treatment depended on the carbonate contents in the shallow till soils. Because of the mineralogic similarities of the Chauvin surface soils to the Caroline area soils, this type of study seemed unsuitable for the Chauvin area. Application of the method based on dissolved hydrocarbon gas has been largely confined to the former Soviet Union or to studies of deep brines. A soil pore hydrocarbon gas study thus appeared to be the most appropriate and feasible for this investigation.

We used the Petrex® method to detect volatile hydrocarbons in the shallow subsurface. The method uses a ferromagnetic curie-point wire coated with activated charcoal and placed in a test tube sampler. The test tube is placed in a shallow hole dug about 1 m deep and left for several weeks to allow emanating hydrocarbon gases to accumulate on the wire. This method eliminates short-term gas fluctuations in the near-surface environment. Upon completion of field sampling, the test tubes are removed from the ground, sealed, and transported to the laboratory. There, hydrocarbon gases are desorbed from the wire, ionized, separated by mass, and counted by an automated mass spectrometer. In this way, a "spectral fingerprint" of atomic mass is obtained for each sample, including hydrocarbons ranging in mass from 15 to 240. This mass range includes the alkanes, alkyl aromatics, and cycloalkanes.

In areas of proven oil production, Petrex employs a set of training samples that are collected over known oil pools throughout the study area. The spectral fingerprints for the training set are then evaluated by a multivariate factor analysis technique to obtain a signature that is indicative of oil. All samples are compared to the training set. Those samples correlating well with the training set are given high factor score values indicating their similarity to the training set.

In this survey, 230 samplers were used, spaced about 400 m apart. The samplers were installed along three offset lines, two running north-south and a third one east-west. The lines were set so that they crossed petroleum accumulations situated beneath both topographic highs

Figure 17—Map of sampling distribution showing total ion counts of benzene plus toluene. Recharge and discharge areas are also shown.

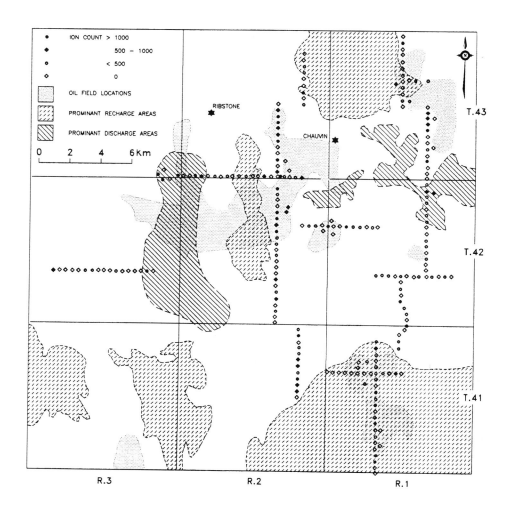

and topographic lows. To obtain background signatures, the lines were also extended into areas where no petroleum accumulations were known.

Field Observations

The set of factor scores provided by Petrex indicated that the strongest, most consistent values occurred in the samples taken over the north part of the Chauvin oil field (T 42–43, R 2). This is a relatively low-lying local area of groundwater discharge. These samples are offset to the east of the high ridge running in a north-south direction. Weaker factor score values occurred over the Chauvin North oil field in T 43, R 1, and over the Hayter oil field in T 41, R 1.

The locations of the soil gas samples and the total benzene plus toluene ion counts recorded at each sampling location are shown in Figure 17. These two compounds were selected for further evaluation for several reasons. Petrex suggested that benzene and toluene were strong indicators of oil in the area and that they constituted a significant part of the oil signature determined from the training set. In addition, benzene and toluene are two hydrocarbons that are least susceptible to microbial

degradation in the subsurface (Hitchon, 1974), while both Hunt (1979) and North (1985) indicate that these compounds have high solubilities in water. It was reasoned, therefore, that they might be more influenced by groundwater flow than other hydrocarbons.

The highest ion counts for benzene and toluene appear to correspond aerially with the highest factor scores, found in the topographic depression in T 42–43, R 2, over the northern part of the Chauvin oil field (Figure 17). This area is offset to the east of the high ridge that serves as a recharge region here. The samples located at the top of this ridge show no, or very low, ion counts for benzene and toluene. Over the Hayter oil field in T 41, R 1, the ion counts for benzene and toluene are again very low. This oil field is situated in a groundwater recharge area on the highest hills in the area. It is therefore inferred that benzene and toluene are transported laterally by groundwater flow to adjacent groundwater discharge areas. Higher ion counts recorded in the samples from the low area to the west of the Hayter field (T 42, R 1) strengthens this hypothesis. These higher ion counts are located over an area where no Mannville oil accumulations exist. Over the Chauvin North oil field, which is also situated beneath a groundwater recharge area, the benzene and toluene ion counts are, again, very low.

Interpretation of Soil Gas Survey Results

In evaluating the factor score distribution, a fundamental question arose about the Petrex analysis method. The collection of a training set over the study area inherently assumes that, over the entire area studied, the subsurface oil accumulations will result in similar spectral signatures at the surface. This assumption holds valid only in regions that are relatively homogeneous, where ascending gases migrate through rock strata with similar adsorptive and transmissive properties and are subjected to similar degrees of biological degradation and similar hydrological conditions. The composition of the oil throughout the study area is also assumed to be uniform. If the spectral fingerprints for the individual samples used in the training set vary as a result of any of these effects, then the oil signature determined by the Petrex method will be a weakened statistical combination of the oil signatures in the areas with varying conditions.

Petrex already acknowledges that petroleum signatures for oil pools associated with faults differ from signatures obtained in the same areas over oil pools with no faulting (Hayes et al., 1984; Villenave and Bloom, 1986). In these situations, both types of signatures indicate the presence of subsurface oil. However, gases migrating up through the faults have been subjected to different conditions than those to which gases migrating straight up through the rock strata are exposed. Different hydrocarbon gases reach the surface in each instance, resulting in different spectral signatures. In these cases, Petrex isolates the training set obtained over the fault from the training set obtained over the other areas. All samples are then compared to either one or both of the training set signatures. We propose that with groundwater recharge and discharge areas, Petrex should be isolating the training sets so that separate petroleum spectral fingerprints are obtained for the two hydraulically different area types. In this way, the spectral fingerprint indicative of petroleum will not be statistically weakened through a combination of possibly different unique fingerprints.

Surficial geochemical signatures obtained through the soil gas survey were not as intense as expected at the outset of the study. We therefore postulate that in the Chauvin area, groundwater descending regionally from the surface to the Mannville Group suppresses the upward movement of hydrocarbon gases, thus preventing the formation of pronounced soil gas anomalies.

The results of this study suggest that groundwater movement can play a significant role in the surface expression of petroleum reservoirs. In the Chauvin area, the distribution of petroleum-related hydrocarbon gases that reach the surface appears to be influenced by localized shallow groundwater movement. Stronger soil gas signatures were detected in the northern part of the Chauvin oil field where subsurface petroleum accumulations are located beneath a local groundwater discharge area. In the case of the Hayter oil field in T 41, R 1, which occurs beneath a topographically high recharge area, the observed soil gas signatures are weak.

SUMMARY AND CONCLUSIONS

Surface geochemical surveys are commonly interpreted in a too simplistic manner. This study intends to draw attention to one mechanism—groundwater flow—which may play a decisive role in the success or failure of such surveys. Specifically, in our view, groundwater flow should be considered in two aspects of surface geochemical exploration: (1) in determining the applicability of the soil gas survey method in the initial assessment of an area and (2) in the interpretation of observational results.

In the Chauvin area of east-central Alberta, groundwater was determined to move downward on a regional scale from the surface to the underpressured Lower Cretaceous Mannville Group. The flow pattern was established by studying the hydrodynamics of this part of the Alberta Basin and by a series of geochemical groundwater analyses. On a local scale, groundwater flow is adjusted to the topography, with highlands serving as local groundwater recharge areas and depressions as local groundwater discharge areas.

Pressure–depth profiles indicate that the vertical pressure gradients are subhydrostatic above and superhydrostatic below the Mannville Group. This pressure pattern suggests downward movement of water from above and upward flow beneath this reservoir. According to the potentiometric surface maps, regional groundwater movement is toward the northeast within the Mannville Group.

This interpretation of the fluid dynamic parameters is supported by the water's chemical properties in the area. Close to the surface, shallow groundwaters are of a calcium bicarbonate type, reflecting meteoric origin. With increased depth, the groundwaters evolve initially to sodium bicarbonate type, through cation exchange, and eventually to a NaCl type, through the incorporation of older formation waters from the low-permeability shaley sediments of the Belly River Formation. Halogen ratios indicate that the increase in chloride concentrations in the Ribstone Creek sandstone waters are not due to the mixing of deeper Mannville brines with shallow meteoric water.

Stable isotope contents of oxygen and hydrogen indicate that all shallow waters examined are derived from recent meteoric sources and not from upward movement of Mannville brines. When age dated using ^{14}C analyses, one Mannville brine sample showed a geologically young age of 24,000 years, also suggesting that fresh meteoric water is moving downward from the surface. Shallow Ribstone Creek water also shows a range in age depending on the depth of the sample from the surface. Analyses show ^{14}C ages ranging from about 2600 to >10,000 years.

The soil gas survey was undertaken once the groundwater flow distribution had been confirmed. The results of the survey suggest that groundwater flow can have a significant effect on the resulting location and intensity of surface geochemical expressions of petroleum accumulations.

The relatively poor intensity and consistency of the observed hydrocarbon anomalies appears to suggest that surface geochemical surveys are unreliable in areas of regional groundwater recharge, that is, where water movement is opposite to the expected movement of gas. Nevertheless, in spite of weak geochemical signatures, factor scores calculated by Petrex, as well as total ion counts for benzene and toluene, were observed to be higher and more consistent where petroleum accumulations were situated beneath local discharge areas. These parameters were considerably weakened where petroleum accumulations were located beneath local groundwater recharge areas.

Geochemical explorationists could modify their investigative and analytical techniques to account for different surface geochemical signatures associated with groundwater recharge and discharge areas. This may serve to refine and improve the success rates of these types of surveys.

As a final conclusion, we suggest that surface geochemical exploration techniques may be better suited to sedimentary basins where there is considerable upward flow of formation waters and extensive regional discharge areas, such as in the Upper Rhine Graben in western Europe (Tóth and Otto, 1993) and in Railroad Valley, Nevada (Reinsborough, 1992). In these situations, areas of ascending formation waters tend to coincide with areas of major petroleum accumulations and with upward movement of hydrocarbon gases. Even in such cases, however, the pattern of groundwater movement between the petroleum reservoirs and the ground surface should be evaluated to ascertain the possible affects of flow on the type, intensity, and areal distribution of the expected geochemical anomalies.

Acknowledgments *The major portion of the operating and analytical expenses of this project were funded from NSERC Strategic Research Grant (Individual) No. G1658 to J. Tóth. Petrex Exploration and Development of Lakewood, Colorado, provided the soil gas collectors, analytical services, and factor analysis at reduced commercial rates. K. Muehlenbachs of the Geology Department, University of Alberta, Edmonton, Canada, made his laboratory available and gave advice on the environmental isotope analyses. We are grateful for all these contributions to our project.*

REFERENCES CITED

Abercrombie, H. J., and E. G. Fullmer, 1992, Regional hydrogeology and fluid geochemistry of the Mannville Group, Western Canada Sedimentary Basin: synthesis and reinterpretation, *in* Y. K. Kharaka and A. S. Maest, eds., Proceedings, International Symposium on Water–Rock Interaction, July 13–18, Park City, Utah: Rotterdam, A.A. Balkhema Publishers, v. 2, p. 1101–1104.

Chave, K. E., 1960, Evidence on history of sea water from chemistry of deeper subsurface waters of ancient basins: AAPG Bulletin, v. 44, p. 357–370.

Christopher, J. E., 1980, The Lower Cretaceous Mannville Group of Saskatchewan—a tectonic overview, *in* L. S. Beck, J. E. Christopher, and D. M. Kent, eds., Lloydminster and beyond: geology of Mannville hydrocarbon reservoirs: Saskatchewan Geological Survey, Special Publication #5, p. 3–32.

Collins, A. G., W. P. Zelinski, and C. A. Pearson, 1967, Bromide and iodide in oilfield brines in some Tertiary and Cretaceous formations in Mississippi and Alabama: U.S. Bureau of Mines, Report of Investigations #6959, 27 p.

Craig, H., 1961, Isotopic variations in meteoric waters: Science, v. 133, p. 1702–1703.

Creaney , S., J. Allen, K. S. Cole, M. G. Fowler, P. W. Brooks, K. G. Osadetz, R. W. Macqueen, L. R. Snowdon, and C. I. Diediger, 1994, Petroleum generation and migration in the Western Canada Sedimentary Basin, *in* G. D. Mossop and I. Sheston, comps., Geological atlas of the Western Canada Sedimentary Basin: Canadian Society of Petroleum Geologists and Alberta Research Council, p. 455–468.

Davis, P. B., 1987, Modeling areal, variable-density, groundwater flow using equivalent freshwater head—analysis of potentially significant errors, *in* Proceedings in solving groundwater problems with models, Conference Exposition, Association of Ground Water Scientists and Engineers and International Water Modeling Center: Halcomb Research Institute, Denver, Colorado, v. 2, p. 888–903.

Debnam, A. H., 1969, Geochemical prospecting for petroleum and natural gas in Canada: Geological Survey of Canada Bulletin, v. 177, 217 p.

Domenico, P. A., and F. W. Schwartz, 1990, Physical and chemical hydrogeology: New York, John Wiley and Sons, 824 p.

Duchscherer, W., 1981, Carbonates and isotope ratios from surface rocks: a geochemical guide to underlying petroleum accumulations, *in* B.M. Gottlieb, ed., Unconventional methods in exploration for petroleum and natural gas II: Dallas, Texas, Southern Methodist University Press, p. 201–218.

Dunning, N. E., H. J. Henley, and A. G. Lange, 1980, The Freemont field: an exploration model for the Lloydminster area, *in* L. S. Beck, J. E. Christopher, and D. M. Kent, eds., Lloydminster and beyond: geology of Mannville hydrocarbon reservoirs: Saskatchewan Geological Survey, Special Publication #5, p. 132–148.

Fetter, C. W., 1994, Applied hydrogeology, 3rd edition: New York, Macmillan College Publishing, 691 p.

Fontes, J. Ch., 1986 Environmental isotopes in groundwater hydrology, *in* P. Fritz and J. Ch. Fontes, eds., Handbook of environmental geochemistry: Amsterdam, Elsevier Scientific Publishing, p. 74–140.

Fontes, J. Ch. and J. M. Garnier, 1979, Determination of the initial ^{14}C activity of the total dissolved carbon: a review of the existing models and a new approach: Water Resources Research, v. 15, p. 399–413.

Freeze, R. A., and J. A. Cherry, 1979, Groundwater: Englewood Cliffs, New Jersey, Prentice Hall, 604 p.

Geyh, A. M., and R. H. Wagner, 1979, Guideline for groundwater sampling for isotope analysis: Laboratory manual, ^{14}C and ^{3}H laboratory, Nieders, Landesamt fur Bodenforschung, Hannover, 4 p.

Hanor, J. S., 1987, Origin and migration of subsurface sedimentary brines: SEPM Short Course Notes #21, 241 p.

Hayes, C., M. M. Thacker, and J. H. Viellenave, 1984, Exploration techniques aided by computer tiering: World Oil, Houston, Texas, Gulf Publishing, 5 p.

Hitchon, B., 1974, Application of geochemistry to the search for crude oil and natural gas, *in* A. A. Levinson, ed., Introduction to exploration geochemistry: Calgary, Alberta, Applied Publishing, p. 509–545.

Hitchon, B., and I. Friedman, 1969, Geochemistry and origin of formation waters in the Western Canadian Sedimentary Basin—I, stable isotopes of hydrogen and oxygen: Geochim. Cosmochim. Acta, v. 33, p. 1321–1349.

Holysh, S., 1989, Petroleum related geochemical signatures and regional groundwater flow, Chauvin area, east-central Alberta: M.Sc thesis, Department of Geology, University of Alberta, Edmonton, 208 p.

Hubbert, M. K, 1940, The theory of groundwater motion: Journal of Geology, v. 48, p. 785–944.

Hunt, J. M., 1979, Petroleum geochemistry and geology: San Fransisco, W.H. Freeman and Company, 617 p.

Jardine, D., 1974, Cretaceous oil sands in western Canada, *in* L. V. Hills, eds., Oil sands—fuel of the future: Canadian Society of Petroleum Geologists Memoir 3, p. 50–67.

Jones, V. T., and R. J. Drozd, 1983, Predictions of oil and gas potential by near-surface geochemistry: AAPG, v. 67, no. 6, p. 932–952.

Laubmeyer, G., 1933, A new geophysical prospecting method, especially for deposits of hydrocarbons: Petroleum, v. 29, p. 1–4.

LeBreton, E. G, 1963, Groundwater geology and hydrology of east-central Alberta: Research Council of Alberta, Bulletin #13, 64 p.

Leckie, D. A., J. P. Bhattacharya, J. Bloch, C. F. Gilboy, and B. Norris, 1994, Cretaceous Colorado/Alberta Group of the Western Canada Sedimentary Basin, *in* G. D. Mossop and I. Sheston, comps., 1994, Geological atlas of the Western Canada Sedimentary Basin: Canadian Society of Petroleum Geologists and Alberta Research Council, p. 335–352.

Lloyd, J. W., K. W. F. Howard, N. R. Pacey, and J. H. Tellam, 1982, The value of iodide as a parameter in the chemical characterization of groundwaters: Journal of Hydrology, v. 57, p. 247–265.

Lusczynski, N. J., 1961, Head and flow of ground water of variable density: Journal of Geophysical Research, v. 66, no. 12, p. 4247–4256.

Mossop, G. D., and I. Sheston, comps., 1994, Geological atlas of the Western Canada Sedimentary Basin: Calgary, Canadian Society of Petroleum Geologists and Alberta Research Council, 510 p.

McCrossan, R. G., N. L. Ball, and L. R. Snowdon, 1972, An evaluation of surface geochemical prospecting for petroleum, Olds–Caroline area, Alberta: Geological Survey of Canada Paper 71-31, 101 p.

North, F. K., 1985, Petroleum geology: London, Allen and Unwin, 607 p.

Philp, R. P., and P. T. Crisp, 1982, Surface geochemical methods used for oil and gas prospecting—a review: Journal of Geochemical Exploration, v. 17, p. 1–34.

Pirson, S. J., 1946, Disturbing factors in geochemical prospecting: Geophysics, v. 11, p. 312–320.

Pirson, S. J., 1969, Geological, geophysical, and chemical modifications of sediments in the environment of oil fields, *in* W. B. Heroy, ed., Unconventional methods in exploration for petroleum and natural gas: Dallas, Southern Methodist University Press, p. 159–186.

Price, L. C., 1980, Utilization and documentation of vertical oil migration in deep basins: Journal of Petroleum Geology, v. 2, no. 4, p. 353–387.

Price, L. C., 1986, A critical overview and proposed working model of surface geochemical exploration , *in* M. J. Davidson, ed., Unconventional methods in exploration for petroleum and natural gas, IV: Dallas, Southern Methodist University Press, p. 245–290.

Putnam, P. E., 1982, Aspects of the petroleum geology of the Lloydminster heavy oil fields, Alberta and Saskatchewan: Canadian Society of Petroleum Geologists Bulletin, v. 30, p. 81–311.

Reinsborough, B. C., 1992, Exploration application of hydrogeology in the Great Basin, Nevada (abs.): AAPG Annual Convention, Program with Abstracts, Calgary, Alberta, Canada, p. 109.

Rittenhouse, G., 1967, Bromide in oil-field waters and its use in determining possibilities of origin of these waters: AAPG Bulletin, v. 51, no. 12, p. 2430–2440.

Schumacher, D., 1990, Surface exploration in mature basins: advances in the eighties, applications for the nineties: American Petroleum Geochemical Explorationists Newsletter, no. 19, p. 1.

Sokolov, V. A., 1933, New prospecting method for petroleum and gas: Technika Bulletin, February, NGRI, no. 1.

Sokolov, V. A., 1936, Gas surveying: Moscow, Gostoptekhizdat, 269 p.

Tóth, J., 1971, Groundwater discharge: a common generator of diverse geologic and morphologic phenomena: Bulletin of International Scientific Hydrology, v. 16, p. 7–24.

Tóth, J., 1978, Gravity induced cross-formational flow of formation fluids, Red Earth region, Alberta, Canada: Canadian Society of Petroleum Geologists Bulletin, v. 27, p. 63–86.

Tóth, J., 1980, Cross-formational gravity flow of groundwater: a mechanism for the transport and accumulation of petroleum (the generalized hydraulic theory of petroleum migration), *in* W. H. Roberts III and R. J. Cordell, eds., Problems of petroleum migration: AAPG Studies in Geology 10, p. 121–167.

Tóth, J., 1988, Ground water and hydrocarbon migration, *in* W. Back, J. S. Rosenshein, and P. R. Seaber, eds., Hydrogeology: GSA Decade of North American Geology, v. O-2, p. 485–502.

Tóth, J., and C. J. Otto, 1993, Hydrogeology and oil deposits at Pechelbronn–Soultz–Upper Rhine Graben: Acta Geologica Hungarica, v. 36, no. 4, p. 375–393.

Vigrass, L. W., 1977, Trapping of oil at intra-Mannville (Lower Cretaceous) disconformity in the Lloydminster area, Alberta and Saskatchewan: AAPG Bulletin, v. 61, p. 1010–1028.

Villenave, J. H., and D. N. Bloom, 1986, The Petrex fingerprint technique as applied to dry gas exploration, Bounde Creek field, Colusa and Glenn counties, California (abs.): AAPG Pacific Coast Section Meeting, Program with Abstracts, April 17, Denver, Colorado.

Wadsworth, Jr., A. H., 1991, President's message: American Petroleum Geochemical Explorationists Newsletter, no. 33, p. 1.

Tóth, J., 1996, Thoughts of a hydrogeologist on vertical migration and near-surface
geochemical exploration for petroleum, *in* D. Schumacher and M. A. Abrams,
eds., Hydrocarbon migration and its near-surface expression: AAPG
Memoir 66, p. 279–283.

Chapter 20

Thoughts of a Hydrogeologist on Vertical Migration and Near-Surface Geochemical Exploration for Petroleum

József Tóth

Department of Earth and Atmospheric Sciences
University of Alberta
Edmonton, Alberta, Canada

Abstract

Near-surface exploration for petroleum is based on the detection and interpretation of a great variety of natural phenomena occurring at or near the land surface or sea floor and attributed, directly or indirectly, to hydrocarbons migrating vertically upward from leaky reservoirs at depth. Development of surface exploration methods began in the early 1930s with the chemical analysis of gaseous hydrocarbons in soil air. It has since expanded to include a wide range of geochemical, geophysical, mineralogic, microbiological, and other types of anomalies. The great advances in the observational and analytical techniques, however, have not been matched with similar improvements in the method's efficiency and effectiveness in terms of new field discoveries. From a hydrogeologic perspective, the inconsistency of the results can be explained, at least in part, by a disregard for the possible effects that groundwater flow may have on the nature and intensity of the anomalies as well as on their positions relative to subsurface sources and accumulations. Since the principles and investigative techniques of regional groundwater flow are well established, introducing a hydrogeologic component of near-surface petroleum exploration should be technically easy and economically feasible. The expected improvement in the results should serve as a strong incentive for purposeful collaboration between near-surface explorationists and hydrogeologists.

INTRODUCTION

Near-surface exploration for petroleum is based on the detection and interpretation of a great variety of natural phenomena occurring at or near the land surface or sea floor and attributed to hydrocarbons migrating vertically upward from leaky reservoirs at depth. Great advances have been made over the last 60 years in the observational and analytical techniques used in near-surface exploration. However, these advances have not resulted in comparable improvements in the method's efficiency and effectiveness in terms of new field discoveries. This dichotomy is not that puzzling to a hydrogeologist. In this paper, I attempt to summarize the reasons why, based on hydrogeologic reasoning, near-surface geochemical exploration for petroleum appears to be a viable and promising approach, but also why the inclusion of hydrogeologic considerations is indispensable if the method is to realize its full potential. Also, I hope to show that the

efficiency and effectiveness of near-surface exploration is bound to improve by considering the role of advective transport of petroleum by moving groundwater. Perhaps these notes will succeed in inducing some dialogue between hydrogeologists and near-surface explorationists. The chances of a breakthrough in the inconsistent performance of near-surface exploration should thereby improve, and as a consequence, the level of confidence of potential users of these methods should increase.

In this paper, I first present a sketch of the evolutionary history and current state of the relevant aspects of near-surface exploration for petroleum. Next, an analysis is given of the principal characteristics of the concept and practice of near-surface exploration, followed by an overview of the relevant hydrogeologic aspects. I conclude the essay by making some suggestions concerning the practical aspects of including hydrogeology in near-surface exploration of petroleum.

NEAR-SURFACE EXPLORATION FOR PETROLEUM TODAY

The phrase *near-surface exploration for petroleum* is used here to denote a variety of observational, sampling, and analytical methods and techniques applied at or near the earth's surface to search for hydrocarbon accumulations at depth and to detect the generation of thermogenic hydrocarbons. These methods are aimed at the detection of (1) gaseous and liquid hydrocarbons that escape from accumulations and move toward the land surface or (2) the various chemical, physical, mineralogic, and biological effects that may accompany their upward migration.

Near-surface exploration for petroleum was started by German and Russian investigators in the early 1930s (e.g., Laubmeyer, 1933; Sokholov, 1935) with the development of methods of detecting trace amounts of hydrocarbon gases, primarily methane, in soil air. During the second half of the 1930s, American explorationists expanded the Russian techniques by analyzing the soil itself for anomalous amounts of absorbed and occluded materials, including heavier hydrocarbons, organic wax, fluorescent substances, and inorganic compounds of carbonates, sulfates, and chlorides (Rosaire, 1938; Horvitz, 1939). As the detection and analytical techniques improved and the range of indicator materials broadened, the possibility of discrepancies between the positions of the near-surface anomalies and their assumed sources at depth had become uncomfortably evident. Referring specifically to ethane emanation intensity, Pirson (1946a, p. 312) wrote,

> A number of factors have been found to be highly disturbing, namely, earth topography, ground water percolation and seepage, barometric pressure variations, etc. These effects result in fluctuations of the rate of escape of hydrocarbons accompanied by horizontal shifts of leakage which give rise to the creation of artificial leakage highs altogether meaningless from the point of view of oil and gas accumulation at depth. Certain quantitative rules of interpretation have been established which permit weeding out the meaningless anomalies provided sufficient information is at hand on the topography and water table movement.

Notwithstanding his reservations, Pirson proceeded to propose several physically based methods for the detection of hydrocarbon migration paths. Recognizing the chemically reducing effect of organic matter, Pirson (1971) suggested that the column of hydrocarbon-rich earth material standing vertically between an oil reservoir and the land surface can act as a "fuel cell," inducing natural electrotelluric currents between the reduced earth column and its relatively oxidized surroundings. The currents are accompanied by anomalies of electrotelluric potential, pH and Eh, which can be measured at the earth's surface and thus be used to map out the reduced column. Pirson (1971) called this reduced column of earth the *geochemical chimney* and associated it with measurably reduced gamma ray activity. Again, he warned of the disturbing effects of various factors, including the hydrodynamics of infiltration.

The geochemical chimney has proved to be a useful concept in near-surface exploration for petroleum. It can be considered as the generator, as well as the reference base in space, of a wide variety of natural processes and phenomena which, by the 1990s, have expanded to include low-frequency electromagnetic induction, soil alteration, electrical resistivity patterns, vegetation changes, and microbiological (bacterial) colonies. Not surprisingly, new detection methods, adapted to the sea floor, sea surface, air, and land surface have also proliferated.

Indeed, over the past few decades, the research and prospecting activities of the near-surface exploration community have focused much more heavily on the development of detection and survey techniques and analytical methods than on the understanding and exploitation of the migration processes that lead to observable phenomena, or on the factors that control them. Consequently, the industry today is sufficiently well equipped to conduct highly sensitive surveys in water, on land, and in air. Nevertheless, in terms of new field discoveries and consistency of predictions, the results of these surveys do not seem to have improved correspondingly since the 1930s, and no breakthroughs in performance can be identified.

The situation today can be characterized by the views of two prominent proponents of near-surface petroleum exploration as expressed in recent issues of the *American Petroleum Geochemical Explorationists Newsletter*. Referring to the large number and variety of prospecting techniques on the market and their competing promoters, Schumacher (1990, p. 1) asks, "Is it any wonder explorationists are confused or at least skeptical?" Wadsworth (1991, p. 1) summarizes his impressions from geologists and geophysicists around the United States in the APGE President's Message as follows: "I am struck by their lack of acceptance of the concept of vertical migration. . . . In my opinion, this is a primary cause for the dismay we all feel about the acceptance of surface prospecting by the exploration community."

HYDROGEOLOGIC PERSPECTIVE

Near-surface exploration for petroleum is really a problem of hydrocarbon migration from deep-seated accumulations to the land surface or the sea floor. Nevertheless, there is virtually no sign of serious and credible attempts in the otherwise extensive literature of near-surface exploration (e.g., Groth and Groth, 1994) to understand the various migration mechanisms, let alone quantify their relative importance and employ this understanding in exploration strategies.

The near-surface explorationist seems to accept the proposition almost without question that hydrocarbons migrate from deposit to surface either by buoyancy or diffusion or both (Klusman, 1993). Observable signatures at the surface are thus expected to be positioned vertically or nearly vertically above the accumulations (Davidson, 1982). This axiomatic reliance on buoyancy and diffusion as the principal mechanisms of hydrocarbon transport is also reflected by the deeply ingrained term *vertical migration*. A "vertical" direction of transport is asserted explicitly, but an "upward" sense of migration is also implied tacitly. The possibility that vertical upward flow of hydrocarbons driven by buoyancy or diffusion may be deflected by lateral advection is usually minimized by unsupported and qualitative judgmental arguments.

For example, in an authoritative work on applied geochemistry, Siegel (1974, p. 231–232) wrote,

> As for the influence of a lateral migration of waters trapped in the rocks overlying the petroleum accumulations on the dispersion of rising gases, such lateral migration is probably slow with respect to the velocity of vertical migration of the gases.

Even if it were supported by quantitative flow rates, such a statement could hardly be generalized in view of the extreme variations in flow-controlling factors such as permeability, hydraulic gradients, fluid densities and viscosities, temperatures, and structure found in actual situations. Indeed, in an exhaustive review of the topic, the conclusions of Philp and Crisp (1982, p. 1) were diametrically opposite to Siegel's position. Broaching the question of the effect of advective transport, Philp and Crisp wrote,

> Onshore geochemical prospecting appears to have more problems associated with it than offshore prospecting due to the more complex migration mechanism of near-surface waters containing dissolved gases. No onshore prospecting studies have been published [that] thoroughly consider this factor, and the success of onshore prospecting remains equivocal. . . . Failures in prospecting to date are attributable to the simplistic manner in which data have been interpreted; insufficient attention has been paid to the hydrological and geological factors which modify the upward migration of indicator species to the surface.

It is telling, perhaps, to compare these words with those of Pirson (1946a) written 36 years earlier!

From a hydrogeologic viewpoint, there is no justification to consider any of the three possible subsurface transport mechanisms—advection, buoyancy, and diffusion—as intrinsically dominant or negligible. However, although relatively little is known about the rates of diffusion and buoyancy other than that they are strongly affected by several intractable and site-specific parameters, the rates of advective groundwater flow are generally known and relatively easy to evaluate. It is thus puzzling that the most tractable of the transport mechanisms remains ignored.

The two most important ways by which advection can lead to erroneous interpretation of near-surface observations are (1) displacement of anomalies by lateral flow with respect to a vertical position above the source deposits—the "artificial leakage highs" of Pirson (1946a)—and (2) prevention of the formation of anomalies by descending groundwaters negating the rise of hydrocarbons that would otherwise form the geochemical chimney and thus the near-surface signatures. In

moderately to relatively highly permeable aquifers, commonly found in the upper 2 km of the earth's crust and used for development of water supplies, lateral flow rates can reach tens or hundreds of meters per year. Although truly meaningful evaluations require rates of diffusion and buoyancy for comparison, it is hard to imagine that these lateral flow rates would not affect the position or even existence of near-surface anomalies. Such possibilities have been convincingly illustrated by the examples of Thrasher et. al. (this volume).

Groundwaters can move downward over extended areas driven by gravitational force gradients, by elastic dilation of pore space consequent upon erosional unloading, or by other mechanisms (Corbet and Bethke, 1992). All these cases are characterized by subhydrostatic pore pressures at some depth. If a zone of underpressuring exists above a source deposit and the downward drive from the surface is sufficiently strong, hydrocarbons possibly escaping from the accumulation may be prevented from reaching the surface. Such situations are common in areas of recent uplift, such as the Western Canada Sedimentary Basin. The possible effect of underpressuring (i.e., downward-driving forces) on anomaly development was stated succinctly by Harnett (1990, p. 4), an independent consulting geologist, based on field observations: "if it is not pressured, it is not going to make it to the surface." Also, the lack of reliability of a soil gas survey has been ascribed by Holysh and Tóth (this volume) to downward flow of groundwaters above oil fields due to underpressuring.

In a recent study of a comparable problem, Stute et al. (1992) compared measured fluxes of helium gas to fluxes calculated by applying Tóth's (1963) model of advective groundwater flow to the Great Hungarian Plains region of the Pannonian Basin. Both the observed and calculated helium concentrations were significantly lower in recharge areas than in discharge areas. The results clearly show the effect that downward and upward water flow can have on the flux of gases migrating upward by buoyancy and diffusion. Also, the effect of the sense and intensity of vertical hydrodynamic flow on the sealing efficiency of leaky cap rocks of petroleum accumulations is well known to the reservoir engineer. A strong downward flow can prevent any escape of hydrocarbons from the pool, while an intensive upward flow can render the cap rock completely inefficient as a seal (e.g., Schowalter, 1979).

If one considers the potentially decisive role that advective transport of hydrocarbons by moving groundwater can have in the formation of near-surface signatures of petroleum accumulations, as well as the relative ease by which regional water flow can be evaluated, the reluctance of near-surface explorationists to include groundwater flow studies in their methodology is puzzling. Can this reluctance be attributed to the fundamental difference between explorationists and hydrogeologists in educational background and professional experience needed to deal with hydrogeologic problems? If this is the case, the gap can be easily bridged by consciously and mutually fostered interdisciplinary collaboration.

The principles and controlling factors of regional groundwater flow, and its potential to interfere with the development of accumulation-indicating signatures, are well understood (Tóth, 1987, 1988). Furthermore, the evaluation of groundwater flow conditions is a matter of daily routine for the regional hydrogeologist and should not represent any major practical problems in most cases. There are three main types of approaches by which regional groundwater flow can be studied: (1) field observations of fluid dynamic parameters, (2) field mapping of natural phenomena (field effects) related to groundwater flow, and (3) mathematical modeling of groundwater flow and potential flow patterns.

Fluid dynamic parameters include hydraulic head, pore pressure–depth profiles, pressure–depth gradients, and dynamic pressure increment (Tóth, 1980). These parameters provide direct information on the actual distribution pattern (direction, depth, and intensity) of groundwater flow. They can be derived from measurements of pore fluid pressures and fluid levels taken in water wells, hydrocarbon boreholes, or test holes dedicated to hydrogeologic studies or drilled for other purposes. Naturally, this approach can be most readily applied in populated areas and mature basins where boreholes already exist and in areas with good access for light drilling equipment.

A wide variety of natural phenomena arise from the physical, chemical, and dynamic interaction between moving groundwater and its physical environment (Tóth, 1984). Regional groundwater flow can have observable effects on a region's hydrology, soil types, mineralogy, vegetation, soil and rock mechanics, geomorphology, economic minerals, and subsurface heat distribution. In addition, these effects tend to be self-organized according to recharge and discharge areas of groundwater flow systems and thus can be diagnostic of the natural flow distribution. It is relevant to note that many of the phenomena indicating natural groundwater discharge, such as quick sand, phreatophytic plants, mud volcanoes, saline soils, algae and bacterial colonies, thermal mineral springs, and soil cementation, are also frequently found associated with near-surface hydrocarbon anomalies.

Mathematical modeling can be used either to give a first approximation of groundwater flow distribution in areas without field data or to verify or refine flow patterns derived from observations made in the field. The most complete knowledge of groundwater flow is obtained from a balanced combination of information provided by all three approaches.

These hydrogeologic studies can yield answers to questions on groundwater flow that are indispensable for the rational evaluation of migration paths from a reservoir to a near-surface anomaly. The following attributes of groundwater flow can be established:

1. Nature—diffuse, through the continuous pore space of the rock, or channeled along faults, unconformities, or other discontinuities of permeability?
2. Direction—upward, downward, or lateral?
3. Intensity—zero, mild, moderate, or high?
4. Zonation—shallow or deep, single or multiple?
5. Relative importance of advection, buoyancy, and diffusion.

Several vital pieces of information for the near-surface explorationist can be garnered from, or assessed with the aid of, this hydrogeologic knowledge. Such information might include the position of anomalies relative to their source (i.e., the attitude, shape, and dimensions of the geochemical chimney) and the expected intensity and pattern of the signatures. Such studies may identify entire regions as being inappropriate for near-surface exploration, such as large areas of significant underpressuring. In contrast, groundwater discharge areas, such as the central parts of most intermontane basins, and other overpressured regions, such as offshore compacting basins, may be particularly rewarding prospects for near-surface exploration.

SUMMARY AND CONCLUSIONS

Near-surface exploration for petroleum was born more than 60 years ago. It started with the field detection and chemical analysis of microseeps of hydrocarbon gases attributed to leakage from subsurface reservoirs of petroleum. Over the years, the approach has been expanded to include chemical analyses for a great variety of organic and inorganic substances, as well as the measurement of a wide range of physical properties of the geochemical chimney, that is, the rocks and soils altered by migrating hydrocarbons above leaky reservoirs. However, notwithstanding the diversity and advanced state of detection and analytical techniques, near-surface exploration still seems to yield inconsistent results not noticeably better than in its early days. As a consequence, the approach lacks credibility among many potential users.

In my view, the main reason for this unsatisfactory and unnecessary state of affairs is that advective transport of hydrocarbons between the reservoired sources and the near-surface anomalies is ignored in interpretation. Near-surface explorationists tend to accept the phrase "vertical migration" literally. They concentrate their intellectual and financial resources on the ever-continuing improvement of the already advanced and diverse techniques of detection and analysis. However, the effect of groundwater flow, which can shift, modify, or obliterate anomalies, is not considered or even recognized by most explorationists despite the sporadically but clearly stated warnings found in the literature.

It is likely that the possible effects of groundwater flow are ignored because near-surface explorationists make no attempt to see them. The natural sciences are replete with examples of discoveries that were there for all to see, but were found only by those who were consciously looking. Hydrogeology is well equipped, both theoretically and technically, to evaluate site-specific situations. In most cases, even existing data can be used and would suffice. What's really needed is for the near-surface exploration community to see, and thus appreciate, the role and importance of groundwater flow in the development of diagnostic anomalies in a number of well-planned and deliberately targeted research projects. The predictable result of such action would be the removal of the stigma of unreliability from near-surface exploration for petroleum.

REFERENCES CITED

Corbet, T. F., and C. M. Bethke, 1992, Disequilibrium fluid pressures and groundwater flow in the Western Canada Sedimentary Basin: Journal of Geophysical Research, v. 95, p. 7203–7217.

Davidson, M. J., 1982, Toward a general theory of vertical migration: Oil and Gas Journal, June 21, p. 288.

Groth, P. K., and L. W. Groth, 1994, Bibliography for surface and near-surface hydrocarbon prospecting methods: Denver, Colorado, Association of Petroleum Geochemical Explorationists, 143 p.

Harnett, R. A., 1990, Technical message: APGE Newsletter, no. 21, p. 3.

Horvitz, L., 1939, On geochemical prospecting: Geophysics, v. 4, p. 210–225.

Klusman, R. W., 1993, Soil gas and related methods for natural resource exploration: Chichester, U.K., John Wiley and Sons, 483 p.

Laubmeyer, G., 1933, A new geophysical prospecting method, especially for deposits of hydrocarbons: Petroleum, v. 29, p. 1–4.

Philp, R. P., and P. T. Crisp, 1982, Surface geochemical methods used for oil and gas prospecting—a review: Journal of Geochemical Exploration, v. 17, p. 1–34.

Pirson, S. J., 1946a, Disturbing factors in geochemical prospecting: Geophysics, v. 11, p. 312–320.

Pirson, S. J., 1946b, Emanometric oil and gas prospecting: The Petroleum Engineer, v. 17, p. 132–142.

Pirson, S. J., 1971, New electric technique can locate gas and oil: World Oil, v. 172, pt. 1, p. 69–72; pt. 2, p. 72–74.

Rosaire, E. E., 1938, Shallow stratigraphic variations over Gulf Coast structures: Geophysics, v. 3, p. 96–115.

Schowalter, T. T., 1979, Mechanics of secondary hydrocarbon migration and entrapment: AAPG Bulletin, v. 63, no. 5, p. 723–760.

Schumacher, D., 1990, Surface exploration in mature basins: advances in the eighties, applications for the nineties: APGE Newsletter, no. 19, p. 1.

Siegel, F. R., 1974, Applied geochemistry, *in* Geochemical prospecting for hydrocarbons: New York, John Wiley and Sons, p. 228–255.

Sokholov, V. A., 1935, Summary of the experimental work of the gas survey: Neftyanoye Khozyastvo, v. 27, no. 5, p. 23–34.

Stute, M., C. Sonntag, J. Deák, and P. Schlosser, 1992, Helium in deep circulating groundwater in the Great Hungarian Plain: flow dynamics and crustal and mantle helium fluxes: Geochimica et Cosmochimica Acta, v. 56, p. 2051–2067.

Tóth, J., 1963, A theoretical analysis of groundwater flow in small drainage basins: Journal of Geophysical Research, v. 68, no. 16, p. 4796–4812. (Reprinted in 1983.)

Tóth, J., 1980, Cross-formational gravity flow of groundwater: a mechanism of the transport and accumulation of petroleum (the generalized hydraulic theory of petroleum migration), *in* W. H. Roberts III and R. J. Cordell, eds., Problems of petroleum migration: AAPG Studies in Geology 10, p. 121–167.

Tóth, J., 1984, The role of regional gravity flow in the chemical and thermal evolution of groundwater, *in* B. Hitchon and E. Wallick, eds., Practical applications of ground water geochemistry: Proceedings of First Canadian/American Conference of Hydrogeology, Banff, Alberta, Canada, p. 1–39.

Tóth, J., 1987, Petroleum hydrogeology: a new basic in exploration: World Oil, v. 205, p. 48–50.

Tóth, J., 1988, Groundwater and hydrocarbon migration, *in* W. Back, J. S. Rosenshein, and P. R. Sieber, eds., The geology of North America: Hydrogeology, v. 0–2, p. 485–502.

Wadsworth, Jr., A. H., 1991, President's Message: APGE Newsletter, no. 33, p. 1.

Price, L. C., 1996, Research-derived insights into surface geochemical hydrocarbon exploration, *in* D. Schumacher and M. A. Abrams, eds., Hydrocarbon migration and its near-surface expression: AAPG Memoir 66, p. 285–307.

Chapter 21

Research-Derived Insights into Surface Geochemical Hydrocarbon Exploration

Leigh C. Price

U.S. Geological Survey
Denver, Colorado, U.S.A.

Abstract

Research studies based on foreland basins (mainly in eastern Colorado) examined three surface geochemical exploration (SGE) methods as possible hydrocarbon (HC) exploration techniques. The first method, microbial soil surveying, has high potential as an exploration tool, especially in development and enhanced recovery operations. Integrative adsorption, the second technique, is not effective as a quantitative SGE method because water, carbon dioxide, nitrous oxide, unsaturated hydrocarbons, and organic compounds are collected by the adsorbent (activated charcoal) much more strongly than covalently bonded microseeping C_1–C_5 thermogenic HCs. Qualitative comparisons (pattern recognition) of C_{8+} mass spectra cannot gauge HC gas microseepage that involves only the C_1–C_5 HCs.

The third method, soil calcite surveying, also has no potential as an exploration tool. Soil calcite concentrations had patterns with pronounced areal contrasts, but these patterns had no geometric relationship to surface traces of established or potential production, that is, the patterns were random. Microscopic examination of thousands of soils revealed that soil calcite was an uncrystallized caliche coating soil particles. During its precipitation, caliche captures or occludes any gases, elements, or compounds in its immediate vicinity. Thus, increased signal intensity of some SGE methods should depend on increasing soil calcite concentrations. Analyses substantiate this hypothesis. Because soil calcite has no utility as a surface exploration tool, any surface method that depends on soil calcite has a diminished utility as an SGE tool. Isotopic analyses of soil calcites revealed carbonate carbon $\delta^{13}C$ values of –4.0 to +2.0‰ (indicating a strong influence of atmospheric CO_2) as opposed to expected values of –45 to –30‰ if the carbonate carbon had originated from microbial oxidation of microseeping HC gases. These analyses confirm a surface origin for this soil calcite (caliche), which is not necessarily related to HC gas microseepage. This previously unappreciated pivotal role of caliche is hypothesized to contribute significantly to the poor and inconsistent results of some SGE methods.

INTRODUCTION

Three different surface geochemical exploration (SGE) methods—microbial soil surveying, integrative adsorption, and soil calcite surveying—were examined in a large cooperative study over a ~300-mi² (~780-km²) area in eastern Colorado, United States. The exact location cannot be disclosed at this time due to legal considerations and agreements with landowners for continued access to the area during the ongoing study. Research concentrated on channel sandstone stratigraphic oil deposits in horizontal, unstructured rocks of a foreland basin. The estimated ultimate recovery of the principal oil field studied is 10–20 million bbl. Producing depths of the oil and occasional gas wells are ~5000–7000 ft (~1500–2100 m).

Interim results of the ongoing microbial soil surveying study have been previously reported (Price, 1993) and are only briefly reviewed here. Integrative adsorption failed in this study, and research was terminated on this technique; however, salient points of existing research are summarized here. Soil calcite surveying was also a disappointment, and research on this method was terminated as well. Nonetheless, results from this method provided significant insight into other SGE methods and are thus discussed in detail.

INTEGRATIVE ADSORPTION METHOD

Introduction and Methods

Concentrations of thermogenic microseeping hydrocarbon (HC) gases in soil gas are affected by different controlling parameters unrelated to HC microseepage (Price, 1986). Consequently, soil gas concentrations can have poor reproducibilities. The integrative adsorption surface method attempts to overcome this problem by "integrating" a HC gas microseepage signal over time by burying a strong adsorbent that collects (adsorbs) microseeping HC gases. This technique was researched by burying gas-monitoring (environmental) screened badges that contained 500 mg of activated coconut charcoal ("Anasorb CA" from SKC West, Inc.).

These badges consisted of a thin cylinder (20 mm wide wide $\times$ 7 mm high) with a screen on one side and a permeable, thin foam rubber backing on the other side covered by a plastic cap. The charcoal was contained between the screen and foam rubber backing. The entire assembly was housed in a sealed metal foil package. For use, the badge was placed in a holder, which was a small (10 cm $\times$ 2.5 cm) lath of thin (1.5 mm) polypropylene with a 20-mm circle cut out near one end of the lath. A cut near the top of the lath into the circle allowed the badge to be snapped into the holder. The holder with the badge was placed upright in an inverted polypropylene bottle (13 cm high $\times$ 5 cm wide) that had a snap-on cap perforated with six 6-mm holes. The inverted bottle with the holder and badge was tightly attached to a 3-ft (0.9-m) surveying stake by baling wire and was buried with dirt in a 2-ft (0.6-m) hole. After a given burial time, the assemblies were recovered and the badges were placed in the metal foil packages whose tops were tightly folded several times and held in place by paper clips. These well-sealed packages prevented any further exposure of the badge charcoal to the atmosphere. Quantitative analysis for the adsorbed C_1–C_4 HC gases was carried out by thermal desorption of the charcoal (see the Chapter Appendix for details).

Results

Surveys were carried out over and around the Bell Creek oil field, Powder River County, Montana, and the eastern Colorado oil field serving as a study area for the microbial surveying study (Price, 1993) (discussed later). In the case of Bell Creek, a traverse was carried out with sample centers spaced 0.1 mi (0.16 km) apart over the field and 0.25 mi (0. 4 km) apart off the field for distances of 6 mi (10 km) from the field edges defined by drilling. There was no significant difference in the average and range of quantitative values for either individual C_1–C_4 HC gases or the sum of C_1–C_4 HC gases for samples from the Bell Creek oil field area. For example, the average of the sums of C_1–C_4 HC gases for sites off the Bell Creek field was 457 ppm gases (by volume in a sealed pipe bomb used for thermal desorption analysis; see Chapter Appendix), with a range of 103–1141 ppm. In contrast, an average of 423 ppm and a range of 37–1032 ppm were obtained for samples from over the field. Plots of values of both individual C_1–C_4 HC gases and the sum of the C_1–C_4 HC gases versus location on the traverse gave no indication of either the location or boundaries of the oil field. Furthermore, the boundaries of Bell Creek are well defined by a large number of dry holes surrounding the field.

In the study area in eastern Colorado, three different surveys were carried out. The first was a traverse with sample centers spaced 0.1 mi (0.16 km) apart over a producing oil field and 0.25 mi (0.4 km) apart for distances of about 6 mi (10 km) on either side of the oil field. The second was a sparse grid with sample centers at intervals of 0.25–0.50 mi (0.4–0.8 km) over an area of about 50 mi^2 (130 km^2) bordering the producing field and shown to be barren of oil deposits by a denser (0.1–0.2 mi, 0.16–0.32 km) sampling grid used for a microbial survey over the same area. The third survey consisted of four closely spaced (7 $\times$ 7) grids with sample centers spaced 4 and 30 ft (1.2 and 9 m) apart with 49 samples each, over and at the border of the producing oil field (see Price, 1993, his figures 19–22).

Ranges and averages of sample values in the traverse for sites over and off the oil field were similar, with the magnitude of values demonstrating no relationship to proximity of the oil field. For example, the range of values for sites off-field was 63–537 ppm versus 83–491 ppm (with one sample at 2126 ppm) over the field, with averages of 242 ppm off field and 277 ppm over the field. The location and boundaries of the producing oil field could not be recognized by the analytical results. Likewise, the range (20–690 ppm) and average (256 ppm) of the samples from sites in the oil-barren sparse regional grid adjacent to the producing field (the second survey above) were similar to the values from sites over the field in the traverse and from the tightly spaced grids over and adjacent to the oil field (the third survey above).

In this latter case, tightly spaced grids both over and areally removed from HC deposits allow an assessment of the reproducibility of a given SGE method (see Price, 1993, his figures 11, 14, 19, 20; and see Figure 9, this paper). In my opinion, such grids are the optimum way to ascertain the reproducibility of many surface methods. However, published results of such grids are rarely available. With this study, four tightly spaced grids demonstrated poor quantitative reproducibility of the integrative adsorption method under research. For example, a closely spaced grid (D grid; see Figures 8, 9) with sample centers 30 ft (9 m) apart located over the study oil field had poor repeatability (Figure 1), with a wide range of values (43–971 ppm) and no defined maximum in the histogram. Furthermore, the distribution of values for the D grid is similar to that of the F grid (Figure 1) which is a 7 $\times$ 7 grid of 49 samples with sample centers 30 ft apart in an area devoid of potential or existing HC production (see Price, 1993, his figure 12). Thus, the integrative adsorption method examined could not quantitatively distinguish oil-bearing from oil-barren areas.

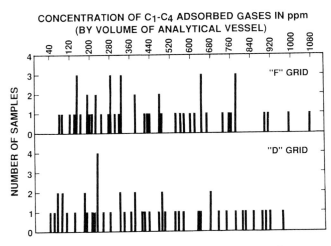

Figure 1—Concentrations of desorbed (C_1–C_4) HC gases from integrative adsorption studies using two closely spaced 7 × 7 grids (30 ft, or 9 m, between sample centers): (a) over oil production (D grid) and (b) over an oil-barren area (F grid). HC concentrations are given in ppm by volume of the closed pressure vessel in which the adsorbed HC gases were thermally desorbed. Plots of only the C_2–C_4 HC gases were similar to plots shown here.

Burial times varied from 3 days to 6 months at each site of the traverse over and adjacent to the oil field. After 1 week, burial time had no measurable effect. Most of the surveys were carried out with burial times of 1 to 2 months.

Discussion

The integrative adsorption method studied allowed absolutely no quantitative assessment of HC gas microseepage. Furthermore, the reasons for this conclusion are not unique to this geologic setting and can thus apply to all integrative adsorption techniques. For example, upon recovery of the buried capsules, they were invariably soaked with soil moisture. In most cases, the aluminum holders containing the charcoal were corroded (oxidized) from the soil moisture, sometimes badly. No attempt was made to measure quantitatively the water on the capsules or the water adsorbed by the charcoal. However, in many cases, the weight of the water probably exceeded that of the charcoal. Upon consideration, this is to be expected. All strong adsorbents have unsatisfied ionic charges by which adsorbents bond with compounds. The more polar (ionic) the compound, the stronger it is adsorbed. Water is by far the most polar and most abundant compound in soils. It is thus taken up in greatest abundance by any buried adsorbent and will literally swamp covalent thermogenic HC gases.

Furthermore, as noted by Price (1986), elevated concentrations of both nitrous oxide and carbon dioxide result from normal microbial activity in soils. Both these compounds are also in higher concentration and are more polar than covalent microseeping thermogenic C_1–C_4 HC

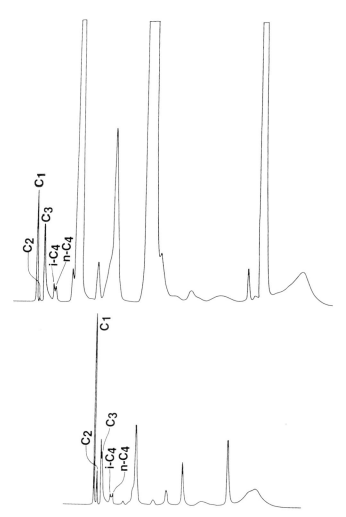

Figure 2—Gas chromatograms of two representative thermally desorbed integrative adsorption samples. C_1 is methane, C_2 ethane, C_3 propane, i-C_4 isobutane, and n-C_4 normal butane. (Analytical conditions are detailed in the Chapter Appendix.)

gases and are thus also adsorbed more strongly than HC gases. However, in this study, no attempt was made to analyze quantitatively either adsorbed carbon dioxide or nitrous oxide.

Price (1986) also noted that a host of light HCs with double or triple carbon–carbon bonds and light organic compounds with nitrogen, oxygen, and sulfur atoms are formed by soil microbial and plant root activity and are more polar and thus adsorbed more strongly than covalent microseeping HC gases. This point is demonstrated by one of the gas chromatograms of two representative samples from this research on integrative adsorption (Figure 2). In the chromatogram in Figure 2a, which is representative of most of the samples in this study, the amount of unidentified organic compounds indigenous to the soil dwarfs the microseeping thermogenic HC gases. Furthermore, a large hump of higher molecular weight compounds resulting from a 30°C/min heating ramp of the gas chromatographic oven (see Chapter Appendix) is not shown in Figure 2. However, the amount of absorbed nonthermogenic HCs was variable, and chromatograms were occasionally obtained in which the thermogenic HCs made up much more of the total peaks by area (Figure 2b). Note in Figure 2b that an unidentified peak elutes before and interferes with propane (C_3).

Clearly all absorbents have a strong preference for water, carbon dioxide, nitrous oxide, alkenes, and oxygen-, nitrogen-, or sulfur-bearing organic compounds compared to covalent thermogenic HC gases. Thus, only a minute and nonreproducible portion of the adsorbent is ever available to bond with the HC gases. Hence, it is difficult to envision how any integrative adsorption method could ever quantitatively integrate a HC gas microseepage signal over time, a conclusion documented by the data of this study.

Because of this flaw with integrative adsorption, some methods qualitatively compare mass spectrograms from C_{8+} compounds thermally desorbed onto a mass spectrometer. Samples from areas of established production provide a representative mass spectrometric pattern from HC-productive sites. Similar or exact patterns from adjacent unproductive areas imply a favorable drilling location. As stressed in Price (1986, 1993), HC gas microseepage involves movement of only the C_1–C_5 thermogenic HCs to the earth's surface; C_{6+} HCs are not involved. The C_{6+} HCs at the earth's surface originate from low levels of HC macroseepage up faults and fractures. Adsorption methods using qualitative comparisons of C_{6+} HCs are not measuring HC gas microseepage, but instead are measuring fault-assisted low-level HC macroseepage not necessarily associated with a HC deposit. It is also a strong possibility that integrative adsorption methods using such qualitative comparisons are simply measuring indigenous C_{6+} organic compounds present in soils from microbial and plant root activity.

In theory, integrative adsorption techniques may work well in true deserts where soil moisture and therefore plant root and soil microbial activity are at a minimum or altogether lacking. However, this possibility was not examined in this study. In contrast to the negative conclusions drawn here concerning integrative adsorption, Potter et al. (this volume) has presented positive results with their use of the PETREX method. However, their report includes insufficient analytical data to allow an independent appraisal of their conclusions.

MICROBIAL SOIL SURVEYING

The microbial surveying technique of this research project has been detailed by Price (1993) and is only briefly reviewed here. Microbial soil surveys were carried out over the same areas in eastern Colorado using the same sample sites as in the integrative adsorption and

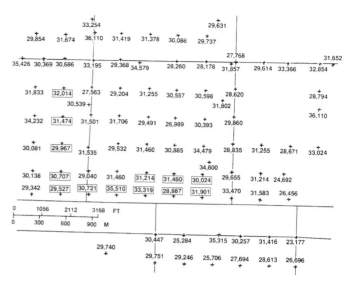

Figure 3—HC gas concentrations (in ppm by volume) after incubation for sites from an area not near existing or potential HC production (see Price, 1993, his figure 5). Samples outlined with boxes were further analyzed in Figure 4 and are explained in text. Vertical and horizontal lines are section lines.

soil calcite studies. This work was pivotal to the present paper because it provided an independent check of the correspondence of calcite soil patterns to surface traces of HC deposits.

In the microbial soil method, soils are incubated in sealed glass vials with nutrient solution and 37,000 ppm (by volume) HC gas concentration (not methane). After given incubation times at constant temperature, the headspace gas in the vials is analyzed by flame ionization detection gas chromatography.

Microbes in soils that are not near existing or potential HC production have no, or small to moderate, propensity to consume HC gas during incubation. In contrast, microbes in soils over existing or potential HC production strongly consume HC gas during incubation, resulting in highly reduced or zero HC gas concentrations. Sample repeatability is 80–100% (most often >90%) for soils both from oil-barren areas and from strong microbial anomalies over oil-bearing areas. However, transition zones of variable width and intensity surround and can be confused with strong microbial anomalies. Sample repeatability is poor in transition zones.

Figures 3 and 4 demonstrate sample reproducibility in oil-barren zones. Figure 3 is from the center of a large (~50 mi², 130 km²) area devoid of existing or potential (by microbial analysis) HC production (Price, 1993). In Figure 3, the lines represent section lines, and no sample has a HC gas concentration of less than 20,000 ppm, with most being greater than 28,000 ppm. In this method, final HC gas concentrations of 2001–20,000 ppm are weakly anomalous, 101–2000 ppm moderately anomalous, 5.01–100 ppm anomalous, and 0–5.0 ppm strongly anomalous. Contiguous samples with HC gas concentrations of 0–5.0

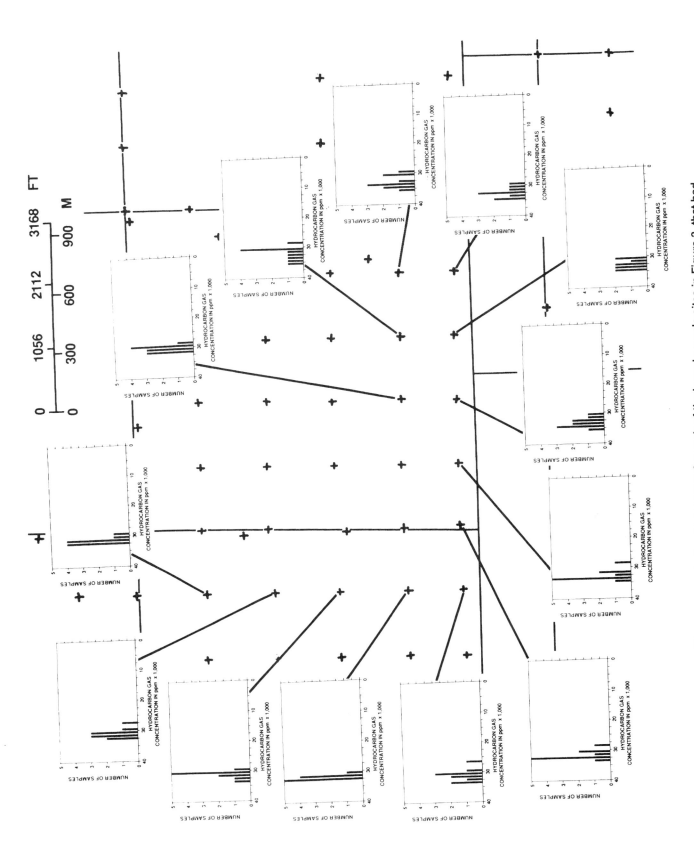

Figure 4—HC gas concentration histograms for each of the boxed sample sites in Figure 3 that had ten repetitive incubations and analyses performed on them.

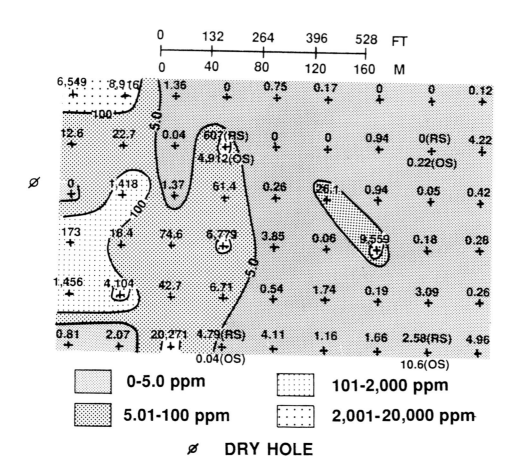

Figure 5—Contour map of HC gas concentrations after incubation (in ppm) showing the results of a closely spaced (132-ft, or 40-m, grid) microbial survey adjacent to an attempted development well (dry hole) at the edge of existing production. The values of 0–5.0 ppm suggest a significant extension to the field that was missed by development drilling. The letters OS and RS represent sample sites sampled twice, once in an earlier survey (OS, original sample) with a 528-ft (161-m) grid and again (RS, resample) with a 132-ft (40-m) grid. Notice the excellent reproducibility of these four sites.

ppm are considered to outline oil-bearing areas. Along the bottom and to the left of the section in the middle of Figure 3 are 13 sites outlined with boxes. Each of these sites had ten replicate incubations and analyses, shown in Figure 4, with HC gas concentration histograms for each site. Only two samples in Figure 4 had HC gas concentrations <28,000 ppm, and no samples had values <26,000 ppm. In this example, sample reproducibility is 100%, characteristic of areas with no potential for HC production.

Figures 5, 6, and 7 demonstrate both the potential utility of the method and the sample reproducibility in strong microbial anomalies. Figures 5 and 7 show closely spaced grids (132 ft, ~40 m) located near attempted development wells around existing production. To reiterate, contiguous sites with HC gas concentrations after incubation of 0–5.0 ppm are thought to outline oil-bearing areas by this method. In Figure 5, a significant oil-bearing area is suggested to the right of the dry hole by the microbial survey. The area between the dry hole and the 0–5.0 ppm (presumed oil-bearing) portion of the microbial anomaly is an oil-barren transition zone always surrounding strong microbial anomalies induced by HC gas microseepage. Figure 5 would have been a useful tool in best positioning the attempted stepout well.

Figure 6 demonstrates sample reproducibility in strong microbial anomalies. Ten replicate analyses each for five random sites inside the microbial anomaly

demonstrated excellent reproducibility, at or near 100% in every case. The reproducibility of Figure 6 mirrors that of microbial anomalies over existing HC production (see Price, 1993). Furthermore, the 50 replicate analyses of Figure 6 were run several years after the first analyses, and the soil microbes would have lost some of their potency because of population decline during storage.

As with Figure 5, Figure 7 would have been useful in best locating the attempted development well drilled in an oil-barren transition zone adjacent to the apparent oil-bearing zone. The two sites in the interior of the strong microbial anomaly (0–5.0 ppm) with values of 18.9 and 39.0 ppm most likely are truly anomalous sites with <100% reproducibility. However, replicate analyses were not performed on these two samples.

Data are preliminary, research is still ongoing, and significant problems exist both with the present method and with microbial surveying in general (Price, 1993). However, my study confirms that this microbial technique appears to have significant potential as an exploration development tool to (1) check prospects generated by other geologic or surface methods, (2) delineate field boundaries after wildcat discoveries, (3) position development wells optimally to maximize recoveries, (4) outline significant field extensions missed by development drilling, and (5) discover by-passed oil in enhanced oil recovery operations in older fields. The technique also may be a useful research tool to examine (1) reservoir het-

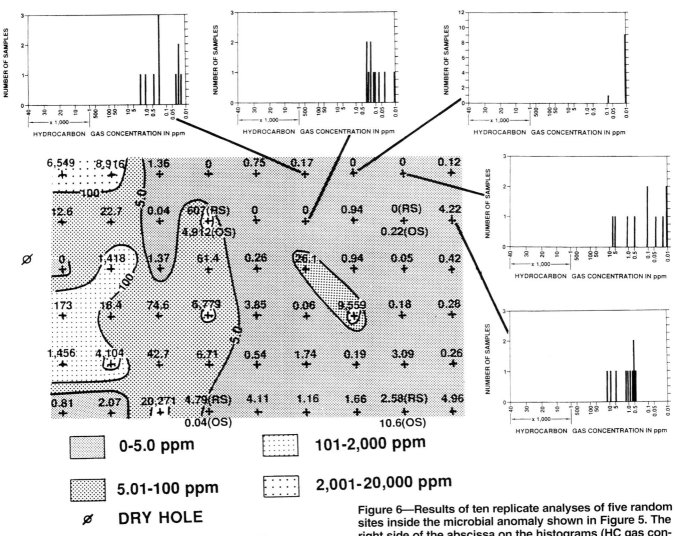

Figure 6—Results of ten replicate analyses of five random sites inside the microbial anomaly shown in Figure 5. The right side of the abscissa on the histograms (HC gas concentration scale) is a logarithmic scale from 0.01 to 1000 ppm. The far-right vertical bars of the histograms represent samples with zero hydrocarbon gas concentration.

erogeneity, (2) distributary patterns of oil-bearing sands, (3) secondary migration patterns, and (4) field growth. However, the results of this microbial study are not an accreditation of microbial surveying in general. As discussed by Price (1993), each method must be judged on its own merit, assuming that enough relevant data are published for a given method to be evaluated. For the purposes of this paper, the present microbial technique is a powerful check on the results of soil calcite surveying.

SOIL CALCITE SURVEYING

Introduction

In a conceptual model, Price (1986) hypothesized that the halo anomaly around HC deposits purportedly detected by different surface methods was due to elevated soil calcite concentrations, in a halo surrounding HC deposits, resulting from erosion and exposure of chimneys at the earth's surface. The chimneys, in turn, had resulted from sulfate-reducing bacteria consuming microseeping gas in rocks above HC deposits over geologic time. Such long-term bacterial consumption of HC gas would result in large amounts of carbon dioxide, leading to precipitation of significant calcite in and around the chimneys. To check this hypothesis, detailed soil calcite surveys were performed on the same samples used for microbial analysis. Over 6000 analyses were performed (see Chapter Appendix for sampling and analytical conditions). Three major conclusions resulted. (1) There was no relationship of soil calcite patterns to established HC production as defined by drilling or microbial anomalies or to potential HC production as

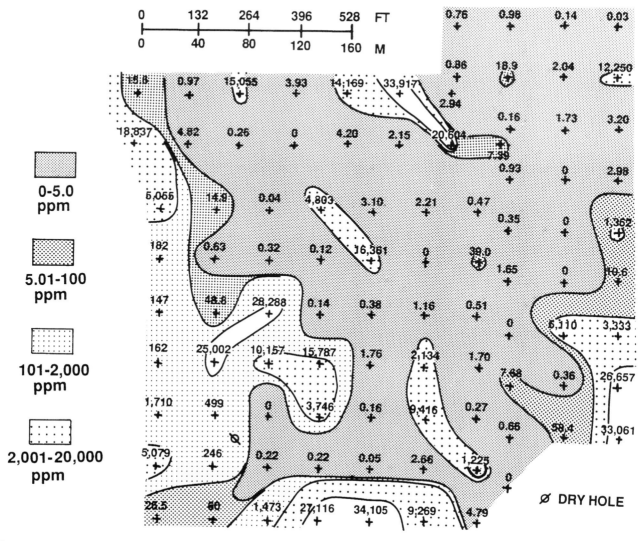

Figure 7—Contour map of HC gas concentrations (in ppm) showing the results of a closely spaced (132-ft, or 40-m, grid) microbial survey around and adjacent to an attempted development well (dry hole) at the edge of existing production. Values of 0–5.0 ppm delineate oil-bearing areas.

defined by microbial anomalies. Soil calcite concentrations were high, low to zero, or highly contrasting over both oil-bearing and oil-barren areas; in other words, they were completely random. (2) Reproducibility of soil calcite concentrations over limited areas was very poor. (3) There was no indication of soil calcite halos around areas of existing or potential HC production.

Oil-Bearing Areas

We surveyed a complete oil field and the area immediately adjacent to it (12 mi², 31 km²) with about 1450 sample centers spaced 528 ft (161 m) apart, with sparser sampling farther from the field. Soil calcite patterns were completely random over, immediately adjacent to, and at a distance from the oil field. Because of the survey's

scope, the results cannot be shown in one figure. One representative portion of the survey is shown in Figure 8 (see Figure 11 for another portion). In Figure 8, that part of the field discussed in figure 18 of Price (1993) is delineated by the square outline (light dashed line). The original area was expanded in the survey shown in Figure 8. The field outlines as delineated by the microbial survey of Price (1993, his figure 18) are shown by the thick dashed lines (hachures inward to the field) and correspond closely to producing wells. One could also draw field boundaries on the basis of producing wells and dry holes. Both the microbial survey and reservoir fluid–pressure relationships show, however, that this single field is actually in two hydraulically separate channel sands. Soil calcite patterns have absolutely no spatial relationship to the shape of the HC deposit as delineated either by the microbial anomalies or by producing versus dry holes.

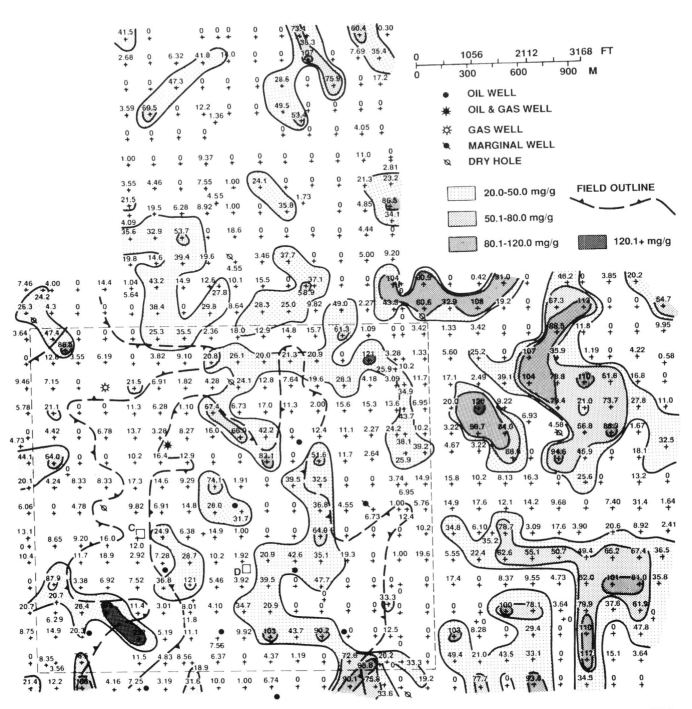

Figure 8—Contour map of soil calcite concentrations (in mg calcite/g soil) showing the results of a soil calcite survey (528-ft, or 161-m, grid) over and adjacent to the edges of a producing oil field. Marginal wells were noncommercial and produced less than 5000 bbl of oil before abandonment. Dashed lines are explained in the text. (After Price, 1993, his figure 18.)

The small square labeled D in Figure 8 is a 7 × 7 grid of 49 samples with sample centers spaced 30 ft (9 m) apart and demonstrates the lack of reproducibility of soil calcite surveys. Based on the 528-ft (161-m) grid (Figure 8), soil calcite concentrations of 0–20 mg/g of soil would be expected. Yet around 70% of the samples have values in

excess of 20 mg/g (Figure 9) with a maximal value of 131 mg/g and a strongly contrasting linear pattern, unexpected from the data in Figure 8. The square in the upper left corner of Figure 9 is another 7 × 7 sample grid with sample centers 4 ft (1 m) apart whose reproducibility is better but still unacceptable (Figure 10). Thus, over an

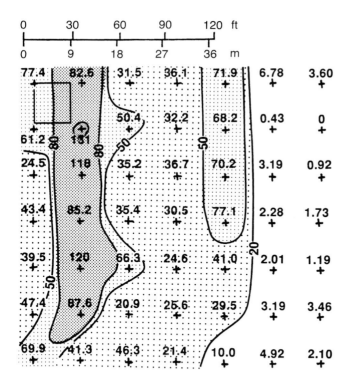

Figure 9—Contour map of soil calcite concentrations (in mg calcite/g soil) for the 7 × 7 sample D grid of Figure 8. Sample centers are spaced 30 ft (9 m) apart.

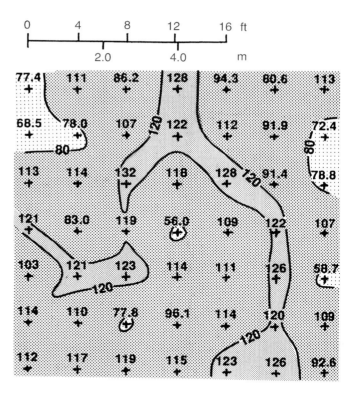

Figure 10—Contour map of soil calcite concentrations (in mg calcite/g soil) for the 7 × 7 sample grid (4 ft, or 1.2 m, sample centers) in the upper left corner of Figure 9.

area of ~575 ft² (~53 m²), soil calcite varies from 56 to 132 mg/g (Figure 10), although the pattern of values corresponds to expectations based on Figure 9. Figures 9 and 10 demonstrate that soil calcite concentrations can have poor reproducibilities over short distances. This possible lack of reproducibility greatly diminishes any utility the method might have as a HC exploration tool.

However, this lack of reproducibility is not universal. For example, the 7 × 7 sample C grid in Figure 8 has a much smaller range of soil calcite concentrations of 4.25–41.6 mg/g, with 44 of the 49 samples less than 21.0 mg/g, which is good reproducibility. Another 7 × 7 sample grid with sample centers 4 ft (1.2 m) apart in the upper right corner of the C grid had equally good reproducibility of 6.00–42.3 mg/g, with 45 of 49 samples between 9.10 and 28.4 mg/g. Likewise, four other 7 × 7 sample grids (see Price, 1993, his figures 10, 12) with 198 total samples had excellent reproducibility with values ranging from 0 to 2.0 mg/g. Good reproducibility is common in areas with low soil calcite concentrations (<5.0 mg/g) that are not adjacent to areas with elevated soil calcite content.

A soil calcite survey over another part of the oil field is presented in Figure 11. Much of this area was covered by a sample grid with centers spaced 528 ft (161 m) apart. A microbial anomaly closely corresponded to the field outline as determined by drilling results (curved dashed line) but was more irregular in shape. The microbial survey showed no potential for significant production outside

the area of the producing field. Strong soil calcite patterns, including halos, that might be interpreted as anomalies are present in Figure 11. However, these patterns have no spatial relationship to the shape of existing production, and strong patterns removed from existing production are not associated with undiscovered HC deposits.

Oil-Barren Areas

Soil calcite concentrations were highly elevated, low, or strongly contrasting over large areas that had no existing production nor any potential for such, according to microbial surveys. For example, weak, spotty soil calcite patterns are present in Figure 12. According to microbial surveying, this area has no potential for significant HC deposits (see Price, 1993, his figure 5). Conversely, strongly contrasting soil calcite patterns shown in Figures 13 and 14, with well-defined partial or complete halos (Figure 14), could be interpreted as SGE anomalies. Yet neither area has existing production nor potential for significant production by microbial surveys. Some sites in Figure 14 were sampled twice during separate field trips. The analytical results were highly disparate (Figure 15), again demonstrating the lack of reproducibility that can occur with this technique. Four sample pairs with concentrations greater than 20 mg/g on or near the diagonal line of Figure 15 did have good agreement, as did 15 sample pairs with zero concentrations. However, many other

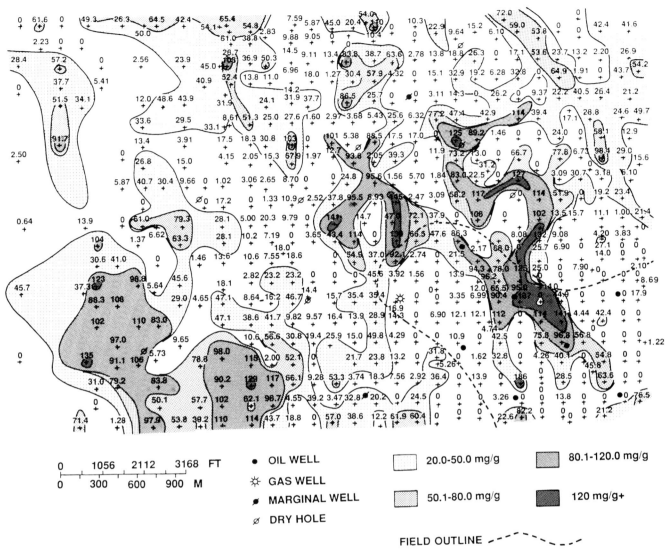

0   1056   2112   3168 FT
0   300   600   900 M

● OIL WELL
☼ GAS WELL
🖋 MARGINAL WELL
∅ DRY HOLE

▢ 20.0-50.0 mg/g   ▨ 80.1-120.0 mg/g
▨ 50.1-80.0 mg/g   ■ 120 mg/g+

FIELD OUTLINE - - - ⌒ - - ⌒ - -

Figure 11—Contour map of soil calcite concentrations (in mg calcite/g soil) showing the results of a soil calcite survey (528-ft, or 161-m, sample centers) over and adjacent to the edges of a producing oil field. This shows another part of the oil field shown in Figure 8. Marginal wells were noncommercial and produced less than 5000 bbl of oil before abandonment. Curved dashed line is explained in the text.

sample pairs in Figure 15 had poor reproducibility. For example, ten sample sites that gave zero values on the first trip had calcite concentrations of 30 mg/g or more on the second trip. In contrast, one first-trip site with a concentration over 30 mg/g had a zero value on the second trip. This difference between the two trips may have been caused by pronounced variations in soil calcite concentrations over short depth intervals (discussed in the Chapter Appendix). In any case, Figure 15 again underscores the lack of reproducibility possible with soil calcite concentrations.

Figure 16 shows soil calcite concentrations for a 16-mi^2 (41-km^2) area adjacent to existing production (note the three dry holes on the left). No HC production exists in the area of Figure 16, nor is there any potential for such

according to microbial surveys. Yet pronounced soil calcite anomalies exist in strongly contrasting patterns throughout the area. Furthermore, several halos are apparent if one ignores the 20.0–50.0 mg/g range of soil calcite concentrations and focuses only on the higher values. As in Figures 13 and 14, strongly contrasting soil calcite patterns (including halos) are not associated with existing or potential HC production.

Soil Calcite Patterns over Marginal Production

Figure 17 shows the results of a soil calcite survey over a 25-mi^2 (65-km^2) area largely devoid of HC production,

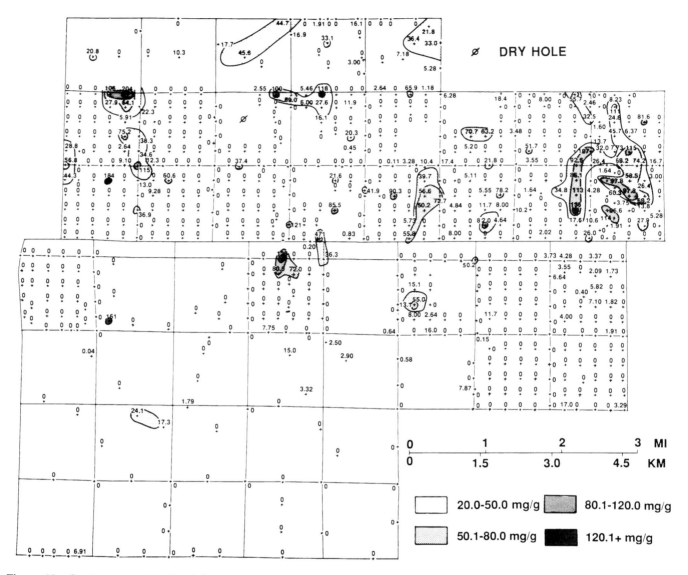

Figure 12—Contour map of soil calcite concentrations (in mg calcite/g soil) over an area with no existing HC production or potential, as determined by microbial surveying. Note the weak, spotty soil calcite patterns. (After Price, 1993, his figure 5.)

with the exception of a single well (center and left in figure) that produced less than 9000 bbl of oil before abandonment. Two dry holes were drilled near the abandoned well. By microbial surveying, most of the area was shown to have no potential for significant deposits. However, weak, spotty microbial anomalies were present around the abandoned well and to the right and below the well in a semi-circle having a radius of ~2100 ft (~640 m). Thus, a closely spaced microbial survey (sample centers at 132 ft, 40 m) was carried out in the area around the well outlined by the box in Figure 17, the results of which are shown in Figure 18.

The spotty (noncontiguous) microbial anomalies of Figure 18 are considered weak anomalies not indicative of significant HC deposits by Price (1993) and by comparison to Figures 5, 6, and 7. As stated earlier, contiguous

sites with final HC gas concentrations after incubation of 0–5.0 ppm delineate oil-bearing areas by this microbial technique. In the weak microbial anomaly associated with the abandoned well, only one site has such a value (0.62 ppm). Furthermore, ten replicate analyses of three of the sample sites in the anomaly around the abandoned well had poor reproducibility (not shown).

Soil calcite analyses were run for sites in the area enclosed by the box in Figure 18, the results of which are shown in Figure 19. Here, strongly contrasting soil calcite anomalies are present, although these anomalies have no spatial relationship to the weak microbial anomaly associated with the abandoned oil well. In fact, in Figure 19, the vertical grain of the calcite anomalies cuts across the lateral grain of the weak microbial anomaly. Soil calcite surveys with both closely spaced (132 ft, 40 m) (Figure 19)

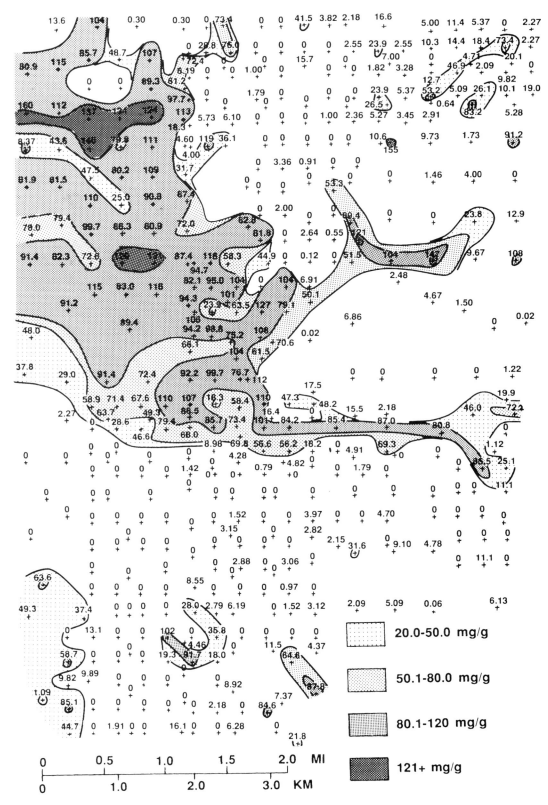

Figure 13—Contour map of soil calcite concentrations (in mg calcite/g soil) over an area with no existing HC production or HC potential, as determined by microbial surveying. Note the strongly contrasting but random patterns.

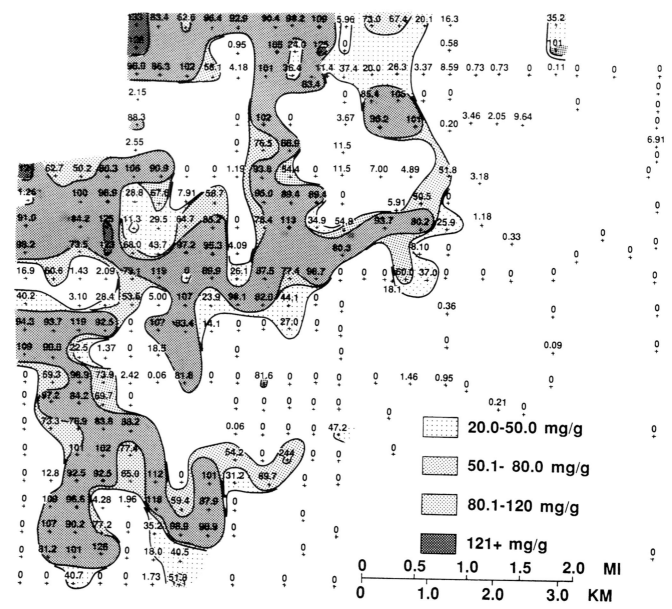

Figure 14—Contour map of soil calcite concentrations (in mg/g soil) over an area with no existing HC production or HC potential, as determined by microbial surveying. Note the partial halo patterns. See Figure 15 for additional analyses of selected sites.

and more widely spaced (528 ft, 91 m) sample centers (Figure 17) give no indication of the insignificant HC production associated with the abandoned well. The soil calcite patterns are completely random.

Returning to Figure 17, large areas are associated with highly contrasting soil calcite patterns that could be interpreted as anomalies. However, as in previous cases, these soil calcite anomalies are completely random and spatially unassociated with existing or potential (by microbial surveys) HC production. Reiterating, on the basis of microbial surveying, most of the area in Figure 17 has no potential for significant HC production.

Carbon Isotope Analysis of Soil Calcite

Carbon isotopic analyses were run on soil calcites from sites over, adjacent to, and removed from existing or potential (by microbial surveys) HC production (Figure 20). The isotopic values in Figure 20 had no relationship to the original sample location. Moreover, these values are much heavier (–4 to +2‰) than the expected values of –20 to –40‰ if the carbonate carbon had originated from microbial oxidation of microseeping HC gas to carbon dioxide in the chimney above the HC deposit according to Price's (1986) model of chimney formation. Micro-

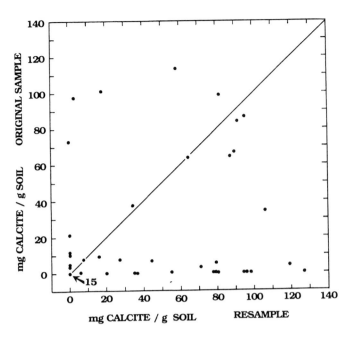

Figure 15—Cross plot of soil calcite concentrations (in mg calcite/g soil) for samples taken at the same sites during separate field trips for selected sites in Figure 14. The dot labeled "15" at the zero point indicates 15 sample pairs from both trips that had zero values. The diagonal line is the focus of perfect agreement between sample pairs.

scopic examination of thousands of soil samples revealed that the soil calcite was caliche, coating soil particles. Caliche is an amorphous (noncrystalline) soil salt usually composed mostly of calcium carbonate and formed from near-surface evaporative processes in soils. The isotopic values of this caliche (Figure 20) demonstrate that a strong equilibration with atmospheric CO_2 has occurred, providing further evidence of a caliche origin from surface soil processes.

The soil calcite may have ultimately originated from HC gas microseepage through a chimney over a HC deposit. However, with the uplift and erosion in the study area, the chimney was exposed to the atmosphere, and the CO_2 in the carbonate molecule of the original calcite exchanged with atmospheric CO_2 such that the original lighter isotopic value of the calcite was lost. In any case, isotopic analyses must be run on calcite from deep (>200 ft, 60 m) core samples unaffected by surface soil processes in order to ascertain if the ultimate origin of this soil calcite is from HC gas microseepage through chimneys above HC deposits. This is beyond the scope of this study.

It is also equally possible that this soil calcite originated from processes totally unrelated to HC gas microseepage because caliche is abundant worldwide in soils derived not just from sedimentary rocks but also from crystalline and volcanic rocks. In either case, soil calcite is distinctly nonutilitarian as a HC exploration tool. Even if soil calcite is originally derived from chimneys over HC deposits, surface soil processes and equilibration with

atmospheric CO_2 alter soil calcite concentration patterns so profoundly that they have no relationship to the original pattern inherited from the chimney.

Dependence of Other SGE Methods on Soil Calcite

Caliche is an amorphous solid with no, or an ill-defined, crystal lattice. As such, when caliche is precipitated during soil evaporative processes, it occludes or captures other entities, such as elements, ions, compounds, and soil gases in its area of precipitation. Thus, Price (1986) hypothesized that the signal strength of some surface methods, such as Horvitz's soil gas acid extraction technique, is dependent on soil calcite concentration. To test this hypothesis, soil samples underwent ΔC analysis (Duchscherer, 1984) and Horvitz analysis. The soils originated from the tightly spaced D grid in Figures 8 and 9 over a producing oil field (indicated by dots in Figures 21, 22), from five other random sites over the oil field (triangles), and from seven sites farther from existing or potential production (crosses). Increasing ΔC signal strength tightly correlates to increasing soil calcite concentration (Figure 21). The tight correlation in Figure 21 is rarely observed in geologic data. Furthermore, ΔC signal intensity is completely invariant to sample location, being no more than a direct measure of the thermal decrepitation of soil caliche. As shown earlier, in my study area, soil calcite has originated from surface soil processes and has no utility as a HC exploration tool. Because of the complete dependence of ΔC signal strength on soil calcite concentration, the conclusion then follows that, in the study area, ΔC also has no utility as a HC exploration tool.

In Figure 22, the two lines enclosing the dots define a moderate correlation of increasing concentration of soil-adsorbed ethane (determined by acid extraction) with increasing soil calcite concentration for samples from the tightly spaced D grid of Figures 8 and 9. It could be argued that a trend exists of decreasing ethane concentration with increasing soil calcite content for off-field samples (crosses) and that the ethane concentration of the on-field samples is invariant versus soil calcite content. However, I do not believe that either of these two conclusions is true for several reasons. First, nine of the twelve samples from both these sets have soil calcite concentrations of ≥ 300 mg/g. As is apparent from Figures 8–17 and 19, soil calcite concentrations usually range from 0 to 150 mg/g, with concentrations of ≥ 300 mg/g being uncommon. Thus, more samples in the range of 0–150 mg/g are needed for valid conclusions concerning these possible trends. Second, populations of five to seven samples are too small for statistically valid conclusions, except in unusual cases (Figure 21). Third, the scatter for the D grid samples in Figure 22 (dots) demonstrates that other (unknown) parameters are also controlling soil-adsorbed ethane concentrations as determined by acid extraction. I consider it probable (although unproven) that the soil processes causing the high (≥ 300 mg/g) soil calcite concentrations also lessen the dependence of caliche-occlud-

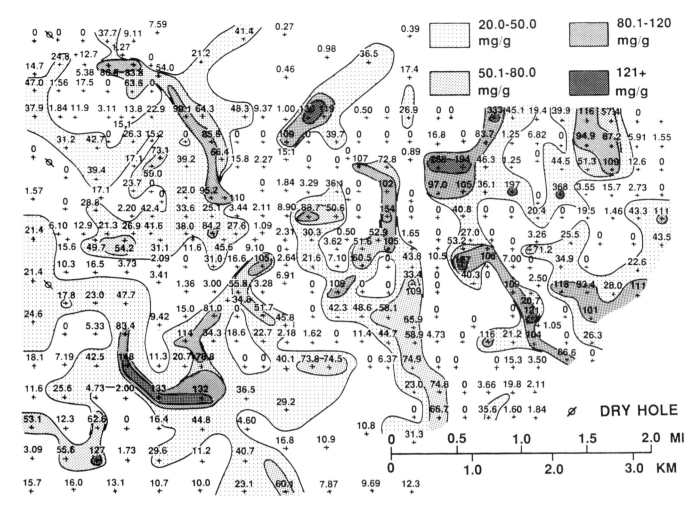

Figure 16—Contour map of soil calcite concentrations (in mg calcite/g soil) over an area with no existing HC production or HC potential, as determined by microbial surveying. Note the strongly contrasting patterns that have no relationship to the occurrence of HCs.

ed ethane on soil calcite. If one considers only samples with soil calcite concentrations of 210 mg/g or less in Figure 22 (which is the range of over 99% of all soil calcites analyzed to date), then the moderate dependence of increasing soil ethane concentrations on increasing soil calcite concentrations persists.

Sample sizes of ≤100 mesh were requested for the ∆C (Duchscherer) analysis and of ≤200 mesh for the Horvitz acid extraction analysis. Both these sample sizes made up variable but small percentages (by weight) of the soils in my study areas. Furthermore, there is a slight to moderate preferential concentration of caliche in the ≤100 mesh size fractions of soils, especially for soils with ≤10 mg/g of calcite. Neither the Duchscherer nor Horvitz laboratories determined the percentage of soils made up by the ≤100 mesh or ≤200 mesh sizes in their normal analytical service. Their analytical results were simply normalized on a per weight basis to the smaller mesh sample sizes. The lack of normalization of the smaller soil mesh sizes to the whole soil sample introduces a large margin of error

in the final analytical results because the percentages of these smaller mesh sizes varies for different soils.

In any case, the utility of acid extraction of soil gas HCs as a surface exploration method is diminished by its moderate dependence on soil calcite concentrations.

Eastern Colorado is typical of other oil-producing foreland basins in the U.S. mid-continent with respect to soil formation processes, moisture content, and climate. Thus, the results from this study area should apply to other oil basins in similar settings. A user of a particular surface method would be well-advised to know the degree of dependence of that method on soil calcite content. Because of the lack of utility of soil calcite surveying as a surface exploration method, other techniques will have their utility diminished according to their degree of dependence on soil calcite content. Some methods can be expected to have significant dependence, while others, such as soil air HCs, have little or no dependence. For example, according to Figure 23, the microbial technique of this study has no dependence on soil calcite content.

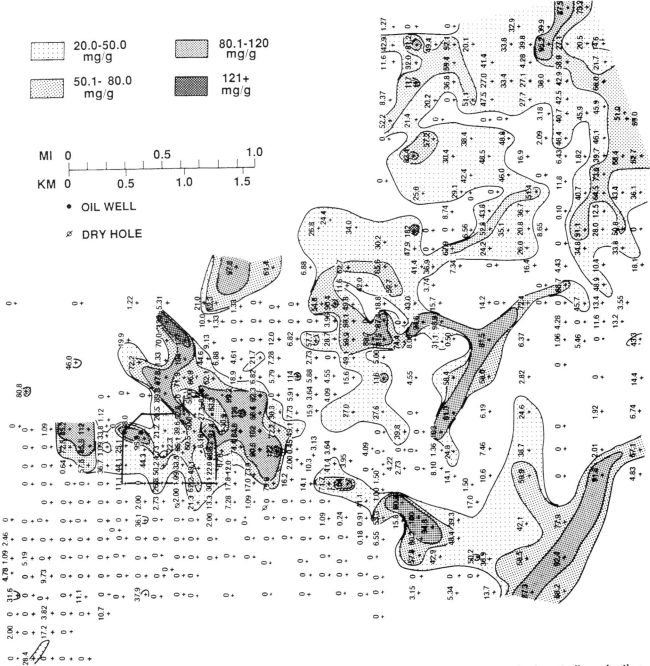

Figure 17—Contour map of soil calcite concentrations (in mg calcite/g soil) over an area with limited past oil production and potential for only small (noncommercial) deposits, according to microbial surveying. Outlined area is detailed in Figure 18 and is explained in the text. (Note that soil calcite concentrations are oriented sideways because of a drafting error.)

In Figure 23, calcite concentrations are plotted on the vertical scale in milligrams of calcite per gram of soil versus post–microbial incubation HC gas concentrations for the same samples. The samples were from the closely spaced (30 ft and 4 ft) C and D grids of Figure 8 (see also Figures 9, 10 of this paper and figures 18–24 of Price, 1993). Both the 7 × 7 C and D grids had 98 samples, 49 with sample centers at 30 ft (9 m) and 49 from smaller

interior grids with sample centers at 4 ft (1.2 m). The C grid samples (crosses, Figure 23) had moderately uniform (5–43 mg/g) calcite concentrations but were from a microbial transition zone adjacent to a strong microbial anomaly. Hence, post–microbial incubation HC gas concentrations were highly variable because of the poor replicability of samples from transition zones (Price, 1993). D grid samples (dots, Figure 23) were from the inte-

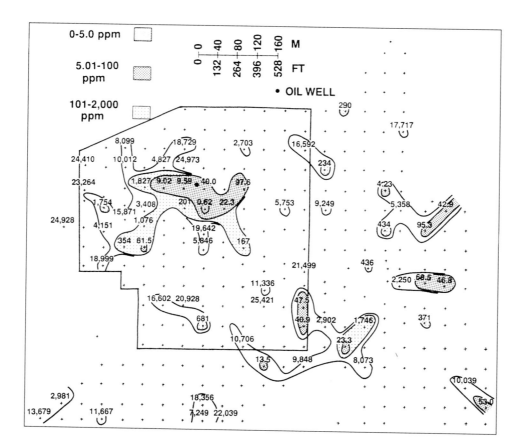

Figure 18—Hydrocarbon concentrations after incubation for a closely spaced (132 ft, or 40.2 m, grid) microbial survey around the abandoned oil well shown in Figure 17. Outlined area is detailed in Figure 19 and explained in the text.

rior of a strong microbial anomaly over the producing oil field. Post–microbial incubation HC gas concentrations were uniformly low, characteristic of the good sample replicability in the middle of strong microbial anomalies (Price, 1993). However, calcite concentrations were highly variable (0–131 mg/g) in D grid samples. Post–microbial incubation HC gas concentrations display absolutely no dependence on soil calcite concentrations for either the C or D grid samples.

THE HALO ANOMALY

The halo anomaly was first popularized in surface exploration by Horvitz's (1939, 1954) classic study around the Hastings salt dome, where he found a pronounced halo around the production surrounding the dome. However, in anomalies associated with production in other areas surveyed by Horvitz (1957, 1965, 1980) that were not associated with salt domes, halos were absent or much less pronounced compared to Hastings. In my study area, soil calcite halos never occurred around either established production (Figures 8, 11) or potential production as defined by microbial anomalies. Furthermore, well-defined partial or total soil calcite halos throughout my study areas were never associated with either established or potential HC production (Figures 12–14, 16–19).

Halos are expected around strong circular or oval geologic features such as salt domes or well-defined anti-

clines. For example, Duchscherer (in Juhlin et al., 1991) found a pronounced ΔC halo around the Siljan Ring in Sweden, which he attributed to microseeping mantle methane. However, this halo anomaly results from upturned Paleozoic carbonates surrounding the fossil crater from the meteor impact (see Castaño, 1993).

It is my premise that the halo anomaly has been overstated with regards to some surface techniques and that this anomaly shape is probably not as common as previously thought. If nothing else, soils and soil formation processes (discussed below) mask or smear out the anomaly shape and thus strongly moderate it.

DISCUSSION

I had much feedback from the Price (1986) article and was shown much data by explorationists attempting to use surface methods and by investigators researching such methods. Although none of their data were ever published, it does provide further insight into HC microseepage and attempted application of surface methods. Thus, some of that feedback and data are discussed here.

Integrative Adsorption

Several explorationists attempting to use integrative adsorption as an exploration technique in different areas

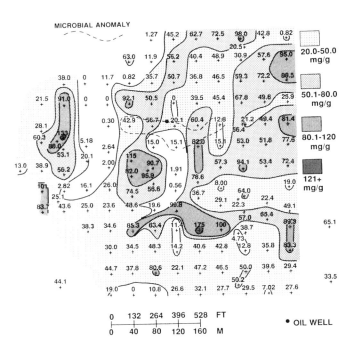

Figure 19—Contour map of soil calcite concentrations (in mg calcite/g soil) for the area outlined in Figure 18.

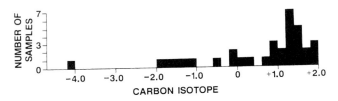

Figure 20—Carbon isotopic analyses for the carbonate carbon of soil calcite from samples over, adjacent to, and away from existing (or potential) HC production.

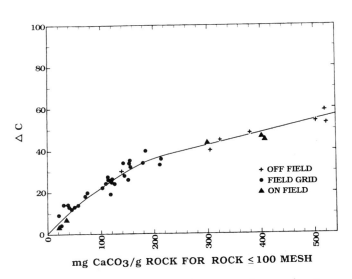

Figure 21—Cross plot of Duchscherer's ΔC versus soil calcite concentration (in mg calcite/g soil) for soil fractions of $\leq$100 mesh. Soil calcite concentrations of the total soil (all size fractions) closely tracks but are slightly less than soil calcite concentrations of the $\leq$100 mesh fraction. Symbols are further explained in the text. (ΔC data supplied by B. Duchscherer.)

drew the same conclusions about the technique as those drawn in this study. However, none of these surveys were carried out in true deserts, thus the possible utility of this technique in desert areas remains unresolved.

Faults

Several explorationists also related the results of surveys run in various faulted terranes using different surface methods where dry holes were drilled on strong anomalies over what later turned out to be faults. Price (1986) maintains that surface exploration methods have their utility significantly diminished in faulted terranes because faults often create strong direct detection anomalies at the earth's surface, whether or not those faults are associated with a HC deposit at depth. Over geologic time, continued movement of HCs along the fault and their microbial consumption at the earth's surface also result in strong indirect detection anomalies. The negative experiences of these explorationists reinforce Price's (1986) contention that the utility of surface exploration methods is significantly reduced in faulted terranes. This point is also supported by Jones and Drozd (1983, their figure 20), which portrays strong soil gas anomalies directly over thrust faults subcropping under Tertiary sediments in the Ryckman Creek field area, Wyoming. These authors noted that "the hydrocarbon seeps occur directly over the subcrop of the thrust faults that reach upward to essentially the bottom edge of the Tertiary" (Jones and Drozd, 1983, p. 947).

Soil Air Hydrocarbons

Price (1986) noted major problems with soil air HC surveys, principally related to diurnal and climatologic breathing of soil and the microbial filter through which microseeping HC gas must pass. Because of these factors, and because of the possible nature of near-surface migration of microseeping HC gases (discussed later), soil air HC gas concentration analyses have not been demonstrated as repeatable. If soil air HC gas concentrations from single sites are not repeatable through time, then data from large regional surveys cannot be cross compared with one another. One explorationist attempted to avoid this problem by taking a daily sample at the same site during large regional surveys. The values for the rest of that day's samples were then normalized to the value for this site, in theory allowing all the samples in multiple-day, regional surveys to be cross comparable. However, this approach appears to be negated by possible poor sample reproducibility of this surface technique

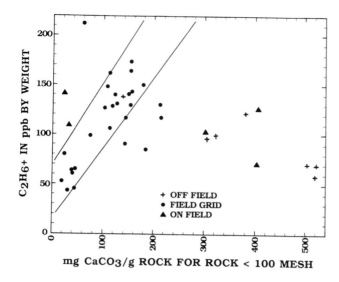

Figure 22—Cross plot of soil ethane (in ppb) by acid extraction analysis for soil fractions of ≤200 mesh versus soil calcite concentration (in mg calcite/g soil) for soil fractions of ≤100 mesh. (Ethane data supplied by P. Horvitz.)

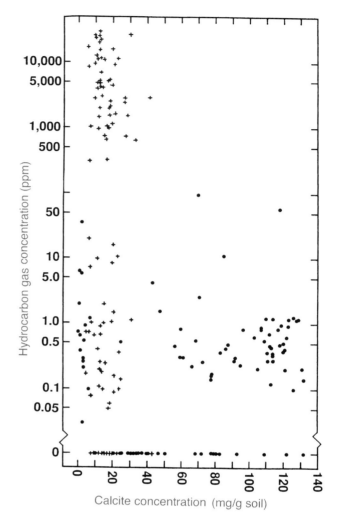

Figure 23—Soil calcite concentrations (vertical arithmetic scale) versus final HC gas concentrations (horizontal logarithmic scale) for incubations carried out on the same samples used for measuring soil calcite concentrations. The samples are the 98 samples of the C grid (crosses) and the 98 samples of the D grid (dots) of Figure 8.

over short distances. I was shown the results of two separate closely spaced grids of 25 samples each with 3 m (9 ft) and 6 m (20 ft) sample centers. These grids were associated with two separate large regional surveys. In both cases, the ranges of values of the grids were as large as those in the regional surveys themselves. Both these grids and surveys were taken in faulted terranes. However, if this lack of reproducibility in soil air HC concentrations is universally true, then the utility of this technique as a HC exploration tool would be limited.

It also remains to be demonstrated if lack of reproducibility of soil air HC analysis is dependent on the analytical technique used; for example, some methods of soil air HC analysis are less reproducible than others. In any case, definitive data from neutral parties are sorely needed here. I am astounded that results from replicability studies using closely spaced grids have never been published for such a widely used surface method. This is telling commentary on the level of basic research carried out on HC microseepage and surface geochemical oil exploration methods in general.

Let's assume that the lack of reproducibility of soil air HC analysis arises from the soil air HCs themselves, as opposed to the analytical technique used to measure them. Then insightful information about the near-surface migration of microseeping HCs would result. HC gas microseepage near the earth's surface can be viewed as occurring in (at least) two ways. The first is a "blanket" emanation of uniform concentration of HC gases, elevated over background values. The second is a "huff and puff" episodic migration related to concentration buildups and threshold escape values at any given site. This second possibility is analogous to the "focusing and

ponding" model of Clayton and Dando (this volume) for HC gas macroseepage. Near-surface HC gas microseepage by an episodic migration would explain a lack of reproducibility of soil air HCs because, at any one site over time, soil air HC gas concentrations would be variable. However, effective microbial soil surveying methods would avoid the problem because soil air HC gas concentrations would be integrated by soil microbes at a given site over time.

In any case, I continue to have strong reservations about the utility of soil air HC analysis as an effective surface exploration tool, even though the technique appears to be the current method of choice among explorationists using surface methods.

Surface Soils and Soil-Formation Processes

The influence of Holocene surficial deposits and soil formation processes on the effectiveness of indirect detection methods is a potential problem. For example, my study area is covered by Holocene loess deposits. Such deposits can obscure or completely mask chimneys (and their accompanying indirect detection signals) exposed at the earth's surface from uplift and erosion. Also, surface soil formation processes seemingly have the capability to smear out such signals, perhaps rendering them meaningless. The extent to which these events occur has not been appraised or even discussed in the literature.

Finally, to echo Price (1986), if a surface method is really an effective tool, its best present use would be in new field development following wildcat discoveries or in enhanced oil recovery operations. After proving itself in these settings, the method would find more ready acceptance by the industry as a possible exploration tool.

CONCLUSIONS

1. Integrative adsorption was not utilitarian as a quantitative SGE method in this study due to strong adsorption of soil moisture, unsaturated HC gases, carbon dioxide, nitrous oxide, and oxygen-, sulfur-, and nitrogen-bearing organic gases. Thus, little adsorptive capacity remains for microseeping thermogenic HCs. The technique may have utility in true deserts, but this remains to be documented. Qualitative comparisons of the mass spectra of C_{8+} compounds are not measurements of HC gas microseepage, which involves C_1–C_5 thermogenic HC gases.
2. Some (but not necessarily all) microbial methods appear to offer significant potential as effective SGE tools.
3. Oil deposits in some reservoir types have irregular boundaries with internal oil-barren zones. Grids with sample centers no greater than 132 ft (40 m) apart are mandatory to delineate drilling sites accurately in such deposits, regardless of the technique used. More widely spaced grids and traverses have increasingly less chance of accurately delineating field boundaries. The general lack of previous application of such tightly spaced grids because of the costs involved is a contributing factor to the inconsistent results of surface methods.
4. In the area studied, soil calcite patterns showed no relationship to the presence or absence of established (or potential) HC production, in other words, these patterns were completely random. Soil calcite concentration analyses also had very poor reproducibilities due to surface soil processes and had no utility in this study as a surface exploration tool.
5. Some surface methods are expected to, and do, have moderate to total dependence on soil calcite and thus have their utility moderately reduced or lost altogether as SGE tools. However, other surface methods are not expected to, and do not, have demonstrable dependence on soil calcite contents.
6. It is probable that the significance of the halo anomaly has been previously overstated.
7. SGE methods lose much of their utility in faulted terranes.
8. Serious unanswered questions remain about (1) the use of soil air HC surveys as a surface technique, and (2) the effect of Holocene surface deposits and surface soil formation processes on masking or completely covering some indirect detection signals from chimneys exposed by previous uplift and erosion.

Chapter Appendix

Analytical Method for Integrative Adsorption

Upon recovery from the field, the buried adsorbent capsule was placed in its foil package (with large tweezers) and most of the air was expressed from the packet as the top was folded over on itself four to five times and held in place by two paper clips. Tests demonstrated no measurable quantitative loss of adsorbed HC gases from the charcoal for at least 3 months at room temperatures when the foil packets were so sealed. However, sealed packets were stored in a freezer at 0°C upon return from the field.

For analysis, the capsules were opened and the charcoal was transferred via a metal funnel to a stainless steel pipe bomb for thermal desorption analysis. These pipe bombs were 20 cm long and were composed of a 15-cm piece of stainless tubing (12 mm o.d., 10 mm i.d.) with a Cajon 316 stainless steel cap silver-soldered onto one end, with the other end sealed by 316 stainless steel Swagelok reducing unions terminated by a cylindrical gas chromatograph septum held in place by a 3/16-in. Swagelok cap. The high content of absorbed water on the activated charcoal complicated the transfer of charcoal to the pipe bomb, necessitating a policing of the charcoal out of the capsule and down the funnel by a metal spatula and bent dissecting needle. After transfer of the charcoal, the pipe portion of the vessel was gingerly tapped with a small open-end wrench to knock all water-soaked charcoal off the pipe walls to the bottom of the pipe bomb and the pipe bomb was sealed with the reducing unions.

The pipe bombs were heated to 280°C in an aluminum six-hole heating block (25-mm-diameter holes) for 25 min on a Scientific Products Tek Pro H2143-1 (120 V/880 watt) heating plate. Previous calibrations with this thermal desorption step demonstrated that with the absolute amounts of C_1–C_4 thermogenic HC gases adsorbed onto the charcoal (about 5–150 µg) in the field samples, within the level of measurement, 100% of the adsorbed HC gases were thermally desorbed into the head space above the charcoal. However, when much larger amounts (5000–7000 µg, or 5–7 mg) of HC gas were exposed to the charcoal in laboratory calibrations, only 50–80% of the adsorbed gases could be thermally desorbed. These higher concentrations were never approached for the adsorbed thermogenic microseepage HC gases in the field samples.

After the 25-min heating period, a headspace gas sample was taken by inserting the needle of a Pressure-Lok gas-tight syringe with a Minneret valve through the septa sealing the pipe bomb. Sample sizes were generally around 200 µL but could vary from 100 µL to 1.0 mL. Sample analysis was by flame ionization detection (FID) gas chromatography on a 6 ft (1.8 m) × 1/4 in. (6 mm) O.D. packed column of Porasil C. Initial oven temperature was 73°C with a 3-min postinjection interval followed by a 30°C/min heating ramp to 300°C with an upper temperature hold of 3 min; FID temperature was 110°C and injection port temperature was 95°C. Large amounts of unidentified compounds of all molecular weights present in the samples were purged from the column during the 30°C/min heating ramp. An attempt to identify the organic compounds eluting with and just after the C_1–C_4 thermogenic HCs by the use of unsaturated HC standards was abandoned because of both difficulty of identification and the academic nature of such an identification. Quantification of the C_1–C_4 HC gases was by peak area analyses using a (mechanical) Disc Integrator. The FID responses of the C_1–C_4 HC gases were normalized to propane, and propane standards were made up using Supelco 250-mL gas bottles by injecting known amounts of propane through the cylindrical septa on the bottles.

Analytical Method for Soil Calcite

Samples of air-dried soil (25 g each) were placed in a 400-mL thick-walled beaker with 50–60 mL of distilled water. A 30-mL beaker with 20 mL of concentrated (12.1 normal) HCl was also placed in the 400-mL beaker using locking forceps and a glass stirring rod. A beginning weight (tare) for the 400-mL beaker (with contents) was obtained using an Arbor model 606 open-pan balance (capacity 10.0 kg, accuracy 0.001 g). Four such samples were prepared. The 30-mL beakers were then tipped over with the glass stirring rods, thus exposing the soil samples to the HCl. After all four 30-mL beakers were tipped, a laboratory timer was set for 6 min and the 20 mL of HCl, 50–60 mL of H_2O, and soils were mixed using the glass

stirring rod. Samples with small amounts of caliche could be quickly and vigorously stirred, whereas samples with high caliche concentrations had to be gingerly and slowly stirred to prevent the reaction from boiling as a slurry over the top of the 400-mL beaker. The samples were again stirred 2 min after tipping and also just before the final weighing at 6 min. All analyses were performed in a fume hood that was not running.

The difference between the final weight and the original weight is due to CO_2 evolution from the sample and a small evaporative loss of water (discussed below). The original amount of $CaCO_3$ in the sample was obtained by multiplying the measured weight loss (in grams) by 2.2742 (based on the atomic weight of $CaCO_3$).

A small evaporative loss of water (0.020–0.040 g/min) occurred from the 400-mL beaker, depending on the laboratory (fume hood) temperature. Because of this evaporative loss, (1) the fume hood was kept off to avoid the complete loss of accuracy and reproducibility that would occur from much higher H_2O evaporative losses with a running fume hood, and (2) measured weight losses had to be adjusted by subtracting results of calcite-barren standards. Standards were soils with zero caliche concentrations as demonstrated by previous analyses. Four such samples were run every 2 hr, and the average weight loss of such caliche-barren samples, generally 0.050–0.120 g depending on lab conditions, was subtracted from the measured weight loss of unknown samples to account for baseline evaporation of water from the unknown samples. Also, unknowns that had no reaction with the HCl, as evidenced by a complete lack of bubbles in the slurry due to CO_2 evolution, served as standards.

Early in the research, it became apparent that a depth dependence existed in soil calcite concentrations. Even in areas with highly elevated calcite concentrations (≥100 mg/g), soils at depths of 0 in. to between 6 and 18 in. (15–45 cm) had zero to low calcite concentrations. Elevated calcite concentrations generally appeared at greater depths, and a uniform sample depth of 2 ft (0.6 m) was arbitrarily chosen early in the study. Several profiles of calcite concentration versus depth (to 4 ft, 1.2 m) also revealed this 6–18 in. depth variation where high calcite concentrations were first observed. Furthermore, once the horizon of elevated calcite concentrations was reached, these concentrations could exhibit significant variation versus depth down to the 4-ft depths examined in this study. Perhaps soil calcite concentrations become more uniform versus depth at greater soil depths. However, the lack of uniformity of soil calcite concentrations versus soil depth observed in this study is a dominant controlling parameter of the lack of reproducibility of soil calcite as a surface exploration method and thus of other surface methods that depend on soil calcite.

REFERENCES CITED

Castaño, J. R., 1993, Prospects for commercial abiogenic gas production: implications from the Siljan Ring area, Sweden, *in* D. G. Howell, ed., The future of energy gases: USGS Professional Paper 1570, p. 138–154.

Duchscherer, W., 1984, Geochemical hydrocarbon prospecting: Tulsa, Oklahoma, PennWell Publishing, 196 p.

Horvitz, L., 1939, On geochemical prospecting: Geophysics, v. 4, p. 210–228.

Horvitz, L., 1954, Near-surface hydrocarbons and petroleum accumulation at depth: American Institute of Mining and Metallurgical Engineers, Transactions, v. 199, p. 1205–1209.

Horvitz, L., 1957, How geochemical analysis helps the geologist find oil: Oil and Gas Journal, no. 45, p. 234–242.

Horvitz, L., 1965, Case histories show results of drilling geochemical prospects: World Oil, November, p. 161–165.

Horvitz, L., 1980, Near-surface evidence of hydrocarbon movement from depth, *in* W. H. Roberts III and R. J. Cordell, eds., Problems of petroleum migration: AAPG Studies in Geology 10, p. 241–263.

Jones, V. T., and R. J. Drozd, 1983, Predictions of oil or gas potential by near-surface geochemistry: AAPG Bulletin, v. 67, p. 932–952.

Juhlin, C., A. A. Aldaham, J. R. Castaño, B. Collini, T. Gorody, and H. Sandstedt, 1991, Scientific summary report of the deep gas drilling project in the Siljan Ring impact structure: Vottenfall Research and Development Report, CI:U688/89-TE30, no. 98460, 278 p.

Price, L. C., 1993, Microbial soil surveying: preliminary results and implications for surface geochemical oil exploration: Association of Petroleum Geochemical Explorationists Bulletin, v. 9, p. 81–129.

Price, L. C., 1986, A critical overview and proposed working model of surface geochemical exploration, *in* M. J. Davidson, ed., Unconventional methods in exploration for petroleum and natural gas, IV: Dallas, Texas, Southern Methodist University Press, p. 245–304.

Abrams, M. A., 1996, Interpretation of methane carbon isotopes extracted from surficial
marine sediments for detection of subsurface hydrocarbons, in D. Schumacher and
M. A. Abrams, eds., Hydrocarbon migration and its near-surface expression: AAPG
Memoir 66, p. 309–318.

Chapter 22

Interpretation of Methane Carbon Isotopes Extracted from Surficial Marine Sediments for Detection of Subsurface Hydrocarbons*

Michael A. Abrams

*Exxon Ventures (CIS), Inc.
Houston, Texas*

**This chapter is a modified version of a paper previously
published in the Association of Petroleum Geochemical
Explorationists Bulletin, v. 5, p. 139–166 (1989).*

Abstract

In surface prospecting for hydrocarbons, it is often assumed that the carbon isotopic composition of methane can provide information on the origin of the gas. This assumption may not be valid for methane carbon isotopes extracted from surficial marine sediments because of (1) mixing of gases with multiple origins and (2) alteration resulting from secondary processes. Examination of a relatively large data set reveals the correlation of methane concentrations with carbon isotopic compositions. Three groups have been defined based on a comparison of molecular concentrations to methane carbon isotopes: *type I*—small concentrations of methane with heavy isotopic compositions; *type II*—large concentrations of methane with very light isotopic compositions; and *type A*—average to above average concentrations of methane with intermediate isotopic compositions.

A large data set from deep core samples must be collected to assess accurately the reliability of methane carbon isotope values to deduce the origin of gas contained in surface sediments. A plot of methane concentration versus $\delta^{13}C$ is used in combination with the hydrocarbon composition data to help determine the significance of the isotopic composition and subsequent interpretation of origin. Samples that contain both type A isotopic compositions and anomalous higher molecular weight hydrocarbons provide the most reliable information on the origin of migrated gas.

INTRODUCTION

Carbon isotopic compositions are presently being used to help distinguish the origin and maturity of light hydrocarbons in Recent sediments (Reitsema et al., 1981; Stahl et al., 1981; Horvitz, 1985; Abrams, 1987). Many explorationists using surface methane isotopic composition data assume that the measured value recorded reflects only the isotopic composition of the source material and the thermal history of the source. However, this assumption may not be valid for many samples extracted from surficial marine sediments.

Examination of nearly 700 methane isotopic compositions in surficial marine sediment samples from several different basins demonstrates a large range of values. Some of this variability is due to differences in the isotopic composition of the source material and thermal histories, but much of it appears to be due to mixing and secondary alteration processes. The purpose of this paper is to document the large range of methane isotopic compositions in Recent marine sediments, to explain why such variations exist, and to offer a technique for possibly distinguishing meaningful data.

DATA COLLECTION AND ANALYSIS

A site-specific approach was used in choosing core locations; locations were chosen on the basis of strong indications of gas (wipe-out zones and bright spots) or indications of actual seepage (gas bubbles) on the reflection records. In addition, core locations were chosen away from areas of potential seepage in order to provide background levels.

Sediment samples were collected using a variety of devices. Gravity cores were obtained with an AE (Abrams-Exxon) open-barrel gravity corer, with the exception of the offshore California samples, which were collected with a modified piston corer (see Horvitz, 1985, for details). The AE open-barrel gravity corer has a modified design that maximizes recoveries in shelfal sediments (Abrams, 1982). Recoveries ranged from 2 to 6 m, with an average of about 3.75 m. Cores were also recovered with a jet core system, which uses high-pressure water through a drill pipe, and with a wireline core system to recover 1-m samples from a maximum depth of 46 m below the water–sediment interface.

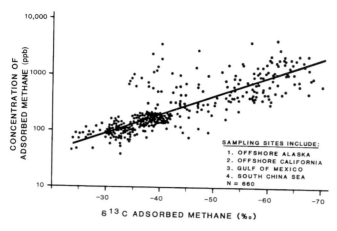

Figure 1—Graph of adsorbed methane gas concentrations versus adsorbed methane isotopic composition.

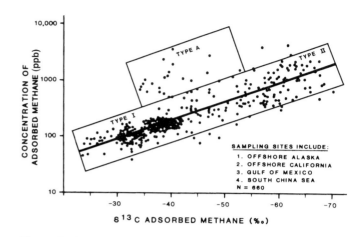

Figure 2—Isotopic classification of adsorbed methane extracted from surficial marine sediments.

Gravity cores were sampled at 1 and 3 m and at the bottom of the core. Jet cores were sampled at 4, 22, and 46 m. For each sample, approximately 400 mL of sediment was placed in a specially modified quart can. The remainder of the can was filled with distilled degassed water and sealed. After 110 mL of water was displaced with chemical grade helium, the can was stored upside-down and kept frozen until the samples were ready for analysis. In addition, 600 mL of sediment from each core was placed in an airtight bag. The air from around the sediment was squeezed out of the bag, and the bagged sample was frozen.

All sediment samples were analyzed for the following:

1. Headspace C_1–C_{5+} hydrocarbon concentration
2. Cuttings C_1–C_{5+} hydrocarbon concentration
3. Adsorbed C_1–C_{5+} hydrocarbon concentration
4. Nonhydrocarbon gases (oxygen, nitrogen, and carbon dioxide)
5. Spectrofluorescence
6. Total organic carbon

When these analyses indicated the presence of high molecular weight hydrocarbons, the following analyses were also performed:

1. Detail gasoline range C_5–C_7
2. Soxhlet extraction detail C_{15+}
3. Biomarker (GC-MS)
4. Whole-oil gas chromotography

Isotopic compositions were analyzed on the adsorbed and headspace methane. Adsorbed gases were removed from a sieved sample (less than 63-μm mesh) using the Horvitz acid extraction process (Horvitz, 1972). Headspace gases were removed from the headspace via a silicone septum with a gas-tight syringe after the can was thawed and subjected to gentle agitation. The carbon dioxide present was removed, and the resulting purified hydrocarbons were converted to carbon dioxide. The carbon isotope ratios were taken from the converted carbon dioxide (see Horvitz, 1985, for details).

Carbon isotope ratios were determined using a Varian mass spectrometer by Horvitz Research Laboratories, Houston, Texas. Results are expressed as deviations in parts per thousand (‰) of the $^{13}C/^{12}C$ ratio of the sample relative to that of a standard. All δ-values are expressed in terms of the standard PDB. The following equation relates the δ-value to the isotope ratios of the sample and standard, respectively:

$$\delta C_{CH_4} = \left(\frac{\dfrac{^{13}C}{^{12}C}_{sample} - \dfrac{^{13}C}{^{12}C}_{standard}}{\dfrac{^{13}C}{^{12}C}_{standard}} \right) \times 10^3 \ (‰)$$

COMPARISON OF METHANE ISOTOPIC COMPOSITIONS AND CONCENTRATIONS

General Observations

Examination of nearly 700 adsorbed methane isotopic compositions from surface marine sediment samples demonstrates a strong negative correlation with adsorbed methane concentration (Figure 1). Samples with small concentrations of methane (<200 ppb) usually have heavy isotopic compositions (heavier than –45‰). Samples with larger concentrations of methane (>1000 ppb) usually have light isotopic compositions (lighter than –55‰).

Figure 2 shows the three categories of surface methane carbon isotopic composition that have been developed

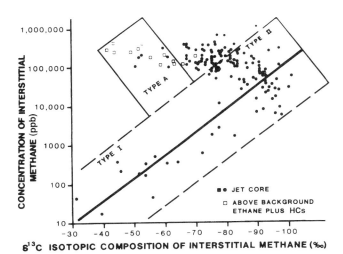

Figure 3—Isotopic composition of interstitial methane extracted from surficial marine sediments in offshore Alaska.

based on the above observations:

Type I are the samples with small concentrations of methane (usually <200 ppb) and relatively heavy isotopic compositions (usually heavier than –45‰, but vary for different areas).

Type II are the samples with large concentrations of methane (usually >1000 ppb) and relatively light isotopic compositions (lighter than –55‰).

Type A are the samples with large concentrations of methane, isotopic compositions in the thermogenic range (–35 to –55‰), and anomalous concentrations of ethane plus hydrocarbons.

Examination of the headspace methane concentration and isotopic composition shows a similar correlation (Figure 3). Samples with small concentrations of methane usually have heavy isotopic compositions, whereas samples with large concentrations of methane usually have light isotopic compositions. There was also a type A group that contained samples with relatively large concentrations of methane, isotopic compositions from –40 to –60‰, and anomalous concentrations of ethane plus hydrocarbons. Therefore, the strong negative correlation between methane concentration and methane isotopic composition is not unique to the adsorbed extraction process.

Variations in Methane Carbon Isotopes in Different Offshore Basin Sediments

Offshore Alaska

The adsorbed methane concentration and corresponding isotopic composition were plotted for surface samples from two distinct areas in offshore Alaska. Three sample groupings were identified in both areas.

Area I had 399 samples taken from 1–45 m below the water–sediment interface (Figure 4). Approximately 32% of the samples contained gas concentrations less than 200 ppb and isotopic compositions heavier than –40‰. These samples are classified as type I. About 60% of the samples contained gas concentrations greater than 200 ppb and isotopic compositions lighter than –50‰. These samples are classified as type II. About 8% of the samples contained gas concentrations greater than 200 ppb and isotopic compositions between –37 and –46‰. In addition, 65% of these samples contained significant amounts of gasoline-range hydrocarbons (concentrations up to 3693 ppb by weight of adsorbed pentane through hexane) (Abrams, 1992).

Area II had 66 samples taken from 3–45 m below the water–sediment interface (Figure 5). About 50% of the samples contained gas concentrations less than 200 ppb and isotopic compositions heavier than –40‰. These samples are classified as type I. Approximately 25% of the samples contained gas concentrations greater than 200 ppb and isotopic compositions lighter than –55‰. These samples are classified as type II. About 25% of the samples contained gas concentrations greater than 200 ppb and isotopic compositions between –43 and –52‰. In addition, several contained above average concentrations of adsorbed wet gases (concentrations >50 ppb by weight of adsorbed ethane through butane) (Abrams, 1992).

Examination of high-resolution seismic profiles throughout both Alaska sampling areas did not reveal any active seepage, such as gas bubbles in the water column (Abrams, 1992). This type of seepage would be classified as passive, as defined by Abrams (Chapter 1, this volume).

Offshore California

The adsorbed methane gas concentrations and corresponding isotopic compositions were plotted for 180 surface samples taken on a 1-mile grid from less than 2 m below the water–sediment interface (Horvitz, 1982, 1984, 1985). Two distinct sample groups were identified (Figure 6). About 95% of the samples contained methane concentrations less than 200 ppb and isotopic compositions heavier than –45‰. These samples are classified as type I. The remaining 5% contained methane concentrations in excess of 200 ppb and isotopic compositions between –48 and –54‰. In addition, most of these samples contained significant amounts of wet gases (concentrations >150 ppb by weight of adsorbed ethane plus hydrocarbons). This type of seepage would be classified as active, as defined by Abrams (Chapter 1, this volume) based on the high-resolution seismic profiles.

South China Sea

The adsorbed methane gas concentrations and corresponding isotopic compositions were plotted for 13 surface samples taken from 5 m below the water–sediment interface. Samples with less than 90 ppb were not analyzed (Figure 7). One sample contained a gas concentration of 95 ppb with a corresponding isotopic composition

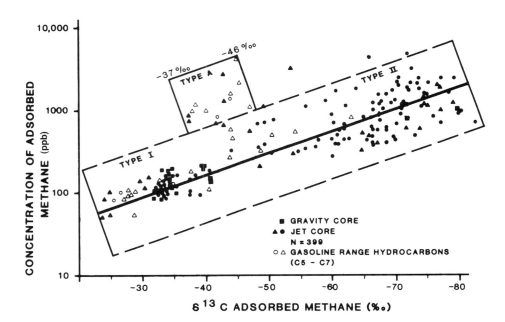

Figure 4—Graph of adsorbed methane gas concentrations versus adsorbed methane isotopic compositions, Navarin Basin, Alaska.

of –32‰. This sample is classified as a type I. The remaining 12 samples contained gas concentrations greater than 100 ppb and isotopic compositions between –34 and –37‰. In addition, over 50% of these samples contained above-background concentrations of wet gases (concentrations >20 ppb by weight of adsorbed ethane through butane).

Examination of high-resolution seismic profiles indicated that ongoing seepage is occurring through gas bubbles in the water column. This type of seepage would be defined by Abrams (Chapter 1, this volume) as active.

OTHER FACTORS THAT CAN AFFECT THE ISOTOPIC COMPOSITION OF METHANE

The isotopic composition of the source material and subsequent thermal history of the sediments may not be the only processes reflected in measured isotopic compositions of methane in surficial sediments (Schoell, 1984). Several other factors can also affect the measured isotopic composition of methane extracted from surface marine sediments.

Migration

Theoretical and experimental evidence as well as empirical data argue against a significant change in the isotopic composition of hydrocarbons during migration (Stahl and Carey, 1975; Fuex, 1977, 1980; Reitsema et al., 1981; Pflaum, 1989). Fuex's (1977) studies indicate that small gas accumulations, gas shows, and seeps may show residual methane depletions.

Microbial Oxidation

Methane-oxidizing bacteria can significantly alter the carbon isotopic composition of methane. Barker and Fritz (1981) and Coleman et al. (1981) showed that bacterial oxidation leaves residual methane enriched in ^{13}C. This results in methane with an isotopic composition identical to that of methane from a mature natural deposit.

Botz's (1981) examination of interstitial methane concentrations and isotopic compositions demonstrates the effect of bacterial oxidation on methane. Compositional and isotopic properties of bacterial gases change their genetic signature to that of thermally derived gases. The depth distribution of methane concentrations on isotopic compositions shown by Botz (1981) and the present study (Figure 8) can be explained by the zones of methane generation in surficial marine sediments, as shown by Bernard (1978). Sediment samples collected within the aerobic or sulfate-reducing zone are more likely to be altered by *in situ* bacterial processes. The depth at which the sulfate is depleted and methane production occurs varies from 1–2 m in the Santa Barbara Channel (Doose, 1980) to 3–4 m in the southern Bering Sea (Figure 8).

Sample Depth and Preparation

As previously mentioned, the depth below the water–sediment interface has been shown to control the concentration and quality of the samples. Samples collected within the aerobic or sulfate-reducing zones may not truly reflect the subsurface leakage.

Faber and Stahl (1983) have shown empirically and in laboratory experiments that degassing leads to preferential loss of methane and enrichment of ^{13}C. Thus, the resulting hydrocarbon gas does not reflect the true hydro-

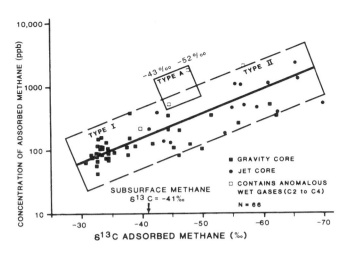

Figure 5—Graph of adsorbed methane gas concentrations versus adsorbed methane isotopic compositions, St. George Basin, Alaska.

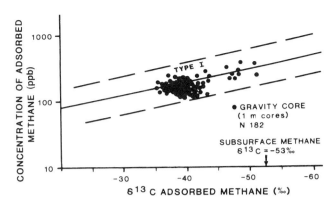

Figure 6—Graph of adsorbed methane gas concentrations versus adsorbed methane isotopic compositions, off-shore California.

carbon or isotopic composition. Stahl's (1983) studies indicate that gas extraction soon after sampling, or freezing immediately after sampling, minimizes post-sampling bacterial activity.

Analytical Techniques

Several different extraction techniques are commonly used with surface sediment samples. The three most common are as follows:

1. Headspace (interstitial)—Gas is sampled through a silicone septum on the top of the can which contains a specified amount of sediment. The can is gently shaken prior to sampling to release interstitial gases contained in the pore space.
2. Cuttings (loosely bound)—A specified amount of sediment is placed in a modified blender. The sediment is broken up, releasing loosely bound gases contained in the sediment.
3. Adsorbed (acid extraction)—The fine-grained portion of the sediment is subjected to heat and phosphoric acid in a partial vacuum.

Examination of the isotopic compositions extracted from the headspace and adsorbed methane revealed very different signatures (Woltemate, 1982; Horvitz, 1985) (Figure 9). The adsorbed methane consistently produced heavier isotopic compositions. Similar results were found when adsorbed samples were subjected to wet sieving to remove the sand fraction (greater than 63-μm mesh) (Figure 10). Horvitz (1985) believed these differences reflected the removal of biologically produced gas present in the interstitial pores through his blending, filtering, and sieving process. Schoell (1984) noted that these processes are not well understood and may reflect fractionation.

Comparison of headspace and adsorbed methane carbon isotopes over a known oil and gas field demonstrates the same phenomenon (Figure 11). These results indicate that the adsorbed methane isotopes track the subsurface isotopic compositions, whereas the headspace methane isotopes tend to be lighter. In this example, one could conclude that the adsorbed extraction technique does indeed reflect subsurface hydrocarbons.

The mechanism of hydrocarbon adsorbing is not well understood. Horvitz (personnel communication, 1984) believed that hydrocarbons are tightly adhered to the clay fraction within surface soils and sediments. Studies by Thompson (1987) indicate that the adsorbed soil and sediment gas is trapped in carbonate minerals, probably as fluid inclusions. Similar conclusions have been reached by Pflaum (1989) during his studies on adsorbed (bound) hydrocarbons. Horvitz and Ma (1988) ground the coarse-grained fraction and extracted the hydrocarbons using the standard adsorbed acid extraction technique. Results were similar to the fine-grained fraction. Both studies suggest that the clay fraction is not the adsorbing mechanism as Horvitz originally believed, but that the hydrocarbons may actually be trapped in carbonate fluid inclusions as Thompson (1987) has suggested.

Mixing

The mass spectrometer is incapable of distinguishing different mixtures of gas within surficial sediments. Subsurface hydrocarbon leakage is difficult to detect isotopically when large volumes of biogenic gas are being produced in surface sediments. Bernard (1978) used the isotopic composition of methane and $C_1/(C_2 + C_3)$ molar ratios to characterize the origin and zones of mixing with light hydrocarbons in marine sediments. He demonstrated that small fractions of biogenic gas can cause a significant change in a predominantly petroleum-related mixture. Similarly, only a small fraction of thermogenic gas can cause a largely biogenic gas to appear to be related to petroleum seepage.

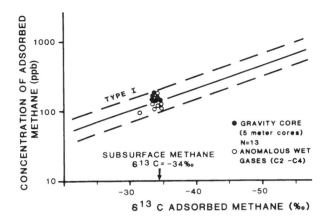

Figure 7—Graph of adsorbed methane gas concentrations versus adsorbed methane isotopic compositions, South China Sea.

Examination of the adsorbed and headspace methane carbon isotopes for the groups found in this study produces similar results (Figure 12). Type II samples fall within the mixing zone, and types I and A lie within the zone of thermocatalytic origin.

INTERPRETATION OF SAMPLE GROUPINGS

Three distinct groups of samples based on the evaluation of adsorbed methane isotopic compositions and concentrations have been defined here. Much of the variability in isotopic composition in this study appears to be due to processes other than the isotopic composition of the source material and the thermal history. Fractionation due to migration is assumed to be negligible based on studies by Fuex (1977, 1980). Fractionation from analytical procedures is also assumed to be negligible because the headspace and adsorbed extraction techniques show the same groupings and because all samples have been frozen to prevent secondary bacterial activity. In addition, examination of lithologies indicates that the groupings are not related to lithology (Figure 13).

The three types are interpreted as follows:

1. **Type I**—The samples with small concentrations and relatively heavy isotopic compositions classified as type I most likely represent methane that has been altered by bacterial oxidation. Schoell (personal communication, 1987) thinks the whole trend shown in Figure 1 represents bacterial oxidation (Figure 14). The extent to which the oxidation proceeds depends on the sulfate-reducing activity. Schoell may be correct. However, in areas where the isotopic composition of the methane in the surficial sediments matches the subsurface methane isotopic composition (e.g., South China Sea), one

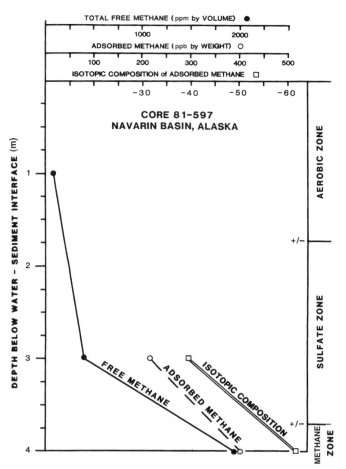

Figure 8—Vertical variability of total free methane, adsorbed methane, and isotopic composition of adsorbed methane in core 81-597 from the Navarin Basin, Alaska.

can only conclude that subsurface seepage can be sampled unaltered in the surface. The extent to which subsurface seeps become altered is not well understood. Further examination of the factors that affect seepage alteration (e.g., seepage activity, localized biological activity, lithology, and sampling depth) must be studied before reliable interpretations of surface methane carbon isotopes can be achieved.

2. **Type II**—The samples with high concentrations and methane isotopic compositions indicative of biogenic gas represent gas that has been derived either exclusively from a biological process or from a mixture of predominantly biogenic gas that has isotopically overwhelmed the presence of migrated thermogenic gas. It would be difficult to determine if thermogenic hydrocarbons are present in these samples.

3. **Type A**—The samples with high concentrations and methane isotopic compositions within the range of thermally derived methane may represent migrated thermogenic hydrocarbons. Examination

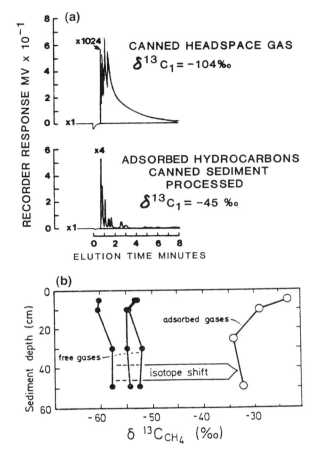

Figure 9—(a) Comparison of free (headspace) and adsorbed gases (from Horvitz, 1985). (b) Graph of methane concentration versus sediment depth showing isotopic shift (from Woltemate, 1982).

(a) Comparison Bulk versus Sieved		
Sample	$\delta^{13}C$ Bulk	$\delta^{13}C$ Sieved
149	-67.6	-40
256	-42.5	-35.1
585	-40.6	-33.3
10	-44.6	-35.6
125	-55.4	-49.6
597L	-47.5	-39.1
597B	-81.1	-62.3
618	-40.5	-33.4
629	-75.3	-68.9

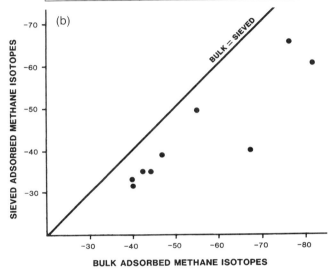

Figure 10—(a) Table of bulk and sieved adsorbed methane carbon isotopes for offshore Alaska surface core samples. (b) Graph comparing bulk versus sieved absorbed methane carbon isotopes.

of other geochemical parameters, such as the presence of significant concentrations of higher molecular weight hydrocarbons, would provide additional evidence that thermogenic seepage has occurred.

DISCUSSION

Over 85% of the 700 methane carbon isotopes collected from Recent marine sediments in this study have been classified as type I or type II. Thus, less than 15% have been interpreted as representing migrated thermogenic hydrocarbons. An examination of the distribution of isotopic classification groups and a comparison of subsurface and surface methane isotopic compositions, seepage activity, and average maximum sampling depths provide some understanding of the problem (Table 1).

In both Alaska areas, geophysical records display numerous acoustic anomalies and surface craters but no actual bubble traces (Abrams, 1992). This indicates that passive seepage is occurring. The large number of type II

samples indicates that high volumes of biogenic gas are actively being produced. The presence of large concentrations of gasoline-range and higher molecular weight hydrocarbons in the surface sediments and the presence of subsurface hydrocarbon accumulations demonstrate thermogenic hydrocarbon generation and subsequent leakage. Therefore, despite passive seepage, biogenic activity, and microbial oxidation, 8% of the samples from area I and 25% from area II did show a type A signature. In addition, the area II type A isotopic compositions were similar to that of the subsurface methane accumulation (Table 1). I believe that the site-specific sampling and deep core recovery (>6 m) provided the opportunity to obtain type A samples within the given constraints. Similar conclusions were suggested by Kvenvolden and Redden (1980) in the southern Bering Sea area.

The California samples were taken on a grid survey with maximum recoveries of 2 m or less. Thus, all samples were collected within the aerobic sulfate-reducing

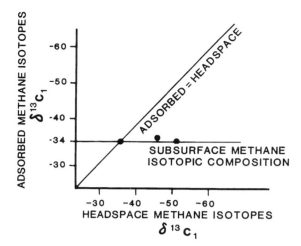

Figure 11—Graph of adsorbed methane versus headspace methane carbon isotopes over a known oil and gas reservoir in the South China Sea.

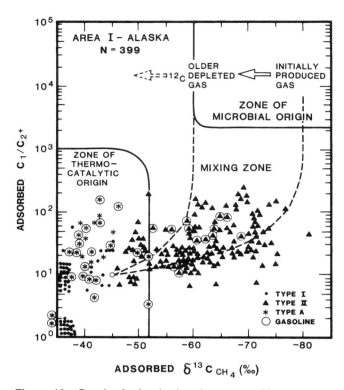

Figure 12—Graph of adsorbed molar composition versus isotopic composition of methane showing zones of origin. (Adapted from Bernard, 1978.)

zone. This is reflected in the large number of type I samples (95%). Light hydrocarbon data from Doose (1980) indicate limited biogenic activity in the surface sediments. This would account for the lack of type II samples. Geophysical records demonstrate active seepage occurring throughout the sampling area. In addition, there are several subsurface hydrocarbon accumulations with methane isotopic compositions of about –52‰. It appears that, despite the shallow sampling, seepage is active enough to overcome *in situ* bacterial alteration.

The South China Sea samples were collected over a major hydrocarbon accumulation where geophysical records display extremely active seepage (gas bubbles in the water column over a surface fault). All but one sample collected had isotopic compositions within 3‰ of the reservoired methane at depth. Samples with less than 90 ppb were not sampled. Therefore, only one smaller concentration sample exhibited bacterial oxidation (type I). Light hydrocarbon compositional data indicate limited biological data, explaining the apparent lack of type II samples. The combination of highly active fault leakage and site-specific sampling provides highly accurate surface methane carbon isotopes.

CONCLUSIONS

The three data sets demonstrate that meaningful surface isotopic compositions can be obtained, but extreme caution should be used in their interpretation. The depth of sampling, seepage activity, biological activity, and sampling approach can modify the data so that erroneous interpretations can easily be made. A large data set from deep site-specific cores must be collected to assess the reliability of surface methane isotopic compositions accurately. An examination of the concentration versus isotopic analysis used in combination with hydrocarbon

composition data helps the interpreter distinguish among biogenic, thermogenic, altered, and mixed gases.

Isotopic compositions measured on adsorbed and interstitial methane show similar trends. Evidence from this study and others (Woltermate, 1982; Horvitz, 1984, 1985) indicates that the adsorbed isotopic compositions provide different signatures. These differences are not well understood. Schoell (1984) thinks these differences are not related to the presence of thermogenic hydrocarbons, whereas Horvitz (1985) believes they reflect a differential adsorption process. Whatever the process, the adsorbed methane does provide more accurate measurements of subsurface methane in the areas sampled in this study. Pflaum's (1989) studies indicate that the adsorbed gas distributions may not be significantly affected by conditions having a profound effect on free gas distribution. If the adsorbed analysis reflects hydrocarbons trapped in carbonate fluid inclusions, as suggested by Thompson (1987), then Horvitz may have been correct but for the wrong reasons. The fluid inclusions may act as a temporary barrier to secondary alterations discussed here.

Isotopic compositional data should never be used alone. Hydrocarbon composition data are also required to determine the origin of seepage. The use of ratios and samples with relatively low concentrations (<50 ppb of adsorbed or <100 ppm of interstitial) does not provide reliable information on the origin. Schoell (1984) noted that methane, ethane, propane, and butane are generated in

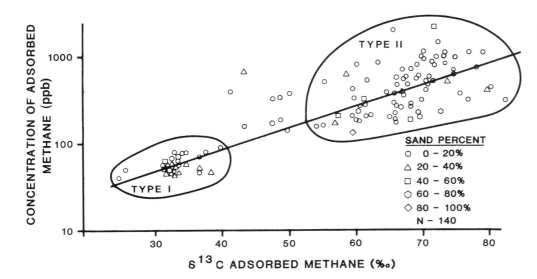

Figure 13—Graph of adsorbed methane concentrations versus adsorbed methane isotopic compositions showing groupings according to sediment type.

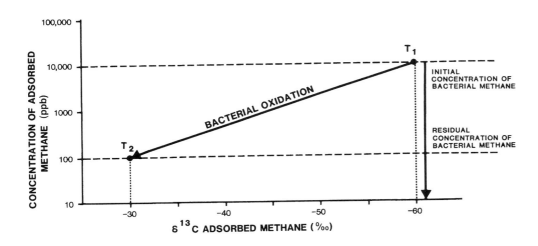

Figure 14—Plot of theoretical bacterial oxidation curve for methane concentration versus and isotopic composition.

Table 1—Distribution of Isotopic Classifications

Area	No. of Samples	Geophysical Seepage Activity	Average Maximum Depth (m)	Subsurface $\delta^{13}C_{CH4}$ (‰)	Type A $\delta^{13}C_{CH4}$ (‰)	Type A (%)	Type I (%)	Type II (%)
Alaska Area I	399	Passive	42	N/A	−37 to −46	8	32	60
Alaska Area II	66	Passive	9.0	−41	−43 to −52	25	25	50
Offshore California	180	Active	1.5	−53	−48 to −54	5	95	—
South China Sea	13	Active	4.0	−34	−34 to −37	92	8	—

sediments in quantities comparable to those observed and used for surface geochemical exploration (5.6–16 ppb of ethane). Only when concentrations exceed 100–3500 ppb (area I in the Alaskan example) for the wet gases and the higher molecular hydrocarbons are present can one feel confident that thermogenic hydrocarbons are present.

Acknowledgments—*The author gratefully acknowledges Exxon Company, U.S.A., and Exxon Production Research Company for funding during this study. Thanks go to several people who provided invaluable assistance in sample collection: Earl Stanford, Philip Bowden, James Hoppie, John Gaither, Chandler Rowell, and the crew of the research vessels. The manuscript was critically reviewed by Al James, Linda Wall, Curtis Koger, Victor Jones, Martin Schoell, Dirk Kettel, and Ray Metter. The author would also like to acknowledge the late Leo Horvitz for his invaluable discussions. The manuscript was typed by Jonelle Glosch and Shevonne Malvo. Figures were drafted by Ray A. Reed and Elizabeth M Brady. The author would also like to thank the Association of Petroleum Geochemical Explorationist for permission to republish this paper.*

REFERENCES CITED

Abrams, M. A., 1982, Modifications for increasing recovery and penetration in an open barrel gravity corer: OCEANS 82 Conference Records 82CH1827.5, p. 661–666.

Abrams, M. A., 1987, Interpretation of surface sediment methane carbon isotopes for detection of subsurface hydrocarbons: AAPG Bulletin, v. 71, no. 5, p. 524.

Abrams, M. A., 1992, Geophysical and geochemical evidence for subsurface hydrocarbon leakage in the Bering Sea, Alaska: Marine and Petroleum Geology Bulletin, v. 9, p. 208–221.

Barker, J. F., and P. Fritz, 1981, Carbon isotope fractionation during microbial methane oxidation: Nature, v. 293, p. 289–291.

Bernard, B. D., 1978, Light hydrocarbons in marine sediments: Ph.D. dissertation, Texas A&M University, College Station, Texas, 144 p.

Botz, R., 1981, Geochemistry Untersuchungen an Spurengasen Aus Wattsedimenten: Ber. Bundes, Geo Sen. Rohstofe Hannover, Arch., p. 95–104.

Coleman, D. D., J. B. Risatti, and M. Schoell, 1981, Fractionation of carbon and hydrogen isotopes by methane-oxidizing bacteria: Geochimica et Cosmochimica Acta, v. 45, p. 1033–1037.

Doose, P. R., 1980, The bacterial production of methane in marine sediments: Ph.D. dissertation, University of California, Los Angeles, p. 240.

Faber, E., and W. Stahl, 1983, Analytic procedure and results of an isotope geochemical surface survey in an area of the British North Sea, *in* J. M. Brooks, ed., Proceedings, British Isles Geological Congress on petroleum geochemistry and exploration of Europe: Geological Society of London, Special Publication No. 11, p. 51–63.

Fuex, A. N., 1977, The use of stable carbon isotopes in hydrocarbon exploration: Journal of Geochemical Exploration, v 7, p. 155–188.

Fuex, A. N., 1980, Experimental evidence against an appreciable isotopic fractionation of methane during migration, *in* A. G. Douglas and J. R. Maxwell, eds., Advances in Organic Geochemistry, 1979: Oxford, Pergamon Press, p. 725–732.

Horvitz, L., 1972, Vegetation and geochemical prospecting for petroleum: AAPG Bulletin, v. 546, p. 925–940.

Horvitz, L., 1982, Upward migration of hydrocarbons from gas and oil deposits: Paper presented at the 183rd American Chemical Society National Meeting, Las Vegas, Nevada, March 28–April 2.

Horvitz, L., 1984, A geochemical survey in part of the Santa Barbara Channel, an area containing subsurface accumulations of heavy oil: Paper presented at the AAPG Research Conference on exploration for heavy crude oil and bitumen, Santa Maria, California, October 28–November 2.

Horvitz, L., 1985, Stable carbon isotopes and exploration for petroleum, *in* Surface and near-surface geochemical methods in petroleum exploration: Rocky Mountain Section AAPG-SEPM-EMD Conference, Denver, Colorado, June 2, p. G1–G14.

Horvitz, E. P., and S. Ma, 1988, Hydrocarbons in near-surface sand, a geochemical survey of the Dolphin field in North Dakota: APGE Bulletin, v. 4, n. 1, p. 30–46.

Kvenvolden, K. A., and G. D. Redden, 1980, Hydrocarbon gases in sediment of the shelf, slope, and basin of the Bering Sea: Geochimica et Cosmochimica Acta, v. 44, p. 1145–1150.

Pflaum, R. C., 1989, Gaseous hydrocarbons bound in marine sediments: Ph.D. dissertation, Texas A&M University, College Station, Texas, p. 155.

Reitsema, R. H., A. J. Kalenback, and F. A. Linberg, 1981, Source and migration of light hydrocarbons indicated by carbon isotopic ratios: AAPG Bulletin, v. 65, p. 1536–1542.

Stahl, W., and B. D. Carey, 1975, Source rock identification by isotope analysis of natural gases from fields in the Val Verde and Delaware basins, West Texas: Chemical Geology, v. 16, p. 257–267.

Stahl, W., E. Faber, B. D. Faber, and D. L. Kirksey, 1981, Near-surface evidence of migration of natural gas from deep reservoirs and source rocks: AAPG Bulletin, v. 65, p. 1543–1550.

Schoell, M., 1984, Recent advances in petroleum isotope geochemistry: Organic Geochemistry, v. 6, p. 645–663.

Thompson, R., 1987, Relationship between mineralogy and adsorbed hydrocarbons in soils and sediments: Paper presented at ISEM–SMU Conference on vertical migration, Ft. Burgwin, New Mexico, October 15–18.

Woltemate, I., 1982, Isotopische Untersuchungen zur bakteriellen Gasbildung in cinem SuBwasseree: Diploma–beit, Universitat Clausthal, Clausthal Zellerfeld, 90 p.

Kettel, D., 1996, A method for processing adsorbed methane stable isotope data
from the near surface based on fractionation, *in* D. Schumacher and M. A.
Abrams, eds., Hydrocarbon migration and its near-surface expression: AAPG
Memoir 66, p. 319–336.

Chapter 23

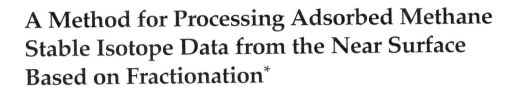

A Method for Processing Adsorbed Methane Stable Isotope Data from the Near Surface Based on Fractionation*

Dirk Kettel

Oil and Gas Consultant
Hannover, Germany

Abstract

Molecular and isotopic fractionation of gases that are controlled by flow processes are observed in nature. Because gas flow in sedimentary basins is highly dynamic, reasonable fractionation occurs on all geochemical parameters that depend on flow. Isotopic ratios of methane adsorbed in near-surface sediments are often measured in geochemical surface exploration. It is apparent, however, that the rough data cannot be used directly to identify hydrocarbon sources in the subsurface because the isotopic composition of methane fractionates during the transition from the dissolved phase in which methane is transported into the adsorbed phase in which it is measured. This implies a knowledge of the physical processes involved in gas migration through near-surface sediments, which is derived here from one of the exceptional cases of prestationary fluid flow found in nature. This physical background is then applied to the more common stationary flow processes, in which isotopic fractionation reaches equilibrium conditions. $\delta^{13}C_1$ fractionates toward the negative and δDC_1 toward the positive. The degree of fractionation depends on the methane flux in the subsurface and on the petrophysical properties of the sampled near-surface host sediment. Determination of the fractionation and the properties of the host sediment allows the methane flux from the subsurface to be calculated. The flux is observed to be stronger over gas accumulations and weaker over oil accumulations. The isotopic data processing method derives hydrocarbon fluxes and identifies source rocks from near-surface measurements and in this way allows the precise identification of hydrocarbon accumulations in the subsurface.

$SURFEX^{TM}$, United States Patent No. 5388456 and international patents.

INTRODUCTION

In the application of near-surface geochemical methods in oil and gas exploration, various physical approaches have been developed and offered during the past few decades. The techniques most frequently used are to measure the heavier hydrocarbon gas (C_{2+}) concentration within the free pore space (interstitial gas) (Sandy, 1989) and to measure the methane and heavier gas (C_2–C_6) concentration, as well as their isotopic composition ($\delta^{13}C$ and δDC), in the adsorbed gas phase that is fixed to the fine particles of the sediment sample (Horvitz, 1972; Faber and Stahl, 1984). However, it has become evident from field measurements that both the molecular and isotopic compositions of free and adsorbed gases in near-surface samples reflect the lithology of the sampled sediment rather than yielding information about the subsurface (e.g., see field survey examples in Faber and Stahl, 1984).

In the present paper, I show that the lithologic parameters controlling the molecular and isotopic fractionations of adsorbed hydrocarbon gases are the permeability and specific surface area of that part of the pore space that participates in the flow process. *Permeability* means fluid flow through porous media. Therefore, fractionation of gas parameters can be modeled by fluid flow equations. An understanding of the fractionation processes offers an opportunity to restore the original isotopic compositions that identify the hydrocarbon source in the subsurface and to calculate the strength of the hydrocarbon flux. This is shown in the following sections.

The analytical procedure followed here is to measure adsorbed methane concentrations and their isotopic ratios, $\delta^{13}C_1$ and δDC_1. Measuring the adsorbed phase excludes any influence of bacterial methane generation, which normally stays within the free pore space. Samples are taken below the local groundwater table. The sample is frozen in liquid nitrogen to avoid secondary alteration during transport and storage. If a new sediment is sampled, a parallel sample is taken for determining petrophysical properties. The coarse fraction (>63 μm) of the bulk sediment is sieved out, thus eliminating the free pore space content. Exact separation of free and adsorbed gases is accomplished by repeated washing procedures. From the remaining fine fraction (<63 μm), the adsorbed gas is extracted by vacuum and acid treatment.

OBSERVATIONS FROM NATURE

Mechanisms of Gas Migration Through Porous Media

When fluid flow through the sedimentary fill of subsiding oil- and gas-generating basins is observed, it can be concluded that normally stationary conditions of flow are reached (Figure 1). *Stationary conditions* are defined as situations in which the input of fluid into a standard volume of porous medium equals the output from the volume over time. *Prestationary processes,* which refer to fluid concentration fronts observed during migration, are rarely found in nature. However, *nonstationary features,* when observed, offer the unique opportunity to determine the physical mechanism and constants responsible for the flow process (Krooss and Leythaeuser, 1988). Therefore, this section on observations from nature uses one of these exceptional cases of prestationary fluid flow observed in nature (B2 well, Figure 2) to derive the physical processes involved in gas migration through near-surface sediments. The following section on secondary fractionation applies the physics thus identified to stationary flow conditions that are likely to be normal in subsiding basins (see Figures 5, 6). Under stationary flow conditions, isotopic fractionation of methane reaches equilibrium. The degree of fractionation can therefore be determined, and from this, the hydrocarbon flux from the subsurface can be derived.

Typical prestationary and semistationary features (as shown in Figure 1) have been observed on adsorbed methane concentration profiles of near-surface sediments in the Münsterland Basin, Germany. These are discussed in more detail in Kettel (in press). Figure 2 shows the B2 well as an example of several identical profiles from this area. The Münsterland Basin sedimentary fill is composed of slightly folded Upper Carboniferous strata containing coal seams and dispersed organic matter that act as gas source rocks, overlain by a 1000-m-thick Upper Cretaceous marl section. The basin has been undergoing uplift since the end of the Cretaceous. The younger sedi-

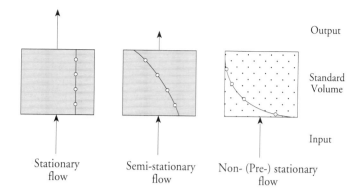

Figure 1—Schematic illustration of three different states of fluid flow in nature. Standard volumes of rock are shown, with the dots indicating virtual data points. The lengths of vertical arrows are proportional to the strength of fluid flow.

mentary record contains glacial drift sediments that are subdivided into a lower claystone representing the Saale glaciation (200,000 yr B.P.) and an upper siltstone representing the Weichsel glaciation (12,000 yr B.P.).

Lommerzheim (1988) published adsorbed methane concentration profiles through the Upper Cretaceous section (reported by Kettel, in press). The methane originates from the Carboniferous coal source rocks, and its concentration profiles show relatively uniform values of hundreds to thousands of parts per billion throughout the section due to the uniform lithology of the sediments. This concentration range is confirmed by sample B2/8 from the B2 well. (Figure 2). The same measurements have been made on samples from the overlying Saale and Weichsel sediments (B2/7–B2/1), with the surprising result that the Saale clay shows a semistationary upward methane flux character and the Weichsel silt a clearly nonstationary flux. The methane concentration peak at the base of the Weichsel silt is particularly unusual for flow parameters normally observed in sedimentary sections.

We attempted to match these characteristics by flow models based on different physical mechanisms, paying special attention to the concentration peak. The model that properly reproduces the measured features (Kettel, in press) is a combined Darcy flow–diffusion model in which the diffusion constant is also a function of the mechanical dispersion of the water flow (Krooss and Leythaeuser, 1988; Bouhroum, 1991). In Figure 3, the present time slice shows the fit of the measured and calculated data. The hydrodynamic interpretation of the model is based on the "Bear equation" (Kinzelbach, 1986; Bear and Corapcioglu, 1987) for tracer flow, in which methane (the tracer) is transported in a water solution and is partially retained by adsorption and desorption (elution) processes on the fine particles of the sediment. The variation of the relative adsorbed methane concentration c through time t and distance x is

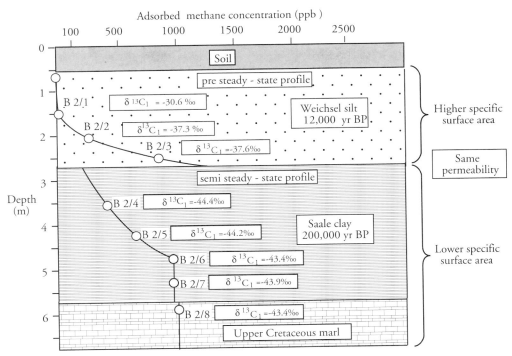

Figure 2—Nonstationary adsorbed methane concentration profiles, B2 well, Münsterland Basin, Germany.

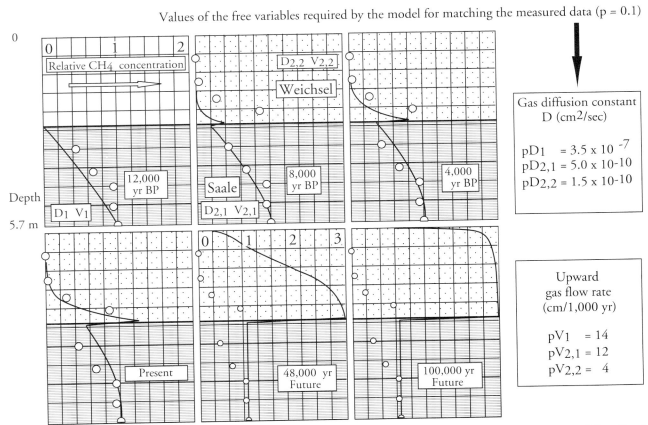

Figure 3—Physical modeling of adsorbed methane concentration profiles in B2 well using the Bear equation (Kinzelbach, 1986; Bear and Corapcioglu, 1987).

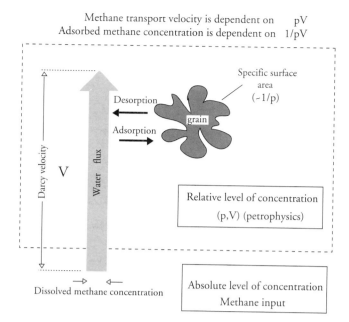

Methane transport velocity is dependent on   pV
Adsorbed methane concentration is dependent on   1/pV

Figure 4—Schematic model of gas adsorption and desorption dynamics in sedimentary rocks.

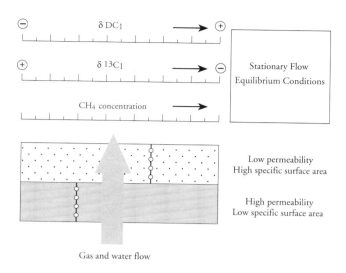

Figure 5—Schematic illustration of molecular and isotopic fractionation occurring at stationary flow equilibrium conditions.

expressed (after the Bear equation) as

$$\frac{\delta[c(t,x)]}{\delta t} = -p\frac{\delta[V(t,x)c(t,x)]}{\delta x} + p\frac{\delta^2[D(t,x)c(t,x)]}{\delta x^2} \quad (1)$$

The free variables for matching the measured profile are products pD and pV, where D is the diffusion constant (cm²/sec), V the Darcy velocity for water (cm/1000 yr), and p the eluted portion of gas molecules relative to the adsorbed portion. (See Table 1 for definitions of all abbreviations and symbols used in this paper.) The eluted portion p is inversely proportional to the adsorption probability q and thus inversely proportional to the specific surface area available to the flow process. During the time of Saale glaciation, the product was $pD(V)_1$ for the Saale layer, and during Weichsel time, the products were $pD(V)_{2,1}$ and $pD(V)_{2,2}$ for the Saale and Weichsel layers, respectively. In Figure 3, the values for the products that match the measured profile are given. The higher diffusion constant needed for Saale glaciation time could coincide with some stronger hydrothermal activity postulated by Lommerzheim (1988) for Tertiary–Quaternary time in this area.

The methane concentration peak at the base of the Weichsel silt in Figure 2 is matched when a sharp reduction in pV (12–4 cm/1000 yr) is introduced for the Weichsel layer relative to the Saale layer. This is realistic because, although the permeabilites are the same in both layers, it coincides with differences in the specific surface areas available to the water flow measured in both layers (see Kettel, in press). The specific surface area is higher by

a factor of three in the Weichsel layer than in the Saale layer. Figure 4 shows schematically that a high adsorbed methane concentration goes along with a reduced water flow $(1/V)$ or a higher specific surface area $(1/p)$. Given the same adsorption probability in time, it is evident that more gas molecules can be adsorbed by the fine particles when the gas transport velocity is low and the specific surface area is high.

The presence of semistationary and nonstationary methane fluxes through glacial drift sediments is due to reduced vertical water flow velocities during that time of basinal uplift. These features have not been the same in the past and will not be the same in the future, as shown in Figure 3. At the beginning of Weichsel deposition at 12,000 yr B.P., the Saale layer would have had a prestationary profile. After deposition of the Weichsel layer, this approached the present semistationary state in a stepwise manner, whereas the nonstationary profile developed through the Weichsel layer. Extrapolation into the future (after 48,000 and 100,000 yr, respectively) shows that the observed nonstationary profiles will be transitional on the way toward a steady-state flux equilibrium condition in which gas concentration is exclusively controlled by changes in the petrophysical properties from layer to layer. This is schematically shown in Figure 5.

The principles of this molecular behavior are also found in the isotopic record (Figure 5). This is to be expected because different isotopes act like different molecules with respect to their adsorption probabilities. Only a fractionation factor must also be introduced into the Bear equation, which expresses the preference in adsorption of heavier over lighter isotopes (see equation 2). Reaching steady-state flux, isotopes exhibit their equilibrium fractionation ($\delta^{13}C_1$ toward the negative and δDC_1 toward the positive, as shown in Figure 5). In that case, changes in isotopic ratios of adsorbed gases through sed-

Table 1—Abbreviations and Symbols Used in This Paper

Symbol	Definition
$\delta^{13}C_1$	Carbon isotope ratio of methane (‰)
δDC_1	Hydrogen isotopic ratio of methane (‰)
C_n	Normal alkanes
ppb	Parts per billion gas weight per rock sample weight (g/g)
$c(t, x)$	Relative adsorbed gas concentration through time t and over distance x
D	Diffusion constant (cm^2/sec)
$V_{2,1}$	Darcy flow velocity for water (cm/1000 yr) (for example 2,1, in Weichsel glaciation time through the Saale layer)
p	Inversely dependent on effective porosity, rock matrix density, and portion of adsorbed gas relative to total gas (adsorbed + dissolved). Describes degree of desorption of gas.
$1/p$	Directly proportional to specific surface area. Describes degree of adsorption of gas q.
q	Gas adsorptivity
A	Geologic factor expressing relationship of gas and water flow through sample multiplied by adsorptivity q. Also expresses q/V. Directly proportional to methane flux rate through sample and its petrophysical properties.
λ	Solubility of gas in water of a certain salinity and under near-surface pressure/temperature conditions (cm^3/cm^3)
α	Fractionation factor between heavier and lighter gas isotope during adsorption from dissolved or free gas phase.
F	Isotopic shift during gas phase transitions (equals delta δC_1).
K	Water permeability of a rock sample (darcy)
i	Vertical water pressure gradient (MPa/m)
R_o	Actual source rock maturity, reflected by the isotopic composition of a 100% free methane.

imentary sections exclusively reflect changes in host rock properties; otherwise they remain constant. Examples from nature are shown by samples B2/8–B2/4 (Figure 2) and by samples from another northern German shallow well (Figure 6). Fractionations of the carbon isotope ratio in the opposite direction (toward the positive), as seen in the nonstationary methane concentration profile through the Weichsel layer (Figure 2, samples B2/3–B2/1), can be explained by short-lived kinetic effects such as those observed by Fuex (1980) in the laboratory. Bacterially mediated oxidation can be excluded for these values from the low content of unsaturated compounds (for further details, see Kettel, in press).

From these observations, it becomes evident that fluid flow through porous media may be the process responsible for gas migration through near-surface sediments. In other words, the upward-directed water flow through the sedimentary column carries gas in solution and guaran-

tees information transfer from the source to the surface. At the site of the sampled sediment, transition of some of the methane occurs from the dissolved phase into the adsorbed phase and a free phase.

Adsorbed Methane Isotopic Fractionation

As explained in the previous section, in the common case of stationary flow, it is observed that the adsorbed methane isotopic ratios and molecular concentration vary according to the petrophysical properties of the host rocks (Figure 5). This observation can only be made where bacterial activity does not interfere and the methane is of purely thermocatalytic origin.

An example is shown in Figure 6 (from Stahl et al., 1984) in which a well drilled down to 40 m penetrated Middle Triassic carbonates and an underlying Lower Triassic siltstone (Upper Bunter). Each lithology is characterized by its own narrow range of $\delta^{13}C_1$ values: the carbonates about –43‰ and the silt about –37‰. Because the gas samples were taken at the same site as a profile, it is assumed that the thermogenic methane migrating from the deeper subsurface upward and penetrating the Triassic strata is the same for all samples and that it carries the same isotopic information about the facies and maturity of its source. This implies that isotopic fractionation is occurring during the transition of methane from the dissolved phase in which it is transported into the adsorbed phase in which it is measured. Consequently, the degree of fractionation is dependent on the lithology of the sampled host rock. If you can imagine these samples taken instead as a survey, sampling various lithologies laterally, you can understand why the rough data of adsorbed or free gases from geochemical surveys must be noisy. The same applies to all geochemical parameters that can be measured in nature and that depend on flow processes.

Figure 6 also shows that, when introducing a $\delta^{13}C_1$ value for free methane of –36.2‰ characterizing the maturity of its humic source at that site, methane in the adsorbed phase is fractionated toward the more negative in the carbon isotope ratio. This observation was made earlier by Bond (1962) in nature and by Fuex (1980) in laboratory experiments. It can also be seen from Figure 6 that fractionation is stronger in low-permeable sediments.

The complete methane isotopic profile through the upper 1 m of the Upper Cretaceous marl and the overlying 3 m of Saale layer is shown in Figure 7. The lithology of both layers are identical. The gaseous content of these strata exhibits isotopic equilibrium fractionation. Even within the same lithology, a small-scale variation in isotopic fractionation is observed: the fractionation is stronger in samples with lower methane yield (samples B2/4 and B2/5).

We also measured the hydrogen δDC_1 ratio here. It was observed that δDC_1 fractionates toward the positive, whereas $\delta^{13}C_1$ fractionates toward the negative. This contrasting behavior of the carbon and hydrogen isotopic ratios has been observed by Fuex (1980) in the laboratory

Adsorbed methane distribution at steady-state equilibrium flow, extrapolated for the B2 well to 100,000 yr into the future

Adsorbed methane concentration →

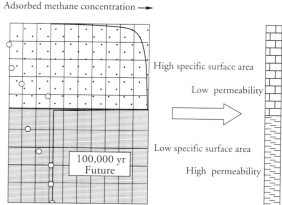

Carbon isotopic record of adsorbed methane, Elm, North Germany

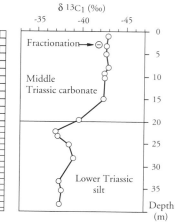

Figure 6—(Left) The theoretical molecular concentration profile reflecting stationary flow conditions, B2 well, Münsterland (taken from Figure 3), compared to (right) adsorbed methane $\delta^{13}C_1$ profile reflecting stationary flow through strata with different lithologies, Reitlingstal well (from Stahl et al., 1984).

during the transition from dissolved gas to free gas. It is also observed in nature where complete isotopic profiles of thermocatalytic methane are measured (e.g., Lommerzheim, 1988; Faber et al., 1990).

SECONDARY FRACTIONATION

Isotopic fractionation of methane during transport as a solute through porous media is defined as *secondary fractionation*. Molecular and isotopic fractionation can only be interpreted in terms of fluid flow strength and petrophysical properties of the conducting rocks when equilibrium of fractionation or stationary flow of fluids is reached. Fractionations are characterized by a fractionation factor between the heavier and lighter isotopes.

Moving along the migration path from deep sources toward the surface, part of the methane leaves the system due to adsorption onto the fine particles and transition into the free phase. Theoretically, a continuous change in the isotopic composition of the dissolved methane should occur during this upward migration. In subsiding basins with strong fluid flow, however, the ratio of adsorbed or free methane to transported methane in the system is so small that the isotopic composition of the dissolved methane remains practically unchanged during migration. This was derived by Lommerzheim (1988) from measurements of adsorbed and free methane concentration profiles through 1000 m of Upper Cretaceous sections in the Münsterland.

This paper explains the reason why the gas flux from a deep-seated gas source and its properties can be determined by analyzing surface sediment or a borehole sample, regardless of the migration path the gas followed. In terms of predicting gas flow, only the last phase transition within the sediment sampled is crucial.

Isotopic fractionation due to phase transitions can be examined theoretically and calibrated *in situ* from observations in nature. Fuex (1980) measured the equilibrium conditions of methane in solution-dissolution experiments. He presented a formula that relates methane con-

centration in the free gas phase, isotopic shift relative to the dissolved phase, and a geologic factor for the special case when equilibrium conditions are reached. Through measurements of these parameters, he determined the fractionation factor for the transition of dissolved methane into free methane for the stable carbon isotope composition $\delta^{13}C_1$. In principle, this can also be applied to the equilibrium between dissolved and adsorbed methane and between free and adsorbed methane. From Fuex's (1980) approach, a mass balance equation can be derived that uses the mole fraction of the lighter isotope, the isotopic composition, a geologic factor A (adsorption), and a fractionation factor α:

$$c = \exp - \left(\frac{\ln \dfrac{1+A}{A}}{\ln \dfrac{1+A}{\alpha+A}} \times \ln \frac{F+1000}{1000} \right) \quad (2)$$

which expresses the mass balance and the theoretical relationship between the relative methane concentration of the adsorbed phase c, the isotopic shift F (equal to delta δC_1 in Figures 8, 9) that the adsorbed phase undergoes relative to the free gas phase after reaching equilibrium, the fractionation factor α of methane between the heavier and lighter isotopes during adsorption, and a geologic factor A. (The derivation of this equation is given in the Chapter Appendix.) Equation (2) is identical to the Bear equation (equation 1) when it is solved for the special case of equilibrium conditions or stationary flow. In equation 2, A is proportional to $1/pV$, c stands for $c(t, x)$, and the isotopic shift F is related to c and A by the fractionation factor.

Fractionation of $\delta^{13}C_1$ and δDC_1

Fractionation factors between ^{13}C and ^{12}C as well as between D and H for the transition of dissolved methane into adsorbed methane starting from the value of a 100% free gas have been determined using case histories with

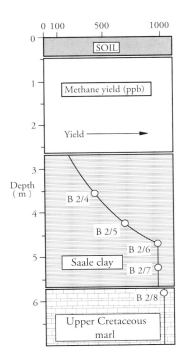

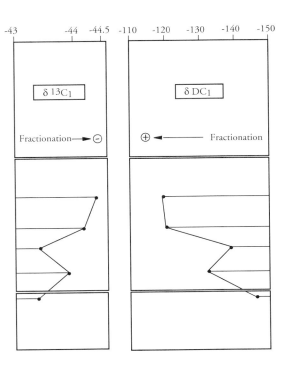

Figure 7—Adsorbed methane isotopic record through strata with the same lithology, B2 well, Münsterland Basin.

data published by Lommerzheim (1988). In these case histories, the methane yield c, the isotopic shift F, and the geologic factor A were known for each sample. With these parameters, equation (2) can be solved for α. Thus, the fractionation factor between ^{13}C and ^{12}C under near-surface temperature conditions has been determined as

$$\alpha_{13C_1/12C_1} = 1.0014493$$

The fractionation factor between D and H under near-surface temperature conditions has been determined as

$$\alpha_{DC_1/HC_1} = 0.9752528$$

A comparison of both fractionation factors shows that fractionation of DC_1 is stronger than that of $^{13}C_1$ by a factor of 18. This is in qualitative agreement with observations from other fractionation processes of methane in nature and in the laboratory (Gunter and Gleason, 1971; Gant and Yang, 1964). DC_1 fractionates toward the more positive, whereas $^{13}C_1$ fractionates toward the more negative. This is also in agreement with laboratory observations (Fuex, 1980).

DATA PROCESSING PROCEDURE: HYDROCARBON FLUX

From the four variables in equation (2), the fractionation factor α has been determined and the methane yield c is known by measurement. Of the remaining two parameters, isotopic shift F and factor A, either of them can be determined if the other is known.

In Figures 8 and 9, equation (2) is shown graphically on a semilogarithmic scale using the fractionation factors for $\alpha_{13C/12C}$ (Figure 8) and $\alpha_{D/H}$ (Figure 9). These are called *fractionation diagrams*. The measured isotope ratios from the profile B2 well (Münsterland) shown in Figure 7 were plotted on these graphs against the respective methane yields. The resulting data points lie close to a regression line, which is then fitted into the set of lines for different factors A. The absolute positions of the data points in Figures 8 and 9 are thus determined, as well as the isotopic shifts and A values. Factor A can be read as about 1.3×10^{-1} for all points on this profile (see Table 3).

Since we now have the value of geologic factor A, we can calculate the actual methane flux from the subsurface at the specific site of the B2 well profile. In relative terms, A is expressed as

$$A = f(q/V) = f(q/\text{Darcy water velocity})$$

where q is the adsorptivity. To determine A in absolute terms, we use the definition of equation (2). In the derivation of equation (2), the geologic factor A equals the relationship between the light isotopes ^{12}C or H for the gas in the free or adsorbed phase and that in the dissolved phase. This means that A is a function of (1) the relationship between light gas and water supply to the sampled sediment over time, (2) the adsorptivity q of the light gas on specific grain surfaces in the sample, and (3) the solubility λ of the light gas in water of a certain salinity and under near-surface pressure and temperature conditions (see Fuex, 1980):

$$A = \frac{q \times \text{gas supply over time}}{\lambda \times \text{water supply over time}} \qquad (3)$$

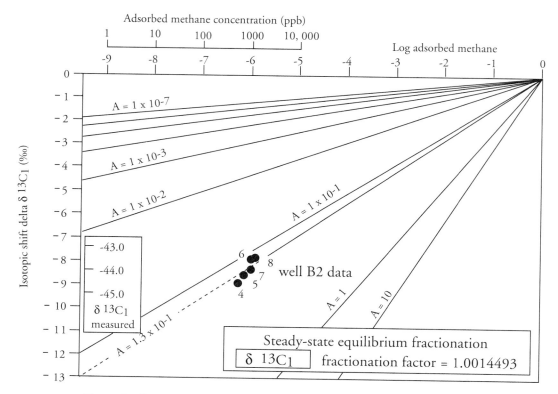

Figure 8—Secondary equilibrium fractionation of δ¹³C₁ with B2 well data.

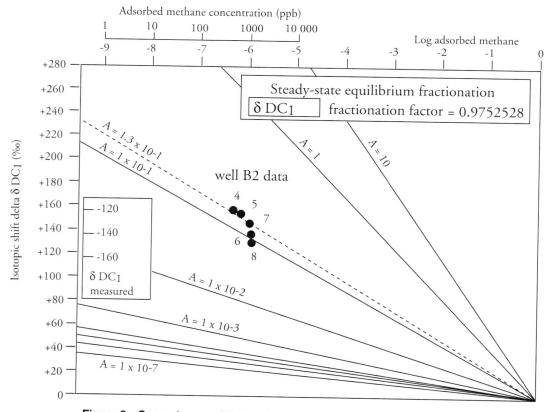

Figure 9—Secondary equilibrium fractionation of δDC₁ with B2 well data.

The gas supply over time to the sampled sediment is the term that we are looking for. It is the vertical methane flux (in standard m^3/km^2) multiplied by the time (in years) in the subsurface at the site where the sample was taken. This parameter determines the absolute level of methane concentration and the isotopic shift, which are then modified by the lithologic properties $1/V$ and $1/p$ (see Figure 4). To calculate this from equation (3) with a known factor A, only the water supply over time to the sample (which equals the water flux V through the sample) and the adsorptivity q (which equals $1/p$) remain to be determined. Doing that, the water permeability K (in darcys) and the specific surface area of the part of the pore space that is available to the flow process need to be determined. The hydraulic pressure gradient i is estimated and kept constant throughout the survey, and time t in years is taken as the depositional age of the sampled sediment. In soft sediments, however, both the permeability and specific surface area are expressed sufficiently well by the content (in %) of the coarse silt fraction (20–63 μm) of the sediment sampled because this contains the part of the pore space that conducts the fluid (Kettel, in press).

From equation (3), with known factor A and a permeability for the 20–63-μm fraction of $K = 2.6 \times 10^{-5}$ d (Table 2), a pressure gradient of $i = 0.01$ MPa/m, $\lambda = 55$ mL/L, and $q = 1$, we can calculate the methane flux from the subsurface for the B2 well site as 2.91×10^4 $m^3/km^2 \times 200{,}000$ yr (see Table 3).

Equation (2) and Figure 8 show that with increasing isotopic fractionation, factor A also increases. According to equation (3), a higher factor A indicates a stronger methane flux, given the same petrophysical properties of the host rock. In this way, a more negative $^{13}C_1$ of adsorbed gas under otherwise equal conditions indicates a stronger methane flux from the subsurface.

PRIMARY FRACTIONATION

Primary fractionation is the gradual isotopic change of gas deliberated from kerogen during heating. It allows identification of the hydrocarbon source within the subsurface according to organic facies and maturity. After determination of the degree of isotopic shift in Figures 8 and 9, the isotopic value of the 100% free phase can be determined and thus the signatures of the gaseous source in the subsurface. From the pair of free gas isotopic signatures, $\delta^{13}C_1$ and δDC_1, obtained in this way, the source can be characterized according to organic facies and maturity applying Figures 10 and 11. Both figures show a correlation between the $\delta^{13}C_1$ (Figure 10) or δDC_1 (Figure 11) of methane and the mean degree of maturity (% R_o) of the source rock interval that actually generates methane directly below.

The data points in Figures 10 and 11 represent case histories grouped according to the organic facies of their source. The data for sapropelic sources are taken from Faber (1987). The data for humic sources are from gas accumulations in the Central European Upper Carbon-iferous Basin reservoired in Lower Permian Rotliegendes sandstones. The resulting characteristics reflect the primary fractionation during methane generation and are in good agreement with those obtained by Chung and Sackett (1979) from pyrolysis experiments. The correlation in these figures once again shows that the degree of fractionation diminishes with decreasing methane generation at higher maturities. Primary fractionation occurs toward the more positive for both isotopic ratios, approaching the ratios of the parent material, which are generally more positive for the humic material and more negative for the sapropelic (Redding et al., 1980). This allows the organic facies of the source to be determined in addition to its maturity.

Figures 10 and 11 are combined in one graph in Figure 12, called the *diagnostic diagram*, which shows $\delta^{13}C_1$ and δDC_1 on the axes (heavy lines represent the facies and thin lines show isomaturity). A set of parallel lines represents the secondary fractionation law between $\delta^{13}C_1$ and δDC_1 according to Figures 8 and 9. Starting from different facies and maturity combinations, each free gas isotopic pair has to fractionate strictly parallel to the fractionation lines (isotopic shift F) during adsorption. Data from one well originating from the same source must fractionate along the same fractionation line. This can be seen by plotting the measured isotopic data pairs from B2 well (Figure 7) onto Figure 12. The measured data points arrange themselves along the same secondary fractionation line.

The data points show different distances from an optional point of origin. According to the inset in Figure 12 (bottom), these distances are controlled by factor A and the methane yield. Since A is the same for all data points (see above), the distances vary as a function of the yields. For each yield and for $A = 1.3 \times 10^{-1}$, the distances are taken from the inset in Figure 12, and the respective data points are shifted back along their secondary fractionation line over these distances until they reach a discrete facies and maturity option. The B2 well data points end up on the humic source line at a maturity of 1.12% R_o. This agrees with the maturity measured for the underlying Upper Carboniferous source rock.

CASE HISTORIES

Single-Source Case History: Lower Saxony Inverted Basin

An example from the Lower Saxony Inverted Basin in Germany shows how isotopic data from the near-surface processed in this way can give a real picture of the gas flow status in the subsurface. Near-surface samples were taken from the Triassic Bunter Formation of sandstones, siltstones, and shales of different permeabilities. They were taken by rotary drilling from a depth of about 5 m. Their adsorbed hydrocarbon gas contents and the bulk permeability for water of each lithology were measured in the laboratory (Table 2).

Table 2—Measured Parameters

Sample	Yield C_1 (ppb)	$\delta^{13}C_1$ (‰)	δDC_1 (‰)	Permeability (d)
B2 Well profile				
B2/4	437	−44.4	−119	2.6×10^{-5}
B2/5	683	−44.2	−120	2.6×10^{-5}
B2/6	960	−43.4	−139	2.6×10^{-5}
B2/7	966	−43.9	−132	3.6×10^{-5}
B2/8	1.014	−43.4	−146	2.0×10^{-5}
Lower Saxony survey				
1	212	−31.6	−131	2.7×10^{-4}
7	268	−34.0	−239	1.1×10^{-4}
8	218	−33.9	−300	1.1×10^{-4}
10	1.464	−33.6	−253	8.8×10^{-5}
13	244	−34.6	−213	1.1×10^{-4}
16	6	−29.0	—	1.1×10^{-4}
18	232	−33.2	−187	1.1×10^{-4}
19	148	−32.2	−152	1.1×10^{-4}
24	49	−30.2	n.m.	1.1×10^{-4}
30	178	−35.2	−300	8.8×10^{-5}
31	272	−37.5	−261	3.8×10^{-4}
32	493	−40.8	−209	3.8×10^{-4}
33	459	−37.3	−155	3.8×10^{-4}
34	831	−37.3	−189	3.8×10^{-4}
42	5	−29.0	—	2.4×10^{-4}
Upper Rhine Graben survey				
5	4.302	−37.2	−134	1.0×10^{-4}
8	1.298	−37.5	−201	1.0×10^{-4}
9	1.879	−36.8	−151	1.0×10^{-4}
10	2.176	−37.0	−137	1.0×10^{-4}
11	4.674	−37.2	−136	1.0×10^{-4}
61	2.784	−38.6	−159	1.0×10^{-4}
62	4.359	−38.2	−126	1.0×10^{-4}
64	2.488	−39.3	−143	1.0×10^{-4}
65	2.797	−37.7	−140	1.0×10^{-4}

In the sample area, the Triassic Bunter is underlain by Upper Permian Zechstein rock salt and Zechstein carbonates, which provide a seal for possible gas accumulations in the Lower Permian Rotliegendes reservoir sandstones. These are in turn underlain by Upper Carboniferous Namurian shales and sandstones that were slightly folded during the Variscan orogeny. Shortly after folding, this area was affected by shear faulting that produced graben zones in which Upper Carboniferous Westfalian coal-bearing humic source rocks were deposited. The whole section reached its maximum burial depth, temperature, and gas generation in Late Cretaceous time (Stahl, 1992) and has undergone uplift since then.

The sampled geochemical line covers part of a graben zone and a structural high for the top of the Rotliegendes reservoir sandstones nearby. The methane yields, both C and H isotopic ratios, and water permeabilities are listed in Table 2 for each sample of the survey. The isotopic data pair $\delta^{13}C_1/\delta DC_1$ for each sample is plotted on the diagnostic diagram in Figure 13. The data points indicate source maturities ranging from 1.05 to 1.45% R_o when they are shifted back along the parallel fractionation lines until they reach a line in the isotopically heavier part of the humic facies. (In this case, it was previously known that the source is strongly humic.) From the distances of backshifting of the data points, the isotopic shifts F of $\delta^{13}C_1$ were then derived and plotted on the fractionation diagram in Figure 14 against the respective methane yields. The resulting factors A vary between 0 and 3×10^{-2}. We then calculated the methane fluxes (in Std. $m^3/km^2 \times 10$ m.y.) using equation (3) and the permeabilities K (in darcys). The value of q is set at 1 and the vertical pressure gradient is set at $i = 0.01$ (MPa/m), which is kept constant throughout the survey.

The methane fluxes are listed in Table 3 and plotted in Figure 15. A higher methane flux is observed over the Westfalian graben subcrop, indicating the presence of a humic source. The nearby seismic structure is associated with a continuous positive gas flux anomaly at the surface, indicating that the structure may be gas filled. The

Table 3—Derived Parameters

Sample	Isotopic Shift delta $\delta^{13}C_1$ (‰)	Gas Source Maturity R_o (%)	Factor A	Methane Flux
B2 Well profile				(Std. $m^3/km^2 \times 200{,}000$ yr)
B2/4	−8.8	1.12	1.3×10^{-1}	2.91×10^4
B2/5	−8.6	1.12	1.3×10^{-1}	2.91×10^4
B2/6	−7.8	1.12	1.2×10^{-1}	2.57×10^4
B2/7	−8.3	1.12	1.3×10^{-1}	5.60×10^4
B2/8	−7.8	1.10	1.2×10^{-1}	1.52×10^4
Lower Saxony survey				(Std. $m^3/km^2 \times 10$ m.y.)
1	−3.2	1.42	1.0×10^{-3}	9.30×10^5
7	−2.0	1.27	2.0×10^{-5}	3.73×10^3
8	0	1.15	0	1.19×10^2
10	−1.4	1.23	1.0×10^{-6}	1.12×10^5
13	−2.9	1.25	6.0×10^{-4}	0
16	−0.6	1.42	0	5.60×10^4
18	−2.7	1.33	3.0×10^{-4}	1.30×10^5
19	−3.0	1.38	7.0×10^{-4}	0
24	−0.2	1.35	0	0
30	−0.4	1.12	0	6.51×10^5
31	−2.7	1.12	3.0×10^{-4}	6.51×10^7
32	−5.6	1.05	3.0×10^{-2}	5.40×10^7
33	−5.6	1.26	2.5×10^{-2}	4.34×10^7
34	−4.8	1.22	2.0×10^{-2}	0
42	−1.0	1.45	0	
Upper Rhine Graben survey				(Std. $m^3/km^2 \times 10{,}000$ yr)
5	−4.8	1.30	2.3×10^{-2}	1.15×10^3
8	−3.2	1.20	2.0×10^{-3}	1.00×10^2
9	−4.1	1.28	1.0×10^{-2}	5.00×10^2
10	−4.7	1.30	1.6×10^{-2}	8.00×10^2
11	−4.8	1.30	2.3×10^{-2}	1.15×10^3
61	−4.8	1.23	2.0×10^{-2}	1.00×10^3
62	−5.4	1.28	4.0×10^{-2}	2.00×10^3
64	−5.4	1.23	3.3×10^{-2}	1.60×10^3
65	−4.9	1.28	2.0×10^{-2}	1.00×10^3

source maturity diagram shows a steady decrease in maturity from a more tectonically stressed area in the south toward a less affected area in the north. This is consistent with geologic field observations.

Double-Source Case History: Upper Rhine Graben Rift Basin

Near-surface samples were taken from an area of the Upper Rhine Graben in Germany. The graben fill shows that increasing subsidence and sedimentation rates have occurred from the early Oligocene until the present. Strong rift-related subsidence was associated with increased heat flow since Pliocene–Pleistocene time (Hoffers, 1981). The rift-related graben fill is underlain by prerift Lower Jurassic sediments. The Jurassic provides a good sapropelic source for the oil accumulations in this region, while the lower Oligocene section bears humic source rock layers.

The geochemical sampling line covered a structure for the top of a sequence of upper Oligocene fluvial reservoir sandstones. The samples were taken from the silty cover of the Rhine River terrace formed during the latest Pleistocene glaciation (10,000 yr B.P.) at an average depth of 3 m. In the laboratory, the <63-μm sediment fraction was sieved out of the samples, and the hydrocarbon gas content adsorbed to this fraction was measured. We also measured the water permeability of this sediment fraction, which was the same for all samples (see previous section on data processing). The portion of the 20–63-μm fraction was also the same for all samples. The laboratory data are listed in Table 2. The isotopic data pair $\delta^{13}C_1/\delta DC_1$ for each sample is plotted on the diagnostic diagram in Figure 13. The resulting data points indicate source maturities between 1.2 and 1.3% R_o when they are shifted back along the parallel fractionation lines until they reach the humic facies line as one of the two source rock facies in the subsurface. The isotopic shifts F of $\delta^{13}C_1$

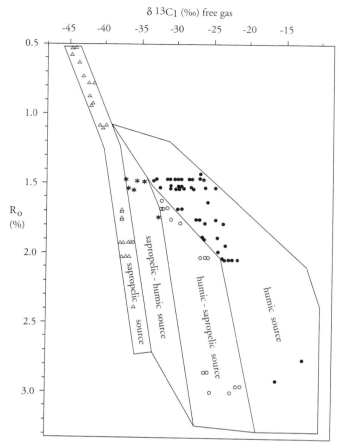

Figure 10—Primary fractionation of $\delta^{13}C_1$ from sources of varying organic facies and maturity (vitrinite reflectance).

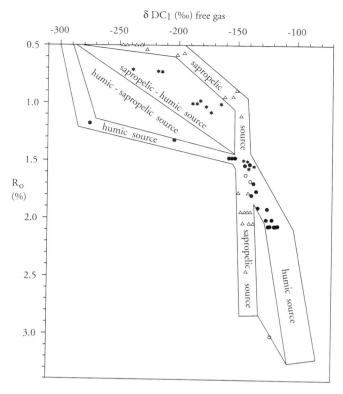

Figure 11—Primary fractionation of δDC_1 from sources of varying organic facies and maturity (vitrinite reflectance).

were derived from the distances of backshifting and were plotted on the fractionation diagram in Figure 14 against the respective methane yields.

The resulting factors A vary between 2.0×10^{-3} and 4.0×10^{-2} and are listed in Table 3. From these, the methane fluxes (in Std. $m^3 / km^2 \times 10,000$ yr) were calculated using equation (3), and λ, q, and i were kept as 55 mL/L, 0.05, and 0.01 MPa/m, respectively. The methane fluxes and source maturities are listed in Table 3 and plotted in Figure 16.

The normal reading away from the structure is a humic gas maturity of 1.3% R_o. This is realistic for the lower Oligocene humic source in this area. It implies that the Jurassic sapropelic source must be overmature now. Therefore, the known oil fields in this area must have been generated earlier while the humic source was immature. However, with the beginning of the Quaternary, the subsidence rate and heat flow increased to such an extent that the lower Oligocene humic source now generates gas.

The structure indicated in this section is associated with a low methane flux anomaly and a low source maturity anomaly at the surface. Assuming that the structure contains oil generated by the Jurassic sapropelic

source, the recently upward-migrating humic gases would be partially dissolved when passing through the oil accumulation. This could easily produce a lower methane flux leaving the structure over the same time period. The low source maturity over the same site may thus indicate an admixture of low-maturity oil associated gases to the high-maturity humic gas. Similar characteristics have often been observed above oil fields and structures in oil-prone regions. It is a typical case history for the superimposition of a passive sapropelic source (oil accumulation dissoved gas) on an active humic gas source which allows the presence of an oil accumulation to be identified.

Discussion of Case Histories

The results of these two case histories point out some aspects of special interest:

1. The methane flux and source maturity resulting from the calculations show continuous trends when plotted over the geologic sections, and no scattering is observed. Therefore, it is likely that the physical process of gas migration and the lithologic influence on the data has been satisfactorily considered and that each data point represents an unambiguous hydrocarbon source characterization.

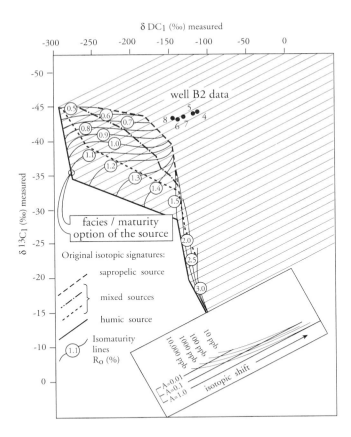

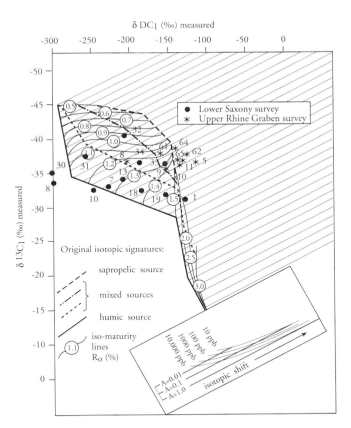

Figure 12—Diagnostic diagram for the determination of facies and maturity of a methane source from near-surface adsorbed gas data (composite of Figures 8, 9, 10, 11 with B2 well data). Heavy lines indicate the facies and maturity options (primary fractionation), and thin parallel lines indicate the secondary fractionation law.

Figure 13—Diagnostic diagram (as in Figure 12) with the case histories data added.

2. The absolute ranges of the methane fluxes calculated from the near-surface isotope data in various geologic settings are in good agreement with those calculated using source rock reaction kinetics and basinal temperature histories (Welte et al., 1984).

3. The methane flux anomalies observed over structures can be positive or negative. In the case of a single source, anomalies are positive because the information from a hydrocarbon-filled structure adds to the information from the source rock when recently active. It is our experience from other surveys that this also occurs when a gas source is above an oil source, where the oil-associated methane mixes with the methane from the gas source. The anomaly is negative when an oil source is above a gas source because the methane from the gas source partially dissolves when migrating through the oil. This process temporarily depresses the strength of the methane flux observed at the surface.

4. In the study of water flow processes in nature, it is important to distinguish between relatively lighter

and heavier hydrocarbon gas molecules. Because methane is the lightest hydrocarbon gas, it is more soluble and far more mobile (Vetsoskiy, 1979; Zhang, 1994). Thus, information transmitted by upward-migrating methane is capable of giving precise images of subsurface structures.

5. Adsorbed methane can easily be attacked by bacterial degradation, which changes the isotopic signature of the gas to such an extent that it must be excluded from further processing. Then why not use ethane or propane, which are thought to be less influenced by bacterial processes? There are two reasons not to use these heavier hydrocarbons. First, yields of adsorbed heavier hydrocarbon gases are sometimes so small that separation problems may arise from the methane during chromatography. Subsequently, the $\delta DC_{2,3}$ ratio cannot be measured with the same accuracy. Without the additional $\delta DC_{2,3}$ parameter, however (referring to the $\delta^{13}C_{2,3}$ value only), there is no contrast between primary and secondary fractionation on which an interpretation scheme can be based.

6. Under special geologic conditions, near-surface sediments may contain only a background gas signature, which means a lack of thermocatalytic

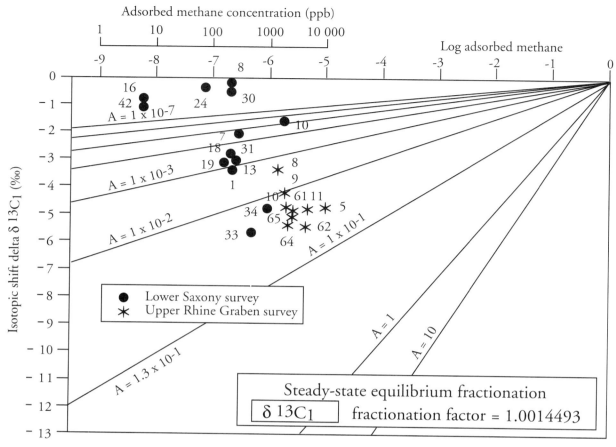

Figure 14—Secondary equilibrium fractionation of δ¹³C₁ (as in Figure 8) with case histories data.

information from the subsurface. From our experience, this is the case with a sampled near-surface layer exhibiting a prestationary adsorbed gas concentration profile. This can occur when deposition of the layer does not coincide with a time of basinal subsidence or, in other words, when the actual sedimentation rate is higher than the vertical upward water flow. We thus recommend that sampling be done on those sediments in which steady-state conditions have been reached during transmission of gas information to the surface by water flow (see B2 well in the Saale layer, Figure 2).

INTERFERING BACTERIAL PROCESSES

It is a widespread opinion that an isotopic ratio of adsorbed methane measured anywhere in near-surface sediments may represent a random value on a mixing line between bacterially produced and bacterially degraded methanes (Abrams, 1989). If this were true, it would exclude any possibility of deriving thermocatalytic information from near-surface sediments.

However, from our experience with many field surveys, this limitation seems to apply only to sediments rich

in organic matter in the range of source rock quality (e.g., Abrams, 1989; Stahl, 1992), where bacterial methane production is so high that it also enters the adsorbed phase. We have observed this repeatedly in our field surveys. When surface sediments are primarily of inorganic components (e.g., clay and siliciclastics), however, it is observed that bacterially generated methane stays within the free pore space (interstitial gas), while the adsorbed gas phase originates exclusively from the thermocatalytic gas input from below.

The problem then becomes to separate these gas phases from each other so as to retain the pure thermocatalytic information. We attain this separation by applying several washing procedures after sieving the fine sedimentary fraction and before acid treatment. That pure thermogenic data can be obtained in this way is proven by published isotopic data that are uniformly distributed over profiles within the same sampled lithology (Figures 6, 7). In the case of bacterial admixing, one would expect these data to be distributed randomly with depth.

Nevertheless, thermocatalytic methane, even in the adsorbed phase, can be affected by bacterial oxidation in some cases. Therefore, the following procedure is recommended to reach pure thermocatalytic information from the subsurface: (1) sample major inorganic sediments

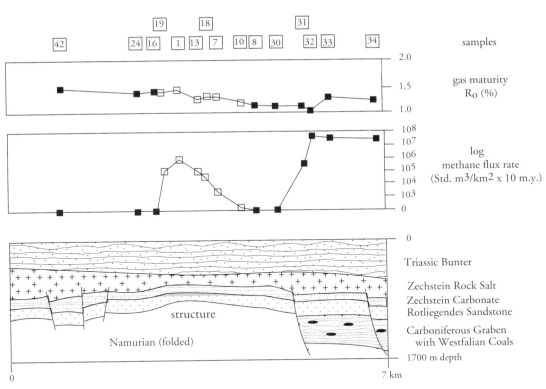

Figure 15—Single-source case history from Lower Saxony Inverted Basin, with a humic source rock (Carboniferous), showing the correspondence of gas maturity and methane flux to the geologic structure and source rocks.

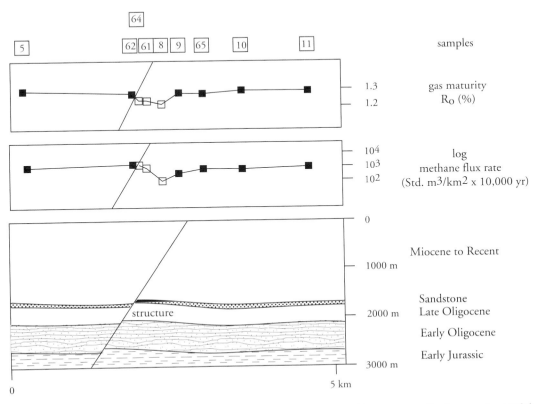

Figure 16—Double-source case history from Upper Rhine Graben Rift Basin, with a sapropelic (Lower Jurassic) and humic (lower Oligocene) source, showing the correspondence of gas maturity and methane flux to the geologic structure and source rocks.

with low organic carbon content, (2) wash the sieved fraction carefully, (3) measure the adsorbed phase of methane, and (4) exclude data from apparently oxidized samples from further processing.

CONCLUSIONS

Molecular and isotopic fractionation of gases that are controlled by flow processes are observed in nature. Because gas flow in sedimentary basins is highly dynamic, reasonable fractionation occurs on all geochemical parameters that depend on flow, even under steady-state conditions. A method is presented here that allows (1) methane fluxes in the subsurface to be calculated from the degree of isotopic fractionation of near-surface adsorbed methane, (2) hydrocarbon sources to be identified, and (3) the maturity and facies of the hydrocarbon sources to be characterized.

The calculated parameters represent the determination of the hydrocarbon flux from the subsurface at every site and show continuous trends and anomalies. The processed parameters do not scatter as do the rough data. It can be derived from *in situ* measurements that, in subsiding basins with strong vertical water flow, the dissolved methane (the phase in which methane is transported) does not change its isotopic composition considerably during migration. For this reason, the generation potential and the properties of a deep-seated gas source can be determined by analyzing a surface sediment or a borehole sample, regardless of the migration path that the gas followed. Only the last phase transition from the dissolved phase to the adsorbed phase in the sampled sediment, and the degree of fractionation produced by this, must be considered to calculate the methane flux from the subsurface at this site.

This method provides a tool that makes sense of near-surface methane isotopic measurements for assessment of the hydrocarbon charge of undrilled structures (Kettel, 1995). Because of the high resolution of the data processing method, this technique is especially useful in mature areas of hydrocarbon exploration.

Acknowledgments—The author thanks BEB Erdgas und Erdoel GMBH and Wintershall AG for permission to publish the results of the two field survey examples. Keith Kvenvolden and Chris Clayton gave helpful comments on the manuscript.

REFERENCES CITED

Abrams, M. A., 1989, Interpretation of methane carbon isotopes extracted from surficial marine sediments for detection of subsurface hydrocarbons: Association of Petroleum Geochemical Explorationists Bulletin, v. 5, no. 1, p. 139–166.

Bear, J., and M. Y. Corapcioglu, eds., 1987, Advances in transport phenomena in porous media: NATO ASI Series E, Applied Sciences, Dordrecht, Nijhoff Publishers, no. 128.

Bond, D. C., 1962, Geochemical process: U.S. patent no. 3,033,287.

Bouhroum, A., 1991, Dispersion und Adsorption in homogenen, porösen Medien: Erdöl-Erdgas-Kohle, v. 107, no. 7/8, p. 318–322.

Chung, H. M., and W. M. Sackett, 1979, Use of stable carbon isotope compositions of pyrolytically derived methane as maturity indices for carbonaceous materials: Geochimica et Cosmochimica Acta, v. 43, p. 1979–1988.

Faber, E., 1987, Zur Isotopengeochemie gasförmiger Kohlenwasserstoffe: Erdöl-Erdgas-Kohle, v. 103, no. 5, p. 210–218.

Faber, E., and W. J. Stahl, 1984, Geochemical surface exploration for hydrocarbons in North Sea: AAPG Bulletin, v. 68, no. 3, p. 363–386.

Faber, E., W. J. Stahl, M. J. Whiticar, J. Lietz, and J. M. Brooks, 1990, Thermal hydrocarbons in Gulf Coast sediments: Ninth Annual Research Conference Proceedings, GCS-SEPM Foundation, October 1, p. 279–307.

Fuex, A. N., 1980, Experimental evidence against an appreciable isotopic fractionation of methane during migration, *in* A. G. Douglas and J. R. Maxwell, eds., Advances in Organic Geochemistry 1979, p. 725–732.

Gant, P. L., and K. Yang, 1964, Chromatographic separation of isotopic methanes: Journal of American Chemical Society, v. 86, p. 5063–5064.

Gunter, B. D., and J. D. Gleason, 1971, Isotope fractionation during gas chromatographic separations: Journal of Chromatographic Science, v. 9, p. 191–192.

Hoffers, B., 1981, A model for hydrothermal convection in the Rhine Graben and its tectonic implications: Tectonophysics, v. 73, p. 141–149.

Horvitz, L., 1972, Vegetation and geochemical prospecting for petroleum: AAPG Bulletin, v. 56, p. 925–940.

Kettel, D., 1995, Procedure in order to detect the gas potential in sedimentary basins and the oil potential obtained from this: U.S. Patent No, 5,388,456.

Kettel, D., in press, Identification of the migration process of methane through the near-surface: 17th International Meeting on Organic Geochemistry, San Sebastian, Spain, 1995.

Kinzelbach, W., 1986, Groundwater modelling: New York, Elsevier.

Krooss, B. M., and D. Leythaeuser, 1988, Experimental measurements of the diffusion parameters of light hydrocarbons in water-saturated sedimentary rocks—results and geochemical significance: Organic Geochemistry, v. 12, no. 2, p. 91–108.

Lommerzheim, A., 1988, Die Genese und Migration von Kohlenwasserstoffen im Münsterländer Becken: Ph.D. dissertation, University of Münster, Germany.

Redding, C. E., M. Schoell, J. Monin, and B. Durand, 1980, Hydrogen and carbon isotopic composition of coals and kerogens, *in* A. G. Douglas and J. R. Maxwell, eds., Advances in Organic Geochemistry 1979, p. 711–723.

Sandy, J., 1989, A pinnacle reef discovery using interstitial soil gas sampling—a case history of the McCrea field, Stephens County, Texas (abs.): AAPG Southwest Section Meeting, Program with Abstracts, San Angelo, Texas.

Stahl, W. J., 1992, Isotope geochemistry of light hydrocarbons adsorbed in Jurassic shales from the Hils syncline, Northwest Germany: Abhandlungen der Braunschweigischen Wissenschaftlichen Gesellschaft, v. 43, p. 103–123.

Stahl, W. J., H. J. Kelch, P. Gerling, and E. Faber, 1984, Gasförmige Kohlenwasserstoffe in Oberflächensedimenten des Elm: Geologisches Jahrbuch, A, v. 75, p. 501–524.

Vetsoskiy, T. V., 1979, Natural gas geology: Moscow, NEDRA Press.

Welte, D. H., R. G. Schaefer, W. Stoessinger, and M. Radke, 1984, Gas generation and migration in the Deep Basin of western Canada: AAPG Memoir 38, p. 35–47.

Zhang, Y., 1994, Factors affecting the dynamic equilibrium of gas accumulations: Journal of Petroleum Geology, v. 17, no. 3, p. 339–350.

Chapter Appendix

Derivation of Equation (2)

Equation (2) in the text describes the isotopic fractionation of gases due to phase transitions, along with variations in concentration controlled by flow processes.

Equations (1)–(7) and (9) in this appendix are from Fuex (1980), while the deductions of equation (8) and (10)–(16) are newly developed for this data processing method to determine precisely the actual isotopic shift F. Definitions of the equations and their derivates are as follows:

$$x_{gi} = {}^{12}C_{gi}/{}^{12}C_0$$

is the mole fraction of ${}^{12}C_0$ in the gas (g) phase after the i-th step.

$$x_{ai} = {}^{12}C_{ai}/{}^{12}C_0$$

is the mole fraction of ${}^{12}C_0$ in the dissolved (a) phase after the i-th step, except for the gas dissolved during the previous i-1th step.

$$\delta = \left(\frac{R_{\text{sample}}}{R_{\text{standard}}} - 1 \right) \times 1000\%_0 \qquad (1)$$

is the definition of the isotopic composition, where

$$R_{\text{sample}} = {}^{13}C/{}^{12}C \; D/H \text{ of methane}$$

and

$$R_{\text{standard}} = {}^{13}C/{}^{12}C \; D/H \text{ of standard}$$

Also,

$$k = x_{gi}/x_{ai} \qquad (2)$$

is a constant dependent on purely geologic parameters.

$$k = {}^{12}C_{gi}/{}^{12}C_{ai} \; H_{gi}/H_{ai} \qquad (3)$$

The fractionation factor α is defined as

$$\alpha = R_{ai}/R_{gi} \qquad (4)$$

Therefore, from equations (1) and (4), we have

$$\alpha = \frac{\delta_{ai} + 1000}{\delta_{gi} + 1000} \qquad (5)$$

The mass balance used for ${}^{13}C$ is

$${}^{13}C_{gi-1} = {}^{13}C_{ai} + {}^{13}C_{gi} \qquad (6)$$

and the mass balance used for ${}^{12}C$ is

$${}^{12}C_{gi-1} = {}^{12}C_{ai} + {}^{12}C_{gi} \qquad (7)$$

or equally applied to D and H.

By mutual substitution of equations (1) through (7), the isotopic shift δ_{gi} of free methane after the i-th step becomes

$$\text{delta } \delta_{gi} = \text{delta } \delta_{g1} \times \frac{\left(\dfrac{1+k}{\alpha+k} \right)^{i-1}}{\dfrac{1-\alpha}{\alpha+k}} \qquad (8)$$

By the same procedure, delta δ_{g1} of the first step becomes

$$\text{delta } \delta_{g1} = \frac{1000(1-\alpha)}{\alpha+k} \qquad (9)$$

Therefore from equations (8) and (9), we obtain

$$\text{delta } \delta_{gi} = \frac{1000(1-\alpha)}{\alpha+k} \times \frac{\left(\dfrac{1+k}{\alpha+k} \right)^{i-1}}{\dfrac{1-\alpha}{\alpha+k}}$$

$$\text{delta } \delta_{gi} = 1000 \times \left(\frac{1+k}{\alpha+k} \right)^{i-1} \qquad (10)$$

When delta δ_{gn} is the sum of isotopic shifts after complete fractionation during n steps ($i = n$), then

$$\text{delta } \delta_{gn} = F \tag{11}$$

From equations (10) and (11), we have

$$F = 1000 \times \left(\frac{1+k}{\alpha + k} \right)^{n-1} \tag{12}$$

$$\frac{F}{1000} + 1 = \left(\frac{1+k}{\alpha + k} \right)^{n}$$

$$\ln \frac{F + 1000}{1000} = n \times \ln \frac{1+k}{\alpha + k}$$

$$n = \frac{\ln \dfrac{F + 1000}{1000}}{\ln \dfrac{1+k}{\alpha + k}} \tag{13}$$

The methane concentration in the gas phase x_{gn} after n steps is

$$x_{gn} = \left(\frac{1+k}{k} \right)^{-n} = \exp - \left(n \times \ln \frac{1+k}{k} \right) \tag{14}$$

Thus, from equations (13) and (14), we have

$$x_{gn} = \exp - \left(\frac{\ln \dfrac{1+k}{k}}{\ln \dfrac{1+k}{\alpha + k}} \times \ln \frac{F + 1000}{1000} \right) \tag{15}$$

Phase transitions from the dissolved or free gas phase into the adsorbed phase were covered in the text. Here we can use the same equation (15), except that factor k is substituted by factor A and x_{gn} by x_{sn}, which are both used for adsorption:

$$x_{sn} = \exp - \left(\frac{\ln \dfrac{1+A}{A}}{\ln \dfrac{1+A}{\alpha + A}} \times \ln \frac{F + 1000}{1000} \right) \tag{16}$$

Equation (16) thus expresses the mass balance and theoretical relationship between the methane concentration of the adsorbed phase x_{sn}, the isotopic shift F that the adsorbed phase undergoes after reaching equilibrium, the fractionation factor α of methane between the lighter and heavier isotopes during adsorption, and geologic factor A.

Thompson,K. F. M., 1996, Postulated generation of bacterial methane from seepage
petroleum in sea floor sediments of the Gulf of Mexico, *in* D. Schumacher and
M. A. Abrams, eds., Hydrocarbon migration and its near-surface expression:
AAPG Memoir 66, p. 331–334.

Chapter 24

Postulated Generation of Bacterial Methane from Seepage Petroleum in Sea Floor Sediments of the Gulf of Mexico

K. F. M. Thompson

Petrosurveys, Inc.
Dallas, Texas, U.S.A.

Abstract

Evidence from large numbers of sediment cores collected on the continental slope of the U.S. Gulf of Mexico suggests that methane is generated by bacteria at depths of about 4 m from seepage petroleum. Marine sediment-dwelling methanogenic bacteria use carbon dioxide generated at shallow depths by aerobic and anaerobic bacteria from available organic material. Until the cores described here were evaluated, it appeared that shallow occurrences of biogenic methane in Pliocene, Pleistocene, and Recent sediments of the Gulf Coast were derived solely from organic detritus, both marine and terrestrial, of the same ages. However, circumstantial evidence indicates that a significant source of organic carbon for methanogenesis is provided by petroleum that frequently seeps into surface sediments. The relative scales of methane generation from these competing sources cannot yet be established.

INTRODUCTION

Seepage of petroleum and natural gas on the U.S. continental slope in the Gulf of Mexico, offshore Louisiana and Texas, is well documented (e.g., Anderson et al., 1983; Kennicutt and Brooks, 1990). This paper provides additional data on biological activity at seepage sites, particularly concerning the generation of light hydrocarbon gases by bacteria and possibly other microbial organisms.

Information has been compiled on microbial generation of olefins and bacterial methane occurring in large numbers of piston cores taken in the uppermost 3–5 m of sediment on the continental slope. The cores were collected and analyzed by the Geochemical and Environmental Research Group (GERG) of Texas A&M University for BP Exploration. Sample coverage extends from the area of the Mississippi Delta, offshore Louisiana, to offshore Brownsville, Texas, at the Mexican border. Sediment samples were sealed in cans and rapidly frozen upon recovery of the core, precluding continuing microbial activity.

EXPERIMENTAL METHODS

On shore, the preserved samples were analyzed for headspace light hydrocarbons by gas chromatography methods. Solvent extracts of the dried sediment were also analyzed; saturated hydrocarbons were analyzed by gas chromatography and polyaromatic hydrocarbons (PAH) by ultraviolet spectroscopy.

Chromatographic analytical precision is indicated by the following information. For example, replicate analyses of a 94-ppm methane mixture yielded a coefficient of variation of 2% and a detector response on the order of 250,000 units. A test of linearity over a wide concentration range had a Pearson correlation coefficient of 0.994 linking response and concentration. Hydrocarbon concentrations below 1 ppm were readily quantified. Accuracy was ensured by running standards to bracket the unknowns. Polyaromatic hydrocarbon levels, quantified by ultraviolet fluorescence spectroscopy, are expressed on a nonlinear scale.

Table 1—Characteristics of a Typical Seepage Site Exhibiting Methanogenesis (Core UGB-142, Upper Garden Banks)

Section	Depth (m)	PAH[a] (UV units)	Methane[b] (ppm)	Ethane (ppm)	Propane (ppm)	n-Butane (ppm)	Ethene (ppm)	Propene (ppm)	Propane/ Propene
14	2.8	6510	31	0.8	0.1	0.1	0.4	0.1	1
18	3.6	4860	13084	25.1	0.6	0.2	0.0	0.2	3
22	4.4	2250	28072	35.7	0.6	0.2	0.0	0.1	6

[a]PAH levels are expressed on a nonlinear, arbitrary, but consistent scale of fluorescence yield measured quantitatively by UV spectroscopy.

[b]Gas concentrations are expressed in ppm by volume of headspace.

RESULTS AND DISCUSSION

Major seepages exhibited extremely high levels of PAH, 30 to 5000 times background levels (Table 1). Saturated hydrocarbons that generated a chromatographically unresolved complex mixture accompanied the PAHs, indicating that the saturated hydrocarbons are biodegraded, that is, bacterially oxidized. A suite of petroleum-like normal alkanes were occasionally observed, believed to characterize the most active seepages.

At sites exhibiting high levels of PAH, light hydrocarbons present in reservoir gases (e.g., ethane, propane, and n-butane) occurred at relatively high concentrations at sub–sea floor depths of 3–4 m and tended to decrease in concentration toward the sediment–water interface, evidently because of increasing biodegradation. These compounds were not generated by biological activity at the sea floor and are therefore always absent from pristine, recent sediments.

Ethene, propene, and 1-butene, compounds with a single double bond (olefins) but otherwise similar to reservoir gases, increase in concentration toward the sea floor at the seepage sites. The mechanism responsible for the generation of olefins has not been identified, but we postulate that it is a microbial biological process, acting upon petroleum hydrocarbons. Olefins, particularly ethene, are commonly detected in soils (Smith and Ellis, 1963), but are rarely reported in deep-water marine sediments. Olefins are generally absent at nonseepage sites in the Gulf of Mexico.

This rationale suggests that the propane/propene concentration ratio in headspace gases provides an index of the degree of bacterial oxidation. The ratio decreases from a value of infinity in fresh reservoir gases where propene is necessarily absent, to values below 0.01 upon intense bacterial and microbial modification of petroleum hydrocarbons in near-surface sediments. The ratio value is an indication of the "freshness" of the seepage hydrocarbons. Not only does the ratio decrease toward the top of a core, but at a given depth, it progressively decreases with time, as biodegradation and alteration proceed.

At high-PAH seepage sites, high concentrations of methane were found in the deepest core sections (Table 1). Carbon isotope ratio analyses representing these occurrences (from 53 high-PAH sites) have $\delta^{13}C$ (methane) values mostly between –70 and –90‰ versus the PDB standard (Figures 1, 2). Both isotopic and compositional data indicate that the methane is bacterial in origin, according to well-established criteria first set forth by Claypool and Kaplan (1974). The histograms in Figure 1 compare $\delta^{13}C$ (methane) values from core gases with 116 analyses of Gulf Coast offshore petroleum reservoir gases, principally from gas and gas-condensate reservoirs, published by Rice and Threlkeld (1983). Only a small proportion of core gas cases could possibly represent escaped reservoir gas, as the two types are largely isotopically distinct.

In summary, although substantial volumes of methane are observed at petroleum seepage sites, the methane is not derived from petroleum reservoirs. An alternative hypothesis is that it is secondarily derived from petroleum carbon by *in situ* bacterial activity, an idea explored below.

Figure 2 provides further distinction between reservoir and core gases. The figure plots methane/ethane ratios based on mole percentages versus $\delta^{13}C$ (methane) compiled from Rice and Threlkeld (1983) and the present core data. Two evolutionary trends are evident in petroleum gases, both starting from a comparatively low maturity condition with an initial $\delta^{13}C$ value of about –50‰ and a methane/ethane ratio on the order of 10. Evolution along the thermogenic pathway proceeds by way of maturation, while that along the alternative pathway involves the mixing of biogenic and thermogenic gases. Figure 2 shows that the core gases represent a potential end-member component for the mixing process and a potential source of the biogenic component of reservoired gases. The core gases possess a median methane/ethane ratio of 2933.

Table 2 presents mean values of PAH and headspace gas concentrations at nonseepage sites, representing cores from an area southeast of the Mississippi Delta. The sampled sediments have total organic carbon concentrations similar to those of the seepage cores, but generally fail to exhibit significant levels of PAH and possess relatively low levels of free gas. The nonzero mean concentrations of ethane, propane, and n-butane suggest that petroleum seepage affects some members of the sample set.

All marine sediments except those of the deep ocean floor exhibit anoxia and methanogenesis at depths ranging from millimeters to several meters below the sediment–water interface (Fenchel and Riedel, 1970), depending on the rate of sedimentation of organic detritus (Volkov and Rozanov, 1983). For these reasons, biogenic

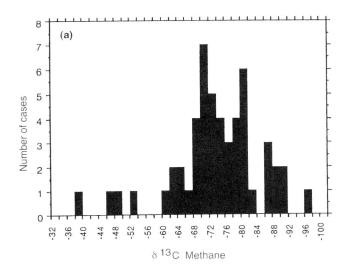

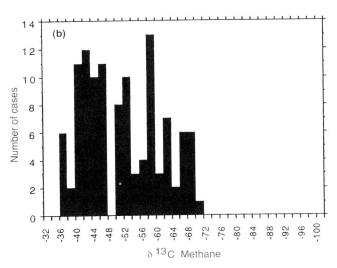

Figure 1—Carbon isotope data from (a) methane in cores (this study) and from (b) methane in petroleum reservoir gases (data from Rice and Threlkeld, 1983).

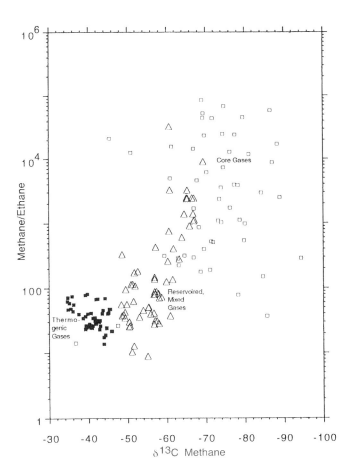

Figure 2—Comparison of two classes of reservoired natural gas, thermogenic and "mixed" (as defined by Rice and Threlkeld, 1983), with core gases. It is suggested that many core gas compositions represent the biogenic end-member in the mixing process.

methane is an expected natural product from Recent or Pleistocene organic matter everywhere in Gulf of Mexico sediments. Large numbers of cores have been taken in slope sediments to depths of 4–5 m without encountering this zone of methanogenesis, as indicated by the data in Table 2. At random slope locations, however, methanogenesis takes place at depths of less than 4 m. This most likely indicates that an abnormally large amount of bacterially accessible organic carbon has been provided at the locations exhibiting shallow methanogenesis. These are the sites where high levels of petroleum-derived PAHs occur.

Spies and Davis (1979) were the first investigators to hypothesize that seepage petroleum provides the basis of an extensive microbial food chain that may also support a megafauna. Examples such as the chemosynthetic com-

munities at seepage sites in the Gulf of Mexico (Mac-Donald et al., 1989, and this volume; Reilly, this volume) are now well known and amply verify the concept. The lowest trophic level of such seepage communities, and of all marine sediment food chains, comprises methanogenic bacteria. The postulate must therefore be allowed that methane at seepage sites is, in part, bacterially derived *in situ* from petroleum.

An examination of the properties of 39 cores interpreted as exhibiting methanogenesis led to the following observations. Methane occurred at concentrations greater than 10,000 ppm in the deepest analyzed sections at core depths of about 4 m, it was isotopically light ($\delta^{13}C = -60$ to $-90‰$), and it was associated with methane at methane/ethane ratios greater than 1000. Methane concentrations decreased continuously upward. It is significant that hydrogen sulfide was detectable in 36 of the 39 cores. It is well known that the zone of methanogenesis invariably occurs immediately below the zone of sulfate reduction (Fenchel and Riedel, 1970). The latter process

Table 2—Mean Concentrations of Sediment PAH and Gaseous Hydrocarbons in Headspace Samples[a]

PAH (UV units)	Methane (ppm)	Ethane (ppm)	Propane (ppm)	*n*-Butane (ppm)	Ethene (ppm)	Propene (ppm)
1330	1131	0.5	0.1	0.02	–	–

[a]Concentrations were averaged over three subsamples of core sediment taken at depths similar to those in Table 1 and over 175 core sites on the continental slope southeast of Mississippi Delta. See Table 1 for explanation of units.

results in the release of hydrogen sulfide to the sediment pore fluid. There is thus adequate reason to believe that the high observed concentrations of methane are presently being formed *in situ* as part of the normal sequence of bacterial activities.

The relative volumetric significance of biogenic methane ultimately generated from seepage petroleum versus that derived from biological detritus remains to be determined. In marine sediments, methane is in fact created by methanogens from carbon dioxide (Whiticar et al., 1986; Faber et al., 1990). Up to the present, only methane generated from biologically created carbon dioxide derived from contemporary organic remains has been implicitly accepted as the source of the biogenic methane of reservoired gases. "Recycled" petroleum carbon should also be considered as a contributory methane precursor.

Regardless of the nature of its carbon source, biogenic methane is generated close to the sediment–water interface and must necessarily be buried, along with its enclosing sediments, by subsequent deposition in order to occur as a deep-reservoired gas component. The data of Rice and Threlkeld (1983) mentioned earlier show that the majority of mixed biogenic-thermogenic gas accumulations occur in Pleistocene reservoirs, but many are found in Pliocene strata and a few in Miocene reservoirs, such as the –60.3‰ methane in well H-9 at 1207 m in Grand Isle Block 47 field, offshore Louisiana. Bacterial methane was generated from sedimentary organic matter throughout the Tertiary and Quaternary, but it is possible that petroleum seepage has taken place intermittently during the same time interval, with partial conversion to methane in near-surface sediments.

Acknowledgments—*Appreciation is expressed to the management of BP Exploration, Inc., for permission to publish this information.*

REFERENCES CITED

Anderson, R. K., R. S. Scalan, P. L. Parker, and E. W. Behrens, 1983, Seep oil and gas in Gulf of Mexico slope sediments: Science, v. 222. p. 69–62.

Claypool, G. E., and I. R. Kaplan, 1974, The origin and distribution of methane in marine sediments, *in* I. R. Kaplan, ed., Natural gases in marine sediments: New York, Plenum Press, p. 99–139.

Faber, E., W. J. Stahl, M. J. Whiticar, J. Lietz, and J. M. Brooks, 1990, Thermal hydrocarbons in Gulf Coast sediments, *in* D. Schumacher and B. Perkins, eds., Gulf Coast oils and gases: Proceedings, SEPM Ninth Annual Research Conference, Gulf Coast Section, p. 297–307.

Fenchel, T. M., and R. J. Riedel, 1970, The sulfide system: a new biotic community underneath the oxidized layer of marine sand bottoms: Marine Biology, v. 7, p. 255–268.

Volkov, I. I., and A. G. Rozanov, 1983, The sulfur cycle in oceans, *in* M. V. Ivanov and J. R. Freney, eds., The global biogeochemical sulfur cycle: Chichester, John Wiley and Sons, p. 357–420.

Kennicutt, M. C., and J. M. Brooks, 1990, Seepage of gaseous and liquid petroleum in the northern Gulf of Mexico, *in* D. Schumacher and B. Perkins, eds., Gulf Coast oils and gases: Proceedings, SEPM Ninth Annual Research Conference, Gulf Coast Section, p. 309–330.

MacDonald, I. R., G. S. Boland, J. S. Baker, J. M. Brooks, M. C. Kennicutt, and R. R. Bidigare, 1989, Gulf of Mexico hydrocarbon seep communities, II. Spatial distribution of organisms at Bush Hill: Marine Biology, v. 101, p. 235–247.

Rice, D. D., and C. N. Threlkeld, 1983, Chemical and isotopic composition of natural gas analyses from selected wells in the Gulf of Mexico: USGS Open File Report 83-152.

Smith, G. H., and M. M. Ellis, 1963, Chromatographic analysis of gases from soils and vegetation, related to geochemical prospecting for petroleum: AAPG Bulletin, v. 47, p. 1897–1903.

Spies, R. B., and P. H. Davis, 1979, The infaunal benthos of a natural oil seep in the Santa Barbara Channel: Marine Biology, v. 50, p. 227–237.

Whiticar, M. J., E. Faber, and M. Schoell, 1986, Biogenic methane formation in marine and freshwater environments: CO_2 reduction vs. acetate fermentation—isotopic evidence: Geochimica et Cosmochimica Acta, v. 50, p. 693–709.

Zyakun, A. M., 1996, Potential of ^{13}C/^{12}C variations in bacterial methane in assessing origin of environmental methane, *in* D. Schumacher and M. A. Abrams, eds., Hydrocarbon migration and its near-surface expression: AAPG Memoir 66, p. 341–352.

Chapter 25

Potential of $^{13}C/^{12}C$ Variations in Bacterial Methane in Assessing Origin of Environmental Methane

A. M. Zyakun

Institute of Biochemistry and Physiology of Microorganisms
Russian Academy of Sciences
Pushchino, Russia

Abstract

Laboratory experimental results measuring carbon isotope fractionation of the methane produced by pure cultures of methanogenic bacteria during their growth on both C_1^- (CO_2, formate, methanol, and methylamine) and C_2^- (acetate) substrates have been carried out. Isotopic effects associated with bacterial methane generation depend on the specific methane production rate, that is, the amount of methane produced by a cell or cell biomass per unit time. In some cases, the isotopic effects also depend on the specific concentration of the substrate, or the amount of bicarbonate per unit of cell biomass or per one bacterial cell.

Biogenic methane has various carbon isotopic compositions both at extremely low ($\delta^{13}C < -55‰$) and comparatively high ($\delta^{13}C > 0‰$) carbon isotope ratios relative to the substrate used. Theoretical calculations show that the carbon isotope content of bacterial methane can be depleted in ^{13}C by $-100‰$ relative to the carbon dioxide in the culture medium.

INTRODUCTION

The carbon isotope content of methane can be used as an indicator of its origin in natural gas (Rice and Claypool, 1981; Whiticar et al., 1986; Tyler et al., 1994). The methane produced by methanogenic bacteria in natural ecosystems is considered to have $\delta^{13}C$ values in the range of -50 to $-70‰$, whereas methane of thermocatalytic genesis is believed to have values higher than $-50‰$ (Collett and Cunningham, 1994; Hansen and Bjoroy, 1994). It is evident that with $\delta^{13}C$ used as a diagnostic parameter of methane origin, the most significant problems are to determine the range of possible values of this parameter and to elucidate the factors affecting carbon isotope ratios in bacterial methane.

It is known that biogenic methane in marine ecosystems has isotopic carbon and hydrogen contents that differ from those in freshwater reservoirs (Whiticar et al., 1986). These differences are believed to be due to substrates used by methanogenic bacteria. Thus, in marine ecosystems, bacterial methanogenesis is associated with the predominant use of $CO_2 + H_2$ as substrates (Kosiur and Warford, 1979; Lein et al., 1981) and in some subsurface marine sediments of methanol and methylated

amines as substrates (King, 1984). In freshwater reservoir materials (e.g., lake silt sediments, waste water settling tank sludge, and peaty soils) acetate is the main precursor for methane formed by methanogens (Cappenberg and Jongejan, 1978).

In laboratory experiments measuring carbon isotope fractionation (CIF) in bacterial methane formation, methanol was used as a substrate to activate methanogenesis by an enrichment culture isolated from reduced deep-water Pacific Ocean silt (Rosenfeld and Silverman, 1959) and lake sediments (Oremland et al., 1982). The methane obtained in those cases was depleted in ^{13}C relative to the substrate up to -70 to $-80‰$. Experiments with enrichment cultures growing on acetate also revealed a significant depletion of ^{13}C in biogenic methane (up to $80‰$) (Nakai, 1960).

However, studies of CIF in pure cultures of methanogenic bacteria grown on $CO_2 + H_2$ and acetate (Games et al., 1978; Krzycki et al., 1987) showed that the differences in the ^{13}C content of bacterial methane relative to the substrate are significantly lower than those observed in marine ecosystems and laboratory experiments with enrichment cultures isolated from these systems and grown on methanol and acetate. In contrast, the determined activities of methanogenic bacteria in marine silt sediments using test substrates of $CO_2 + H_2$ and acetate showed that, in these systems, bacterial methanogenesis results predominantly from the consumption of $CO_2 + H_2$ (Kosiur and Warford, 1979; Lein et al., 1981; Senior et al., 1982). The difference in the isotopic composition of methane and CO_2 carbon in the marine sediments assayed was –60 to –80‰ (Claypool and Kaplan, 1974; Alekseev and Lebedev, 1975).

The question is to what extent laboratory experiments on CIF in bacterial methanogenesis reflect the processes occurring in natural ecosystems and which factors affect them. This paper is an attempt to specify the range of possible $\delta^{13}C$ values for the methane produced by methanogenic bacteria, as well as to reveal the factors that effect fractionation of carbon isotopes by these microorganisms.

THEORETICAL CONSIDERATIONS

Scheme of Bacterial Methane Production

Recent years have witnessed a breakthrough in the understanding of the biochemical pathways of bacterial methanogenesis (Keltjens and van der Drift, 1986). In a general scheme of the carbon dioxide reduction to methane (Figure 1), several stages can be identified in which transfers of metabolites containing C_1 units from one cycle to another are accompanied by CIF of methane.

It has been shown that methanogenic bacteria growing on H_2 and CO_2 use carbon dioxide as one of the forms dissolved in water (H_2CO_3, HCO_3^-, CO_3^{2-}, and CO_2) (Fuchs et al., 1979). Carbon dioxide is a C_1 unit that reacts with methanofuran (MF) and reduces to formyl-MF. According to Leigh et al. (1985), formyl-MF is readily oxidized to produce CO_2 and MF under $H_2 + CO_2$ atmosphere if 5,6,7,8-tetrahydromethanopterin (H_4MPT) is absent. The formyl group bound to MF is enzymatically transferred to H_4MPT in an ATP-independent reaction (Keltjens et al., 1986). Furthermore, the reduction of formyl-H_4MPT to methenyl-H_4MPT is reversible. Methenyl-H_4MPT is reduced to methylene-H_4MPT under H_2 atmosphere, but in N_2, the latter compound is oxidized to methenyl-H_4MPT with evaluation H_2 (Keltjens and Vogels, 1981).

The next step of this process is the reduction of methylene-H_4MPT to methyl-H_4MPT catalyzed by the methylene-H_4MPT reductase (Keltjens et al., 1986). The H_3C group is transferred from methyl-H_4MPT to HS–CoM. The terminal step of bacterial methanogenesis involves a reduction of $H_3C–S–CoM$ to CH_4 and HS–CoM, although this mechanism has not yet been adequately investigated (Keltjens and van der Drift, 1986). Thus, reversibility of enzyme-catalyzed processes in C_1 unit reduction is typical and must be dealt with. It should be noted that all C_1 units of various stages in the bacterial reduction of a substrate are partially transformed from stage to stage.

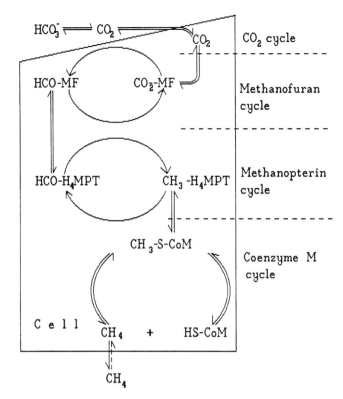

Figure 1—The steps of carbon isotope fractionation by methanogenic bacteria growing on $CO_2 + H_2$ (a biochemical scheme of HCO_3^- reduction adapted from Keltjens and van der Drift, 1986). See the text for values of α.

Isotope Fractionation Steps

The transfer of C_1 units from one cycle to another is considered as a process with ramified stages (Melander and Saunders, 1980), as shown in Scheme 1:

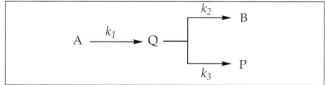

Scheme 1

where A is the initial compound, Q the intermediate compound, B and P the products, and k the rate of formation. If Q possesses high energy and is available in small concentrations, steady-state conditions can be applied. Then, the observed constant of the formation rate of product B from A (k_{AB}) can be written as

$$k_{AB} = k_1 k_2 / (k_2 + k_3) \tag{1}$$

It is assumed that k_1 and k_2 are dependent on isotopic effects in the systems containing ^{12}C and ^{13}C atoms, yet the isotopic substitution does not affect the k_3 value. In this case, a ratio of the rate constants of formation of

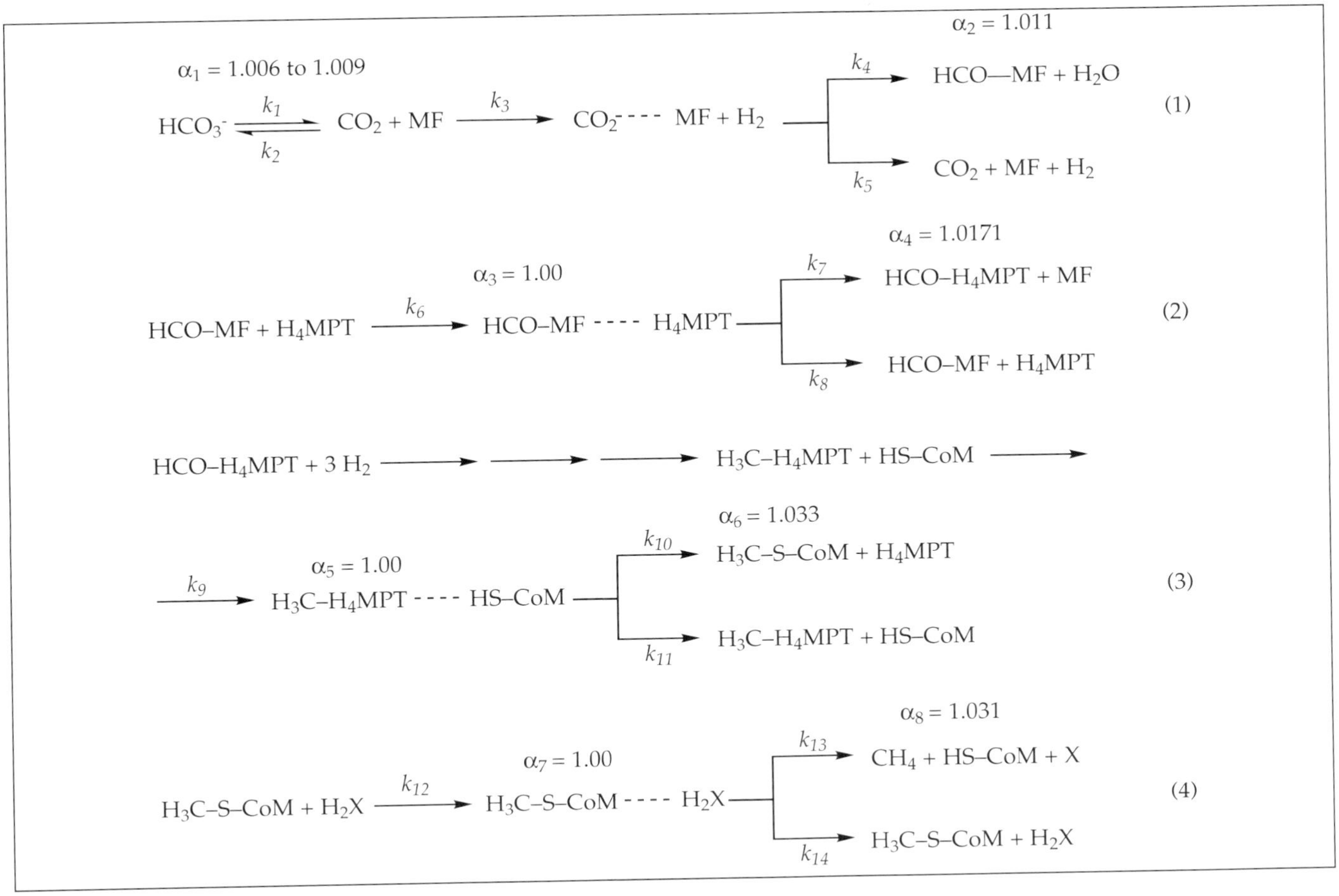

$\alpha_1 = 1.006\ to\ 1.009$

$\alpha_2 = 1.011$

$$HCO_3^- \underset{k_2}{\overset{k_1}{\rightleftharpoons}} CO_2 + MF \xrightarrow{k_3} CO_2 \text{ - - - - } MF + H_2 \begin{cases} \xrightarrow{k_4} HCO-MF + H_2O \\ \xrightarrow{k_5} CO_2 + MF + H_2 \end{cases} \quad (1)$$

$\alpha_3 = 1.00$

$\alpha_4 = 1.0171$

$$HCO-MF + H_4MPT \xrightarrow{k_6} HCO-MF \text{ - - - - } H_4MPT \begin{cases} \xrightarrow{k_7} HCO-H_4MPT + MF \\ \xrightarrow{k_8} HCO-MF + H_4MPT \end{cases} \quad (2)$$

$$HCO-H_4MPT + 3\ H_2 \longrightarrow \longrightarrow \longrightarrow H_3C-H_4MPT + HS-CoM \longrightarrow$$

$\alpha_5 = 1.00$

$\alpha_6 = 1.033$

$$\xrightarrow{k_9} H_3C-H_4MPT \text{ - - - - } HS-CoM \begin{cases} \xrightarrow{k_{10}} H_3C-S-CoM + H_4MPT \\ \xrightarrow{k_{11}} H_3C-H_4MPT + HS-CoM \end{cases} \quad (3)$$

$\alpha_7 = 1.00$

$\alpha_8 = 1.031$

$$H_3C-S-CoM + H_2X \xrightarrow{k_{12}} H_3C-S-CoM \text{ - - - - } H_2X \begin{cases} \xrightarrow{k_{13}} CH_4 + HS-CoM + X \\ \xrightarrow{k_{14}} H_3C-S-CoM + H_2X \end{cases} \quad (4)$$

Scheme 2

isotopically distinguishable products B from A or the coefficient of CIF (α_{AB}) can be given as

$$\alpha_{AB} = k_{AB}/k'_{AB} = \frac{k_1 k_2}{k'_1 k'_2} \times \frac{k'_2 + k'_3}{k_2 + k_3} \quad (2)$$

where k_{AB} and k'_{AB} are rate constants for formation of product B from A containing isotopes ^{12}C and ^{13}C, respectively; k'_1, k'_2, and k'_3 are rate constants for the formation of products ^{13}C-Q, ^{13}C-B, and ^{13}C-P, respectively; and $k_3 = k'_3$.

If the process occurring in Scheme 1 is accompanied mainly by the formation of product P, such that $k_2 << k_3$, then the coefficient of CIF (α_{AB}) can be derived from the following equation (product B is obtained from A):

$$\alpha_{AB} = (k_1/k'_1)(k_2/k'_2) = \alpha_1\alpha_2 \quad (3)$$

where α_1 and α_2 are coefficients of CIF at the stages of Q and B formation, respectively.

If product B is mainly formed ($k_2 >> k_3$), the coefficient of CIF can be determined by the first stage of a reaction:

$$\alpha_{AB} = k_1/k'_1 = \alpha_1 \quad (4)$$

It can be shown that if a small amount of C_1 metabolite (product B) is transferred from one cycle to another, the total coefficient of CIF (α) is multiplication of α_i coefficients for separate n cycles:

$$\alpha^* = \prod_{i=1}^{n} \alpha_i \quad (5)$$

Carbon Dioxide Reduction

In line with Keltjens and van der Drift (1986), the main steps of CO_2 reduction are written as shown in Scheme 2 (above). The steps of the C_1 units reduction (1–4, Scheme 2) are represented as remified reactions of intermediate complexes catalyzed by the enzymes. The isotope effect is believed to be related to the processes of formation and disruption of bonds between the C_1 unit and its carrier complex, as well as with a transfer of the C_1 unit from one carrier to another. Since reversible reactions of high molecular weight substances are observed, the isotope effects of binding and disrupting processes of the C_1 unit complex with the subsequent C_1 carrier are taken to be insignificant. The transfer of the C_1 unit from one C_1 carrier to another must be accompanied by the *kinetic isotopic effect* (KIE), mainly determined by masses of isotopically different C_1 units. Thus, the isotope discrimination coefficients (α) of $^{13}C_1$ units relative to $^{12}C_1$ units can be calculated as

$$\alpha = \sqrt{m_2 / m_1} \qquad (6)$$

where m_1 and m_2 are the masses of $^{12}C_1$ and $^{13}C_1$ units, respectively.

Cycle I (the CO_2 cycle) (Figure 1) is associated with carbon dioxide transport from a liquid medium into a bacterial cell and vice versa. The CIF in this case can be determined by two factors: the *thermodynamic isotopic effect* (TIE) (α_1, Scheme 2) in the system HCO_3^-–CO_2, and the KIE (α_2, Scheme 2) during the binding of CO_2 to methanofuran (transfer of CO_2 to methanofuran cycle; step 1, Scheme 2). It has been shown that the CIF between gaseous CO_2 and HCO_3^- varies from -9.2 to $-6.8‰$ over a temperature range of $0°$–$30°C$ (Deuser and Degens, 1967). The TIE can then be defined as $\alpha_1 = 1.0068$ to 1.0092 for the equilibrium reaction between HCO_3^- and CO_2. The KIE of CO_2 transfer to the methanofuran cycle can be estimated, according to equation 6, as

$$\alpha_2 = \sqrt{45 / 44} = 1.011$$

If $k_5 \gg k_4$ (step 1, Scheme 2), then the total isotopic effect of carbon dioxide bound with methanofuran is $\alpha_{1-2} = \alpha_1\alpha_2 = 1.0179$ to 1.0203.

In the methanofuran cycle, CO_2 is reduced to the formyl group (C_1 unit), which is transferred to the following methanopterin cycle (step 2, Scheme 2). It is assumed that the isotopic fractionation coefficient for binding of HCO–MF and H_4MPT produced as an intermediate complex is insignificant ($\alpha_3 = k'_6/k_6 = 1.0$). If $k_8 \gg k_7$ (Scheme 2), then KIE in the transfer of the formyl group (HCO$^-$) is calculated according to equation 6 as

$$\alpha_4 = \sqrt{30 / 29} = 1.0171$$

During the next cycle (step 3, Scheme 2), the C_1 unit is reduced to the methyl group, which is then transferred to HS–CoM (2-mercaptoethanesulfonic acid, or coenzyme M). As in step 2, the coefficient $\alpha_5 = k'_9/k_9$ is equal to 1.0 (step 3, Scheme 2). If $k_{11} \gg k_{10}$, the transfer of the methyl group to HS–CoM is accompanied by the KIE according to equation 6 as

$$\alpha_6 = \sqrt{16 / 15} = 1.033$$

In the case where CH_3S–CoM (2-methylthioethanesulfonic acid, or methyl coenzyme M) is reduced to CH_4 irreversibly (such that $k_{13} \gg k_{14}$; step 4, Scheme 2), the CIF of methane relative to HCO_3^- in the medium can be assessed as

$$\alpha_{1-6} = \prod_{i=1}^{6} \alpha_i = 1.069 \text{ to } 1.072$$

If, however, some of the methane produced diffuses from the cell ($k_{13} \ll k_{14}$; step 4, Scheme 2), the KIE of methane is also observed in this step. The value of the fractionation coefficient at the stage of methane diffusion from the cell according to equation 6 is $\alpha_8 = 1.031$.

Consequently, the maximum assessment of the CIF during bacterial methane generation on $CO_2 + H_2$ can be defined by

$$\alpha_{max} = \prod_{i=1}^{8} \alpha_i = 1.102 \text{ to } 1.105$$

Otherwise, the carbon isotope content of bacterial methane will be depleted in ^{13}C by more than $100‰$ relative to the bicarbonate substrate. The realization of the maximum CIF during carbon dioxide reduction to methane by methanogenic bacteria is feasible only if the transfer of respective carbon forms of the substrate (C_1 units) used is minor from cycle to cycle, such that $k_4 \ll k_5$, $k_7 \ll k_8$, $k_{10} \ll k_{11}$, and $k_{13} \ll k_{14}$ (steps 1–4, Scheme 2). In the case where these rate constants hold opposite values, such that $k_4 \gg k_5$, $k_7 \gg k_8$, $k_{10} \gg k_{11}$, and $k_{13} \gg k_{14}$, the coefficients from α_3 to α_8 must be about 1.0. Thus, the minimum CIF during a bacterial CO_2 reduction is attained by $\alpha_{min} = \alpha_1\alpha_2 = 1.018$ to 1.020.

C_1 Organic Substrate Reduction

The methylotrophic methanogen *Methanosarcina barkeri* is able to grow on CO_2 and H_2, as well as on the H_3C group of organic substrates—methanol, methylamines, and acetate (Balch et al., 1979). In general pathways of C_1 unit metabolism of *M. barkeri*, the H_3C group of organic substrates reacts with B_{12} (5-hydroxybenzimidazol cobamide). This is metabolized in two directions: a reduction to CH_4 via H_3C–S–CoM and oxidation to CO_2 via H_3C–H_4MPT, as shown in Scheme 3 (based on principles described by Keltjens and van der Drift, 1986). In parellel with these metabolic pathways, it should be noted that methanol oxidase participates in the direct oxidation of the methanol H_3C group. An unknown carrier of the methanol C_1 unit (such as X–CH_2OH) is transferred to H_4MPT and produces methylene-H_4MPT or methenyl-H_4MPT, which are oxidized to CO_2 (Blaut et al., 1985).

There are some difficulties in estimating the carbon isotope fractionation of the C_1 units for every metabolic stage of the bacterial C_1 organic substrate reduction to methane, as shown in Scheme 3 (next page). The literature provides insufficient data on the CIF for the initial stages of metabolic pathways of bacterial methanogenesis using methanol and methylamines. Thus, our reasoning is based on the general idea of the CIF of C_1 units and has been modified by laboratory experiments.

The first step of CIF is determined by a diffusion of methanol to the bacterial cell from a culture medium. In this case, the α_1 value (Scheme 3) is varied from 0 at equilibrium between the methanol content of the cell and the medium to 1.0155 at methanol flux to the cell. In the next

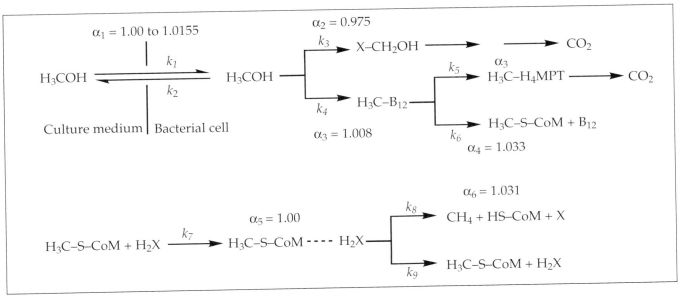

Scheme 3

step, the cell methanol fluxes are divided in two directions: oxidation of the H_3C group to CO_2 and its reduction to CH_4. The ratio of the H_3C quantities, which is oxidized to CO_2 and reduced to CH_4, is 1:3 (as determined in our experiments). It is assumed that the H_3C group of the cell methanol transformed to B_{12} is depleted in ^{13}C, whereas the residual H_3C of methanol oxidized to CO_2 must be enriched in ^{13}C with respective to the cell methanol pool. According to mass isotopic balance calculations, $\delta^{13}C$ of CO_2 and CH_4 must be about +25‰ ($\alpha_2 = 0.975$) and −8‰ ($\alpha_3 = 1.008$), respectively, to the cell methanol. This means that metabolic CO_2 is enriched in ^{13}C relative to methanol.

Consequently, the maximum CIF of methane by *M. barkeri* grown on methanol is determined by $\alpha_{max} = \alpha_1\alpha_3\alpha_4\alpha_6$, which ranges from 1.072 to 1.090 at the low methane generation rate (such that $k_1 << k_2$, $k_3 = (1/3)k_4$, $k_6 << k_5$, and $k_8 << k_9$, Scheme 3). The minimum value is estimated by $\alpha_{min} = \alpha_1\alpha_3 = 1.008$ to 1.024 at the high methane generation rate (such that $k_1 >> k_2$, $k_3 = (1/3)k_4$, $k_6 >> k_5$, and $k_8 >> k_9$).

Acetate Reduction

Studies dealing with cells of *M. barkeri* grown on acetate have shown that a reduction of H_3C of acetate to CH_4 and oxidation of $COOH$ to CO_2 accrue to a large extent. Acetate activates to acetyl-CoA, which is then cleaved into two enzyme-bound C_1 units (H_3C and CO). The CO dehydrogenase oxidizes the bound CO to CO_2. According to Keltjens and van der Drift (1986), the utilization stages of acetate by *M. barkeri* can be written as shown in Scheme 4 (next page).

The KIE of acetate transport to a cell is assumed to be determined by $\alpha_1 = 1.0$ if $k_1 >> k_2$ and $\alpha_1 = 1.008$ if $k_1 << k_2$. De Niro and Epstein (1977) have shown that a ^{13}C depletion in the acetaldehyde produced by an enzymatic

decarboxylation of pyruvate is concentrated in the carbonyl carbon of acetaldehyde. The difference between the $\delta^{13}C_{CO}$ values of the 100% and 1.9% acetaldehyde production in this reaction is about 14.1‰. Therefore, the CIF of the carbonyl carbon of acetate can be carried out during the bonding of acetate and HS–CoA (step 1, Scheme 4). By analogy with CIF results in the above reaction (DeNiro and Epstein, 1977), the carbonyl carbon of H_3C–CO–S–CoA should be depleted in ^{13}C relative to the $COOH$ group of the residual acetate: $\alpha_{2\,CO} = 1.014$ if $k_3 << k_4$. A split of the acetyl-CoA and B_{12} complex into two C_1 units ($k_6 = k_7$) makes it possible to assume that the carbon isotope content of CO_2 corresponds to CO of acetyl-CoA.

Metabolic pathways of H_3C–B_{12} diverge into two directions to produce H_3C–H_4MPT (k_8, step 3a) and H_3C–S–CoM (k_9, step 3b). If the binding of H_3C and B_{12} determines a reaction rate for step 2 (Scheme 4), the CIF of the H_3C group will take place ($\alpha_3 = 1.033$). In conformity with Scheme 4, the CIF of the methane produced by *M. barkeri* grown on acetate at the low methane generation rate (such that $k_1 << k_2$, $k_3 << k_4$, $k_6 = k_7$, $k_9 << k_8$, $k_{12} << k_{13}$, and $k_{15} << k_{16}$) can be determined by some steps of the H_3C group transfer: from acetyl-CoA to B_{12} ($\alpha_3 = 1.033$), from B_{12} to HS–CoM ($\alpha_7 = 1.033$), and by reduction of H_3C–S–CoM to CH_4 diffused from cells ($\alpha_9 = 1.031$). Thus, the maximum CIF of the methane produced by methanogens grown on acetate can be determined by the α value, which is about $\alpha_{max} = \alpha_1\alpha_3\alpha_7\alpha_9 = 1.097$ to 1.105.

In an assessment of the minimum CIF of methane produced by methanogens grown on acetate, one consideration should be taken into account. It has been shown (Krzycki et al., 1982) that about 15% of CO_2 is produced during oxidation of H_3C by methanogens grown on acetate. That is, about 15% of H_3C in H_3C–B_{12} (Scheme 4) can be oxidized to CO_2 and a 85% residual part of H_3C in H_3C–B_{12} is reduced to CH_4. Let us assume that the α_6

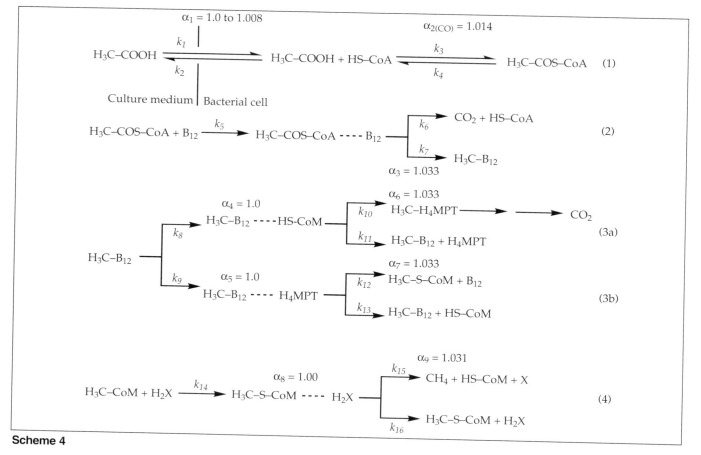

Scheme 4

value for the transfer of the H_3C group to H_4MPT is about 1.033. The CH_4 enrichment of $\delta^{13}C$ at about +5‰ relative to the H_3C group of acetate has been calculated.

Thus, at the high methane generation rate of methanogens grown on acetate, the CIF of the methane realized can be as high as the $\delta^{13}C$ values of about +5‰ or more relative to the H_3C group of acetate. According to these statements, one could conclude that the carbon isotope content of the CO_2 produced at the high methane generation rate must be depleted in ^{13}C and that, at the low rate, is enriched in ^{13}C relative to CH_4. Consequently, based on theoretical considerations, the CIF of methane by methanogens relative to a substrate depends on the amount of methane produced by the bacterial cell per unit time, that is, on the *specific methanogenesis rate* (SMR).

It was shown earlier that the SMR depends on the temperature, pH of the medium, and concentration of NaCl (Wolkin et al., 1983). Hence, it is possible to regulate SMR by changing the conditions of the medium where methane is produced by bacteria grown on $CO_2 + H_2$ and to study CIF by methanogens.

EXPERIMENTAL METHODS

The methodology used to change the SMR by bacteria grown on different C_1 and C_2 substrates has been described by Ivanov et al. (1983). Briefly, it envisages variability in the culture medium parameters: pH values from 5.7 to 8.6, NaCl contents from 0.9 to 80 g/L, and temperatures from –2°C to +37°C.

The rate of methanogenesis was measured as milliliters of CH_4 produced by 1 mg dry biomass or by one bacterial cell per hour. The total number of bacteria were determined by direct count using membrane filters. The quantity of methane and CO_2 was determined by gas chromatography. To prepare samples for mass spectrometric analysis, methane was purified from CO_2 by passing it through alkali and then burnt in oxygen to CO_2 in a closed device. The carbon isotope composition of CO_2 was measured with a Varian-MAT CH-7 double-channeled mass spectrometer by a comparative technique. Data are reported as $\delta^{13}C$ for CO_2 gas relative to the PDB standard as defined by

$$\delta^{13}C = (R_{sa}/R_{st} - 1)1000‰ \qquad (7)$$

where R_{sa} and R_{st} are the ratios of $[^{13}C]/[^{12}C]$ for the CO_2 sample and standard, respectively. The determination error of $\delta^{13}C$ values was ±0.2‰. The coefficient of CIF of products relative to the residual substrate was estimated as follows:

$$\alpha = (\delta^{13}C_{subst} + 1000)/(\delta^{13}C_{prod} + 1000) \qquad (8)$$

RESULTS AND DISCUSSION

Carbon Dioxide Reduction

In our experiments, SMR varied within two orders of magnitude from 10^{-3} to 10^{-1} mL CH_4/mg dry biomass per hour, or from 10^{-11} to 10^{-9} mL CH_4/cell per hour (Ivanov et al., 1983). Figures 2 and 3 demonstrate the evaluation of the isotopic composition of methane produced by pure cultures of *Methanobacterium formicicium* str. Kuznetcov (Figure 2) and *Methanosarcina barkeri* str. MS (Figure 3) growing on CO_2 and H_2 relative to carbon dioxide depending on SMR.

The maximum CIF by *M. formicicum* ($\alpha = 1.066$, Figure 2) was obtained at low SMR attained with values of about 0.32×10^{-2} mL CH_4/mg dry biomass per hour. In this case, the growth conditions of bacteria were as follows: temperature $-2°C$, NaCl content 45 g/L, and pH 7.5. In contrast, at a high SMR of about 22.4×10^{-2} mL CH_4/mg dry biomass per hour, the CIF between CH_4 (product) and CO_2 (substrate) was decreased ($\alpha = 1.022$, Figure 2). The bacteria were grown at the following conditions: temperature $37°C$, NaCl content 0.9 g/L, and pH 7.3.

The CIF of *M. barkeri* grown on CO_2 and H_2 versus SMR (Figure 3) was generally identical to the abundance curve in the previous case. The maximum CIF of these bacteria ($\alpha = 1.057$) was obtained under their growth at a temperature of $37°C$, a NaCl content of 47 g/L, and a pH of 7.0 and at a low SMR of about 3.7×10^{-2} mL CH_4/mg dry biomass per hour. The minimum CIF of *M. barkeri* ($\alpha = 1.024$, Figure 3) was also exhibited at a high SMR of about 12.7×10^{-2} mL CH_4/mg dry biomass per hour at a temperature of $37°C$, a NaCl content of 0.9 g/L, and a pH of 6.8.

Studies of CIF of methane produced by thermophilic methanogens have shown the same tendency for a change in $\delta^{13}C$ values versus SMR (Figure 4, curve 3). The maximum CIF of methane produced by *Methanobacterium thermoautotrophicum* str. ΔH ($\alpha = 1.032$) was obtained under their growth at a temperature of $55°C$, a NaCl content of 0.9 g/L, and a pH of 7.2 and a low SMR for this culture of about 1.3 mL CH_4/mg dry biomass per hour (Ivanov et al., 1985). The minimum CIF of methane produced by this microbial culture ($\alpha = 1.015$) was exhibited at a high SMR of about 8.0 mL CH_4/mg dry biomass per hour and at a temperature of $66°C$, a NaCl of content 0.9 g/L, and a pH of 7.2. It is remarkable that the SMR of these thermophilic bacteria was about 100 times higher than the observed rate of mesophilic methanogenic bacteria.

Thus, in all cases of methanogenesis by mesophilic and thermophilic bacteria growing on CO_2 and H_2, the difference in values between the carbon isotopic composition of CO_2 and the methane produced increased as the SMR decelerated. The results are in good agreement with those predicted by our theoretical considerations for the CIF of methane produced by methanogens grown on CO_2 and H_2 (see Scheme 2).

Along with consumption of CO_2 by methanogenic bacteria, there is also a release of metabolic CO_2 during

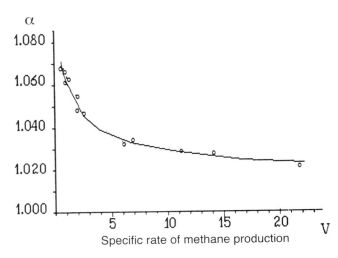

Figure 2—Fractionation of carbon isotopes by *Methanobacterium formicicum* growing on CO_2 + H_2 as dependent on the specific rate of methane production, V = 10^{-2} mL CH_4/mg dry biomass per hour. (Data are from Figure 4.)

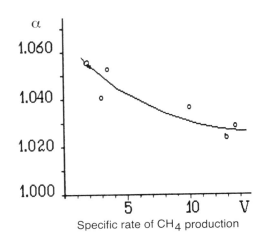

Figure 3—Fractionation of carbon isotopes by *Methanosarcina barkeri* growing on CO_2 + H_2 as dependent on the specific rate of methane production, V = 10^{-2} mL CH_4/mg dry biomass per hour. (Data are from Figure 4.)

the biosynthesis of cell biomass, such as in isocitrate decarboxylation (Zeikus, 1983) and during methane "oxidation" by methanogens (Zehnder and Brock, 1979). It was assumed that metabolic CO_2 from bacterial cells is depleted in ^{13}C relative to exogenous CO_2 and that its mixture with exogenous CO_2 also affects the difference in the carbon isotope composition of CO_2 (substrate) and methane (product).

Figure 5 shows the dependence of the difference in the carbon isotopic composition of methane and CO_2 versus the specific concentration of CO_2 calculated as ratios of quantities of carbon dioxide (mg of carbon in CO_2) to the cell biomass (mg of organic carbon) for *M. formicicum*

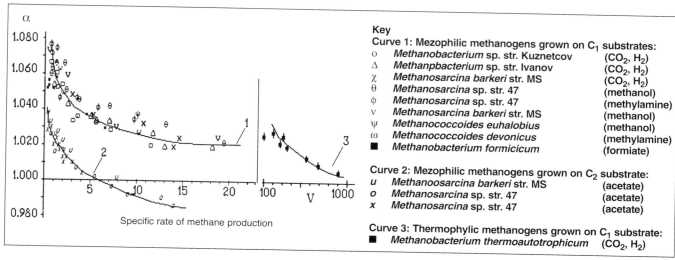

Figure 4—Fractionation of carbon isotopes by methanogenic bacteria as dependent on the specific rate of biogenic methane production, $V = 10^{-10}$ mL CH_4/cell per hour. (From Zyakun et al., 1988.)

grown on $CO_2 + H_2$ (Zyakun et al., 1988). In this experiment, the specific carbon dioxide concentration varied from 6.5 to 190 mg of CO_2 carbon/mg of organic carbon, whereas the specific concentration of methane was within 14–20 mg of CH_4 carbon/mg of organic carbon. These CO_2 concentrations occur in the culture medium when differences between $\delta^{13}C$ values of CH_4 and CO_2 decrease with a diminution of the specific CO_2 concentration during bacterial methanogenesis. This means that the increasing share of metabolic carbon dioxide in the CO_2 pool of the culture medium significantly decreases the CIF, in other words, the difference between $\delta^{13}C$ values of biogenic methane and residual carbon dioxide diminishes.

Thus, according to theoretical considerations and experimental data, possible $\delta^{13}C$ values of methane produced by bacteria relative to CO_2 (substrate) vary within a wide range (–19 to –100‰), depending on the SMR of $CO_2 + H_2$ and methanogenic species, in particular, thermophilic and mesopilic bacterial genera. In some instances, at low specific concentrations of CO_2 and high specific concentrations of methane, the differences in the $\delta^{13}C$ values of bacterial methane and carbon dioxide can be as much as –10‰.

C_1 Organic Substrate Fermentation

To further elucidate the factors that influence the CIF in bacterial methanogenesis, we assessed the effect of organic substrates (formate, methanol, methylamine, and acetate) on the observed carbon isotopic composition of methane produced (Ivanov et al., 1983). A range of cultures using organic substrates (except formate) was confined to representatives of the genera *Methanosarcina* and *Methanococcoides*. In the case of formate, a representative of the genus *Methanobacterium* was used (*M. formicicum*) (Figure 6).

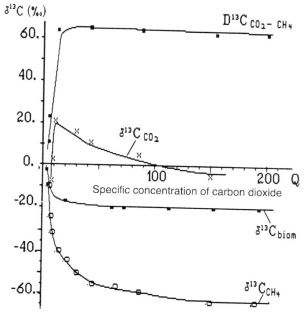

Figure 5—Fractionation of carbon isotopes by *Methanobacterium formicicum* grown on CO_2 and H_2 as dependent on the specific concentration of carbon dioxide, Q = mg of CO_2 carbon/mg of organic carbon. $D^{13}C_{CO_2-CH_4}$ = $\delta^{13}C_{CO_2} - \delta^{13}C_{CH_4}$; $\delta^{13}C_{CO_2}$, $\delta^{13}C_{biom}$, and $\delta^{13}C_{CH_4}$ are the carbon isotope ratios in carbon dioxide, cell biomass, and biogenic methane, respectively. (From Zyakun et al., 1988.)

The growth of *M. formicicum* on formate was accompanied by generation of ^{13}C-depleted methane and carbon dioxide enriched in this isotope relative to formate. In this case, as during the growth on $CO_2 + H_2$, we observed a tendency toward decreased CIF with increased SMR. Thus, at a SMR of 0.9×10^{-2} to 5.2×10^{-2} mL CH_4/mg dry

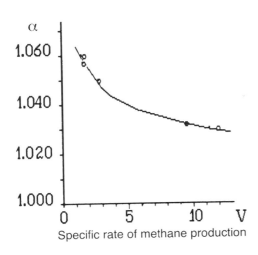

Figure 6—Fractionation of carbon isotopes by *Methanobacterium formicicium* growing on formate as dependent on the specific rate of methane production, $V = 10^{-2}$ mL CH_4/mg dry biomass per hour. (Data are from Figure 4.)

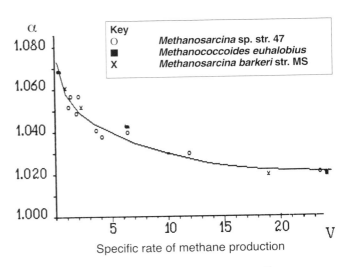

Figure 7—Carbon isotope fractionation by methagens growing on methanol as dependent on the specific rate of methane production, $V = 10^{-2}$ mL CH_4/mg dry biomass per hour. (Data are from Figure 4.)

biomass per hour, the $\delta^{13}C$ difference between the substrate and the methane produced was 44–61‰, and at a SMR of 9×10^{-2} to 12×10^{-2} mL CH_4/mg dry biomass per hour, it was 31–34‰.

The assays on CIF during bacterial methane generation on methanol were carried out using the cultures *Methanosarcina* sp. str. 47, *Methanosarcina barkeri* str. MS, and *Methanococcoides euchalobius* (Figure 7). As during methanogenesis on $CO_2 + H_2$, the CIF by bacteria grown on methanol depended on SMR. Thus, SMR of 1×10^{-2} to 3×10^{-2} mL CH_4/mg dry biomass per hour, which is the difference in the $\delta^{13}C$ values of methanol and methane produced, was 50–57‰, while at 25×10^{-2} mL CH_4/mg dry biomass per hour, it was only 21.4‰ (Figure 7). A similar CIF dependence of methane was noted in the case of methanogenic bacteria grown on methylamine.

All our experiments on bacterial methanogenesis with C_1 organic substrates featured the generation of biogenic carbon dioxide enriched in ^{13}C relative to the substrate by 15–19‰. Thus, the experimental results of carbon isotope content for both methane and carbon dioxide produced by methanogens grown on C_1 organic substrates are in agreement with the previous theoretical predictions (see Scheme 3).

Acetate Fermentation

Of great interest are the results obtained in Scheme 4 from the CIF of CH_4 and CO_2 produced by growth of *Methanosarcina barkeri* str. MS and *Methanosarcina* sp. str. 47 on acetate versus SMR (Ivanov et al., 1984). Similar to CIF of methane in bacterial methanogenesis on C_1 substrates ($CO_2 + H_2$, methanol, methylamine, and formate), we observed a dependence of CIF of methane produced by methanogenic bacteria grown on acetate (C_2 substrate) relative to SMR (Figure 4). At low SMR, the carbon iso-

topic content of methane is significantly depleted in ^{13}C compared to the acetate methyl which is its source (Krzycki et al., 1987). At increased SMR, the α values of methane decreased relative to this methyl group.

However, in keeping with theoretical considerations (Scheme 4) and in contrast with results from the C_1 substrate methanogenesis experiments, at maximum SMR by bacteria grown on acetate, the carbon isotopic composition of biogenic methane was enriched in ^{13}C relative to the methyl group of acetate consumed. These experiments (Figure 4) show that the biochemical mechanism of acetate transformation by methanogenic bacteria may lead to an isotopic effect resulting not in a decrease of ^{13}C of methane (as observed earlier) but in an increase of its content relative to the substrate consumed (an apparent reverse isotopic effect). For instance, at a high SMR of 10×10^{-2} to 12×10^{-2} mL CH_4/mg biomass per hour, the methane produced was enriched in ^{13}C by 10–20‰, whereas at lower rates of 0.1×10^{-2} to 0.4×10^{-2} mL CH_4/mg dry biomass per hour, the carbon isotopic composition of methane was depleted in ^{13}C by 30–33‰ relative to the substrate.

To better understand the controls on the observed CIF, experiments were conducted with complete acetate exhaustion by methanogenic bacteria. Figure 8 shows the dependence of the carbon isotopic composition of residual acetate in the culture liquid and the products formed (methane and carbon dioxide) for *Methanosarcina* sp. str. 47 grown on acetate versus the acetate share used (F) at two specific rates of methanogenesis: slow, 1.8×10^{-2} mL CH_4/mg biomass per hour, and fast, 6.9×10^{-2} mL CH_4/mg biomass per hour. In the former case (Figure 8A), while acetate is being consumed, its remaining part is enriched in ^{13}C (the effect of Raleigh exhaustion). In the latter case (Figure 8B), the carbon isotopic composition of residual acetate does not change during the assay.

As might be expected from theoretical conciderations (Scheme 4), the significant changes in the values of $\delta^{13}C_{CO_2}$ and $\delta^{13}C_{CH_4}$ have been shown to depend on SMR and the extent of acetate exhaustion. At low SMR and in the initial stages of substrate consumption, the carbon isotope content of metabolic CO_2 is depleted in ^{13}C relative to the COOH group of acetate by about $-10\%o$ and CH_4 relative to the H_3C group by about $-22\%o$ (Figure 8A). At high SMR (Figure 8B), metabolic carbon dioxide is depleted in ^{13}C by about $-13\%o$ relative to the COOH group, whereas the biogenic methane is enriched in ^{13}C relative to the H_3C group by about $+6\%o$. As the substrate (acetate) amount in the medium decreases in both cases (Figures 8A, B), the ^{13}C content of CO_2 has a tendency to increase relative to its initial values (e.g., at low substrate consumption, F < 0.2).

At low SMR (Figure 8A), the carbon isotopic composition of methane is depleted in ^{13}C and alters in accordance with a change in the ^{13}C content of acetate. In other words, as acetate is being consumed, the carbon isotopic compositions of methane and residual acetate are enriched in ^{13}C relative to their initial portions. At high SMR, however, the isotopic composition of methane is enriched in ^{13}C relative to the substrate consumed, but it becomes depleted in ^{13}C as the reaction proceeds (e.g., $\delta^{13}C_{CH_4} = -22\%o$ at F < 0.1 to $\delta^{13}C_{CH_4} = -34\%o$ at F < 0.9).

Thus, during methane generation by bacteria grown on C_2 substrate (acetate), the carbon isotopic composition of methane produced can change relative to the substrate consumed both toward the depletion of isotope ^{13}C and its enrichment.

The determining factor in the CIF of methane is the biochemical pecularities of the methanogens' growth on acetate that are specified by different SMRs. According to theoretical considerations (Scheme 4), the CIF of methane is determined by a share of the H_3C group of metabolite H_3C-B_{12} consumed for CH_4 production. The carbon isotopic composition of metabolic CO_2 is formed at the stage of acetyl-CoA formation and the oxidation of its CO group to CO_2, as well as during the oxidation of part of H_3C of metabolite H_3C-B_{12} to CO_2.

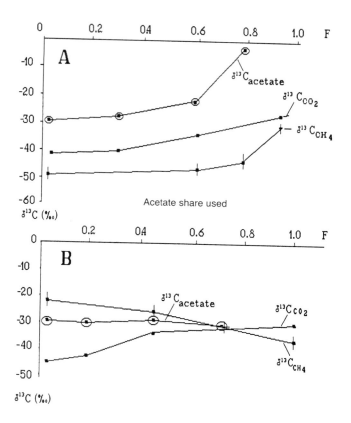

Figure 8—Carbon isotope composition of acetate ($\delta^{13}C_{acetate}$), methane ($\delta^{13}C_{CH_4}$), and carbon dioxide ($\delta^{13}C_{CO_2}$) during *Methanosarcina barkeri* growth on acetate versus the acetate share used (F). (A) Specific rate of methanogenesis is 1.8×10^{-2} mL CH_4/mg biomass per hour. (B) Specific rate of methanogenesis is 6.9×10^{-2} mL CH_4/mg biomass per hour. The following carbon isotope contents were used in the initial acetate: $\delta^{13}C_{CH_3} = -27.8\%o$, $\delta^{13}C_{COOH} = -32\%o$. (From Zyakun et al., 1988.)

CONCLUSIONS

Analysis of laboratory experiments on the CIF of methane and CO_2 during bacterial methanogenesis reveals that biogenic methane can have variable isotopic compositions of carbon both with a very low ($\delta^{13}C < -55\%o$) and comparatively high ($\delta^{13}C > 0\%o$) content of isotope ^{13}C relative to the substrate used. The carbon isotopic composition of biogenic methane is primarily determined by the isotopic composition of the substrate used, the biochemical pathways of bacterial methane generation, the specific methanogenesis rate (SMR), and the specific substrate concentration. At low SMR, the carbon isotopic composition of methane formed is significantly depleted in ^{13}C relative to carbon dioxide or the organic substance (methanol, methylamine, or acetate) consumed

as the carbon source by methanogenic bacteria. One can assume that methane with a $\delta^{13}C$ value lower than $-55\%o$ is of bacterial origin.

Methane with higher ^{13}C content detected in natural gas reservoirs can be either of bacterial or thermogenic origin. In this case, to characterize the origin of methane, one should use microbiological, biogeochemical, and other data in addition to methane isotopic data.

An important conclusion should be made when comparing the CIF of methane produced by methanogens growing on C_1 and C_2 substrates (Figure 4). At similar SMRs by methanogens grown on C_1 substrates, the carbon isotopic composition of methane is systematically depleted in ^{13}C by nearly $30\%o$ relative to methane produced by these bacteria grown on C_2 substrates. Thus, when analyzing the CIF of methane produced by methanogens in natural ecosystems, it is necessary to take into account the C_1 and C_2 substrates used, in addition to the SMR.

Acknowledgments—*I would like to thank D. Schumacher of ESRI, University of Utah, and M. Abrams of Exxon Ventures (CIS) for reviewing the manuscript and for useful advice and comments. I am also grateful to T. Zhuravleva and E. Chervyakova for their help in preparing the English version of this paper.*

REFERENCES CITED

Alekseev, F. A., and V. S. Lebedev, 1975, The carbon isotope composition of carbon dioxide and methane in the bottom sediments of the Black Sea (in Russian), *in* Dispersed gases and biochemical conditions of sediments and rock: Moscow, VNIIYGG, p. 49–53.

Balch, W. E., G. E. Fox, L. J. Magrum, C. R. Woese, and R. S. Wolfe, 1979, Methanogens: reevaluation of a unique biological group: Microbiology Review, v. 43, p. 260–296.

Blaut, M., V. Muller, K. Fiebig, and G. Gottschalk, 1985, Sodium ions and an energized membrane required by *Methanosarcina barkeri* for the oxidation of methanol to the level of formaldehyde: Journal of Bacteriology, v. 164, p. 95–101.

Cappenberg, Th. E., and E. Jongejan, 1978, Microenvironments for sulfate reducing and methane production in freshwater sediments, *in* W. E. Krumbein, ed., Environmental biogeochemistry and geomicrobiology: Michigan, Ann Arbor Science Publishers, p. 129–139.

Claypool, G. E., and I. R. Kaplan, 1974, The origin and distribution of methane in marine sediments, *in* I. R. Kaplan, ed., Natural gases in marine sediments: New York, Plenum Press, p. 99–139.

Collett, T. S., and K. I. Cunningham, 1994, Geologic controls on gas migration within northern Alaska (abs.), *in* Near-surface expressions of hydrocarbon migration: AAPG Hedberg Research Conference Abstracts, April 24–28, Vancouver, British Columbia.

De Niro, M. J., and S. Epstein, 1977, Mechanism of carbon isotope fractionation associated with lipid synthesis: Science, v. 197, p. 261–263.

Deuser, W. G., and E. T. Degens, 1967, Carbon isotope fractionation in the system $CO_{2(g)}$–$CO_{2(aq)}$–HCO_3^-: Nature, v. 215, no. 5105, p. 1033–1035.

Fuchs, G., R. Thauer, H. Ziegler, and W. Stichler, 1979, Carbon isotope fractionation by *Methanobacterium thermoautotrophicum*.: Archives of Microbiology, v. 120, p. 135–139.

Games, L. M., J. M. Hayes, and R. P. Gunsalus, 1978, Methane-producing bacteria: natural fractionation of the stable carbon isotopes: Geochimica et Cosmochimica Acta, v. 42, p. 1295–1297.

Hansen, G., and M. Bjoroy, 1994, Correlation of carbon isotope composition of adsorbed gas and nature of extractable hydrocarbons in shallow core (abs.), *in* Near-surface expressions of hydrocarbon migration: AAPG Hedberg Research Conference Abstracts, April 24–28, Vancouver, British Columbia.

Ivanov, M. V., S. S. Beljaev, A. M. Zyakun, V. A. Bondar, K. S. Laurinavichus, and O. V. Shipin, 1983, Carbon isotope fractionation by methane-producing bacteria growing on different substrates (in Russian), *in* Isotopen in der Natur, 3: Arbeitstagung von November 15–18, Leipzig, v. 1, p. 139–158.

Ivanov, M. V., S. S. Beljaev, A. M. Zyakun, V. A. Bondar, K. S. Laurinavichus, and O. V. Shipin, 1984, Carbon isotope

fractionation by methane-producing sarcine (in Russian): Dokl. Akad. Nauk SSSR, v. 277, no. 1, p. 225–229.

Keltjens, J. T., and G. D. Vogels, 1981, Novel coenzymes of methanogens, *in* H. Dalton, ed., Micrbial growth on C_1 compounds: Heyden, London p. 152–158.

Keltjens, J. T., and C. van der Drift, 1986, Electron transfer reactions in methanogenesis: FEMS Microbiology Review, v. 39, p. 259–303.

Keltjens, J. T., G. C. Caerteling, C. van der Drift, and G. D. Vogels, 1986, Methanopterin and the intermediary steps of methanogenesis: Systematic Applied Microbiology, v. 7, p. 370–375.

King, G. M., 1984, Utilization of hydrogen, acetate and "non-competitive" substrates by methanogenic bacteria in marine sediments: Geomicrobiology Journal, v. 3, p. 275–306.

Kosiur, D. R., and A. L. Warford, 1979, Methane production and oxidation in Santa Barbara basin sediments: Estuarine and Coastal Marine Science, v. 48, p. 379–385.

Krzycki, J. A., R. H. Wolkin, and J. G. Zeikus, 1982, Comparison of unitrophic and mixotrophic substrate metabolism by an acetate-adapted strain of *Methanosarcina barkeri*: Journal of Bacteriology, v. 149, p. 247–254.

Krzycki, J. A., W. R. Kenealy, M. J. De Niro, and J. G. Zeikus, 1987, Stable carbon isotope fractionation by *Methanosarcina barkeri* during methanogenesis from acetate, methanol or carbon dioxide hydrogen: Applied Environmental Microbiology, v. 53, no. 10, p. 2597–2599.

Leigh, J. A., K. L. Rinehart, Jr., and R. S. Wolfe, 1985, Methanofuran (carbon dioxide reduction factor), a formyl carrier in methane production from carbon ioxide in *Methanobacterium*: Biochemistry, v. 24, p. 995–999.

Lein, A. Yu., B. B. Namsaraev, V. Yu. Trotsyuk, and M. V. Ivanov, 1981, Bacterial methanogenesis in Holocene sediments of the Baltic Sea: Geomicrobiology Journal, v. 2, no. 4, p. 299–317.

Melander, L., and W. H. Saunders, Jr., 1980, Reaction rates of isotopic molecules: New York, Wiley-Interscience, 344 p.

Nakai, N., 1960, Geochemical studies on the formation of natural gases: Journal of Earth Sciences, Nagoya University, v. 8, no. 2, p. 174–180.

Oremland, R. S., L. M. Marsh, and D. J. Des Mereis, 1982, Methanogenesis in Big Soda Lake, Nevada: an alkaline, moderately hypersaline desert lake: Applied Environmental Microbiology, v. 43, no. 2, p. 462–468.

Rice, D. D., and G. E. Claypool, 1981, Generation, accumulation, and resource potential of biogenic gas: AAPG Bulletin, v. 61, p. 5–25.

Rosenfeld, W. D., and S. R. Silverman, 1959, Carbon isotope fractionation in bacterial production of methane: Science, v. 130, p. 1658–1659.

Senior, E., E. B. Lindstrom, T. M. Banat, and O. W. Nedwell, 1982, Sulfate reduction and methanogenesis in the sedimentary salt marsh on the east coast of the United Kingdom: Applied Environmental Microbiology, v. 43, no. 5, p. 987–996.

Tyler, S. C., G. W. Brailsford, K. Yagi, K. Minami, and R. J. Cicerone, 1994, Seasonal variations in methane flux and $\delta^{13}CH_4$ values for rice paddies in Japan and their implications: Global Biogeochemistry Cycles, v. 8, no. 1, p. 1–12.

Whiticar, M. J., E. Faber, and M. Schoell, 1986, Biogenic methane formation in marine and fresh water environments: CO_2 reduction vs. acetate fermentation isotopic evidence: Geochimica et Cosmochimica Acta, v. 50, p. 693–709.

Wolkin, R., S. Belyaev, and G. Zeikus, 1983, The isolation and characterization of methanogenic bacteria from the Bonduzhskoye oil field: Applied Environmental Microbiology, v. 45, no. 2, p. 691.

Zehnder, A. J., and T. D. Brock, 1979, Methane formation and methane oxidation by methanogenic bacteria: Journal of Bacteriology, v. 137, no. 1, p. 420–432.

Zeikus, J. G., 1983, Metabolism of one-carbon compounds by chemotrophic anaerobes: Advances in Microbial Physiology, v. 25, p. 215–299.

Zyakun, A. M., V. A. Bondar, K. S. Laurinavichus, O. V. Shipin, S. S. Beljaev, and M. V. Ivanov, 1988, Carbon isotope fractionation during the growth of methane-producing bacteria on different substrates (in Russian): Microbiology Journal, v. 50, no. 2, p. 16–22.

Tucker, J., and D. Hitzman, 1996, Long-term and seasonal trends in the response of hydrocarbon-utilizing microbes to light hydrocarbon gases in shallow soils, *in* D. Schumacher and M. A. Abrams, eds., Hydrocarbon migration and its near-surface expression: AAPG Memoir 66, p. 353–357.

Chapter 26

Long-Term and Seasonal Trends in the Response of Hydrocarbon-Utilizing Microbes to Light Hydrocarbon Gases in Shallow Soils

James Tucker

Daniel Hitzman

Geo-Microbial Technologies
Ochelata, Oklahoma, U.S.A.

Abstract

The use of population variations in hydrocarbon-utilizing microorganisms as a surface geochemical exploration technique for oil and gas has been vigorously studied since the 1940s. The Microbial Oil Survey Technique (MOST) was developed by Phillips Petroleum Company in the 1950s and held proprietary until 1985. In the development of the methodology, Phillips periodically sampled a particular test traverse over known production near Bartlesville, Oklahoma, beginning in 1957. Geo-Microbial Technologies (GMT) continued to sample the test traverse beginning in 1986. In addition to the yearly sampling, the test traverse was sampled monthly in 1993 by GMT to determine if seasonal variations affect the specific suite of microorganisms measured in MOST. Although slight changes in the absolute microbial concentrations did occur through the seasons, after factoring out oil production effects, the trend in microbial highs and lows across the test traverse remained constant. In the monthly surveys throughout 1993, of all the seasonal and weather factors tracked, only soil moisture content seemed to correlate with the monthly change in absolute microbial counts. Microbial lineplots covering this 37-year period demonstrate the continuous nature of hydrocarbon microseepage over a known hydrocarbon accumulation and the reproducibility of associated microbial prospecting.

INTRODUCTION

In the 1950s, Phillips Petroleum Company developed surface geochemical exploration technology called the Microbial Oil Survey Technique (MOST) to aid in oil and gas exploration efforts (Hitzman, 1959). This technology successfully identifies the populations of a specific suite of shallow soil microorganisms that use light hydrocarbons (Beghtel et al., 1987). To evaluate the technique, a test traverse 1.6 km (1 mile) in length over known production near Bartlesville, Oklahoma, was sampled periodically to determine the applicability and reproducibility of microbial prospecting to identify hydrocarbon microseepage. In 1986, Geo-Microbial Technologies (GMT) continued to sample the Phillips test traverse periodically using the same MOST technology. Between the Phillips and GMT surveys, a database covering over 37 years of microbial surveys of the same test traverse has been produced.

Microbial prospecting involves collecting shallow soil samples and screening the soils for a specific suite of microorganisms that indicate the presence of light hydrocarbons. In screening, a set weight of the collected soil sample is dispersed in water and diluted by a series of dilutions to 1:40,000. The diluted sample is then plated on petri dishes with agar gel that has been spiked with butanol as the only carbon source for microbial growth. Only those microorganisms capable of oxidizing light hydrocarbons (particularly butane) will grow. After one week incubation, the microbial colonies on the plates are counted and reported.

The advantages of microbial prospecting are the ease of collecting samples, the speed of sample collection in the field, and the low cost of sample analysis (Hitzman, 1961). This allows for many samples to be collected economically over a short period of time to build a consistent sample set and help eliminate many of the unknown

Table 1—Statistical Microbial Values for 1957–1993

Sample Site	1957	1958	1986	August 1993	1993 Average[a]
17	34	38	41	40	52.4
16	36	44	50	59	57.2
15	41	52	58	74	62.9
14	44	64	69	93	68.6
13	53	69	64	96	68.2
12	57	62	57	89	65.8
11	54	51	48	84	63.8
10	40	42	47	74	62.3
9	31	42	48	77	63.8
8	29	38	54	85	69.5
7	33	36	60	93	77.9
6	37	27	64	88	82.4
5	40	19	59	75	78.2
4	44	17	53	65	71.6
3	48	28	48	60	63.3
2	48	44	46	55	58.3
1	45	55	45	48	55.3

[a]The average value for the 13 months of the 1993 monthly surveys is presented as the 1993 average.

noise factors associated with geochemical exploration. In addition, GMT uses a five-point moving-average smoothing function to smooth the extreme values, both high and low, within the data set. The microbial values presented in this paper are the smoothed data. The raw data are available to interested researchers upon request.

GMT initiated a year-long monthly sampling program in 1993 to sample the same test traverse to determine if seasonal variations in the microbial activity could be identified. The monthly surveys continued to identify the same microbial trends with varying magnitudes in absolute microbial activity. The microbial trends were compared to measured weather factors to determine if seasonal variations affected microbial growth.

YEARLY SURVEYS 1957–1993

The 1.6-km test traverse was sampled approximately every 100 m in 1957 and 1958 by Phillips Petroleum Company and in 1986 and 1993 by GMT (Table 1). The statistical smoothed microbial values are presented as lineplots for the surveys in each of these years (Figure 1).

The greatest variations occur between the yearly surveys. A double-peaked microbial microseepage signature occurs along the test traverse, one in the west and one in the east. The microbial high on the western end of the traverse is present in all the surveys (samples 15–11). The microbial high on the eastern end appears to have migrated from the extreme eastern end (samples 1–3) in 1957 and 1958 westward to its position in the 1986 and 1993 surveys (samples 4–8).

A possible explanation for the apparent anomalous migration may be the amount of oil production through the years. The test traverse was originally chosen to sample an area of known production and then into background. Production is from the Bartlesville Sandstone at a depth of about 800 ft. The western end has never been produced, while the eastern end has undergone active production since before the initial traverse. Since GMT has been conducting the yearly surveys, production from the field has stopped except for two periodic producers (Figure 2).

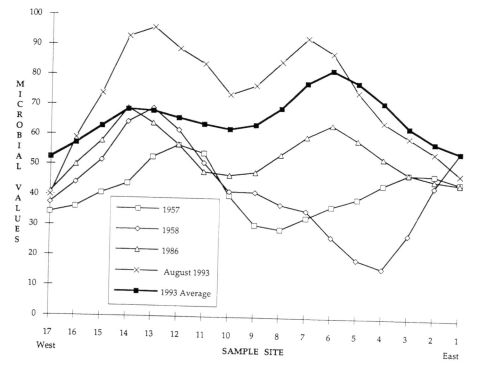

Figure 1—Statistical microbial lineplots for samples collected over the same traverse from 1957 to 1993. The microbial values on the left for 1957 and 1958 are lower than those for 1986 and 1993 due to oil production from the area in 1957–58. In 1986 and 1993, production stopped, allowing the reservoir to repressurize and hydrocarbon microseepage to resume.

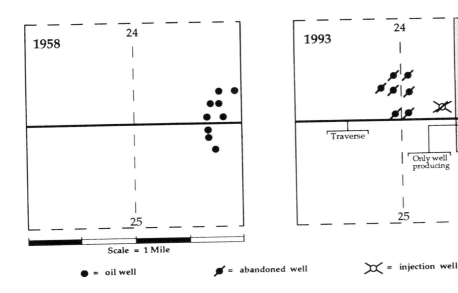

Figure 2—Maps showing the test traverse and the status of production and injection wells in 1958 and in 1993. The number of active producers in 1958 on the eastern end of the traverse may have reduced the microseepage signature identified here during this time. By 1993, the reservoir had been repressurized by the continuing waterflood from 1958 onward, which reinitiated the elevated microseepage identified in 1986 and 1993 on the eastern end of the traverse.

● = oil well   = abandoned well   = injection well

Since production of oil and gas has been shown to have a direct effect on decreasing microbial counts at the surface (Tucker et al., 1994), we suggest that the reason for the reduced counts in the eastern half of the test traverse is the intensive production that occurred in 1957 and 1958. Conversely, the anomaly on the western end (an area that has never been produced) has been identified consistently in all the yearly surveys. The microbial populations have increased in the eastern part of the traverse since the initial sampling because production has virtually stopped, and the reservoir has been repressurized by a waterflood and perhaps by continuing hydrocarbon recharge.

MONTHLY SURVEYS, 1993

The 1993 monthly sampling program of the same test traverse was conducted to determine if seasonal weather variations affected the microbial populations identified in the test traverse. Except for May, the monthly surveys continued to identify the same microbial trends as identified in the 1986 survey, with only varying magnitude in absolute microbial activity (Table 2). This suggests that some factors affect the absolute population values but not the locations of elevated and low microseepage. This also suggests that the MOST technique is highly reproducible

Table 2—Statistical Microbial Values for Each Month from January 1993 to January 1994.

Sample Site	13-Month Statistical Average	Jan 93	Feb	Mar	Apr	May	June	July	Aug	Sept	Oct	Nov	Dec	Jan 94
17	52.4	61	44	41	57	63	50	71	40	62	50	39	46	57
16	57.2	62	47	50	69	65	50	76	59	77	50	45	37	57
15	62.9	60	53	58	82	65	55	80	74	90	56	49	37	59
14	68.6	64	60	69	82	70	60	81	93	96	60	52	51	54
13	68.2	70	62	64	69	68	65	76	96	90	61	47	68	50
12	65.8	79	65	57	57	67	63	72	89	86	54	45	75	47
11	63.8	85	63	48	55	62	61	67	84	81	54	43	77	50
10	62.3	77	60	47	57	64	55	69	74	75	54	45	78	55
9	63.8	71	54	48	58	67	51	75	77	69	64	48	87	60
8	69.5	70	63	54	60	70	54	91	85	72	74	50	97	63
7	77.9	92	73	60	65	71	61	100	93	78	86	54	109	71
6	82.4	102	79	64	75	72	69	104	88	84	90	55	113	76
5	78.2	100	73	59	74	76	68	94	75	78	76	55	108	80
4	71.6	99	66	53	71	77	64	85	65	76	57	48	93	77
3	63.3	89	54	48	60	82	59	71	60	68	47	41	74	70
2	58.3	77	52	46	57	77	55	61	55	67	51	38	59	63
1	55.3	62	50	45	53	79	54	56	48	65	60	39	51	57

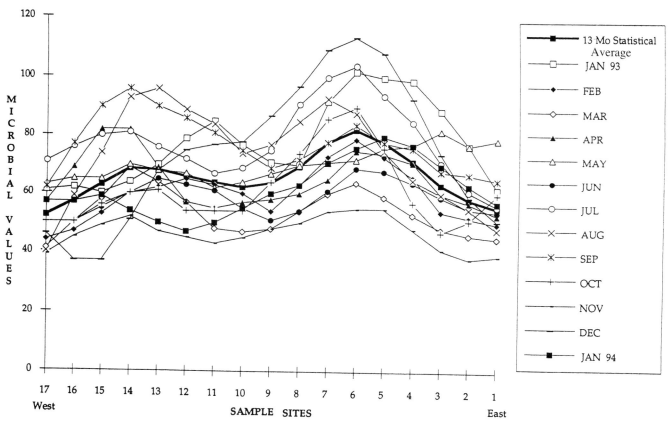

Figure 3—Statistical microbial lineplots for samples collected over the same traverse each month from January 1993 to January 1994. Note how the microbial trends do not change throughout the year, although the absolute microbial populations do vary seasonally.

in identifying elevated and low hydrocarbon microseepage trends regardless of season (Figure 3).

The microbial trends from the May survey do not demonstrate the same double-peaked microbial signature as identified in the other months. The featureless microbial pattern from May is probably related to an exceptionally long wet period just prior to sample collection. At the time of collection, there was abundant water standing in all the low areas and ditches. However, the surface soil moisture content was not reflected in the amount of standing free water. Either the amount of water flushing through the shallow soil or the saturated soil below the sampling interval affected the microbial growth rate.

The microbial trends were compared to measured weather factors to determine if seasonal variations affected microbial growth. The microbial trend for each month is the microbial average for the particular survey. Measured environmental factors included high and low air temperature, air temperature difference (Δ air temperature), high and low soil temperatures, soil temperature difference (Δ soil temperature), precipitation, and soil moisture content. Soil moisture was measured directly from the collected samples, while the other parameters

were obtained from the National Weather Bureau in Tulsa, Oklahoma. The value for each of the obtained parameters used in this study is the average parameter value for five days prior to actual sample collection.

Table 3 lists the correlation coefficients between the measured environmental factors and the microbial average for each month. No significant positive or negative correlation can be identified to control the microbial population changes. This lack of correlation indicates that transient environmental factors do not affect the hydrocarbon-utilizing microbial populations found in the shallow soils.

CONCLUSIONS

In all surveys, the microbial samples were collected every 100 m along the existing road on the 1.6-km test traverse. Two clusters of elevated microbial activity on the east and west ends of the test traverse have continually been identified through the years. Slight shifts in the exact location of the microbial signatures as well as shifts in the magnitude of the activity are identified.

Table 3—Correlation Coefficients Between Measured Environmental Factors and Microbial Average for Each Month[a]

Variables	Microbes/g Dry Soil	Microbial Average	% Soil Moisture	Max Air Temp	Min Air Temp	Δ Air Temp	Precipi- tation	Max Soil Temp	Min Soil Temp	Δ Soil Temp
Microbes/g Dry Soil	1.000	.966	−.041	.429	.506	−.466	.445	.445	.458	.124
Microbial Average	.966	1.000	−.298	.564	.602	−.375	.357	.594	.605	.280
% Soil Moisture	−.041	−.298	1.000	−.572	−.444	−.248	.244	−.651	−.646	−.587
Maximum Air Temp	.429	.564	−.572	1.000	.967	−.288	.057	.957	.947	.884
Minimum Air Temp	.506	.602	−.444	.967	1.000	−.523	.287	.950	.948	.761
Δ Air Temp	−.466	−.375	−.248	−.288	−.523	1.000	−.888	−.371	−.394	.097
Precipitation	.445	.357	.244	.057	.287	−.888	1.000	.147	.175	−.360
Maximum Soil Temp	.445	.594	−.651	.957	.950	−.371	.147	1.000	.999	.776
Minimum Soil Temp	.458	.605	−.646	.947	.948	−.394	.175	.999	1.000	.750
Δ Soil Temp	.124	.280	−.587	.884	.761	.097	−.360	.776	.750	1.000

[a]No significant positive or negative correlation can be identified to control the microbial population changes. This lack of correlation indicates that transient environmental factors do not affect the hydrocarbon utilizing microbial populations found in the shallow soils.

The various microbial surveys recorded by Phillips Petroleum and GMT over a 37-year time span indicate several empirical observations:

- There is a consistent trend in high and low microbial populations throughout the years over the same test traverse.
- The microbial populations depend on light hydrocarbons, indicating that a consistent source reaches the shallow soils.
- Although the microseepage trend is consistent, both the yearly and monthly surveys show variation in the absolute microbial values.
- Correlation of the environmental parameters indicate that transient environmental factors have no significant control on the absolute microbial population trends.
- Although the microseepage trends are constant,

the factors that control the hydrocarbon microseepage flux appear to be the ultimate controlling factors in absolute microbial populations.

REFERENCES CITED

Beghtel, F. W., D. O. Hitzman, and K. R. Sundberg, 1987, Microbial oil survey technique (MOST) evaluation of new field wildcat wells in Kansas: Association of Petroleum Geochemical Explorationists Bulletin , v. 3, p. 1–14.

Hitzman, D. O., 1959, Prospecting for petroleum deposits: U.S. Patent 3,880,142, March 31, 1959.

Hitzman, D. O., 1961, Comparison of geomicrobial prospecting methods used by various investigators: Developments in Industrial Microbiology, v. 2, p. 33–42.

Tucker, J. D., and D. C. Hitzman, 1994, Detailed microbial surveys help improve reservoir characterization: Oil & Gas Journal, v. 92, no. 23, p. 65–68.

Barwise, T., S. Hay, and J. Thrasher, 1996, Contamination of shallow cores: a common problem, *in* D. Schumacher and M. A. Abrams, eds., Hydrocarbon migration and its near-surface expression: AAPG Memoir 66, p. 359–362.

Chapter 27

Contamination of Shallow Cores: A Common Problem

Tony Barwise

BP Exploration
BP Research and Engineering Centre,
Sunbury on Thames, U.K.

Steve Hay

BP–Statoil R+D Alliance
Trondheim, Norway

Present address:
BP Exploration Company (Europe) Ltd.
Dyce, Aberdeen, Scotland

Jane Thrasher

BP Exploration
BP Research and Engineering Centre
Sunbury on Thames, U.K.

Present address:
Sir Alexander Gibb & Partners
Reading, U.K.

Abstract

Detecting migrated petroleum in shallow cores requires techniques that reliably indicate the presence of thermogenic petroleum in a core extract. Often, the level of petroleum in the extract is very low, and in our experience, many surveys have been prone to some form of contamination. It is important to be able to recognize these contaminants to avoid ascribing contaminants to actual petroleum in the core.

The most common form of contamination in our coring surveys has been distilled vacuum pump oil in the extracts. This arises during the freeze-drying process when the wet cores are subjected to a high vacuum to dry them prior to extraction. In many of our shallow-coring studies, this material was mistakenly interpreted as "condensate" largely because of its fluorescence characteristics. However, GC-MS analysis shows that this material is of low thermal maturity and has a similar composition in coring studies from widely differing parts of the world where very different source characteristics are expected. This constancy in composition is strongly suggestive of contamination. Despite attempts to remove this pump oil with a series of filters in the freeze-drying apparatus, it has proved to be difficult to prevent it from reaching the cores. We now air-dry all our cores to prevent this contamination and have found that our data more clearly show the presence of true petroleum in cores. Several other contaminants have been detected during coring surveys, mainly from contamination in the laboratory. It has proved necessary to carefully monitor each coring survey and the data produced by contractors to achieve reliable interpretations.

INTRODUCTION

Shallow coring involves detection of small quantities of petroleum in cores taken from the sea bottom. In recent years, BP has undertaken a large number of coring surveys from around the world (Figure 1) and has gained extensive experience in interpretation of data from shallow coring. In basins such as the Gulf of Mexico, the level of seepage is high and thus large quantities of petroleum are commonly observed in cores. However, in most basins, we have found that seepage levels are low, such that the levels of petroleum found in core extracts are very small. This means that geochemical techniques for detecting petroleum are potentially prone to problems with background contamination either during the core collection phase or in the laboratory. It has been our experience that most of BP's coring surveys have experienced some sort of contamination problem.

Figure 1—A map showing the location of recent BP coring surveys throughout the world.

British Petroleum has used a number of contractors for its shallow coring work in recent years. All of the contractors used have had problems with contamination in one form or another, and so no one particular contractor can be blamed for these problems. The types of contamination we have experienced include

1. Freeze-drier pump oil distilled into cores in the laboratory.
2. Contamination from impure solvents.
3. Contamination from plastic gloves used in the extraction stage.
4. Carry-over of high-performance liquid chromatography (HPLC) standards into core extracts during saturate/aromatic separation prior to gas chromatography–mass spectrometry (GC-MS) analysis.
5. Contamination of gas chromatogram samples by septum caps on sample bottles.

All these contaminants are common problems during geochemical analysis and are well-known to most geochemists. However, one contaminant is unique to shallow coring—freeze-drier pump oil. This oil has been the largest single contamination problem we have experienced from shallow coring. It can be difficult to interpret and to eradicate.

FREEZE-DRIER PUMP OIL

Freeze-drying has been used by several contractors to dry cores prior to extraction with solvent in the laboratory. The process involves subjecting cores to a vacuum in which water in the cores is separated in a cold trap. The vacuum is created by a rotary pump that often contains a petroleum-based oil. It appears that under these condi-

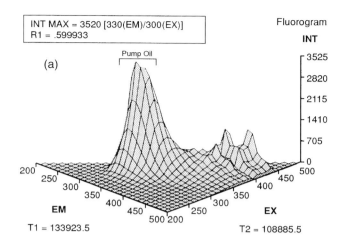

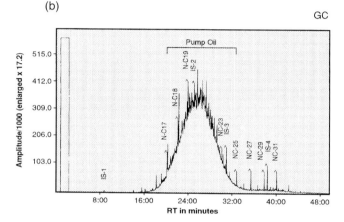

Figure 2—(a) A fluorogram and (b) a gas chromatogram of a core extract contaminated with pump oil.

tions, and despite several filters placed in-line on the vacuum system, a small but significant quantity of pump oil gets into the cores. This material effectively distills into the core. The characteristics of the distillate most likely depend on the original pump oil, the vacuum used, and the temperature of the pump.

Fluorescence has been one of the main tools for detecting petroleum in core extracts (see Barwise and Hays, this volume, for an explanation of the parameters measured from fluorescence). Pump oil fluoresces intensely, and so core extracts contaminated with pump oil have a characteristic fluorogram with a Max Ex/Max Em at about 300/330 nm ± 10 nm, as shown in the fluorogram in Figure 2a. Using fluorescence alone, it is difficult to spot the presence of pump oil in core extracts. Gas chromatographic analysis, however, clearly shows the presence of the pump oil. A superficial look at such a gas chromatogram hump may lead to an interpretation of the presence of biodegraded oil. However, close inspection reveals that the pump oil distillate forms a symmetrical hump atypical of biodegradation, which usually results

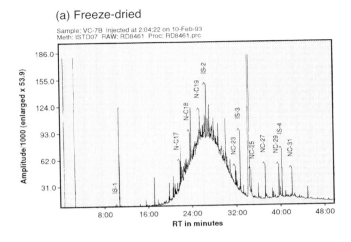

(a) Freeze-dried

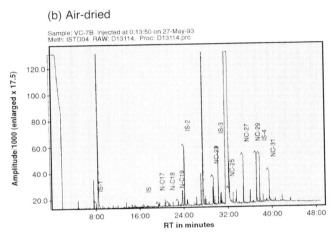

(b) Air-dried

Figure 3—A comparison of two gas chromatograms of extracts from the same core in which (a) freeze-drying and (b) air-drying had been used.

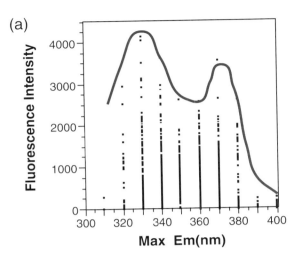

(a)

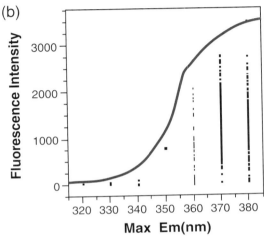

(b)

Figure 4—A comparison of the distribution of maximum fluorescence intensity versus wavelength of maximum emission (Max Em) for a series of cores that had been (a) freeze-dried and (b) air-dried.

in an asymmetric hump. The hump is relatively devoid of normal alkanes and has a maximum intensity at about the same retention time as C_{20} *n*-alkane. The lack of *n*-alkanes occurs because of the denormalization process used in producing lubricating oils in a refinery.

The only way we have found to avoid the pump oil contamination is to air-dry core samples instead of freeze-drying them. The example depicted in Figure 3 shows the difference between the two techniques. The gas chromatogram in Figure 3a shows a core that was originally freeze-dried and has a large hump. The gas chromatogram in Figure 3b shows the same core, but this time the sample was air-dried. As can be seen, the sample on the top was massively contaminated compared to the sample on the bottom.

The effect of pump oil contamination on our coring surveys has been significant. Because fluorescence was used to interpret many surveys, the pump oil was easy to misidentify as mature petroleum condensate. For exam-

ple, in one survey (Figure 4a), a plot of fluorescence intensity versus maximum emissions (Max Em) revealed a bimodal plot for freeze-dried samples. The most intense fluorescence observed from the core extracts was about 4000 units at a Max Em of 330 nm. A smaller secondary maximum was seen at a Max Em of 370 nm. To examine the effect of the pump oil contamination, aliquots of the cores were then air-dried and reextracted. Fluorescence data on the fresh extracts (Figure 4b) showed that the 330 nm peak had almost completely disappeared, indicating that this peak was entirely due to pump oil contamination. The second peak at 370 nm was still evident, but there was a subtle shift in the average Max Em to between 370 and 380 nm. This indicates that even when there was less contamination in some of the originally freeze-dried cores, there was still enough to affect the average Max Em value.

The geochemical interpretation of such contaminated data is radically altered because of the pump oil contamination, as shown in comparative maps in Figure 5. If a

(a)   Freeze Dried Cores
Contaminated with Lube Oil
Mixed low and high R1

Latitude

(b)   Oven-Dried Cores
No Lube Oil contamination
Mainly high R1

Latitude

◇	R1 < 1.5 "High API fluids"
△	R1 > 1.5 "Low API fluids"
▪	No Fluorescence data

fluorescence parameter R_1 value of 1.5 or less is taken to mean the presence of thermally mature light oil in cores, then the interpretation of the contaminated dataset would be very different from the same cores that had been air dried. Since the fluorescence technique is unable to distinguish thermogenic oil from contaminant, other techniques such as gas chromatogram and GC-MS must be used to confirm the presence of petroleum. This shows that extreme care must be taken to obtain quality analytical data in surface geochemical studies to avoid expensive mistakes in interpretation.

CONCLUSIONS

Coring surveys demand good quality analytical control, otherwise the data are worthless. Sample blanks must be used throughout the analytical procedure and must be checked diligently. Over-reliance on an analytical technique such as fluorescence, which cannot distinguish contamination from real petroleum, is dangerous.

Figure 5—A comparison of the mapped distributions of R_1 values in a survey for cores that had been either (a) freeze-dried or (b) air-dried.

Barwise, T., and S. Hay, 1996, Predicting oil properties from core fluorescence, *in* D. Schumacher and M. A. Abrams, eds., Hydrocarbon migration and its near-surface expression: AAPG Memoir 66, p. 363–371.

Chapter 28

Predicting Oil Properties from Core Fluorescence

Tony Barwise

BP Exploration
BP Research and Engineering Centre
Sunbury on Thames, U.K.

Steve Hay

BP–Statoil R+D Alliance
Trondheim, Norway

Present address:

BP Exploration Company (Europe) Ltd.
Dyce, Aberdeen, Scotland

Abstract

One method for detecting the presence of oil in shallow cores uses fluorescence of core extracts. Synchronous scanning of the extracts is performed using excitation wavelengths of 250–450 nm monitoring emission intensity over the same wavelengths. Several parameters are recorded that describe the fluorescence characteristics. First, the intensity of maximum emission is recorded along with the wavelengths at which this maximum was obtained: Max Ex is the excitation wavelength and Max Em is the emission wavelength. Second, a ratio (R_1) of emission intensity at 320 and 360 nm arising from excitation at 270 nm is recorded. To interpret these measurements in terms of oil density and maturity, fluorescence spectra have been compiled from about 130 oils representing a wide range of source rock types and thermal maturities from basins around the world. A large data set has been gathered for North Sea oils (about 56) which all have a marine siliciclastic source but vary greatly in thermal maturity. Several naturally biodegraded oils from the North Sea and other basins are included in the study. Bulk oil data and molecular parameter data have been acquired for many of the oils.

The prime effect on oil fluorescence characteristics is the thermal maturity of an oil. Oil fluorescence depends on the quantity and type of aromatics in oils, and since aromatics decrease with increasing thermal maturity, oil fluorescence intensity also decreases substantially. A light 50°–60° API oil from the North Sea is about 10 times less fluorescent than a 20°–30° API oil from the North Sea. There appears to be little effect of source type on oil fluorescence properties, but biodegradation may alter fluorescence. Shallow cores appear to have a constant background fluorescence signature with a Max Ex/Max Em of 320/370 nm. We attribute this to recent organic matter, but further work is needed to be certain of its origin. This background makes fluorescence an unreliable tool for detecting microseepage in shallow cores.

INTRODUCTION

A major goal of seepage interpretation is to be able to predict the likely phase and composition of reservoired petroleum at depth in a basin based on extracts from cores taken at the surface. During seabed coring surveys, large numbers of cores are often taken, and a cheap, reliable tool is required for screening. The first step in the screening process involves identifying which cores may contain thermogenic petroleum. Several analytical techniques have been applied to core extracts to achieve this goal, with varying degrees of success. One technique that has been used in recent years to estimate the quantity and type of petroleum in core extracts is fluorescence (e.g.,

Brooks et al., 1983, 1986). Cores are dried and extracted with solvent, and the extracts are analyzed using fluorimetry. The resultant fluorograms must be interpreted to determine petroleum type and quantity, and thus it is necessary to have a well-understood and calibrated technique to do this successfully.

The purpose of our study is to attempt to establish the relationships between oil properties and oil fluorescence such that the fluorescence technique can be calibrated and used with confidence. Fifty-six oils from the North Sea have been analyzed using the fluorescence technique, and data derived from fluorescence have been compared with basic oil properties, such as oil density, and with molecular maturity data. This North Sea dataset comprises oils

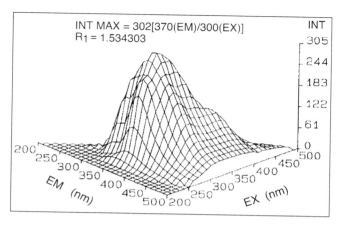

Figure 1—A three-dimensional fluorogram from a typical oil.

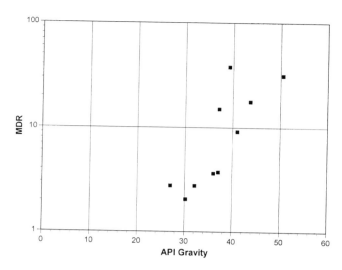

Figure 2—Plot of API gravity versus 4-methyl/1-methyl dibenzothiophene (MDR) parameter for selected North Sea oils.

derived largely from one source type, and hence the differences in properties are thought to be mainly due to variation in the average temperature of expulsion from the source rock. Several of the oils chosen were known to have undergone varying degrees of biodegradation.

An additional series of oils were chosen for examination because they represent a broad spectrum of source types, making it possible to investigate whether fluorescence has any dependence on source type.

METHODS AND SAMPLES

All oils were analyzed by diluting them to a standard concentration (10 ppm oil by weight in hexane). A standard concentration was chosen because it was found that fluorescence intensity and characteristics are sensitive to concentration effects. Too high a concentration leads to quenching of the fluorescence and a shift in the wavelengths of maximum emission. Four standard parameters were calculated for each oil: intensity, Max Ex, Max Em, and R_1. These parameters have been described previously (e.g., Brooks et al., 1986). The oil samples were briefly irradiated with light scanning from 250 to 500 nm at 10-nm intervals in a spectrometer. The fluorescence emission spectrum wes recorded for each excitation wavelength, again scanning from 250 to 500 nm. In this way, a three-dimensional spectrum is built up; a typical spectrum from an oil is shown in Figure 1. The maximum intensity of emission is recorded along with the emission wavelength (Max Em) and the excitation wavelength that caused this emission (Max Ex). R_1 is the ratio of emission at 360 nm to emission at 320 nm when excitation at 270 nm is used. This parameter attempts to show variation in the shape of fluorescence spectra. The aromatics content is determined using high-performance liquid chromatography (HPLC) in which oils were separated into three fractions: saturates, aromatics, and polars. The aromatics content is expressed in weight percent (wt. %) of the total oil.

As stated above, 56 oils from the North Sea were chosen for this study. Table 1 lists the basic properties of these oils and the fluorescence data obtained from them. Several of the oils were analyzed using gas chromatography–mass spectrometry (GC-MS) to measure molecular maturity parameters. Four biodegraded oils from the North Sea were also investigated, and data from these are given in Table 1. Data for oils from a worldwide oil survey are shown in Table 2.

NORTH SEA OILS

Maturity

The undegraded oils from the North Sea vary in density from about 30° API to as high as 55° API. This variation results mainly from the varying temperatures of expulsion of the oils from their source rock. At higher temperatures, oil to gas cracking causes a change in oil composition and a lowering of average oil density. While other nonthermal effects such as phase separation could cause variation in oil density, the maturity effect can be illustrated by plotting a thermal maturity parameter such as the methyl dienzothiophene (MDR) ratio versus oil gravity. MDR increases with increasing thermal maturity for sediment extracts and oils (Radke, 1987). A plot of API gravity versus MDR for several North Sea oils confirms that the high API gravity oils are indeed higher in thermal maturity than the low API gravity oils (Figure 2).

This change in oil properties with increasing thermal maturity results in a variation in the relative amount of aromatic material in oils. Low thermal maturity oils with low API gravity have much more aromatic material than high thermal maturity oils with high API gravity

Table 1—Fluorescence Properties and Basic Oil Data for Oils from the North Sea

Sample No.	Oil Class	API Gravity	Sulfur (wt. %)	Aromatics[a] (wt. %)	Max Intensity (per µg oil)	Max Ex (nm)	Max Em (nm)	R_1	A3[b]	MDR[b]
N1	B		0.3	34	57.1	300	360	2.26	0.48	
N2	B	27	0.8	43	53.6	300	370	2.56	0.42	2.74
N3	B	54	0	25	6.3	290	340	0.95		
N4	B	36	0.1	27	53.7	300	360	2.10	0.77	
N5	B	30.2	0.6	41	81.6	330	380	3.15	0.22	2.03
N6	B	32.1	0.4	31	49	300	370	2.46	0.38	2.73
N7	B	10.6	1.1		59.4	330	390	4.62		
N8	B	51	0	18	12.4	290	340	0.76	0.88	
N9	B	44.3	0.1		17.3	290	350	1.53	0.80	
N10	B	46.6	0.1	1	13.6	290	350	1.35	0.69	
N11	B	20	0.5		79.5	330	380	2.68		
N12	B	35	0.2		46.6	300	370	2.38	0.40	
N13	B	35.5	0.2	29	43.1	300	370	1.92		
N14	B	36.3	0.3	30	52.7	320	380	2.30	0.65	
N15	B	38	0.2		60.1	300	380	2.27	0.70	
N16	B	40	0.2	30	35.9	300	360	1.87	0.37	
N17	B	36	0.3		60.5	300	370	2.35	0.68	
N18	B	42.5			26.3	290	350	1.40	0.95	
N19	B	45.8	0	25	15	290	350	1.48		37.6
N20	B	39.2	0.1	15	37.3	300	370	1.63		
N21	B	41	0.1	22	22.9	290	350	1.17		
N22	B	48	0	18	11.1	290	340	1.43	0.84	
N23	B	52	0.2	13	7.4	290	340	1.28	0.84	
N24	B	41	0.1	19	24.4	290	360	2.00	0.52	
N25	B	60	2.3		12.7	280	330	0.79		
N26	B	44.2	0.1	30	59.3	300	370	2.36		
N28	B	19		34	69.1	300	370	2.29	0.49	
N29	B	37	0.3	35	41.8	300	380	2.32	0.58	3.71
N30	B	34	0.4	40	63.1	330	390	2.76	0.82	
N31	B	36		32	58.5	320	380	2.41		
N32	B	50.5	0	20	27.5	290	340	0.94	1.00	31.2
N33	B	36	0.2		36.8	300	370	2.28	0.66	
N34	B	38	0.1	25	30.7	300	360	1.74	0.66	
N35	B	26	0.2	27	45.5	300	370	2.26	0.75	
N36	B	40	0.1	27	38.4	300	370	2.00	0.56	
N37	B	20.2	0.4	41	113	300	360	2.23	0.41	
N38	B	14.7		38	92.3	320	380	3.13	0.31	
N39	B	47.7		18	16.2	290	330	1.04	0.42	
	B	49.7	0	39					0.77	
N40	B	36	0.16	28	75	320	370	2.36		
N41	B	37.1	0.28	26.2	49	300	370	1.84		3.6
N42	B		0.62		98	330	390	2.95		15
N43	B		0.18		50	300	380	2.02		3.83
N44	B		0.13		29	300	370	2.11		12.9
N45	B	41	0.05	12	22	300	350	1.39		9.12
N46	B	44	0.14		30	300	380	2.08		
N47	B		0.27	37.6	56	300	370	2.96		7.6
N48	B	22.5	0.13		40	310	370	2.28		
N49	B		0.13		29	290	380	1.91		8.5
N50	B		0.04	8	10	300	370	1.83		
N51	B	45	0.05	10.5	20	300	370	1.67		
N52	B	43	0.03	10.6	21	300	370	1.46		17.6
N53	B	43.7	0.04		22	300	370	1.58		
N54	B				102	300	360	2.43		
N55	B	46.6		17.7	29	300	370	2.46		
N60	B				92	300	370	2.13		

[a]Aromatics content is defined by HPLC and is in wt. % of oil.

[b]A3 and MDR are molecular parameters measured by GC-MS. A3 is the ratio of C_{20} triaromatic sterane/(C_{20} + C_{28} triaromatic sterane). MDR is the ratio of 4-methyl/1-methyl dibenzothiophene.

Table 2—Fluorescence Properties and Basic Oil Data for Oils from a Variety of Source Rock Types[a]

Sample No.	Country/Location	Oil Class	API Gravity	Sulfur (wt. %)	Aromatics (wt. %)	Max Int. (per μg oil)	Max Ex (nm)	Max Em (nm)	R_1	A3	MDR
E1	Egypt	A	13.7	5.20	54.4	67.4	340	400	4.62	0.23	0.67
E2	Egypt	A	21.1	2.70		28.2	330	370	3.38		
E3	Egypt	A	36.9	7.00		52.7	330	380	2.72		
E4	Egypt	A	21.6	7.00	41.8	96.2	330	390	3.99	0.31	
E5	Egypt	A	17.8	3.80	37.4	71.9	330	380	4.10	0.22	1.95
E6	Egypt	A	29.5	1.60	47.4	74.7	330	390	3.43	0.22	
E7	Egypt	A	30.5	1.70	45.1	79	330	390	3.46		
E14	Egypt	A	32	0.90						0.50	2.29
E8	Egypt	A	34	0.50		60.4	290	360	2.96	0.47	
E9	Egypt	A	12.3			85.4	330	380	3.47	0.40	
E10	Egypt	A	19.8	2.50		54.5	330	390	3.45	0.23	
E11	Smackover, U.S.A.	A					290	340	0.90		40.3
E12	Smackover, U.S.A.	A					370	420	3.81		1.07
E13	Smackover, U.S.A.	A					290	360	0.93		218
I1	Indonesia/Malaysia	C/D/E	30.5			66.2	290	340	1.19		
I2	Indonesia/Malaysia	C/D/E				153	300	360	2.43		
I3	Indonesia/Malaysia	C/D/E				65.6	300	360	1.87		
I4	Indonesia/Malaysia	C/D/E	19.5			47	320	380	2.14		
I5	Indonesia/Malaysia	C/D/E				45.3	300	360	1.63		
I6	Indonesia/Malaysia	C/D/E				22.6	300	360	1.57		
I7	Indonesia/Malaysia	C/D/E	30.8			69.3	330	390	3.28		
I8	Indonesia/Malaysia	C/D/E	38.1			42	290	340	1.69	0.41	4.81
I9	Indonesia/Malaysia	C/D/E				35.8	300	410	3.09		
I10	Indonesia/Malaysia	C/D/E				98.4	290	350	1.34		
I11	Indonesia/Malaysia	C/D/E				87.6	290	350	1.90		
I12	Indonesia/Malaysia	C/D/E				56.1	290	340	1.72		
I13	Indonesia/Malaysia	C/D/E			54	75.6	290	340	1.41		
I14	Indonesia/Malaysia	C/D/E	28		26	76	320	380	2.64	0.35	5.07
I15	Indonesia/Malaysia	C/D/E	42			86.6	330	370	2.66	0.35	5.10
I16	Indonesia/Malaysia	C/D/E				30.2	300	370	1.53		
I17	Indonesia/Malaysia	C/D/E				73.2	290	350	1.45		
I18	Indonesia/Malaysia	C/D/E				54.2	290	340	1.43		
I19	Indonesia/Malaysia	C/D/E		0.10	43	66.8	290	330	1.44	0.45	6.30
A1	Gippsland Basin, Australia	C/D/E	53.3	0.10	27	11.9	290	340	1.52	0.51	2.16
A2	Gippsland Basin, Australia	C/D/E	41.1	0.10	15	68.9	290	340	1.33	0.62	7.70
A3	Gippsland Basin, Australia	C/D/E	40.3	0.10	9	39.6	290	330	1.41	0.60	5.00
A4	Gippsland Basin, Australia	C/D/E	44.6	0.10	13	36.7	290	340	0.97	0.74	4.30
A5	Gippsland Basin, Australia	C/D/E	42.3	0.10	12	26.3	290	330	0.78	0.65	4.10
A6	Gippsland Basin, Australia	C/D/E	42.7	0.10	12	61.1	290	340	1.21	0.77	7.70
T1	Thailand	C/D/E				7.8	280	340	0.86		
T2	Thailand	C/D/E	23.4	0.30	28	48.7	330	390	3.34		
T3	Thailand	C/D/E	26	0.20	26	81	330	390	4.08	0.84	1.44
T4	China	C/D/E				153	330	390	4.73		
T5	China	C/D/E	29.6	0.20	21	60.7	330	390	3.02	0.20	
T6	China	C/D/E				15.3	290	350	1.04		
T7	China	C/D/E				209	330	380	5.13		
C1	China	C/D/E			22	55.2	300	380	2.97		
C2	China	C/D/E				49.6	300	370	2.23		
C3	China	B	46.9	0.00		14.3	280	330	0.57		
C4	China	C/D/E	47.8		11	10.3	280	340	1.43	1.00	
C5	China	C/D/E	35.7	0.10	21	20.1	300	350	1.96		
C6	China	C/D/E	43.3	0.00	15	23.9	300	350	1.86		
C7	China	C/D/E	28	0.20		45.4	300	370	2.53		
C8	China	B				8.5	300	370	2.43		
C9	China	C/D/E	25.1	0.10	12	31	300	380	3.25		
C10	China	B				116	310	370	2.49		
C11	China	B	44.2			34.4	290	340	1.18		
C12	China	C/D/E	20.2	0.20		39.7	330	390	4.07		
C13	China	C/D/E				40.4	300	380	2.94		
C14	Nigeria	C/D/E	28.9	0.20	27.2		290	340	0.99		
C15	Nigeria	C/D/E	20.2	0.40	43.7		290	360	1.52		
C16	Nigeria	C/D/E	20.8	0.30	42.3		300	360	1.52		
C17	Nigeria	C/D/E	36.4	0.10	27.6		290	330	0.82		

[a]See Table 1 for definitions.

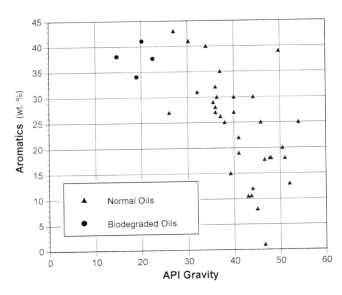

Figure 3—Aromatics content (wt. %) plotted against API gravity for normal and biodegraded North Sea oils.

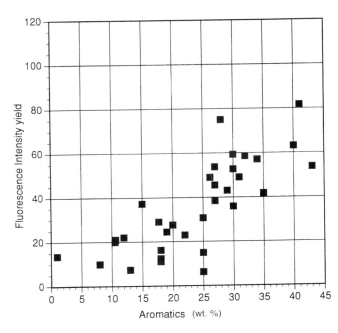

Figure 4—Aromatics content (wt. %) plotted against fluorescence intensity yield (units/µg oil) for North Sea oils.

(Figure 3), and the fluorescence yield from an oil is directly proportional to its aromatics content (Figure 4). A consequence of this variation is that 20° API gravity oils exhibit about a tenfold greater oil fluorescence intensity yield per unit weight of oil than a light 60° API oil mainly because the latter has far less aromatics than the former. Heavy oils are therefore more likely to be detected than light oils using fluorescence.

Both Max Ex and Max Em vary with increasing oil gravity and increasing thermal maturity, although significant scatter is seen. For example, Max Em plotted against API gravity (Figure 5) shows that the average Max Em drops about 50 nm as gravity increases from 20° to 60° API. However, there is considerable scatter: oils with a Max Em of 370 nm range from 25° to 45° API. Therefore, Max Em is not a very accurate or reliable indicator of an oil's API gravity.

A much better indicator is R_1 (Figure 6). R_1 drops from about 3 to 0.5 between 25° and 60° API. This plot has much less scatter than the other plots, and a least-squares plot through the data gives the following predictive relationship for API gravity with an r^2 fit of 0.66:

$$\text{API gravity} = -10.1 \times R_1 + 59.8$$

The fact that R_1 is dependent on maturity can be seen by crossplotting R_1 against a molecular maturity parameter, A_3, which is based on aromatized steranes (Figure 7). The value of A_3 increases from about 0.3 at low maturity to 1.0 at high thermal maturity (Mackenzie, 1984) and is a useful oil maturity parameter. R_1 clearly decreases with increasing values of A_3. However, A_3 can be affected by physical separation of oils in the subsurface, which could account for some of the scatter seen in Figure 7. The maturity dependence is also demonstrable from a crossplot of

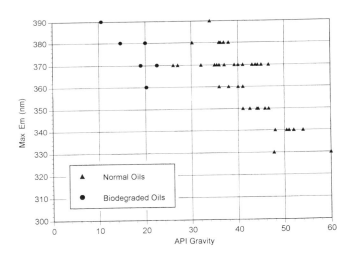

Figure 5—API gravity plotted against fluorescence parameter Max Em (nm) for North Sea oils.

the dibenzothiophene parameter MDR versus R_1 (Figure 8). MDR increases and R_1 decreases with increasing thermal stress, as clearly shown in this crossplot.

Biodegradation

While the above relationships between oil fluorescence and oil properties look encouraging, it is necessary to consider that most petroleum reaching the sediment surface undergoes varying degrees of biodegradation.

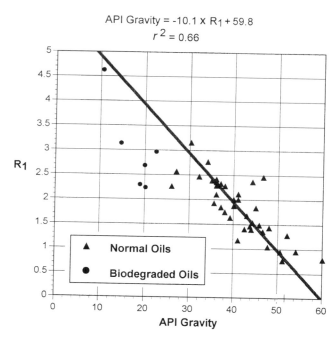

API Gravity = -10.1 × R_1 + 59.8

$r^2 = 0.66$

Figure 6—API gravity plotted against fluorescence parameter R_1 for North Sea oils.

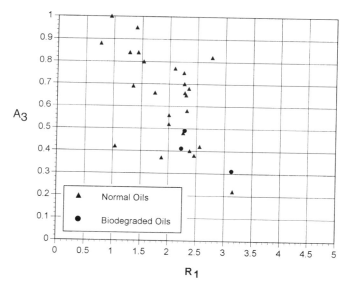

Figure 7—Fluorescence parameter R_1 plotted against A_3 (C_{20} triaromatic sterane/C_{20} + C_{28} triaromatic sterane) for normal and biodegraded North Sea oils.

This causes chemical and physical alteration of oil composition and can alter oil fluorescence properties. Four oils from subsurface accumulations in the North Sea that have undergone varying degrees of degradation were examined. The most degraded oil has undergone extreme alteration, leaving a 10° API fluid in the reservoir. Plots of Max Em and R_1 versus API gravity for the degraded oils (Figures 5, 6) show that degradation appears to alter both oil gravity and fluorescence properties. The most degraded oil exhibits the highest R_1 value and has the highest Max Em value seen for all North Sea oils. This implies that biodegradation may alter the aromatic composition of oils, which is known to be the case for these oils. GC-MS analysis of this most degraded oil shows that many of the aromatics have indeed been altered, with removal of parent aromatics such as naphthalene and phenanthrene and several of the alkylated homologs of these parent aromatics.

One difficulty with using the previously described crossplots is that it is difficult to estimate what the original thermal maturity of the undegraded oils would have been prior to the degradation process. It is thus impossible to know what the original fluorescence properties of the degraded oils would have been before degradation. If the Max Em and R_1 values are to be believed, then the oils are of low thermal maturity. Alternatively, biodegradation may have caused an increase in R_1 and Max Em, making the oils appear less mature. Evidence that the biodegraded oils may be of low thermal maturity comes from the A_3 versus R_1 crossplot (Figure 7), which shows that they are toward the low end of the maturity scale.

We conclude that the fluorescence characteristics of biodegraded seeps should be used cautiously when attempts are made to relate them to unaltered reservoired petroleum in the subsurface.

MATURITY OF WORLDWIDE OIL TYPES

A number of oils from a variety of source types (both marine and nonmarine) were examined. These oils vary from high-sulfur marine sources, such as from Egypt and the Gulf of Mexico, to nonmarine lacustrine and deltaic sourced oils from Thailand, Malaysia, China, Nigeria, and Indonesia. As those from the North Sea, some of these oils have been biodegraded, resulting in alteration of their physical properties. Table 2 presents a compilation of data from these oils.

Compositionally, the nonmarine oils in this dataset have a lower aromatics content than marine oils of the same oil density (Figure 9). Because of their lower thermal maturity, the high-sulfur carbonate oils have the highest aromatics contents, reaching up to 50%. The marine oils essentially form a continuum of decreasing aromatics content with increasing oil gravity, such that a 50° API gravity oil has about one-fifth the aromatics content of a high-sulfur 10° API gravity carbonate-derived oil.

Overall, it appears that oils derived from nonmarine and marine source rocks have similar yields of fluorescence per unit weight of oil. The nonmarine oils show a similar decrease in aromatics content with increasing gravity, but at any particular gravity, they have a slightly lower aromatics content than oils from marine source

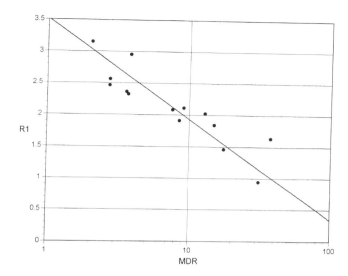

Figure 8—Fluorescence parameter R₁ plotted against molecular maturity parameter MDR for North Sea oils.

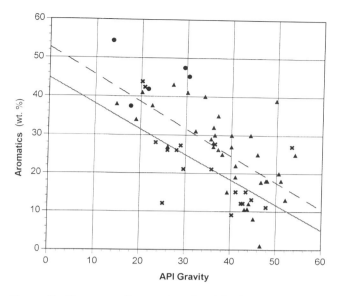

Figure 9—The aromatics content (wt. %) of oils from a worldwide survey plotted against API gravity. The dashed line shows a least-squares fit through the North Sea oils and the solid line shows the same for nonmarine oils. Key: ▲ = North Sea oils, ✖ = nonmarine oils, ● = marine carbonate-sourced oils.

rocks. A least-squares fit for both the nonmarine and North Sea oils is shown in Figure 9. This lower abundance of aromatics for nonmarine oils seems to be counterbalanced by a higher fluorescence intensity yield per unit weight of aromatics, as shown in Figure 10. The nonmarine oils may be more fluorescent because of their less abundant higher molecular weight polars and asphaltenes compared to marine oils. This may result in less internal quenching and therefore a higher fluorescence yield.

As with the North Sea oils, it can be stated that the densities of most of the oils in the global dataset are primarily a result of the temperatures at which the oils were expelled from their source, although nonthermal effects such as phase separation and biodegradation play an important secondary role. There is a good general fit between the R_1 parameter measured from the oils and oil density (Figure 11). High-sulfur carbonate-derived oils exhibit the highest R_1 values because of their low thermal maturity, as seen in the R_1/API gravity crossplot. The marine high-sulfur and medium-sulfur oils essentially form a maturity continuum, and a reasonably good correlation exists between R_1 and API gravity following the line defined from Figure 6.

Nonmarine oils follow a trend similar to that of the marine oils for the API versus R_1 correlation, although there are a few exceptions. However these exceptions are largely due to the effects of biodegradation. As with the North Sea oils, the biodegraded oils have much lower gravities than normal oils, but their R_1 values are affected less than oil gravity. It is tempting to assign original oil gravity from the fluorescence data, and indeed, the minimum gravity should be assignable using R_1 since biodegradation should only increase R_1. Therefore, R_1 is potentially a good indicator of subsurface oil gravity and oil properties for marine and nonmarine oils, providing the oils are not too heavily biodegraded.

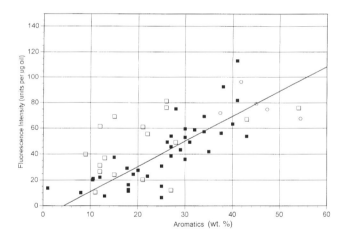

Figure 10—The fluorescence intensity yield (units/μg oil) plotted against aromatics content (wt. %) for oils from a worldwide survey. A solid line shows a least-squares fit through the North Sea oils. Key: ■ = North Sea oils, ○ = high-sulfur oils, ❏ = nonmarine oils.

SEDIMENT CORES

The previous oil data suggests that fluorescence should be a useful technique for estimating oil density and inferring something about its thermal maturity. However, for core extracts taken from the near subsurface, several other factors must be considered before such a tool can be used with confidence.

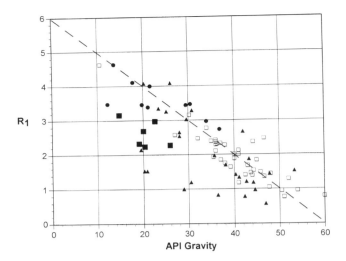

Figure 11—A crossplot of API gravity against fluorescence parameter R_1 for oils derived from a variety of source types. The dashed line is the North Sea API gravity versus R_1 correlation for nondegraded oils from Figure 6. Key: ● = carbonate-sourced oil, ▢ = North Sea clastic marine shale, ▲ = nonmarine lacustrine/deltaic/paralic sources, ■ = biodegraded North Sea oils.

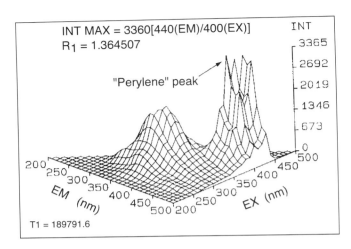

Figure 12—A fluorogram from background recent organic matter (ROM) in a typical shallow core.

The primary consideration is that oils change in composition both physically and chemically during migration from a reservoir to the surface. Physical separation of fluids occurs as the petroleum migrates vertically to lower temperature and pressure regimes, resulting in gas separation and an altered molecular weight distribution in the fluids. A large bacterial alteration effect is usually seen resulting from extensive biodegradation as the petroleum migrates through low-temperature sediments. Unless a seep is very fresh and has travelled quickly to the surface from a subsurface accumulation, surface seeps are invariably biodegraded. This degradation varies in extent, but extensive degradation can make it difficult to be confident of the original fluid composition in the subsurface. Such a drastic alteration possibly changes fluorescence properties, as shown in the oil examples already discussed.

A secondary consideration in using core extracts is that most recent sediments contain significant amounts of indigenous organic matter in yields that are comparable and often exceed the levels of microseepage found in surface cores. This organic matter contains aromatic material, and in our experience with coring surveys in basins from around the world, there appears to be a constant fluorescence background in most of the cores even where other analyses indicate the lack of thermogenic petroleum in the core. This background may be attributable to recent organic matter (ROM) in the core extracts. An example of this background fluorescence is shown in Figure 12. In addition to the perylene fluorescence seen above 400 nm, fluorescence with a Max Ex/Max Em of about 320/370 nm is invariably seen in most of the core extracts we have

examined. In fine-grained sediments such as muds, the ROM signature can be intense and the fluorescence intensity can reach up to 6000 intensity units. In these ROM-dominated extracts, there is invariably little evidence for any thermogenic petroleum. In contrast, we have encountered several cores that have low fluorescence yields but do contain mainly thermogenic petroleum.

The origin of this fluorescence remains unknown at present, but it could come from several sources. One possibility is that ROM contains significant reworked mature organic matter due to erosion and reworking of older sediments into present-day sediments. However, the universal appearance of this signature in all of the basins we have examined and the association with fine-grain muds leads us to conclude that it is due to indigenous organic matter in the sediments themselves. More work is required to be certain of this.

Since a Max Ex/Max Em of 320/370 nm is similar to that from a low-maturity oil, this background fluorescence implies that one cannot distinguish ROM from low-maturity oil in core extracts using the fluorescence technique. This is especially true for microseepage where the levels of petroleum in cores is at or below the level of ROM in the same cores. Other techniques such as gas chromatography or GC-MS must be used to confirm or deny the presence of thermogenic petroleum in a core extract. Fluorescence can only be used with confidence when a core is heavily oil-impregnated, which can swamp out the ROM effect. It is not recommended for detection of microseepage.

CONCLUSIONS

Calibration of oil fluorescence with data from North Sea oils shows that there is a consistent relationship between oil composition and oil fluorescence. These changes are largely the result of variation in the thermal maturity of the oils. The best parameter to use for predicting oil gravity is R_1. Biodegradation possibly causes an alteration in oil properties and in oil fluorescence, such that biodegraded oils appear thermally less mature from the fluorescence properties than they really are. The relationship between oil composition and fluorescence characteristics appears to be largely independent of source rock type. The combined effects of subsurface degradation and the ubiquitous presence of background fluorescence implies that the fluorescence technique can only be used with confidence when cores contain significant oil (macroseeps) and not when low levels of seepage (microseeps) are present.

REFERENCES CITED

Brooks, J. M., M. C.Kennicutt, L. A. Barnard, G. J. Denoux, and B. D. Carey, 1983, Applications of total scanning fluorescence to exploration geochemistry: Proceedings of the 15th Offshore Technology Conference, Houston, Texas, v. 3, p. 393–400.

Brooks, J. M., M. C.Kennicutt, and B. D. Carey, 1986, Offshore surface geochemical exploration: Oil and Gas Journal, October, p. 66–72.

Mackenzie, A. S., 1984, Application of biological markers in petroleum geochemistry, *in* J. Brooks and D. Welte, eds., Advances in petroleum geochemistry: London, Academic Press, v. 1, p. 115–214.

Radke, M., 1987, Organic geochemistry of aromatic hydrocarbons, *in* J. Brooks and D. Welte, eds., Advances in petroleum geochemistry: London, Academic Press, v. 2, p. 141.

Thrasher, J., D. Strait, and R. Alvarez Lugo, 1996b, Surface geochemistry as an exploration tool in the South Caribbean, *in* D. Schumacher and M. A. Abrams, eds., Hydrocarbon migration and its near-surface expression: AAPG Memoir 66, p. 373–384.

Chapter 29

Surface Geochemistry as an Exploration Tool in the South Caribbean

Jane Thrasher

BP Exploration
Research and Engineering Centre
Sunbury-on-Thames, U.K.

Present address:
Sir Alexander Gibb & Partners
Reading, U.K.

David Strait

BP Exploration Company (Colombia) Ltd.
Santafe de Bogota, Colombia

Present address:
BP Exploration (Alaska)
Anchorage, Alaska, U.S.A.

Ricardo Alvarez Lugo

BP Exploration Company (Colombia) Ltd.
Santafe de Bogota, Colombia

Abstract

The risk on petroleum charge in acreage in the South Caribbean, offshore Colombia, has been assessed using integrated seepage detection techniques, including seabed coring. Eight offshore wells have been drilled but have only discovered limited volumes of dry gas. Oil seepage is well known onshore in the basin, and mud diapirs can be seen on seismic profiles offshore. Seepage detection and sampling techniques can be used to better define exploration risk by determining if the oil seepage extends into the offshore area, and if so, what is the most likely source of the oil.

An airborne survey located two main areas of possible offshore oil seepage and was followed by a shallow coring survey that retrieved oil-bearing cores from an offshore mud volcano. Other cores collected on mud diapirs and shallow faults identified from seismic did not retrieve readily identifiable oil. The oil extracted from the offshore mud volcano core was correlated with oil samples collected from onshore seepage sites. Biomarker analysis led to identification of two oil families in onshore seeps, one mainly from a Cretaceous marine carbonate and the other from a Tertiary mixed marine–terrestrial source. Tertiary oils are common in northwestern South America, but are seldom volumetrically important; a Cretaceous marine carbonate source is likely to have much more potential. The offshore seep correlated with the Tertiary oil.

The other shallow cores did not contain readily identifiable oil, but appeared to have abundant recent organic material (ROM) masking any potential thermogenic signature on gas chromatography or gas chromatography–mass spectrometry (GC-MS) results. Even cores with sufficient oil to leave a slick on the sea surface during retrieval were overprinted by ROM on the GC-MS hopane and sterane traces. High resolution GC-MS, however, did show that the cores contained aromatic biomarkers in the relative proportions expected in thermogenic oils but not in ROM. The thermal maturities derived from these biomarkers made a good match with the crudely modeled maturity of a middle Tertiary source rock. The extensive ROM overprint is probably due to the present-day depositional environment. Most of the mud diapirs and shallow faults identified on seismic profiles do not break the surface, but are covered by a veneer of recent sediments; any petroleum leaking into this is diluted by the abundant ROM.

The results of the seabed core analysis alone were ambiguous, but once integrated with geologic models and other seepage data, they improved the definition of the exploration risk on petroleum charge. Cores with a wide geographic spread offshore do contain seeped thermogenic petroleum, but this is most likely from a Tertiary source rather than the potentially more prolific Cretaceous source identified onshore.

INTRODUCTION

Between 1991 and 1993, British Petroleum held an Association Contract to evaluate the "Sur del Caribe" contract area of Colombia. The conditions of the contract required the acquisition of new seismic data and reprocessing of existing seismic profiles and the acquisition of surface geochemical data. The decision at the end of the contract period was either to drill an exploration well or to relinquish all interest in the acreage. The company strategy was to look for a significant new oil discovery with the potential for the area to be a future profit center.

The Sur del Caribe contract area covers the entire South Caribbean Basin of northwestern Colombia, from onshore out to the 1000-m water depth contour, with a total area of approximately 21,000 km² (Figure 1).

Geologic Setting

The South Caribbean has a complex tectonic setting, having been an active plate margin at the junction of several plates since the Cretaceous. The boundaries of the South Caribbean Basin were most likely strike-slip up to Eocene time, when the predominant structural style became compressional. The basin is now bounded on the west by the frontal thrust belt, the westernmost limit of tectonism, and on the east by the Montañas de Maria, which separate the South Caribbean Basin, floored by oceanic crust, from the Lower Magdalena Basin, floored by continental crust.

A schematic structural cross section of the basin (Figure 2) shows the Miocene (and older) deposits thrust westward, with Pliocene sediments filling basins created on the backs of the thrust sheets. Pleistocene sediments prograde over the structures and basins. The South Caribbean Basin is characterized by extensive mud

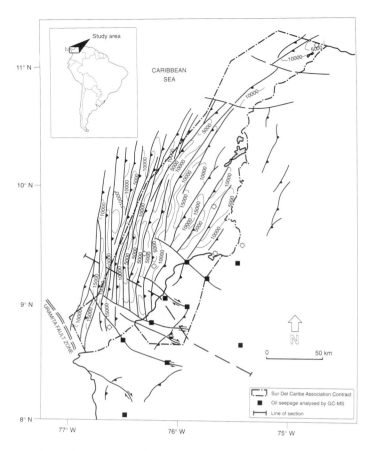

Figure 1—Map showing the location of the Sur del Caribe association contract area in northwestern Colombia. Onshore seepage sites analyzed by GC-MS are shown. Offshore contours indicate structure of top Miocene (?) reflector (contour interval is 5000 ft). Seabed cores were collected throughout the offshore area of the association contract.

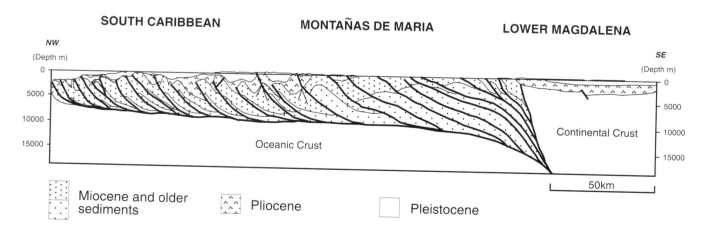

Figure 2—Schematic geologic cross section through the Sur del Caribe association contract area. See Figure 1 for location of line of section.

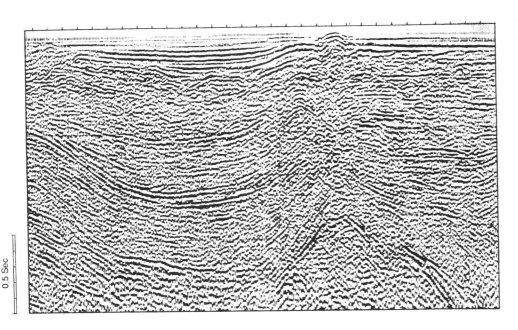

0.5 Sec

Figure 3—Seismic section showing surface-breaking mud diapir in the Sur del Caribe. Airborne surveys detected oil slicks down-wind of this structure on two occasions several weeks apart. (Section is about 8 km wide.)

diapirism, often formed on the footwalls of the thrust sheets. The mud appears to originate from the Miocene or older sections, perhaps at the level of detachment.

Potential play types in the offshore basin area include thrust-related structures in the Miocene with deep-water clastic reservoirs; structural-stratigraphic traps with Pliocene shelf and slope clastic reservoirs; and stratigraphic traps with slope sand reservoirs.

Source Risk

Key risks in the basin include reservoir presence and quality and source presence and quality. Historically, many oil and gas seeps have been recorded in the onshore area (e.g., Hedberg, 1980; Hovland and Judd, 1988), and gas-related features are known in the northern offshore area (Shephard, 1973; Vernette, 1989). Prior to this study, the extension of a working oil source into the offshore area and the nature of this source were unknown. Low-potential Tertiary marine siliciclastic source rocks are widespread in South America. However, to be commercially viable, it was considered that an oil province in the area would require the presence of a world-class source rock, such as the Cretaceous La Luna marine carbonate, which is the source of most of the oil in the Venezuelan Maracaibo Basin.

Surface geochemistry has the potential to reduce source risk in underexplored areas such as the South Caribbean ahead of the drill. In addition to proving the existence of a working petroleum system, geochemical analysis of seep oils can provide invaluable information about the likely origin and maturity of the seep, which can then be used to constrain geologic models.

A surface geochemical program, incorporating a regional offshore airborne survey, onshore seepage sampling, and seabed coring, was therefore carried out.

AIRBORNE SURVEY

The airborne survey was flown over a period of a few weeks in September and October 1991. An oil slick was identified at similar locations on two separate sorties, several weeks apart. The visible slick extended for about 6 km and was located downwind from a major shallow mud diapir structure observed on regional seismic profiles (Figure 3). Evidence for oil slicks was identified at other locations, but the observations were not repeated on subsequent flights. The presence of a repeating visible oil slick at a site where seepage is most likely to occur is good evidence for the presence of a working oil source rock in at least that part of the basin. Without samples, however, it is impossible to confirm that the slick is in fact a seepage slick or to infer the likely source of the oil.

ONSHORE SEEPAGE SAMPLING

Numerous hydrocarbon seeps have been documented in the onshore part of the Sur del Caribe area, with many gassy mud volcanoes recorded as well as black oil seeps. Samples were collected at a number of these seepage sites, and 10 oil samples were selected for detailed geochemical analysis. The positions of the seeps within the stratigraphic column are shown in Figure 4. Two seeps were collected in the Paleocene–Eocene section, and the other seeps are all from the upper Miocene or Pliocene. The geographic location of the seeps is shown in Figure 1. Gas chromatography (GC) showed most of the oils to be biodegraded, with few normal alkanes visible on the GC traces. The biomarker distributions, however, appeared little affected by biodegradation, and the oils can be grouped into two families.

Figure 4—Stratigraphic section of the onshore sequence, showing the stratigraphic location of seepages that were analyzed geochemically. (Stratigraphy from Geotec Ltda., 1991.)

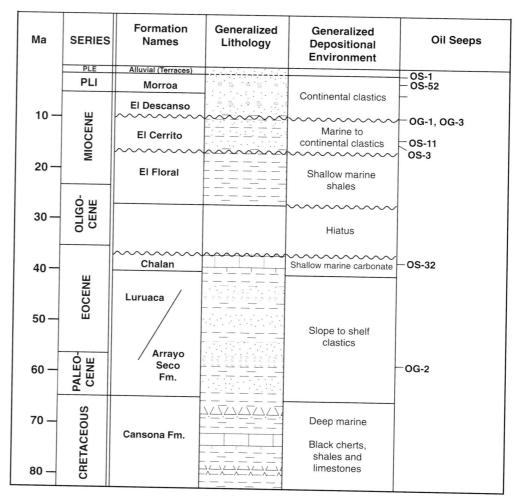

Ma	SERIES	Formation Names	Generalized Lithology	Generalized Depositional Environment	Oil Seeps
	PLE	Alluvial (Terraces)			OS-1
	PLI	Morroa		Continental clastics	OS-52
10	MIOCENE	El Descanso			OG-1, OG-3
		El Cerrito		Marine to continental clastics	OS-11
20		El Floral		Shallow marine shales	OS-3
30	OLIGOCENE			Hiatus	
40	EOCENE	Chalan		Shallow marine carbonate	OS-32
50		Luruaca		Slope to shelf clastics	
60	PALEOCENE	Arrayo Seco Fm.			OG-2
70	CRETACEOUS	Cansona Fm.		Deep marine	
80				Black cherts, shales and limestones	

The oils collected from seeps in the lower Tertiary (Figure 5) are rich in tricyclic terpanes and extended hopanes, with few diasteranes, typical of oils from marine carbonate source rocks. X-ray fluorescence elemental analysis on one of these oils showed it to be rich in vanadium and also relatively rich in sulfur and nickel. There is good correlation with the Cretaceous La Luna Formation oils of the Venezuelan Maracaibo Basin (Talukdar et al., 1986).

In contrast, the oils collected in seeps in the upper Miocene and Pliocene (Figure 6) are characteristic of oils derived from marine siliciclastic source rocks, with a major land plant input. They all contain high proportions of the biomarker oleanane, considered indicative of angiosperm land plant input (and thus unlikely to be older than Late Cretaceous), and relatively high levels of diasteranes, indicating siliciclastic sediment. The sulfur content and C_{27}/C_{29} sterane ratio are too high for a true delta-top source rock, suggesting significant marine influence. The oils correlate well with the terrestrial sourced oils of the Maracaibo Basin (Talukdar et al., 1986). They also correlate well with oils from the Lower Magdalena Valley.

These data suggest that the onshore seepage oils in the South Caribbean are derived from two primary sources, a Cretaceous marine carbonate and a Tertiary (possibly Oligocene–Miocene) marine siliciclastic source rock with major land plant input. Analogy with other northwestern South American basins suggests that only the Cretaceous marine carbonate has the potential to charge a major petroleum province.

SEABED CORES

A seabed coring survey was carried out in the South Caribbean, with the objective of sampling offshore oil seepage to allow oil typing and to high grade the most prospective parts of the basin.

Coring Procedure and Descriptions

Core locations were largely based on regional seismic lines, focusing on shallow mud diapirs, near-surface faults, and shallow bright spots. In the deeper water,

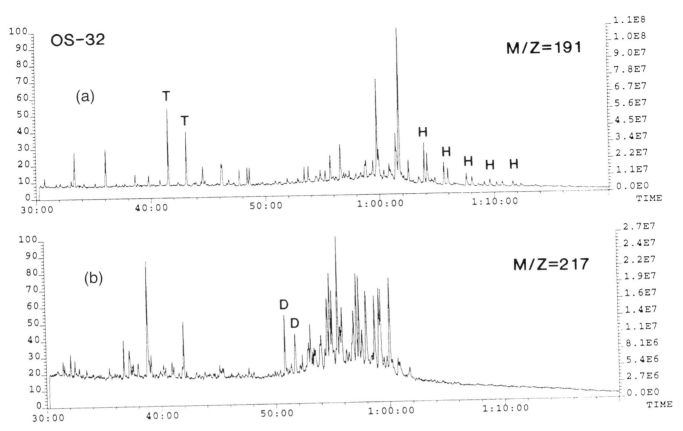

Figure 5—HR GC-MS mass fragmentograms of (a) *m/z* 191 (terpanes) and (b) *m/z* 217 (steranes) for onshore oil seepage OS-32 typical of oil seeps in the lower Tertiary. The relatively high proportion of tricyclic terpanes and extended hopanes and low proportion of diasteranes suggest that this oil comes from a marine carbonate source rock. The presence of oleanane suggests some mixing with a Tertiary source. T, tricyclic terpane; H, extended hopane; D, diasterane.

some cores were also targeted at areas of possible bottom-simulating reflectors (BSR), which may indicate gas hydrate.

Cores were collected using a piston corer, which achieved relatively good penetration in most cases, with average core recovery greater than 3 m. A vibrocorer was used at several shallow-water locations where there was poor piston corer penetration.

Most cores retrieved consisted of dark greenish clay, with some sand in very shallow nearshore areas. Most cores were free of visible debris, but in a few locations, large pieces of wood and other vegetation were identified in cores.

No clearly visible oil staining was identified in any of the cores, but cores from one location were highly gas charged and left an oil slick on the sea surface after retrieval. At another location, the core lithology was mud breccia comprising clasts of lithified mud embedded in a mud matrix, visually similar to cores from oil-bearing mud volcanoes in the Black Sea (R. Crisp, personal communication, 1993).

Geochemical Results

Gas Analysis

Headspace gas concentrations were measured on sections from all cores. Levels of headspace gas in nearly all cores were very low, less than 10 ng/g sediment. The highest headspace gas concentrations, up to 62,000 ng/g (62 μg/g), were measured in the cores that left an oil slick at the sea surface. This gas was very dry (C_1/C_1–C_4 = 0.997) and isotopically light ($\delta^{13}C$ methane –60‰). Although this composition is typical of biogenic gas, it is also normal for headspace gas compositions of gas-rich cores from established oil seepage areas, such as the Gulf of Mexico (BP data). Lower but significant headspace gas concentrations were found in cores containing mud breccia (up to 3000 ng/g). This gas was also very dry (C_1/C_1–C_4 = 0.995) but isotopically heavier ($\delta^{13}C$ methane –44‰), more typical of oil-associated gases. Headspace gas anomalies with concentrations of 80–300 ng/g were identified in several other cores that were not notably different from the rest of the cores in other respects.

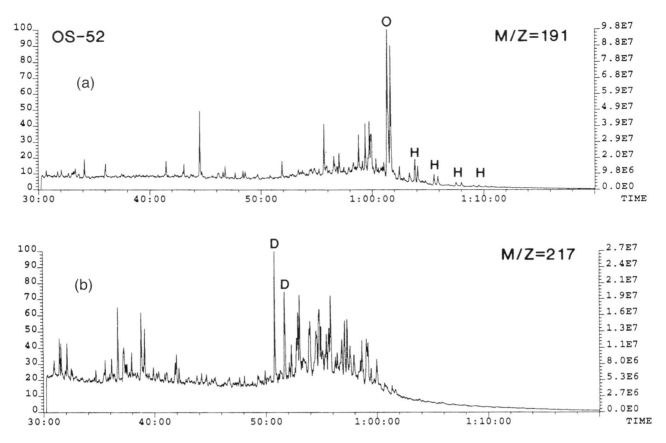

Figure 6—HR GC-MS mass fragmentograms of (a) *m/z* **191 (terpanes) and (b)** *m/z* **217 (steranes) for onshore oil seepage OS-52 typical of oil seeps in the upper Miocene and Pliocene. The presence of oleanane and high proportion of diasteranes are among the features suggesting that this oil comes from a marine siliciclastic source rock with major land plant input. O, oleanane; H, extended hopane; D, diasterane.**

Soluble Extract Analysis

Sections from all cores were oven dried and soxhlet extracted with hexane by the Geochemical and Environmental Research Group (at Texas A&M) prior to analysis by total scanning fluorescence (TSF) and whole-extract gas chromatography (GC).

Cores from only one location, those containing mud breccia, contained clear evidence for oil in the GC or TSF, with high extract yields, very high fluorescence intensity, and a large unresolved complex mixture (UCM) hump on the GC trace (Figure 7a). A section of one of these cores was selected for detailed analysis by gas chromatography–mass spectrometry (GC-MS).

Extract yields and fluorescence intensities in the other cores were all relatively low compared to cores from the Gulf of Mexico slope (BP data). The GC traces showed no clear evidence for migrated oil but abundant evidence for recent organic material. Some features of the GC traces (Figure 7b–f), including elevated baselines (UCM) and a broad molecular weight range of *n*-alkanes with little odd-even preference, can be interpreted as indicating the presence of thermogenic petroleum in the cores. Sections from nine cores considered representative on geographic

and geochemical grounds, and with a reasonable possibility of containing thermogenic petroleum, were selected for detailed analysis by high-resolution (HR) GC-MS.

HR GC-MS

Samples from 10 cores were oven dried and soxhlet extracted in hexane at BP's Sunbury Geochemistry Laboratory prior to whole-extract analysis by HR GC-MS. This technique eliminates the separation steps from sample preparation, reducing the likelihood of sample contamination with very small sample sizes.

The extract from only one core, the mud breccia, contained a clearly recognizable oil signature in *m/z* 191 and *m/z* 217 traces (Figure 8). This shows a good correlation with the oleanane-rich seep oils, which were interpreted as most likely derived from a Tertiary siliciclastic source rock with major land plant input. Aromatic and saturated hydrocarbon maturity parameters suggest that this oil is of relatively low maturity.

The extracts from all the other cores analyzed by GC-MS showed the *m/z* 191 and *m/z* 217 traces to contain predominantly unidentified functionalized or unsaturated terpenoids and steroids derived from recent organic

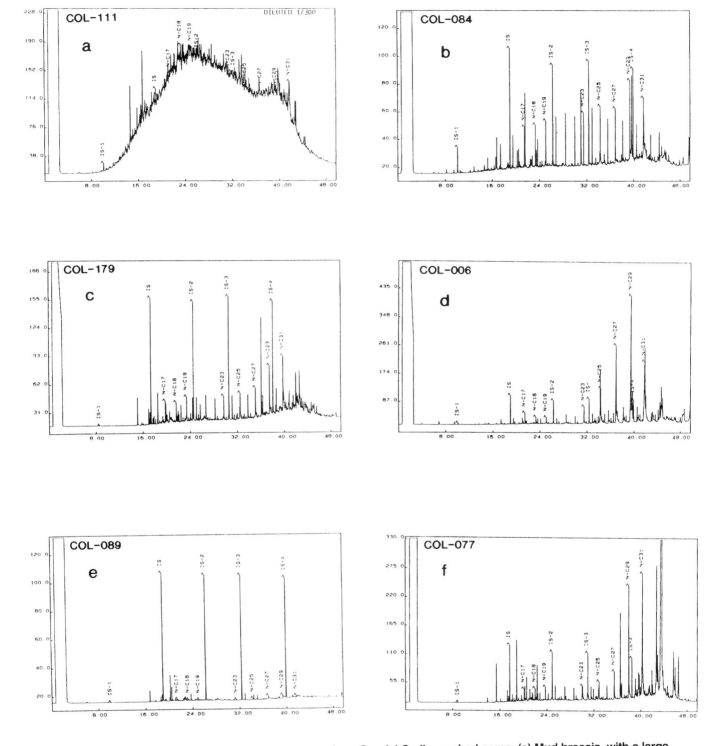

Figure 7—Gas chromatograms of typical whole extracts from Sur del Caribe seabed cores. (a) Mud breccia, with a large amount of unresolved complex mixture. (b–f) More typical core extracts, with very low extract concentrations (note the preponderance of the internal standard peaks IS, IS-2, IS-3, and IS-4), especially in sample (e). Samples (b) and (c) show some elevated baseline and broad distribution of *n*-alkanes, which may indicate low concentrations of thermogenic material.

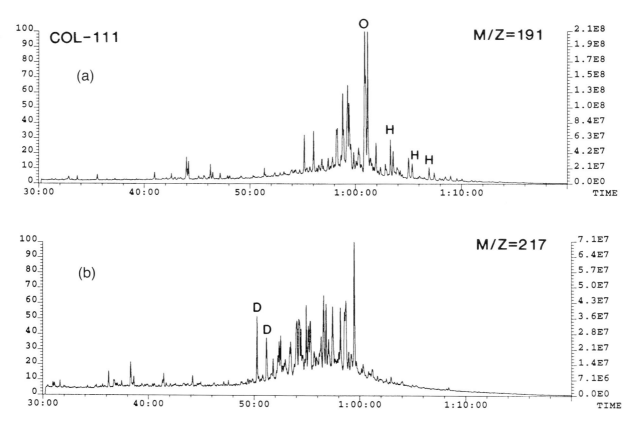

Figure 8—HR GC-MS mass fragmentograms of (a) *m/z* 191 (terpanes) and (b) *m/z* 217 (steranes) for an extract from seabed core COL-111, the mud breccia. This extract correlates well with the onshore oil seeps interpreted as most likely having a Tertiary marine siliciclastic source (see Figure 6). O, oleanane; H, extended hopane; D, diasterane.

matter (Figure 9). The GC-MS does, however, show the cores to contain aromatic biomarkers common in petroleum, including phenanthrenes, dibenzothiophenes, biphenyls, and aromatic steroids (Figure 10). The molecular ratios of these aromatic compounds are typical of mature petroleum. Moreover, the maturity ratios co-vary consistently in different samples (Figure 11), suggesting that the cores contain some migrated petroleum, with maturity varying between cores. The traces of mature petroleum are too weak to identify the likely oil source rock.

Integration of Seabed Coring Results with Geologic Models

The thermal histories of Cretaceous and Oligocene–Miocene potential source rocks in the offshore South Caribbean have been roughly modeled using the sparse well data and seismic interpretations (British Petroleum Colombia, 1993). These suggest that even in the shallowest parts of the basin, a Cretaceous oil source rock would have expelled most of its oil before the Miocene traps could have formed. An Oligocene source, however, could still be generating oil in some places and could have charged traps elsewhere in the past 10 m.y.

Middle Tertiary maturity can be approximated by a map of interpreted depth to the top Miocene (?) seismic reflector (Figure 12). This shows a remarkably good correlation with the thermal maturities estimated from the aromatic biomarker maturities in the seabed cores. The most mature signatures in the cores are in the deepest part of the basin, and the lowest maturity oil signatures are in areas where the Oligocene–Miocene could still be generating. This correlation increases confidence in the interpretation of the seabed cores as containing migrated oil from a Tertiary source.

Discussion

Although many mud diapir structures were identified on seismic sections and targeted by shallow cores, cores from only one diapir (Figure 13) proved to be clearly oil-bearing. As previously described, the lithology of these cores was an unusual mud breccia, which looked like the pure extrusion product of a mud volcano. Other cores retrieved only mud without lithified clasts, which looked like detrital sediment. It is suggested that this detrital mud could be rich enough in recent organic material to mask the small amounts of petroleum seeping to the surface. For example, the very gassy core, which left an oil

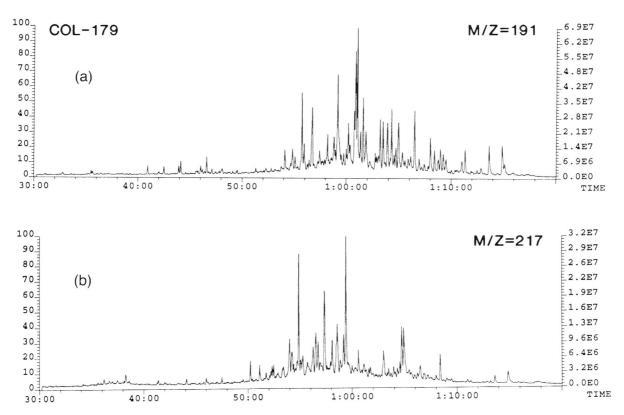

Figure 9—HR GC-MS mass fragmentograms of (a) *m/z* 191 (terpanes) and (b) *m/z* 217 (steranes) for an extract from seabed core COL-179 (see Figure 7c). Major peaks appear to be functionalized or unsaturated terpenoids and steroids derived from recent organic matter.

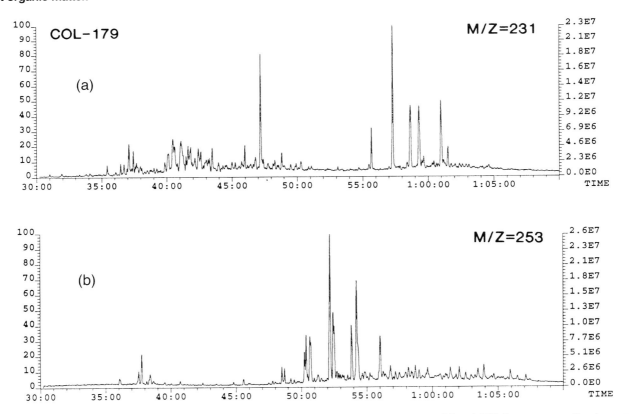

Figure 10—HR GC-MS mass fragmentograms of (a) *m/z* 231 (triaromatic steranes) and (b) *m/z* 253 (monoaromatic steranes) for an extract from seabed core COL-179. Biomarker distributions are typical of moderate maturity oils.

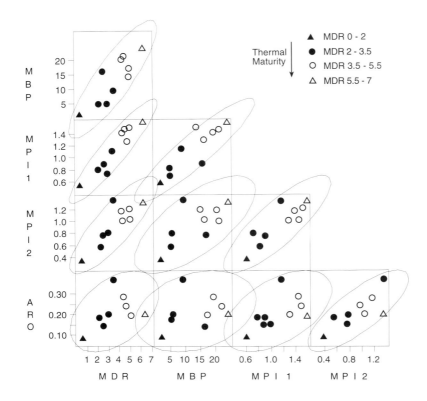

Figure 11—Crossplot of aromatic molecular maturity parameters calculated from HR GC-MS of Sur del Caribe seabed core extracts. Symbols show grouping of cores based on maturity. Computer-generated density ellipses show good cross-correlation of most parameters (e.g., MPI1 with MDR). Key: MDR, 4-methyl dibenzothiophene/1-methyl dibenzothiophene; MBP, 3-methyl biphenyl/2-methyl biphenyl; MPI1, methyl phenanthrene (ME) index 1 [(3ME + 2ME)/(9ME + 1ME)], where the numbers indicate the position of the methyl group; MPI2, methyl phenanthrene index 2 [(3ME + 2ME)/(phenanthrene + 9ME + 1ME)] × 1.5; ARO, C_{20} triaromatic sterane/(C_{20} triaromatic sterane + C_{28} 20(R)-triaromatic sterane).

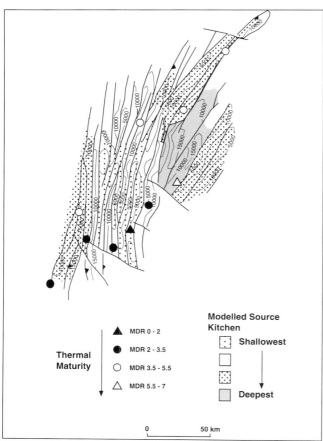

Figure 12—Map of core extract maturities determined from molecular maturity parameters superimposed on approximate depth to source kitchen.

slick at the sea surface but did not contain geochemically traceable oil , was made up of unlithified detrial mud. It is proposed that only cores that hit the target migration pathway exactly and retrieved freshly extruded material had a chance of containing enough seeped petroleum to be readily detectable. Cores collected in the vicinity of the mud diapir thought to have sourced the oil slick seen in the airborne survey probably missed the target. Close inspection of regional seismic data suggests that many of the targeted mud diapirs were older structures and draped in recent sediment cover (Figure 14). This would also have the effect of diluting any seepage signal with recent organic matter.

Summary of Seabed Core Results

Cores from one location, which appears to be the site of a submarine mud volcano, contained good evidence of oil seepage. The oil was most likely derived from a marine siliciclastic source rock with major land plant input of Tertiary age, possibly Oligocene–Miocene.

Cores from a second location, close to a deep water mud volcano seismic feature, contained high concentrations of gas and evidence of traces of oil, with an oil slick left at the sea surface around the core barrel.

The GC-MS evidence for oil in this gassy core and other cores from the South Caribbean is based on the presence of aromatic biomarker molecules. The more typical terpenoid and steroid saturated biomarkers are overprinted by abundant recent organic matter. There is reasonable evidence for traces of migrated petroleum in all cores selected for GC-MS analysis.

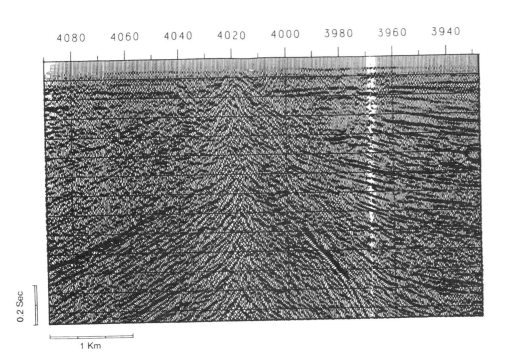

Figure 13—Regional seismic profile across mud diapir from which oil-bearing mud breccia cores were collected. Although the seabed is poorly resolved, the mud volcano does appear to break the surface.

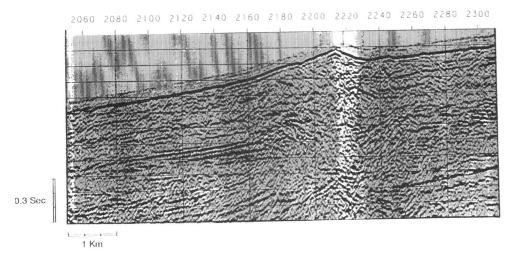

Figure 14—Regional seismic profile across a deep water mud volcano feature. An oil slick was seen when seabed cores collected here reached the sea surface, but geochemical analysis suggests that any thermogenic petroleum is largely overprinted by recent organic material. The continuous reflector across the mud volcano structure suggests that the feature may be draped by recent sediment.

CONCLUSIONS

The evidence for an oil charge to the offshore part of the South Caribbean Basin would be fairly tenuous if based on only one of the surface geochemistry techniques described here. When taken together, however, confidence in the interpretation increases, and the evidence for a widespread oil source in the southern offshore area becomes more compelling, providing information that can influence exploration decisions.

Acknowledgments—A large number of people contributed to the background work described in this paper, including the South Caribbean Exploration team at BP Colombia, particularly Scott Opdyke, Andrew Evans, and Jaime Martínez. The airborne survey was carried out by Alan Williams' team. Geomapa collected the onshore seep samples. Russell Crisp supervised the seabed coring, carried out by GERG, who also did the screening geochemistry. HR GC-MS was undertaken by Malcolm Dee at BP in Sunbury. Thanks are also due to Nigel Goodwin for help with the GC-MS interpretation and to Andy Fleet for support when preparing the manuscript. BP XTP Graphics drafted the figures. Finally, we thank BP Colombia for permission to publish.

REFERENCES CITED

British Petroleum Colombia, 1993, Reporte final Contrato de Asociación "Sur del Caribe": Unpublished Ecopetrol report, 82 p.

Geotec Ltda., 1991, The hydrocarbon potential of the Caribbean basin—northwest Colombia: Geotec Ltda. report, Bogotá, Colombia.

Hedberg, H. D., 1980, Methane generation and petroleum migration, *in* W. H. Roberts III and R. J. Cordell, eds., Problems of petroleum migration: AAPG Studies in Geology No. 10, p. 179–206.

Hovland, M., and A. G. Judd, 1988, Seabed pockmarks and seepages: impact on geology, biology and the marine environment: London, Graham and Trotman, 293 p.

Shephard, F. P., 1973, Sea floor off Magdalena Delta and Santa Marta area, Colombia: GSA Bulletin, v. 84, p. 1955–1972.

Talukdar, S., O. Gallango, and M. Chin-a-Lien, 1986, Generation and migration of hydrocarbons in the Maracaibo Basin, Venezuela: an integrated basin study, *in* Advances in Organic Geochemistry 1985: Organic Geochemistry, v. 10, p. 261–279.

Vernette, G., 1989, Examples of diapiric control on shelf topography and sedimentation patterns on the Colombian Caribbean continental shelf: Journal of South American Earth Sciences, v. 2., p. 391–400.

Piggott, N. and M. A. Abrams, 1996, Near-surface coring in the Beaufort and
Chukchi Seas, Northern Alaska, in D. Schumacher and M. A. Abrams, eds.,
Hydrocarbon migration and its near-surface expression: AAPG Memoir 66,
p. 385–399.

Chapter 30

Near-Surface Coring in the Beaufort and Chukchi Seas, Northern Alaska

N. Piggott

BP Exploration Operating Company
Uxbridge, Middlesex, U.K.

M. A. Abrams

Exxon Ventures (CIS) Inc.
Houston, Texas, U.S.A.

Abstract

Hydrocarbon charge systems on the Arctic shelf were evaluated using 450 cores from five geologic provinces at different states of exploration maturity: Hanna Trough (Chukchi), western Beaufort rift margin, Camden Bay Basin, Demarcation Basin, and Canadian Beaufort-Mackenzie Basin. Each core was screened for seeping hydrocarbons, and any samples with high hydrocarbon content were also characterized by biomarker content and isotope-mass spectrometry for oil–source correlation.

The Chukchi and western Beaufort areas are potentially charged from the North Slope upper Paleozoic–Mesozoic source rocks. Timing of generation and migration was mainly due to Cretaceous burial. Hence, present-day seepage is likely to be limited. Passive seepage was anticipated and minimal evidence of seepage was detected in only 2% of the cores. However, poor core penetration and recovery means these areas were essentially not evaluated by this study. In contrast, deeper site-specific samples collected from shallow rotary boreholes in an earlier survey detected thermogenic liquid hydrocarbons below 10 m. These can be correlated with reservoired oils tested in several exploration wells. This result shows how critical it is to use more site-specific deep coring acquisition techniques here, such as shallow rotary boring or jet-coring. The three Tertiary depocenters farther east have deeply buried the upper Paleozoic–Mesozoic source rocks to postmature levels. No correlation between the seeps and subsurface oils could be made. Fluorescence was due to reworked early Cretaceous source rocks eroded from outcrop and redeposited unoxidized in the Arctic climate. This result shows the importance of ground truthing seeps identified from screening techniques, which can identify hydrocarbons of variable origin.

The absence of significant evidence for a seep in 450 near-surface cores suggests that seepage is not reaching the seabed. Shallow seismic evidence shows that this may be due to a hydrate (or even relict permafrost in the near-shore) barrier preventing leakage to the sediment–water interface. Future seepage detection should focus on obtaining 10–20 m penetration by jet-core or rotary boring acquisition techniques to get through this barrier and also reach pre-Holocene sediments where reworked HRZ should not be such a problem.

INTRODUCTION

Piston, gravity, and vibrocoring for seepage detection have been applied as a useful technology for exploration in a wide range of settings around the world. In the marine environment, it has the advantage (over more remote techniques such as airborne and sniffer surveys) of providing a large enough physical sample to be fully characterized by conventional oil geochemistry analysis. In this way it provides insight into the petroleum system ahead of the drill bit, allowing constraints on critical source presence and effectiveness risks which need to be understood for optimal exploration business decisions.

For instance, if a retrieved seep can be typed and a likely source rock stratigraphic position defined, then the maturity level of the oil can be estimated and some perception of oil plumbing in the subsurface deduced since a seep is essentially a migration pathway terminus. The technique has been particularly successful in basins experiencing rapid recent sedimentation promoting active seepage, such as in the deep-water Gulf of Mexico. A coring program to test this technology in the Arctic Ocean, offshore North America, was mounted by BP and Exxon in the summer of 1990. Acquisition and operator of the cruise was the Geochemical Environmental Research Group (GERG) of Texas A&M University.

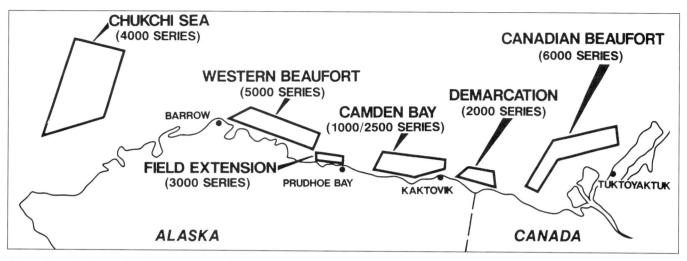

Figure 1—Study area location map showing the five geologic provinces surveyed in 1990 Beaufort and Chukchi seas shallow coring program.

The program acquired 450 cores from the Chukchi Sea in the west to Tuktoyaktuk in the east. Five geologic provinces in varying states of exploration maturity were evaluated: Hanna Trough (Chukchi), western Beaufort rift margin, Camden Bay Basin, Demarcation Basin, and Canadian Beaufort-Mackenzie Basin (Figure 1).

GEOLOGIC AND GEOCHEMICAL FRAMEWORK

The Alaska North Slope is a relatively mature hydrocarbon province that includes the giant Prudhoe Bay and Kuparuk oil fields and several smaller peripheral pools producing oil close to the shoreline adjacent to the Beaufort Sea. As such, the North Slope geologic framework (Hubbard and Edrich, 1987) and petroleum system are fairly well understood. Reservoirs and source rocks range in age from late Paleozoic to Mesozoic within a tripartite plate sequence subdivision of the northern Alaska stratigraphy (Figure 2). The first of these, the Ellesmerian plate sequence, was deposited on the passive margin of North America. The basinal distal sediment-starved Ellesmerian (Kuna-Otuk) sequence is a rich source rock in the southwestern part of the North Slope and in the Hanna Trough. At the end of the Ellesmerian, source deposition occurred up onto the shelf (via upwelling?) with development of the Triassic Shublik Formation, an important source rock. The Jurassic–Neocomian Beaufort plate sequence developed as northern Alaska rifted away from its original North American position and the Arctic Ocean opened. Early rifting events failed, and in this low-energy setting, the Early Jurassic Kingak source rock was developed.

Eventual successful break-up in the Neocomian was followed by collision of northern Alaska with exotic Alaska to the south along the Gobuk suture with folding and thrusting forming the Brooks Range and totally switching the basin polarity. A series of foreland basin sediment wedges comprise the third plate sequence deposited in response to flexural subsidence. The first (Torok) early Brookian foreland basin formed in the southwestern part of the North Slope and filled the Colville Trough foredeep to the northeast and the Hanna Trough to the north.

Condensed pelagic sedimentation distal to the foreland wedge led to the third important source rock: the Aptian–Albian highly radioactive zone (HRZ). The filling of the Colville Trough was effectively axial, and the second (middle Brookian) foreland basin sediment wedge prograded ever farther to the northeast. Condensed pelagic sedimentation distal to the middle Brookian continued, and the Cenomanian–Campanian Shale Wall source rock was deposited. In the eastern North Slope on the coastal plain of the Arctic Natural Wildlife Refuge (ANWR), continuous pelagic sedimentation has caused the HRZ and Shale Wall section to be undifferentiated, and it is referred to lithostratigraphically as the middle–Upper Cretaceous Hue Shale (Magoon et al., 1987). Similarly, in the Mackenzie delta area, this is the Boundary Creek Formation source rock.

The Eocene–Recent late Brookian wedge sedimentation occurs offshore in the U.S. eastern Beaufort and Canadian Beaufort seas. A range of poorly constrained Tertiary age source developments occur in the late Brookian in what appear to be predominantly paralic terrestrially influenced settings. The upper Brookian sediments are as much as 8 km thick.

To summarize, there are essentially four major source rocks:

- The basinal Ellesmerian, which is restricted to the west
- The Shublik and lower Kingak, which due to stratigraphic proximity can be considered as one system

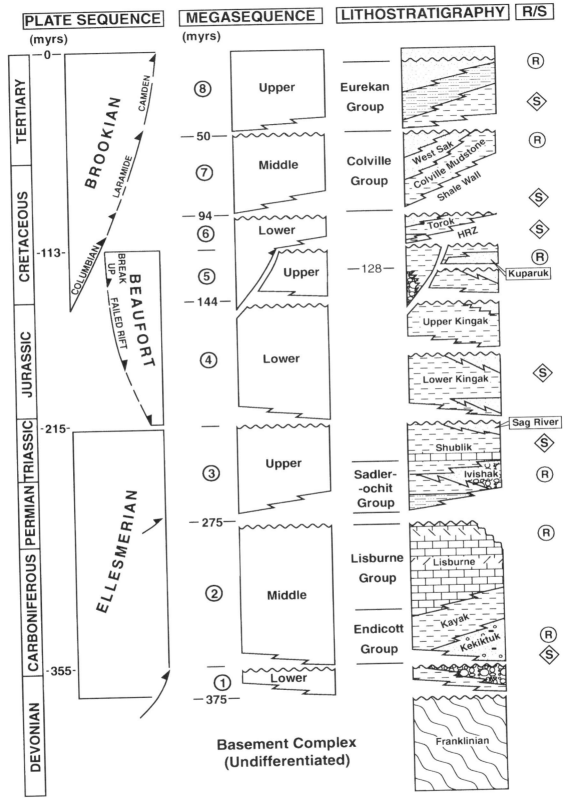

Figure 2—Stratigraphic terminology of the Alaskan North Slope area showing the three-part plate sequence subdivision (first column). (After Hubbard and Edrich, 1987.)

- The condensed distal pelagic equivalents of the lower–middle Brookian foreland basin sequences, HRZ and Shale Wall. These are lithostratigraphically grouped together and referred to as the Hue Shale (in U.S.) or the Boundary Creek (in Canada). The distal correlative conformity between the lower and middle Brookian means this is effectively one system offshore in the Beaufort Sea.
- Tertiary paralic source rocks restricted to late Brookian depocenters

Within the geologic framework, it is the foreland basin wedge sedimentation and burial that drives generation and maturation of the source rocks, as summarized in Figure 3. Thus, in the Chukchi and Beaufort seas, the Hanna Trough and western Beaufort rift margin received most sedimentation in early–middle Brookian (middle–Late Cretaceous) time, which matured mainly older Paleozoic and Mesozoic source systems (distal Ellesmerian and Shublik-Kingak). In contrast, Camden Bay, Demarcation, and Canadian Beaufort basins received most sedimentation in late Brookian (Tertiary–Recent) time, which has matured younger source systems. The extreme thickness of the upper Brookian strata means that north of the shoreline the HRZ, Shale Wall, Hue Shale, and Boundary Creek system has been buried well beyond the oil floor in the basinal areas of the three depocenters, resulting in only the Tertiary paralic source system being effective.

This pattern has been corroborated by exploration results to date. Oil shows and discoveries to date in the Chukchi and western Beaufort rift margin carry a distal Ellesmerian or Shublik-Kingak source signature (e.g., Mukluk-1, Figure 4), while oil from Camden Bay, Demarcation, and Canadian Beaufort-Mackenzie basins has a Tertiary source signature (e.g., Tarsiut well, Figure 5) (Brooks, 1986). Active fresh coastal oil seepage (26.7° API) is seen onshore from Camden Bay at Manning Point (Anders et al., 1987), again generated from a Tertiary source rock.

Thus, ahead of the seepage detection coring program, a low-intensity passive leakage seep signal for the distal Ellesmerian or Shublik-Kingak sourced oil was anticipated for the Chukchi and western Beaufort rift margin segments of the survey. It was hoped a higher intensity active seepage signal for the Tertiary sourced oil would be observed in the three eastern segments in Camden Bay, Demarcation, and Canadian Beaufort-Mackenzie basins.

CORING SURVEY

Coring locations were selected by explorers, predominantly from a regional two-dimensional seismic grid, focusing on areas having evidence for shallow gas charge or indications of faults coming to the surface. This was supplemented locally with shallow penetrative analog seismic data acquired for site-surveying purposes. In terms of coring practice, the survey leaned heavily on experience from the contracting company acquired from years of successful coring in the deep water of the Gulf of Mexico using a piston coring device. With its past experience in geotechnical coring in the Beaufort Sea, Exxon perceived that a potentially hard substrate was present on the Beaufort shelf. It was thus advised that other acquisition technology would be needed to get sufficient penetration and sample recovery from deeper than any shallow disturbed zone, where there is a better chance of seepage signal preservation (Abrams, 1992). A vibrocoring device was therefore also mobilized in case this proved to be a problem.

The cores retrieved were initially subjected to a hydrocarbon seepage screening protocol that relied on headspace gas analysis of canned samples for the presence of gas and light hydrocarbons. Analysis also included solvent extraction followed by total scanning fluorescence (TSF) (Brooks et al., 1985) and whole extract gas chromatography (GC) for the presence of higher molecular weight hydrocarbons (oil). In addition, any samples containing sufficient hydrocarbons present to permit analysis were further characterized by separation, gas chromatography–mass spectrometry, and isotope mass spectrometry. (Experience from the Gulf of Mexico indicated that it is rare to obtain sufficient extract yield for further geochemical characterization if the TSF intensity is <2000 units in the contractor's screening analysis.) This additional analysis was done first to "truth" any seepage from possible pollutants (anthropomorphic in origin) and, second, to permit oil–source correlation with established regional source correlations, which was then incorporated into a complete understanding of the petroleum system. This detailed geochemical characterization was to prove to be a critical step in understanding the survey results.

SURVEY RESULTS

The results of each leg of the survey are summarized in Figures 6 through 11, which show histograms of core length (indicating penetration and recovery) and the screening TSF data used to detect oil seepage. The first obvious result is that poor penetration was achieved, with core lengths of only about 0.5 m being typical. This was a disappointment, even if expected based on earlier Exxon experience. The vibrocore was deployed and failed to achieve significantly better penetration and recovery than the simpler and faster piston core, so it was not used extensively. These results attest to the presence of a surface layer of stiff clays that severely inhibited coring device penetration.

Penetration and recovery were particularly poor in the Chukchi, western Beaufort, and Field Extension segments (Figures 6, 7, 8) of the cruise, areas where only a weak passive signal was anticipated anyway due to the burial history of this part of the margin. This meant that the section and any source rocks present may not be at maximum burial any longer and therefore not actively

(a) EARLY BROOKIAN OIL GENERATION

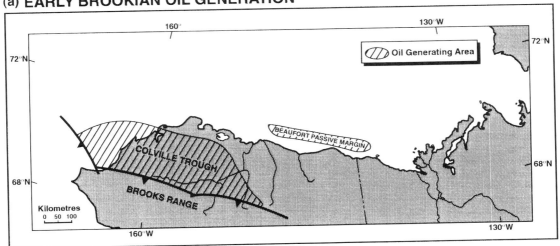

(b) MID - BROOKIAN OIL GENERATION

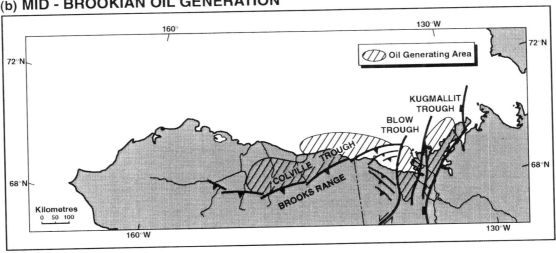

(c) LATE BROOKIAN OIL GENERATION

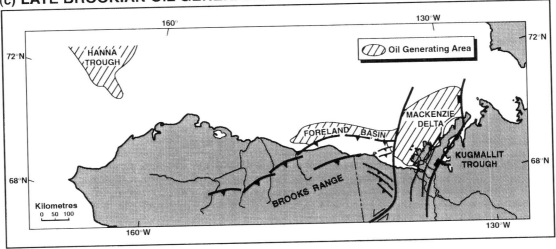

Figure 3—Oil kitchens and generation history of source rocks in the Alaskan North Slope region during (a) early, (b) middle, and (c) late Brookian oil generation. (After Hubbard and Edrich, 1987.)

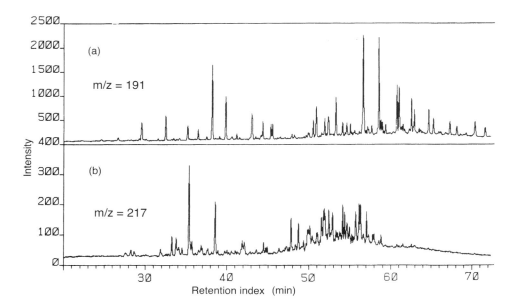

Figure 4—Gas chromatograph showing biomarker distribution for Mukluk-1 well oil show. (a) Triterpanes; (b) steranes.

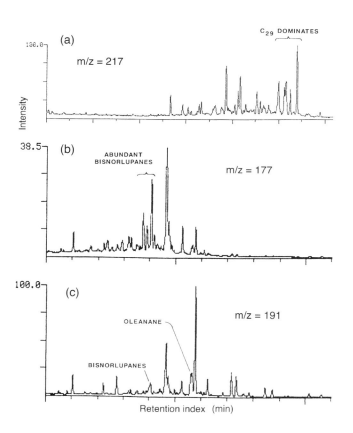

Figure 5—Gas chromatographs showing saturate biomarker distributions for the Tarsiut oil field, Canadian Beaufort Sea. (a) Steranes; (b) demethylated triterpanes; (c) triterpanes.

generating and expelling oil and gas. Nevertheless, some significant live oil shows were reported in several wells along with evidence of seepage in deep rotary drilled boreholes (to 50 m or deeper) in these areas. Seal Island/North Star and Sandpiper, unitized oil and gas discoveries, are also located in the Field Extension segment, and there were no signs of hydrocarbons leaking to surface in any of these positions.

Only three cores from the western Beaufort segment fluoresced with sufficient intensity to yield extractable oil for further geochemical characterization. Only 2% of the 174 cores contained anomalous quantities of hydrocarbons in the passive seepage area. Better core penetration and recovery may have significantly improved seepage detection, and it is felt that seepage detection in these segments was essentially unevaluated in this study.

To the east, in the eastern Beaufort (Camden Bay), Demarcation, and Canadian Beaufort segments, areas where active seepage was anticipated, an occasional longer core was retrieved (Figures 9, 10, 11). Even in these segments, however, the average core length remained low. The poor core recovery was particularly disappointing because an initial interpretation indicated that in situations where longer cores are retrieved, higher fluorescence intensities are achieved.

While the TSF is only a semi-quantitative screening tool, it seems that longer cores have a better chance of detecting oil seepage than do cores of <1 m length. This is borne out by inspection of Figures 6–11 and from cross-plotting core length against fluorescence intensity for all 42 cores from the Canadian Beaufort segment of the survey, where the best core penetration and the highest fluorescence were both achieved (Figure 12). A total of 35 cores in the active seepage segments gave fluorescence intensities high enough for extraction and further geo-

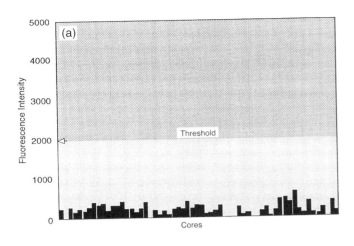

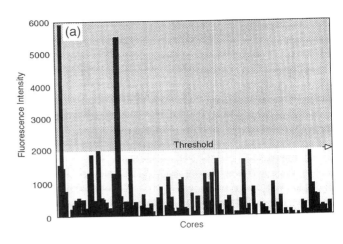

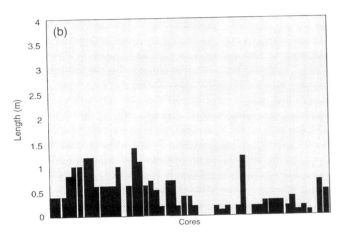

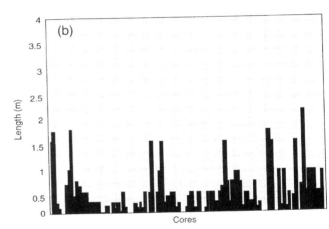

Figure 6—Histograms showing relationship between (a) fluorescence intensity and (b) core penetration depth for Chukchi Sea cores. Note the very low average core lengths (0.45 m) for this area (51 cores total).

Figure 7—Histograms showing relationship between (a) fluorescence intensity and (b) core penetration depth for western Beaufort Sea cores. Average core length (103 cores) is 0.4 m.

chemical characterization, suggesting that seepage may have been detected in up to 12% of cores taken. (This would still have been poor given the seismic focus on optimal locations, but still reasonable given the low penetration depths.)

Whole-oil GC data for the core extracts yielded promising *n*-alkane envelopes typical of a thermogenically derived petroleum, not derived solely from recent surficial organic matter nor from a narrow range envelope typical of an artificial distilled product pollutant (Figure 13). Not one visible oil stain was observed nor petroliferous odors noted on any of the cores retrieved, a result not typical in a coring survey of this extent carried out over a so-called active seepage area.

The other screening technique for seepage focuses on light hydrocarbons. However, few headspace gas concentrations were noted at a significant level. In population of cores with higher TSF intensities, it was found that only

one of the oil seeps could be corroborated with gas seepage (Figure 14), which is unusual because gas seepage typically accompanies oil seepage. Furthermore, only 18 cans from the entire cruise contained gas in sufficient concentration within the headspace to permit isotopic determination of the methane present. More than half of the values obtained were for dry gases and proved to be isotopically light and of predominantly bacterial origin. Although bacterial methane concentrations associated with visible oil seeps can often be high (biogenic generation from the oil), they usually also contain appreciable wet C_{2-4} components, a feature not observed here.

Thus, an initial interpretation of the coring survey based on screening criteria indicated that there was some evidence for the occurrence of oil seepage, but that this was only rarely detected due to short core lengths. Furthermore, there was a disturbing lack of support for the fluorescence data from the headspace gas data.

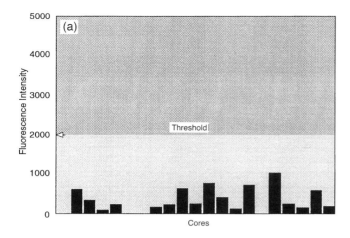

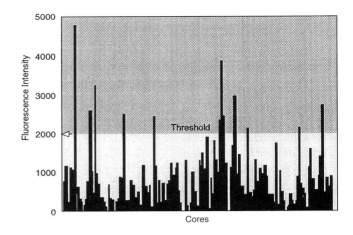

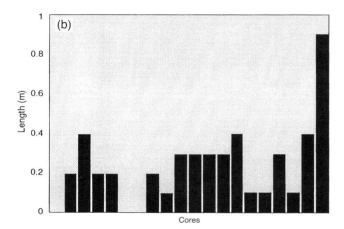

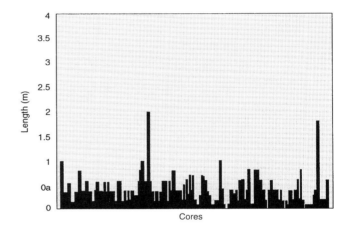

Figure 8—Histograms showing relationship between (a) fluorescence intensity and (b) core penetration depth for cores from the Field Extension area of the Western Beaufort Sea. Average core length (20 cores) is 0.24 m.

Figure 9—Histograms showing relationship between (a) fluorescence intensity and (b) core penetration depth for eastern Beaufort Sea (Camden Bay) cores. Average core length (169 cores) is 0.9 m.

GEOCHEMICAL SIGNIFICANCE OF HIGHER FLUORESCENCE

Biomarker and isotope mass spectrometry techniques pointed to a different interpretation for the highly fluorescent cores obtained from the active seepage segments. As for the GC traces, the sterane and triterpane distributions at first sight seemed to indicate genuine petroleum seepage, if slightly immature to early mature in its aspect. However, the biomarkers highlighted a major problem: the "seepage" was not derived from Tertiary source rocks and did not match the subsurface oil discoveries found offshore in the Canadian Beaufort Sea or Camden Bay. For instance, oleanane and bisnorlupanes were missing (Figure 15). The biomarker and isotopic signatures (Figure 16) were closest to oils derived from the Upper Cretaceous HRZ, Hue Shale, and Boundary Creek source system, observed in some onshore fields. This was a difficult result to rationalize in terms of maturity level (with-

in the gas window beneath the thick Tertiary sediments), timing of oil generation (modeled to occur well before the Holocene sediments cored would have been deposited), and migration pathway (circumventing subsurface reservoirs but reaching the surface!).

An insight into this apparent conundrum was obtained by looking at the numeric ratios for maturity sensitive biomarkers (Figure 17). A remarkable uniformity of the various ratios (almost at the limit of analytical precision) was obtained across the Canadian Beaufort, Demarcation, and Camden Bay segments. None of the biomarkers at equilibration level suggested homogeneous immaturity. This is astonishing for seepage obtained over a wide area where source rocks can be expected to be generating and expelling oil from a range of different burial depths, giving rise to a range of isomer ratios. This is what is observed in deep-water offshore Texas where a similar population of cores with seepage indications (often visible oil staining) have been characterized by this technique (Figure 18).

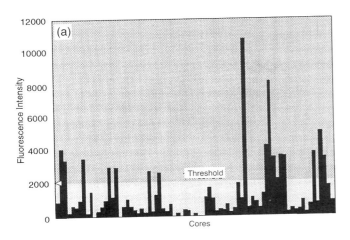

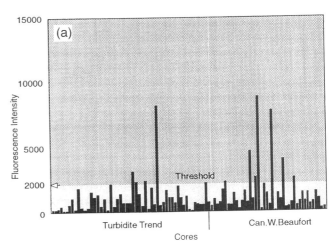

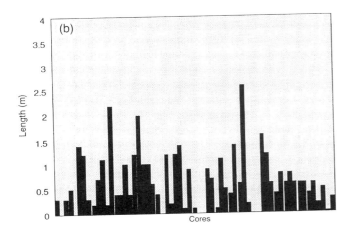

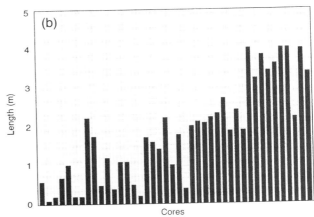

Figure 10—Histograms showing relationship between (a) fluorescence intensity and (b) core penetration depth for Demarcation Bay cores. Average core length (61 cores) is 0.7 m.

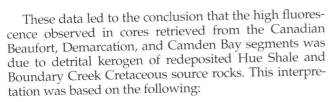

Figure 11—Histograms showing relationship between (a) fluorescence intensity and (b) core penetration depth for Canadian Beaufort Sea cores. Average core length (42 cores) is 1.9 m.

These data led to the conclusion that the high fluorescence observed in cores retrieved from the Canadian Beaufort, Demarcation, and Camden Bay segments was due to detrital kerogen of redeposited Hue Shale and Boundary Creek Cretaceous source rocks. This interpretation was based on the following:

- Lack of associated headspace gas anomalies to support the seepage as bona fide migrated oil reaching the surface
- Absence of a recent organic matter or artificial distilled product (pollutant) signature, which can sometimes explain elevated fluorescence levels
- Lack of correlation of core extracts with offshore subsurface oil discoveries, which carry a strong Tertiary source signature
- Remarkable geochemical uniformity of all extracts both isotopically and in biomarker content.

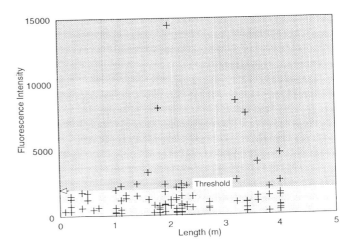

Figure 12—Graph of fluorescence intensity versus core length for Canadian Beaufort Sea segment cores.

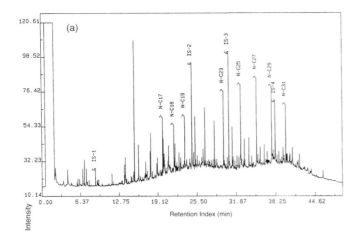

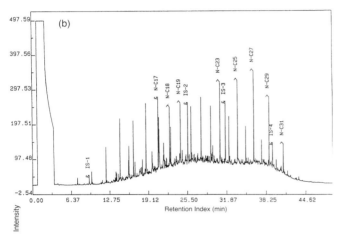

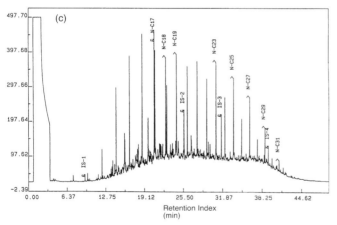

Figure 13—Representative gas chromatograph traces for core extracts from cores with high fluorescence intensity in the three eastern segments: (a) Camden Bay, (b) Demarcation Basin, and (c) Canadian Beaufort Sea. The back-end odd carbon number dominance is derived from recent organic matter, but the underlying envelope is oil-like. IS = internal standard; N = normal alkane.

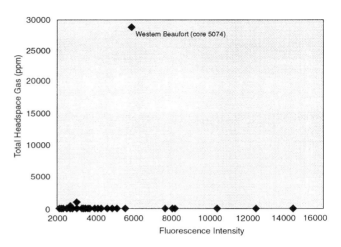

Figure 14—Graph of headspace gas concentration versus fluorescence intensity for 48 cores with >2000 units fluorescence.

- Correlation with immature to early mature Cretaceous (Hue Shale and Boundary Creek) outcrop materials
- Pyrolysis of fluorescent core materials showing moderate (redeposited and diluted) source potentials, with S_2 values of 1.5–4.5 kg/t and total organic carbon (TOC) contents of 0.8–1.5%.
- A large population of Late Cretaceous palynomorphs revealed by age dating of the fluorescent cores and abundance correlating broadly with fluorescence and TOC content.

Given this weight of evidence, it is clear that the seepage relates to redeposited Cretaceous detrital kerogen, which is not being fully oxidized during outcrop erosion, river transport along the Mackenzie system and smaller rivers draining the ANWR (such as the Canning, Katakturuk, and Niguanak rivers), and then subsequent redeposition on the Beaufort shelf. It may be significant that this model is consistent with the hydrology of the Mackenzie delta and Beaufort shelf, which argues that the best organic matter (detrital kerogen) preservation potential would exist distally on the outer shelf beyond storm reworking effects (Figure 19). This correlates with the distribution of most fluorescent cores, which were retrieved from deeper water coring sites. Thus, although core length correlates well with fluorescence intensity, it is actually a correlation with water depth that is the true control.

The longer cores were recovered from locations having depositional environments in which the sediment–water interface conditions were the lowest energy and thus the chances for preservation of the reworked kerogen signal were the greatest. It is not clear whether this is a unique problem associated with the Arctic climate (slowing the rate of oxidation of organic matter during transport and redeposition) or whether it is caused by the hydrology of the Beaufort shelf, which has a narrow near-shore high-energy zone due to pack ice restricting dominant wave

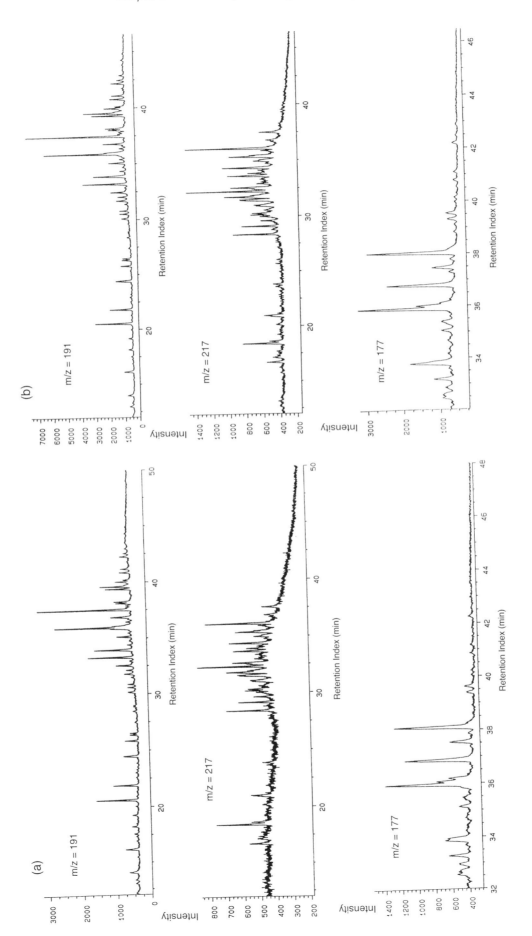

Figure 15—Typical sterane and triterpane signatures for core extracts from high-fluorescence cores from the eastern study segments: (top) triterpanes, (middle) steranes, and (bottom) demethylated triterpanes. (a) Camden Bay cores; (b) Demarcation Basin cores.

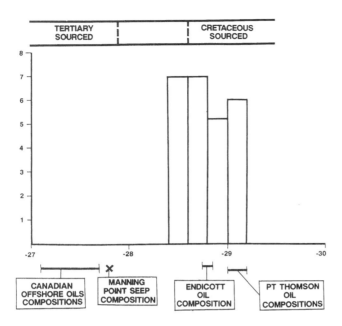

Figure 16—Bar chart of carbon isotopic compositions of core extracts indexed to known oil compositions derived from Tertiary or Cretaceous sources.

fetch in the Arctic Ocean. This effect, along with freshwater "capping" by river run-off leading to local water column stratification, leads to conditions having enhanced organic matter preservation potential. Whatever the major control, the strongest reworked Cretaceous kerogen signal is recorded in distal shelf locations where the low density organic debris is concentrated by being the last to drop out of suspension from the sediment load. This problem may be a major potential pitfall for near-surface gravity (piston) and vibrocoring in Arctic areas where immature to early mature source rocks outcrop in the hinterlands of rivers feeding the prospective shelf.

It should be noted that no reworked kerogen problem was encountered in the three western segments of this cruise. Here, few indications of a seepage signal in the form of high fluorescence intensity were noted. However, the three anomalous cores were actually associated with slightly elevated headspace gas contents and upon extraction and full characterization yielded seepage having isotopic and biomarker compositions consistent with subsurface discoveries and shows. Also significantly, they differed from one another and from the uniform biomarker ratios obtained for the core extracts in the three eastern segments (Figure 17). The western locations are removed from the area impacted by rivers with immature to early mature Hue Shale and Boundary Creek strata in their drainage hinterlands. The Colville River system farther west contains only postmature HRZ Cretaceous source rock in its hinterland, matured beneath the early–middle Brookian foreland basin wedges during Cretaceous time. These cores were the only three indications of true seepage recovered in the entire cruise.

PLANNING FUTURE SEEPAGE SURVEYS IN BEAUFORT SEA

Clearly, the message from the western segments of this cruise in the western Beaufort and Chukchi seas is that a subtle passive seepage signal has poor chances of detection in cores averaging 0.5 m in length. Site-specific deep coring techniques such as jet-coring and rotary boring must be used for future seepage detection efforts in these geologic settings, where near-surface coring by gravity (piston) and vibrocore technology simply cannot provide sufficient penetration and recovery.

But what of the eastern segments of the study area where an active seepage signal was anticipated? Is there a way around the reworked kerogen problem encountered? This cannot be assessed with certainty given the existing results, but there are indications that seepage is being inhibited from reaching the near surface. First, from deep seismic records, there are indications of impressive hydrocarbon leakage, such as near the Immiugak A-06 well in the Canadian sector (Figure 20). Amoco experienced operational difficulties in drilling this feature due to shallow gas.

Second, the U.S. Geological Survey (Wolf et al., 1987) notes that there is evidence for mappable horizons within the shallow seismic records of the eastern Beaufort shelf at depths 10–20 m below the seabed. The continuity and lack of structure on these horizons indicates they may be bottom-simulating reflectors corresponding to hydrate layers at that depth range. The low temperatures of the eastern Beaufort shelf would cause near-surface conditions to fall in the hydrate stability field. We speculate that these hydrates may act as a barrier to vertically leaking hydrocarbons, preventing oil and gas from reaching the top ~10 m of the sediment column and seabed. Core penetration and recovery in this study were insufficient to test such a hypothesis. All the same, the 270 cores of up to 4 m penetration could not get through the reworked kerogen signal to see if active seepage terminates deeper in the sediment pile.

The only solution to this problem seems to be deeper acquisition techniques, as demonstrated to be more effective for seepage detection farther west in the Chukchi. In the east, however, the impact of the reworked kerogen signal must be "screened out." Only with a deep boring down to >30 m is it possible to test whether a strong active seepage signal of vertically migrating or leaking oil and gas (possibly concentrated by a hydrate layer) could overwhelm the detrital source rock input at depth. These future acquisition strategies are summarized in Figure 21, which emphasizes the need for careful choice of acquisition techniques to detect seepage in this environment.

CONCLUSIONS

This study highlights the importance of detailed geochemical analysis of seeps to authenticate hydrocarbon indications within a coherent geologic framework and

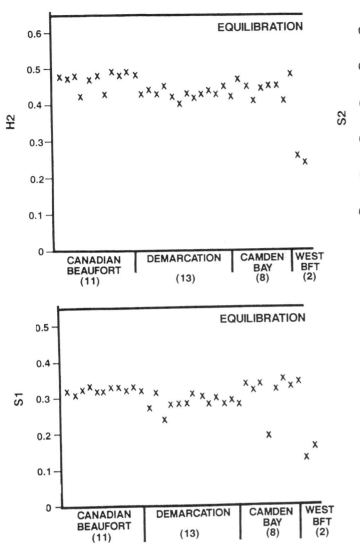

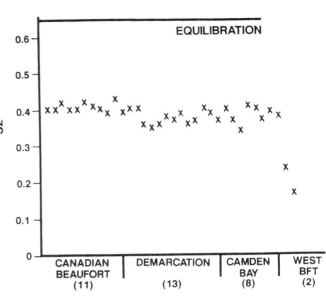

H2 = C$_{31}$ S/C$_{31}$ R + S Hopane.

S2 = C$_{29}$ $\alpha\beta\beta$/C$_{29}$ $\alpha\alpha\alpha$ + α ßß Sterane

S1 = C$_{29}$ S/C$_{29}$R + S Sterane.

Figure 17—Numeric values for three maturity-sensitive sterane and hopane biomarker ratios (H2, S2, and S1) for Arctic core extracts. The core extracts show remarkable uniformity, with the exception of the western Beaufort cores.

petroleum charge system. Recovery of a physical sample is crucial within this process, and a more remote technique for seepage detection that does not retrieve a sample is prone to misinterpretation in many ways. In this area, reworked detrital kerogen is a potential false indicator of seepage within near-surface cores. This problem may be particularly acute in Arctic shelf environments due to slow oxidation during erosion and transport of source rocks. It also depends on an appreciation of source presence and maturity within drainage hinterlands.

The reworked kerogen problem in the Canadian Beaufort, Demarcation, and Camden Bay segments of this survey, plus poor core penetration and recovery to the west where low-level passive seepage was anticipated, mean that piston, gravity, and vibrocore acquisition technology is inappropriate for successful seepage detection. Deeper, more penetrative acquisition using highly site-specific jet-coring or rotary boring is recommended for any future seepage detection in this area.

Acknowledgments—The authors would like to thank BP Exploration and Exxon Exploration for their support and permission to publish. In particular, Helmer Fucke, Alan Ross, and Dan Hughes provided assistance in execution of the project.

REFERENCES CITED

Abrams, M. A., 1992, Geophysical and geochemical evidence for subsurface hydrocarbon leakage in the Bering Sea, Alaska: Marine and Petroleum Geology, v. 9, p. 208–224.

Anders, D. E., L. B. Magoon, and S. C. Lubek, 1987, Geochemistry of surface oil shows and potential source rocks: USGS Bulletin, v. 1778, p. 181–198.

Brooks, J. M., M. C. Kennicutt, and B. D. Carey, 1986, Offshore surface geochemical exploration: Oil & Gas Journal, v. 84, no. 42, p. 66–72.

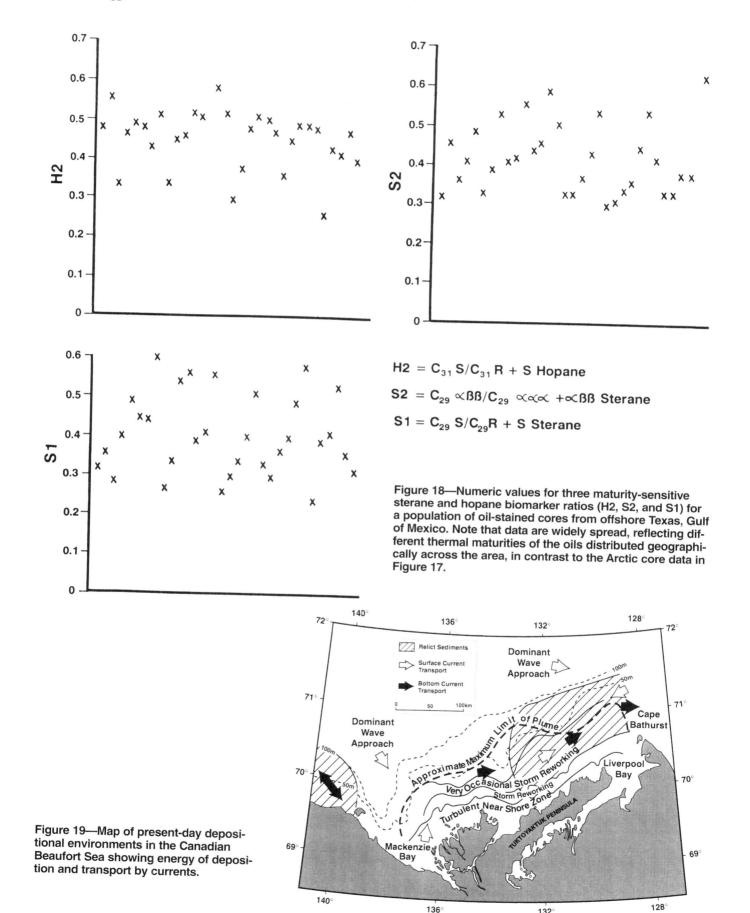

$$H2 = C_{31} S/C_{31} R + S \text{ Hopane}$$

$$S2 = C_{29} \alpha\beta\beta/C_{29} \alpha\alpha\alpha + \alpha\beta\beta \text{ Sterane}$$

$$S1 = C_{29} S/C_{29} R + S \text{ Sterane}$$

Figure 18—Numeric values for three maturity-sensitive sterane and hopane biomarker ratios (H2, S2, and S1) for a population of oil-stained cores from offshore Texas, Gulf of Mexico. Note that data are widely spread, reflecting different thermal maturities of the oils distributed geographically across the area, in contrast to the Arctic core data in Figure 17.

Figure 19—Map of present-day depositional environments in the Canadian Beaufort Sea showing energy of deposition and transport by currents.

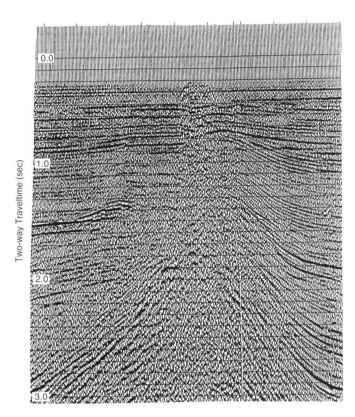

Figure 20—Seismic profile showing evidence of gas leakage in the subsurface close to the Immiugak feature in the Canadian Beaufort Sea.

Brooks, P. W., 1986, Biological marker geochemistry of oils from the Beaufort-Mackenzie region, Arctic Canada: Canadian Petroleum Geology Bulletin, v. 34, no. 4, p. 490–505.

Magoon, L. B., P. V. Woodward, A. C. Banet, S. B. Griscom, and T. A. Daws, 1987, Thermal maturity, richness, and type of organic matter of source rock units: USGS Bulletin, v. 1778, p. 127–180.

Hubbard, R. J., and S. P. Edrich, 1987, Geologic evolution and hydrocarbon habitat of the Arctic Alaska microplate: Marine and Petroleum Geology, v. 4, p. 1–34.

Wolf, S. C., P. W. Barnes, D. M. Rearic, and E. Reimnitz, 1987, Quaternary seismic stratigraphy of the inner continental shelf north of the Arctic National Wildlife Refuge: USGS Bulletin, v. 1778, p. 61–78.

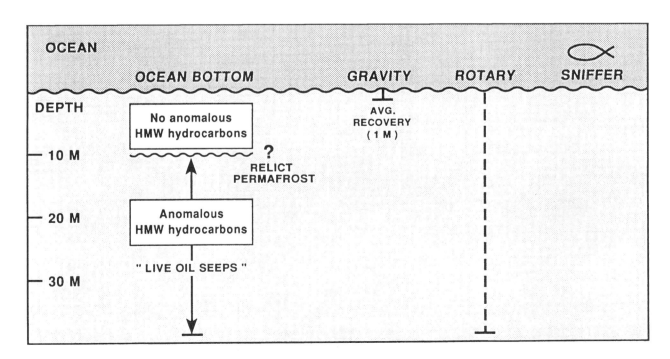

Figure 21—Summary of offshore Alaska surface geochemical studies using various acquisition techniques showing shallow impermeable barriers and depth distribution of high molecular weight hydrocarbons.

Mello, M. R., F. T. T. Gonçalves, N. A. Babinski, and F. P. Miranda, 1996,
Hydrocarbon prospecting in the Amazon rain forest: application of surface
geochemical, microbiological, and remote sensing methods, *in* D. Schumacher
and M. A. Abrams, eds., Hydrocarbon migration and its near-surface expression: AAPG Memoir 66, p. 401–411.

Chapter 31

Hydrocarbon Prospecting in the Amazon Rain Forest: Application of Surface Geochemical, Microbiological, and Remote Sensing Methods

M. R. Mello

F. T. Gonçalves

N. A. Babinski

F. P. Miranda

Petrobrás Research & Development Center
Rio de Janeiro, Brazil

Abstract

Geochemical, microbiological, and remote sensing investigative techniques were used to reduce exploration risk in the difficult task of finding oil in the Amazon rain forest. These nonconventional methods were used successfully in ranking exploratory leads close to a major gas occurrence in the Juruá area, Solimões Basin, Brazil. Near-surface soil and sediment samples were collected along seismic lines and analyzed by geochemical and microbiological methods. The results showed anomalous concentrations of hydrocarbons and hydrocarbon-consuming bacteria concordant with structural traps and aligned with the trace of a major reverse fault. The remote sensing survey presented an anomalous spectral response of vegetation over the geochemical and microbial anomalies which may represent the response of plants to long-term anaerobic conditions due to gas leakage from subsurface reservoir rocks. Further exploration on potential prospects proved the existence of gas and gas-condensate accumulations. The use of surface geochemical, microbiological, and remote sensing techniques thus proved to be cost-effective in constraining exploration risk.

INTRODUCTION

Surface geochemistry is one of the oldest and most successful surface exploration methods. Surface seeps and microseeps occur because processes and mechanisms such as diffusion, effusion, and buoyancy allow hydrocarbons to escape from reservoirs and migrate to the surface where they may be retained (absorption or adsorption) in the sediments and soils or diffuse into the atmosphere or water columns. Historically, the search for subsurface hydrocarbon accumulations was oriented toward finding macroscopic surface deposits of hydrocarbons, often associated with specific geological conditions (Link, 1952). Geochemical techniques for hydrocarbon prospecting were first applied by Sokolov et al. (1936) in Russia and by Horvitz (1939) in the United States. Link (1952) reported that a large number of giant oil accumulations in onshore basins around the world are associated with or located near oil or gas seeps (e.g., Maracaibo oil field, Venezuela; San Juan oil field, California; and Persian Gulf region).

Today, the application of surface geochemistry to hydrocarbon exploration has improved greatly as a result of the availability of new analytical instrumentation (e.g., gas chromatograph, capillary columns, and computer systems), which provide better identification of individual hydrocarbon components at higher sensitivity. Such advances allow the identification and characterization of previously undetected microseeps, thus offering the petroleum geologist a powerful auxiliary tool to rank exploration priority among prospects identified in a basin.

Surface exploration methods include the direct detection and quantification of hydrocarbons that are present in surface soils and sediments as well as the geomicrobial methods, which are based on the distribution of oil-consuming bacteria. Remote sensing methods, based on the spectral behavior of soils or vegetation, can also reveal features associated with hydrocarbon microseeps. Therefore, the ideal surface exploration approach is one that combines these methods with other techniques. This approach greatly increases the diagnostic value of surface exploration as an investigative tool to reduce exploration risk.

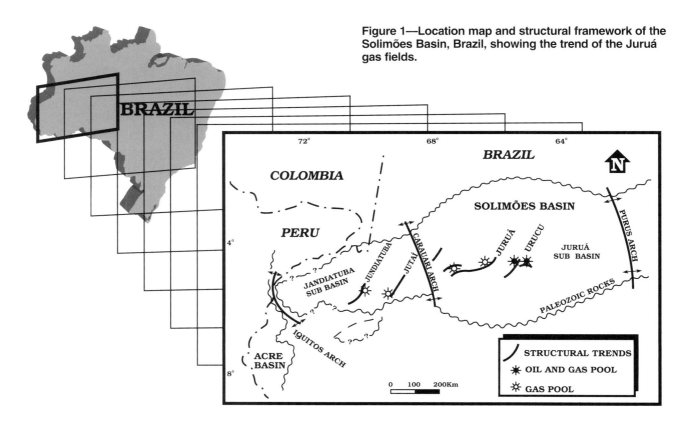

Figure 1—Location map and structural framework of the Solimões Basin, Brazil, showing the trend of the Juruá gas fields.

This study embraces a specific area in the Solimões Basin, Brazil, in which the Juruá gas accumulation is located (Figure 1). This gas accumulation was discovered in 1978 by Petrobrás on the basis of interpreted seismic data. To evaluate the occurrence of hydrocarbon microseeps in and around the Juruá gas field and to rank the exploration priority of the two structures to be drilled, a surface geochemical survey was carried out by Rezende et al. (1980). The results were so promising that a more detailed surface exploration study was carried out that included geochemical, microbiological, and remote sensing detailed surface exploration methods.

GEOLOGY AND PETROLEUM SYSTEM

The Solimões Basin is a wide interior sag (600,000 km²) located in northern Brazil and covered by the Amazon rain forest. It is separated from the Acre Basin by the Iquitos Arch on the west and from the Amazonas Basin by the Purus Arch on the east (Figure 1). The Solimões Basin is the only Brazilian Paleozoic basin with commercial hydrocarbon production. To the present, more than 110×10^9 m³ of gas and 70×10^6 m³ of oil and condensate (in-place) have been discovered (Murakami et al., 1992).

The sedimentary evolution of the Solimões Basin is characterized by transgressive and regressive cycles that are limited by wide regional unconformities (Figure 2). The approximately 4000 m of deposits consist mainly of Devonian and Carboniferous strata intruded by diabase sills and covered by Cretaceous and Tertiary rocks (Mosmann et al., 1986; Murakami et al., 1992). The Upper Devonian (late Frasnian–early Fammenian) marine black shales (Grahn, 1992) of the Jandiatuba Formation (Figure 2) comprise the hydrocarbon source rocks. The main reservoirs are Pennsylvanian eolian sandstones of the Juruá Formation, and the seal is provided by an overlying thick evaporitic layer of the Carauari Formation (Figure 2).

During the opening of the North Atlantic Ocean, the Paleozoic section of the Solimões Basin was intruded by Triassic–Jurassic (200 ± 20 Ma) diabase dikes and sills (Mizusaki et al., 1992), which are emplaced at three main distinct levels (Figure 2). The first level of sills occurs at the Fonte Boa Formation, the second level at the top of the Carauari Formation, and the third level at the base of the Carauari, close to reservoir and source rocks.

Normal faults of Devonian and Carboniferous age have controlled sedimentation and were associated with large, smooth structural highs. An intense Early Jurassic compressive tectonic event is responsible for the formation of an array of transpressional structures, particularly northeast-trending reverse faults associated with assymetrical anticlines, which also affected the diabase sills. Such structures comprise the main hydrocarbon field trends, such as the Juruá (Figures 1, 3, 4).

Upper Devonian source rocks are characterized by a thick section (up to 80 m) with total organic carbon (TOC) contents ranging from 2 to 12%. The low hydrocarbon source potential values and hydrogen indices (S_2 and HI from Rock-Eval analysis), combined with the observed vitrinite reflectance values ($R_o > 1.0\%$), indicate that the source rocks have reached an overmature stage of ther-

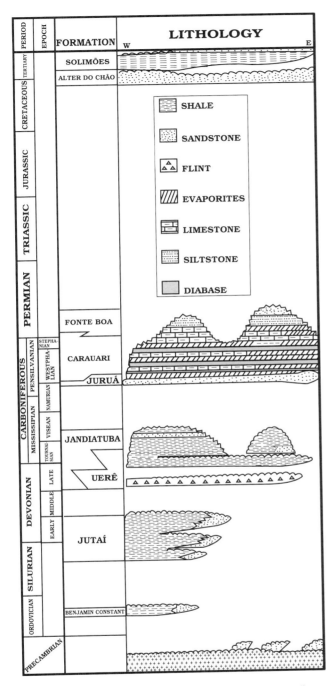

Figure 2—Schematic stratigraphic chart of the Solimões Basin. (Modified from Eiras et al., 1994).

mal evolution (Figure 5). The vitrinite reflectance data suggest that the heat effect of the diabase intrusions has strongly impacted the thermal evolution of both the source rock and the petroleum accumulations. In fact, regional mapping of maturation data (Figure 6) shows that the third level of diabase sills close to the reservoir and source rocks in the Juruá area is associated with the overmature stage, where mostly gas accumulations

occur. The absence of these intrusions in the Rio Urucu area is related to the highly mature stage that controls the presence of the gas and light oil fields (Rodrigues et al., 1990).

The most remarkable features of the analyzed oils and condensates are API gravities higher than 40°, saturate contents higher than 70%, $\delta^{13}C$ (PDB) values of about −28.0‰, and dominance of low molecular weight *n*-alkanes together with pristane greater than phytane. The biomarker and isotopic data of the light oils and organic extracts of the Upper Devonian shaly section show good correlation (Rodrigues et al., 1990). The geochemical data from the analyzed oils and the isotopic data ($\delta^{13}C$) of methane from gas samples, ranging from −34 to −38‰, are consistent with a highly mature stage of source rock thermal evolution.

In summary, in the Juruá area, the hydrocarbons sourced by the Upper Devonian marine black shales migrated during the Jurassic–Triassic and initially accumulated in Pennsylvanian eolian sandstone reservoirs structured by Paleozoic and Jurassic–Triassic tectonism. The thermal effect of the Jurassic–Triassic igneous intrusions cracked the source rock remnant and the trapped petroleum to light oil, condensate, and gas. Jurassic postmagmatic tectonism remobilized the previous accumulations, with the loss of large volumes of petroleum. Geological and geochemical data qualify the Solimões Basin as a condensate- and gas-prone area.

SURFACE GEOCHEMICAL AND MICROBIOLOGICAL STUDIES

To evaluate the occurrence of hydrocarbon microseeps in and around the Juruá gas field, a surface geochemical, microbial, and remote sensing study was carried out. For this purpose, surface soil and sediment samples were collected along and concomitantly with seismic lines acquired in the area. The samples were analyzed using geochemical and microbial techniques. After statistical treatment, the resulting data were mapped in order to correlate the microseep anomalies with the Juruá gas field and two other structural prospects outlined in the area. The objective was to rank two prospects using this auxiliary exploration tool along with available geological data.

Sampling Procedures

A total of 377 surface geochemistry samples were collected along 180 km of seismic lines in special drill holes. Holes were drilled every 500 m using a manual auger; the sampling depth ranged from 1.2 to 1.5 m. This procedure avoided hydrocarbons derived from biological processes. Samples were stored in sealed metal cans closed with appropriate lids. The amount of soil sample was about half the volume of the can.

Microbiological samples were collected at 40 previously selected points over and adjacent to the seismically

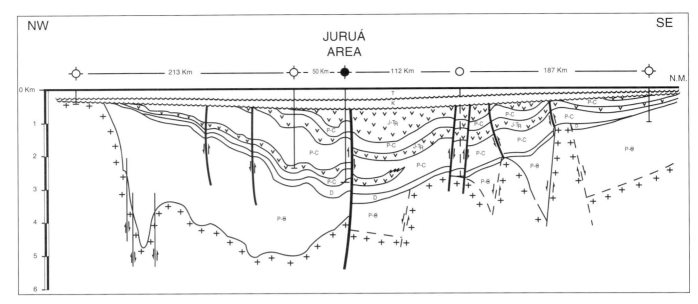

Figure 3—Schematic northwest–southeast structural cross section of the Solimões Basin, Brazil.

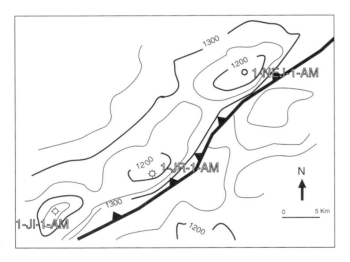

Figure 4—Structural map of the top of the Juruá Formation reservoirs drawn on the basis of seismic data. Locations of three wells are shown. Contours are in two-way traveltime in milliseconds.

mapped structures. To calibrate this survey with a known field, nine soil samples were collected over the Juruá gas field. Samples (200 g) were collected from 1-m-deep trenches using a sterilized spatula and transferred into sterilized marked plastic bags. These samples were refrigerated until arrival at the laboratory in Rio de Janeiro.

The surface geochemistry samples were analyzed by high-resolution gas chromatography, and after statistical treatment, the resulting data were mapped to locate microseep anomalies. Microbiological samples were incubated and analyzed by the Petrobrás ethanol method, and the results were reported as number of organisms per gram of soil.

Analytical Procedures

The surface geochemistry samples were submitted to high-resolution gas chromatography analysis. The cans were first agitated in a multidirectional mechanical agitator for 5 min. Then, they were subjected to ultrasonic homogenization at 40°C for 15 min and left to rest for 12 hr. The gaseous fraction, homogeneously dispersed in the headspace, was then sampled with a 500-µL syringe.

The analyses were carried out on an HP-5890A gas chromatograph equipped with a flame ionization detector and hydrogen as carrier gas. A 50-m-long fused silica column with Al_2O_3/KCl was used as the stationary phase. Samples were injected in split mode (split ratio = 1:50) at 80°C. The column was programmed to 180°C at a rate of 6°C/min, and the oven temperature remained at 180°C for 13 min. As a result, gas chromatograms with a distribution of linear, branched, and cyclic alkanes and alkenes up to seven carbon atoms were obtained. Data acquisition was done by the HP-3350A Laboratory Automation System which integrates, identifies, and quantifies all hydrocarbon peaks and reports the chemical composition of the samples in parts per million.

The microbiological samples were distributed in specially designed Petri plates. Each sample was separated into five equal portions with different dissolution patterns. After a 28-day period of incubation at 38°C, samples were analyzed by the Petrobrás NMP ethanol method. The results were reported as the number of microorganisms per gram of dry sample.

Statistical Treatment

An important factor in the application of surface organic geochemistry as an exploration tool is the definition of background and anomalous values. The problem

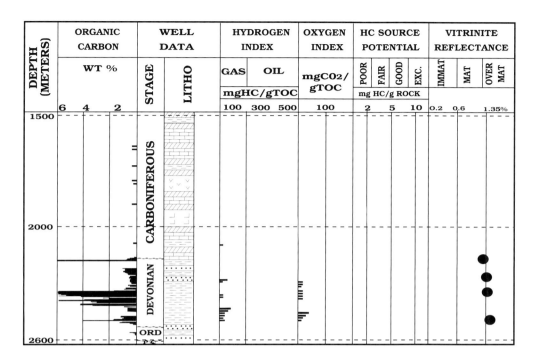

Figure 5—Typical geochemical well log showing the Upper Devonian organic-rich sediments of the Jandiatuba Formation. (After Murakami et al., 1992).

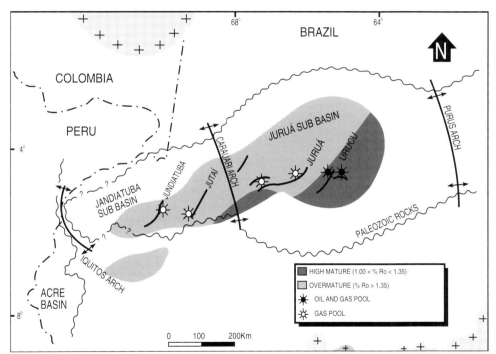

Figure 6—Map of thermal maturity of the Upper Devonian source rocks showing the distribution of highly mature and overmature stages. (From Rodriques et al., 1990.)

is how to synthesize the enormous amount of data into simple maps indicating the zones of related relevant targets. Isovalues maps are not adequate by themselves because high values might be associated with secondary concentrations of organic matter or with dispersion due to topographic influence. The problem can be considered as a monovariant or multivariant one. In the monovariant case, the usual procedure is to define an anomalous cut-off value from the statistical distribution curve before mapping.

For statistical analysis of the geochemistry data points around the Juruá gas field, a statistical graphic system (Statigraphics, version 4.0) was used based on previous experience (Rezende et al., 1980; Mello et al., 1992). The statistical procedures were as follows:

1. To obtain an overview and to study the shape of the distribution grades of all variables, a frequency histogram of the total data was prepared. The graphs were used to ensure a log-normal

Figure 7—Map of methane anomalies in the Juruá area.

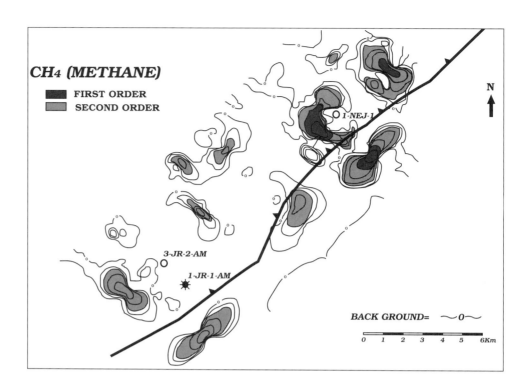

distribution for data descriptions and to reveal the presence of possible outliers.

2. Using the natural log of the observed data, normal probability plots were processed for each variable by the Statigraphics program. In each case, after analyzing the histogram, a visual estimate was made and only one population was considered. Individual histograms, normal probability plots, and summary statistics were obtained.

3. Regardless of the geological setting, the results were interpreted as a whole anomaly, and values greater than 1, 2, or 3 standard deviations from the mean (M) characterized first, second, or third order anomalies. These values were converted to parts per million before mapping.

4. Interpretation of one of the populations was straightforward. The background was M, and anomalous values were grouped in M+1, M+2, and M+3 standard deviations, respectively.

Results and Discussion

The prime objective of the geochemical prospecting study in the Juruá area was the detection and quantification of free, absorbed, and adsorbed hydrocarbon gases chemically or physically connected to the soils, sediments, and rocks. The method included direct sampling and mechanically or chemically induced release prior to concentration.

The light hydrocarbon gas data and the statistical results were used to make a series of maps of concentrations of methane, ethane, propane, butane, pentane, ethane-pentane, hexane, gases heavier than hexane, and

gases heavier than ethane. The maps were made with the GEOMAP geological mapping package, available in Petrobrás' IBM 3090 computer. A grid for each gas was made over the studied area with cell dimensions of 500 × 500 m. Shading from light to dark gray was used to indicate increasing concentrations of hydrocarbons.

Care was taken in interpreting anomalies because of the grid parameters used to make the maps. The sampling points did not constitute a regular network, and thus some anomalies were better defined than others according to the number of points they represented. Grid parameters were defined to make long-distance extrapolations and not to leave "blank" areas. For this reason, some anomalies may look exceedingly large. This effect can be minimized by considering only those points that actually define the anomaly and their distribution in space.

The methane concentration map (Figure 7) shows that most of the area contains values higher than the mean value of the population. Major anomalies are associated with a northeast-southwest trend that is related to a coincident set of faults, which probably served as conduits for hydrocarbon migration. Locally, the higher anomalies are associated with structure 1-NEJ-1 situated northeast of the trend. As a whole, the methane concentrations decrease gradually to the southwest (Figure 7). The anomalies within the seismically mapped structure form a longitudinal or linear northeast-southwest anomaly. This elongate anomaly type is generally related to surface expressions of major fault systems (Sokolov, 1970).

The anomalous concentrations of C_{2+} hydrocarbons detected in soil samples of the area are also concordantly aligned with the trace of the major northeast-southwest reverse fault system present in the area (Figure 8).

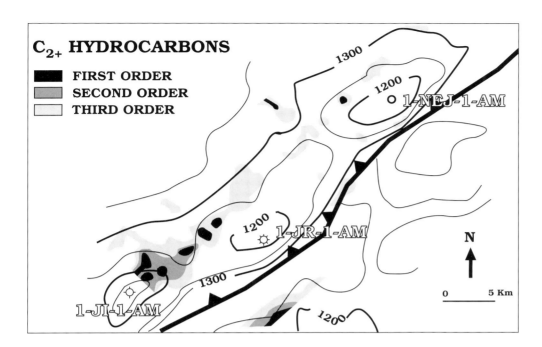

Figure 8—Map of anomalies of hydrocarbons heavier than methane (C$_{2+}$) and structural contours of the top of the Juruá Formation based on seismic data.

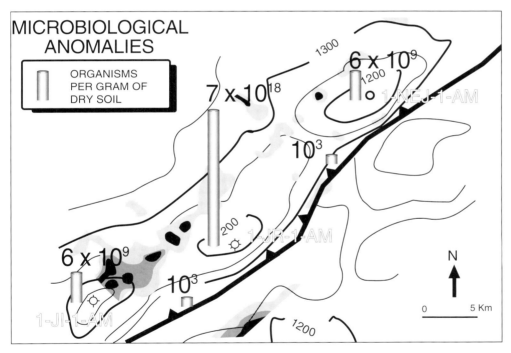

Figure 9—Map of microbiological and C$_{2+}$ anomalies and structural contours of the top of the Juruá Formation based on seismic data.

Compared to the methane anomalies shown in Figure 7, the concentrations of C$_{2+}$ compounds present a strong anomaly in the southeastern area associated with the 1-JR-1 structure. The anomaly increases greatly from northeast to southwest (Figure 8). A few other anomalies are located in the northwestern area of the block, but due to their lack of consistency, they must be viewed with caution.

The C$_{2+}$ concentration map (Figure 8) summarizes the results of all the gas maps because it is a summation of their individual values. The first, second, and third order anomalies shown in Figure 8 represent the contours of

M+1, M+2, and M+3 standard deviation values for C$_{2+}$ gases. As can be seen, the stronger anomalies (first and second order) show a consistent longitudinal configuration in the southwestern area with a high concentration of C$_{2+}$ hydrocarbons. Such anomaly systems are typically the result of migration from a possible underlying liquid hydrocarbon accumulation with major fault control. Based on these factors, the presence of a well-controlled structural trap (e.g., 1-JR-1) below the hydrocarbon surface anomalies is considered of prime importance in ranking exploration targets in the area.

Figure 10—Map of morphostructural anomalies (dome) in the Juruá gas field area and basement structural contours based on seismic data. Contours are in two-way traveltime (20-msec intervals). Curved and straight dashed lines indicate annular and radial drainage, respectively. (After Miranda and Boa Hora, 1986.)

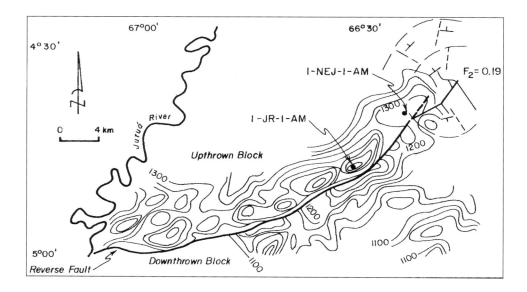

REMOTE SENSING STUDIES

Morphostructural Analysis

The prospective Paleozoic section in the Solimões Basin is unconformably covered by Mesozoic and Cenozoic continental sediments up to 500 m thick (Figure

By extending the geochemical prospecting survey in the area, a preliminary geomicrobial survey was applied over the previously drilled structures around the 1-JR-1, 1-JU-1, and 1-NEJ-1 wildcat wells. Microbiological samples were collected at 40 previously selected points over and adjacent to the seismically mapped structures. To calibrate this survey with known features, three soil samples were collected over each of the structures. Although not very consistent (a small number of samples and the lack of a regular grid), the analytical results suggest the occurrence of medium to high microbial populations ($>10^4$ organisms/g dry soil) associated with all mapped structures (Figure 9). In addition, the negative to low results ($<10^3$ organisms/g dry soil) from the microbiological survey are in agreement with the absence of surface hydrocarbon anomalies.

The results of the microbiological survey were compared to the light gas anomalies (Figure 9) to check the consistency of the geochemical results and to help rank the more prospective areas. As can be seen, the best structure to be tested is the 1-JR-1 because it was analyzed by two different techniques, each of which measures different physical-chemical and microbiological parameters. Such an approach greatly increases the diagnostic value of prospecting geochemistry as a predictive hydrocarbon tool in the area. The integration of these data within the geological, hydrological, and environmental framework of the area allows a better understanding of the real significance of hydrocarbon anomalies related to the major northeast-southwest reverse fault system.

2). In addition, jungle vegetation and intensive weathering cause many problems for photogeological interpretations. At first glance, these factors seem to complicate the detection of features of the Juruá structural trend using remote sensing techniques. However, a significant contribution to regional studies has been achieved in this area by morphostructural analysis of Landsat MSS and SLAR imagery at the 1:250,000 scale (Cunha and Carneiro, 1978; Miranda, 1984). The main objective of these studies was to define the morphostructural anomalies, which can be indicative of hydrocarbon traps.

Morphostructural anomalies, as defined by Soares et al. (1981), are based on the simultaneous occurrence of annular, radial, and asymmetric drainage. It is thought that such configurations represent drainage networks controlled by underlying structures such as domes and structural depressions. It is assumed that the underlying structures are reflected at the surface by tectonic reactivation or differential compaction, which makes the identification of such features feasible with the aid of morphostructural analysis.

To characterize the morphostructural anomalies in an objective manner, Soares et al. (1981) proposed the use of two quantitative parameters: the similarity factor and the confidence factor. The *similarity factor (F_2)* estimates the extent to which the configuration of a given morphostructural anomaly differs from the ideal model of a dome. The *confidence factor (F_1)* indicates how sure the interpreter is about the determination of the dip slope and the plane of surface bedding with the aid of the drainage network. The characterization of a morphostructural anomaly according to both F_1 and F_2 is discussed in detail by Miranda and Boa Hora (1986).

By using these parameters, the drainage network was quantitatively evaluated in the study area. This analysis led to the detection of a morphostructural anomaly representative of a dome in the vicinity of the gas-producing well 1-NEJ-1-AM (Figure 10). Furthermore, topographic maps at 1:100,000 scale show that the gas-producing

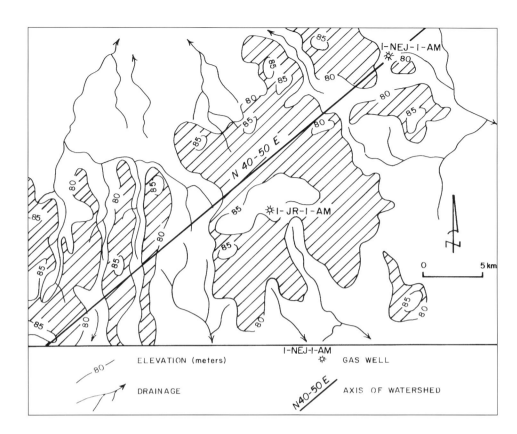

Figure 11—Morphologic outline of the Juruá area. (Modified from Cunha, 1982.)

wells of the Juruá field are located over a 80-m-high watershed (Cunha, 1982). This morphological feature is aligned with the northeast-southwest structures of the Juruá trend (Figure 11).

Digital Analysis of Landsat MSS Data

A digital analysis of Landsat MSS data was performed over the Juruá area (scene 001-63, acquired July 13, 1976). The output of the Landsat MSS is a set of brightness measurements corresponding to bands 4, 5, 6, or 7. As suggested by Swain (1978), it is convenient to think of the four measurements as defining a four-dimensional space, which is referred to as the *measurement space*. Any point in the measurement space can be represented by a measurement vector. The "decision maker" or "classifier" ascribes each measurement vector to one specific class according to a given classification approach. Classification of the MSS data for the Juruá gas field was performed in an unsupervised fashion by using the K-MEANS clustering algorithm. The unsupervised classifier divides a given amount of remote sensing data into categories in a nonsubjective manner according to a preestablished criterion (a clustering algorithm), without making use of a training phase.

The use of an unsupervised classification proved to be the most effective approach to identify the existence of a selectively distributed spectral class in the vegetation cover. It shows that one spectral class is aligned with the Juruá structural trend (Figure 12) (Cunha, 1979; Miranda

and Cunha, 1981). There are several geological factors that may account for such a coincidence:

1. The topographic map of the Juruá gas field shows a tabular watershed aligned with the northeast-southwest structural trend (Figure 11). This morphological feature corresponds to an erosional surface of Pleistocene age associated with weathering products such as bauxite and laterite (Oliveira et al., 1977). The soil in this area is impermeable and low in nutrients (Cunha, 1982).
2. The anomalous spectral behavior of the vegetation over the gas and condensate fields may represent the response of plants to long-term anaerobic soil conditions which, in turn, are related to gas leakage from the Paleozoic reservoir (Figures 7, 8, 9).

In any case, Cunha (1982) reported that the vegetation near the gas- and condensate-producing wells (e.g., 1-JR-1-AM) is not as dense as the forest outside their limits.

CONCLUSIONS

The combined use of surface geochemical, microbiological, and remote sensing methods with full technological and scientific support, integrated with geological and geophysical data, allowed the priority ranking of exploration leads mapped in the Solimões Basin, Amazon rain forest, Brazil. The integration of all data indicates that the

Figure 12—Unsupervised classification (K-MEANS algorithm) performed in the Juruá area (Landsat MSS scene 001-63), assuming four initial centers (K = 4). Note the spectral cluster aligned with the Juruá structural trend. (After Cunha, 1979.)

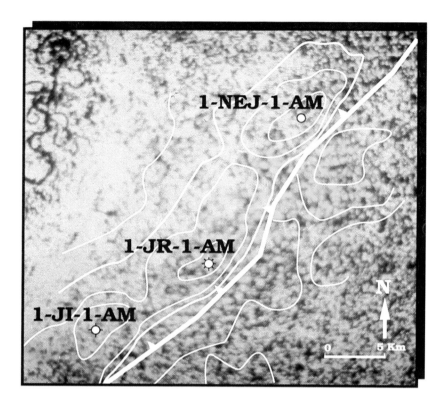

1-JR-1 prospect in the southwestern area of the block has the most consistent microbiological and hydrocarbon anomalies, which range from C_2 to C_{6+}. The anomalies are in and around the seismically mapped structures forming a northeast-southwest longitudinal or linear anomaly. Such an elongate anomaly is generally related to surface expression of major fault systems (Kartsev et al., 1959; Sokolov, 1970). Indeed, the anomalous concentrations of hydrocarbons and hydrocarbon-consuming bacteria detected in the soil are concordantly aligned with the trace of a major reverse fault. A few other anomalies are located in the northwestern area of the block, but due to their lack of consistency they must be interpreted with caution.

The remote sensing survey (Landsat MSS data) showed an anomalous spectral response of the vegetation over the geochemical anomalies, which may represent the response of plants to long-term anaerobic conditions due to the gas leakage from subsurface reservoir rocks.

Integration of all geochemical analyses with geological and geophysical data allowed the recognition of areas that were more likely to contain hydrocarbon accumulations in the subsurface. Further exploration in the area proved the success of the 1-JI-1 prospect (southwestern anomaly) with the discovery of a large gas-condensate accumulation. The drilling of the northeastern anomaly (structure 1-NEJ-1), linked with the higher values of methane anomalies, proved to contain a significant dry gas acumulation.

An integration of all data allowed the recognition of structural leads which are more likely to contain hydro-carbon accumulations in the subsurface. Further exploration proved the existence of gas and gas-condensate acummulations. Therefore, the use of surface geochemistry, microbiological, and remote sensing techniques proved to be cost effective and contributes to constrain exploration risk.

Acknowledgments—We would like to thank the Geochemistry Section of Petrobrás Research and Development Center (CENPES) for all of the analyses. C. B. Eckardt, L. A. F. Trindade, R. Sassen, A. G. Requejo, and Regina C. R dos Santos for their helpful comments and revision of the text, and Ricardo M. P da Silva for editing the figures. We also thank Petrobrás for permission to publish this paper.

REFERENCES CITED

Cunha, F. M. B., 1979, Estudo preliminar das microexsu-dações do Rio Juruá–AM por imagens de satélite: Petrobrás, Internal Report, 17 p.

Cunha, F. M. B., 1982, Significado das reflectâncias anômalas registradas na vegetação da Bacia do Alto Amazonas: II Simpósio Brasileiro de Sensoriamento Remoto, v. 1, p. 169–177.

Cunha, F. M. B., and R. G. Carneiro, 1978, Interpretação morfológica preliminar da área do Rio Juruá: Petrobrás, Internal Report, 6 p.

Eiras, J. F., C. R. Becker, E. M. Souza, F. G. Gonzaga, J. G. F. Silva, L. M. F. Daniel, N. F. Matsuda, and F. J. Feijó, 1994, Bacia do Solimões (Resumo da revisão estratigráfica): Boletim de Geociências da Petrobrás, Rio de Janeiro, v. 8, p. 17–46.

Grahn, C. Y., 1990, Evaluation of the chitinozoan biostratigraphy of the Solimões Basin: Petrobrás, Internal Report, 34 p.

Horvitz, L., 1939, On geochemical prospecting, I: Geophysics, v. 4, p. 210–218.

Link, W. K., 1952, Significance of oil and gas seeps in world oil exploration: AAPG Bulletin, v. 36, p. 1505–1540.

Mello, M. R., N. A. Babinski, H. L. B. Penteado, and M. F. G. Meniconi, 1992, Surface geochemistry survey in the NC-162 Block, Ghadames Basin, Libya: Petrobrás, Internal Report, 36 p.

Miranda, F. P., 1984, Significado geológico das anomalias morfo-estruturais da Bacia do Alto Amazonas: II Symposium Amazônico, v. 1, p. 103–116.

Miranda, F. P., and F. M. B. Cunha, 1981, Automatic analysis of Landsat multispectral data as a source of additional information for hydrocarbon exploration in the Alto Amazonas Basin: Paper presented at the COGEODATA-IAMG Meeting for South America, Rio de Janeiro, Brazil.

Miranda, F. P., and M. P. P. Boa Hora, 1986, Morphostructural analysis as an aid to hydrocarbon exploration in the Amazonas Basin, Brazil: Journal of Petroleum Geology, v. 9, p. 163–178.

Mizusaki, A. M. P., J. R. Wanderley Fo, and J. R. Aires, 1992, Caracterização do magmatismo básico das bacias do Solimões e Amazonas: Petrobrás, Internal Report, 57 p.

Mosmann, R., F. U. H. Falkenhein, A. Gonçalves, and F. Nepomuceno F°, 1986, Oil and gas potential of the Amazon Paleozoic basins: AAPG Memoir 40, p. 207–241.

Murakami, C. Y., F. T. T. Gonçalves, J. F. Eiras, C. R. Becker, M. P. Lima, and L. M. F. Daniel, 1992, Habitat of petroleum in the Solimões Basin, Brazil: Third Latin American Congress on Organic Geochemistry, Manaus, Brazil, Extended Abstracts, p. 113–115.

Oliveira, A. A. B., I. H. L. Pitthan, and M. G. L. Garcia, 1977, Folha SB. 19 Juruá—II Geomorfologia: Projeto RADAMBRASIL, MME-DNPM, v. 15, p. 91–142.

Rezende, W. M., M. R. Mello, and C. Bettini, 1980, Detecção de hidrocarbonetos por prospecção geoquímica terrestre (projeto experimental): Proceedings of the 31st Brazilian Congress of Geology, Camboriú, Santa Catarina, v. 1, p. 431–443.

Rodrigues, R., J. A., Trigüis, C. V. Araújo, and I. R. Brazil, 1990, Geoquímica e faciologia orgânica dos sedimentos da Bacia do Solimões: Petrobrás, Internal Report, 115 p.

Soares, P. C., J. T. Mattos, et al., 1981, Análise estrutural integrada com imagens de Radar e Landsat na Bacia do Paraná: Paulipetro-Consórcio IPT-CESP Internal Report, 21 p.

Sokolov, V. A., 1936, New method of surveying oil and gas formations: Tekhnnefti, Gas surveying, Moscow, Gostoptekhizdat, v. 16, 269 p.

Sokolov, V. A., 1970, The theoretical foundation of geochemcial prospecting for petroleum and natural gas and the tendencies of its development: International Geochemical Exploration Symposium, Toronto, Canada, p. 544–549.

Swain, P. H., 1978, Fundamentals of pattern recognition in remote sensing, *in* P. H. Swain and S. M. Davis, eds., Remote sensing: a quantitative approach: New York, McGraw Hill, p. 136–153.

.

Kornacki, A. S., 1996, Petroleum geology and geochemistry of Miocene source rocks and heavy petroleum samples from Huasna Basin, California, *in* D. Schumacher and M. A. Abrams, eds., Hydrocarbon migration and its near-surface expression: AAPG Memoir 66, p. 413–430.

Chapter 32

Petroleum Geology and Geochemistry of Miocene Source Rocks and Heavy Petroleum Samples from Huasna Basin, California

Alan S. Kornacki

Shell Offshore, Inc.
New Orleans, Louisiana, U.S.A.

Abstract

Only subcommercial accumulations of heavy oil have been discovered in the Huasna Basin, which contains numerous active oil seeps and several source rocks. The petroleum geochemistry of a heavy oil sample from the only field in the basin and of five tar samples obtained at seeps has been determined and compared to the chemistry of core samples of Monterey and Rincon shales. The bulk chemistry and mineralogy of these Miocene shales can be used to identify several lithofacies, including phosphatic and siliceous shales containing variable amounts of detrital clay and carbonate minerals. Pyrolysis-FID and visual kerogen analysis demonstrate that Monterey and Rincon shales are good quality, oil-prone source rocks in the Huasna Basin. Thermal maturity modeling calibrated using bottom-hole temperature data and vitrinite reflectance measurements in several wells indicate that Monterey and Rincon source rocks are thermally mature in the axial syncline of the Huasna Basin.

The tarry petroleum obtained at seeps is more severely biodegraded than a heavy oil sample from the Huasna field. The chemistry of these petroleum samples, especially their V/Ni ratios and sulfur concentrations, can be used to classify them into distinct groups. High-sulfur samples contain >4 wt. % sulfur and are enriched in V relative to Ni. Low-sulfur samples contain <2 wt. % sulfur and are enriched in Ni relative to V. One tar sample exhibits an intermediate chemistry. Sour oil and tars from the western flank of the Huasna Basin were generated by phosphatic Monterey shales or clay-poor siliceous Monterey shales containing sulfur-rich (Type II-S) kerogen. The sweeter petroleum seeping along the eastern margin of the basin was probably generated by more proximal, clay-rich Miocene shales containing Type II kerogen. Monterey lithofacies variations in the Huasna Basin and the distribution of subsurface oil and gas support this interpretation, as does the organic geochemistry of bitumen samples extracted from mature shales. Thus, inferred differences in Miocene oil quality within the Huasna Basin seem to have been influenced primarily by source effects (with a significant overprint of biodegradation) rather than by thermal maturity effects alone.

INTRODUCTION

The presence of oil macroseeps or tar outcrops in a sedimentary basin is direct evidence that a petroleum system is present. Oil seeps and tar outcrops do not indicate hydrocarbon charge volumes, but their geochemistry can be used to perform oil–source rock correlations and to constrain independent geologic models of source rock maturity and depositional environments. Analysis of petroleum seepage can therefore play an important role in evaluating the exploration potential of a basin (Macgregor, 1993).

Unfortunately, the petroleum in most oil seeps or outcrops has generally been altered by such secondary

processes as biodegradation, water washing, and evaporative loss of volatile compounds. The effects of these transformation processes compromise the utility of some standard geochemical methods used to characterize the source and maturity of petroleum. For this reason, the biomarker chemistry of seep samples and their sulfur, nickel, and vanadium concentrations are commonly used to characterize seeps (Miiller et al., 1987).

The Huasna Basin is a small, fault-bounded basin in central California that lies adjacent to the prolific Santa Maria Basin (Figure 1). Both of these basins contain thick deposits of Neogene sediments, including the oil-prone Monterey Formation (Hall and Corbato, 1967; Crawford, 1971). In contrast to the Santa Maria Basin, however, only

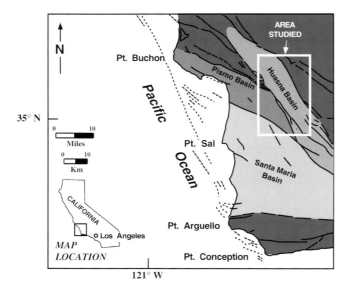

Figure 1—Index map of the Huasna Basin in coastal California (adapted from Hall, 1978). This small, elongate basin lies northeast of the prolific Santa Maria Basin.

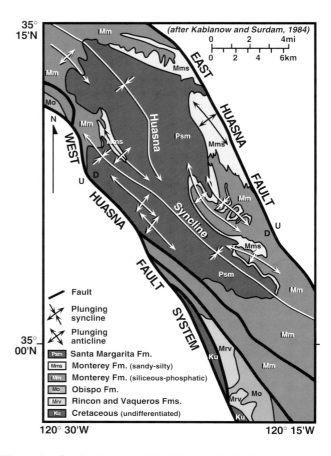

Figure 2—Geologic map of the Huasna Basin (adapted from Hall and Corbato, 1967; Hall, 1973a,b; Kablanow and Surdam, 1984). Neogene sediments outcrop in this narrow, fault-bounded basin. The Huasna syncline is the dominant structural feature.

subcommercial oil accumulations have been discovered in the Huasna Basin (King, 1943; Bell, 1951). Kablanow and Surdam (1984) concluded that Monterey source rocks are thermally mature in the Huasna Basin and their conclusion is supported by the presence of numerous oil seeps (Hodgson, 1980). The purpose of this study was to perform a detailed evaluation of the petroleum charge in the Huasna Basin, including identification and characterization of source rocks and petroleum systems, determination of source rock thermal maturity, and geochemical analysis and correlation of oil, tar seep, and bitumen samples.

GEOLOGY OF HUASNA BASIN

The Huasna Basin is an elongate, doubly plunging, curvilinear synclinorium (Figure 2). It is bounded by the steeply dipping East Huasna and West Huasna faults, which exhibit both strike-slip and dip-slip components. Fold axes in the Huasna Basin are generally subparallel to these boundary faults. Johnson and Page (1976) and McLean (1991) have described the development and style of folding in the Huasna Basin in detail.

Figure 3 summarizes the stratigraphy in the Huasna Basin. In the Santa Maria Basin, the middle Miocene Point Sal and Monterey formations generally rest unconformably on Franciscan basement rocks or isolated remnants of Mesozoic clastic sedimentary rocks (Gray, 1980; Hall, 1982). But in the Huasna Basin, correlative shales are underlain by the Oligocene–lower Miocene Sespe, Vaqueros, and Rincon formations (Taliaferro, 1943). This transgressive marine sequence resembles the pre-Monterey stratigraphy in the Ventura Basin. In addition,

the Huasna Basin contains deposits of early middle Miocene Obispo volcaniclastic sedimentary rocks. The Monterey Formation also undergoes a significant lithologic change across the Huasna Basin, grading from pelagic siliceous and phosphatic shales along the western, southern, and southeastern flanks of the basin to more proximal siltstone and fine-grained sandstone along the northeastern flank (Figure 4) (Kablanow and Surdam, 1984).

SOURCE ROCKS: CHEMISTRY, MINERALOGY, AND PETROLEUM POTENTIAL

The Huasna Basin contains Cretaceous, lower Miocene, and middle Miocene shales that could be petroleum source rocks. The mineralogy, inorganic chemistry, and organic richness of these possible source rocks were determined by analyzing whole-core or cuttings samples from 11 wells identified in Table 1. The locations of these wells are shown in Figure 5.

Siliceous Monterey Shales

Siliceous Monterey shales in the Huasna Basin consist largely of biogenic silica that was initially deposited as a diatomaceous ooze (Figure 6). The amorphous opal comprising the tests of these diatoms has generally been converted to opal-CT or diagenetic quartz in the Huasna Basin (Kablanow and Surdam, 1984). Clay-poor ("clean") siliceous shales commonly contain ≥80 wt. % SiO_2 and have low concentrations of Al_2O_3 (≤5.0 wt. %), Fe_2O_3 (≤2.0 wt. %), and P_2O_5 (≤0.50 wt. %) (Table 2). As would be expected, clay-rich siliceous Monterey shales are more enriched in Al_2O_3, Fe_2O_3, and some trace elements (e.g., barium and zirconium). Dolomitic siliceous shales and siliceous dolomites, in contrast, contain more carbonate (typically ≥5 wt. % CO_2) and lower concentrations of SiO_2, Al_2O_3, Fe_2O_3, P_2O_5, and uranium (due to dilution by diagenetic dolomite). Kablanow et al. (1984) have described the chemistry and origin of these dolomites in detail.

Siliceous Monterey shales in the Huasna Basin are good oil-prone source rocks, typically containing ~1.4–4.5 wt. % total organic carbon (TOC) and ≥0.70 wt. % total hydrocarbon (HC) yield (Table 2; Figure 7). Core samples from several deep wells exhibit vitrinite reflectance (R_o) values of ≥0.5%, a level of thermal maturity at which the sulfur-rich kerogen common to this lithofacies probably has entered the oil window (Baskin and Peters, 1992). This suggests that the original HC yield of some siliceous shale samples was probably greater than that measured. Visual kerogen analysis of several samples of siliceous Monterey shale confirms that this lithofacies contains Type II kerogen; >85% oil-prone macerals (mainly structureless organic matter, SOM) and <10% gas-prone macerals were identified (Figure 8).

Phosphatic Monterey and Point Sal Shales

Bulk samples of these lithofacies, which can be characterized as phosphatic marls (Figure 6), contain relatively large amounts of phosphorus (≥1.0 wt. % P_2O_5) and uranium (≥10 ppm). Phosphatic shales from the Monterey and informally named "Point Sal" generally contain more aluminum, iron, and sulfur than siliceous Monterey shales and highly variable amounts of carbonates (Table 2). These shales are excellent oil-prone source rocks, containing >3 wt. % TOC and >1.5 wt. % HC. Visual kerogen analysis demonstrates that their organic matter is comprised of SOM-rich Type II kerogen with <10% gas-prone macerals (Figure 8).

Rincon Shales and Cretaceous Shales

Rincon shales are calcareous claystones that contain high concentrations of aluminum (~10–12 wt. % Al_2O_3) and iron (~4–5 wt. % Fe_2O_3), moderate amounts of carbonates and uranium, and low concentrations of phos-

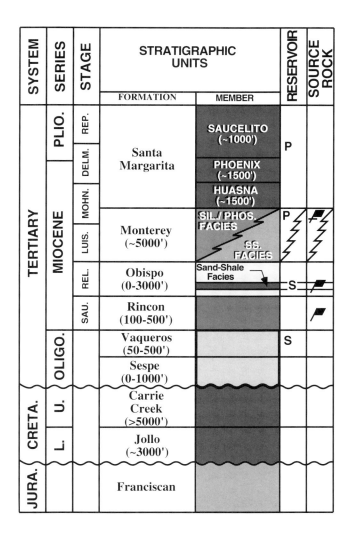

Figure 3—Stratigraphy of the Huasna Basin (age and thickness of rock units adapted from Hall and Corbato, 1967; Hall, 1973a,b). Proven reservoirs (P), potential reservoirs with good petroleum shows (S), and identified source rocks (flag symbols) are also shown.

phorus (generally <0.5 wt. % P_2O_5) (Table 2). Although not as enriched in organic matter as phosphatic Monterey shales, Rincon shales are nevertheless good petroleum source rocks in the Huasna Basin (commonly containing ~2.5–3.5 wt. % TOC and ~1.5–2.0 wt. % HC) (Figure 7). Visual kerogen analysis indicates that the kerogen in Rincon shales is a mixture of oil-prone and gas-prone macerals (Figure 8). Correlative lower Miocene shale outcrops along the Santa Barbara Channel exhibit similar source rock potential (Stanley et al., 1993). However, Cretaceous shales sampled at three locations in the Huasna Basin are poor petroleum source rocks, containing only <0.10 wt. % HC.

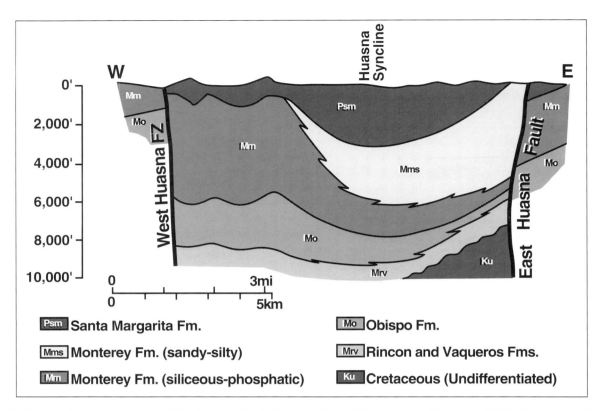

Figure 4—Schematic cross section of the Huasna Basin (adapted from Kablanow and Surdam, 1984). The sandy-silty facies of the Monterey Formation is well developed along the eastern margin of the basin.

Table 1—Inventory of Source Rock and Petroleum Samples from Huasna Basin

Map No.	Well Name	Location	Number and Type of Samples
Source Rock Samples			
1	Hunt Tar Springs Ranch 1	S25-T32S-R14E	2 Monterey and 1 Rincon core sample
2	Hancock Sherer-Dickes 1	S30-T12N-R33W	11 Monterey core samples
3	W. Gulf Huasna Community 1	S35-T31S-R14E	7 Monterey core samples
4	Standard Jessup 1	S18-T32S-R15E	2 Monterey core samples
5	Standard Porter 1-1	S21-T32S-R15E	9 Monterey core samples
6	Phillips Kerckhoff A-2	S14-T32S-R15E	5 Monterey, 4 Rincon, and 1 Cretaceous well cuttings
7	Danciger Twitchell 1	S35-T32S-R15E	5 Monterey core samples
8	Phillips Porter D-2	S11-T11N-R33W	13 Monterey well cuttings
9	Honolulu Rancho Suey 1	S22-T11N-R33W	5 Rincon and 1 Cretaceous core sample
10	Brit.-Amer. Rancho Suey C-1	S5-T10N-R32W	4 Monterey core samples
17	Honolulu Sunray-Sisquoc 1	S5-T9N-R31W	6 Monterey and 1 Cretaceous core sample
Heavy Oil Sample			
11	Unidentified well in the LaVoie-Hadley area of the Huasna field	S30-T12N-R33W	Collected from bottom of well casing
Tar Seep Samples			
12	Tar Springs Ranch	S14-T32S-R14E	Seep #3-46 in Hodgson (1980)
13	Arroyo Grande Creek, Lopez Lake	S35-T31S-R14E	Seep #3-40 in Hodgson (1980)
14	Rust Ranch	S34-T32S-R15E	Seep #3-56 in Hodgson (1980)
15	Carrie Creek, Stephens Canyon	S35-T32S-R15E	Seep #3-57 in Hodgson (1980)
16	Huasna River	S14-T11N-R33W	Seep #3-61 in Hodgson (1980)

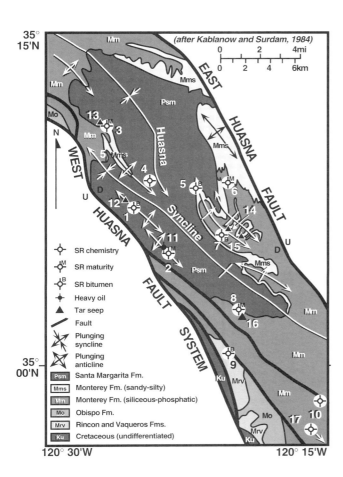

Figure 5—Location map of wells and seeps in the Huasna Basin at which source rock, bitumen, tar, and heavy oil samples were obtained. The wells and seeps are further identified in Table 1.

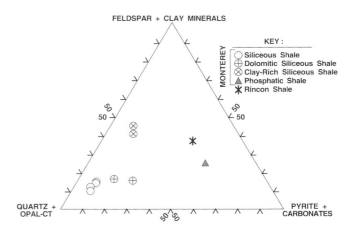

Figure 6—Mineralogy of representative samples of Monterey shale and Rincon shale in the Huasna Basin. Different facies of siliceous Monterey shale contain variable amounts of biogenic silica, clastic minerals (clay, silt, and feldspar), and diagenetic carbonate minerals (principally dolomite).

THERMAL MATURITY OF MIOCENE SOURCE ROCKS

Geothermal Gradients

The present geothermal gradient in the Huasna Basin was determined by analyzing bottom-hole temperature (BHT) data from all wells deeper than 2000 ft (~600 m). Unfortunately, Horner corrections could be applied to BHT data from only two recent wells (Phillips Kerckhoff A-2 and Phillips Porter D-2). These data yielded disparate geothermal gradients of ~1.5° and ~2.25°F/100 ft (~27° and ~41°C/km), respectively. BHT data from the much larger set of older wells in the basin that were corrected empirically suggest that the current geothermal gradient in the Huasna Basin is ~2.0°F/100 ft (~36°C/km).

The geothermal gradient in the Huasna Basin has been influenced by previous episodes of volcanism, uplift, and subsidence. Paleogeothermal gradients prevailing during the burial of Miocene source rocks were estimated by empirically correcting the present geothermal gradient to

account for the thermal effects of geologic processes (using proprietary methods developed at Shell Oil Company). For example, the thick blankets of cold Monterey and Santa Margarita sediments that were deposited from middle Miocene to early Pliocene time would have cooled the basin and temporarily lowered the geothermal gradient by ~0.5°F/100 ft (~9°C/km). In addition, the presence of volcanic sediments in this basin records a period of anomalously high heat flow during the early middle Miocene (Heasler and Surdam, 1985).

Source Rock Thermal Maturity Measurements and Modeling

The thermal maturity of Monterey and Rincon oil-prone source rocks in the axial syncline of the Huasna Basin was determined after first calibrating their modeled maturity to their measured maturity in four key wells (Figure 5). Good matches were obtained in the calibration wells between the measured and modeled maturity of Monterey and Rincon shales using reasonable geothermal gradients and corrections for the inferred erosion of younger strata at the well sites that are currently preserved in the axis of the Huasna syncline (Figure 9). The presence of diagenetic quartz in shallow Monterey sediments near some of these wells demonstrates that a significant amount of erosion has occurred (Kablanow and Surdam, 1984).

Thermal maturity parameters calibrated in the key wells were subsequently used to model the maturity of Miocene source rocks in the elongate "cooking pot" straddling the axial syncline of the Huasna Basin (Figure 10). These results indicate that Rincon shales have attained a moderately high level of thermal maturity ($R_o \approx 0.9\%$) in the axis of the Huasna syncline and that Monterey source

Table 2—Selected Bulk Chemical Properties of Source Rock Core Samples from Huasna Basin

Sample No.	Well No.	HC (wt. %)	TOC (wt. %)	SiO_2 (wt. %)	Al_2O_3 (wt. %)	Fe_2O_3 (wt. %)	P_2O_5 (wt. %)	CO_2 (wt. %)	S (wt. %)	U (ppm)	Zr (ppm)	Ba (ppm)
Siliceous Monterey Shale												
S-2408	17	1.51	3.48	83.5	3.79	1.70	0.32	0.16	1.27	6.1	20	260
S-2410	17	3.30	4.46	80.3	3.97	1.76	0.28	0.02	1.52	12.3	40	280
S-3673	1	0.94	1.56	80.1	4.96	2.06	0.51	1.06	n.d.[a]	7.6	42	292
S-4095	5	0.43	1.22	73.1	11.6	2.12	0.31	0.40	n.d.	4.6	152	687
S-2409	17	2.09	4.61	72.7	6.86	2.36	0.40	0.15	1.90	16.2	60	380
S-4088	7	1.17	1.83	71.4	10.9	2.95	0.30	0.82	n.d.	5.0	126	501
Dolomitic Siliceous Monterey Shale												
S-4096	5	0.75	1.47	73.9	8.04	2.20	0.29	2.83	n.d.	4.7	71	410
S-4097	5	0.73	1.85	70.4	7.04	2.69	0.36	4.52	n.d.	6.5	59	351
S-3678	5	0.96	2.05	69.8	6.65	2.47	0.28	5.94	n.d.	4.1	66	274
S-3679	5	0.80	1.39	65.9	5.51	2.12	0.55	8.81	n.d.	4.2	117	376
S-4089	7	1.33	1.89	64.9	3.58	1.81	0.89	9.42	n.d.	4.9	20	235
S-3676	4	0.71	1.44	58.2	1.77	1.08	0.35	15.6	n.d.	3.7	<10	114
Siliceous Monterey Dolomite												
S-4093	5	0.26	0.65	48.6	1.36	3.48	0.18	21.5	n.d.	1.1	<10	108
S-4091	7	0.58	1.75	31.1	5.15	2.64	0.15	25.6	n.d.	5.9	43	238
S-2368	10	1.57	2.46	30.6	1.54	0.94	0.24	29.7	0.70	1.9	10	170
S-2369	10	0.63	1.29	27.2	1.88	1.14	0.18	29.8	0.97	2.8	<10	140
S-4092	5	0.82	2.29	17.4	1.68	5.90	0.16	35.1	n.d.	1.0	<10	94
Phosphatic Monterey Shale or Point Sal Shale												
S-2413	5	3.51	3.98	40.2	9.20	3.44	7.23	6.56	2.60	20.2	70	260
S-2414	5	5.26	8.25	48.1	8.88	3.43	5.28	0.58	3.49	23.5	100	390
S-3674	1	1.95	3.34	33.4	4.81	1.96	4.92	20.2	n.d.	10.1	48	112
S-4090	7	0.95	3.32	54.3	12.7	6.50	2.06	0.11	n.d.	10.4	123	344
S-2370	10	1.69	6.61	62.4	8.11	3.96	1.95	1.21	3.17	12.9	70	480
S-3680	5	0.47	5.96	57.0	7.91	3.68	1.07	7.48	n.d.	12.0	82	545
S-3677	7	0.46	1.73	52.0	8.62	2.37	0.80	12.1	n.d.	8.3	137	390
Rincon Shale												
S-2405	9	1.52	3.25	60.1	11.9	4.19	0.38	2.97	1.05	7.1	120	690
S-2404	9	1.64	3.20	54.0	10.1	3.94	0.43	7.38	1.12	7.8	90	690
S-2403	9	1.95	3.67	51.5	9.93	4.41	0.60	7.70	1.23	7.3	60	740
S-2402	9	0.83	2.64	48.9	11.6	4.74	0.23	8.30	1.05	6.6	90	550
S-3675	1	1.94	3.41	46.9	10.1	4.25	0.32	11.0	n.d.	8.8	63	703
S-2406	9	1.52	3.03	45.5	9.96	3.81	0.32	11.4	1.00	8.4	70	460

[a]n.d. = not determined.

rocks also generated petroleum (at an assumed $R_o \approx 0.5\%$) where they once were buried below a depth of ~7000 ft. Additionally, these Miocene source rocks first entered their respective oil windows about 4–5 Ma. Kablanow and Surdam (1984) concluded independently that Monterey and Rincon source rocks have attained thermal maturity in the Huasna Basin, although their modeling suggests that Rincon shales have reached an even higher level of thermal maturity. Vitrinite reflectance is commonly suppressed in source rocks containing Type II kerogen (Price and Barker, 1985). Thus, the modeled maturity of Miocene source rocks in this study may be systematically low since modeling parameters were calibrated in large part using vitrinite reflectance measure-

ments obtained on SOM-rich Monterey samples whose kerogen may exhibit suppressed reflectance.

OIL PRODUCTION AND PETROLEUM SHOWS IN HUASNA BASIN

The Huasna Basin contains only one small abandoned oil field, which lies along the western flank of the basin (Figure 11). About 20,000 bbl of very heavy (≤10° API) oil were produced from a few wells in the LaVoie-Hadley area of the Huasna field and about 11,500 bbl of heavy (16° API) oil were produced from a single well in the Tar

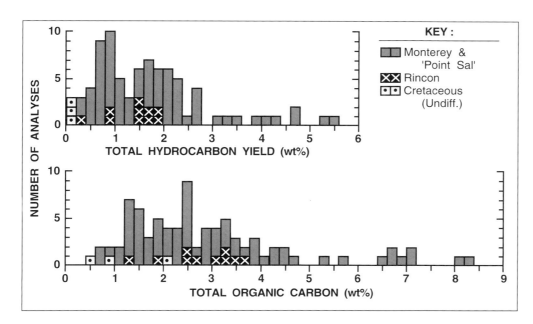

Figure 7—Organic richness of Monterey, Rincon, and Cretaceous shales in the Huasna Basin. Only the Miocene shales are rich enough to serve as good petroleum source rocks. Total hydrocarbon yields were determined using a pyrolysis flame ionization detector (PFID).

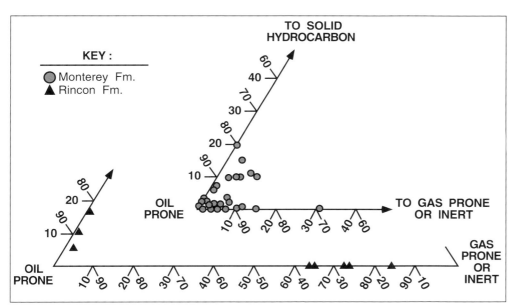

Figure 8—Visual kerogen analysis of Miocene source rocks in the Huasna Basin. Monterey shales are greatly enriched in oil-prone kerogen (principally structureless organic matter). Some samples of Rincon shale contain more gas-prone than oil-prone macerals.

Springs area of this field. All of these wells produced from shallow (<3000 ft, or <900 m) Monterey and/or Santa Margarita pay zones.

Significant oil shows in Monterey reservoirs were reported in several other exploration wells (Table 3). Two deep wells in the vicinity of the Huasna field (Hunt Tar Springs Ranch 1 and Verde Union Dickes 1) encountered light oil and/or free gas in the Monterey Formation. However, in several other exploration wells, shows of only medium-gravity (15–27° API) oil were reported in the Monterey. Good oil and gas shows have been reported in pre-Monterey reservoirs as well; for example, free gas and light oil (≥28° API) were encountered in the Obispo and Vaqueros formations in 10 wells.

The areal distribution of oil and gas shows in the Huasna Basin demonstrates that oil-prone source rocks have generated and expelled petroleum across this basin. The widespread occurrence of oil seeps in the basin supports this notion as well. In addition, the common occurrence of light oil or free gas in pre-Monterey reservoirs and the presence of generally heavier oil in Monterey and younger reservoirs suggest that the basin may contain two petroleum systems: (1) a Monterey–Santa Margarita system in which mainly low- to medium-gravity oil was generated by Monterey source rocks at a relatively low level of thermal maturity, and (2) a Vaqueros-Rincon-Obispo system in which Rincon source rocks generated lighter oil and free gas at a higher level of thermal

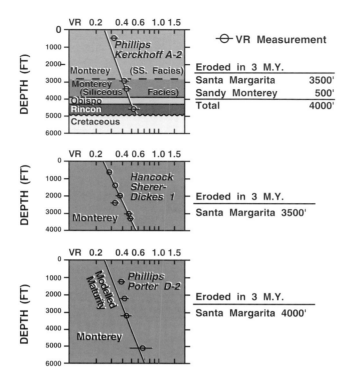

Figure 9—Thermal maturity modeling of three calibration wells in the Huasna Basin. A good fit was obtained between the modeled and measured maturity of Miocene Monterey and Rincon source rocks. The geothermal gradient varied through time in response to episodes of Obispo volcanism, rapid deposition of Monterey and Santa Margarita sediments, and recent uplift and erosion. Geothermal gradients (in °F/100 ft) are as follows: present = 2.1, post-Sta. Margarite = 1.9, post-Monterey = 2.1, post-Obispo = 2.4, post-Rincon = 2.2.

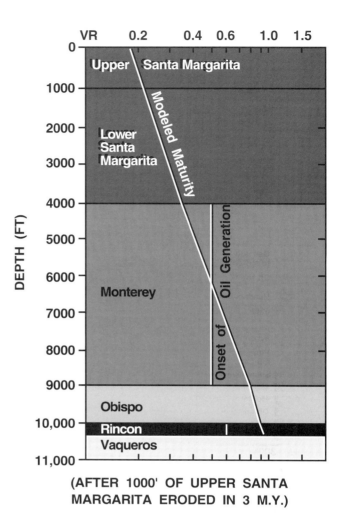

(AFTER 1000' OF UPPER SANTA MARGARITA ERODED IN 3 M.Y.)

Figure 10—Thermal maturity modeling of Monterey and Rincon source rocks in the "cooking pot" astride the axial syncline of the Huasna Basin. A significant volume of Monterey source rocks and the entire Rincon Formation appear to be thermally mature in this cooking pot (which has undergone only a moderate amount of uplift and erosion).

maturity. Furthermore, these distinct petroleum systems may be separated by a regional seal comprised of impermeable Obispo volcanics or Point Sal shales. Oil quality in the cool, shallow reservoirs within the Monterey–Santa Margarita petroleum system may have been influenced by biodegradation as well.

These hypotheses, which in general terms are supported by the source rock thermal maturity modeling described earlier, can be tested by studying and correlating the chemistry of oil and source rock samples from the Huasna Basin. A sample of heavy oil oozing from the bottom of the casing of an abandoned well in the LaVoie-Hadley area of the Huasna field was collected, and tar samples were obtained at five of the active oil seeps identified by Hodgson (1980) (Table 1). In addition, bitumen was extracted from well-characterized core samples of Monterey and Rincon shales obtained in four exploration wells. The locations of these petroleum samples are shown in Figure 5.

BULK PROPERTIES OF PETROLEUM SAMPLES

Biodegradation and other transformation processes generally cause the relative amount of sulfur in residual petroleum to increase. However, Monterey source rocks commonly contain sulfur-rich kerogen that generates heavy, sour oil at a relatively low level of thermal maturity ($R_o \leq 0.5\%$) (Petersen and Hickey; 1987; Baskin and Peters, 1992). Orr (1986) designated this type of organic matter as Type II-S kerogen. Thus, it is important to determine if a sample of heavy crude from a basin containing Monterey source rocks consists of unaltered, low-maturity petroleum generated by sulfur-rich kerogen, or if it

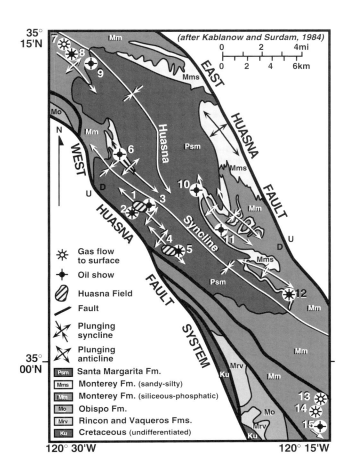

Figure 11—Location map of the modest oil production in the Huasna Basin and exploration wells that reported good oil or gas shows. The fields and wells are further identified in Table 3.

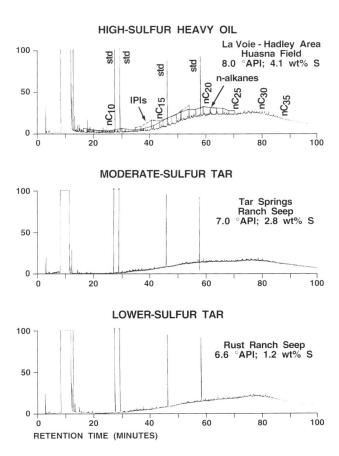

Figure 12—Gas chromatograms of heavy oil and tar samples from the Huasna Basin. The presence of C_{15+} normal alkanes (*n*-alkanes) and isoprenoid isoalkanes (IPIs) in the sour oil sample from the Huasna field indicates that it is less severely biodegraded than the sweeter, tarry petroleum samples obtained at the Tar Springs Ranch and Rust Ranch seeps. The prominent peaks are standards that were added during the analysis.

consists largely of residual, biodegraded oil that may have been generated at a higher level of thermal maturity by Monterey (or possibly other) source rocks.

Gas-liquid chromatography demonstrates that the tarry petroleum samples obtained at oil seeps in the Huasna Basin are more severely biodegraded than the heavy oil sample from Huasna field (Figure 12). Normal alkanes and isoprenoid isoalkanes (IPIs), which are readily metabolized by bacteria, are absent from the tar samples. The heavy oil sample from the Huasna field, in contrast, contains low concentrations of IPIs and normal alkanes heavier than C_{15}, indicating that it has undergone only a moderate amount of biodegradation.

Sulfur Chemistry of Heavy Oil and Tar Samples

Sulfur, nickel, and vanadium concentrations can be used to correlate transformed petroleum samples and to infer the depositional environment of the source rocks that generated them. Sulfur concentrations in oil and tar

samples from the Huasna Basin were used to classify them into three groups (Table 4). The first group, composed of very sour (>4.0 wt. % S) petroleum samples obtained along the western flank of the Huasna Basin, includes the heavy oil from the Huasna field and a tar seep near Lopez Lake (Figure 5). The second group, which contains significantly sweeter (<1.6 wt. % S) petroleum, is comprised of three tar samples obtained at active seeps (Rust Ranch, Carrie Creek, and Huasna River) along the eastern and southern flanks of the Huasna Basin. A tar sample obtained at a seep on Tar Springs Ranch along the western flank of the basin, which contains an intermediate amount of sulfur, is the sole member of the third group.

However, the amount of sulfur in the oil and tar samples from the Huasna Basin is not simply related to their degree of biodegradation. For example, the moderately transformed oil sample from Huasna field contains significantly more sulfur than the more severely transformed tar samples from the Tar Springs Ranch or Rust

Table 3—Petroleum Production and Shows in Huasna Basin

Map No.	Well Name	Show Description[a]
Production/Shows in Santa Margarita or Monterey Reservoirs		
6	Occidental Glaser 1	508 ft of gassy oil from Monterey (DST)
10	Chevron Porter 1-1	One pint of 18° API oil from Monterey (JFT)
3	Shell Tar Springs Ranch 1A	Pumped 19 BOPD (27° API) from Monterey
2	Hunt Tar Springs Ranch 1	400 ft of 33° API oil and gas to surface from Monterey (JFT)
1	Texas Pacific Trustees 1[b]	Pumped 11,500 bbl of 16° API oil from Monterey
5	Verde Union Dickes 1	Gas to surface from Monterey (DST)
4	Homestake LaVoie-Hadley wells[c]	Pumped 19,800 bbl of 8–10° API oil from Santa Margarita
12	Pan Petroleum Williams 1	560 ft of 15–18° API gassy oil and gas to surface from Monterey
15	Colgrove Fleisher 16	Pumped 10–30 BOPD (20° API) from Monterey
Production/Shows in Obispo or Vaqueros Reservoirs		
7	Tidewater USL 18-9	Flowed 500–650 mcfd of gas from Obispo
9	Tidewater USL 18-15	Pumped 3–10 BOPD (32° API) from Obispo
8	Gross Ida B-1	Trace of 28° API oil and 120–260 mcfd of gas from Vaqueros
6	Occidental Glaser 1	100 ft of 28–38° API oil from Obispo (DST)
2	Hunt Tar Springs Ranch 1	Gas to surface from Obispo (JFT)
11	Danciger Twitchell 1	30° API oil from Obispo (JFT)
5	Verde Union Dickes 1	Trace of 32° API oil from Vaqueros (DST)
12	Pan Petroleum Williams 1	Gas to surface from Vaqueros (DST)
13	British-American Rancho Suey C-1	150 mcfd of gas from Obispo and from Vaqueros (JFT)
14	British-American Rancho Suey B-1	275 mcfd of gas from Obispo (JFT)

[a]DST = drillstem test, JFT = Johnston formation test.
[b]This well was the only producing well in the Tar Springs area of the Huasna field.
[c]These five wells were the producing wells in the LaVoie-Hadley area of the Huasna field.

Ranch seeps (Figure 12). This observation supports the proposal that biodegradation causes the sulfur content of residual oils in the California coastal basins to increase by only a moderate amount (Magoon and Isaacs, 1983; Baskin and Jones, 1993). Furthermore, it leads to the notion that the sulfur content of the Huasna Basin seeps largely reflects source or maturity effects related to the source rocks that generated them.

Vanadium and Nickel Chemistry of Oil and Tar Samples

Oil and tar samples from the Huasna Basin contain moderately high concentrations of nickel and vanadium (Table 4). The high-sulfur oil and tar samples are enriched in vanadium relative to nickel (V/Ni > 1), while sweeter tar samples are enriched in nickel relative to vanadium (Ni/V ≈ 2–3) (Figure 13). The V/Ni ratio in petroleum samples is controlled by the depositional environment of the source rocks that generated them (Lewan and Maynard, 1982). In particular, high V/Ni ratios (and high sulfur concentrations) typically occur in oils generated by clay-poor marine source rocks that formed under conditions of intense anoxia (Lewan, 1984).

Sulfur, Vanadium, and Nickel Chemistry of Bitumen Samples

The chemical composition of bitumens extracted from various Monterey and Rincon lithofacies supports the notion that the sulfur, vanadium, and nickel chemistry of the heavy oil and tar samples in this basin was controlled primarily by source effects rather than maturity effects. Thermally mature ($R_o \geq 0.55\%$) phosphatic Monterey and Point Sal shales contain sulfur-rich bitumen with high V/Ni ratios, while the bitumen in thermally mature ($R_o \geq 0.55\%$) clay-rich siliceous Monterey and Rincon shales has less sulfur and lower V/Ni ratios (Table 4; Figure 13). Kablanow (1986) also documented the presence of high-sulfur and low-sulfur Monterey bitumens in this basin. Odermatt and Curiale (1991) measured V/Ni ratios of ≥1 in bitumens extracted from core samples of phosphatic and siliceous Monterey shale from the Santa Maria Basin. However, all of their siliceous shale samples may contain Type II-S kerogen.

Lewan (1986a) reported that immature bitumens in proximal Monterey shales containing mud and silt have less sulfur than bitumens extracted from more pelagic Monterey shales. The Type II kerogen in clay-rich Miocene

Table 4—Chemistry of Oil, Tar, and Bitumen Samples from Huasna Basin

Seep or Well	Sample No.	Gravity (°API)	S (wt. %)	V (ppm)	Ni (ppm)	$\delta^{13}C$ [a] (PDB)
High-Sulfur Oil and Tar Samples						
Lopez Lake seep	O-710	n.d. [b]	4.30	120	106	−22.34
LaVoie-Hadley oil	O-642	8.0	4.11	102	77	−22.02
Moderate-Sulfur Tar Sample						
Tar Springs Ranch seep	O-641	7.0	2.81	60	96	−21.57
Lower Sulfur Tar Samples						
Huasna River seep	S-5244	n.d.	1.55	53	103	−21.19
Carrie Creek seep	S-5245	n.d.	1.55	42	137	n.d.
Rust Ranch seep	O-643	6.6	1.21	21	62	−21.08
High-Sulfur Bitumen Samples from Phosphatic Monterey or Point Sal Shales						
Hunt Tar Springs Ranch 1	S-3674	n.d.	5.0	303	162	n.d.
Danciger Twitchell 1	S-4091	n.d.	2.4	158	40	n.d.
Lower Sulfur Bitumen Sample from Clay-Rich Siliceous Monterey Shale						
Standard Porter 1-1	S-4095	n.d.	1.5	34	146	n.d.
Lower Sulfur Bitumen Samples from Rincon Shales						
Honolulu Rancho Suey 1	S-2406	n.d.	2.0	n.d.	205	−23.04
Hunt Tar Springs Ranch 1	S-3675	n.d.	1.6	141	352	n.d.

[a] Carbon isotopic composition of the C_{15+} fraction of the petroleum sample.

[b] n.d. = not determined.

shales in the Huasna Basin contains relatively low concentrations of organic sulfur, probably because bacterial sulfide preferentially reacted with iron available on clay minerals, forming authigenic sulfide minerals. Phosphatic marls, in contrast, commonly contain both sulfur-rich (Type II-S) kerogen and moderate amounts of authigenic sulfide minerals. This may be because this pelagic lithofacies was deposited during a highstand period of slow sedimentation when the rate of bacterial sulfide production was high enough to eventually overwhelm sources of clay-bound reactive iron. Clay-poor siliceous Monterey shales, which were deposited in an iron-poor environment dominated by biogenic silica sedimentation, also typically contain Type II-S kerogen (Orr, 1986).

CARBON ISOTOPE CHEMISTRY

A distinctive feature of Monterey oil samples is the presence of isotopically heavy carbon ($\delta^{13}C > -24\%_0$, relative to PDB standard) (Lewan, 1986b). The isotopic composition of the carbon in oil and tar samples from the Huasna Basin ranges from about −21.1‰ to −22.3‰ (Table 4). In addition, a good relationship exists between the carbon isotopic composition of these samples and the amount of sulfur they contain (Figure 14).

Because maturity effects can influence the sulfur content of a suite of related oils, it might seem reasonable to interpret the modest systematic changes in the carbon isotopic composition of these petroleum samples as a maturity effect as well, especially because the "high-maturity," low-sulfur tars are more enriched in isotopically heavy carbon than the "low-maturity," sour oil and tar samples (as would be expected for a maturity effect) (Chung et al., 1981; Lewan, 1983). However, V/Ni ratios in these oil and tar samples indicate they were not simply generated by the same source rock at different levels of thermal maturity. Similarly, the systematic change in the carbon isotope chemistry of these samples is not readily explained by secondary transformation because the carbon isotopic composition of oil is not significantly influenced by biodegradation (Stahl, 1980).

Alternatively, the presence of slightly heavier carbon in sweeter petroleum samples from the Huasna Basin may be a source effect. It has not been demonstrated that systematic differences exist between the carbon isotopic composition of the kerogen in various Miocene source rocks in California, or in different organic facies of the same source rock. But in the nearby Cuyama Basin, bitumen in the Soda Lake Shale Member of the Vaqueros Formation, and the sweet oils this lower Miocene source rock generated, contain carbon that is about 1–2‰ heavier than the carbon in Monterey bitumen (Kornacki, 1988).

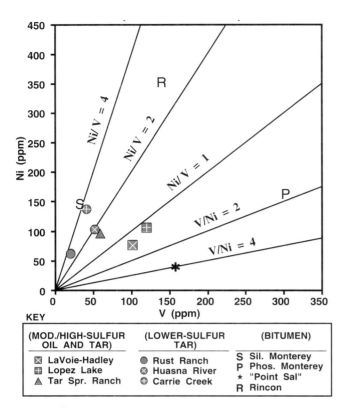

Figure 13—Vanadium and nickel concentrations in petroleum samples from the Huasna Basin. Vanadium is more abundant than nickel in the sour oil sample from the LaVoie-Hadley area of the Huasna field, the sour tar sample from the Lopez Lake seep, and the sulfur-rich bitumens extracted from phosphatic Monterey and Point Sal source rocks. However, nickel is more abundant than vanadium in clay-rich source rocks containing lower sulfur bitumen and in low-sulfur tar samples.

SATURATE BIOMARKER CHEMISTRY

Biomarkers are geochemical trace fossils that can be used to determine the source and maturity of petroleum (Seifert and Moldowan, 1986). Because some types of saturate and aromatic biomarkers are relatively resistant to bacterial transformation, these compounds are commonly used to correlate biodegraded oil and tar samples (Seifert et al., 1984). Only saturate biomarkers were used in this study, however. Steranes are generally less resistant to bacterial transformation than pentacyclic triterpanes (hopanes), which in turn are less resistant than tricyclic diterpanes. For example, the sterane chemistry of most tar samples obtained at seeps in the Huasna Basin has been severely disturbed by biodegradation. But this is not the case for the moderately transformed heavy oil sample from the Huasna field nor for the bitumens extracted from Monterey and Rincon source rocks.

Furthermore, some biomarker maturity indicators are sensitive to source effects as well. Differences in the sulfur, vanadium, and nickel chemistry of oil and tar samples from the Huasna Basin indicate that they were generated by different types of kerogen (and possibly by different source formations). For this reason, the value of biomarker maturity parameters in petroleum samples that contain different amounts of sulfur should be interpreted with caution.

Sterane Biomarker Chemistry

The low relative abundance of C_{29} normal steranes in all petroleum samples from the Huasna Basin mirrors the chemistry of Monterey oil and bitumen samples from the Santa Maria Basin (Figure 15) (Curiale et al., 1985). The sour oil sample from the Huasna field and sulfur-rich bitumens extracted from phosphatic shales exhibit higher C_{27}/C_{29} normal sterane ratios than the lower sulfur bitumens extracted from clay-rich source rocks, supporting the interpretation that the high sulfur content of this heavy oil sample is a source effect. For some unknown reason, most bitumen samples from the Huasna Basin contain higher concentrations of C_{28} normal steranes than the heavy oil sample.

The relative abundance of rearranged steranes in petroleum samples from this basin reflects source and transformation processes. Rearranged steranes are more resistant to bacterial transformation than normal steranes or isosteranes. This explains the high proportion of rearranged steranes in tar samples from the Lopez Lake and Rust Ranch seeps (Table 5). The transformed tar sample obtained at the Huasna River seep and the sour oil sample from the Huasna field are also enriched in residual rearranged steranes, but to a lesser degree. The relative abundance of rearranged steranes in bitumen samples, in contrast, appears to have been influenced primarily by source effects. Bitumens extracted from Rincon source rocks contain more rearranged steranes than bitumens extracted from phosphatic shales. The bitumen in a sample of clay-rich siliceous Monterey shale contains an intermediate amount of rearranged steranes.

The proportionality between the 20(S) and 20(R) isomers of C_{29} normal sterane is a standard biomarker maturity parameter that attains an equilibrium value of ~0.55 in many petroleum systems (Seifert and Moldowan, 1986). Values close to this occur in bitumens extracted from thermally mature source rock samples from the Huasna Basin, but the sour oil sample from the Huasna field exhibits a significantly lower value (Figure 16d). Monterey oils commonly exhibit anomalously low values of this maturity parameter (Curiale et al., 1985; Zumberge, 1987).

Hopane Biomarker Chemistry

The high abundance of hopanes relative to steranes and the low abundance of hopanes relative to tricyclic diterpanes in tar samples from the Huasna Basin are secondary transformation effects that render these biomarker parameters useless for correlation purposes (Figures 16b, c). Curiale and Odermatt (1989) observed that phos-

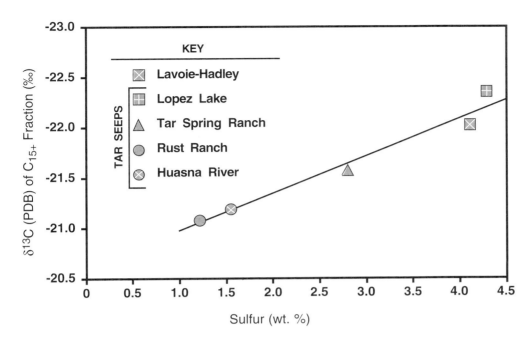

Figure 14—Carbon isotopic composition of oil and tar samples from the Huasna Basin. The carbon in sour petroleum samples is slightly lighter than the carbon in lower sulfur petroleum samples.

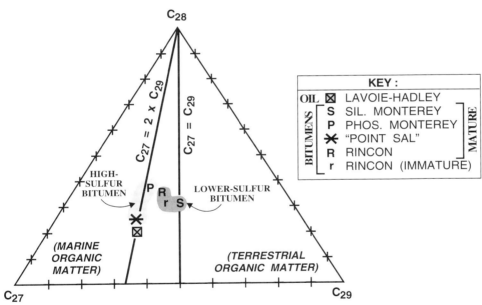

Figure 15—Normal sterane chemistry of oil and bitumen samples from the Huasna Basin. The sour oil sample from the LaVoie-Hadley area of the Huasna field and sulfur-rich bitumens extracted from phosphatic Monterey and Point Sal shales are more enriched in C_{27} normal steranes (relative to C_{29}) than are lower sulfur bitumens extracted from clay-rich source rocks.

phatic Monterey shales contain higher concentrations of bacterially derived hopanes relative to steranes than siliceous Monterey shales that are at the same level of thermal maturity. In the Huasna Basin, the proportionality of hopanes to steranes does not vary systematically among bitumens extracted from Rincon and Monterey source rocks (Figure 16b).

The values of a hopane "maturity" indicator that is relatively resistant to bacterial alteration (Ts/Tm) illustrate the importance of source effects in interpreting the biomarker chemistry of oils and source rocks. A sample of mature Rincon shale has a higher Ts/Tm ratio than a sample of immature Rincon shale (Figure 16e). But the value of this parameter is significantly higher in the

immature Rincon shale than in a sample of phosphatic Monterey shale that is thermally mature. Similarly, the observation that the Ts/Tm ratio is different in sour petroleum samples from the Huasna field and Lopez Lake, and in sweeter tar samples from seeps at Rust Ranch and Huasna River as well, indicates that this hopane maturity parameter has only limited utility in the Huasna Basin.

The ratio of the unusual triterpane compound 28,30-bisnorhopane to C_{30} hopane (a common triterpane) generally increases with increasing sulfur content in unaltered Monterey oils from the Santa Maria Basin (Curiale et al., 1985). The highest concentrations of 28,30-bisnorhopane in Monterey core samples have been found in phosphatic shales, leading to the suggestion that this ratio

Table 5—Biomarker Chemistry of Oil, Tar, and Bitumen Samples from Huasna Basin

Seep Location or Well No.[a] & Source Rock Maturity	Sample No.	Hopane[b] (ppm)	Sterane Parameters[c]							Pentacyclic Triterpane Parameters[d]					
			A	B	C	D	E	F	G	H	I	J	K	L	M
High- to Moderate-Sulfur Oil and Tar Samples															
Lopez Lake	O-710	n.d.	n.d.	n.d.	n.d.	n.d.	n.d.	0.88	0.90	0.53	n.d.	n.d.	n.d.	n.d.	0.26
LaVoie-Hadley well [e]	O-642	1394	0.53	0.21	0.27	0.38	0.28	0.38	0.77	0.29	0.08	0.29	0.80	0.74	0.63
Tar Springs Ranch	O-641	15	n.d.	n.d.	n.d.	n.d.	n.d.	n.d.	>0.99	0.33	15.5	46.7	2.22	70.2	0.26
Lower Sulfur Tar Samples															
Rust Ranch	O-643	n.d.	n.d.	n.d.	n.d.	n.d.	n.d.	n.d.	>0.99 0.85	0.34	n.d.	n.d.	n.d.	n.d.	0.20
Huasna River	S-5244	18	n.d.	n.d.	n.d.	n.d.	0.22	0.49	0.89	0.53	14.8	27.7	0.16	34.1	0.27
High-Sulfur Bitumen Samples Extracted from Phosphatic Monterey or Point Sal Shales															
Well #1 ($R_o \approx 0.55$)	S-3674	1349	0.41	0.37	0.22	0.56	0.28	0.03	0.50	0.19	0.09	0.46	0.92	1.18	0.72
Well #7 ($R_o \approx 0.70$)	S-4091	490	0.51	0.25	0.24	0.52	0.29	0.06	0.30	0.44	0.15	0.34	<0.01	0.88	0.53
Lower Sulfur Bitumen Sample Extracted from Clay-Rich Siliceous Monterey Shale															
Well #5 ($R_o \approx 0.55$)	S-4095	938	0.35	0.31	0.34	0.49	0.62	0.08	0.31	0.42	0.10	0.24	0.76	0.73	0.64
Lower Sulfur Bitumen Samples Extracted from Rincon Shales															
Well #9 ($R_o \approx 0.45$)	S-2406	n.d.	0.38	0.31	0.31	0.12	0.91	0.16	0.13	0.44	0.09	0.21	0.01	0.66	0.86
Well #1 ($R_o \approx 0.65$)	S-3675	1389	0.38	0.35	0.27	0.52	0.59	0.13	0.47	0.56	0.13	0.24	<0.01	0.92	0.74

[a]Wells can be identified by referring to the Map No. in Table 1.

[b]Concentration of C_{30} hopane in the C_{15+} fraction of the saturate component.

[c]Key to C_{27}–C_{29} sterane parameters: **A** = C_{27} normal/C_{27}-C_{29} normal (20R), **B** = C_{28} normal/C_{27}-C_{29} normal (20R), **C** = C_{29} normal/C_{27}-C_{29} normal (20R), **D** = 20S/(20S+20R) C_{29} normal steranes, **E** = C_{27}-C_{29} normal/(normal + iso), **F** = rearranged steranes/total steranes, **G** = triterpanes/(triterpanes + steranes). n.d. = not determined (too lean in steranes).

[d]Key to C_{27}–C_{35} pentacyclic triterpane parameters: **H** = Ts/Tm, **I** = Ts/hopane, **J** = Tm/hopane, **K** = 28,30-bisnorhopane/hopane, **L** = 30-norhopane/hopane, **M** = triterpanes/(triterpanes + diterpanes). n.d. = not determined (too lean in hopanes).

[e]A heavy oil sample collected from the well casing of an unidentified abandoned well in the LaVoie-Hadley area of the Huasna field.

is largely controlled by source effects (Curiale and Odermatt, 1989; Brenner, 1994). The sour oil sample from Huasna field is moderately enriched in 28,30-bisnorhopane, but sweeter tar samples obtained at two seep sites contain only trace amounts of this compound (Figure 16a). The bitumen extracted from Rincon shales also contains only trace amounts of 28,30-bisnorhopane, evidence that these seeps may have been generated by this lower Miocene source rock.

However, the high 28,30-bisnorhopane/hopane ratio in a moderately sour tar sample obtained at a seep on Tar Springs Ranch is interpreted to be a secondary transformation effect (Figure 16a). Under some circumstances, the C_{27}-C_{29} hopanes (Ts, Tm, 28,30-bisnorhopane, and 30-norhopane) are apparently more resistant against bacterial alteration than the C_{30} hopane molecule (Requejo and Halpern, 1989). Similar trends are observed in tar samples from the Huasna Basin. For example, the C_{15+} saturate fraction of the severely biodegraded petroleum from the Tar Springs Ranch and Huasna River seeps contains <20 ppm hopane, compared to ≥500 ppm hopane in source

rock bitumens and in the heavy oil sample from the Huasna field (Table 5). Furthermore, the unreasonably high values of hopane ratios, such as Ts/hopane, Tm/hopane, and 30-norhopane/hopane, in these tar samples are unlikely to be primary features.

Secohopanes and Demethylated Hopanes

The sour oil sample from the Huasna field and the Rust Ranch tar sample contain C_{29}–C_{33} secohopanes. These tetracyclic biomarkers have a modified hopane structure: the third (C) ring of the hopane skeleton is open. The origin of secohopanes is uncertain, but it has been suggested that ring C of the hopane structure can be opened by bacteria during severe biodegradation (Rüllkotter and Wendisch, 1982) or by acids associated with clay catalysis during thermal maturation (Schmitter et al., 1982). The severely transformed petroleum in seeps from the Huasna Basin does not contain demethylated hopanes, which Volkman et al. (1983) suggest form during severe biodegradation.

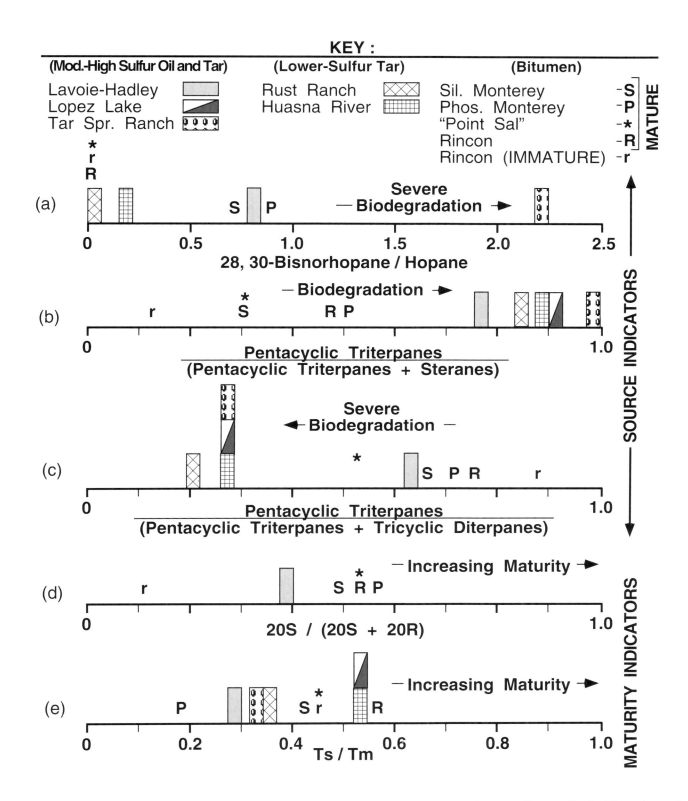

Figure 16—Saturate biomarker source and maturity indicators in petroleum samples from the Huasna Basin. The values of sterane and hopane parameters in the tarry petroleum obtained at seeps are commonly disturbed by biodegradation.

Figure 17—Kerogen organic facies in the Huasna Basin. The heavy, sour petroleum that occurs along the western margin of this basin was probably generated by the sulfur-rich (Type II-S) kerogen in siliceous and phosphatic Monterey shales. Clay-rich source rocks that contain lower sulfur (Type II) kerogen probably generated sweeter crudes and gas.

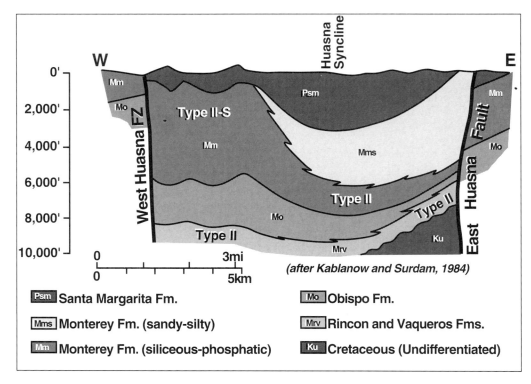

(after Kablanow and Surdam, 1984)

Psm Santa Margarita Fm.	**Mo** Obispo Fm.
Mms Monterey Fm. (sandy-silty)	**Mrv** Rincon and Vaqueros Fms.
Mm Monterey Fm. (siliceous-phosphatic)	**Ku** Cretaceous (Undifferentiated)

CONCLUSIONS

Several oil-prone source rocks are present in the Huasna Basin: siliceous Monterey shales containing variable amounts of clay, phosphatic Monterey and Point Sal shales, and Rincon claystones. The bitumen in clay-rich lithofacies contains less organic sulfur than the bitumen in phosphatic shales. A heavy, sour oil sample from the Huasna field is moderately transformed and enriched in vanadium relative to nickel. But five tar samples obtained at active seeps are so severely transformed that not even their saturate biomarker chemistry can be used to correlate them.

Nevertheless, the concentration of sulfur in these seep samples, their V/Ni ratios, and their carbon isotopic composition can be used to classify them into three groups. Sour petroleum samples from the western margin of the Huasna Basin were probably generated at low levels of thermal maturity by pelagic Monterey source rocks (e.g., phosphatic shales and "clean" siliceous shales) containing Type II-S kerogen. Sweeter petroleum samples from the southern and eastern margins of this basin were probably generated at higher levels of thermal maturity by clay-rich source rocks (e.g., clay-rich Monterey shales and Rincon shales) that contained Type II kerogen (Figure 17). Source rock thermal maturity modeling and the distribution of subsurface petroleum shows support the notion that more than one Miocene petroleum system has been active in this small basin.

Acknowledgments—The concepts and interpretations presented in this paper were stimulated by enlightening discussions with many Shell Oil Company scientists, especially John Castaño, Win Goter, Henry Halpern, Pat Knigge, Frank Mango, Bob McNeil, John Smith, and Joe Westrich. Organic geochemical analyses of source rock samples were performed by Jennifer Thompson, Janet Rush, Mary Lou Weiss, and Linda Peacock; T. P. Fan performed biomarker analyses of oil, tar, and bitumen samples. Inorganic geochemical analyses of source rock samples were performed by X-Ray Assay Laboratories, Don Mills, Ontario, Canada. Several seep samples were obtained by the Geochemical and Environmental Research Group at Texas A&M University. I thank Renee Ford and Cheryl Ryan for preparing figures, and the management of Shell Oil Company and its affiliates for permission to publish this paper.

REFERENCES CITED

Baskin, D. K., and K. E. Peters, 1992, Early generation characteristics of a sulfur-rich Monterey kerogen: AAPG Bulletin, v. 76, p. 1–13.

Baskin, D. K., and R. W. Jones, 1993, Prediction of oil gravity prior to drill-stem testing in Monterey Formation reservoirs, offshore California: AAPG Bulletin, v. 77, p. 1479–1487.

Bell, G. G., 1951, Geology and exploration for oil in the Huasna district, San Luis Obispo County (abs): AAPG Bulletin, v. 35, p. 2630.

Brenner, J., 1994, Lithologic variation of the rare biomarker compound 28,30-bisnorhopane within the Miocene Monterey Formation in the Texaco Anita-14 well, Santa Barbara County, California (abs.): AAPG Bulletin, v. 78, p. 658.

Chung, H. M., S. W. Brand, and P. L. Grizzle, 1981, Carbon isotope geochemistry of Paleozoic oils from Big Horn Basin: Geochimica et Cosmochimica Acta, v. 45, p. 1803–1815.

Crawford, F. D., 1971, Petroleum potential of Santa Maria Province, California, *in* I. H. Cram, ed., Future petroleum provinces of the United States—their geology and potential: AAPG Memoir 15, p. 316–328.

Curiale, J. A., D. Cameron, and D. V. Davis, 1985, Biological marker distribution and significance in oils and rocks of the Monterey Formation, California: Geochimica et Cosmochimica Acta, v. 49, p. 271–288.

Curiale, J. A., and J. R. Odermatt, 1989, Short-term biomarker variability in the Monterey Formation, Santa Maria Basin: Organic Geochemistry, v. 14, p. 1–13.

Gray, L. D., 1980, Geology of Mesozoic basement rocks in the Santa Maria Basin, Santa Barbara, and San Luis Obispo counties, California: Master's thesis, San Diego State University, San Diego, California, 81 p.

Hall, Jr., C. A., 1973a, Geologic map of the Morro Bay South and Port San Luis quadrangles, San Luis Obispo County, California: USGS Miscellaneous Field Studies Map MF-511, 1:24,000.

Hall, Jr., C. A., 1973b, Geology of the Arroyo Grande 15' quadrangle, San Luis Obispo County, California: California Division of Mines and Geology, Map Sheet 24, 1:48,000.

Hall, Jr., C. A., 1978, Origin and development of the Lompoc–Santa Maria pull-apart basin and its relation to the San Simeon–Hosgri strike-slip fault, western California: California Division of Mines and Geology, Special Report 137, p. 25–31.

Hall, Jr., C. A., 1982, Pre-Monterey subcrop and structure contour maps, western San Luis Obispo and Santa Barbara counties, south-central California: USGS Miscellaneous Field Studies Map MF-1384, 1:62,500.

Hall, Jr., C. A., and C. E. Corbato, 1967, Stratigraphy and structure of Mesozoic and Cenozoic rocks, Nipomo quadrangle, southern Coast Ranges, California: GSA Bulletin, v. 78, p. 559–582.

Heasler, H. P., and R. C. Surdam, 1985, Thermal evolution of coastal California with application to hydrocarbon maturation: AAPG Bulletin, v. 69, p. 1386–1400.

Hodgson, S. F., 1980, Onshore oil and gas seeps in California: California Division of Oil and Gas Publication no. TR26, 97 p.

Johnson, A. M., and B. M. Page, 1976, A theory of concentric, kink, and sinusoidal folding and of monoclinal flexuring of compressible, elastic multilayers—VII. Development of folds within Huasna syncline, San Luis Obispo County, California: Tectonophysics, v. 33, p. 97–143.

Kablanow II, R. I., 1986, Influence of the thermal and depositional histories on diagenesis and hydrocarbon maturation in the Monterey Formation—Huasna, Pismo, and Salinas basins, California: Ph.D. dissertation, University of Wyoming, Laramie, Wyoming, 264 p.

Kablanow II, R. I., and R. C. Surdam, 1984, Diagenesis and hydrocarbon generation in the Monterey Formation, Huasna Basin, California, *in* R. C. Surdam, ed.,

Stratigraphic, tectonic, thermal, and diagenetic histories of the Monterey Formation, Pismo and Huasna basins, California: SEPM Guidebook 2, p. 53–68.

Kablanow II, R. I., R. C. Surdam, and D. Prezbindowski, 1984, Origin of dolomites in the Monterey Formation: Pismo and Huasna basins, California, *in* R. C. Surdam, ed., Stratigraphic, tectonic, thermal, and diagenetic histories of the Monterey Formation, Pismo and Huasna basins, California: SEPM Guidebook 2, p. 38–49.

King, V. L., 1943, Huasna area development: California Department of Natural Resources Division of Mines Bulletin, v. 118, p. 448–449.

Kornacki, A. S., 1988, Provenance of oil in southern Cuyama Basin, California (abs.): AAPG Bulletin, v. 72, p. 207.

Lewan, M. D., 1983, Effects of thermal maturation on stable organic carbon isotopes as determined by hydrous pyrolysis of Woodford Shale: Geochimica et Cosmochimica Acta, v. 47, p. 1471–1479.

Lewan, M. D., 1984, Factors controlling the proportionality of vanadium to nickel in crude oils: Geochimica et Cosmichimica Acta, v. 48, p. 2231–2238.

Lewan, M. D., 1986a, Organic sulfur in kerogens from different lithofacies of the Monterey Formation (abs): 192nd National Meeting of the American Chemical Society, Abstracts of Papers, Anaheim, California, Sept. 7–12.

Lewan, M. D., 1986b, Stable carbon isotopes of amorphous kerogens from Phanerozoic sedimentary rocks: Geochimica et Cosmochimica Acta, v. 50, p. 1583–1591.

Lewan, M. D., and J. B. Maynard, 1982, Factors controlling enrichment of vanadium and nickel in the bitumen of organic sedimentary rocks: Geochimica et Cosmochimica Acta, v. 46, p. 2547–2560.

Macgregor, D. S., 1993, Relationships between seepage, tectonics, and subsurface petroleum reserves: Marine and Petroleum Geology, v. 10, p. 606–619.

Magoon, L. B., and C. M. Isaacs, 1983, Chemical characteristics of some crude oils from the Santa Maria Basin, California, *in* C. M. Isaacs and R. E. Garrison, eds., Petroleum generation and occurrence in the Miocene Monterey Formation, California: SEPM Pacific Section Special Publication, p. 201–211.

McLean, H., 1991, Distribution of disharmonic en echelon folds in siliceous beds of the Miocene Monterey Formation east of San Luis Obispo, California (abs.), AAPG Bulletin, v. 75, p. 374.

Miiller, D. E., A. G. Holba, and W. B. Hughes, 1987, Effects of biodegradation on crude oils, *in* R. F. Meyer, ed., Exploration for heavy crude oil and natural bitumen: AAPG Studies in Geology 25, p. 233–241.

Odermatt, J. R., and J. A. Curiale, 1991, Organically bound metals and biomarkers in the Monterey Formation of the Santa Maria Basin, California: Chemical Geology, v. 91, p. 99–113.

Orr, W. L., 1986, Kerogen/asphaltene/sulfur relationships in sulfur-rich Monterey oils: Organic Geochemistry, v. 10, p. 499–516.

Petersen, N. F., and P. J. Hickey, 1987, California Plio-Miocene oils: evidence of early generation, *in* R. F. Meyer, ed., Exploration for heavy crude oil and natural bitumen: AAPG Studies in Geology 25, p. 351–359.

Price, L. C., and C. E. Barker, 1985, Suppression of vitrinite reflectance in amorphous rich kerogen—a major unrecognized problem: Journal of Petroleum Geology, v. 8, p. 59–84.

Requejo, A. G., and H. I. Halpern, 1989, An unusual hopane biodegradation sequence in tar sands from the Pt. Arena (Monterey) Formation: Nature, v. 342, p. 670–673.

Rüllkotter, J., and D. Wendisch, 1982, Microbial alteration of 17α-(H)-hopanes in Madagascar asphalts: removal of C-10 methyl group and ring opening: Geochimica et Cosmochimica Acta, v. 46, p. 1545–1553.

Schmitter, J. M., W. Sucrow, and P. J. Arpino, 1982, Occurrence of novel tetracyclic geochemical markers: 8,14-seco-hopanes in a Nigerian crude oil: Geochimica et Cosmochimica Acta, v. 46, p. 2345–2350.

Seifert, W. K., J. M. Moldowan, and G. R. Demaison, 1984, Source correlation of biodegraded oils: Organic Geochemistry, v. 6, p. 633–643.

Seifert, W. K., and J. M. Moldowan, 1986, Use of biological markers in petroleum exploration, *in* R. B. Johns, ed., Biological markers in the sedimentary record: Methods in Geochemistry and Geophysics 24, Amsterdam, Elsevier, p. 261–290.

Stahl, W. J., 1980, Compositional changes and $^{13}C/^{12}C$ fractionations during the degradation of hydrocarbons by bacteria: Geochimica et Cosmochimica Acta, v. 44, p. 1903–1907.

Stanley, R. G., Z. C. Valin, and M. J. Pawlewicz, 1993, Rock-Eval pyrolysis and vitrinite reflectance results from lower Miocene strata in the onshore Santa Maria Basin and Santa Barbara coastal area, California (abs.): AAPG Bulletin, v. 77, p. 716–717.

Taliaferro, N. L., 1943, Geology of Huasna area: California Department of Natural Resources Division of Mines Bulletin, v. 118, p. 443–447.

Volkman, J. K., R. Alexander, R. I. Kagi, and G. W. Woodhouse, 1983, Demethylated hopanes in crude oils and their application in petroleum geochemistry: Geochimica et Cosmochimica Acta, v. 47, p. 785–794.

Zumberge, J. E., 1987, Terpenoid biomarker distributions in low maturity crude oils: Organic Geochemistry, v. 11, p. 479–496.

Potter II, R. W., P. A. Harrington, A. H. Silliman, and J. H. Viellenave, 1996,
Significance of geochemical anomalies in hydrocarbon exploration, *in* D.
Schumacher and M. A. Abrams, eds., Hydrocarbon migration and its near-sur-
face expression: AAPG Memoir 66, p. 431–439.

Chapter 33

Significance of Geochemical Anomalies in Hydrocarbon Exploration: One Company's Experience

Robert W. Potter II

Santa Fe Minerals Inc.,
Dallas, Texas, U.S.A.

Present address:

Vintage Petroleum
Tulsa, Oklahoma, U.S.A.

Paul A. Harrington

Alan H. Silliman

James H. Viellenave

NERI LLC
Lakewood, Colorado, U.S.A.

Abstract

An independent evaluation of alternate technologies for petroleum exploration undertaken by Santa Fe Minerals selected the PETREX soil gas geochemical method. A subsequent exploration program involving 139 geochemical surveys led to drilling 141 wells in previously undrilled prospects. A total of 43 wells were drilled in negative geochemical anomalies and 41 of these encountered no hydrocarbons. A total of 98 wells were drilled in positive geochemical anomalies and 37 were commercial successes. The geochemical success rate was even higher, accurately predicting the presence of hydrocarbons 92% of the time. Surface soil gas hydrocarbon signatures were shown to be very similar to those of volatile gases from producing hydrocarbons in four new wells ranging in depth from 2000 to 15,000 ft. This evidence supports the vertical migration of reservoir hydrocarbons. Geochemical anomalies are apical and lateral shifts of anomalies are rare. A general trend of increasing anomaly strength with increasing porosity times reservoir thickness has been observed. The integration of seismic data and PETREX geochemical data results in prospect evaluation that is better than that available using either method alone.

INTRODUCTION

In 1986, Santa Fe Minerals Inc. began an investigation of various types of alternate exploration technologies in order to develop a program to reduce the risk of drilling exploratory wells. In addition to geochemical methods, various geophysical methods were used to evaluate the same prospect and known analog fields. Tests were conducted on a variety of plays to evaluate both the methods and the plays to be pursued by Santa Fe Minerals. These early analog studies focused our attention on two geophysical methods and three surface geochemical methods. Early results over the analog fields showed that the PETREX methodology appeared better than other geochemical methods, particularly in prediction of dry holes (Figure 1). The analog studies further indicated that, while many of the predicted producers from the other geochemical methods were in fact producers, the production came from zones other than the target horizon. In contrast, because of the modeling capability of the PETREX method, the results were specific to the horizon of interest. This modeling capability greatly facilitated the

integration of geochemical data with geophysical data allowing for a truly integrated exploration program.

Analog studies do not address the significance of geochemical anomalies in hydrocarbon exploration, although they give one some confidence that surface methods are applicable. Analog studies fail in this regard because the survey grid reflects the known field; hence, the study is biased from the onset. Further complications arise from the possibility of leaking wells and depletion of reservoir pressure due to production. Because of the limited scope of analog studies, certain issues cannot be adequately resolved. For example, can anomalies be present where no hydrocarbons occur? Can hydrocarbon accumulations be found where no geochemical anomaly exists? These questions can only be answered with results from exploration programs using geochemical methods where there are no existing fields.

The purpose of this paper is to present the results of the consistent use of the PETREX geochemical method integrated with geophysical data in hydrocarbon exploration for the period 1986–1993. During this period, a total of 139 geochemical surveys consisting of 33,100 samples

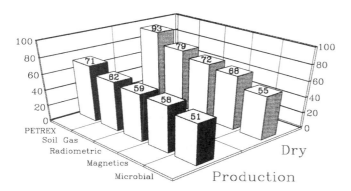

Figure 1—Success rates (in %) of selected exploration methods.

were conducted. In the surveyed areas, 141 wells were later drilled either by Santa Fe Minerals and its partners or by other companies, for which detailed log and test data are available. These exploration activities were conducted in mature to frontier basins in a wide variety of environments: frozen ground, farm land, deserts, mountains, swamps, and offshore. The targets ranged in depth from 1000 to 15,000 ft and covered the complete spectrum of trap type. Survey areas ranged from as small as a few hundred acres to regional programs covering 1000 sq mi. The results of these integrated exploration programs, as defined by the 141 drilled wells, form the basis for the conclusions reported here about the significance of geochemical anomalies and hence surface geochemistry in petroleum exploration.

GEOCHEMICAL METHOD

The PETREX method involves three steps: sample acquisition, sample analysis, and data interpretation (Hickey, 1986). Acquisition of hydrocarbon gas is done using the PETREX sampler (Figure 2). The sampler consists of activated charcoal placed on a ferromagnetic wire and inserted into a glass tube. The sampler is buried at a shallow depth and stays in the ground usually 2 to 3 weeks. The charcoal gradually adsorbs gases migrating to the surface and reaches equilibrium with the soil gases during its resident period. This method of acquisition alleviates the effects of short-period variation and meteorologic interference on the sample. After retrieval, the sampler is sealed and sent to the laboratory for analysis.

Analysis is performed by thermal desorption mass spectrometry (Figure 2). The mass spectrometer analyzes each sample over a mass range that includes hydrocarbons from C_2 to C_{19}. The instrument can effectively detect alkanes, cycloalkanes, alkenes, cycloalkenes, naphthalenes, and aromatic and substituted aromatic compounds. Methane is excluded because it has rarely been a discriminating compound due to its various mechanisms of formation, many of which are not related to hydrocarbon accumulation.

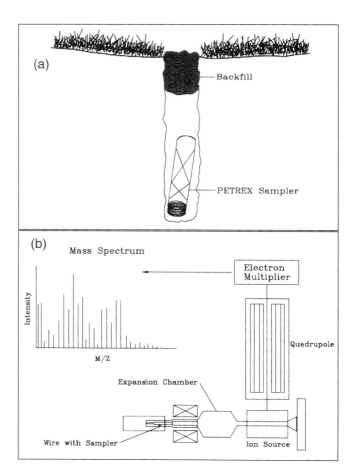

Figure 2—(a) PETREX sampler in place. (b) PETREX mass spectrometer.

Data interpretation involves the evaluation of the chemical composition of each sample. The mass spectral data of a typical PETREX survey may include 100–150 mass spectral peaks (mass measurements), resulting from the fragmentation of perhaps many incident volatile or semivolatile compounds. No single compound is likely to be diagnostic of a petroleum; however, mixtures of compounds may be diagnostic. A variety of statistical pattern recognition techniques can be used to elucidate the structure of such multivariate data (Klusman, 1993).

The particular mass spectral pattern of a sample is referred to as its "fingerprint." The presence of compounds and their relative abundances determine the appearance of the fingerprint. A specific geochemical source yields a consistent fingerprint in resulting mass spectral data. Principal component analysis (PCA) is used to identify fingerprints in the mass spectral data and to describe them (geochemical sources) with "factor" constructs to facilitate the interpretation of the geochemical survey data. An instructive discussion of the applications of the PCA technique is given by Davis (1986). The PCA factor data can be processed using several techniques in order to classify samples and identify anomalous areas.

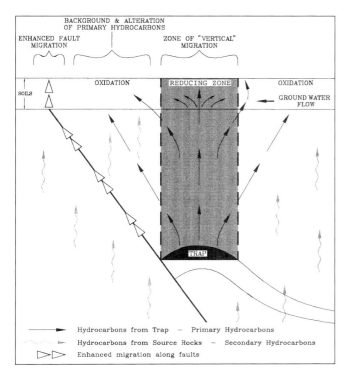

Figure 3—Schematic diagram of hydrocarbon leakage from a trap, showing zone of vertical migration.

The general problem of sample, or observation, classification is discussed by James (1985). Such techniques as cluster analysis, discriminant analysis, and rotation of factors are common. Each of these techniques can be applied to a set of geochemical survey data, depending on the type of geochemical modeling to be performed.

In nonfrontier regions, specific soil gas geochemical models are constructed to identify the fingerprint of each known production type. Samples are placed across fields from known background to production (taking care to avoid contamination near producing wells). Background and production types are characterized separately to develop specific fingerprint models. The models are then compared to the grid sample data.

Either discriminant analysis or rotation of factors is used in the classification of types and evaluation of site-specific geochemical models. Grid data are assessed for outliers (e.g., fault character), and reprocessing is done if necessary. Maps are prepared that show anomalous areas (those similar in character to the fingerprint model). Traditional geostatistical techniques are used to guide isopleth contouring.

In areas where no production data are available, the survey objective is usually limited to finding petroliferous character in the survey data that might be related to hydrocarbons at depth. Cluster analysis allows for identification of geochemical groupings in the data. The samples of each grouping are then examined for petroliferous character and can be used in subsequent modeling efforts.

VERTICAL MIGRATION

Since the inception of surface geochemical methods, their proponents have needed to explain vertical migration to justify their applicability. Figure 3 is a schematic diagram of the migration of hydrocarbons to the earth's surface. Simplistically, there are two sources of migrating hydrocarbons: source rocks and hydrocarbons contained within a reservoir. Reservoir hydrocarbons reach the surface via migration along faults or other enhanced permeability conduits and via lateral migration and vertical migration through rock matrices.

If vertical migration occurs, then the migration paths depicted in Figure 3 would require that soil gas compositions in the zone of vertical migration be very similar to the hydrocarbons at depth. This is because the only process that affects this composition is the preferential adsorption of heavy hydrocarbons over light hydrocarbons. Although Santa Fe Minerals conducted no original studies of migration mechanisms, to verify empirically the occurrence of vertical migration, we compared surface soil gas compositions with compositions of volatile gases from produced oils or gases encountered in new field wildcats. A comparison of the composition of soil gas at 1 ft depth to the composition of volatile gases from produced hydrocarbons in four different new field wildcats ranging from 2000 to 15,000 ft in depth was revealing (Figure 4). As can be seen from these examples, the gas composition at 1 ft depth in the soil is strikingly similar to the produced hydrocarbons from reservoirs as deep as 15,000 ft. The only differences are some shifts in the ratios of heavy hydrocarbons to light hydrocarbons, as predicted. It is important to note that hydrocarbons as heavy as the C_{12} range are present in the soils and are migrating there from the reservoirs at depth.

Another consequence of the migration paths depicted in Figure 3 is that one would predict a change in composition of soil gas due to such processes as oxidation and bacterial activity mixing with background gases as they migrate nonvertically out of the reservoir to the surface. When the soil gas composition of the vertical zone is compared to the adjacent zones charged primarily by lateral migration, a strong difference in composition is detected (Figure 5).

Thus, we conclude that while nonvertical and vertical migration is occurring, the vertical component is naturally accentuated and can be further accentuated by careful modeling of the compositional data. The migration mechanism of the hydrocarbons is still not clear. The presence of significant C_{6+} hydrocarbons in soil gas, which is clearly from deep subsurface sources, argues against simple diffusion. Based on the types of heavy hydrocarbons observed in soil gases, we suggest that the migration mechanism might involve a "carrier" gas (or fluid), such as CO_2 or CH_4.

Migration along faults and shear zones is also well demonstrated by oil seeps and in geochemical surveys as well. Fault zones are enhanced migration pathways and allow significant increase in heavy hydrocarbons at the surface (Figure 6). The PETREX methodology, specifically

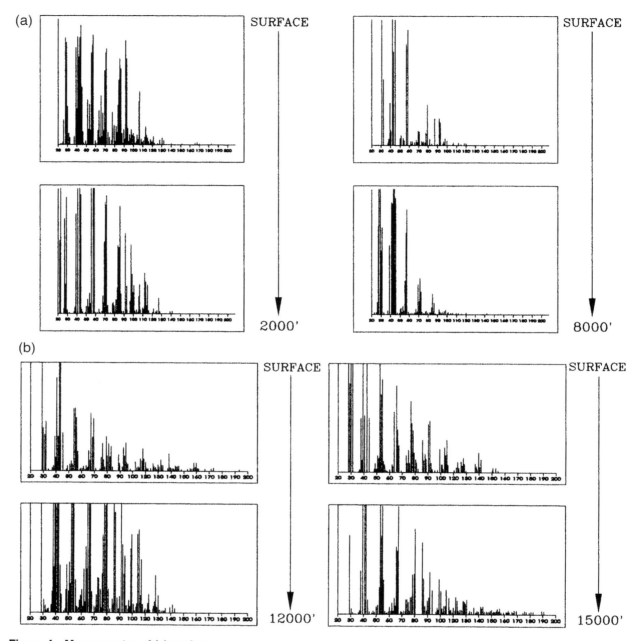

Figure 4—Mass spectra of (a) surface samples near producers and (b) headspace samples of producing oils.

its approach to geochemical modeling, tends to treat such fault signatures with difficulty, as do other geochemical methods. Fault samples in effect represent geochemical outliers that may distort the results of any statistical processing. The fault samples are often classified at the extremes of either the negative (background) or positive (production) models, depending on the type of hydrocarbon being modeled. Oil and wet gas models usually classify these fault signatures as production, while dry gas models classify them as background (Figure 7). This example illustrates a critical reason to integrate geochemical interpretations with geophysical interpretations to understand the geochemical data properly.

How vertical is vertical migration, or alternatively, do geochemical anomalies shift? Compositional anomalies detected with surface geochemical data using the PETREX approach were found to be apical in nature, as one would expect from the paths depicted in Figure 3. We undertook some detailed comparisons between the surface extent of geochemical anomalies and the subsurface extent of hydrocarbon accumulations as defined by geophysical and subsequent well control. In general, anomalies were apical, but a tendency was noted for shifts to occur in the up-dip direction. The maximum observed shift was about 700 ft (Figure 8). More frequently, shifts were not demonstrable or were less than 500 ft.

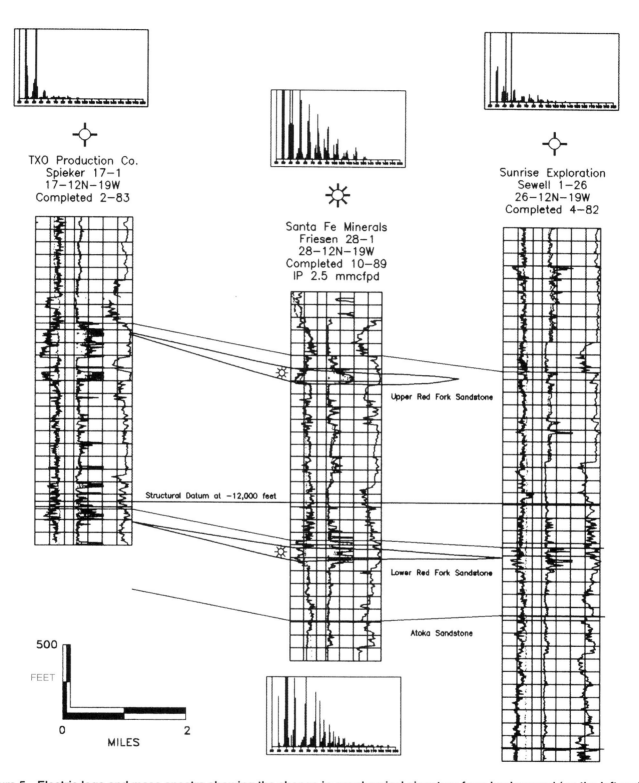

Figure 5—Electric logs and mass spectra showing the change in geochemical signature from background (on the left and right) to production (in the middle), from the Red Fork prospect, Oklahoma.

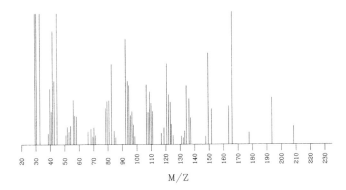

Figure 6—A mass spectrum showing a typical geochemical signature detected within a fault zone. Vertical scale is relative response.

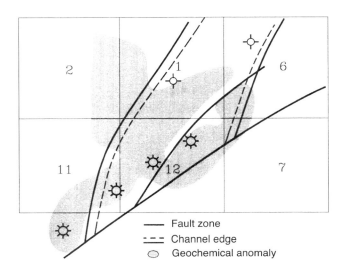

Figure 7—Map of a gas field in the Coalgate, Oklahoma, area showing the effect of a fault zone on geochemical anomalies.

ABSENCE OF HYDROCARBONS

Although the purpose of a geochemical survey is to find evidence of hydrocarbons, many surveys demonstrated that an exploration prospect area simply had no indications of any hydrocarbons being present. That is, the modeled values for the sample grid all reflected background geochemical signatures. Early in our use of geochemistry, we would drill such prospects—a practice we soon began to regret. Within negative anomalies delineated by PETREX, a total of 43 wells have been drilled, and of these, 41 were immediately plugged and abandoned with no significant hydrocarbons detected (Figure 9). Two wells encountered significant hydrocarbons, but only one of these wells was completed. Both were oil wells that had been drilled near the oil–water contact. The

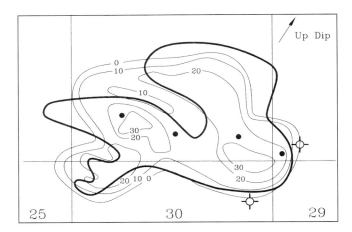

Figure 8—Map showing an apical PETREX surface geochemical anomaly (heavy outline), with an interpreted shift of about 700 ft in the up-dip direction relative to the prospect at depth. The light contours are sand isopachs (in feet).

single completed well watered out prior to recovering sufficient oil to pay out the completion costs.

These results indicate that when no hydrocarbons are indicated with geochemical data and PETREX modeling techniques, subsequent drilling has proven 95% of the time that no accumulations of hydrocarbons are present in the prospect. All of the prospects drilled by the 43 wells were valid prospects and most were correctly defined structures or traps that had sand and closure but contained water, not hydrocarbons.

These results have great significance to hydrocarbon exploration. One often faces the question of whether a frontier area has potential, or a recently mapped but undrilled structure in an established trend has potential. A geochemical survey can be used as a screening tool to answer this question quickly and at relatively low cost. For example, for a frontier area in Chile that covered 1000 sq mi (Figure 10), we were able to determine that a significant part of the subbasin had little or no hydrocarbon potential and thus were able to avoid expensive seismic and drilling commitments to obtain the block (Wensley and Viellenave, 1992).

PRESENCE OF HYDROCARBONS

The desired result of a geochemical survey is a positive anomaly that indicates hydrocarbons are present and, when drilled, yields a commercial discovery. As of December 1993, a total of 98 wells were drilled in exploration prospects that were indicated to be hydrocarbon-bearing by surface geochemistry. The results can be broken down into four categories, illustrated in Figure 11.

The first category consists of eight wells that did not encounter any hydrocarbons. Three of the wells had been drilled on combined amplitude variation with offset

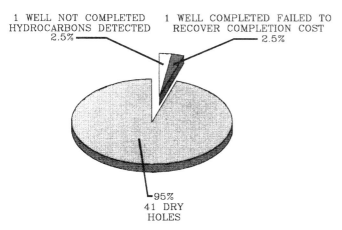

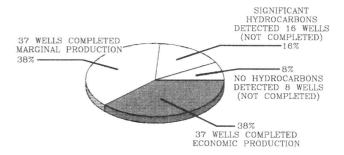

Figure 11—Result of 98 wells drilled in areas of positive geochemical anomalies.

Figure 9—Results of 43 wells drilled in areas of negative geochemical anomalies.

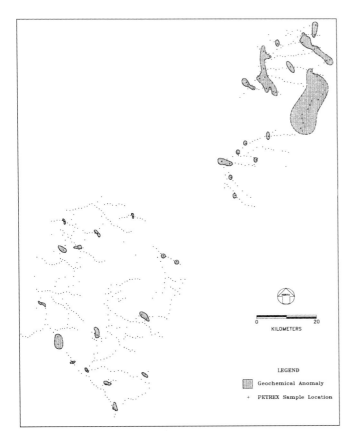

Figure 10—Map showing interpreted geochemical anomalies over a frontier area in northern Chile.

(AVO) and geochemical anomalies, and one well drilled through a thick fault-repeated coal section. However, these eight dry holes must be viewed as a failure of the interpretation or the method.

The second category consists of 16 wells that encountered significant hydrocarbons but in quantities such that completion of the wells was not warranted. These wells are typified by the Pasquale 31-1 (Figure 12), which encountered 4 ft of gas on water. These wells can be counted as geochemical successes but economic failures. The results of this category caused us to scrutinize the geochemistry and its integration with geophysical data to weed out such prospects.

The third category comprises 37 wells that were completed but lacked sufficient flow rates to pay out the cost of the prospect (G&G, leasehold, and drilling). The majority of these wells will pay out their completion costs and part or all of their drilling costs. In this group were 9 wells that encountered significant oil columns (in one case, 225 ft of oil-bearing rock), but which only yielded subcommercial flows of 20 BOPD or less. Another 10 wells had limited drainage areas, either for stratigraphic reasons or proximity to the hydrocarbon–water contact. The final group of 18 wells was completed in sandstones that lacked sufficient permeability for commercial productivity.

The fourth category consists of 37 wells that were commercial successes. The results of these wells require little explanation.

When the entire group of 98 wells is analyzed, some interesting observations are noted. The overall success rate for predicting commercial wells appears to vary with depth (Table 1). When the percentage of commercial success rates are plotted against the mean depth of each depth class, we obtain a correlation coefficient of 0.99. This may be caused by the enhanced response of geochemical methods to hydrocarbons at shallower depths, increasing the likelihood of detecting subcommercial quantities of hydrocarbons. No significant differences could be found with respect to hydrocarbon type (e.g., oil versus gas) and success rates. For wells with hydrocarbons present, there was a general trend of increasing anomaly strength with increasing values of porosity (ϕ) times thickness (h) of the reservoir body. This trend is particularly striking in the Anadarko Basin, Oklahoma (Figure 13). However, one should note that a high value for ϕh can be obtained from a thick section of tight reservoir rock that does not yield hydrocarbons at a commercial rate.

Table 1—Number and Rates of Successful Predictions (Out of 98 Total) of Commercial Wells at Various Depths

	No. of Successful Predictions (and Rate)		
Well Status	At 0 to 4999 ft	At 5000 to 9999 ft	At ≥10,000 ft
Dry hole	6 (15%)	2 (6%)	0 (0%)
HC present (not completed)	10 (25%)	4 (11%)	2 (9%)
Marginal	14 (35%)	15 (43%)	8 (35%)
Commercial	10 (25%)	14 (40%)	13 (56%)
Totals	40	35	23

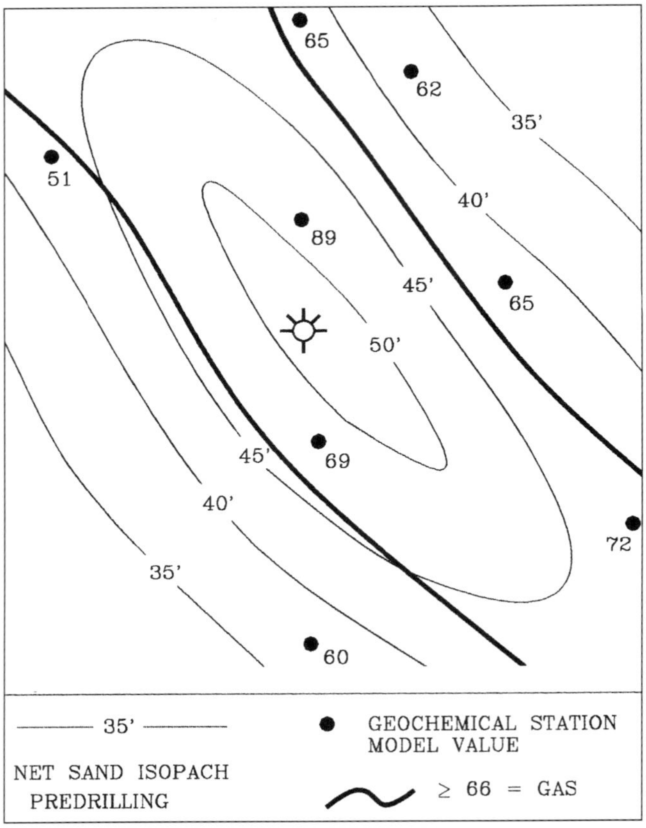

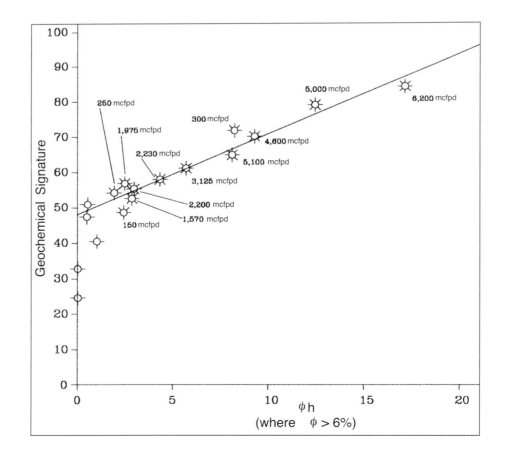

Figure 12—Map of the area around the Pasquali 31-1 well, Oklahoma, showing an example of a geochemical success yet an economic failure. →

Figure 13—Graph of the relationship between porosity-thickness (ϕh) and geochemical anomaly strength for the Anadarko Basin, Oklahoma.

CONCLUSIONS

The results of wells drilled on geochemical surveys conducted by Santa Fe Minerals clearly indicate that geochemical anomalies detected by the PETREX method are a significant tool in hydrocarbon exploration. The data collected demonstrate vertical migration of hydrocarbons from the reservoir to the surface. Since the presence and absence of specific hydrocarbon compositions in the soil directly relates to hydrocarbons present or absent in reservoirs, then by inference, other geochemical methods may be useful to hydrocarbon exploration (albeit with different rates of success). PETREX geochemical anomalies predict the presence of hydrocarbons accurately 92% of the time and the absence of hydrocarbons 95% of the time.

The empirical demonstration of vertical migration, and the ability to distinguish vertical from nonvertical migration, implies that geochemical techniques can be used as a reconnaissance method. Since geochemical costs are lower than seismic costs, geochemical surveys should be used to site seismic surveys, particularly in high-cost seismic regions. Furthermore, the use of surface geochemical data is ideal for screening frontier areas for hydrocarbon potential. Our experience indicates that integration of seismic data and geochemical data yields greater definition of exploration targets than provided by either method separately. Because of the lower cost of geochemical surveys, surface geochemical data can be used to infill between seismic lines and constrain the mapping of AVO/amplitude anomalies between seismic lines.

The primary application of surface geochemistry is to reduce risk. In our experience, *commercial* production was correctly predicted 38% of the time; 95% of the time obvious dry holes could be avoided. Thus, surface geochemistry was able to predict the correct commercial outcome 55% of the time. This success rate for true commercial success is substantially higher than that obtained by classic exploration strategies. Most oil companies would like geochemical data to indicate the presence of commercial hydrocarbon accumulations. However, surface geochemical methods only indicate the presence of hydrocarbons and not whether the reservoir is economically viable.

REFERENCES CITED

Davis, J. C., 1986, Statistics and data analysis in geology, 2nd edition: New York, John Wiley & Sons, 646 p.

Hickey, J. C., 1986, Preliminary investigation of an integrative gas geochemical technique for petroleum exploration: Master's thesis, Colorado School of Mines, Golden, Colorado.

James, M., 1985, Classification algorithms: New York, John Wiley & Sons, 209 p.

Klusman, R. W., 1993, Soil gas and related methods for natural resource exploration: New York, John Wiley & Sons, 483 p.

Wensley, J. R., and Viellenave, J. H., 1992, Enhancement of concession offerings and focusing exploration activities using PETREX surface geochemical techniques: Presentation to the Institute of Americas Conference, Caracas, Venezuela.

Index

◆

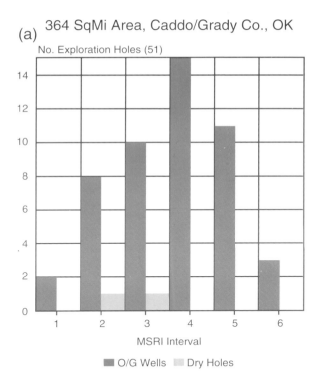

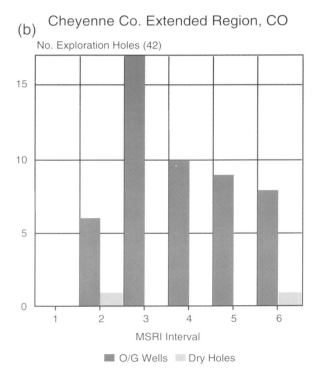

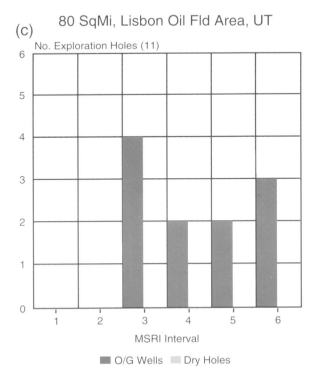

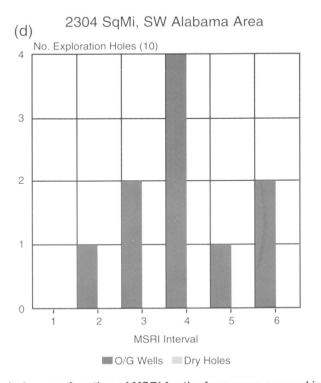

Figure 2—Histograms of the number of in-field exploration holes as a function of MSRI for the four areas covered in this study: (a) Caddo and Grady counties, Oklahoma; (b) Cheyenne County, Colorado; (c) Lisbon oil field area, San Juan County, Utah; (d) southwestern Alabama. Note that the legend for wells and dry holes applies to figures 2, 3, and 4.

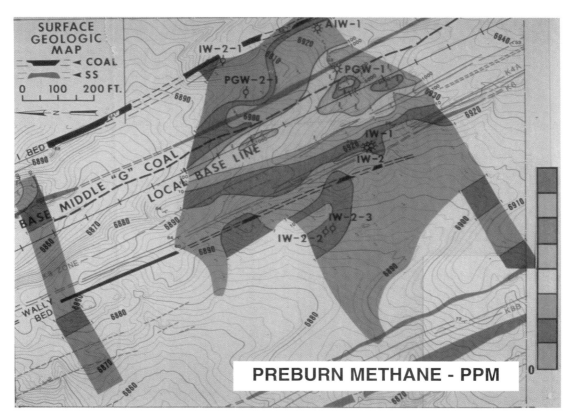

Figure 10—Color contour map of methane concentrations measured at permanent stations installed over an underground coal bed retort. Data were averaged over preburn, prepressure time window, July 22 to August 16, 1981.

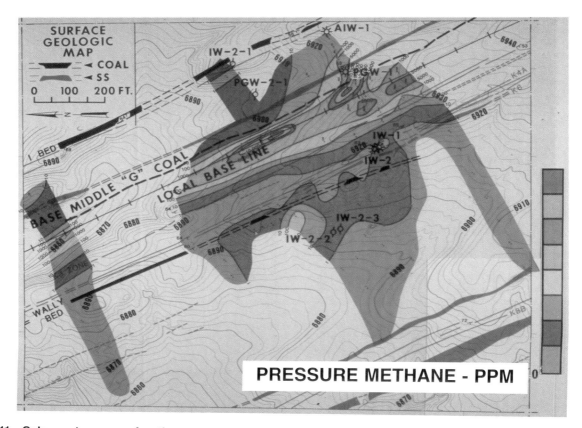

Figure 11—Color contour map of methane concentrations measured at permanent stations installed over an underground coal bed retort. Data were averaged over pressure time window, August 17 to 23, 1981.

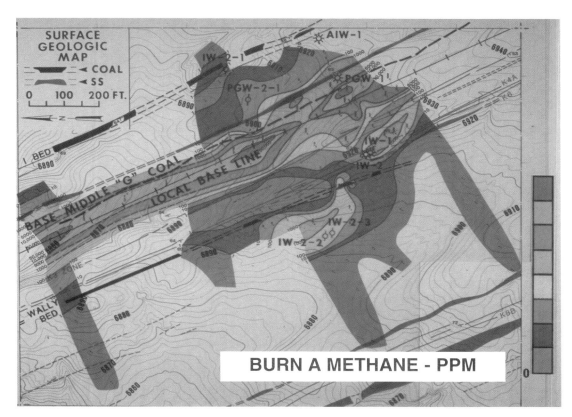

Figure 12—Color contour map of methane concentrations measured at permanent stations installed over an underground coal bed retort. Data were averaged over burn (burn A) time window, August 24 to November 10, 1981.

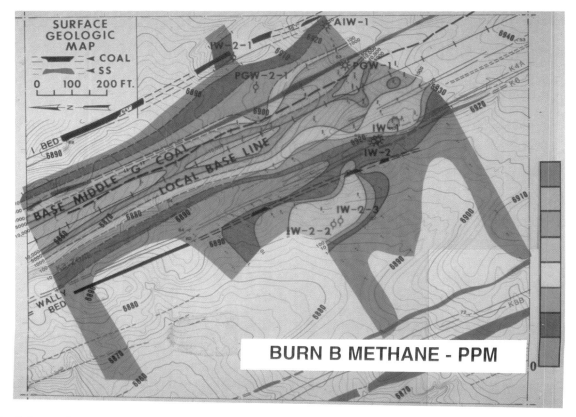

Figure 13—Color contour map of methane concentrations measured at permanent stations installed over an underground coal bed retort. Data were averaged over postburn (burn B) time window, November 11 to December 12, 1981.